Biocatalysis for Practitioners

Biocatalysis for Practitioners

Techniques, Reactions and Applications

Edited by

Gonzalo de Gonzalo
Iván Lavandera

Editors

Prof. Gonzalo de Gonzalo
Universidad de Sevilla
Dpto. de Química Orgánica
c/ Profesor García González 2
41012 Sevilla
Spain

Prof. Iván Lavandera
Universidad de Oviedo
Dpto. de Química Orgánica e Inorgánica
Avenida Julián Clavería 8
33006 Oviedo
Spain

Library of Congress Card No.: applied for

British Library Cataloguing-in-Publication Data
A catalogue record for this book is available from the British Library.

Bibliographic information published by the Deutsche Nationalbibliothek
The Deutsche Nationalbibliothek lists this publication in the Deutsche Nationalbibliografie; detailed bibliographic data are available on the Internet at <http://dnb.d-nb.de>.

Print ISBN: 978-3-527-34683-7
ePDF ISBN: 978-3-527-82444-1
ePub ISBN: 978-3-527-82445-8
oBook ISBN: 978-3-527-82446-5

Cover Design Adam-Design, Weinheim, Germany
Typesetting SPi Global, Chennai, India
Printing and Binding CPI Group (UK) Ltd, Croydon, CR0 4YY

Printed on acid-free paper

C097608_240321

Contents

Foreword

The application of biocatalysis to chemical processes has had an exponential growth in the past 20 years. Biotransformations provide the central tools in industrial biotechnology because they address the need for processes with less environmental impact in terms of energy, raw materials, and waste production. Over the past few years, the use of enzymes as biocatalysts for the introduction of enantiopure active compounds has become an established manufacturing process in the specialty and pharmaceutical industry. It has demonstrated that synthetic and computational tools can be exploited to generate new biocatalysts with novel structure and chemical properties. The employment of biocatalysts is attractive for synthetic organic chemists for producing optically active molecules.

In the present book "*Biocatalysis for Practitioners*," the authors have attempted to cover some of the most challenging areas in practical enzymatic catalysis through 16 interesting chapters by different specialists of recognized prestige in their field. In my opinion, it is important to recognize the work that the editors have carried out to cover in five sections the most current trends in the field of biocatalysis, which makes this work very useful not only for researchers in this field but also for students who are beginning to understand biocatalysis, so it can be a good textbook for some careers.

The first section consists of four chapters where different techniques to improve and discover new biocatalysts are described. Thus, purification, modification, immobilization, and compartmentalization techniques are treated in each of the chapters with great success, updating the reader on the progress of the different methodologies.

The second section also consists of four chapters of varied themes whose general title is *Enzymes Handling and Applications*. The fifth chapter tries to explain formal aspects of the so-called catalytic promiscuity, where processes that in principle were difficult to imagine can be nowadays catalyzed by different biocatalysts. Enzymes applied to the synthesis of amines, some applications of oxidoreductases, and the use of glycosyltransferases for the preparation of glycosides are the other chapters of this part.

Recently, many efforts are aimed to optimize the different biocatalytic processes by modifying the reaction medium or improving the recycling of the cofactor. Thus, in Section 3, entitled *Ways to Improve Enzymatic Transformations*, three very interesting reviews are described on topics of great importance, such as Application of nonaqueous media in biocatalysis, non-conventional cofactor regeneration systems, and Biocatalysis under continuous flow conditions.

A vision of current trends and evolution in synthetic aspects of biotransformations can be found in Section 4 with three chapters, where we can verify the usefulness of photobiocatalysis, multienzymatic transformations, and finally an interesting subject as chemoenzymatic sequential protocols.

To end this interesting book, in the fifth part, two chapters that deal with the increasing possibilities that biocatalysis has in the industry make a very successful closing of this work, on the one hand check the possibilities and the growth of enzymatic catalysis in industrial processes to end with a review of the editors of the book with a very practical chapter, where researchers and industrial sectors can go to check which biocatalysts are commercially available.

Finally, for the subscriber of this prologue, it is an honor to be able to verify how two excellent researchers and people such as Drs Lavandera and de Gonzalo, for whom I was their supervisor of their doctoral theses, have evolved in an extraordinary way with great work since they started their studies, with postdoctoral stays and today with a permanent position at the Universities of Oviedo and Seville. They are doing a great job. For me, it is a great satisfaction to be able to see how the disciples have overcome the teacher.

Vicente Gotor
Professor Emeritus and
Ex-Rector of the University of Oviedo (Spain)

Part I

Enzyme Techniques

1

Techniques for Enzyme Purification

Adrie H. Westphal[1] and Willem J. H. van Berkel[1,2]

[1] *Wageningen University & Research, Laboratory of Biochemistry, Stippeneng 4, 6708WE Wageningen, The Netherlands*
[2] *Wageningen University & Research, Laboratory of Food Chemistry, Bornse Weilanden 9, 6708WG, Wageningen, The Netherlands*

1.1 Introduction

Biocatalysis is the chemical process through which enzymes or other biological catalysts perform reactions between organic components. Biocatalysis gives an added dimension to synthetic chemistry and offers great opportunities to prepare industrial useful chiral compounds [1, 2]. Depending on the goal of the chemical conversion and the costs involved, biocatalyst-driven reactions are performed using whole cell systems or isolated enzymes, either in free or immobilized form [3–5].

Initially, industrial applications utilizing isolated enzymes were mainly developed with amylases, lipases, and proteases [6–8]. These hydrolytic enzymes were usually applied in a partially purified form, also because crude enzyme preparations are often more stable than the purified ones. However, for obtaining highly pure products, especially in the pharmaceutical industry, the purity of the enzyme preparation can be a critical factor.

Many enzyme purification methods have been developed over the years. Traditional purification procedures make use of the physicochemical properties of the enzyme of interest. These procedures were developed during the twentieth century for elucidating enzyme mechanisms and solving protein three-dimensional structures but also appeared to be valuable for the preparation of highly pure biocatalysts. Yet, progress in the preparation of biocatalysts has been given the biggest boost by the amazing developments in recombinant DNA technology and the accompanying revolutionary changes in enzyme production, enzyme purification, and enzyme engineering [9].

Here, we describe our experiences with the contemporary techniques for enzyme purification. For more information about the practical issues of enzyme purification, the reader is referred to the "Guide to Protein Purification" in Methods in Enzymology 463 [10].

Biocatalysis for Practitioners: Techniques, Reactions and Applications, First Edition.
Edited by Gonzalo de Gonzalo and Iván Lavandera.

1.2 Traditional Enzyme Purification

Before summarizing the traditional enzyme purification methods, it is important to note that the purification of enzymes is made easier by the fact that they are such specific catalysts. This enables the determination of the amount of a given enzyme in units (where 1 unit [U] of enzyme activity is defined as the amount of enzyme that catalyzes the conversion of 1 μmol substrate per minute) and its specific activity (in $U\,mg^{-1}$) in crude extracts and after each purification step. The specific activity is a good indication of the purity and quality of the enzyme preparation, especially if the specific activity of the pure enzyme under defined conditions is known. During enzyme purification, the improvement in specific activity and the yield of the enzyme after each purification step can be summarized in a purification scheme. The purification factor (specific activity obtained after a purification step divided by that of the starting material) provides an insight into the "efficiency" of each step. If a pure enzyme is obtained, it also indicates the relative amount of that enzyme present in the starting material. A theoretical example of a purification scheme, comprising three purification steps, is shown in Table 1.1.

Enzymes that are used for biocatalysis are typically purified from microbial cells or from culture media after or during growth of microorganisms (in case of excreted proteins). The enzyme purification generally starts with a cleared cell extract in which the enzyme is present in a soluble form. If the enzyme to be purified is excreted into the culture medium, it is usually sufficient to remove the cells from the medium by centrifugation (for small-scale purifications) or by filtration (for large-scale industrial purifications). In the case of an intracellular enzyme, cells should be broken first to release the protein into solution. Depending on the type of cells, different techniques are employed. The microbial cells are first harvested from the culture medium by centrifugation and resuspended in a small amount of buffer. The cells can be broken using a variety of techniques, e.g. by treatment with enzymes that digest cell walls (e.g. lysozyme), followed by osmotic shock, by using lysis buffers containing detergents, by exposure to ultrasound using sonicators, by pushing cells under high pressure through a small orifice using a pressure cell system, or by grinding frozen cells in liquid nitrogen. Extracts thus obtained are cleared from unbroken cells and large, insoluble particles by centrifugation or filtration. To prevent enzyme inactivation during these treatments, and also in the following purification steps, the temperature

Table 1.1 Imaginary traditional enzyme purification scheme.

Step	Volume (ml)	Activity (U)	Protein (mg)	Specific activity ($U\,mg^{-1}$)	Yield (%)	Purification factor
CE	500	3000	15 000	0.2	100	1.0
AS	100	2400	4000	0.6	80	3.0
IEC	45	1440	500	2.9	48	14.5
GF	50	1000	125	8.0	33	40.0

Steps: CE, cell extract; AS, ammonium sulfate fractionation; IEC, ion exchange chromatography; GF, gel filtration.

of the enzyme solution is usually kept around 4 °C. Proteolytic degradation of the enzyme to be purified can be precluded by adding a protease inhibitor cocktail during breaking of the cells.

Once a cell-free extract has been obtained, several methods can be employed for further purification of the desired enzyme. These separation methods can be roughly divided into the following categories: (i) selective precipitation, (ii) separation based on charge, (iii) separation based on molecular size, (iv) separation based on bio-affinity, and (v) separation based on adsorption principles. Except for the first category, all these methods generally make use of column chromatography, with column sizes depending on the scale of the sample volumes and protein concentrations.

The strategy applied during enzyme purification is such that separation methods belonging to different categories are carried out in a logical order until the goal is reached. A good purification results in the recovery of most of the enzyme activity (i.e. a high yield) and in removal of many "contaminating" proteins and other types of (bio)molecules (i.e. a strong increase in specific activity). An often-experienced phenomenon during purification is the inactivation and/or aggregation of the enzyme (Figure 1.1). Because of increased enzyme concentration in the final steps of purification, aggregation can occur. If proteases are still present, the enzyme becomes more and more the only target for the protease, which can lead to proteolysis. In addition, wrong physical conditions (pH, temperature, and ionic strength) can lead to (partly) unfolding, followed by aggregation and/or proteolysis. Changing the type of buffer, pH, and/or ionic strength and the addition of protecting agents may alleviate these processes.

The purity of the final enzyme preparation can be tested in several ways. The most common methods used are sodium dodecyl sulfate polyacrylamide gel electrophoresis (SDS-PAGE) (Figure 1.2), analytical gel filtration, and mass spectrometry [11].

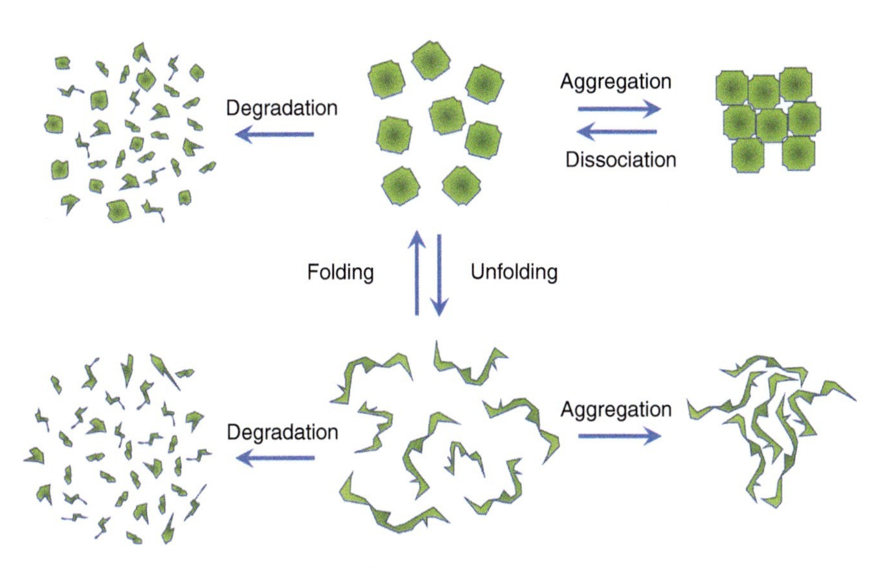

Figure 1.1 Enzyme aggregation and proteolytic degradation processes.

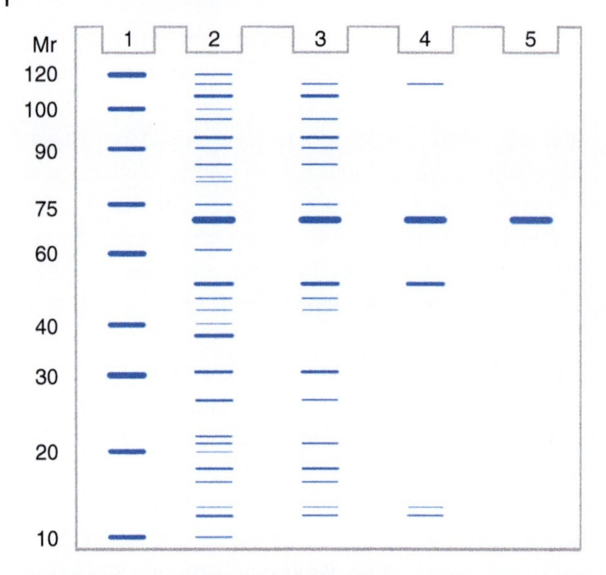

Figure 1.2 Example of an SDS-PAGE gel. (1) Molecular mass markers, (2) cell extract, (3) sample after ammonium sulfate fractionation, (4) sample after ion exchange chromatography, and (5) sample after gel filtration. Mr., relative molecular mass (kDa).

Traditional enzyme purification procedures many times start with an ammonium sulfate fractionation. This type of fractionation makes use of the fact that individual proteins precipitate at different concentration ranges of ammonium sulfate [12].

To make an estimation of the fractionation range, a small-scale pilot experiment can be performed. For such an experiment, different amounts of ammonium sulfate (from 0% to 90% saturation) are added to small samples of cell extract (usually, 1 ml). After dissolving the ammonium sulfate and removal of the formed protein precipitates by centrifugation, enzyme activity of the supernatants is measured (Figure 1.3). Such an analytical pilot experiment tells us at which saturation value the enzyme starts to precipitate (in our pilot, around 30%) and at which degree of saturation precipitation of the enzyme is more or less complete (in our pilot, around 65%). Once these values have been determined, the bulk of the cell extract is fractionated using these percentages and the precipitate obtained after the second addition of ammonium sulfate is used for further purification. If desired, removal of ammonium sulfate can be accomplished by dialysis, ultrafiltration, or gel filtration (e.g. with desalting columns).

Ammonium sulfate fractionation has been used in our group for the purification of several oxidoreductases. For the purification of vanillyl alcohol oxidase from *Penicillium simplicissimum* [13], the ammonium sulfate fractionation of the cell extract from 30% to 60% saturation gave a yield of 85%. Although few protein impurities were removed (as judged from SDS-PAGE and from the rather low purification factor of 1.2), this step appeared to be advantageous for the subsequent purification using a Phenyl Sepharose column, especially because ammonium sulfate removal could be omitted before this hydrophobic interaction chromatography (HIC) step. A similar experience was made with the purification of catalase peroxidase from *P. simplicissimum* [14] and 4-hydroxybenzoate 3-hydroxylase from

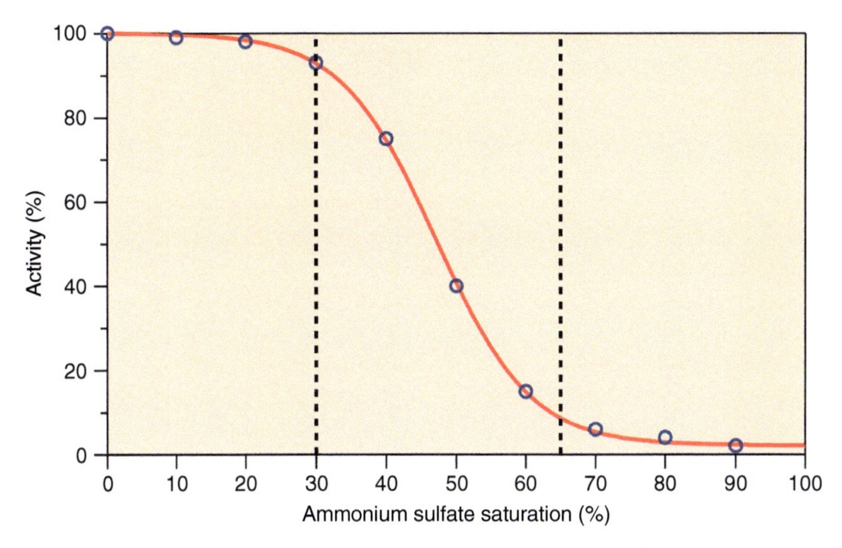

Figure 1.3 Ammonium sulfate fractionation. In this example, a pilot experiment is performed on small cell extract samples. Next, 30% saturation is used on the total extract sample and the precipitates formed are removed by centrifugation. Then, the supernatant is brought to 65% saturation and the precipitate, which contains most of the enzyme activity, is collected by centrifugation.

Rhodococcus opacus 557 [15]. With hydroquinone dioxygenase from *Pseudomonas fluorescens* ACB [16], the cell-free extract was adjusted to 25% ammonium sulfate saturation before loading onto a Phenyl Sepharose column. Ammonium sulfate can also be used to concentrate the solution during enzyme purification. In the case of 4-hydroxybenzoate 1-hydroxylase from *Candida parapsilosis* [17], the enzyme fraction obtained after Q-Sepharose ion exchange chromatography (IEC) was adjusted to 70% saturation with pulverized ammonium sulfate and the resulting precipitate, collected by centrifugation, was dissolved in a small amount of buffer.

1.2.1 Ion Exchange Chromatography

IEC is one of the most widely used methods for enzyme purification. It separates protein molecules according to their differences in charge [18]. The stationary phase (matrix) in IEC carries charged functional groups fixed by chemical bonds. The fixed groups are associated with exchangeable counterions. In anion exchange chromatography, the fixed groups have positive charges and in cation exchange chromatography, these groups are negatively charged. As a rule of thumb, proteins bind to an anion exchanger at pH values above their isoelectric point (pI) and to a cation exchanger at pH values below the pI. Protein IEC usually involves the following steps:

(1) *Equilibration:* The ion exchange resin is equilibrated with a low-salt buffer that allows binding of the enzyme of interest.
(2) *Sample application and adsorption*: Protein molecules with a proper charge displace counterions and bind reversibly to the matrix. The ionic strength of the buffer in which

the protein sample is loaded should be low, as a high concentration of salt usually prevents binding. Any volume of sample can be applied as long as the total amount of protein does not exceed the binding capacity of the matrix. Yet, a large sample volume having a low concentration of precipitates may eventually clog the column. Proteins in the sample not bound by the matrix can be washed from the column using a loading buffer. To follow elution of nonbinding proteins, the absorbance of the column effluent can be monitored at 214 or 280 nm in the case of a low amount of protein in the sample or at 305 nm in the case of high protein concentrations.

(3) *Desorption of bound proteins*: A stepwise increase of salt concentration or, in most cases, a gradual increase of the salt concentration (gradient) of the elution buffer is used. Again, the elution of proteins can be monitored by measuring the absorbance of the column effluent at 214 or 280 nm and, in addition, at a visible wavelength in the case of colored proteins. Elution with a shallow continuous gradient has the advantage that proteins with small differences in pI values are better separated and elute from the column in sharp, symmetrical peaks. For some enzymes, activity may be lost at high salt concentrations (e.g. because of dissociation of subunits). In that case, an elution can be attempted using a pH change step or a pH change gradient.

(4) *Cleaning of the column*: Proteins and other substances that are bound very strongly to the column are removed. This is usually done by "cleaning-in-place," using 2 M NaCl or 0.5 M NaOH solutions, followed by washing with water/buffer and 20% ethanol for storage.

IEC is a very powerful (preparative) purification method because (i) the high binding capacity of ion exchange columns allows elution of proteins in a very concentrated form and (ii) a proper choice of elution conditions results in separation of the bound proteins at high resolution.

For many years, IEC was included in almost every enzyme purification procedure, both on lab scale and at industrial level. Although this picture has changed after the introduction of the recombinant DNA technology providing the use of affinity tags, IEC remains a superior technology for enzyme purification because of its large resolving power and high recovery of enzyme activity.

In our experience, the IEC technique appeared to be crucial for the purification of a wide range of oxidoreductases, including monooxygenases, oxidases, dioxygenases, peroxidases, reductases, and dehydrogenases (Table 1.2).

A specific application of IEC involved the separation of native and oxidized forms of the flavoenzyme 4-hydroxybenzoate 3-hydroxylase from *P. fluorescens* [21]. The sensitivity of this dimeric enzyme to air oxidation resulted in different isoforms, which could be separated on a preparative scale with a DEAE-Sepharose column (Figure 1.4). Further analysis with an analytical Mono-Q column and isoelectric focusing experiments revealed the 10 different isoforms possible, assigned to combinations of the sulfhydryl, sulfenic acid, sulfinic acid, and sulfonic acid state of the surface-accessible Cys116 of each subunit. Mixing a native enzyme and a fully oxidized enzyme resulted in extremely slow formation of hybrid dimers with one native and one fully oxidized subunit, pointing to the high stability of the enzyme dimer.

Table 1.2 Ion exchange chromatography of oxidoreductases.

Enzyme family	Enzyme	References
Flavoprotein hydroxylases	4-Hydroxybenzoate 3-hydroxylase	[14, 19–22]
	4-Hydroxybenzoate 1-hydroxylase	[16]
	3-Hydroxyphenylacetate 6-hydroxylase	[23]
	Hydroquinone hydroxylase	[24]
	Phenol hydroxylase (PheA1)	[25]
	3-Hydroxybenzoate 6-hydroxylase	[26]
Baeyer–Villiger monooxygenases	4-Hydroxyacetophenone monooxygenase	[27]
Copper-dependent monooxygenases	Polyphenol oxidase (tyrosinase)	[28]
	Lytic polysaccharide monooxygenase	[29]
Flavoprotein oxidases	Vanillyl alcohol oxidase	[13]
	Eugenol oxidase	[30]
Multicopper oxidases	Laccase-like multicopper oxidase	[31, 32]
Non-heme iron dioxygenases	Hydroquinone dioxygenase	[33]
Heme-dependent peroxidases	Catalase peroxidase	[12]
	Cationic peroxidase	[34]
Reductases	NADH reductase	[25, 35]
	Flavin reductase (PheA2)	
Nicotinamide-dependent dehydrogenases	Alcohol dehydrogenase	[36]
	Carveol dehydrogenase	[37]
Flavin-dependent dehydrogenases	Galactonolactone dehydrogenase	[38, 39]
	Proline dehydrogenase	[40]

Most of the listed enzymes were purified with several traditional separation methods described in this review. See references for details.

1.2.2 Gel Filtration

In gel filtration, also referred to as molecular sieve or size exclusion chromatography (SEC), sample molecules do not bind to the column but are fractionated based on their relative size and shape [41]. The liquid phase in such a column (total volume, V_t) has two measurable volumes: external or "void" volume, consisting of the liquid between the beads (V_0), and the internal volume (V_i), constituted by the liquid within the pores of the beads. Molecules being too large to enter the pores cannot equilibrate with V_i and therefore emerge first from the column, while small molecules can equilibrate with V_i and therefore elute later.

The most important parameters in SEC are (i) the diameter of the pores allowing access to the internal volume of the beads, (ii) the total internal volume of the beads, (iii) the hydrodynamic diameter of the sample molecules, (iv) the flow rate of the liquid phase, and (v) the operation temperature and viscosity of the buffer used.

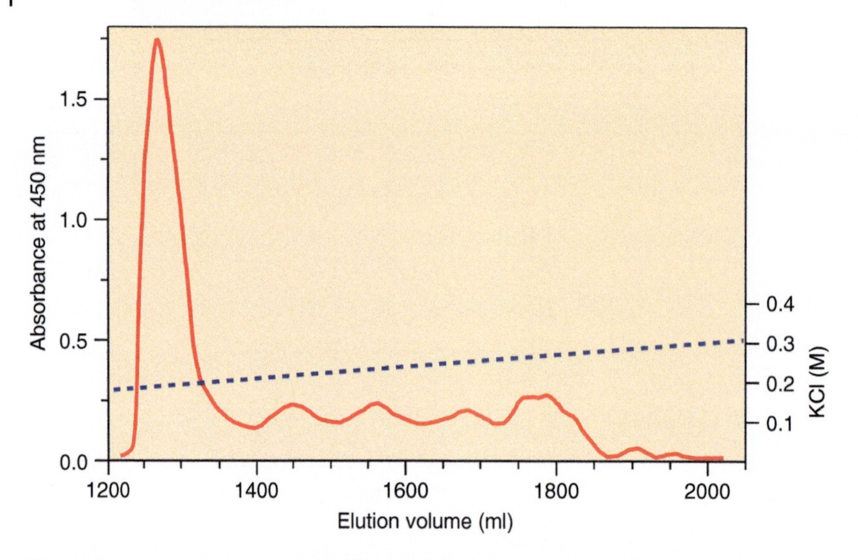

Figure 1.4 IEC of isoforms of highly pure 4-hydroxybenzoate 3-hydroxylase (PHBH) from *Pseudomonas fluorescens*. Preparative separation on DEAE-Sepharose CL-6B (650 mg protein) using a gradient elution. The small peaks after the main peak contain differently oxidized forms of PHBH. Source: Modified from van Berkel and Müller [21].

Elution volumes of fractionated molecules should be intermediate between V_0 and V_t. The elution volume (V_e, Figure 1.5a) relates to the accessibility of the molecule to the pores of the beads: $V_e = V_0 + K_{AV} * V_i$ (where the partition coefficient $K_{AV} = (V_e - V_0)/(V_t - V_0)$). A semi-logarithmic plot illustrating the relation between K_{AV} and protein molecular weight (M_r) is given in Figure 1.5b. The separation of proteins according to M_r is greatest in the central, linear region of the sigmoidal curve, spanning K_{AV} values between 0.2 and 0.8. This span is described as the fractionation range of a size exclusion matrix. A steep slope of the sigmoidal curve indicates a large resolving power of a matrix for a certain molecular weight range.

Next to being a suitable purification step [28, 29, 35, 40, 42], SEC is extremely useful to get information about the molecular weight of the native protein and its possible subunit composition [43]. By using this technique, we established that 4-hydroxybenzoate 3-hydroxylase from *P. fluorescens* is a homodimer, both in its holo and apo form [19, 44]. For lipoamide dehydrogenase from *P. fluorescens*, we experienced that nicotinamide adenine dinucleotide reduced (NADH) binding strongly stimulates flavin adenine dinucleotide (FAD)-induced dimerization [45].

For vanillyl alcohol oxidase from *P. simplicissimum*, we found that the holoenzyme favors the octameric state [13, 46], whereas the apoenzyme [47] mainly exists as a dimeric species. The octamer–dimer equilibrium of the holoenzyme varied with the ionic strength of the buffer solution, with kosmotropic salts stimulating the octameric state [48, 49]. More recently, it was established that a single loop at the protein surface is essential for the octamerization of vanillyl alcohol oxidase [30].

For hydroquinone dioxygenase from *P. fluorescens* ACB, we obtained strong indications from gel filtration that this non-heme, iron-dependent enzyme is an $\alpha 2\beta 2$

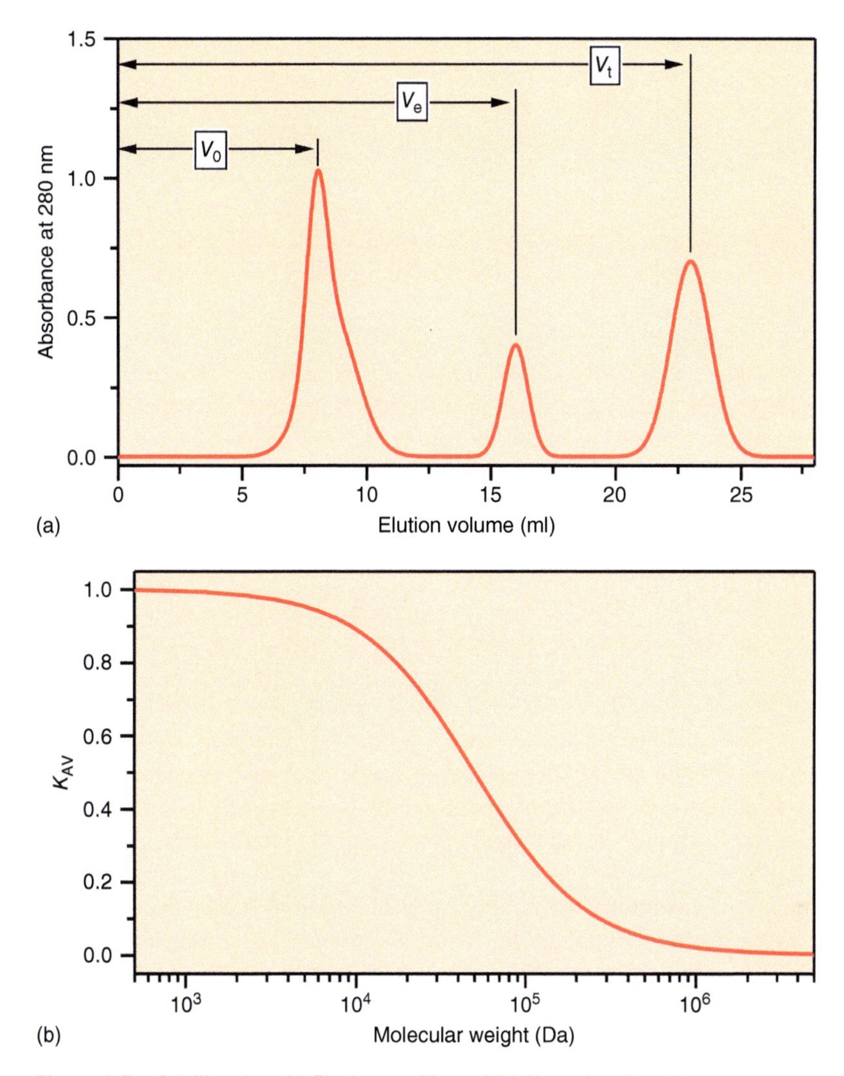

Figure 1.5 Gel filtration. (a) Elution profile and (b) K_{AV} vs log M_r plot.

heterotetramer [33]. With proline dehydrogenase from *Thermus thermophilus*, we showed that the native enzyme is a homotetramer [40, 50] and that dimerization of the protein subunits strongly increases the enzyme thermostability [51]. For 3-hydroxybenzoate 6-hydroxylase from *Rhodococcus jostii* RHA1, we obtained evidence that the monomer–monomer contact of the dimer is stabilized by the binding of a phosphatidylinositol ligand [20].

1.2.3 Bio-affinity Chromatography

Bio-affinity chromatography is one of the most powerful procedures in protein purification. This method has a very high selectivity as it utilizes the specific, reversible interactions between biomolecules [52]. Classic enzyme affinity chromatography mainly focused on methods that made use of the specific interactions of enzymes with ligands, such as substrates, coenzymes, inhibitors, and activators. Through immobilization of such ligands on suitable matrices, enzymes can be selectively bound to these resins (see example of old yellow enzyme given below). Preferably, the dissociation constant (K_d) of an enzyme-immobilized ligand complex should not change substantially compared to that of the enzyme–ligand complex free in solution. The dissociation constant range of the complex may vary from micromolar (enzyme–coenzyme complexes) to nanomolar (enzyme inhibitor complexes).

In bio-affinity chromatography, proteins to be purified are brought onto a column containing the immobilized ligand. After application of sample to the column, nonbinding proteins are washed out. Protein(s) that are retained on the column by their specific interaction with the ligand are removed by changing the elution conditions. The most specific way is by using a soluble ligand, which is competitive for the matrix-bound ligand with which the enzyme is associated. Bio-specific elution is not always possible, in which case elution can be stimulated by, for example, using a gradient of increasing salt concentration or by changing the pH-value of the elution buffer.

Because bio-affinity chromatography makes use of specific interactions, it may result in a high degree of purification. In some cases, using this technique, an enzyme can be obtained from a crude extract almost completely pure in a single chromatographic step (see example of Old Yellow Enzyme given below). However, commercially available bio-affinity resins are often very costly and have a limited choice in coupled ligands. In addition, these resins are also not easily prepared "at home," especially when expensive biomolecules must be used as immobilized ligands. Bio-affinity columns containing such ligands are usually more difficult to clean than, e.g., ion exchangers; therefore, the lifetime of a bio-affinity column is often limited.

We applied traditional bio-affinity chromatography for the purification of a number of enzymes, ranging from oxidoreductases to transferases. For 4-hydroxybenzoate 3-hydroxylase, we developed a Cibacron Blue dye affinity matrix, which appeared to be very useful for increasing the specific activity and yield of the enzyme, as isolated from different microbial sources [14, 15, 19–21]. Glutathione *S*-transferase isoenzymes from rat liver were purified using *S*-hexylglutathione affinity chromatography, followed by chromatofocusing on a Mono-P column [53]. A novel branched-chain alcohol dehydrogenase was purified from *Saccharomyces cerevisiae* using a Procion Red dye affinity column, which was selected based on its capacity to bind to a wide range of nicotinamide adenine dinucleotide phosphate (NADP)-dependent enzymes [36].

Old Yellow Enzyme from *Saccharomyces carlsbergensis* is the canonical member of a large family of ene reductases [54]. These flavoenzymes catalyze the asymmetric *trans*-hydrogenation of alkenes, resulting in industrially relevant chiral products [55]. Because Old Yellow Enzyme strongly interacts with phenolic compounds that act as competitive inhibitors, the enzyme was purified originally in high yield from brewer's bottom yeast by affinity chromatography using *N*-(4-hydroxybenzoyl)aminohexyl agarose [56–58]. This affinity matrix was prepared from agarose in four steps [56]:

(1) Agarose beads were equipped with an aminohexyl spacer arm by activating the beads with cyanogen bromide in the presence of 1,6-diaminohexane.
(2) The resulting aminohexyl agarose was reacted with 4-acetoxybenzoic acid to give *N*-(4-acetoxybenzoyl)aminohexyl agarose.
(3) Remaining free amino groups were acetylated with acetic anhydride.
(4) The protecting acetoxy group was removed from the ligand by incubation with imidazole, yielding the *N*-(4-hydroxybenzoyl)aminohexyl agarose affinity matrix (Figure 1.6).

The purification of Old Yellow Enzyme from brewer's bottom yeast then went as follows [56]:

(1) 350 g dried yeast was suspended in 1 l of demineralized water, containing 10 μM phenylmethylsulfonylfluoride to inactivate serine proteases.
(2) The suspension was homogenized for 30 seconds at the high-speed setting of a Waring Blendor.
(3) *Autolysis*: the mixture was transferred to a glass beaker and mechanically stirred for 4 hours at 37 °C. All subsequent operations were performed at 0–4 °C.
(4) The extract was clarified by centrifugation and precipitated with solid ammonium sulfate to 78% saturation.
(5) *Enzyme reduction to remove phenolic ligands*: the precipitate was collected by centrifugation and dialyzed overnight against 6 l of 0.1 M Tris–HCl pH 8.0, containing 0.1 M ammonium sulfate, 10 μM phenylmethylsulfonylfluoride, and 10 mM sodium dithionite.
(6) *Enzyme reoxidation*: dialysis with the same buffer, omitting sodium dithionite, continued for another 6 hours with one additional buffer change.
(7) Centrifugation to remove a white precipitate and stirring the clarified yeast extract for another 30 minutes to ensure reoxidation.
(8) A column of *N*-(4-hydroxybenzoyl)aminohexyl agarose (Figure 1.6, bed volume 20 ml) was washed with 0.1 M Tris–HCl pH 8.0, containing 0.1 M ammonium sulfate and 10 μM phenylmethylsulfonylfluoride.
(9) The clarified yeast extract was applied on the column and the column was extensively washed with buffer (about 2 l) until the absorbance at 280 nm is lower than 0.2.
(10) The Old Yellow Enzyme was eluted with 400 ml of washing buffer, which was degassed, flushed with oxygen-free nitrogen, and supplemented with 3 mM sodium dithionite.

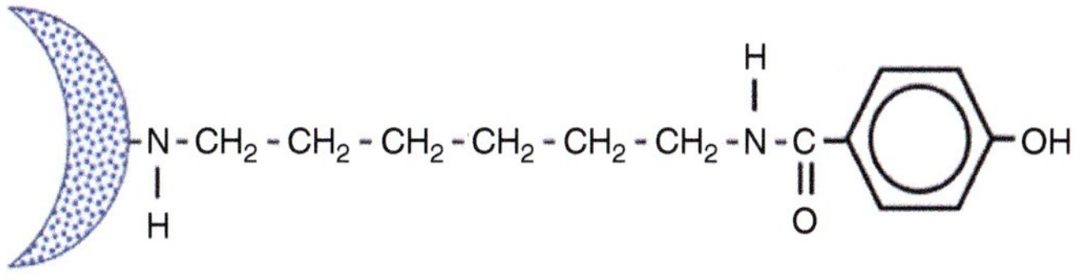

Figure 1.6 *N*-(4-Hydroxybenzoyl)aminohexyl agarose affinity matrix.

(11) The enzyme eluted directly upon flavin reduction and turned bright yellow after reoxidation by air.
(12) The collected enzyme (about 100 ml) was concentrated by ultrafiltration and stored frozen in 1 ml aliquots.
(13) Regeneration of the affinity matrix was accomplished by washing the agarose beads with 0.2 M acetate buffer pH 5.0, containing 6 M GuHCl.
(14) Storage of the gel in 10% ethanol with 1 mM sodium azide to prevent microbial damage.

SDS-PAGE showed that the Old Yellow Enzyme from *S. carlsbergensis* (relative subunit molecular mass = 49 kDa) was obtained in pure form. Absorption spectral analysis confirmed that the dimeric enzyme contained one tightly bound molecule of flavin mononucleotide (FMN) per subunit. The yield of enzyme was about 85% (130 mg) and its specific activity (turnover number) was slightly higher than the value obtained for the enzyme purified by conventional procedures.

Besides IEC, there are other chromatographic separation methods that make use of the properties of a protein's surface for adsorption to a specific chromatographic resin. The most commonly applied methods are described below.

1.2.4 Hydrophobic Interaction Chromatography

HIC is a very useful technique for the fractionation of proteins [59]. In proteins, some hydrophobic groups, or clusters of hydrophobic groups, can occur at the surface of a protein and thus contribute to the surface hydrophobicity. The surface hydrophobicity allows a protein to undergo hydrophobic interactions not only with other proteins but also with column materials carrying hydrophobic groups.

Hydrophobic interactions between nonpolar compounds are enhanced by a polar environment and are energetically favorable because of a gain in entropy on forming. It is the liberation of ordered water molecules in contact with hydrophobic surfaces that drives clustering of hydrophobic groups. It follows that hydrophobic interactions will be affected if the structure of water is changed by dissolved salts or organic solvents. Kosmotropic salts (e.g. ammonium sulfate) tend to favor the strength of hydrophobic interactions, whereas chaotropic salts (e.g. sodium thiocyanate) disrupt the structure of water and thus tend to decrease the strength of hydrophobic interactions. Organic solvents are also commonly used to alter the polarity of water.

Although the mechanisms of hydrophobic interactions are complicated, chromatographic techniques based on hydrophobic interactions are easy to use. The most common resins for HIC are substituted with *n*-butyl, *n*-octyl, or phenyl groups. For an uncharacterized protein, as a start, phenyl-substituted resin is usually the best choice because strongly hydrophobic proteins are not easily eluted from the highly hydrophobic octyl-substituted resins. The phenyl ligand is intermediate in hydrophobicity between *n*-butyl and *n*-pentyl and will bind to aromatic amino acids through π–π interactions.

A salt concentration just below that used for salting out of a protein is normally used for binding the protein to a hydrophobic matrix. A common procedure is to start purification of a protein from a crude extract with an ammonium sulfate precipitation at a concentration of ammonium sulfate that leaves the protein of interest just in solution, followed by removal of proteins that precipitate at this salt concentration by centrifugation, and loading the clarified extract onto an HIC column.

Proteins bound to an HIC column are eluted by reducing the concentration of kosmotropic salt (e.g. ammonium sulfate) in the buffer using a negative gradient. This successively releases proteins from the column in order of hydrophobicity. Proteins that are tightly bound such that they do not elute at zero salt concentration can be eluted using a positive gradient of a polarity-reducing organic solvent (usually up to 50% ethylene glycol).

We applied the HIC technology in quite some traditional enzyme purifications, usually after ammonium sulfate fractionation, and often in combination with IEC and an additional chromatographic step. This enabled us to characterize the catalytic properties of several oxidoreductases [12–14, 23, 24, 27, 33, 37, 39, 60, 61] (see also Table 1.2).

A special case of HIC concerns the purification of the recombinant forms of lipoamide dehydrogenase from *A. vinelandii* [62] and *P. fluorescens* [63]. Both these enzymes, overproduced in *Escherichia coli*, could be purified in a single chromatographic step by binding them to a Sepharose 6B gel filtration column, equilibrated in 0.1 M potassium phosphate buffer pH 7.0, containing ammonium sulfate at 50% saturation. After washing with equilibration buffer, the lipoamide dehydrogenases were obtained in pure form by eluting with the same buffer at 25% ammonium sulfate saturation.

We also applied the HIC technology for the reversible removal of the flavin cofactor in a number of flavoproteins, including lipoamide dehydrogenase, glutathione reductase, mercuric reductase, and butyryl-CoA dehydrogenase [64, 65]. Figure 1.7 presents a schematic overview of the procedures. Each flavoprotein is bound to Phenyl Sepharose in a high-salt buffer. After changing to low pH, the FAD cofactor is released from the bound protein. Reconstitution of holoenzyme with natural FAD, chemically modified FAD, or isotopically enriched FAD (FAD*) is performed on-column at neutral pH. Next, the reconstituted protein is released from the column with 50% ethylene glycol. Alternatively, the apoprotein is released from the column with 50% ethylene glycol and the holoenzyme is reconstituted in solution. The HIC technology appeared to be superior to classical methods for the reversible dissociation of the FAD cofactor of these enzymes, especially because the reversible immobilization procedure gave excellent yields and could be applied at large scale.

1.2.5 Hydroxyapatite Chromatography

Another adsorption chromatography method for enzyme purification concerns hydroxyapatite (HAP) chromatography [66]. HAP is a chromatographic support consisting of calcium phosphate crystals.

The amino groups of proteins adsorb to HAP primarily as a result of nonspecific electrostatic interactions between their positive charges and negative charges on the HAP column when the column is equilibrated with phosphate buffer. The carboxyl groups in proteins bind specifically by complexation to the calcium sites on the column. It was also found that basic proteins are eluted either as a result of normal Debye–Hückel charge screening or by a specific displacement with Ca^{2+} and Mg^{2+} ions, which

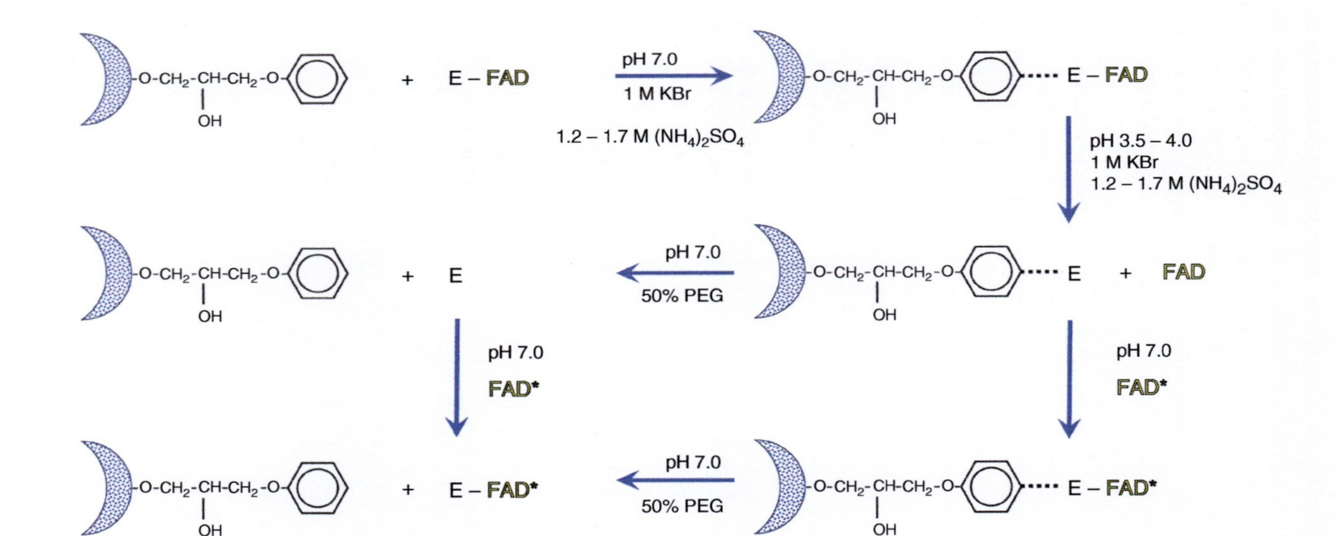

Figure 1.7 Preparation and reconstitution of apo-flavoproteins on Phenyl Sepharose. FAD* is isotopically enriched (^{13}C, ^{15}N)-FAD. Source: Modified from van Berkel et al. [64].

form a complex with column phosphates and neutralize their negative charges. Acidic proteins are eluted by displacement of their carboxyl groups from calcium sites by ions that form stronger complexes with calcium than do carboxylate groups, e.g. fluoride or phosphate. The ineffectiveness of chloride as an eluent of acidic proteins is due to the fact that it does not form a complex with Ca^{2+} and, thus, cannot compete with the calcium carboxylate complexes. The ability of $CaCl_2$ and $MgCl_2$ to strengthen the interaction of acidic proteins with HAP is due to the formation of additional bridges between protein carboxyl groups and column phosphate sites. In practice, proteins are often applied in 5–10 mM phosphate buffer at pH 7.0 and eluted with an increasing gradient of phosphate buffer.

We successfully applied HAP chromatography for the purification of vanillyl alcohol oxidase [13, 60], NADH reductase [35], catalase peroxidase [12], 4-hydroxybenzoate 3-hydroxylase [20], 4-hydroxybenzoate 1-hydroxylase [61], carveol dehydrogenase [37], flavin reductase PheA2 [25], and galactonolactone oxidoreductase [39] (see also Table 1.2). In most cases, HAP chromatography was introduced at the final stage of purification, after IEC or HIC, and before SEC.

1.3 Example of a Traditional Enzyme Purification Protocol

We already discussed the theoretical aspects and some specific applications of the traditional enzyme purification methods. In this paragraph, we describe the purification of 3-hydroxyphenylacetate 6-hydroxylase from *Flavobacterium* JS-7 as an example of a traditional enzyme purification protocol [23].

(1) 50 g of *Flavobacterium* JS-7 cells, grown with 0.1% phenylacetic acid as a sole source of carbon and energy, was suspended in 10 mM potassium phosphate at pH 7.0, containing 0.5 mM ethylenediaminetetraacetic acid (EDTA) (for complexation of heavy metal ions), 0.5 mM dithiothreitol (DTT) (to prevent cysteine oxidation), 1.5 mM 3-hydroxyphenylacetic acid (binding of this aromatic substrate in the active site might protect the enzyme from inactivation), and 10 μM FAD (to prevent apoenzyme formation). Next, 1 mg of DNase was added and cells were broken in a precooled French Press. The resulting cell extract was clarified by centrifugation at 4 °C. All further operations were performed between 0 and 4 °C using the abovementioned starting buffer omitting the aromatic substrate. After each chromatographic step, the enzyme solution was concentrated by ultrafiltration and desalted through either dialysis or gel filtration before performing the next step.
(2) Cell extract, containing about 3 g of protein, was loaded onto a DEAE-Sepharose CL-6B column (17 × 2.4 cm; bed volume 77 ml). After washing with five volumes of starting buffer, the enzyme was eluted with a 0–0.5 M KCl gradient in 1000 ml of starting buffer. Pooled active fractions eluting between 0.10 and 0.15 M KCl were loaded onto a Phenyl Sepharose CL-4B column (2.4 × 9 cm; bed volume 41 ml), equilibrated with 15% ammonium sulfate. After washing with five volumes of equilibration buffer, the enzyme was eluted with a decreasing gradient of 15 to 0% ammonium sulfate in 500 ml starting buffer. Pooled active fractions eluting between 8.4 and 0% ammonium sulfate were

Table 1.3 Purification of 3-hydroxyphenylacetate 6-hydroxylase from *Flavobacterium* JS-7.

Step	Activity (U)	Protein (mg)	Spec. activity ($U\,mg^{-1}$)	Yield (%)
CE	1050	3080	0.3	100
IEC	799	250	3.2	76
HIC	712	60	11.9	68
BAC	510	28	18.2	49

Source: Adapted from Van Berkel and van den Tweel [23].
Steps: CE, cell extract; IEC, ion exchange chromatography; HIC, hydrophobic interaction chromatography; BAC, bio-affinity chromatography.

loaded onto a Reactive Red 120 agarose column (2.4×9 cm; bed volume 41 ml). After washing with five volumes of starting buffer, the enzyme was eluted with a linear gradient of 0–1 M KCl in 500 ml starting buffer. Pooled active fractions eluting between 0.16 and 0.46 M KCl were concentrated in 50 mM starting buffer, and 1 ml aliquots with a protein concentration of $2\,mg\,ml^{-1}$ was distributed over Eppendorf tubes, flash frozen in liquid nitrogen, and stored at −70 °C. The results of this purification procedure are summarized in Table 1.3.

1.4 Purification of Recombinant Enzymes

Developments in recombinant DNA technology and genome sequencing at the turn of the millennium have led to new procedures for the discovery and production of biocatalysts (see also Chapter 2). Nowadays, most biocatalysts are produced in heterologous hosts through the expression of synthetic genes. Many strategies for the expression of these recombinant proteins exist [67–69]. Many new biocatalysts are produced as fusion proteins, either to allow their rapid and efficient purification, enhance their proper folding and solubility, or facilitate cofactor regeneration [70].

One of the most widely applied methods for the purification of recombinant proteins concerns immobilized metal affinity chromatography (IMAC). This versatile method is described in the following section.

1.4.1 Immobilized Metal Affinity Chromatography

IMAC is a specialized adsorption chromatography technique that has turned into a powerful tool for single-step purification of recombinant proteins into which a metal binding site ("His-tag") has been manufactured by genetic modification [71]. In IMAC, transition metal ions such as cobalt, nickel, or zinc act as electron acceptors (Lewis acids) for groups with electron-donating atoms (Lewis bases). The imidazole nitrogen atoms of His-tags introduced at the N- or C-terminal ends of proteins appear to be perfect electron donors for these metal ions.

In order to utilize this interaction for chromatographic purposes, the metal ion must be immobilized onto an insoluble support. The most widely applied chelating group for that is nitrile-triacetic acid (NTA), which has four bases for binding a nickel ion.

Binding of proteins to IMAC columns is usually best between pH 6.0 and pH 8.0 where the imidazole groups of the histidine residues are deprotonated. Chelating agents such as EDTA or citrate, but also Tris buffer, can reduce the binding strength. In order to prevent a metal affinity column to also function as an ion exchanger, buffers for loading, washing, and eluting proteins usually contain a relatively high ionic strength.

His-tags can be inserted at the N- or C-terminal ends of proteins or within exposed loops. The tag is added by modification of the gene encoding the protein to include codons for 6 to 10 consecutive histidine residues. These codons can be added by site-directed mutagenesis or by cloning the gene directly, in the correct reading frame, into a vector that contains these codons, resulting in a fusion protein. It is of major importance that the inserted histidine residues do not prohibit proper folding of the protein nor interfere with its enzyme activity. Site-directed mutagenesis also facilitates removal of His-tags after purification of the protein. This is commonly done by insertion of a recognition site for a highly specific endopeptidase between His-tag and native protein.

Ni-NTA metal affinity columns have a high binding capacity (5–10 mg protein per ml packed gel). His-tagged proteins bound to Ni-NTA are eluted from the column by inclusion of a high-concentration (250–500 mM) imidazole in the elution buffer. Before eluting His-tagged proteins, it is worth washing the column with a low concentration of imidazole (10–50 mM) in order to remove weakly interacting proteins.

We used the His-tag technology for the purification of several oxidoreductases including galactonolactone dehydrogenase from *Arabidopsis thaliana* [72], styrene monooxygenase from *R. opacus* 1CP [73, 74], 3-hydroxybenzoate 6-hydroxylase from *Rhodococcus jostii* RHA1 and *Pseudomonas alcaligenes* [26, 75], ene reductase from *R. opacus* CP1 [76], pyranose 2-oxidase from *Arthrobacter siccitolerans* [77], styrene monooxygenase reductase from *R. opacus* 1CP [78], vanillyl alcohol oxidase from *P. simplicissimum* [79], eugenol oxidase from *Rhodococcus jostii* RHA1 [80], 5-(hydroxymethyl)furfural oxidase from *Methylovorus* sp. strain MP688 [81, 82], and 4-hydroxybenzoate hydroxylase from *Cupriavidus necator* [83].

The hyperthermostable laminarinase from *Pyrococcus furiosus* (LamA) was also purified using the His-tag technique. The cell-fee extract, obtained after breaking overexpressing *E. coli* cells and subsequent centrifugation, was subjected to Ni-NTA affinity chromatography, in which the LamA was eluted with a linear gradient of 0–0.5 M imidazole [42]. Next to highly active monomers, a minor part of the soluble enzyme constituted less active native-like oligomers and non-native monomers.

It is important to mention here that almost 50% of the overproduced LamA protein ended up in the pellet fraction that was obtained after initial centrifugation of the lysed cells. Notably, a large fraction of the soluble and insoluble LamA aggregates could be recovered in fully active form by incubating them for several hours at 80 °C in the presence of 3 M guanidine hydrochloride (GuHCl). Similar chaotropic heat treatment protocols might be useful for increasing the yield of other hyperthermostable enzymes that tend to aggregate during production in *E. coli*.

Obviously, this method is not applicable to less thermotolerant enzymes that initially are produced in inclusion bodies. For such His-tagged proteins, a soluble enzyme exhibiting the desired activity might be obtained by solubilizing them with a proper unfolding agent and subsequent purifying and refolding using a Ni-NTA affinity matrix [84].

Care must be taken regarding the correct unfolding and refolding conditions. For instance, for aryl-alcohol oxidase from *Pleurotus eryngii* overproduced in *E. coli* (without His-tag), the washed inclusion bodies were first solubilized in 20 mM Tris–HCl buffer at pH 8.0, containing 2 mM EDTA, 30 mM DTT, and 8 M urea. Next, optimization of the refolding conditions yielded a decently active enzyme after incubating the recombinant protein at 0.4 mg ml^{-1} for 80 hours at 16 °C and pH 9.0 in the presence of 35% glycerol, 80 μM FAD, 0.6 M urea, 1 mM DTT, 2.5 mM oxidized glutathione, and 1.25 mM reduced glutathione [85].

Recently, Fraaije and coworkers used the IMAC technology for the purification of a self-sufficient cytochrome P450 monooxygenase from *Thermothelomyces thermophila* [86]. They also developed experimental protocols for screening of thermostable Baeyer–Villiger monooxygenases by purifying the His-tagged proteins with Ni-NTA beads in a 96-well plate format [87]. Robust cyclohexanone monooxygenases, stabilized by applying mutations found by computational and experimental library design, were characterized after purification by metal affinity chromatography [88].

A special application of Ni-NTA affinity chromatography concerns the deflavinylation and reconstitution of flavoproteins [89]. This generic method, originally developed for the NifL PAS domain from *A. vinelandii* [89], also turned out to be very useful for the reversible resolution of flavin and pterin cofactors of His-tagged *E. coli* DNA photolyase [90].

1.4.2 Affinity Chromatography with Protein Tags

Next to IMAC, affinity chromatography procedures are available for enzyme purification that make use of a specific interaction of protein tags with accompanying resins. Well-known examples of such fusion tags are thioredoxin (Trx), glutathione transferase (GST), maltose binding protein (MBP), and small ubiquitin modifier (SUMO) [70]. These so-called solubility tags fold rapidly into a stable and highly soluble protein upon translation and fusion to the N-terminus of the enzyme of interest might stimulate the expression and solubility of the target enzyme. Again, removal of the tag from the purified protein can be facilitated by insertion of a recognition site for a highly specific endopeptidase between the tag and the native protein.

Dijkman and Fraaije used the SUMO technology for the production of 5-(hydroxymethyl)furfural oxidase (HFMO) from *Methylovorus* sp. strain MP688 [81]. To that end, they designed a His_6-SUMO-HFMO gene construct, which, after ligation in a pET-SUMO vector, could be expressed in *E. coli* BL21(DE3) cells. After production, the His_6-SUMO-HMFO fusion was cleaved with His_6-SUMO protease. Subsequently, a Ni-NTA column was used to capture both the His_6-SUMO tag and His_6-SUMO protease, yielding purified HMFO in the flow through. A similar strategy was applied for the production of *Thermocrispum municipale* cyclohexanone monooxygenase (*Tm*CHMO) variants, with the aim of selecting for variants of *Tm*CHMO with changed stereo- and regioselectivity [91, 92] and substrate preference [93].

The gene coding for polycyclic ketone monooxygenase (PockeMO) from *Thermothelomyces thermophila* was fused into a His_6-SUMO-containing pET vector (yielding pET-His_6-SUMO-PockeMO) and also into a His_6 cofactor-recycling phosphite dehydrogenase (PTDH) containing the pBAD vector (yielding pBAD-His_6-PTDH-linker-PockeMO) by Fraaije and coworkers [94]. For the expression of the SUMO fusion, *E. coli* BL21(DE3) was used and for expression of the PTDH fusion, *E. coli* NEB10β cells were used. Both constructs yielded high expression levels and the fusion proteins were easily purified with Ni-NTA affinity chromatography by exploiting the N-terminal His_6-tags. PockeMO is active with bulky ketones and can perform enantioselective oxidations on steroids. The PTDH-PockeMO protein showed an increased activity compared to the native enzyme. After removal of the SUMO tag, applying the method described above for *Tm*CHMO, the protein could be crystallized and its structure solved. PockeMO exhibits the typical Bayer–Villiger monooxygenase organization with an FAD domain, an NADP domain, and a helical domain [94].

Fraaije and coworkers also showed that the flavoprotein alditol oxidase (AldO) from *Streptomyces coelicolor* is expressed at extremely high levels in *E. coli* when the enzyme is fused to MBP [95]. The extreme overproduction of the fused protein allowed for a single-step purification procedure using a Q-Sepharose ion exchanger. To find out if fused MBP-AldO protein behaves similarly to native AldO, the MBP tag was cleaved off with trypsin. Using amylose affinity chromatography, the MBP tag was removed. Electrospray ionization mass spectrometry analysis of free AldO established that cleavage had occurred at the expected site and that a homogeneous AldO preparation was obtained. AldO turned out to be active with the same range of substrates as found for fused MBP-AldO protein. AldO accepts various polyols, with xylitol and sorbitol being the preferred substrates. Intriguingly, AldO also oxidizes thiols such as DTT to the corresponding thiocarbonyls [82].

Incorporation of an N-terminal MBP solubility tag strongly increased the yield of recombinant form of the membrane-associated proline dehydrogenase from *T. thermophilus* (*Tt*ProDH) [37]. Expressed MBP-*Tt*ProDH comprised about 50% of the total protein content of expressing E. coli cells, and by applying amylose affinity chromatography, more than 250 mg of active enzyme was obtained per liter of culture. Native *Tt*ProDH was obtained by cleaving the fusion protein with trypsin in the presence of the detergent *n*-octyl β-D-glucopyranoside. Analytical gel filtration suggested that both native and MBP-fused *Tt*ProDH form tetramers that are prone to aggregation through non-native self-association. Site-directed mutagenesis (F10E/L12E variant) then showed that the hydrophobic N-terminal helix of ProDH is responsible for the self-association process [96]. Truncation of the N-terminal arm of *Tt*ProDH (ΔA and ΔAB variants) resulted in highly active tetramers [50], while selective disruption of two ion pairs in the dimerization interface of the enzyme (D205K/E207K variants) resulted in monomer formation (Figure 1.8) [51]. The newly created *Tt*ProDH monomer showed excellent catalytic properties but a significant lower thermal stability than the tetramer. Finally, using a riboflavin auxotrophic *E. coli* strain and MBP as a solubility tag, we also succeeded in producing the apoprotein of *Tt*ProDH. Reconstitution experiments together with structural studies and flavin content analysis led to the surprising conclusion that *Tt*ProDH does not discriminate between FAD and FMN as a cofactor [97].

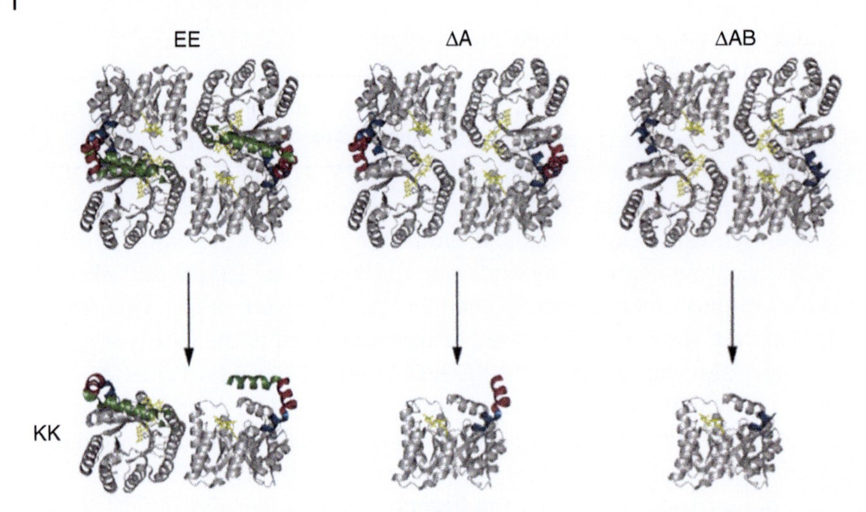

Figure 1.8 Oligomeric state of *Tt*ProDH and its site-directed mutants. EE: F10E/L12E variant. ΔA: *Tt*ProDH without the N-terminal helix αA. ΔAB: *Tt*ProDH without the N-terminal helices αA and αB. KK: D205K/E207K variants of EE, ΔA and ΔAB. The flavin cofactor is shown in yellow, helix αA in green, helix αB in red, and helix αC in blue. Source: Modified from Huijbers et al. [51].

1.5 Column Materials

A small selection of commercially available column materials is shown in Table 1.4. These bulk resins can be used to construct columns of various sizes in accordance with the type and volume of samples applied. In addition, a large number of prepacked columns for direct use are also available from these companies.

Table 1.4 Selection of various commercially available chromatography media.

Ion exchange	Matrix	Functional group	Capacity (mg ml^{-1})	Brand
Capto Q	Highly cross-linked agarose with a dextran surface extender	Quaternary amine	High	1
Capto S	Highly cross-linked agarose with a dextran surface extender	Sulfonate group	High	1
Source 15Q	Rigid polystyrene – a divinyl benzene polymer	Quaternary amine	High	1
Source 15S	Rigid polystyrene – a divinyl benzene polymer	Sulfonate group	High	1
Macro-Prep High Q	Methacrylate copolymer	Quaternary amine	37 BSA	2

Table 1.4 (Continued)

Ion exchange	Matrix	Functional group	Capacity (mg ml^{-1})	Brand
Macro-Prep High S	Methacrylate copolymer	Sulfonate group	49 IgG	2
Macro-Prep DEAE	Methacrylate copolymer	Diethylamino-ethyl	40 BSA	2
Macro-Prep CM	Methacrylate copolymer	Carboxy methyl	40 BSA	2
Gel filtration	**Matrix**	**Separation range (kDa)**		**Brand**
Sephacryl S-100 HR	Copolymer of allyl dextran and *N,N'*-methylene bisacrylamide	1–100		1
Sephacryl S-200 HR	Copolymer of allyl dextran and *N,N'*-methylene bisacrylamide	5–250		1
Sephacryl S-500 HR	Copolymer of allyl dextran and *N,N'*-methylene bisacrylamide	40–20 000		1
Superdex 75 Prep Grade	Composite of cross-linked agarose and dextran	3–70		1
Superdex 200 Prep Grade	Composite of cross-linked agarose and dextran	10–600		1
Bio-Gel P–30	Copolymerization of acrylamide and *N,N'*-methylenebisacrylamide gel	2.4–40		2
Bio-Gel P–100	Copolymerization of acrylamide and *N,N'*-methylenebisacrylamide gel	5–100		2
Bio-Gel A 1.5 m	Agarose	10–1500		2
(Bio)-affinity	**Matrix**	**Functional group**	**Capacity (mg ml^{-1})**	**Brand**
HiPrep Heparin FF	Cross-linked 6% agarose	Heparin	High	1
Ni Sepharose HP	Highly cross-linked agarose	Ni-charged	40	1
Glutathione Sepharose 4 FF	Highly cross-linked 4% agarose	Glutathione	10	1
Strep-Tactin® XT Sepharose	Rigid cross-linked agarose	Strep-Tactin	10	1
Blue Sepharose 6 Fast Flow	Cross-linked 6% agarose	Cibacron blue 3G	High	1
Nuvia™ IMAC	Inert hydrophilic beads	Nitrilotriacetic acid	>40	2

(Continued)

Table 1.4 (Continued)

Ion exchange	Matrix	Functional group	Capacity (mg ml^{-1})	Brand
Affi-Gel® Blue Gel	Cross-linked agarose	Cibacron blue F3GA	>11	2
Hydrophobic interaction	**Matrix**	**Functional group**	**Capacity (mg ml^{-1})**	**Brand**
Phenyl Sepharose 6 FF	Cross-linked 6% agarose	Phenyl	High	1
Butyl Sepharose 4 FF	Cross-linked 4% agarose	Butyl	High	1
Octyl Sepharose 4 FF	Cross-linked 4% agarose	Octyl	High	1
Macro-Prep® Methyl HIC	Methacrylate	Methyl	12	2
Macro-Prep *t*-butyl HIC	Methacrylate	*t*-Butyl	12	2
Hydroxyapatite interaction	**Matrix**	**Functional group**	**Capacity (mg g^{-1})**	**Brand**
Bio-Gel® HT Hydroxyapatite	Calcium phosphate	Calcium phosphate	10 BSA	2

1. GE Healthcare Life Sciences/Cytiva, VWR International B.V., Amsterdam, The Netherlands; 2. Bio-Rad Laboratories B.V., Veenendaal, The Netherlands.

1.6 Conclusions

Traditional enzyme purification procedures, whereby different chromatographic steps are required in order to obtain a pure protein, have played an eminent role in the development of the biocatalysis field. Nowadays, these methods have been replaced for the most part by affinity chromatographic techniques in which recombinant proteins fused with a specific tag can be effectively purified in a single step. Regarding the latter, it should be kept in mind that in many cases, an additional polishing step is required to remove critical impurities.

In this review, we have summarized our longstanding experience with the purification and characterization of redox enzymes. Purification procedures for these proteins do not significantly differ from the purification methods used for, e.g., hydrolases or transferases. However, with redox enzymes, and, in general, with proteins containing cofactors, potential cofactor dissociation during purification steps should be taken into account. Thus, addition of excess cofactor during purification can be essential for keeping the enzyme stable and active. We show that conventional chromatographic purification techniques can also be used for the large-scale preparation of apoproteins and for gaining insights into the molecular and hydrodynamic properties of the enzymes. Combining the specific opportunities of the different chromatographic techniques with the still growing DNA-recombinant toolbox will make it possible in the future to identify and purify new exciting biocatalysts.

References

1 Kohls, H., Steffen-Munsberg, F., and Höhne, M. (2014). Recent achievements in developing the biocatalytic toolbox for chiral amine synthesis. *Curr. Opin. Chem. Biol.* 19: 180–192.

2 Zmijewski, M. (2000). *Stereoselective biocatalysis*. New York: American Chemical Society.

3 Faber, K. (2018). *Biotransformations in organic chemistry*, 1–434. Heidelberg: Springer-Verlag.

4 Sheldon, R.A. and Brady, D. (2018). The limits to biocatalysis: pushing the envelope. *Chem. Commun.* 54: 6088–6104.

5 Woodley, J.M. (2019). Reaction engineering for the industrial implementation of biocatalysis. *Top. Catal.* 62: 1202–1207.

6 de Souza, P.M. and de Oliveira Magalhães, P. (2010). Application of microbial α-amylase in industry: a review. *Braz. J. Microbiol.* 41 (4): 850–861.

7 Bornscheuer, U.T. and Kazlauskas, R.J. (2006). *Hydrolases in organic synthesis: regio- and stereoselective biotransformations*, 1–355. Weinheim: Wiley-VCH Verlag GmbH and Co. KGaA.

8 Filho, D.G., Silva, A.G., and Guidini, C.Z. (2019). Lipases: sources, immobilization methods, and industrial applications. *Appl. Microbiol. Biotechnol.* 103 (18): 7399–7423.

9 Bornscheuer, U.T., Huisman, G.W., Kazlauskas, R.J. et al. (2012). Engineering the third wave of biocatalysis. *Nature* 485 (7397): 185–194.

10 Burgess, R.R. and Deutscher, M.P. (eds.) (2009). *Guide to Protein Purification*. Methods in Enzymology 463 San Diego: Academic Press.

11 Rhodes, D.G. and Laue, T.M. (2009). Determination of protein purity. In: *Methods in Enzymology* (eds. R.R. Burgess and M.P. Deutscher), 463: 677–689. San Diego: Academic Press.

12 Burgess, R.R. (2009). Protein precipitation techniques. In: *Methods in Enzymology* (eds. R.R. Burgess and M.P. Deutscher), 463: 331–342. San Diego: Academic Press.

13 de Jong, E., van Berkel, W.J.H., van der Zwan, R.P. et al. (1992). Purification and characterization of vanillyl-alcohol oxidase from *Penicillium simplicissimum*. *Eur. J. Biochem.* 208 (3): 651–657.

14 Fraaije, M.W., Roubroeks, H.P., Hagen, W.R. et al. (1996). Purification and characterization of an intracellular catalase-peroxidase from *Penicillium simplicissimum*. *Eur. J. Biochem.* 235: 192–198.

15 Jadan, A.P., van Berkel, W.J.H., Golovleva, L.A. et al. (2001). Purification and properties of *p*-hydroxybenzoate hydroxylases from *Rhodococcus* strains. *Biochem. Mosc.* 66 (8): 898–903.

16 Moonen, M.J.H., Kamerbeek, N.M., Westphal, A.H. et al. (2008). Elucidation of the 4-hydroxyacetophenone catabolic pathway in *Pseudomonas fluorescens* ACB. *J. Bacteriol.* 190 (15): 5190–5198.

17 Eppink, M.H.M., Schreuder, H.A., and van Berkel, W.J.H. (1998). Interdomain binding of NADPH in *p*-hydroxybenzoate hydroxylase as suggested by kinetic, crystallographic and modeling studies of histidine 162 and arginine 269 variants. *J. Biol. Chem.* 273 (33): 21031–21039.

18 Jungbauer, A. and Hahn, R. (2009). Ion-exchange chromatography. In: *Methods in Enzymology* (eds. R.R. Burgess and M.P. Deutscher), 463: 349–371. San Diego: Academic Press.

19 Müller, F., Voordouw, G., van Berkel, W.J.H. et al. (1979). A study of *p*-hydroxybenzoate hydroxylase from *Pseudomonas fluorescens*. *Eur. J. Biochem.* 101 (1): 235–244.

20 Seibold, B., Matthes, M., Eppink, M.H.M. et al. (1996). 4-Hydroxybenzoate hydroxylase from *Pseudomonas* sp. CBS3. Purification, characterization, gene cloning, sequence analysis and assignment of structural features determining the coenzyme specificity. *Eur. J. Biochem.* 239 (2): 469–478.

21 van Berkel, W.J.H. and Müller, F. (1987). The elucidation of the microheterogeneity of highly purified *p*-hydroxybenzoate hydroxylase from *Pseudomonas fluorescens* by various biochemical techniques. *Eur. J. Biochem.* 167 (1): 35–46.

22 van Berkel, W.J.H., Westphal, A.H., Eschrich, K. et al. (1992). Substitution of Arg214 at the substrate binding site of *p*-hydroxybenzoate hydroxylase from *Pseudomonas fluorescens*. *Eur. J. Biochem.* 210 (2): 411–419.

23 van Berkel, W.J.H. and van den Tweel, W.J.J. (1991). Purification and characteriation of 3-hydroxyphenylacetate 6-hydroxylase: a novel FAD-dependent monooxygenase from a *Flavobacterium* species. *Eur. J. Biochem.* 201 (3): 585–592.

24 Eppink, M.H.M., Cammaart, E., van Wassenaar, D. et al. (2000). Purification and properties of hydroquinone hydroxylase, a FAD-dependent monooxygenase involved in the catabolism of 4-hydroxybenzoate in *Candida parapsilosis* CBS604. *Eur. J. Biochem.* 267 (23): 6832–6840.

25 Kirchner, U., Westphal, A.H., Müller, R. et al. (2003). Phenol hydroxylase from *Bacillus thermoglucosidasius* A7, a two-protein component monooxygenase with a dual role for FAD. *J. Biol. Chem.* 278 (48): 47545–47553.

26 Montersino, S. and van Berkel, W.J.H. (2012). Functional annotation and characterization of 3-hydroxybenzoate 6-hydroxylase from *Rhodococcus jostii* RHA1. *Biochim. Biophys. Acta* 1824 (3): 433–442.

27 Kamerbeek, N.M., Moonen, M.J.H., van der Ven, J.G.M. et al. (2001). 4-Hydroxyacetophenone monooxygenase from *Pseudomonas fluorescens* ACB. *Eur. J. Biochem.* 268 (9): 2547–2557.

28 Kuijpers, T.F.M., van Herk, T., Vincken, J.-P. et al. (2014). Potato and mushroom polyphenol oxidase activities are differently modulated by natural plant extracts. *J. Agric. Food Chem.* 62 (1): 214–221.

29 Frommhagen, M., Sforza, S., Westphal, A.H. et al. (2015). Discovery of the combined oxidative cleavage of plant xylan and cellulose by a new fungal polysaccharide monooxygenase. *Biotechnol. Biofuels* 8 (101): 12.

30 Ewing, T.A., Gygli, G., and van Berkel, W.J.H. (2016). A single loop is essential for the octamerization of vanillyl-alcohol oxidase. *FEBS J.* 283 (13): 2546–2559.

31 Tamayo-Ramos, J.A., van Berkel, W.J.H., and de Graaff, L.H. (2012). Biocatalytic potential of laccase-like multicopper oxidases from *Aspergillus niger*. *Microb. Cell Factories* 11: 165.

32 Ferraroni, M., Westphal, A.H., Borsari, M. et al. (2017). Structure and function of *Aspergillus niger* laccase McoG. *Biocatalysis* 3 (1): 10.

33 Moonen, M.J.H., Synowsky, S.A., van den Berg, W.A.M. et al. (2008). Hydroquinone dioxygenase from *Pseudomonas fluorescens* ACB: a novel member of the family of nonheme-iron(II)-dependent dioxygenases. *J. Bacteriol.* 190 (15): 5199–5209.

34 Dicko, M.H., Gruppen, H., Hilhorst, R. et al. (2006). Biochemical characterization of the major sorghum grain peroxidase. *FEBS J.* 273 (10): 2293–2307.

35 Weber, F.J., van Berkel, W.J.H., Hartmans, S. et al. (1992). Purification and properties of the NADH reductase component of alkene monooxygenase from *Mycobacterium* strain E3. *J. Bacteriol.* 174 (10): 3275–3281.

36 van Iersel, M.F., Eppink, M.H.M., van Berkel, W.J.H. et al. (1997). Purification and characterization of a novel NADP-dependent branched-chain alcohol dehydrogenase from *Saccharomyces cerevisiae*. *Appl. Environ. Microbiol.* 63 (10): 4079–4082.

37 van der Werf, M.J., van der Ven, C., Barbirato, F. et al. (1999). Stereoselective carveol dehydrogenase from *Rhodococcus erythropolis* DCL14: A novel nicotinoprotein belonging to the short chain dehydrogenase/reductase superfamily. *J. Biol. Chem.* 274 (37): 26296–26304.

38 Leferink, N.G.H., Heuts, D.P.H.M., Fraaije, M.W. et al. (2008). The growing VAO flavoprotein family. *Arch. Biochem. Biophys.* 474 (2): 292–301.

39 Kudryashova, E.V., Leferink, N.G.H., Slot, I.G.M. et al. (2011). Galactonolactone oxidoreductase from *Trypanosoma cruzi* employs a FAD cofactor for the synthesis of vitamin C. *Biochim. Biophys. Acta* 1814 (5): 545–552.

40 Huijbers, M.M.E. and van Berkel, W.J.H. (2015). High yields of active *Thermus thermophilus* proline dehydrogenase are obtained using maltose-binding protein as a solubility tag. *Biotechnol. J.* 10 (3): 395–403.

41 Stellwagen, E. (2009). Gel filtration. In: *Methods in Enzymology* (eds. R.R. Burgess and M.P. Deutscher), 463: 373–385. San Diego: Academic Press.

42 Westphal, A.H., Geerke-Volmer, A.A., van Mierlo, C.P.M. et al. (2017). Chaotropic heat treatment resolves native-like aggregation of a heterologously produced hyperthermostable laminarinase. *Biotechnol. J.* 12 (6): 11.

43 Rhodes, D.G., Bossio, R.E., and Laue, T.M. (2009). Determination of size, molecular weight, and presence of subunits. In: *Methods in Enzymology* (eds. R.R. Burgess and M.P. Deutscher), 463: 691–723. San Diego: Academic Press.

44 Müller, F. and van Berkel, W.J.H. (1982). A study on *p*-hydroxybenzoate hydroxylase from *Pseudomonas fluorescens*. *Eur. J. Biochem.* 128 (1): 21–27.

45 van Berkel, W.J.H., Benen, J.A.E., and Snoek, M.C. (1991). On the FAD-induced dimerization of apo-lipoamide dehydrogenase from *Azotobacter vinelandii* and *Pseudomonas fluorescens*. *Eur. J. Biochem.* 197 (3): 769–779.

46 Fraaije, M.W., van den Heuvel, R.H.H., van Berkel, W.J.H. et al. (1999). Covalent flavinylation is essential for efficient redox catalysis in vanillyl-alcohol oxidase. *J. Biol. Chem.* 274 (50): 35514–35520.

47 Tahallah, N., van den Heuvel, R.H.H., van den Berg, W.A.M. et al. (2002). Cofactor-dependent assembly of the flavoenzyme vanillyl-alcohol oxidase. *J. Biol. Chem.* 277 (39): 36425–36432.

48 Fraaije, M.W., Mattevi, A., and van Berkel, W.J.H. (1997). Mercuration of vanillyl-alcohol oxidase from *Penicillium simplicissimum* generates inactive dimers. *FEBS Lett.* 402: 33–35.

49 van Berkel, W.J.H., van den Heuvel, R.H.H., Versluis, C. et al. (2000). Detection of intact megaDalton protein assemblies of vanillyl-alcohol oxidase by mass spectrometry. *Protein Sci.* 9 (3): 435–439.

50 Huijbers, M.M.E., van Alen, I., Wu, J.W. et al. (2018). Functional impact of the N-terminal arm of proline dehydrogenase from *Thermus thermophilus*. *Molecules* 23 (1): 184.

51 Huijbers, M.M.E., Wu, J.W., Westphal, A.H. et al. (2019). Dimerization of proline dehydrogenase from *Thermus thermophilus* is crucial for its thermostability. *Biotechnol. J.* 14 (5): 1800540.

52 Urh, M., Simpson, D., and Zhao, K. (2009). Affinity chromatography: general methods. In: *Methods in Enzymology* (eds. R.R. Burgess and M.P. Deutscher), 463: 417–438. San Diego: Academic Press.

53 Vos, R.M.E., Snoek, M.C., van Berkel, W.J.H. et al. (1988). Differential induction of rat hepatic glutathione S-transferase isoenzymes by hexachlorobenzene and benzyl isothiocyanate: comparison with induction by phenobarbital and 3-methylcholanthrene. *Biochem. Pharmacol.* 37 (6): 1077–1082.

54 Williams, R.E. and Bruce, N.C. (2002). New uses for an old enzyme: the old yellow enzyme family of flavoenzymes. *Microbiology* 148 (6): 1607–1614.

55 Scholtissek, A., Tischler, D., Westphal, A.H. et al. (2017). Old yellow enzyme-catalysed asymmetric hydrogenation: linking family roots with improved catalysis. *Catalysts* 7 (5): 130.

56 Abramovitz, A.S. and Massey, V. (1976). Purification of intact old yellow enzyme using an affinity matrix for the sole chromatographic step. *J. Biol. Chem.* 251 (17): 5321–5326.

57 Eweg, J.K., Müller, F., and van Berkel, W.J.H. (1982). On the enigma of old yellow enzyme's spectral properties. *Eur. J. Biochem.* 129 (2): 303–316.

58 Stott, K., Saito, K., Thiele, D.J. et al. (1993). Old yellow enzyme. The discovery of multiple isozymes and a family of related proteins. *J. Biol. Chem.* 268 (9): 6097–6106.

59 McCue, J.T. (2009). Theory and use of hydrophobic interaction chromatography in protein purification applications. In: *Methods in Enzymology* (eds. R.R. Burgess and M.P. Deutscher), 463: 405–414. San Diego: Academic Press.

60 Benen, J.A.E., Sánchez-Torres, P., Wagemaker, M.J.M. et al. (1998). Molecular cloning, sequencing, and heterologous expression of the *vaoA* gene from *Penicillium simplicissimum* CBS 170.90 encoding vanillyl-alcohol oxidase. *J. Biol. Chem.* 273 (14): 7865–7872.

61 Eppink, M.H.M., Boeren, S.A., Vervoort, J.M. et al. (1997). Purification and properties of 4-hydroxybenzoate 1-hydroxylase (decarboxylating), a novel flavin adenine dinucleotide-dependent monooxygenase from *Candida parapsilosis* CBS604. *J. Bacteriol.* 179 (21): 6680–6687.

62 Westphal, A.H. and de Kok, A. (1988). Lipoamide dehydrogenase from *Azotobacter vinelandii*. Molecular cloning, organization and sequence analysis of the gene. *Eur. J. Biochem.* 172 (2): 299–305.

63 Benen, J.A.E., van Berkel, W.J.H., van Dongen, W.M.A.M. et al. (1989). Molecular cloning and sequence determination of the *lpd* gene encoding lipoamide dehydrogenase from *Pseudomonas fluorescens*. *Microbiology* 135 (7): 1787–1797.

64 van Berkel, W.J.H., van den Berg, W.A.M., and Müller, F. (1988). Large-scale preparation and reconstitution of apo-flavoproteins with special reference to butyryl-CoA dehydrogenase from *Megasphaera elsdenii*. *Eur. J. Biochem.* 178 (1): 197–207.

65 Hefti, M.H., Vervoort, J.M., and van Berkel, W.J.H. (2003). Deflavination and reconstitution of flavoproteins: Tackling fold and function. *Eur. J. Biochem.* 270 (21): 4227–4242.

66 Cummings, L.J., Snyder, M.A., and Brisack, K. (2009). Protein chromatography on hydroxyapatite columns. In: *Methods in Enzymology* (eds. R.R. Burgess and M.P. Deutscher), 463: 387–404. San Diego: Academic Press.

67 Brondyk, W.H. (2009). Selecting an appropriate method for expressing a recombinant protein. In: *Methods in Enzymology* (eds. R.R. Burgess and M.P. Deutscher), 463: 131–147. San Diego: Academic Press.

68 Zerbs, S., Frank, A.M., and Collart, F.R. (2009). Bacterial systems for production of heterologous proteins. In: *Methods in Enzymology* (eds. R.R. Burgess and M.P. Deutscher), 463: 149–168. San Diego: Academic Press.

69 Cregg, J.M., Tolstorukov, I., Kusari, A. et al. (2009). Expression in the yeast *Pichia pastoris*. In: *Methods in Enzymology* (eds. R.R. Burgess and M.P. Deutscher), 463: 169–189. San Diego: Academic Press.

70 Malhotra, A. (2009). Tagging for protein expression. In: *Methods in Enzymology* (eds. R.R. Burgess and M.P. Deutscher), 463: 239–258. San Diego: Academic Press.

71 Block, H., Maertens, B., Spriestersbach, A. et al. (2009). Immobilized-metal affinity chromatography (IMAC): a review. In: *Methods in Enzymology* (eds. R.R. Burgess and M.P. Deutscher), 463: 439–473. San Diego: Academic Press.

72 Leferink, N.G.H., van den Berg, W.A.M., and van Berkel, W.J.H. (2008). L-Galactono-γ-lactone dehydrogenase from *Arabidopsis thaliana*, a flavoprotein involved in vitamin C biosynthesis. *FEBS J.* 275 (4): 713–726.

73 Tischler, D., Eulberg, D., Lakner, S. et al. (2009). Identification of a novel self-sufficient styrene monooxygenase from *Rhodococcus opacus* 1CP. *J. Bacteriol.* 191 (15): 4996.

74 Riedel, A., Heine, T., Westphal, A.H. et al. (2015). Catalytic and hydrodynamic properties of styrene monooxygenases from *Rhodococcus opacus* 1CP are modulated by cofactor binding. *AMB Express* 5 (1): 112.

75 Montersino, S., te Poele, E., Orru, R. et al. (2017). 3-Hydroxybenzoate 6-hydroxylase from *Rhodococcus jostii* RHA1 contains a phosphatidylinositol cofactor. *Front. Microbiol.* 8 (1110): 11.

76 Riedel, A., Mehnert, M., Paul, C.E. et al. (2015). Functional characterization and stability improvement of a 'thermophilic-like' ene-reductase from *Rhodococcus opacus* 1CP. *Front. Microbiol.* 6: 1073.

77 Mendes, S., Banha, C., Madeira, J. et al. (2016). Characterization of a bacterial pyranose 2-oxidase from *Arthrobacter siccitolerans*. *J. Mol. Catal. B* 133: S34–S43.

78 Heine, T., Scholtissek, A., Westphal, A.H. et al. (2017). N-terminus determines activity and specificity of styrene monooxygenase reductases. *Biochim. Biophys. Acta* 1865 (12): 1770–1780.

79 Ewing, T.A., van Noord, A., Paul, C.E. et al. (2018). A xylenol orange-based screening assay for the substrate specificity of flavin-dependent *para*-phenol oxidases. *Molecules* 23 (164): 18.

80 Nguyen, Q.T., de Gonzalo, G., Binda, C. et al. (2016). Biocatalytic properties and structural analysis of eugenol oxidase from *Rhodococcus jostii* RHA1: a versatile oxidative biocatalyst. *ChemBioChem* 17 (14): 1359–1366.

81 Dijkman, W.P. and Fraaije, M.W. (2014). Discovery and characterization of a 5-hydroxymethylfurfural oxidase from *Methylovorus* sp. strain MP688. *Appl. Environ. Microbiol.* 80 (3): 1082.

82 Ewing, T.A., Dijkman, W.P., Vervoort, J.M. et al. (2014). The oxidation of thiols by flavoprotein oxidases: a biocatalytic route to reactive thiocarbonyls. *Angew. Chem. Int. Ed.* 53 (48): 13206–13209.

83 Westphal, A.H., Tischler, D., Heinke, F. et al. (2018). Pyridine nucleotide coenzyme specificity of *p*-hydroxybenzoate hydroxylase and related flavoprotein monooxygenases. *Front. Microbiol.* 9 (3050): 10.

84 Schlager, B., Straessle, A., and Hafen, E. (2012). Use of anionic denaturing detergents to purify insoluble proteins after overexpression. *BMC Biotechnol.* 12: 95.

85 Ruiz-Dueñas, F.J., Ferreira, P., Martínez, M.J. et al. (2006). *In vitro* activation, purification, and characterization of *Escherichia coli* expressed aryl-alcohol oxidase, a unique H_2O_2-producing enzyme. *Protein Expr. Purif.* 45 (1): 191–199.

86 Fürst, M.J.L.J., Kerschbaumer, B., Rinnofner, C. et al. (2019). Exploring the biocatalytic potential of a self-sufficient cytochrome P450 from *Thermothelomyces thermophila*. *Adv. Synth. Catal.* 361 (11): 2487–2496.

87 Fürst, M.J.L.J., Martin, C., Lončar, N. et al. (2018). Experimental protocols for generating focused mutant libraries and screening for thermostable proteins. In: *Methods in Enzymology* (ed. N. Scrutton), 608: 151–187. San Diego: Academic Press.

88 Fürst, M.J.L.J., Boonstra, M., Bandstra, S. et al. (2019). Stabilization of cyclohexanone monooxygenase by computational and experimental library design. *Biotechnol. Bioeng.* 116 (9): 2167–2177.

89 Hefti, M.H., Milder, F.J., Boeren, S.A. et al. (2003). A His-tag based immobilization method for the preparation and reconstitution of apoflavoproteins. *Biochim. Biophys. Acta* 1619 (2): 139–143.

90 Xu, L., Zhang, D., Mu, W. et al. (2006). Reversible resolution of flavin and pterin cofactors of His-tagged *Escherichia coli* DNA photolyase. *Biochim. Biophys. Acta* 1764 (9): 1454–1461.

91 Li, G., Fürst, M.J.L.J., Mansouri, H.R. et al. (2017). Manipulating the stereoselectivity of the thermostable Baeyer–Villiger monooxygenase *Tm*CHMO by directed evolution. *Org. Biomol. Chem.* 15 (46): 9824–9829.

92 Li, G., Garcia-Borràs, M., Fürst, M.J.L.J. et al. (2018). Overriding traditional electronic effects in biocatalytic Baeyer–Villiger reactions by directed evolution. *J. Am. Chem. Soc.* 140 (33): 10464–10472.

93 Fürst, M.J.L.J., Romero, E., Gómez Castellanos, J.R. et al. (2018). Side-chain pruning has limited impact on substrate preference in a promiscuous enzyme. *ACS Catal.* 8 (12): 11648–11656.

94 Fürst, M.J.L.J., Savino, S., Dudek, H.M. et al. (2017). Polycyclic ketone monooxygenase from the thermophilic fungus *Thermothelomyces thermophila*: a structurally distinct biocatalyst for bulky substrates. *J. Am. Chem. Soc.* 139 (2): 627–630.

95 Heuts, D.P.H.M., van Hellemond, E.W., Janssen, D.B. et al. (2007). Discovery, characterization, and kinetic analysis of an alditol oxidase from *Streptomyces coelicolor*. *J. Biol. Chem.* 282 (28): 20283–20291.

96 Huijbers, M.M.E. and van Berkel, W.J.H. (2016). A more polar N-terminal helix releases MBP-tagged *Thermus thermophilus* proline dehydrogenase from tetramer-polymer self-association. *J. Mol. Catal. B* 134: 340–346.

97 Huijbers, M.M.E., Martínez-Júlvez, M., Westphal, A.H. et al. (2017). Proline dehydrogenase from *Thermus thermophilus* does not discriminate between FAD and FMN as cofactor. *Sci. Rep.* 7: 43880.

2

Enzyme Modification

Antonino Biundo[1], Patricia Saénz-Méndez[2]*, and Tamas Görbe[3]*

[1] *University of Bari, Department of Biosciences, Biotechnology and Biopharmaceutics, via Orabona, 4, Bari, 70125, Italy, and REWOW srl, Via Ciasca 9, Bari, 70124, Italy*
[2] *KTH Royal Institute of Technology, School of Engineering Sciences in Chemistry, Biotechnology, and Health Science for Life Laboratory, Gamma building floor 5, Tomtebodavägen 23, Solna, 171 65, Sweden, and Karlstads Universitet, Faculty of Health, Science and Technology, Universitetsgatan 2, 65188 Karlstad, Sweden*
[3] *COO and Co-Founder of Menten AI, 2225 E Bayshore Road, Suite 200, Palo Alto, CA, 94303, USA*

2.1 Introduction

The chemical society is witnessing an increasing use of enzymes as biocatalysts for sustainable production of pharmaceuticals, chemical building blocks, biofuels, additives, and materials. The vast range of biocatalytic reactions and substrate scope allows to set up such processes to fulfill industrial goals (see also Chapter 15). This trend is a clear consequence of both environmental concerns (e.g. depletion of fossil fuels and transition metals, greenhouse effect, and marine environment contamination) and the tremendous advances in enzyme engineering or the ability to tailor native enzymes to display new and improved catalytic properties. Currently, it is possible to exploit the ability of enzymes to selectively catalyze a plethora of reactions under unusual mild conditions. However, the natural features of enzymes still maintain several limitations for the broad range of industrial applications: (i) limited scope of substrates, (ii) inadequate activity, (iii) incorrect stereo- and/or regioselectivity, and (iv) low stability in the desired reaction environment [1, 2].

Clearly, to generate robust biocatalytic processes, the catalytic performance of enzymes must be improved to adapt them to desired regio- and enantioselectivity, activity, substrate scope, stability, etc.

To this end, enzyme engineering strategies can be employed to deliver the desired biocatalyst and at least four possible pathways can be followed: (i) random, (ii) *de novo*, (iii) semi-rational, and (iv) rational design [3, 4].

The **random** approach, best known as directed evolution (DE), imitates Darwinian's survival of the fittest evolutionary theory by performing repeated cycles of randomized gene mutagenesis, expression of the variants, and high-throughput screening of the

---

*Both authors contribute- equally to this chapter.

Biocatalysis for Practitioners: Techniques, Reactions and Applications, First Edition.
Edited by Gonzalo de Gonzalo and Iván Lavandera.

mutants against the desired feature (e.g. thermostability and substrate scope). This cyclic process is repeated until the desired improvement is reached [5–7]. The clear advantage of DE is that it does not require any previous knowledge about the structure of the enzyme or how the reaction takes place, but the preparation of a good-quality library and the screening process are the most challenging steps [8, 9]. The term DE was first suggested by Hansche in 1972 and further improved by different groups around the world [10]. It has become a powerful tool for both enzyme engineering for industrial purposes and for understanding the relationship between protein sequence, structure, and function [11].

Totally contrast to DE is ***de novo*** enzyme design, based on the structure of a calculated transition state for the reaction in hands, which is computationally inserted into different protein backbones until finding a potentially active enzyme based on the natural occurring structural motifs [12–14].

The **rational** enzyme design requires at least the three-dimensional (3D) structure of the biocatalyst. In addition, if the structure contains a substrate bound to the active site, detailed information about binding interactions might help in the design. Molecular modeling calculations (i.e. active site analysis, molecular docking, molecular dynamics (MDs) simulations, free energy calculations, etc.) can assist in the identification of key hot spots to mutate to improve the desired property of the enzyme [15–18].

The combination of rational design and DE can have synergistic effects in producing small and "smart" libraries, which could have greater hit's rate and lower number of clones. This is referred to as **semi-rational design** or focused DE and includes the use of site saturation mutagenesis (SSM) or focused random mutagenesis over a certain region of the enzyme rather than the entire coding sequence. This approach requires a less labor-intensive and more time-efficient way for the creation of the "smart" library, which contains a higher proportion of beneficial mutations [19].

2.2 Practical Approach: Experimental Information, Analytical Methods, Tips and Tricks, and Examples

2.2.1 Directed Evolution

DE is a powerful tool, if any sequence that shows a detectable activity for a chemical transformation of interest is known or discovered [5]. The DE of enzymes can exploit the ability of certain microorganisms to produce chemicals and to perform novel reactions that can be used in industries. The power of DE relies on the introduction of random mutations that do not require structure knowledge and it is not based on the mutagenesis of specific residues in the biocatalyst, but the entire gene is taken into consideration. Proteins are readily evolvable, and through DE, certain properties can be improved, such as stability on harsher environments, activity on non-natural substrates, thermostability, and changes on enantioselectivity. Interestingly, the critical mutations can appear far from the active site and, in certain cases, on the enzyme surface, which still have dramatic effects during the catalysis [20, 21]. DE is based on the principle of natural evolution processes such as non-recombining (or random mutagenesis) and recombining methods (or genetic recombination) [22].

Random mutagenesis approaches are based on the introduction of mutations in a single gene sequence encoding the biocatalyst of interest. The mutagenesis of the specific gene sequence can be obtained chemically, *in vivo* or *in vitro*. Chemical mutagenesis uses chemical substances (e.g. nitrous acid) or irradiations (e.g. ultraviolet radiation), which can induce mismatching in the double-stranded DNA (dsDNA) [23]. *In vivo* approaches are performed through the aid of a mutator strain, such as *Escherichia coli* XL1-red [24], yeast orthogonal replication [25], or phage-assisted continuous evolution (PACE), which relies on the presence of genetic circuits that link the biocatalyst function with phage infectivity [26]. Polymerase chain reaction (PCR) is the main technique used to introduce non-matching nucleotides in the DNA sequence in *in vitro* approaches. Error-prone PCR (epPCR) is one of the most common methods used during the years to insert random mutations in the DNA sequence. Specifically, a polymerase enzyme lacking the proofreading activity can be applied to introduce errors during the PCR cycles. The introduction of random mutations can be induced by the presence of nonoptimal concentration of Mg^{2+} and Mn^{2+} or an unequal amount of nucleotides (dNTPs) [21, 27].

The *in vivo* approach of the **genetic recombination** is based on the reassortment of mutations to access beneficial combinations of mutations through homologous or nonhomologous recombination to form chimeras of different genes. Homologous recombination methods such as DNA (or gene) shuffling [28] and RAndom CHImeragenesis on Transient Templates (RACHITTs) [29] are based on the fragmentation of a gene and further PCR reaction using the DNA fragments as primers. In the case of a DNA shuffling of orthologs identified through a phylogenetic analysis, the DNA family shuffling [30] is carried out to use the naturally occurring genetic diversity of family genes. The staggered extension process (StEP) uses normal PCR primers but shorter elongation steps producing shorter extension products creating recombination of multiple DNA templates into one amplicon [31]. Assembly of Designed Oligonucleotides (ADOs) uses synthetic overlapping primers, which extend one another, and it has become a preferred strategy [32].

Non-homologous recombination is based on the random recombination of genes lacking the requirement of sequence similarity. Certain techniques, such as Sequence Homology-Independent Protein RECombination (SHIPREC) [33], Incremental Truncation of the Creation of Hybrid enzYmes (ITCHYs) [34], and a combination of ITCHY and DNA shuffling (SCRATCHY) [35] for multiple crossovers, are based on the truncation of the genes without sequence homology. The method Sequence-Independent Site-Directed Chimeragenesis (SISDC) [36] with sticky ends uses Type IIb restriction enzymes for the recombination. The computational algorithm, SCHEMA, is used in enzyme engineering to identify fragments of proteins that can be recombined in chimeras without disturbing the integrity of the protein 3D structure. The identification is based on the calculation of interaction between residues [37]. Certain guidelines are present for practitioners who enter the field of DE, such as RAndom MUtagenesis Strategy flowchart (RAMUS) and keep it simple and smart (KISS) [38]. Finally, acknowledging to the low cost of sequencing, quick quality control (QCC) at the level of the DNA sequence should be carried out when possible to identify the statistics of the DE experiment in order to remove biases.

2.2.1.1 (Ultra)High-Throughput Screening and Selection

Full sampling of a gene sequence is impossible, because the construction of a library of a polypeptide of 40 amino acid residues built with 20 proteinogenic amino acids and containing only a single molecule of each possible sequence would exceed several orders of magnitude the mass of the Earth [39]. Moreover, enzymes are way larger than only 40 residues; thus, clever approaches are needed to generate libraries that contain a maximum number of useful variants. Screening for biocatalyst libraries is the most time-consuming and labor-intensive step during enzyme engineering. The need for a fast and accurate screening is continuously required [40]. Techniques based on **fluorescence** and **absorbance** are the most common methods of choice. Fluorescence-activated cell sorting (FACS) allows the daily screening of up to 10^8 variants. It has been used together with cell surface display to evolve different enzymes [41, 42]. Alternatively, *in vitro* compartmentalization (IVC) methods and, especially fluorescence-activated droplet sorting (FADS) can be linked to microfluidic-based screening using water-in-oil droplets as the reaction environment [43, 44]. Co-compartmentalization with cells expressing the enzyme of interest and a fluorogenic substrate in the droplet can link the genotype and phenotype. Absorbance-based readouts can dramatically increase the application of this technology. The microcapillary single-cell analysis and laser extraction (μSCALE) can employ fluorescence imaging to assay the enzyme activity with spatially segregating single cells within a microcapillary array with further recovery of active clones through laser extraction [45]. Specifically, carboxyl esterases can be used in the "Quick *E*" assay to determine the enantioselectivity on *para*-nitrophenyl esters spectrophotometrically [46] or through the change of pH caused by the proton release [47, 48]. Surface-enhanced resonance Raman scattering (SERRS) is based on the absorption of released products on silver nanoparticles and on their selective and sensitive detection, thus enabling the quantification of enzymatic activity [49]. Selection techniques are higher throughput than screening methods but not always possible to implement because they are based on agar plate methodologies and rely on a direct correlation between cell survival and the desired enzyme property, without interfering with cellular metabolism [50, 51].

2.2.1.2 Applications of Directed Evolution Methodology

Industrial conditions are usually harsher than natural environments and biocatalysts cannot withstand them [39, 52–54]. Protein engineering for improved stability must be capitalized to enhance their activity on different environments. Temperature, solvent, and pH are the main factors that can destabilize a biocatalyst because of the non-natural and harsher industrial conditions, and a wide range of methods have been applied for increasing the biocatalyst's tolerance to these factors [55].

DE approaches have been used to improve the stability of biocatalysts in different surroundings. For instance, **thermostability** of enzymes was improved through different techniques of DE, both for hot and cold applications. Approaches using a single [56–59] or multiple [60, 61] techniques of DE have been reported. **Solvent** tolerance has been addressed through DE in order to stand different solvents and their high concentration, which is important in organic synthesis because of the low water solubility of most of the organic compounds [62–64]. In contrast to thermostability and solvent tolerance, the **pH-dependent inactivation** of enzymes has not been as widely studied. This mechanism may be regulated by the introduction of certain amino acid residues with ionizable side chains. However, it is still debated which regions (surface or core) are more important for the biocatalyst stability and DE can shed the light on this mechanism, especially for enzymes that could find applications in environments with a certain pH [65].

Enzyme engineering for desired specificities is a challenging task that plays an important role especially in the fields of chemoenzymatic and synthetic biotechnology. Enantioselectivity of biocatalysts plays a key factor for the production of different chemicals that are used in several fields [66, 67]. DE can improve the selectivity of enzymes in order to accept different substrates than the natural ones [68–70]. Most of the organic synthesis reactions use compounds that are not the natural substrates for the enzyme. For this reason, DE alone or in a synergistic approach together with an improved biocatalyst stability and selectivity can increase the enzyme activity on natural [71–74] and non-natural [75–77] substrates being able to bring to light novel reactions [78, 79] that are not found in biology.

2.2.2 Semi-rational Design

The semi-rational design approach, which is a combination of rational design and DE, has increased its value through the development of powerful computational algorithms. It can simplify the decision making and permits the screening of libraries *in silico* for the identification of beneficial mutations.

Similarly, to the DE method, a semi-rational design can be performed on a single gene or on a series of orthologs from the same family based on phylogenetic analysis. A restriction of DE is the complete random nature of the inserted mutations. When structural and/or functional information exists, the engineering could be directly focused where the mutations can be very beneficial and more effective, in order to produce smaller libraries, especially in the case of lacking high-throughput screening methods [19, 80]. The generation of a "smart" library would then enhance the value of lower-throughput screening methods [81]. Saturation mutagenesis on more residues can create a number of variants that, due to degeneracy of the genetic code, will result in unbalanced library. In order to reduce the number of variants, the use of variations of NNN codons (64 codons) are of need. For reducing the codon redundancy, a method called 22c-trick uses a mixture of two primers with degeneracy codons and one primer containing a TGG codon. This mixture does not contain any stop codon and only two redundant codons for Val and Leu [82]. The OSCARR methodology (One-pot Simple methodology for CAssette Randomization and Recombination) bridges the gap between site-directed mutagenesis and full randomization by making use of carefully designed mutagenic cassettes and an optimized one-pot megaprimer PCR. This method is especially suited to construct libraries of up to ten randomized codons for focused DE, which exhibits up to 97% efficiency in the amplification of mutated over wild-type products. This method is sufficiently versatile to allow mutagenesis and recombination of several cassettes within the same gene [83].

Structure-based enzyme redesign exploits 3D structure and biochemical data of biocatalysts and the identification of specific regions that are taken into consideration for the methods. SSM is the method of choice and especially the iterative saturation mutagenesis (ISM), which reduces drastically the molecular biology work and the screening efforts focusing on rationally chosen hot spot regions, which are consecutively mutated [84]. The combinatorial approaches, such as combinatorial saturation mutagenesis (CSM) and hydroclassified CSM (HCSM), which divides amino acid residues present in certain regions depending on their hydrophobicity and size, reduce significantly the size of the library because of the simultaneous mutagenesis of different residues. The combinatorial active site saturation test

(CASTing), which derives from ISM, identifies amino acid residues of specific regions based on the 3D structures found in databases, such as the commercial 3DM database [85], with the further randomization of these residues and their adjacent ones. This method enables the introduction of probabilistic elements with synergistic activity around the active site [86].

Sequence-based enzyme engineering uses the evolutionary information and prediction to identify hot spots in the sequence of the biocatalysts. Multiple sequence alignment (MSA) and phylogenetic analysis have been widely used for the discovery of conserved amino acid residues and ancestral relationships among groups of homologous protein sequences [87]. The HotSpot Wizard server [88] groups information from extensive structure and sequence database searches with functional data to create a map of the mutations for the protein of interest. Meanwhile, the 3DM database integrates information from GenBank and RSCB Protein Data Bank (PDB) with literature tracking to create alignments of protein superfamilies. Methods using the evolutionary history of a biocatalyst are, for instance, consensus design, which uses the alignment of all sequences after the calculation of their conservation and, thus, identifies the most frequent residues at each position [89]. Ancestral sequence reconstruction (ASR) uses phylogenetic hierarchy to go back in evolutionary time to build statistically predicted ancestors [90, 91]. Several databases are now present to collect samples. The most commonly used are Pfam, PROSITE, and SMART [92–94]. Other methods go beyond the amino acid analysis to focus on DNA sequences, such as the reconstructing evolutionary adaptive path (REAP) methods that exploit the sequence data from ancestral proteins to create focused and functional-enriched libraries [95].

In order to increase the power of semi-rational engineering based on protein design algorithms, **computational methods** were developed either to perform a virtual screening of a vast library or to redesign the active sites of enzymes. The ultrahigh throughput of these methodologies allows the *in silico* screening of enormous numbers of virtual candidates. For instance, the computational screening can analyze libraries of 10^{80} variants to eliminate mutations that could destabilize the protein fold [96]. The protein design automation (PDA) can predict the optimal sequence for a desired fold and the sequence with the lowest conformational energy is chosen for iteration through Monte Carlo simulated annealing [97].

Structure–activity relationship has also been studied through **machine learning** techniques such as the heuristic establishment of quantitative structure–activity relationship (QSAR) algorithm. This technique is based on a statistical modeling to establish a causal relationship between the structures of interacting molecules and measurable properties of scientific and commercial interest. Meanwhile, the PROtein Structure Activity Relationship (ProSAR) algorithm [98] screens a small number of samples from a combinatorial library and is able to build a statistical model that correlates the sequences with the measured activities and identify hot spots for the improved activity building the next generation of library by incorporating the most valuable residues [99, 100].

2.2.2.1 Applications of Semi-rational Design Methodology

Semi-rational design approaches have been performed to increase the enantioselectivity of biocatalysts and to enlarge the substrate acceptance of the enzyme in combination with DE and rational design approaches.

Structure-based redesign: The target of the active site residues to increase the substrate range of biocatalysts have been widely used to increase the promiscuity of an enzyme, which can rationally be further modified to change its substrate specificity [101–103] and to improve the enantioselectivity [104–106] of the biocatalyst. A combination of rational design and semi-rational engineering can be very successful for increasing the substrate scope of enzymes [107]. Moreover, the usage of both DE approaches and semi-rational design can introduce novel activity [108], change the substrate specificity [109–111], and improve the enantioselectivity of biocatalysts [112–114].

Sequence-based redesign: Other characteristics such as tolerance to temperature and pH or resistance to specific agents (e.g. proteases) can also be tackled through MSA or phylogenetic analysis to identify specific hot spots [115]. The enantioselectivity of enzymes can also be improved by alignment with further saturation mutagenesis studies [116]. Furthermore, the identification of ancestor through phylogenetic analysis enabled the increase of Tm of the spiroviolene synthase from *Streptomyces violens* (a terpene cyclase) with increased substrate promiscuity [117].

2.2.3 *De Novo* Enzyme Design

Proteins fold to their lowest free energy state [118]. This is the main principle followed by *de novo* enzyme design. The design of novel protein structures needs to produce amino acid sequences with lowest energy state in the prescribed structure, leading to two main problems: producing a protein fold that is designable and finding a sequence with its lowest energy state in the produced structure [119]. Interestingly, *de novo* protein design removes the dependence on naturally evolved scaffolds, leading to a deeper understanding of the contribution that every side chain of the residues make toward the structure, stability, and function of the biocatalyst [120]. Three main approaches are present in the state of the art for *de novo* protein design.

Minimalist approach is based on the design of binary patterns of polar (p) and hydrophobic (h) residues. The latter is usually built through α-helices because of their natural behavior to fold and assemble with "hpphppp" sequence pattern forming amphipathic helices encoded by single polypeptide chains or self-associating peptides. Indeed, the production of four-helix bundles has been thoroughly studied [121, 122]. However, the lack of high-resolution structures for many of these constructs emphasizes the need to consider the stereochemical arrangement of side chains to produce well-ordered protein cores and better-defined 3D structures. Moreover, this approach requires very little computational power and usually the structures are built hierarchically from an initial simplified model with the addition of optimized elements.

The second approach that has been used for *de novo* protein design is the **rational parametric design of functional assemblies** using selected protein folds, which are described mathematically with a minimal number of parameters. The latter has also been achieved with four-helix bundles [123]. Coiled-coil proteins have also been used for parameterization because of their relatively straightforward sequence and structures [124, 125].

The last approach is the **fragment-based computational design**. This method is based on the power of computers that can produce thousands of designs, which can be analyzed *in silico*. Shortly, libraries of fragments or motifs are selected from structural databases, and algorithms are developed to combine these fragments to assemble target structures. In order to assess the assembled structures and sequences that best fit onto them, scoring functions are used. The perfect example is the Rosetta suite for computational protein design [126].

The process leads to the formation of a specific type of model for biological catalysis, using the knowledge of modern quantum mechanical (QM) methods and programs. Specifically, it builds an assay of functional groups, the so-called theoretical enzyme or theozyme, in a geometry predicted by theory to provide transition-state stabilization [127]. Similarly, mimicking the natural combination of larger fragments, a related approach has proven successful for the generation of *de novo* proteins [128, 129]. Following the same approach, *de novo* biocatalysts have been designed incorporating cavities with potential for catalysis using RosettaRemodel, which was developed for repeat proteins [130]. The rational design and the computational design approaches can be synergistically incorporated into the design of novel biocatalysts, thus reducing the number of models that need to be built and scored.

2.2.3.1 Applications of *De Novo* Enzyme Design Methodology

The use of *de novo* design has been used to introduce novel functions that are also thought to be lacking in nature, such as the Kemp elimination reaction, which involves the deprotonation of the ligand substrate 5-nitrobenzisoxazole by a catalytic base with the corresponding electronic rearrangement. Moreover, the Kemp elimination reaction is simple and goes through a single transition state.

Minimalistic approaches: This method has been widely used to introduce a base in a hydrophobic cavity, which would increase the pK_a for Kemp elimination [131, 132]. Most of the structures used for this purpose were based on the TIM barrel, which is a ubiquitous fold constituted of eight α-helices and eight β-strands arranged in tandem. This structure has also been used to introduce hydrolase and lactonase activity [133]. A second structure broadly used for the introduction of specific residues to introduce specific catalysis is the calmodulin (CaM) structure, which contains a hydrophobic pocket ideal to introduce the substrate binding cavity [134].

Rational parametric design: Through this approach, coiled-coil structures have been used to introduce specific activities. The heptametic coiled-coil scaffold CC-Hept has been utilized to introduce hydrolase activity through the incorporation of the catalytic triad Cys-His-Glu [135].

Fragment-based computational design: TIM barrel structures have also been used for this approach. Particularly, *de novo* four-fold symmetric $(\beta\alpha)_8$-barrels can be designed with potential for catalysis [130].

2.2.4 Rational Enzyme Design

Computer-aided enzyme design requires the high-quality 3D structure of a protein. One clear limitation of rational design is that it fully relies on the availability of protein structures and sufficient knowledge about the enzymatic mechanism. However, with the introduction of good-quality homology modeling, it is possible to generate 3D structures based on known homologous templates [136–138]. Rational enzyme design can take enormous advantage of the exponential increase in protein coordinates deposited in the RSCB PDB [139]. Since 1976 (13 structures available), the repository has been continuously growing, thanks to the analytical advancements in X-ray crystallography, protein NMR techniques, and lately the cryo-EM technology, containing today almost 145 000 protein

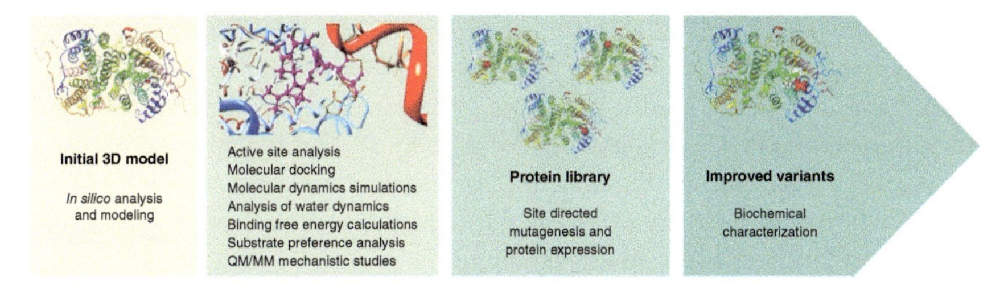

Figure 2.1 Overview of the rational enzyme engineering strategy.

structures (http://www.rcsb.org/stats/growth/protein (accessed November 2019)). By analyzing the structures, it is possible to detect hot-spots for mutations. Further computational modeling using state-of-the-art algorithms makes possible to calculate the binding energy, the reaction pathways, water/solvent dynamics, and the energy gain or loss of single and multiple mutations. Selected rational mutations can be experimentally introduced by site-directed mutagenesis. After the expression of the variant, experimental assays for testing the enzyme must be performed (Figure 2.1) [3, 4].

The increasing computer power and the continuous development of theoretical algorithms have notably improved the accuracy of computational enzyme engineering.

Molecular docking methods aim to predict the experimental binding modes and affinities of substrates within the active site of a particular protein target. The "best pose" is selected by "sampling" the internal conformational space of the substrate and by "scoring" (i.e. ranking) the predicted binding poses [16]. Structural analysis helps to identify the residues that may lead to an improved ligand affinity when substituted by a particular amino acid. However, the analysis of "static" structures of a highly dynamic system (i.e. proteins) has clear limitations [140]. **Ensemble docking** alleviates this problem by allowing to dock substrates into multiple receptor conformations that can be in turn obtained through MD simulations [141, 142].

MD simulations can be employed for generating multiple conformations and for analyzing the motions of a macromolecular system. MD simulation is an empirical method based on solving Newton's equations of motion, which allows simulating the movement of solvated proteins [16, 143].

However, modeling of an enzymatic reaction involves electronic changes during catalysis. **QM** methods must be employed when chemical reactions need to be described (i.e. electronic structure of molecules). Nonetheless, the large size of biological systems hampers the use of such methods. Therefore, to model the catalytic mechanism of an enzyme, QM/molecular mechanics (MM) techniques can be employed. **QM/MM** means only a preselected part of the system involved in the chemical reaction are modeled using QM (active site residues, substrate, and participating water molecules if any), and for the remaining parts of the protein, MM is used [142].

2.2.4.1 Applications of Rational Design Methodology

Many successful examples of rational enzyme design are available, and some of them will be summarized in the following section [53, 144–146].

Protein stability is usually related to resistance to unfolding [17]. Stable proteins are desirable for the industrial implementation of enzymatic processes but also to use them under different conditions than the wild-type analogs (higher temperatures, higher concentration of substrates, toleration of organic solvents, etc.). In addition, more stable proteins are better starting points for further enzyme engineering, considering that mutations aiming to modify protein native structure and function usually result in lowering the stability of the system by each step of mutagenesis [147].

Engineering more stable proteins involves making mutations that usually decrease the flexibility of the unfolded form (thus favoring the folded one) by adding salt bridges, H-bonding interactions, disulfide cross-links, or introducing proline residues [148, 149]. It is well known since the 1980s that each disulfide bond contributes to the thermodynamic stability of proteins in about 2.2–5.2 $kcal\,mol^{-1}$ [150, 151]. Moreover, a disulfide bond restricts the motion of the unfolded state in a greater extent than that of the folded one; the main reason of increasing stability is due to the loss of conformational entropy of the unfolded state [152].

Early attempts to engineer disulfide bonds were based on the distance of potential residue pairs to be mutated to cysteine. However, these too simple criteria do not take into account the introduction of strain in the folded form [153]. The web-based program SSBOND predicts locations in the protein where disulfide bonds could be introduced without adding strain in the folded form (http://129.128.191.54/forms/ssbond.html (accessed November 2019)) [154]. It only requires the PDB file of the protein of interest and identifies pairs of residues with proper distances (Cβ–Cβ c. 2.9–4.6 Å), excluding pairs if the sulfur atoms are not properly oriented. It includes a final energy minimization, which allows calculating the energy cost of deviation from the ideal geometry.

Another web-based tool to design disulfide bonds is disulfide by design (DbD) (http://cptweb.cpt.wayne.edu/DbD2/ (accessed November 2019)) [155–157]. The program estimates the torsion angle around $C\beta_1$–$S\gamma\gamma_2$–$C\beta_2$. This torsion angle is between −87° and +97°. The energy of each possible disulfide bond is calculated, and possible mutations are ranked. Finally, the PDB file for those selected disulfide bonds can be created.

A different strategy to reduce the flexibility of the protein and thus increasing the stability is the substitution of some amino acid residues by another one less flexible. With glycine being the most flexible amino acid and proline the least flexible one (with the other 18 natural amino acids in between), replacing any residue with proline is expected to reduce the flexibility and to stabilize the folded protein (in about 1.2 $kcal\,mol^{-1}$) [158]. The most flexible regions of the protein can be pointed out by running a MD simulation [159] or analyze the B-factors in the X-ray crystal structure [160, 161]. B-FITTER is a free-to-download program that identifies the residues having the highest B-factors [160] (https://www.kofo.mpg.de/en/research/biocatalysis (accessed November 2019)). Starting from this short analysis, it is possible to introduce mutations within the flexible region aiming to decrease flexibility. After identifying flexible regions, several different techniques can be combined, such as disulfide bond design, MD simulations, and rational analysis of predicted variants. A library of variants for increasing haloalkane dehydrogenase stability was created, identifying 17 substitution and one disulfide bond that enhanced thermostability. The best variant showed a 23 °C increase in apparent melting temperature and about 200-fold longer half-time at 60 °C [162].

Another approach might be the introduction of amino acid residues that are highly conserved in homologous proteins, assuming that evolution tends to conserve beneficial residues, such as those increasing stability. This consensus approach is based on the sequence analysis and the comparison does not require any structural knowledge [163]. However, this is a highly efficient approach because it allows to introduce multiple mutations at the same time. Lehmann et al. [164] determined a consensus sequence of fungal phytases employing 13 different sequences, resulting in a protein 15–26 °C more thermostable than any of its parent sequences.

Besides stability, many other properties can be modified/optimized, such as substrate specificity, stereoselectivity, activity, and the introduction of novel promiscuous activities. In the following section, some significant works are briefly discussed to provide the reader with an overview of the state of the art in computational enzyme design.

Regarding **introducing new chemical activities**, Korendovych et al. [165] redesigned calmodulin, a regulatory calcium binding protein, into an allosterically controlled Kemp eliminase (AlleyCat), which is activated upon binding to Ca^{2+}, by introducing a single nonpolar residue at the bottom of a hydrophobic cavity, as shown in Figure 2.2.

The Kemp elimination (i.e. the irreversible deprotonation of benzisoxazole to form cyanophenoxides) has been employed in biocatalyst design, as a benchmark model for C—H bond proton abstraction [12, 166, 167]. It is known that dehydrated carboxylates are good catalysts for Kemp elimination because of their basicity toward protons in carbon atoms [168]. Starting from this point, when placing a single carboxylate in a hydrophobic pocket, the anionic form is destabilized and thus more basic and able to deprotonate C—H bonds [169]. Calmodulin has a cavity that binds the aromatic side chains of peptides and it was screened to find a site where Glu or Asp would fit. The model substrate was computationally docked, and several rotamers were tested. Only one position turned out to

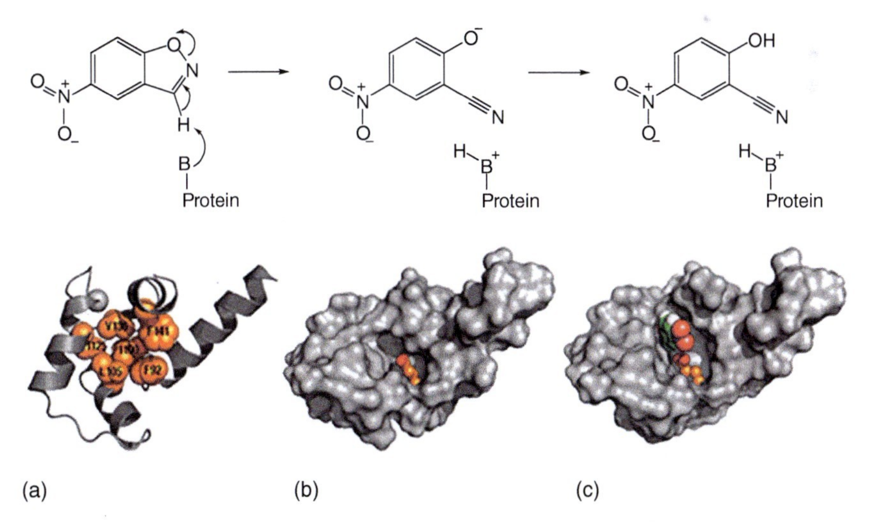

Figure 2.2 Kemp elimination reaction and summery of the computational approach. (a) The identified mutation points, where a catalytic residue will be introduced into the enzyme structure. (b) Initial modeling of the Glu residue in the identified position of the substrate. (c) The docking of the substrate's transition state. Source: Korendovych et al. [165].

accommodate an acidic residue, with Glu being the best option in terms of potential energy. Thus, the mutant Phe92Glu (so called AlleyCat) was experimentally tested exhibiting Kemp elimination activity.

In the same field of designing a new catalytic function in proteins, Korendovych et al. also converted calmodulin into an ester hydrolase, by designing an enzyme bearing a catalytically active histidine (Figure 2.3) [170].

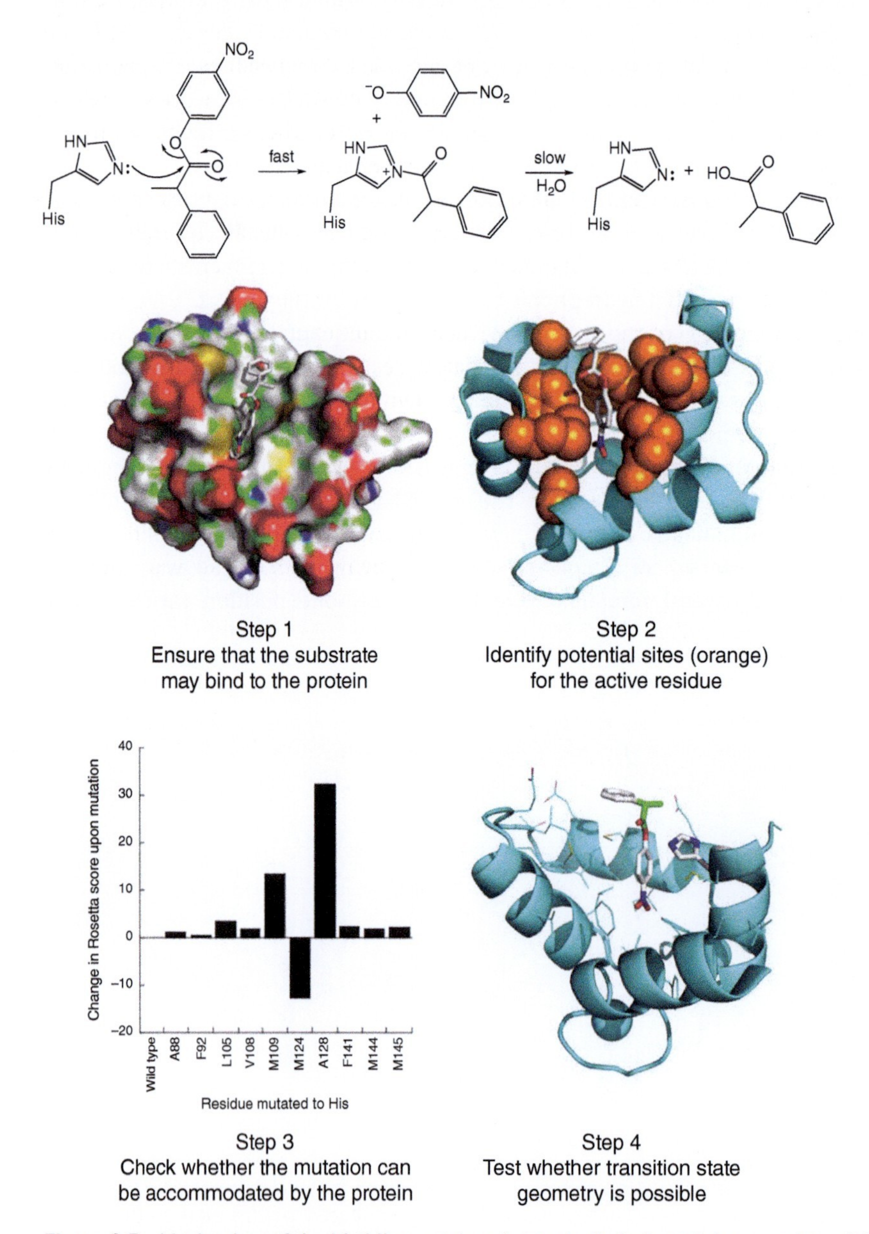

Figure 2.3 Mechanism of the histidine-catalyzed ester hydrolysis and the overview of the design methodology. Source: Moroz et al. [170].

By docking the substrate *p*-nitrophenyl-2-phenylpropanoate, the residues facing the substrate were identified and analyzed the ability to accommodate histidine in those positions without generating clashes. Finally, the possibility of Michaelis–Menten complex formation was assessed, and one variant (Met144His) was able to hydrolyze the *R* enantiomer of the substrate.

The rational design of metalloproteins is extremely challenging because the metal-binding sites of the protein display higher variability in terms of ligand donors, preferred geometry, and metal ion oxidation states. Yeung et al. succeeded to rationally design nitric oxide reductase (NOR), a metalloprotein, by introducing three histidine residues and one glutamate into the distal pocket of myoglobin. Remarkably, the crystal structure of the variant revealed that the protein contains a heme/non-heme Fe_B center, and it is highly similar to the computer model and also displays nitric oxide reduction activity [171].

The **synthesis of enantiomerically pure** chiral building blocks is of utmost interest of both academia and industry. To this regard, rational design of enzymes generating pure stereoisomers has been the topic of many research studies employing several approaches.

Alcohol dehydrogenases have been extensively employed for the asymmetric synthesis of enantiomerically pure alcohols. Besides the successfully engineered dehydrogenases through DE, rational approaches have also been employed. The natural enantiopreference (Prelog) of a secondary alcohol dehydrogenase from *Thermoanaerobacter ethanolicus* was switched into an anti-Prelog specificity, by a single-point mutation (I86A) [172]. The Prelog's rule (Figure 2.4) depends on the relative sizes of the two substituents of the prochiral ketone substrate.

Many alcohol dehydrogenases display a Prelog stereopreference and thus reverting it is of great interest. The crystal structure of the alcohol dehydrogenase showed that the enzyme has two hydrophobic pockets, one larger than the other. Thus, enlarging the size of the small pocket indeed affected the stereospecificity. The resulting mutant (I86A) reduced

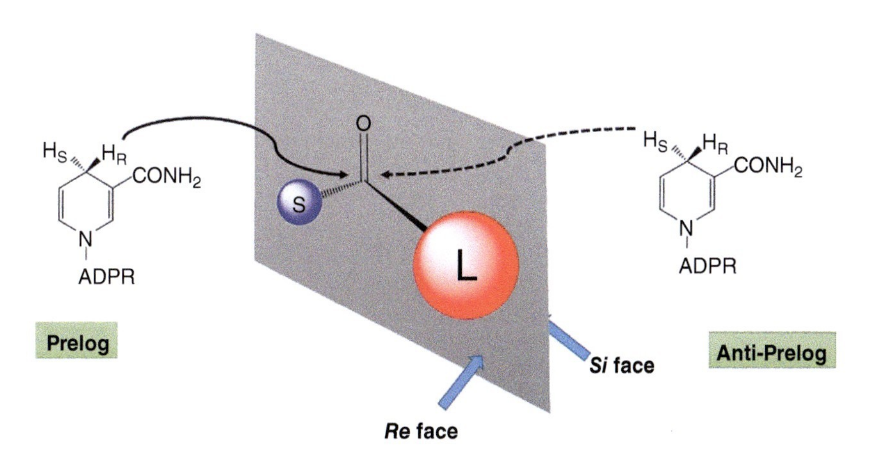

Figure 2.4 Prelog's rule for predicting the outcome of an alcohol dehydrogenase-mediated ketone reduction. ADPR is adenosine diphosphoribose. L and S stand for large and small groups. The L group is more sterically demanding and has a higher Cahn-Ingold-Prelog priority than the S group.

the substrates to the corresponding anti-Prelog alcohols, with moderate to good yields and high enantioselectivities.

Optically pure epoxides are highly valuable building blocks for the production of pharmaceuticals, and thus, the preparation of the latter products using epoxide hydrolases is of great interest. In 2010, Lonsdale et al. studied the selectivity of the soluble epoxide hydrolase (sEH), which catalyzes the epoxide ring opening to give the corresponding vicinal diols. The authors rationalized the observed selectivity of the enzyme over two different substrates, where one contained equivalent epoxide carbon atoms (both regio- and stereoselectivity) [173]. The authors suggested that the regioselectivity of the nucleophilic attack of sEH is mostly due to the different relative electrophilicity of the two carbon atoms when nonequivalent. However, the presence of equivalent carbon atoms showed a more regioselectivity caused by the nucleophile present in the enzyme–substrate complex. A similar and more recent study in 2017 was performed on the microsomal epoxide hydrolase (mEH). In fact, the homology modeling of human mEH based on the fungal (*Aspergillus niger*) EH crystallographic template (PDB ID: 1QO7) is the only recent theoretical work available on the literature. The results allowed to identify the key residues for substrate binding, stereoselectivity, and intermediate stabilization during the reaction [174]. These insights provide useful information for the design of epoxide hydrolase variants.

Because of the fact that lipases are the most commonly used enzymes in industrial applications, several engineered hydrolases are available [175, 176]. Subtilisin Carlsberg (SC), a serine protease, has been rationally engineered toward increased enantioselectivity of (*S*)-1-phenylethanol substrate. Based on the extensive literature search and molecular modeling, two mutation points have been identified, which were G165 and M221. A few point mutants were designed, expressed, and screened for the selectivity of the transesterification reaction of 1-phenylethanol. The synergistically successful mutations were G165L/M221F, where the M221F mutation, next to the beneficial opening of the S_1' pocket, was also responsible for a further increase in the stability of the protein (Figure 2.5). The

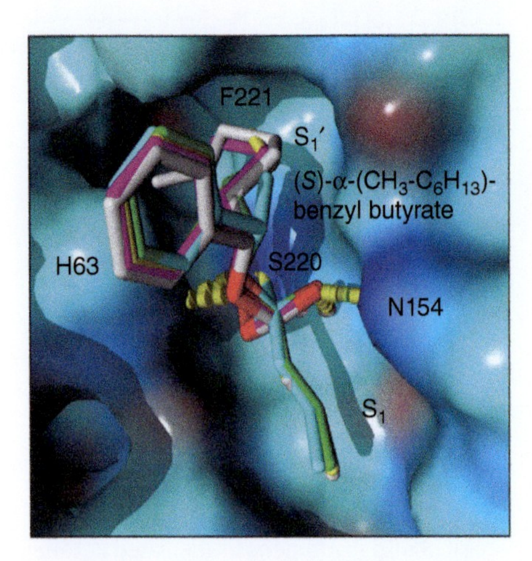

Figure 2.5 Structure of Subtilisin Carlsberg is shown in surface mode. The tetrahedral intermediates with different lengths of alkyl chains are coordinated into the active site of SC G165L/M221F double mutant. Source: Dorau et al. [175].

screening was performed in tetrahydrofuran (THF) solvent under inert conditions and showed an E value of >100 toward (*S*)-1-phenylethanol. By further studying the models of the enzyme, it has been theorized and tested that the G165L mutations would allow to convert longer alkyl chain-containing *sec*-alcohols, which were not possible with the wild-type enzyme [177].

Extending the **substrate scope** is also very interesting as this approach allows the use of enzymatic catalysis on novel substrates. A very useful enzyme type that has been extensively studied and modeled are Baeyer–Villiger monooxygenases (BVMOs) [178]. These enzymes introduce one oxygen atom into their substrates, allowing the transformation of ketones in esters or lactones with high regio- and stereoselectivity. Phenylacetone monooxygenase (PAMO) from *Thermobifida fusca* possesses high stability regarding both temperature and organic solvents. Thus, it is a very interesting candidate for applications in organic synthesis. Fraaije's group computationally designed PAMO variants, selecting several active-site residues [179]. The M446G variant showed activity toward aromatic ketones, amines, and sulfides, which was not shown by the PAMO wild-type. Moreover, the variant was able to convert indole into indigo blue, a completely novel reaction for a BVMO.

Cytochrome P450 monooxygenases catalyze a broad range of reactions, although with low specificity and narrow substrate scope. Protein engineering of several P450 enzymes has been extensively performed employing DE [5]. Computational approaches can be employed not only for rational design of enzymes but also to explain the rationality beyond the successful substitutions during the DE process. Butler et al. presented the first crystal structure of the *Bacillus megaterium* P450 BM3 enzyme, showing how the substitutions alter the landscape of the active site. These new insights may assist in a faster identification of variants with desired activities [180].

Polyethylene terephthalate (PET), the most commonly produced thermoplastic, is difficult to degrade and thus tends to accumulate in the environment. To overcome this problem, hydrolases, in particular esterases and cutinases, are being increasingly discussed in the recent literature, regarding their application in plastic and textile recycling [181–183]. Several rational approaches led to very interesting results [184], such as the recent work by Austin et al. [185], describing the successful modification of the so-called PETase (PET-digesting enzyme). In this work, the authors present the crystal structure of the PETase from *Ideonella sakaiensis*, a bacterium that is able to grow on PET as a major source of carbon and energy. Moreover, the computational analysis of the enzyme allowed for designing a variant with improved PETase activity, introducing two simultaneous mutations, i.e. S238F/W159H. The result was obtained by trying multiple orientations of PET employing induced fit docking, indicating that both the substrate and the region of the protein close to the ligand are treated as flexible, while the rest of the protein (more than 8 Å from the ligand) is treated as a rigid body. The carbonyl of PET is placed within a proper distance from the nucleophilic Ser160 for the attack to take place. At the same time, His237 and Asp206 are at an ideal distance to activate Ser160. Thus, the computational model of the designed mutant predicted complexes having PET in proper conformations to be cleaved.

Currently, only 1% of all the polymers produced are generated from renewable resources [186], being then produced from fossil resources, negatively impacting the environment. One interesting concept, analogous to the chemical retrosynthesis developed by Corey [187], is the retro-bio-synthesis one, suggested by Turner and O'Reilly [188, 189]. In a recent work, a retro-biosynthesis approach was employed to design a chemoenzymatic route to generate pinene-derived polyesters through lactone **3** (Scheme 2.1) [190], in which the pinene starting materials are readily available from direct natural resources [191].

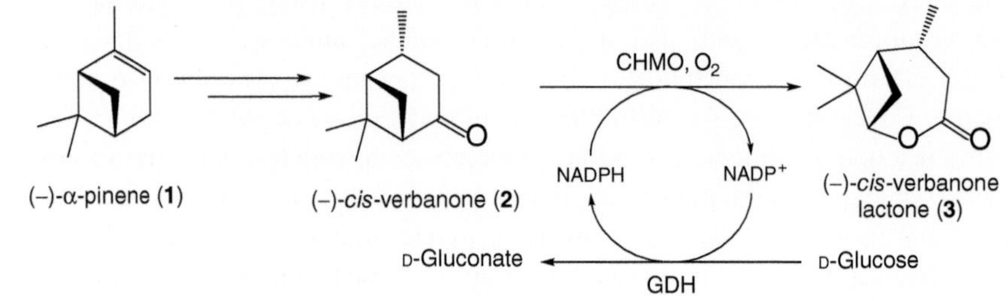

CHMO, O_2
NADPH
$NADP^+$
(−)-α-pinene (**1**)
(−)-*cis*-verbanone (**2**)
(−)-*cis*-verbanone lactone (**3**)
D-Gluconate
D-Glucose
GDH

Scheme 2.1 Chemoenzymatic route of lactone **3** from (−)-α-pinene **1**. This lactone is a precursor of pinene-derived polyesters.

To allow the conversion of **2** into the bulky bicyclic lactone **3**, cyclohexanone monooxygenase (CHMO) from *Acinetobacter calcoaceticus* was rationally engineered by employing molecular modeling. Eight amino acids were selected as hot spots for mutations based on the proximity of the substrate and lactone product models docked into the active site. Twenty-six variants were built, including single, double, and triple mutants. One of the variants, L426A, showed activity toward the oxidation of **2** into **3** (39% conversion after 24 hours). L426 is placed in the binding pocket and mutating it into a smaller amino acid allowed the productive binding of the bulkier substrate **2** and its conversion into the lactone **3**.

In a different approach, Bornscheuer's group has successfully applied the 3DM dataset to guide protein engineering of hydrolases, thus creating "smart libraries." The approach involves the design of mutant libraries by only taking into account amino acids frequently occurring at a given position [85, 192]. Employing this methodology, the activity of an esterase from *Pseudomonas fluorescens* was increased 240-fold, at the same time that enantioselectivity was notably improved ($E_{mutant} = 80$, $E_{wild\text{-}type} = 3.2$) by mutating only four residues close to the active site [193].

The same group rationally designed mutants that were computationally predicted to act as transaminases. By further searching in the protein databases for proteins carrying these mutations, 17 (*R*)-selective amine transaminases producing (*R*)-amines with up to >99% enantiomeric excess were discovered [194].

A completely different strategy to identify mutational hot spots for the generation of tailored biocatalyst is the **identification of tunnels** and cavities, which are present in a high number of proteins and have an important role in function and substrate recognition. If the active site is buried within the enzyme, the first step of the recognition is indeed the passage of the substrate through the tunnel and then the actual binding in the active site. In fact, tunnels are present in all six EC classes of enzymes [195]. Hence, by modifying the size, physicochemical properties, and tunnel dynamics, a protein might be engineered and their properties are improved (i.e. activity, substrate scope, and stability). Engineering of tunnels may improve the access of desired substrates, while avoiding the access of non-preferred ones, modifying the substrate preference. The recently proposed "keyhole-lock-key" model reflects the importance of the substrate passage through tunnels to reach the active site and react [196].

Several computational tools have been developed for analyzing and engineering tunnels and cavities in proteins, such as CAVER [197], MOLE [198], and BetaCavityWeb [199]. All these tools are based on computational geometry methods using the Voronoi diagrams [200]. Cavities and tunnels play an essential role in biocatalysis, allowing the entrance of the substrate and water molecules to the active site, the exchange of protons, and the release of products. Moreover, the enzyme specificity is not only determined by protein–substrate interactions, but also highly influenced by the selectivity of these access tunnels. Interestingly, CAVER allows to analyze the tunnels in an ensemble of protein conformations (e.g. MD simulation snapshots), avoiding to neglect transient tunnels [201].

CAVER has been applied in several protein engineering studies (Figure 2.6) [202], such as cytochrome P450 and carbamoyl phosphate synthase enzymes. The computational design of variants of haloalkane dehalogenases (DhaA) with modified tunnels has also been performed, showing 32 times higher activity toward the non-natural substrate 1,2,3-trichloropropane. Haloalkanes dehydrogenases catalyze the hydrolysis of haloalkanes to the corresponding alcohol, showing preference for primary carbon–halogen bonds. Pavlova et al. identified residues in those tunnels connecting the surface with the active site and then the mutants were built by randomizing those residues through DE. The results showed that identifying hot spot residues padding the access tunnels by computer modeling using CAVER combined with DE is a valuable and generally applicable methodology for engineering biocatalysts with buried active sites [203].

2.3 Expectations and Perspectives

Enzyme engineering has been successfully applied in the generation of protein variants with improved properties, such as activity, thermostability, toleration of organic solvents, modified substrate specificity, modified stereoselectivity, and even creation of completely novel activities. Either we can learn from nature and employ DE or to recall into the

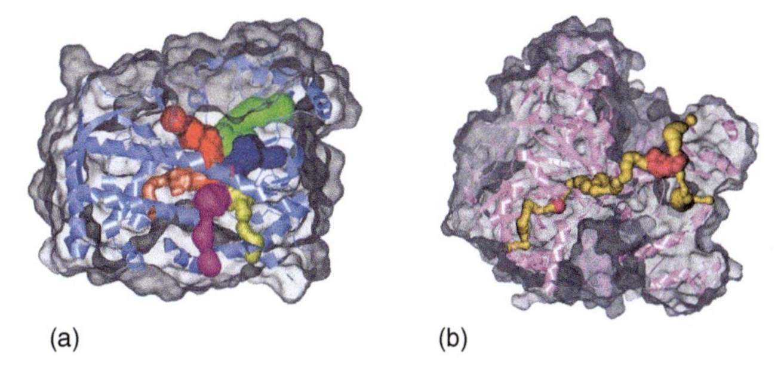

Figure 2.6 The detectable access tunnels in the crystal structures of cytochrome P450 and carbamoyl phosphate synthase enzymes. (a) Various colors show access tunnels leading to the active site of a human cytochrome P450 CYP3A4 (PDB 5VCC). (b) The long yellow tunnel connecting two active sites (magenta) in human carbamoyl phosphate synthase (PDB 1A9X). Source: Kokkonen et al. [202].

knowledge base of rational design or even, more recently, to create enzymes "from scratch" using *de novo* design.

The continuous improvements in DE allow the focus on more complex systems such as the assembly of sequentially acting enzymes into synthetic pathways to produce high-value chemicals following the 12 principles of Green Chemistry. The Golden Gate gene assembly strategy is routinely used to assemble DNA sequences for **synthetic biology** applications, bringing together biocatalysts from different sources in the same expression plasmid to create hybrid pathways [204]. Several engineering pathways focus on the amplification of the flux of key biosynthetic precursors in *E. coli* to improve the production of certain compounds, such as terpenes [205, 206]. Moreover, adaptive laboratory evolution (ALE) experiments can engineer complex phenotypes using DE *in vivo* through the natural evolution of microbial cultures grown for hundreds of generations in defined selective conditions [207–209]. Furthermore, the discovered tool for synthetic biology given by the Clustered Regularly Interspaced Short Palindromic Repeats (CRISPRs) system can increase diversity by altering the expression levels in genome-wide loss-of-function and gain-of-function screens [210, 211].

The combination of computational *de novo* enzyme design and laboratory evolution is a successful strategy for the development of biocatalysts with non-natural functions. Chemical activities that are unknown in nature might be shown and incorporated in biochemical processes of living organisms that can increase the importance of production through biotechnology. In order to be able to interact with the cell metabolism, the *de novo* protein must be biocompatible.

In this chapter, we have shown the success of rational design, highlighting some examples. However, it is also possible to find even more examples of failed attempts of rational protein engineering. Nonetheless, there is still room for improvements, mainly related to the structural and mechanistic complexities of biocatalysts. The use of a sole approach may not account for all features of an enzymatic system. Thus, the use of tunnel analysis combined with MDs simulations and docking experiments may give a better overview of the problem in hands. The best approach undoubtedly will be the combination of DE, rational, and *de novo* design to take advantage of the best features of each methodology and avoid the pitfalls of one approach with another.

2.4 Concluding Remarks

This chapter has addressed the diverse methods available for enzyme modification, providing a comprehensive state of the art of directed evolution, rational, semi-rational, and *de novo* design. Undoubtedly, directed evolution techniques have been extensively developed and can be applied in many cases. However, depending on the particular enzymatic system and on the available information (3D structure, enzymatic mechanism, etc.), other techniques such as rational design are also becoming stronger in the field. With the simultaneous development of software packages and the improvement of hardware, computational approaches may reduce the cost and time of the random experiments. By focusing on smart libraries, protein engineering can be further improved. Thus, depending on factors that are selected on a case basis, all approaches are presently being successfully employed, although not or rarely at industrial level. As a final message, enzyme engineering is continuously gaining importance in many applications, and it is expected to play a main role in catalysis in the coming years, as the different methodologies allow to customize enzymes.

References

1 Faber, K. (2011). *Biotransformations in Organic Chemistry*, 6e. Heidelberg: Springer.
2 Reetz, M.T. (2013). Biocatalysis in organic chemistry and biotechnology: past, present, and future. *J. Am. Chem. Soc.* 135 (34): 12480–12496.
3 Bornscheuer, U.T. (2013). Protein engineering as a tool for the development of novel bioproduction systems. *Adv. Biochem. Eng. Biotechnol.* 137: 25–40.
4 Bornscheuer, U.T., Huisman, G.W., Kazlauskas, R.J. et al. (2012). Engineering the third wave of biocatalysis. *Nature* 485 (7397): 185–194.
5 Arnold, F.H. (2018). Directed evolution: bringing new chemistry to life. *Angew. Chem. Int. Ed.* 57 (16): 4143–4148.
6 Zhang, R.K., Chen, K., Huang, X. et al. (2019). Enzymatic assembly of carbon–carbon bonds via iron-catalysed sp(3) C–H functionalization. *Nature* 565 (7737): 67–72.
7 Cho, I., Jia, Z.-J., and Arnold, F.H. (2019). Site-selective enzymatic C–H amidation for synthesis of diverse lactams. *Science* 264: 575–578.
8 EMJ, G., Copp, J.N., and Ackerley, D.F. (eds.) (2014). *Directed Evolution Library Creation. Methods and Protocols*, 2 ed. Totowa, NJ: Humana Press, Springer.
9 Currin, A., Swainston, N., Day, P.J., and Kell, D.B. (2015). Synthetic biology for the directed evolution of protein biocatalysts: navigating sequence space intelligently. *Chem. Soc. Rev.* 44 (5): 1172–1239.
10 Francis, J.C. and Hansche, P.E. (1972). Directed evolution of metabolic pathways in microbial populations. I. Modification of the acid phosphatase pH optimum in *S. cerevisiae*. *Genetics* 70 (1): 59–73.
11 Porter, J.L., Rusli, R.A., and Ollis, D.L. (2016). Directed evolution of enzymes for industrial biocatalysis. *ChemBioChem* 17 (3): 197–203.
12 Rothlisberger, D., Khersonsky, O., Wollacott, A.M. et al. (2008). Kemp elimination catalysts by computational enzyme design. *Nature* 453 (7192): 190–195.
13 Kries, H., Blomberg, R., and Hilvert, D. (2013). De novo enzymes by computational design. *Curr. Opin. Chem. Biol.* 17 (2): 221–228.
14 Kiss, G., Celebi-Olcum, N., Moretti, R. et al. (2013). Computational enzyme design. *Angew. Chem. Int. Ed.* 52 (22): 5700–5725.
15 García-Guevara, F., Avelar, M., Ayala, M., and Segovia, L. (2016). Computational tools applied to enzyme design – a review. *Biocatalysis* 1 (1): 109–117.
16 Genheden, S., Reymer, A., Saenz-Méndez, P., and Eriksson, L.A. (2017). Computational chemistry and molecular modelling basics. In: *Computational Tools for Chemical Biology*. Chemical Biology (ed. S. Martín-Santamaría), Chapter 1, 1–38.
17 Kazlauskas, R. (2018). Engineering more stable proteins. *Chem. Soc. Rev.* 47 (24): 9026–9045.
18 Davey, J.A., Damry, A.M., Goto, N.K., and Chica, R.A. (2017). Rational design of proteins that exchange on functional timescales. *Nat. Chem. Biol.* 13 (12): 1280–1285.

19 Chica, R.A., Doucet, N., and Pelletier, J.N. (2005). Semi-rational approaches to engineering enzyme activity: combining the benefits of directed evolution and rational design. *Curr. Opin. Biotechnol.* 16 (4): 378–384.
20 Arnold, F.H. and Aa, V. (1999). Directed evolution of biocatalysts. *Curr. Opin. Chem. Biol.* 3: 54–59.
21 Cadwell, R.C. and Joyce, G.F. (1992). Randomization of genes by PCR mutagenesis. *PCR Methods Appl.* 2: 28–33.
22 Packer, M.S. and Liu, D.R. (2015). Methods for the directed evolution of proteins. *Nat. Rev. Genet.* 16: 379–394.
23 Myers, R., Lerman, L., and Maniatis, T. (1985). A general method for saturation mutagenesis of cloned DNA fragments. *Science* 229: 242–247.
24 Greener, A., Callahan, M., and Jerpseth, B. (1997). An efficient random mutagenesis technique using an *E. coli* mutator strain. *Mol. Biotechnol.* 7: 189–195.
25 Ravikumar, A., Arrieta, A., and Liu, C.C. (2014). An orthogonal DNA replication system in yeast. *Nat. Chem. Biol.* 10: 175–177.
26 Esvelt, K.M., Carlson, J.C., and Liu, D.R. (2011). A system for the continuous directed evolution of biomolecules. *Nature* 472: 499–503.
27 Vanhercke, T., Ampe, C., Tirry, L., and Denolf, P. (2005). Reducing mutational bias in random protein libraries. *Anal. Biochem.* 339: 9–14.
28 Stemmer, W.P. (1994). Rapid evolution of a protein in vitro by DNA shuffling. *Nature* 370: 389–391.
29 Coco, W.M., Levinson, W.E., Crist, M.J. et al. (2001). DNA shuffling method for generating highly recombined genes and evolved enzymes. *Nat. Biotechnol.* 19 (4): 354–359.
30 Crameri, A., Raillard, S.A., Bermudez, E., and Stemmer, W.P. (1998). DNA shuffling of a family of genes from diverse species accelerates directed evolution. *Nature* 391: 288–291.
31 Zhao, H., Giver, L., Shao, Z. et al. (1998). Molecular evolution by staggered extension process (StEP) in vitro recombination. *Nat. Biotechnol.* 16: 258–261.
32 Zha, D., Eipper, A., and Reetz, M.T. (2003). Assembly of designed oligonucleotides as an efficient method for gene recombination: a new tool in directed evolution. *ChemBioChem* 4: 34–39.
33 Sieber, V., Martinez, C.A., and Arnold, F.H. (2001). Libraries of hybrid proteins from distantly related sequences. *Nat. Biotechnol.* 19: 456–460.
34 Ostermeier, M., Shim, J.H., and Benkovic, S.J. (1999). A combinatorial approach to hybrid enzymes independent of DNA homology. *Nat. Biotechnol.* 17: 1205–1209.
35 Lutz, S., Ostermeier, M., Moore, G.L. et al. (2001). Creating multiple-crossover DNA libraries independent of sequence identity. *Proc. Natl. Acad. Sci. U. S. A.* 98: 11248–11253.
36 Hiraga, K. and Arnold, F.H. (2003). General method for sequence-independent site-directed chimeragenesis. *J. Mol. Biol.* 330: 287–296.
37 Voigt, C.A., Martinez, C., Wang, Z.-G. et al. (2002). Protein building blocks preserved by recombination. *Nat. Struct. Biol.* 9: 553–558.
38 Wong, T., Zhurina, D., and Schwaneberg, U. (2006). The diversity challenge in directed protein evolution. *Comb. Chem. High Throughput Screen.* 9: 271–288.
39 Zeymer, C. and Hilvert, D. (2018). Directed evolution of protein catalysts. *Annu. Rev. Biochem.* 87: 131–157.

40 Otten, L.G., Hollmann, F., and Arends, I.W.C.E. (2010). Enzyme engineering for enantioselectivity: from trial-and-error to rational design? *Trends Biotechnol.* 28: 46–54.
41 Varadarajan, N., Cantor, J.R., Georgiou, G., and Iverson, B.L. (2009). Construction and flow cytometric screening of targeted enzyme libraries. *Nat. Protoc.* 4: 893–901.
42 Chen, I., Dorr, B.M., and Liu, D.R. (2011). A general strategy for the evolution of bond-forming enzymes using yeast display. *Proc. Natl. Acad. Sci. U. S. A.* 108: 11399–11404.
43 Baret, J.-C., Miller, O.J., Taly, V. et al. (2009). Fluorescence-activated droplet sorting (FADS): efficient microfluidic cell sorting based on enzymatic activity. *Lab. Chip.* 9: 1850–1858.
44 Agresti, J.J., Antipov, E., Abate, A.R. et al. (2010). Ultrahigh-throughput screening in drop-based microfluidics for directed evolution. *Proc. Natl. Acad. Sci. U. S. A.* 107: 4004–4009.
45 Chen, B., Lims, S., Kannan, A. et al. (2016). High-throughput analysis and protein engineering using microcapillary arrays. *Nat. Chem. Biol.* 12: 76–81.
46 Janes, J.L., Kazlauskas, R.J., and Quick, E. (1997). A fast spectrophotometric method to measure the enantioselectivity of hydrolases. *J. Org. Chem.* 62 (14): 4560–4561.
47 Janes, L.E., Löwendahl, A.C., and Kazlauskas, R.J. (1998). Quantitative screening of hydrolase libraries using pH indicators: identifying active and enantioselective hydrolases. *Chem. – A Eur. J.* 4: 2324–2331.
48 Baumann, M., Hauer, B.H., and Bornscheuer, U.T. (2000). Rapid screening of hydrolases for the enantioselective conversion of 'difficult-to-resolve' substrates. *Tetrahedron: Asymmetry* 11: 4781–4790.
49 Moore, B.D., Stevenson, L., Watt, A. et al. (2004). Rapid and ultra-sensitive determination of enzyme activities using surface-enhanced resonance Raman scattering. *Nat. Biotechnol.* 22: 1133–1138.
50 Boersma, Y.L., Dröje, M.J., van der Sloot, A.M. et al. (2008). A novel genetic selection system for improved enantioselectivity of *Bacillus subtilis* lipase A. *ChemBioChem* 9: 1110–1115.
51 Reetz, M.T., Höbenreich, H., Soni, P., and Fernández, L. (2008). A genetic selection system for evolving enantioselectivity of enzymes. *Chem. Commun.* 43: 5502–5504.
52 Engqvist, M.K.M. and Rabe, K.S. (2019). Applications of protein engineering and directed evolution in plant research. *Plant Physiol.* 179: 907–917.
53 Bottcher, D. and Bornscheuer, U.T. (2010). Protein engineering of microbial enzymes. *Curr. Opin. Microbiol.* 13 (3): 274–282.
54 Jemli, S., Ayadi-Zouari, D., Hlima, H.B., and Bejar, S. (2016). Biocatalysts: application and engineering for industrial purposes. *Crit. Rev. Biotechnol.* 36: 246–258.
55 Ebrahimi, M., Lakizadeh, A., Agha-Golzadeh, P. et al. (2011). Prediction of thermostability from amino acid attributes by combination of clustering with attribute weighting: a new vista in engineering enzymes. *PLoS One* 6: e23146.
56 Ness, J.E., Welch, M., Giver, L. et al. (1999). DNA shuffling of subgenomic sequences of subtilisin. *Nat. Biotechnol.* 17: 983–986.
57 Taguchi, S., Ozaki, A., and Momose, H. (1998). Engineering of a cold-adapted protease by sequential random mutagenesis and a screening system. *Appl. Environ. Microb.* 64: 492–495.

58 Zhang, N., Suen, W.C., Windsor, W. et al. (2003). Improving tolerance of *Candida antarctica* lipase B towards irreversible thermal inactivation through directed evolution. *Protein Eng.* 16: 599–605.

59 Jochens, H., Hesseler, M., Stiba, K. et al. (2011). Protein engineering of α/β-hydrolase fold enzymes. *ChemBioChem* 12: 1508–1517.

60 Giver, L., Gershenson, A., Freskgard, P.O., and Arnold, F.H. (1998). Directed evolution of a thermostable esterase. *Proc. Natl. Acad. Sci. U. S. A.* 95: 12809–12813.

61 Zhao, H. and Arnold, F.H. (1999). Directed evolution converts subtilisin E into a functional equivalent of thermitase. *Protein Eng. Des. Sel.* 12: 47–53.

62 Hao, J. and Berry, A. (2004). A thermostable variant of fructose bisphosphate aldolase constructed by directed evolution also shows increased stability in organic solvents. *Protein Eng. Des. Sel.* 17: 689–697.

63 Zumárraga, M., Bulter, T., Shleev, S. et al. (2007). In vitro evolution of a fungal laccase in high concentrations of organic cosolvents. *Chem. Biol.* 14: 1052–1064.

64 Moore, J.C. and Arnold, F.H. (1996). Directed evolution of a *para*-nitrobenzyl esterase for aqueous-organic solvents. *Nat. Biotechnol.* 14: 458–467.

65 Suplatov, D., Panin, N., Kirilin, E. et al. (2014). Computational design of a pH stable enzyme: understanding molecular mechanism of penicillin acylase's adaptation to alkaline conditions. *PLoS One* 9: e100643.

66 Reetz, M.T. (2008). Directed Evolution as a Means to Engineer Enantioselective Enzymes. In: *Asymmetric Organic Synthesis with Enzymes* (eds. V. Gotor, I. Alfonso and E. García-Urdales), 21–63. Wiley-VCH Verlag GmbH & Co. KGaA.

67 Reetz, M.T., Puls, M., Carballeira, J.D. et al. (2007). Learning from directed evolution: further lessons from theoretical investigations into cooperative mutations in lipase enantioselectivity. *ChemBioChem* 8: 106–112.

68 Reetz, M.T., Zonta, A., Schimossek, K. et al. (1997). Creation of enantioselective biocatalysts for organic chemistry by in vitro evolution. *Angew. Chem. Int. Ed.* 36: 2830–2832.

69 Bornscheuer, U.T., Altenbuchner, J., and Meyer, H.H. (1998). Directed evolution of an esterase for the stereoselective resolution of a key intermediate in the synthesis of epothilones. *Biotechnol. Bioeng.* 58: 554–559.

70 Kirschner, A. and Bornscheuer, U.T. (1999). Directed evolution of a Baeyer–Villiger monooxygenase to enhance enantioselectivity. *Appl. Microbiol. Biotechnol.* 81: 465–472.

71 Kumamaru, T., Suenaga, H., Mitsuoka, M. et al. (1998). Enhanced degradation of polychlorinated biphenyls by directed evolution of biphenyl dioxygenase. *Nat. Biotechnol.* 16: 663–666.

72 Machielsen, R., Leferink, N.G., Hendriks, A. et al. (2008). Laboratory evolution of *Pyrococcus furiosus* alcohol dehydrogenase to improve the production of (2S,5S)-hexanediol at moderate temperatures. *Extremophiles* 12: 587–594.

73 Virus, C. and Bernhardt, R. (2008). Molecular evolution of a steroid hydroxylating cytochrome P450 using a versatile steroid detection system for screening. *Lipids* 43: 1133–1141.

74 Koch, D.J., Chen, M.M., van Beilen, J.B., and Arnold, F.H. (2009). In vivo evolution of butane oxidation by terminal alkane hydroxylases AlkB and CYP153A6. *Appl. Environ. Microb.* 75: 337–344.

75 Yano, T., Oue, S., and Kagamiyama, H. (1998). Directed evolution of an aspartate aminotransferase with new substrate specificities. *Proc. Natl. Acad. Sci. U. S. A.* 95: 5511–5515.

76 Bartsch, S., Kourist, R., and Bornscheuer, U.T. (2008). Complete inversion of enantioselectivity towards acetylated tertiary alcohols by a double mutant of a *Bacillus subtilis* esterase. *Angew. Chem. Int. Ed.* 47: 1508–1511.

77 Sawayama, A.M., Chen, M.M., Kulanthaivel, P. et al. (2009). A panel of cytochrome P450 BM3 variants to produce drug metabolites and diversify lead compounds. *Chemistry* 15: 11723–11729.

78 Brandenberg, O.F., Fasan, R., and Arnold, F.H. Exploiting and engineering hemoproteins for abiological carbene and nitrene transfer reactions. *Curr. Opin. Biotechnol.* 47: 102–111.

79 Hammer, S.C., Knight, A.M., and Arnold, F.H. (2017). Design and evolution of enzymes for non-natural chemistry. *Curr. Opin. Green Sustain. Chem.* 7: 23–30.

80 Morley, K.L. and Kazlauskas, R.J. (2005). Improving enzyme properties: when are closer mutations better? *Trends Biotechnol.* 23 (5): 231–237.

81 Lutz, S. (2010). Beyond directed evolution--semi-rational protein engineering and design. *Curr. Opin. Biotechnol.* 21 (6): 734–743.

82 Kille, S., Acevedo-Rocha, C.G., Parra, L.P. et al. (2013). Reducing codon redundancy and screening effort of combinatorial protein libraries created by saturation mutagenesis. *ACS Synth. Biol.* 2: 83–92.

83 Hidalgo, A., Schließmann, A., and Bornscheuer, U.T. (2014). One-pot simple methodology for CAssette randomization and recombination for focused directed evolution (OSCARR). *Methods Mol. Biol. (Clifton, NJ)* 1179: 207–212.

84 Reetz, M.T. and Carballeira, J.D. (2007). Iterative saturation mutagenesis (ISM) for rapid directed evolution of functional enzymes. *Nat. Protoc.* 2: 291–903.

85 Kuipers, R.K., Joosten, H.J., van Berkel, W.J. et al. (2010). 3DM: systematic analysis of heterogeneous superfamily data to discover protein functionalities. *Proteins* 78 (9): 2101–2113.

86 Reetz, M.T., Bocola, M., Carballeira, J.D. et al. Expanding the range of substrate acceptance of enzymes: combinatorial active-site saturation test. *Angew. Chem. Int. Ed.* 44: 4192–4196.

87 Thornton, J.W. (2004). Resurrecting ancient genes: experimental analysis of extinct molecules. *Nat. Rev. Genet.* 5: 366–375.

88 Pavelka, A., Chovancova, E., and Damborsky, J. (2009). HotSpot Wizard: a web server for identification of hot spots in protein engineering. *Nucleic Acids Res.* 37: W376–W383.

89 Steipe, B., Schiller, B., Plückthun, A., and Steinbacher, S. (1994). Sequence statistics reliably predict stabilizing mutations in a protein domain. *J. Mol. Biol.* 240: 188–192.

90 Gumulya, Y. and Gillam, E.M.J. (2017). Exploring the past and the future of protein evolution with ancestral sequence reconstruction: the 'retro approach to protein engineering. *Biochem. J.* 474: 1–19.

91 Merkl, R. and Sterner, R. (2016). Ancestral protein reconstruction: techniques and applications. *Biol. Chem.* 379: 1–21.

92 Finn, R.D., Bateman, A., Clements, J. et al. (2014). Pfam: the protein families database. *Nucleic Acids Res.* 42: D2222–D2230.

93 Hulo, N., Bairoch, A., Bulliard, V. et al. (2006). The PROSITE database. *Nucleic Acids Res.* 34: D227–D230.

94 Letunic, I., Doerks, T., and Bork, P. (2015). SMART: recent updates, new developments and status in 2015. *Nucleic Acids Res.* 43: D257–D260.

95 Chen, F., Gaucher, E.A., Leal, N.A. et al. (2010). Reconstructed evolutionary adaptive paths give polymerases accepting reversible terminators for sequencing and SNP detection. *Proc. Natl. Acad. Sci. U. S. A.* 107: 1948–1953.

96 Dahiyat, B.I. (1999). In silico design for protein stabilization. *Curr. Opin. Biotechnol.* 10: 387–390.

97 Hayes, R.J., Bentzien, J., Ary, M.L. et al. (2002). Combining computational and experimental screening for rapid optimization of protein properties. *Proc. Natl. Acad. Sci. U. S. A.* 99: 15926–15931.

98 Fox, R.J., Davis, C., Mundorff, E.C. et al. (2007). Improving catalytic function by ProSAR-driven enzyme evolution. *Nat. Biotechnol.* 25: 338–344.

99 Fox, R. (2005). Directed molecular evolution by machine learning and the influence of nonlinear interactions. *J. Theor. Biol.* 234: 187–199.

100 Fox, R., Roy, A., Govindarajan, S. et al. (2003). Optimizing the search algorithm for protein engineering by directed evolution. *Protein Eng. Des. Sel.* 16: 589–597.

101 Hill, C.M., Li, W.-S., Thoden, J.B. et al. (2003). Enhanced degradation of chemical warfare agents through molecular engineering of the phosphotriesterase active site. *J. Am. Chem. Soc.* 125: 8990–8991.

102 Wise, E.L., Yew, W.S., Akana, J. et al. (2005). Evolution of enzymatic activities in the orotidine 5′-monophosphate decarboxylase suprafamily: structural basis for catalytic promiscuity in wild-type and designed mutants of 3-Keto-l-gulonate 6-phosphate decarboxylase. *Biochemistry* 44: 1816–1823.

103 Wu, S., Acevedo, J.P., and Reetz, M.T. (2010). Induced allostery in the directed evolution of an enantioselective Baeyer-Villiger monooxy. *Proc. Natl. Acad. Sci. U. S. A.* 107: 2775–2780.

104 Kourist, R., Bartsch, S., and Bornscheuer, U.T. (2007). Highly enantioselective synthesis of arylaliphatic tertiary alcohols using mutants of an esterase from *Bacillus subtilis*. *Adv. Synth. Catal.* 349: 1393–1398.

105 Park, S., Moreley, K.L., Horsman, G.P. et al. (2005). Focusing mutations into the *P. fluorescens* esterase binding site increases enantioselectivity more effectively than distant mutations. *Chem. Biol.* 12: 45–54.

106 Kotik, M., Stepánek, V., Kyslík, P., and Maresová, H. (2007). Cloning of an epoxide hydrolase-encoding gene from *Aspergillus niger* M200, overexpression in *E. coli*, and modification of activity and enantioselectivity of the enzyme by protein engineering. *J. Biotechnol.* 132: 8–15.

107 Rui, L., Cao, L., Chen, W. et al. (2004). Active site engineering of the epoxide hydrolase from *Agrobacterium radiobacter* AD1 to enhance aerobic mineralization of *cis*-1,2-dichloroethylene in cells expressing an evolved toluene *ortho*-monooxygenase. *J. Biol. Chem.* 279: 46810–46817.

108 Peimbert, M. and Segovia, L. (2003). Evolutionary engineering of a β-lactamase activity on a D-Ala D-Ala transpeptidase fold. *Protein Eng. Des. Sel.* 16: 27–35.

109 Sio, C.F., Riemens, A.M., van der Laan, J.-M. et al. (2002). Directed evolution of a glutaryl acylase into an adipyl acylase. *Eur. J. Biochem.* 269: 4495–4504.

110 Geddie, M.L. and Matsumura, I. (2004). Rapid evolution of β-glucuronidase specificity by saturation mutagenesis of an active site loop. *J. Biol. Chem.* 279: 26462–26468.

111 Savile, C.K., Janey, J.M., Mundorff, J.C. et al. (2010). Biocatalytic asymmetric synthesis of chiral amines from ketones applied to sitagliptin manufacture. *Science* 329: 305–309.

112 Reetz, M.T. (2004). Asymmetric catalysis special feature part II: controlling the enantioselectivity of enzymes by directed evolution: practical and theoretical ramifications. *Proc. Natl. Acad. Sci. U. S. A.* 101: 5716–5722.

113 Horsman, G.P., Liu, A.M.F., Henke, E. et al. (2003). Mutations in distant residues moderately increase the enantioselectivity of *Pseudomonas fluorescens* esterase towards methyl 3-bromo-2-methylpropanoate and ethyl 3-phenylbutyrate. *Chem. – A Eur. J.* 9: 1933–1939.

114 May, O., Nguyen, P.T., and Arnold, F.H. (2000). Inverting enantioselectivity by directed evolution of hydantoinase for improved production of L-methionine. *Nat. Biotechnol.* 18: 317–320.

115 Ehren, J., Govindarajan, S., Morón, B. et al. (2008). Protein engineering of improved prolyl endopeptidases for celiac sprue therapy. *Protein Eng. Des. Sel.* 21: 699–707.

116 Reetz, M.T. and Wu, S. (2009). Laboratory evolution of robust and enantioselective Baeyer–Villiger monooxygenases for asymmetric catalysis. *J. Am. Chem. Soc.* 131: 15424–15432.

117 Hendrikse, N.M., Charpentier, G., Nordling, E., and Syren, P.O. (2018). Ancestral diterpene cyclases show increased thermostability and substrate acceptance. *FEBS J.* 285 (24): 4660–4673.

118 Epstein, C.J., Goldberger, R.F., and Anfinsen, C.B. (1963). The genetic control of tertiary protein structure: studies with model systems. *Cold Spring Harb. Symp. Quant. Biol.* 28: 439–449.

119 Koepnick, B., Flatten, J., Husain, T. et al. (2019). De novo protein design by citizen scientists. *Nature* 570: 390–394.

120 Dawson, W.M., Rhys, G.G., and Woolfson, D.N. (2019). Towards functional de novo designed proteins. *Curr. Opin. Chem. Biol.* 52: 102–111.

121 Degrado, W.F., Summa, C.M., Pavone, V. et al. (1999). De novo design and structural characterization of proteins and metalloproteins. *Annu. Rev. Biochem.* 68: 779–819.

122 Moffet, D.A. and Hecht, M.H. (2001). De novo proteins from combinatorial libraries. *Chem. Rev.* 101 (10): 3191–3204.

123 Grigoryan, G. and Degrado, W.F. (2011). Probing designability via a generalized model of helical bundle geometry. *J. Mol. Biol.* 405: 1079–1100.

124 Harbury, P.B., Zhang, T., Kim, P.S., and Alber, T. (1993). A switch between two-, three-, and four-stranded coiled coils in GCN4 leucine zipper mutants. *Science* 262: 1401–1407.

125 Thomas, F., Boyle, A.L., Burton, A.J., and Woolfson, D.N. (2013). A set of de novo designed parallel heterodimeric coiled coils with quantified dissociation constants in the micromolar to sub-nanomolar regime. *J. Am. Chem. Soc.* 135: 5161–5166.

126 Leaver-Fay, A., Tyka, M., Lewis, S.M. et al. (2011). ROSETTA3: an object-oriented software suite for the simulation and design of macromolecules. *Methods Enzymol.* 487: 545–574.

127 Tantillo, D.J., Chen, J., and Houk, K.N. (1998). Theozymes and compuzymes: theoretical models for biological catalysis. *Curr. Opin. Chem. Biol.* 2: 743–750.

128 Höcker, B. (2014). Design of proteins from smaller fragments-learning from evolution. *Curr. Opin. Struct. Biol.* 27: 56–62.

129 Jacobs, T.M., Williams, B., Williams, T. et al. (2016). Design of structurally distinct proteins using strategies inspired by evolution. *Science* 352: 687–690.

130 Huang, P.-S., Feldmeier, K., Parmeggiani, F. et al. (2016). De novo design of a four-fold symmetric TIM-barrel protein with atomic-level accuracy generated the cluster map. *Nat. Chem. Biol.* 12 (1): 29–34.

131 Malisi, C., Kohlbacher, O., and Höcker, B. (2009). Automated scaffold selection for enzyme design. *Proteins* 77: 74–83.

132 Zanghellini, A., Jiang, L., Wollacott, A.M. et al. (2006). New algorithms and an in silico benchmark for computational enzyme design. *Protein Sci.* 15: 2785–2794.

133 Lapidoth, G., Khersonsky, O., Lipsh, R. et al. (2018). Highly active enzymes by automated combinatorial backbone assembly and sequence design. *Nat. Commun.* 9: 1–9.

134 Marshall, L.R., Zozulia, O., Lengyel-Zhand, Z., and Korendovych, I.V. (2019). Minimalist de novo design of protein catalysts. *ACS Catal.* 9: 9265–9275.

135 Burton, A.J., Thomson, A.R., Dawson, W.M. et al. (2016). Installing hydrolytic activity into a completely de novo protein framework. *Nat. Chem.* 8: 837–844.

136 Krieger, E., Nabuurs, S.B., and Vriend, G. (2003). Homology modeling. In: *Structural Bioinformatics* (eds. P.E. Bourne and H. Weissig), 507–521. Wiley-Liss, Inc.

137 Krieger, E., Joo, K., Lee, J. et al. (2009). Improving physical realism, stereochemistry, and side-chain accuracy in homology modeling: four approaches that performed well in CASP8. *Proteins* 77: 114–122.

138 Xiang, Z. (2006). Advances in homology protein structure modeling. *Curr. Protein Pept. Sci.* 7 (3): 217–227.

139 Berman, H., Henrick, K., and Nakamura, H. (2003). Announcing the worldwide protein data bank. *Nat. Struct. Biol.* 10 (12): 980.

140 Eisenmesser, E.Z., Bosco, D.A., Akke, M., and Kern, D. (2002). Enzyme dynamics during catalysis. *Science* 295: 1520–1523.

141 Amaro, R.E., Baudry, J., Chodera, J. et al. (2018). Ensemble docking in drug discovery. *Biophys. J.* 114 (10): 2271–2278.

142 Horberg, J., Saenz-Mendez, P., and Eriksson, L.A. (2018). QM/MM studies of Dph5 – a promiscuous methyltransferase in the eukaryotic biosynthetic pathway of diphthamide. *J. Chem. Inf. Model.* 58 (7): 1406–1414.

143 Childers, M.C. and Daggett, V. (2017). Insights from molecular dynamics simulations for computational protein design. *Mol. Syst. Des. Eng.* 2 (1): 9–33.

144 Davids, T., Schmidt, M., Bottcher, D., and Bornscheuer, U.T. (2013). Strategies for the discovery and engineering of enzymes for biocatalysis. *Curr. Opin. Chem. Biol.* 17 (2): 215–220.

145 Frushicheva, M.P., Mills, M.J., Schopf, P. et al. (2014). Computer aided enzyme design and catalytic concepts. *Curr. Opin. Chem. Biol.* 21: 56–62.

146 Damborsky, J. and Brezovsky, J. (2014). Computational tools for designing and engineering enzymes. *Curr. Opin. Chem. Biol.* 19: 8–16.

147 Bloom, J.D., Labthavikul, S.T., Otey, C.R.A., and Frances, A. (2006). Protein stability promotes evolvability. *Proc. Natl. Acad. Sci. U. S. A.* 103 (15): 5869–5874.

148 Bommarius, A.S. and Broering, J.M. (2005). Established and novel tools to investigate biocatalyst stability. *Biocatal. Biotransform.* 23 (3–4): 125–169.

149 Polizzi, K.M., Bommarius, A.S., Broering, J.M., and Chapparo-Riggers, J.F. (2007). Stability of biocatalyst. *Curr. Opin. Chem. Biol.* 11 (2): 220–225.

150 Pace, C.N., Grimsley, G.R., Thomson, J.A., and Barnett, B.J. (1988). Conformational stability and activity of ribonuclease TI with zero, one, and two intact disulfide bonds. *J. Biol. Chem.* 263 (24): 11820–11825.

151 Tidor, B. and Karplus, M. (1993). The contribution of cross-links to protein stability: a normal mode analysis of the configurational entropy of the native state. *Proteins: Struct., Funct., Genet.* 15: 71–79.

152 Dombkowski, A.A., Sultana, K.Z., and Craig, D.B. (2014). Protein disulfide engineering. *FEBS Lett.* 588 (2): 206–212.

153 Pabo, C.O. and Suchanek, E.G. (1986). Computer-aided model-building strategies for protein design. *Biochemistry* 25 (20): 5987–5991.

154 Hazes, B. and Dijkstra, B.W. (1988). Model building of disulfide bonds in proteins with known three-dimensional structure. *Protein Eng.* 2 (2): 119–125.

155 Dombkowski, A.A. and Crippen, G.M. (2000). Disulfide recognition in an optimized threading potential. *Protein Eng.* 13 (10): 679–689.

156 Dombkowski, A.A. (2003). Disulfide by design: a computational method for the rational design of disulfide bonds in proteins. *Bioinformatics* 19 (14): 1852–1853.

157 Craig, D.B. and Dombkowski, A.A. (2013). Disulfide by design 2.0: a web-based tool for disulfide engineering in proteins. *BMC Bioinformatics* 14 (346): 1–6.

158 Matthews, B.W., Nicholson, H., and Becktel, W.J. (1987). Enhanced protein thermostability from site-directed mutations that decrease the entropy of unfolding. *Proc. Natl. Acad. Sci. U. S. A.* 84: 6663–6667.

159 Pikkemaat, M.G., Linssen, A.B.M., Berendsen, H.J.C., and Janssen, D.B. (2002). Molecular dynamics simulations as a tool for improving protein stability. *Protein Eng.* 15 (3): 185–192.

160 Reetz, M.T., Carballeira, J.D., and Vogel, A. (2006). Iterative saturation mutagenesis on the basis of B factors as a strategy for increasing protein thermostability. *Angew. Chem. Int. Ed.* 45 (46): 7745–7751.

161 Jochens, H., Aerts, D., and Bornscheuer, U.T. (2010). Thermostabilization of an esterase by alignment-guided focussed directed evolution. *Protein Eng. Des. Sel.* 23 (12): 903–909.

162 Floor, R.J., Wijma, H.J., Colpa, D.I. et al. (2014). Computational library design for increasing haloalkane dehalogenase stability. *ChemBioChem* 15 (11): 1660–1672.

163 Yu, H., Yan, Y., Zhang, C., and Dalby, P.A. (2017). Two strategies to engineer flexible loops for improved enzyme thermostability. *Sci. Rep.* 7: 41212.

164 Lehmann, M., Pasamontes, L., Lassen, S.F., and Wyss, M. (2000). The consensus concept for thermostability engineering of proteins. *Biochim. Biophys. Acta* 1543: 408–415.

165 Korendovych, I.V., Kulp, D.W., Wu, Y. et al. (2011). Design of a switchable eliminase. *Proc. Natl. Acad. Sci. U. S. A.* 108 (17): 6823–6827.

166 Hollfender, F., Kirby, A.J., and Tawfik, D.S. (1997). Efficient catalysis of proton transfer by synzymes. *J. Am. Chem. Soc.* 119: 9578–9579.

167 Casey, M.L., Kemp, D.S., Paul, K.G., and Cox, D.D. (1973). Physical organic chemistry of benzisoxazoles. I. Mechanism of the base-catalyzed decomposition of benzisoxazoles. *J. Org. Chem.* 38: 2294–2301.

168 Kemp, D.S., Cox, D.D., and Paul, K.G. (1997). Physical organic chemistry of benzisoxazoles. IV. Origins and catalytic nature of the solvent rate acceleration for the decarboxylation of 3-carboxybenzisoxazoles. *J. Am. Chem. Soc.* 97: 7312–7318.

169 Pey, A.L., Rodriguez-Larrea, D., Gavira, J.A. et al. (2010). Modulation of buried ionizable groups in proteins with engineered surface charge. *J. Am. Chem. Soc.* 132 (4): 1218–1219.

170 Moroz, Y.S., Dunston, T.T., Makhlynets, O.V. et al. (2015). New tricks for old proteins: single mutations in a nonenzymatic protein give rise to various enzymatic activities. *J. Am. Chem. Soc.* 137 (47): 14905–14911.

171 Yeung, N., Lin, Y.W., Gao, Y.G. et al. (2009). Rational design of a structural and functional nitric oxide reductase. *Nature* 462 (7276): 1079–1082.

172 Musa, M.M., Lott, N., Laivenieks, M. et al. (2009). A single point mutation reverses the enantiopreference of *Thermoanaerobacter ethanolicus* secondary alcohol dehydrogenase. *ChemCatChem* 1 (1): 89–93.

173 Lonsdale, R., Hoyle, S., Grey, D.T. et al. (2012). Determinants of reactivity and selectivity in soluble epoxide hydrolase from quantum mechanics/molecular mechanics modeling. *Biochemistry* 51 (8): 1774–1786.

174 Saenz-Mendez, P., Katz, A., Perez-Kempner, M.L. et al. (2017). Structural insights into human microsomal epoxide hydrolase by combined homology modeling, molecular dynamics simulations, and molecular docking calculations. *Proteins* 85 (4): 720–730.

175 Kourist, R., Brundiek, H., and Bornscheuer, U.T. (2010). Protein engineering and discovery of lipases. *Eur. J. Lipid Sci. Technol.* 112 (1): 64–74.

176 Shu, Z.-Y., Jiang, H., Lin, R.-F. et al. (2010). Technical methods to improve yield, activity and stability in the development of microbial lipases. *J. Mol. Catal. B Enzym.* 62 (1): 1–8.

177 Dorau, R., Görbe, T., and Svendahl Humble, M. (2018). Improved enantioselectivity of subtilisin carlsberg towards secondary alcohols by protein engineering. *ChemBioChem* 19 (4): 338–346.

178 Bermudez, E., Ventura, O.N., Eriksson, L.A., and Saenz-Mendez, P. (2014). Improved homology model of cyclohexanone monooxygenase from *Acinetobacter calcoaceticus* based on multiple templates. *Comput. Biol. Chem.* 49: 14–22.

179 Pazmiño, D.E.T., Snajdrova, R., Rial, D.V. et al. (2007). Altering the substrate specificity and enantioselectivity of phenylacetone monooxygenase by structure-inspired enzyme redesign. *Adv. Synth. Catal.* 349 (8–9): 1361–1368.

180 Butler, C.F., Peet, C., Mason, A.E. et al. (2013). Key mutations alter the cytochrome P450 BM3 conformational landscape and remove inherent substrate bias. *J. Biol. Chem.* 288 (35): 25387–25399.

181 Yoshida, S. and Hiraga, K. (2016). A bacterium that degrades and assimilates poly(ethylene terephthalate). *Science* 351 (6278): 1196–1199.

182 Biundo, A., Ribitsch, D., and Guebitz, G.M. (2018). Surface engineering of polyester-degrading enzymes to improve efficiency and tune specificity. *Appl. Microbiol. Biotechnol.* 102 (8): 3551–3559.

183 Joo, S., Cho, I.J., Seo, H. et al. (2018). Structural insight into molecular mechanism of poly(ethylene terephthalate) degradation. *Nat. Commun.* 9 (1): 382.

184 Liu, B., He, L., Wang, L. et al. (2018). Protein crystallography and site-direct mutagenesis analysis of the poly(ethylene terephthalate) hydrolase PETase from *Ideonella sakaiensis*. *ChemBioChem* 19: 1471–1475.

185 Austin, H.P., Allen, M.D., Donohoe, B.S. et al. (2018). Characterization and engineering of a plastic-degrading aromatic polyesterase. *Proc. Natl. Acad. Sci. U. S. A.* 115 (19): E4350–E4357.

186 Zhang, X., Fevre, M., Jones, G.O., and Waymouth, R.M. (2018). Catalysis as an enabling science for sustainable polymers. *Chem. Rev.* 118 (2): 839–885.

187 Corey, E.J. (1988). Retrosynthetic thinking-essentials and examples. *Chem. Soc. Rev.* 17: 111–133.

188 Turner, N.J. and O'Reilly, E. (2013). Biocatalytic retrosynthesis. *Nat. Chem. Biol.* 9 (5): 285–288.

189 Honig, M., Sondermann, P., Turner, N.J., and Carreira, E.M. (2017). Enantioselective chemo- and biocatalysis: partners in retrosynthesis. *Angew. Chem. Int. Ed.* 56 (31): 8942–8973.

190 Stamm, A., Biundo, A., Schmidt, B. et al. (2019). A retro-biosynthesis-based route to generate pinene-derived polyesters. *ChemBioChem* 20 (13): 1664–1671.

191 Gandini, A. (2011). *Monomers and Macromonomers from Renewable Resources*, 1–33. Weinheim: Wiley-VCH.

192 Kourist, R., Jochens, H., Bartsch, S. et al. (2010). The alpha/beta-hydrolase fold 3DM database (ABHDB) as a tool for protein engineering. *ChemBioChem* 11 (12): 1635–1643.

193 Jochens, H. and Bornscheuer, U.T. (2010). Natural diversity to guide focused directed evolution. *ChemBioChem* 11 (13): 1861–1866.

194 Hohne, M., Schatzle, S., Jochens, H. et al. (2010). Rational assignment of key motifs for function guides in silico enzyme identification. *Nat. Chem. Biol.* 6 (11): 807–813.

195 Gora, A., Brezovsky, J., and Damborsky, J. (2013). Gates of enzymes. *Chem. Rev.* 113 (8): 5871–5923.

196 Damborsky, J. and Brezovsky, J. (2009). Computational tools for designing and engineering biocatalysts. *Curr. Opin. Chem. Biol.* 13 (1): 26–34.

197 Chovancova, E., Pavelka, A., Benes, P. et al. (2012). CAVER 3.0: a tool for the analysis of transport pathways in dynamic protein structures. *PLoS Comput. Biol.* 8 (10): e1002708.

198 Sehnal, D., Svobodova Varekova, R., Berka, K. et al. (2013). MOLE 2.0: advanced approach for analysis of biomacromolecular channels. *J. Cheminform.* 5: 1–13.

199 Kim, J.K., Cho, Y., Lee, M. et al. (2015). BetaCavityWeb: a webserver for molecular voids and channels. *Nucleic Acids Res.* 43 (W1): W413–W418.

200 Brezovsky, J., Kozlikova, B., and Damborsky, J. (2018). Computational analysis of protein tunnels and channels. In: *Protein Engineering Methods and Protocols* (eds. U. Bornscheuer and M. Höhne), 25–85. Springer.

201 Brezovsky, J., Chovancova, E., Gora, A. et al. (2013). Software tools for identification, visualization and analysis of protein tunnels and channels. *Biotechnol. Adv.* 31 (1): 38–49.

202 Kokkonen, P., Bednar, D., Pinto, G. et al. (2019). Engineering enzyme access tunnels. *Biotechnol. Adv.* 37 (6): 107386.

203 Pavlova, M., Klvana, M., Prokop, Z. et al. (2009). Redesigning dehalogenase access tunnels as a strategy for degrading an anthropogenic substrate. *Nat. Chem. Biol.* 5 (10): 727–733.

204 France, S.P., Hepworth, L.J., Turner, N.J., and Flitsch, S.L. (2017). Constructing biocatalytic cascades: in vitro and in vivo approaches to de novo multi-enzyme pathways. *ACS Catal.* 7: 710–724.

205 Krieg, T., Sydow, A., Faust, S. et al. (2018). CO_2 to terpenes: autotrophic and electroautotrophic α-humulene production with *Cupriavidus necator*. *Angew. Chem. Int. Ed.* 57: 1879–1882.

206 Leonard, E., Ajikumar, P.K., Thayer, K. et al. (2010). Combining metabolic and protein engineering of a terpenoid biosynthetic pathway for overproduction and selectivity control. *Proc. Natl. Acad. Sci. U. S. A.* 107: 13654–13659.

207 Cobb, R.E., Sun, N., and Zhao, H. (2013). Directed evolution as a powerful synthetic biology tool. *Methods* 60: 81–90.

208 Dragosits, M. and Mattanovich, D. (2013). Adaptive laboratory evolution – principles and applications for biotechnology. *Microb. Cell Fact.* 12: 64.

209 Tenaillon, O., Rodríguez-Verdugo, A., Gaut, R.L. et al. (2012). The molecular diversity of adaptive convergence. *Science* 335: 457–461.

210 Shalem, O., Sanjana, N.E., Hartenian, E. et al. (2014). Genome-scale CRISPR-Cas9 knockout screening in human cells. *Science* 343: 84–87.

211 Konermann, S., Brigham, M.D., Trevino, A.E. et al. (2015). Genome-scale transcriptional activation by an engineered CRISPR-Cas9 complex. *Nature* 517: 583–588.

3

Immobilization Techniques for the Preparation of Supported Biocatalysts: Making Better Biocatalysts Through Protein Immobilization

Javier Rocha-Martín[1], Lorena Betancor[2], and Fernando López-Gallego[3,4]

[1] *Institute of Catalysis and Petrochemistry (ICP), Department of Biocatalysis, CSIC Campus UAM, Cantoblanco, 28049 Madrid, Spain*
[2] *Universidad ORT Uruguay, Biotechnology Department, Faculty of Engineering, Mercedes 1237, 11100, Montevideo, Uruguay*
[3] *Basque Research and Technology Alliance (BRTA), Heterogeneous Biocatalysis Laboratory, Center for Cooperative Research in Biomaterials (CIC biomaGUNE), Paseo de Miramon 182, 20014, Donostia-San Sebastián, Spain*
[4] *IKERBASQUE, Basque Foundation for Science, Maria Diaz de Haro 3, 48013 Bilbao, Spain*

3.1 Introduction

One of the major challenges of sustainable chemistry is expanding the palette of bio-based fine chemicals. In this context, industrial microbiology and biocatalysis are already key enabling technologies to efficiently and sustainably synthesize chemicals. Nowadays, biotransformations rely on two main approaches: microbial fermentations [1, 2] and cell-free systems [3]. The former approach dominates current chemical biomanufacturing projects using native and engineered pathways to mainly produce alcohols, acids, or amino acids. By using microbes as microfactories, we can assemble highly complex synthetic schemes, but the product purity is limited because of unwanted side reactions and reproducibility issues that jeopardize system robustness. In the past decade, cell-free multienzyme systems come up as an attractive alternative to whole cell biotransformations [4, 5]. By using isolated soluble enzymes, one can assemble a biosynthetic pathway with high selectivity but suffering from the instability issues underlying soluble enzymes. Furthermore, isolated enzymes increase the complexity of downstream processing once the reaction is completed because the soluble biocatalysts are more difficult to separate from the reactants than the whole cells.

Unfortunately, isolated soluble enzymes are yet underutilized at industrial level, finding important hurdles for industrial applications because of their **low stability/robustness** under nonphysiological conditions. For this reason, scientists have devoted enormous efforts engineering enzymes to overcome their limitations as industrial catalysts. Protein engineering pursues adapting enzymes to industry needs [6]. Unfortunately, this technique by itself cannot address the enzyme solubility issue that hampers biocatalyst reusability

Biocatalysis for Practitioners: Techniques, Reactions and Applications, First Edition.
Edited by Gonzalo de Gonzalo and Iván Lavandera.

and workability in flow processes. Complementary to protein engineering, protein immobilization makes enzymes more suitable for industrial processes (i.e., by simplifying the downstream processing) [7–9]. Immobilization entails the heterogenization of the enzyme, giving rise to heterogeneous biocatalyst easily reusable during several operation cycles and ready to be integrated into plug-flow reactors for continuous operation. In some particular cases, enzyme immobilization may also enhance other enzyme properties such as stability, activity, and selectivity [10].

3.2 General Aspects to Optimize Enzyme Immobilization Protocols

The use of heterogeneous biocatalysts has paved the way for applied biocatalysis at industrial scale [11]. Hence, enzyme immobilization seems to be a safe path to the industrialization of enzymatic processes. Nevertheless, enzyme immobilization is not trivial and so several parameters must be considered to optimally immobilize enzymes on solid carriers [12].

3.2.1 Carrier Nature

The nature of the carriers may affect the physicochemical properties of the enzymes and consequently impact their functionality. For example, highly hydrophobic carriers promote the interfacial activation of lipases; a type of enzymes that naturally hydrolyzes the ester bonds from the triacylglycerols, but with more than 50 different applications in industrial biocatalysis. By immobilizing lipases on these hydrophobic surfaces, the intrinsic activity of the immobilized lipases increases 2–10 times depending on the source of the lipase. These immobilized preparations are also able to work under anhydrous conditions [13].

3.2.2 Immobilization Chemistry

Depending on the immobilization chemistry, enzymes can be reversibly or irreversibly attached to the solid material, which determines the ultimate application of the heterogeneous biocatalyst. For example, in those processes where the enzyme lixiviation issues must be avoided, irreversible immobilization is mandatory. On the contrary, when carrier costs are the main contribution to the process cost, reversible chemistry will be demanded to remove the inactivated enzymes and replenish the carrier with fresh ones to increase the useful life of the costly materials.

3.2.3 Protein Orientation

Many enzymes, when attached to the carriers, acquired orientations that limit their functionalities [14]. When the optimization target is retaining the maximum molar activity of the immobilized enzymes, control of the enzyme orientation is crucial. This aspect becomes a necessity in electrobiocatalytic and sensing systems because the enzyme attached to the electrodes must work at their maximum performance to achieve high chemical conversions. In particular, enzyme orientation has experienced the major advances in the biosensing field [15]. For example, the orientation of flavin-dependent oxidases (i.e. glucose oxidase) on electrodes needs to be controlled to achieve the highest direct electron transfer efficiency from the substrates (i.e. glucose) to the electrode, aiming at detectable current even with extremely low analyte concentrations.

3.2.4 Multivalence of the Protein Attachment

For some applications, enzyme stability matters to a higher extent than activity, and for this reason, more robust immobilized proteins must be attained. It is widely accepted that the more intense the attachment, the higher the enzyme stability [16]. Those processes where enzyme production may limit their economic feasibility demand reusable biocatalysts that ensure long operational uses of the costly enzyme. For example, the synthesis of semisynthetic β-lactam antibiotics currently relies on penicillin G acylase (PGA) immobilized on porous microparticles through a covalent and irreversible multivalent attachment. The heterologous production of this enzyme presents some difficulties that are necessary to stabilize the protein for its industrial applications [17]. Immobilized PGA on Eupergit® C can work for more than 800 operational cycles retaining more than 60% of its initial activity [18].

3.2.5 Chemical and Geometrical Congruence

The molecular interaction between the enzyme and the carrier relies on the reactivity and density of chemical groups at both enzyme and carrier surfaces to immobilized enzymes [19]. Such reactivity relies on ready-to-react groups, ideally without adding a catalyst that speeds up the immobilization process. The complementarity between the reactivity and density of enzyme and carrier groups is defined as chemical congruence and determines not only the immobilization kinetics but also the multivalency of the final attachment. Besides their reactivity and density, exposure of reactive groups is fundamental for a proper interaction between enzyme and support. Herein, the geometrical congruence between enzymes and carrier surfaces is fundamental for immobilization purposes and depends on physicochemical features such as pore size and mesh architecture of porous materials, tertiary and quaternary structures of the enzymes, and enzyme molecular weight [20]. For example, carriers with a high density of reactive groups but having pore sizes lower than the protein size would result in poorly efficient immobilizations because the vast majority of the carrier-reactive groups will remain inaccessible for most of the enzyme molecules. On the contrary, carriers with a high reactive group density and suitable pores for the enzyme size would be inefficient to immobilized enzymes with low reactive amino acids (i.e. lysines).

3.2.6 Enzyme Spatial Organization

The spatial organization is often an unaccountable parameter to optimize the fabrication of heterogeneous biocatalysts [21]. However, there is a line of experimental evidence that indicates that some enzymes located at the inner regions of porous carriers perform significantly worse than those located at the outermost surface. The effect of spatial distribution is rather significant for those enzymes that work on substrates that undergo transport restrictions to diffuse into the porous carriers. Of course, spatial organization issues are do not exist when using nonporous materials, where the enzymes can only be located at the material surface that is exposed to the bulk. When preparing heterogeneous biocatalysts without controlling the spatial organization, a fraction of immobilized enzymes can be wasted if the substrate never reaches the enzyme active sites located at the inner regions of the materials [22, 23]. For each process, the maximum loading capacity that provides the maximum apparent activity must be assessed, accounting for the enzyme spatial distribution to avoid its wasting.

3.3 Type of Carriers for Immobilized Proteins

The selection of a suitable material for enzyme immobilization is key to ultimately optimizing the performance of the heterogeneous biocatalysts. However, a universal material has not been described for all different applications of immobilized enzymes. Ideally, the material should be inert, resistant to physicochemical degradation, biocompatible, resistant to microbial attack, and inexpensive [24–26]. Porous carriers are mostly chosen because the high surface area enables a higher enzyme loading and enzymes are protected from the reaction bulk. Nanostructures have also got momentum in virtue of their quick synthesis, large surface area, and fluent movement. When selecting one material for enzyme immobilization, we need to consider its nature (type), its geometry, its dimensions, and its reactive groups.

3.3.1 Types of Materials

3.3.1.1 Organic Materials

These green carriers are commonly subclassified into natural and synthetic polymers according to their origin [12]. Natural polymers possess interesting properties such as high protein affinity, nontoxicity, hydrophilic nature, and biodegradability for food industry, pharmaco-medical applications, and agricultural procedures. The most common natural polymers are cellulose [27], agarose [28], pectins, chitin, chitosan, proteins (gelatin, albumin, and collagen), and alginate.

On the other hand, synthetic polymers are sometimes preferred as a result of their low cost, high stability, and resistant to biodegradation. Polymethacrylate [29], polyacrylamide, and silicone are some examples of synthetic polymers for enzyme immobilization.

3.3.1.2 Inorganic Materials

Although these materials can also be subdivided into natural (silica, bentonite, and activated carbon) and synthetic (glass, zeolite [30], and ceramics), the most remarkable inorganic material for enzyme immobilization is silica [31, 32]. This material can be easily synthesized and activated with different reactive groups. Because silica is quite stable to high temperatures, inert, and less expensive, this material has been widely used for industrial manufacturing and also for biomedical purposes [33].

3.3.2 Geometry

3.3.2.1 Beads

The majority of the commercially available carriers for enzyme immobilization are porous microbeads based on the types of materials described above. This type of geometry is preferred because of their large surface area that allows high enzyme loads, resulting in heterogeneous biocatalysts with high volumetric activities. Moreover, the use of beads eases the implementation of enzymes in both batch and packed-bed reactors for discontinuous and continuous processes. Immobilized glucose isomerase is one of the most productive enzyme currently exploited at large scale for the production of fructose syrups from saccharose. This immobilized biocatalyst is able to produce 11 000 kg of product per 1.0 kg of immobilized enzyme.

3.3.2.2 Monoliths

The geometry of monoliths finds useful applications in flow biocatalysis where the enzyme is immobilized across the surface and length of monolithic porous materials integrated into plug-flow columns. Using this geometry, the biotransformation takes place flushing a solution of substrates and additives across the monolith at a suitable flow rate that assures the optimal residence times to achieve their full conversion. Monoliths can be readily coupled with downstream steps like product extraction or crystallization. This geometry can not only be applied for both large-scale reactors as well as in conventional chemistry but also exploited in microreactors for more modern applications. Monoliths normally present a better response to high flow rates as pressure back issues are less dramatic than using packed beds of porous beads.

3.3.2.3 Membranes

Porous membranes are usually coupled to the outlet of a plug-flow column. These membrane reactors are normally thinner than the reactors based on monoliths and are mainly exploited for those biotransformations where the starting materials are gases [34]. For example, the conversion of CO_2 to methanol has been performed by a four-enzyme system immobilized on functionalized membranes [35, 36].

3.3.3 Dimensions

The dimensions of the carriers are very important because the particle size impacts on the external reaction diffusion of substrates. Hence, carriers formed by smaller particles (<10 μm) pose lower external diffusion restriction to the substrates than larger particles (>100 μm). Nevertheless, larger particles are more easily recovered or integrated to packed-bed reactors. On the other hand, pore size determines the internal diffusion restriction of the substrates, which impacts on the apparent catalytic efficiency of the immobilized enzymes. The size of the immobilized enzymes is also restricted by the pore size. Likewise, the pore size affects the accessibility of the substrates to the enzyme active sites. Nowadays, hierarchical porous materials are gaining momentum because they contain both large and small pores that allow to assemble more complex architectures. Using these materials with no uniform porosity, small and large enzymes can be co-immobilized as well as biotransformations using both small and bulky substrate can be performed. Finally, large surface areas are desired in carriers for enzyme immobilization because they can load high amount of proteins to achieve highly active heterogeneous biocatalysts.

3.3.4 Commercially Available Porous Carriers for Enzyme Immobilization

Nowadays, several companies, such as Agarose Bead Technologies (ABT), Purolite Ltd., Enginzyme AB, Resindion Srl, and Merck KGaA among others, supply a variety of carriers for enzyme immobilization with different types, geometries, and dimensions (particle and pore size). Table 3.1 shows the most used commercially available carriers when preparing heterogeneous biocatalysts.

3.4 Immobilization Methods and Manners

Hitherto, no immobilization protocol has proved to be universal for the enhancement of both activity and stability of the immobilized enzymes. In consequence, enzyme

Table 3.1 Properties of commercial carriers for protein immobilization.[a)]

Carrier (supplier)	Material	Particle size (μm)	Pore size (nm)	Geometry	Website address
ABT™ (ABT)	Agarose	20–350	50–260	Beads	https://www.abtbeads.es/size-exclusion-chromatography/plain-agarose-beads
Purolite™ *(Purolite)*	Methacrylic polymers	100–300	30–180		https://www.purolite.com/ls-product-type/enzyme-immobilization-resins
Relizyme™ *(Resindion)*		100–500	20–40		https://www.resindion.com/index.php?option=com_virtuemart&page=shop.browse&category_id=30&Itemid=53&TreeId=46
CPG-Silica *(Merck)*	Silica	75–125	170		https://www.sigmaaldrich.com/catalog/search?term=MFCD00163370&interface=MDL%20No.&N=0+&mode=partialmax&lang=es®ion=ES&focus=product
EziG™ *(EnginZyme)*		50–150	30–50	Amorphous	http://enginzyme.com/wp-content/uploads/2017/03/EziG%E2%84%A2-Product-Data-Sheet.pdf

a) Data have been obtained from the company's websites. For more information, see the references.

immobilization is still an empirical exercise based on trial-and-error experiments rather than a rational task. Underlying disciplines as materials science, protein structure, enzymology, and reactor engineering must converge during the fabrication of heterogeneous biocatalysts [37].

The main immobilization chemistries between the enzyme and the carrier can be divided into reversible and irreversible immobilization (Figure 3.1). On the one hand, enzymes can be reversibly immobilized by hydrophobic, electrostatic, or van der Waals interactions with the surface of the carriers. Reversible immobilization can be easily reverted by changing the media conditions. This is an attractive point to tackle the high-cost issues of the carriers because when the enzymatic activity decays, the carrier can be regenerated and reloaded with fresh enzyme. On the other hand, irreversible immobilization involves covalent interactions to irreversibly bind the enzyme on the carrier surface. Therefore, the enzyme cannot be physically detached from the carrier. Although the carrier cannot be regenerated once the enzyme is inactive, irreversible immobilization is frequently adopted to prolong the stability of the biocatalyst. Multivalent attachments foment the rigidification of the enzyme and consequently the stabilization of the enzyme [38]. Subsequently, the operational lifetime of the biocatalyst increases as long as the enzyme is not lixiviated from the solid carrier. Nonetheless, the severe rigidification of enzymes must be balanced because it may cause a decrease of the protein activity.

Basically, the most used immobilization chemistries are as follows:

- *Physical adsorption*: This is the most straightforward and oldest technique used to immobilize proteins. The physical adsorption is achieved by weak and nonspecific forces such as hydrophobic or ionic interactions [39]. Hydrophobic adsorption has been used since more than a century ago to nowadays [40], but the interaction can be very easily disrupted. In contrast, ionic adsorption is usually preferred because the interaction between the enzyme and the carrier is stronger, although the enzyme can be still removed by modifying the ionic strength or the pH [41, 42]. This methodology exploits the interaction between the positive (basic residues) and negative (acidic residues) charges of enzymes with the positively or negatively charged groups of the carrier surface.

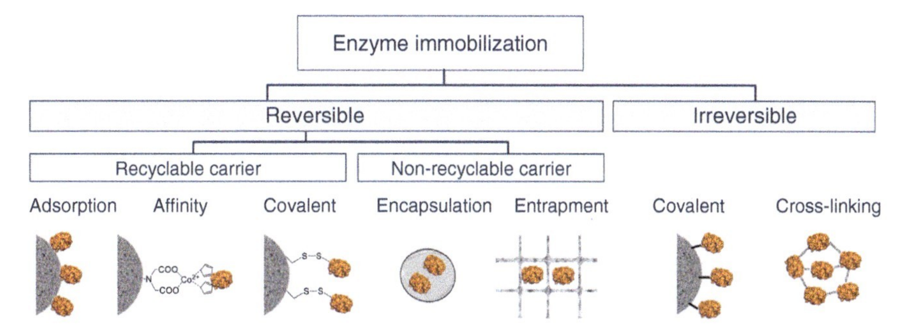

Figure 3.1 Scheme of main enzyme immobilization chemistries.

- *Affinity immobilization*: This type of reversible immobilization relies on the selectivity and complementarity of biomolecules [43], e.g. antibodies and antigens, avidin and biotin, polyhistidine tag and metal ions (Co^{2+}, Cu^{2+}, Zn^{2+}, or Ni^{2+}), hormones and receptors, lectins and glycosilated molecules, and DNA and DNA-binding protein domains, among others. The protein is usually genetically modified and the carrier is activated with the corresponding group before the immobilization process. The minimal conformational changes and the control of protein orientation during affinity immobilization results in a high enzymatic activity of the biocatalyst [44]. Unfortunately, its stability is low upon immobilization because of potential lixiviation of the enzyme.
- *Covalent binding*: Reversible (usually imine or disulfide [45] bonds) and irreversible covalent binding are applied to attach proteins to carriers, although reversible covalent bonds (Schiff's bases) are frequently reduced to become irreversible. The side chains of lysine, cysteine, tyrosine, histidine, arginine, and glutamic/aspartic acid can perform a nucleophilic attack to the reactive groups (epoxy, aldehyde, vinyl, and cyanogen bromide groups) of the carrier for the covalent binding [38, 46–48]. Afterward, the enzyme remains bound to the carrier by multipoint covalent attachment, which is a strong linking but does not orient the enzyme [38]. Covalent binding offers benefits in terms of biocatalyst stability [46] and reuse, which are crucial for industrial biocatalytic processes. However, the irreversible nature of the immobilization does not allow the recycling of the expensive carriers for a new enzyme immobilization.
- *Entrapment/encapsulation*: Unlike the previous immobilization strategies, the entrapment and encapsulation are not carried out on a prefabricated matrix (carrier). Typically, entrapment strategies need the presence of enzymes during the synthesis of the carrier. The most common carriers are polymeric matrices [49, 50], hollow fibers [51], and microcapsules [52]. Entrapment generally requires a chemical contact between the enzyme and the carrier, while encapsulated enzymes are confined within microcapsules but can freely move inside the microcapsules without continuous chemical interaction with the carrier. For this reason, this type of immobilization can be reversible or irreversible. Although this technique offers many possibilities for the design of the biocatalyst, the carriers cannot be reused.
- *Cross-linking*: This carrier-free immobilization methodology relies on the irreversible intermolecular cross-linkages between a cross-linking agent (commonly, glutaraldehyde) and the lysine residues of enzymes [53]. Cross-linked enzyme aggregates (CLEAs) are primarily prepared by precipitation of enzymes with acetone, ethanol, butanol, or ammonium sulfate, followed by the polymerization step. The great advantage of cross-linking is the lack of expensive carriers to immobilize enzymes [54]. The stability of the biocatalyst and the low enzyme release are other benefits of cross-linking [55]. Nevertheless, loss of activity and mass transfer limitations are the main disadvantages of the high enzyme crowding on CLEAs.

3.5 Evaluation of the Enzyme Immobilization Process

Despite considerable efforts over decades by researchers to predict the behavior of an immobilized biocatalyst and to perfect it, it remains a challenging task. The optimization of immobilized enzymes in terms of activity and stability continues to be approached mainly empirically.

In the following sections, we will give some tips and tricks for a successful enzyme immobilization as well as indications on the minimum number of data required to define an enzymatic immobilization process and thus understand the properties of immobilized enzymes.

3.5.1 Considerations Before Immobilization

3.5.1.1 Preparation of the Enzymatic Solution to Be Immobilized

When using lyophilized enzymes, it is important to dissolve the desired amount of enzyme in a buffer that does not interfere with the immobilization process. For example, if we have selected an immobilization protocol by ionic adsorption, we should not use a buffer immobilization with high ionic strength [39]. If we are going to immobilize an enzyme on supports activated with glyoxyl groups (short aliphatic aldehydes), we should not use buffers containing amine compounds such as tris(hydroxymethyl)aminomethane (Tris) [56].

In many other occasions, we use commercial enzymatic solutions in which the manufacturer does not provide information on the composition in which the soluble enzyme is found. In this case, it is advisable to dialyze or gel filtrate the enzymatic solution before immobilization in order to remove the compounds that could interfere with the selected immobilization protocol.

Next, we must check that the enzyme is completely soluble in the immobilization buffer. We should also check that the pH of the enzyme solution is as required by the selected immobilization protocol, and if necessary, adjust it accordingly.

3.5.1.2 Stability of the Soluble Enzyme Under Immobilization Conditions

In an ideal scenario, the soluble enzyme should be completely stable for the entire immobilization process under immobilization conditions. However, some immobilization protocols require the immobilization process to be developed at a particular pH (e.g. the use of glyoxyl-activated supports requires the immobilization process to run at pH 10 [56] or the addition of detergents to prevent the formation of enzymatic aggregates particularly when we work with lipases [57]). Therefore, it is advisable to evaluate the stability of the soluble enzyme under immobilization conditions. For this purpose, an enzymatic solution must be incubated under the same immobilization conditions without adding support or replacing the activated support by the inert support.

In general, it is advisable to select immobilization protocols where the soluble enzyme is fully stable. When this is not possible, stabilizing additives (such as sugars and polyols) can be used [58–60]. In addition, in some cases, immobilization is very rapid and allows stabilization of the immobilized enzyme while the control solution, which contains the soluble enzyme, is inactivated. In this particular case, the activity of the immobilized enzyme decreases much less than the control solution.

3.5.2 Parameters Required to Define an Immobilization Process

In the following section, we describe the key parameters that we need to quantify when reporting an immobilization protocol.

3.5.2.1 Immobilization Yield

Immobilization yield (IY) provides information about the amount of enzyme incorporated to the carrier upon the immobilization process. This parameter allows us to determine the percentage of the immobilized enzyme on the carrier as long as the enzyme is active and stable under immobilization conditions. To calculate it, the following equation can be used:

$$\text{IY}(\%) = \frac{(A_i - A_s)}{A_i} * 100 \tag{3.1}$$

where A_i is the activity of the solution offered for immobilization and A_s is the activity in the supernatant at the end of the immobilization process.

Another parameter that it is advisable to include in any immobilization report is the activity of the suspension (it includes supernatant activity plus the activity of the immobilized enzyme). Besides, it is advisable to conduct a parallel experiment incubating the control solution (soluble enzyme or soluble enzyme plus inert carrier) under the exact immobilization conditions to compensate for enzyme inactivation if necessary.

The most recommendable graphical representation of the time course of an immobilization process includes three activity curves: suspension, supernatant, and control solution. Thus, the *X*-axis represents the activity and on the *Y*-axis the immobilization time. The information shown in this plot helps to understand the immobilization process and to look for solutions if any problem arises during the process.

Sometimes, researchers show the performance of immobilization in terms of protein instead of activity. When the IY is less than 100%, it can lead to error, especially when the solution of the enzyme to be immobilized is not pure and is, thus, composed of a mixture of proteins. In this case, we would have an average measure of the immobilized total protein because the different enzymes may have different IYs. It should also be noted that many immobilization protocols can selectively immobilize the target enzymes. Therefore, the measurement of the activities of control solution, suspension, and supernatant provides more accurate information on the immobilization process. However, monitoring protein concentration in the supernatant and enzyme activity can sometimes be very useful to determine the enzyme load of an immobilized biocatalyst or to discard any inactivation of the soluble form.

3.5.2.2 Expressed Activity or Apparent Activity

The expressed or apparent activity (EA), also known as activity recovery or recovered activity, represents the activity in the immobilized preparation. This parameter goes beyond the intrinsic activity of the immobilized enzymes because it also accounts for the steric hindrances and mass transfer issues of the reactants to access the active site of the heterogeneous biocatalyst. This parameter is calculated from the activity in the enzyme control solution and the IY:

$$\text{EA}(\%) = \frac{A_{\text{immobilized}}}{(A_i - A_s)} * 100 \tag{3.2}$$

where $A_{\text{immobilized}}$ is the activity of the immobilized preparation (measurement of the activity of the immobilized biocatalyst washed and suspended in the activity assay buffer), A_i is the activity of the solution offered for immobilization, and A_s is the activity in the supernatant at the end of the immobilization process. This parameter provides information about the effect of immobilization protocol on the enzyme activity.

3.5.2.3 Specific Activity of the Immobilized Biocatalyst

The specific activity of an immobilized biocatalyst (SA), also known as global enzyme activity yield, is the amount of enzyme activity expressed per unit mass (or unit volume) of the biocatalyst and its dimension is μmol/(min × g support):

$$\text{SA}(\%) = \frac{\text{EA}}{M_{\text{biocatalyst}}} \tag{3.3}$$

where EA is the expressed activity and $M_{\text{biocatalyst}}$ is the mass or volume of the biocatalyst. This parameter relates the final activity observed in the immobilized biocatalyst with the initial solution offered for immobilization.

The SA of a heterogeneous biocatalyst is defined as the product of two factors: enzyme loading (quantity) and the intrinsic SA of the immobilized enzyme (quality) [37]. The loss of intrinsic enzyme activity as a result of the attachment to the solid surface is due to reversible or irreversible alterations in the native structure of the protein. The unproductive binding of the enzyme in a too rigid conformation (affecting its efficiency) or in an inadequate orientation also leads to a decrease in the intrinsic activity. For a deeper study about the possible causes responsible for the loss of activity during the immobilization process, we recommend a recently published review by Boudrant et al. [61].

3.6 Applied Examples of Immobilized Enzymes

A commercial glycerol dehydrogenase (GlyDH) from *Cellulomonas* sp. (Sigma-Aldrich Co., St. Louis, IL, USA) is immobilized by ionic adsorption on monoaminoethyl-*N*-aminoethyl (MANAE)-agarose carrier (Figure 3.2) [62]. About 1.0 g of the carrier is incubated in an enzyme solution containing a total of 10 international units (IU) of GlyDH activity and 0.3 mg of enzyme. After one hour, the beads are filtered and washed. The total GlyDH activity left in supernatant upon immobilization is 0 IU, and the total leftover enzyme concentration is 0 mg.

The washed immobilized biocatalyst is assayed for activity and the total activity of the biocatalyst is found to be 7.6 IU. In this case, the IY would be 100%, the expressed activity is 76%, and the SA of the immobilized biocatalyst is 7.6 U g^{-1} (Table 3.2 and Figure 3.3). As shown in Figure 3.3, the soluble enzyme is completely stable under the conditions of immobilization (control solution). Because enzyme was not eluted during the washings (with low ionic strength buffer), the decrease in the EA attains for the effect of immobilization.

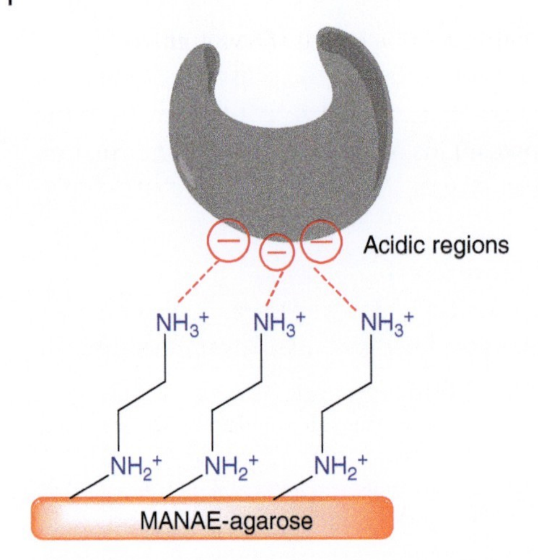

Figure 3.2 Immobilization chemistry of enzymes on MANAE-agarose. Dashed lines depict the ionic interaction between the most acidic regions of the proteins and the positive surface of the carrier.

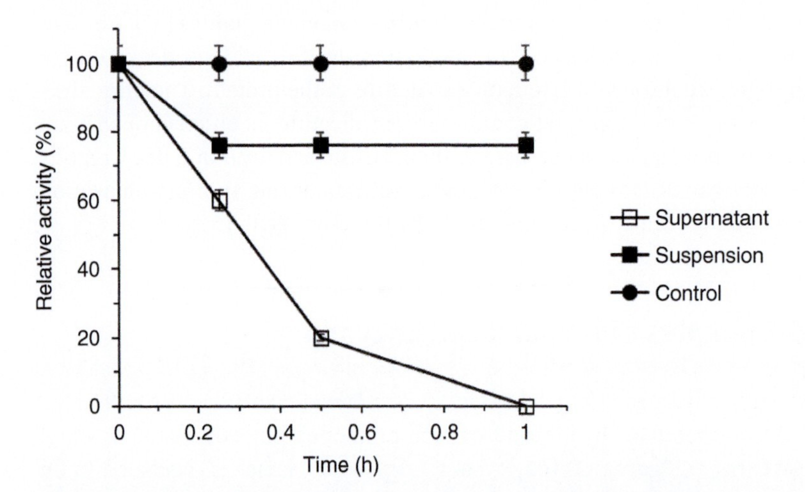

Figure 3.3 Graphical representation of the time course immobilization of the GlyDH enzyme on MANAE agarose carrier at pH 7.0 for 1 hour. Control (●) is the soluble GlyDH incubated under immobilization conditions, suspension (■) includes the supernatant plus the immobilized enzyme activity, and supernatant is the activity of the soluble fraction not attached to the support (□).

3.6.1 Characterization of the Immobilized Biocatalyst

3.6.1.1 Determination of the Catalytic Activity of the Final Immobilized Biocatalyst and Maximum Protein Loading Capacity

For industrial and economically viable use, heterogeneous biocatalysts should have the highest possible SA. This assessment is also done determining the immobilization effectiveness (η).

Table 3.2 Experimental data obtained in the immobilization time course of GlyDH enzyme on MANAE agarose carrier.

Time (h)	Supernatant (IU)	%	Suspension (IU)	%	Control (IU)	%
0	10	100	10	100	10	100
0.25	6	60	7.6	76	10	100
0.5	2	20	7.6	76	10	100
1	0	0	7.6	76	10	100

Supernatant represents a liquid sample of the immobilization bulk after separating the carrier through either filtration or centrifugation. **Suspension** represents a sample that includes the immobilization bulk plus the carrier particles. **Control** is defined as the soluble enzyme (liquid sample) incubated under the same immobilization conditions replacing the carrier volume by the same volume of buffer. It is mandatory to withdraw and measure the three types of samples at each analysis time.

$$\eta = \frac{\text{Immobilized SA}}{\text{Soluble SA}} \tag{3.4}$$

When this effectiveness tends to 1, we conclude that the immobilization negligibly affects the enzyme performance, neither by causing negative structural distortions nor posing diffusion barriers. Ideally, we tend to load as much enzyme as possible to maximize the volumetric activity of the heterogeneous biocatalyst [63]. However, higher loads dramatically decrease η, making the reaction limited by diffusion instead by enzyme kinetics. The conventional characterization of heterogeneous biocatalysts implies the determination of the load capacity of the support used to perform the immobilization. This parameter shows the amount of protein loaded on the carrier, usually from an end point equilibrium with the liquid phase after the attachment has reached an apparent equilibrium [64]. Then, the SA of the heterogeneous biocatalyst is determined.

However, many effects can mask the real SA, complicating the biocatalyst optimization. Typically, there is no linear relationship between the enzyme load and the specific activity of a heterogeneous biocatalyst [65]. The maximum enzyme loading and the time required to reach the maximum load vary considerably according to different immobilization conditions [66].

In the case of a porous support, the interaction between the enzyme and the solid surface, the spatial distribution and density of the immobilized enzymes are critical parameters when selecting a suitable immobilization protocol [26].

Under conditions of nonhomogeneous enzyme distribution, effects such as enzyme aggregation and pore-clogging (even when the support is not completely saturated) could affect the general activity and, therefore, the kinetics of the reaction [65]. The use of confocal laser scanning microscopy (CLSM) to determine the spatial distribution and the density of enzymes across solid porous surfaces is gaining momentum to better characterize the heterogeneous biocatalysts [67, 68].

In other occasions, protein–protein interactions have been described when enzyme-saturated supports are used, which can alter not only the activity but also the stability of heterogeneous biocatalysts. For example, Zaak et al. [69] have recently described how in the case of *Thermomyces lanuginosus* lipase, enzyme concentration is a critical variable when porous supports are used. Highly loaded immobilized biocatalysts were less stable than low-load biocatalysts. The same effect has also been described by Fernández-López et al. [70] for *Candida antarctica* lipase B. The biocatalyst immobilized at a maximum load showed less stability than other preparations with less enzymatic load. Therefore, the crowding effect can affect the final performance of the immobilized enzyme altering substrate diffusion, enzyme structure flexibility, and catalytic activity. Bolivar et al. have recently listed some of the emerging techniques for the direct quantification of the amount of protein immobilized on a solid surface [37].

3.6.1.2 Apparent Kinetic Parameters of the Immobilized Enzyme

The determination of the apparent kinetic parameters (K_m, K_{cat}, and V_{max}) is very useful to evaluate the effect that immobilization has on enzyme performance, comparing the values obtained with the immobilized enzyme and the soluble form.

An important issue to take into account is the existence of mass transfer limitations, which can affect the kinetic behavior of the immobilized enzyme. When enzymes are immobilized in a solid carrier, mass transfer limitations are commonly referred to as diffusional restrictions because the transport of the product and substrate taking place within the particle (internal diffusional restrictions) or in the stagnant layer adjacent to the carrier–liquid interface (external diffusional restrictions) will occur by molecular diffusion [71]. Therefore, mass transfer rates can severely restrict the catalytic potential of an enzyme. In the case of porous carriers, external diffusional restrictions are less important than internal diffusional restrictions because diffusion occurs within the porous matrix. In contrast, external diffusion restrictions may be significant for enzymes immobilized on the surface of a nonporous carrier.

The kinetic behavior of immobilized enzyme under the influence of mass transfer limitations is termed as apparent, and it does not represent the real behavior of the enzyme. To calculate the apparent kinetic parameters, we measure the initial reaction rates at different substrate concentrations under steady-state conditions and fit the experimental data to the Michaelis–Menten equation. Typically, the intrinsic kinetic parameters of an immobilized enzyme do not correspond to the parameters of its soluble form because of a combination of structural changes driven by the immobilization and mass restriction occurred during the process. Only in the absence of mass transfer limitations, we can compare the intrinsic kinetic parameters of both soluble and immobilized enzymes.

Diffusional problems can be identified from Lineweaver–Burk (or double reciprocal) plot that represents $1/V$ versus $1/[S]$ (V = initial rate and $[S]$ = substrate concentration). When diffusional problems exist, there is no proportionality between the enzyme load and observed activity, which means that the substrate consumption rate is higher than the substrate diffusion rate within the porous microstructure of the carriers. Therefore, the substrate concentration decreases across the pores and the substrate may not be available for the inner enzyme molecules. There are some simple tricks to determine whether the decrease in activity is caused by diffusional problems. One strategy is to mill the catalyst to reduce particle size. This will reduce the diffusion limitation of the substrate and increase the activity. Another option is to prepare biocatalysts with different enzymatic loads. For example, if a biocatalyst contains 20 mg of immobilized lipase per gram of support, the activity should be exactly 20-fold higher than a heterogeneous biocatalyst of the same enzyme loaded with 1 mg of lipase per gram of support. In this case, the activity would not be affected by diffusion problems. A detailed analysis about heterogeneous enzyme kinetics with immobilized enzymes can be consulted by the interested reader [63].

3.6.1.3 Biocatalyst Stability

To evaluate the stability of the immobilized enzyme, it is required at least to determine the activity (or residual activity [RA]) of the soluble and immobilized enzyme at time intervals under simulated operational conditions. From the data collected, a RA versus time plot is prepared. RA is defined as:

$$\mathrm{RA}(\%) = \frac{\mathrm{SA}_{\mathrm{time}}}{\mathrm{SA}_{\mathrm{initial}}} \tag{3.5}$$

where $\mathrm{SA}_{\mathrm{time}}$ is the specific activity at a given time and $\mathrm{SA}_{\mathrm{initial}}$ is the initial specific activity. Stabilities at different temperatures, pH and/or in the presence of solvents, can also be determined following the same logic. Figure 3.4 shows an example of the graphical representation of a time course inactivation.

The inactivation profile of the enzyme is modeled based on the deactivation theory proposed by Henely and Sadana [72]. Most of the inactivation mechanisms used are single-stage, first-order mechanisms and two-stage series mechanisms with no RA. Models derived from these inactivation mechanisms are compared to the experimental data. After the corresponding statistical analysis, the one that best fits the experimental data is thus selected. The reader can refer Chapter 5 from the book "Enzyme Biocatalysis: Principles and Applications" for more detailed information [63].

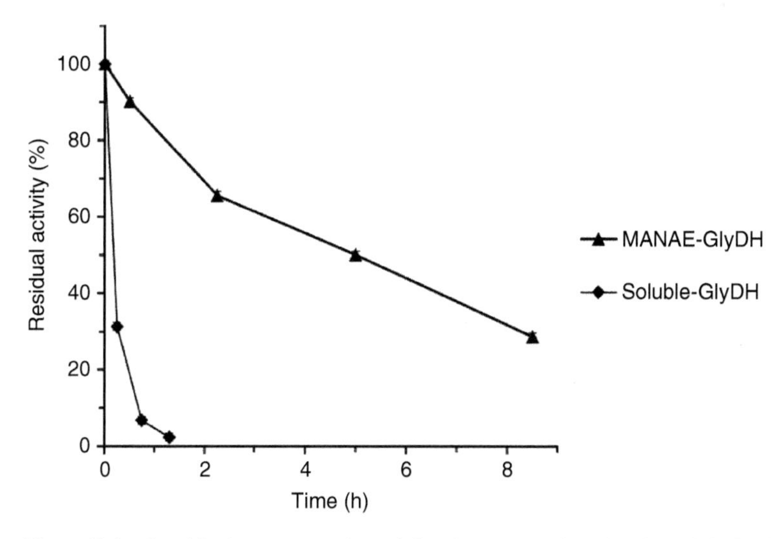

Figure 3.4 Graphical representation of the time course inactivation of GlyDH immobilized on MANAE agarose carrier at pH 9.0 and 55 °C. Symbols: (◆) soluble GlyDH and (▲) GlyDH immobilized on MANAE agarose support.

Parameters such as half-life time ($t_{1/2}$) and the total turnover number (TTN) and stabilization factor (SF) are often used to show the stability of an immobilized biocatalyst:

3.6.1.3.1 The Half-life Time of Biocatalysts

$t_{1/2}$ indicates the time period needed for the residual enzymatic activity to decrease to half of its initial value. When the inactivation process occurs in one step, the inactivation constant (k_d) is calculated as a first-order kinetic constant;

$$\frac{e}{e_0} = \exp\left(-k_d \cdot t\right) \tag{3.6}$$

and $t_{1/2}$ is expressed as:

$$t_{1/2} = \frac{\ln 2}{k_d} \tag{3.7}$$

where k_d is the deactivation constant [time^{-1}].

When the enzyme inactivation involves more than one inactivation step (Figure 3.5), a series mechanism is proposed. Herein, the native (active) enzyme structure (E) is deactivated through to intermediate enzyme conformations (E_1) with specific activity lower specific activity than the native enzyme, until reaching a completely inactive state (E^*).
In this scenario, k_1 and k_2 are the first-order inactivation constants that define the transition rate between E and E_1 and between E_1 and E^*, respectively. They are calculated from the following equation:

$$\frac{e}{e_0} = \left[1 + A \cdot \frac{k_1}{k_2 - k_1}\right] \cdot \exp\left(-k_1 \cdot t\right) - \left[A \cdot \frac{k_1}{k_2 - k_1}\right] \cdot \exp\left(-k_2 \cdot t\right) \tag{3.8}$$

where A is the specific activity ratio of E_1 and E. Values of the inactivation parameters (k_1, k_2, and A) are determined by nonlinear regression of the experimental data to Eq. (3.7) and the coefficient of determination (R^2) will reflect how well the model fits the experimental data. Finally, $t_{1/2}$ can be determined from the model described by Eq. (3.7), considering $e/e_0 = 0.5$.

The half-life time depends on the reaction conditions (temperature, pH value, concentration of substrate and product, cosolvents, etc.), and it is a limiting factor for the TTN (or TON) [73, 74].

The Total Turnover Number (TTN) is defined as the total number of product moles produced by enzyme moles until the biocatalyst is deactivated completely. Therefore, the accurate determination of the TTN involves the measurements until the catalyst activity is completely lost, which indicates a measure of the efficiency of a biocatalyst. The calculation of the TTN will depend on our experimental setup. For example, in batch reactors,

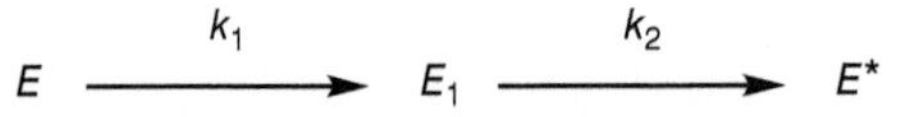

$$E \xrightarrow{k_1} E_1 \xrightarrow{k_2} E^*$$

Figure 3.5 Enzyme inactivation with more than one step.

a filtration unit must be used at the end of each batch to recover the biocatalyst. If we operate a continuous reactor, the biocatalyst must also be retained through a filter integrated into a plug-flow column.

Therefore, the value of TTN should be as high as possible to minimize the impact of the biocatalyst in the final costs of the product and ease the application of enzyme-catalyzed reactions in industrial processes. Two general considerations about the information provided by the *TTN* must be taken into account [64]:

i) When a heterogeneous biocatalyst has too low TTN caused either by its low activity or stability, it does not make sense to immobilize it because the additional costs of modification and recycling facilities limit the process benefits;
ii) When the stability and activity of the soluble biocatalyst is very high (high *TTN* value) and/or the product has a high-added value (the total cost of the biocatalyst is <0.05% of the product value), there would be no economic need to reuse the biocatalyst. However, the improvement of the quality and purity of the product could be a good argument for enzyme immobilization.

The SF is a parameter used to compare quantitatively the stability of the immobilized biocatalysts with their soluble counterpart. It is defined as:

$$\mathrm{SF} = \frac{t_{\frac{1}{2},\text{immobilized enzyme}}}{t_{\frac{1}{2},\text{soluble enzyme}}} \tag{3.9}$$

where $t_{1/2,\text{immobilized enzyme}}$ is the half-life of the immobilized enzyme and $t_{1/2,\text{soluble enzyme}}$ is the half-life of the soluble form.

3.7 Challenges and Opportunities in Enzyme Immobilization

The investigation of new enzyme-immobilized systems is still increasingly driven by economic and environmental constraints that fuel the development of alternatives to conventional synthetic methods. Arguably, the major driven force for the development of novel immobilization approaches is the search for robustness and specific activity that outperform the existing alternatives. Moreover, new biosynthetic applications may also encourage scientists to ample and explore novel possibilities for enzyme immobilization.

One such line of development in the field is the use of new carriers [75]. Metal–organic frameworks (MOFs), for instance, were long studied before applied as a support for enzyme immobilization. From a subtle first sole mention in the literature in 2011 [76] for the immobilization of a peroxidase to a well-documented application today, the past five years have seen a pile of reports on the advantages of this material as an enzymatic carrier [77, 78]. MOFs are organic–inorganic hybrid materials with a crystalline structure such as zeolites whose superior flexibility allows for novel topologies and versatile three-dimensional structures [79]. The structural changes achieved with atomic level precision and the uniform physicochemical environment within the MOF matrix make them highly amenable platforms to immobilize diverse biocatalysts [80].

Using an origami-like approach, Li et al. [81] recently proved that optimization of angles, geometry, and chemical structures of the carrier builders created an interconnected hierarchical channel system for the immobilization of L-lactate dehydrogenase (l-LDH) and diaphorase. Varying pore, aperture, and window sizes of the tailored support facilitated a substrate channeling effect that surpassed the solution regeneration rate of nicotinamide adenine dinucleotide reduced (NADH). This work demonstrated the potential of the carrier for immobilization of the new trend cell-free enzymatic systems as it allows for different interactions of enzymes to the support and precise synthetic control of its architecture that may improve the coupling of biocatalytic tandems.

Another research line in the search of improved carriers is the combination of different materials, each providing added advantages to the composite material for a better heterogeneous biocatalyst.

Silica, although a classic material for enzyme immobilization, has experimented a new age when combined with organic molecules or other inorganic materials. In a recent report, Correa et al. [82] demonstrated that a chemical modification of previously integrated polyethyleneimine molecules in the matrix allowed for a three-dimensional immobilization of an entrapped peroxidase that resulted ~20 times more stable compared to the solely entrapped enzyme in biomimetic silica and ~ 300 times more stable than the soluble enzyme. Addition of magnetic nanoparticles to the composite not only ordered the architecture of the immobilized biocatalyst but also improved the stability and facilitated its separation from reaction media.

In a similar manner but using the well-known biodegradable and environmentally friendly cellulose for enzyme immobilization, Je et al. [83] designed a heterogeneous biocatalyst composed of different materials. The strategy involves a three-step approach of enzyme precipitate coating, consisting of covalent enzyme attachment, enzyme precipitation, and cross-linking. These two last steps were combined with magnetic nanoparticles that facilitated the separation of the biocatalyst but also significantly improved the loading and stability of an α-chymotrypsin.

A deeper comprehension of support material influences on the activity/stability of enzymes has undoubtedly demonstrated that functionality and catalytic behavior are influenced by slight differences in the material composition of the carriers. The ability of certain carriers to be carefully tuned in terms of flexibility, hydrophilicity, or particle size make them extremely attractive to potentiate the properties of enzymes that would not be otherwise applicable [78, 84].

An interesting approach that has been recently explored to improve immobilized enzymes is the integration of costly cofactors to the immobilization matrixes [4]. The leakage of cofactors from immobilized enzymes or poor accessibility to inner portions of the heterogeneous biocatalyst from the media has motivated researchers to find effective solutions. Cofactors can be covalently attached to supports [85] or reversibly bound [86] with the additional advantage of being selectively eluted and reloaded in the heterogeneous biocatalysts extending its half-life time.

Aligned to the concept of self-sufficient catalysts and following the trend of utilizing multienzyme systems in biocatalysis [4, 87], strategies for immobilization must evolve to accommodate multiple enzymes within the same support. Co-immobilization provides benefits such as enhancement of the efficiency of one of the enzymes via *in situ* generation of its substrate, simplification of the process that is conventionally carried out in several steps, elimination of non-desired by-products of an enzymatic reaction, or cofactor regeneration [88]. The continuous discovery of non-natural cascades or adaptation of natural cycles able to work *in vitro* for the preparation of biotechnological valuable products is a constantly challenging investigator for new immobilization designs and architectures [89, 90].

Additionally, to the above-mentioned tendencies in enzyme immobilization, there are efforts to find the best orientation to integrate enzymes on supports [91]. Moreover, methods for the analysis of orientation of immobilized enzymes are increasingly available. The approach confers an element of rationality to the immobilization design and the possibility to arrive to supported conclusions instead of providing hypothesis upon observation of the mean behavior of populations of randomly oriented enzymes on the support. As orientation may provide stabilization, improved catalytic efficiency, and/or selectivity, there is a constant need on studying this parameter as new biocatalysts–support pairs appear [92]. Orientation may be achieved via a specific support functionalization that targets one or many defined amino acids on the surface of an enzyme [93]. Enzymes can be genetically modified to introduce specific tethers or amino acids that serve as an anchor for an oriented integration of the protein to the support [94]. This concept has been exceptionally explored by Benítez-Mateos et al. [95], where they constructed plasmids that fused the N-terminus of superfolded green fluorescent protein (sGFP) with different peptide-tags (poly-(6X)Cys, poly-(6X)His, and poly-(6X)Lys). These tags drove the immobilization of the protein on different suitable activated supports: agarose beads with different functionalities, gold nanorods, and silica nanoparticles. They also achieved the incorporation of non-natural amino acids coupling the enzyme synthesis using a commercially available S30 cell-free extract from *Escherichia coli* with a cyclooctyne/azide click immobilization in a one-pot concurrent process that, as expected, provided an oriented immobilization.

Finally, advanced analytical tools should be deployed to better understand how enzymes are attached to carriers and how they perform within the confined spaces of the solid materials. In the past years, there are many review articles [37, 96] that detail the battery of techniques to study the concentration (density), the conformation, and even the *in operando* activity of the proteins within the carriers. Herein, microscopic techniques coupled to spectroscopic measurements can elicit not only the density of the enzymes immobilized across the surface of the solids but also the conformation changes undergone by them after their utilization with spatio–temporal resolution.

3.8 Conclusions

Immobilization is undoubtedly a powerful tool to improve biocatalysts and enable or expand their applied use. Although it has been used for decades, its utilization has not decayed responding to the constant advent of new enzymes and new biotransformation processes or configurations. Researchers can nowadays exploit the advantages of numerous materials, mixing, and matching the best support for a particular enzyme while walking through the multiple options to optimize the immobilization protocol to obtain highly active and stable heterogeneous biocatalysts. To achieve this, future efforts should include a rigorous study of the immobilization parameters that match the underlying criteria of protein technologists and chemical engineers. The basic parameters to characterize immobilized preparations are not sophisticated and are in constant evolution. Unfortunately, the standardization of immobilization protocols is unmet need, likely because of the lack of consensus about what parameters shall be measured and reported (i.e. IY, specific activity, and half-life time). This chapter pursues providing non-expert researchers with fundamental notions and standard parameters to rapidly select one immobilization protocol over the other. The dynamic nature of this field and the leaning toward a more sustainable chemical industry are propelling a universal use of immobilization when working with enzymes. Latest trends focus on orientation of enzymes on support and sophisticated architectures to accommodate multiple enzymes working in tandem where sciences such as protein engineering and materials science converge. The future will dictate what other science fields will contribute to enzyme immobilization to continue contributing to the improvement of biocatalysts.

List of Abbreviations

η	Immobilization effectiveness
CLEAs	cross-linked enzyme aggregates
EA	Expressed or apparent activity
GlyDH	glycerol dehydrogenase from *Cellulomonas* sp.
IU	International unit
IY	Immobilization yield
k_d	Inactivation constant
L-LDH	L-lactate dehydrogenase
MANAE	monoaminoethyl-*N*-aminoethyl
MOFs	Metal–organic frameworks
NADH	Nicotinamide adenine dinucleotide reduced
PGA	penicillin G acylase
RA	Residual activity
SA	Specific activity of an immobilized biocatalyst
SF	Stabilization factor
sGFP	Superfolded green fluorescent protein
$t_{1/2}$	Half-life time
Tris	Tris(hydroxymethyl)aminomethane
TTN	Total turnover number

References

1 Chen, G.G. and Jewett, M.C. (2016). Editorial: transforming biotechnology with synthetic biology. *Biotechnol. J.* 11 (2): 193–194.

2 Zhang, F., Rodríguez, S., and Keasling, J.D. (2011). Metabolic engineering of microbial pathways for advanced biofuels production. *Curr. Opin. Biotechnol.* 22 (6): 775–783.
3 Dudley, Q.M., Karim, A.S., and Jewett, M.C. (2015). Cell-free metabolic engineering: biomanufacturing beyond the cell. *Biotechnol. J.* 10 (1): 69–82.
4 López-Gallego, F., Jackson, E., and Betancor, L. (2017). Heterogeneous systems biocatalysis: the path to the fabrication of self-sufficient artificial metabolic cells. *Eur. J. Chem.* 23 (71): 17841–17849.
5 Fessner, W.D. (2015). Systems biocatalysis: development and engineering of cell-free "artificial metabolisms" for preparative multi-enzymatic synthesis. *N. Biotechnol.* 32 (6): 658–664.
6 Otte, K.B. and Hauer, B. (2015). Enzyme engineering in the context of novel pathways and products. *Curr. Opin. Biotechnol.* 35: 16–22.
7 Rehm, F., Chen, S., and Rehm, B. (2016). Enzyme engineering for in situ immobilization. *Molecules* 21 (10): 1370.
8 Godoy, C.A., Bdl, R., Grazú, V. et al. (2011). Glyoxyl-disulfide agarose: a tailor-made support for site-directed rigidification of proteins. *Biomacromolecules* 12 (5): 1800–1809.
9 Abian, O., Grazu, V., Hermoso, J. et al. (2004). Stabilization of penicillin G acylase from *Escherichia coli*: site-directed mutagenesis of the protein surface to increase multipoint covalent attachment. *Appl. Environ. Microbiol.* 70 (2): 1249–1251.
10 García-Galan, C., Berenguer-Murcia, Á., Fernández-Lafuente, R., and Rodrigues, R.C. (2011). Potential of different enzyme immobilization strategies to improve enzyme performance. *Adv. Synth. Catal.* 353 (16): 2885–2904.
11 DiCosimo, R., McAuliffe, J., Poulose, A.J., and Bohlmann, G. (2013). Industrial use of immobilized enzymes. *Chem. Soc. Rev.* 42 (15): 6437–6474.
12 Cantone, S., Ferrario, V., Corici, L. et al. (2013). Efficient immobilisation of industrial biocatalysts: criteria and constraints for the selection of organic polymeric carriers and immobilisation methods. *Chem. Soc. Rev.* 42 (15): 6262–6276.
13 Zaks, A. and Klibanov, A.M. (1985). Enzyme-catalyzed processes in organic solvents. *Proc. Natl. Acad. Sci. U. S. A.* 82 (10): 3192–3196.
14 Godoy, C.A., de las Rivas, B., Grazu, V. et al. (2011). Glyoxyl-disulfide agarose: a tailor-made support for site-directed rigidification of proteins. *Biomacromolecules* 12 (5): 1800–1809.
15 Putzbach, W. and Ronkainen, N.J. (2013). Immobilization techniques in the fabrication of nanomaterial-based electrochemical biosensors: a review. *Sensors* 13 (4): 4811–4840.
16 Gregurec, D., Velasco-Lozano, S., Moya, S.E. et al. (2016). Force spectroscopy predicts thermal stability of immobilized proteins by measuring microbead mechanics. *Soft Matter* 12 (42): 8718–8725.
17 Li, K., Mohammed, M.A.A., Zhou, Y. et al. (2020). Recent progress in the development of immobilized penicillin G acylase for chemical and industrial applications: a mini-review. *Polym. Adv. Technol.* 31 (3): 368–388.
18 Kallenberg, A.I., van Rantwijk, F., and Sheldon, R.A. (2005). Immobilization of penicillin G acylase: the key to optimum performance. *Adv. Synth. Catal.* 347 (7–8): 905–926.
19 Mateo, C., Palomo, J.M., Fuentes, M. et al. (2006). Glyoxyl agarose: a fully inert and hydrophilic support for immobilization and high stabilization of proteins. *Enzyme Microb. Technol.* 39 (2): 274–280.

20 Hidalgo, A., Betancor, L., Mateo, C. et al. (2004). Purification of a catalase from *Thermus thermophilus* via IMAC chromatography: effect of the support. *Biotechnol. Prog.* 20 (5): 1578–1582.

21 Roberts, C.C. and Chang, C.E. (2015). Modeling of enhanced catalysis in multienzyme nanostructures: effect of molecular scaffolds, spatial organization, and concentration. *J. Chem. Theory Comput.* 11 (1): 286–292.

22 Do, D.D., Clark, D.S., and Bailey, J.E. (1982). Modeling enzyme immobilization in porous solid supports. *Biotechnol. Bioeng.* 24 (7): 1527–1546.

23 Ladero, M., Santos, A., and García-Ochoa, F. (2001). Diffusion and chemical reaction rates with nonuniform enzyme distribution: an experimental approach. *Biotechnol. Bioeng.* 72 (4): 458–467.

24 Datta, S., Christena, L.R., and Rajaram, Y.R.S. (2013). Enzyme immobilization: an overview on techniques and support materials. *3 Biotech.* 3 (1): 1–9.

25 Mohamad, N.R., Marzuki, N.H.C., Buang, N.A. et al. (2015). An overview of technologies for immobilization of enzymes and surface analysis techniques for immobilized enzymes. *Biotechnol. Biotechnol. Equip.* 29 (2): 205–220.

26 Sheldon, R.A. and van Pelt, S. (2013). Enzyme immobilisation in biocatalysis: why, what and how. *Chem. Soc. Rev.* 42 (15): 6223–6235.

27 Klein, M.P., Scheeren, C.W., Lorenzoni, A.S.G. et al. (2011). Ionic liquid-cellulose film for enzyme immobilization. *Process Biochem.* 46 (6): 1375–1379.

28 Zucca, P., Fernández-Lafuente, R., and Sanjust, E. (2016). Agarose and its derivatives as supports for enzyme immobilization. *Molecules* 21 (11): 1577.

29 Palomo, J.M., Muñoz, G., Fernández-Lorente, G. et al. (2002). Interfacial adsorption of lipases on very hydrophobic support (octadecyl–Sepabeads): immobilization, hyperactivation and stabilization of the open form of lipases. *J. Mol. Catal. B: Enzym.* 19–20: 279–286.

30 Díaz, J.F. and Balkus, K.J. (1996). Enzyme immobilization in MCM-41 molecular sieve. *J. Mol. Catal. B: Enzym.* 2 (2): 115–126.

31 Blanco, R.M., Terreros, P., Fernández-Pérez, M. et al. (2004). Functionalization of mesoporous silica for lipase immobilization: characterization of the support and the catalysts. *J. Mol. Catal. B: Enzym.* 30 (2): 83–93.

32 Carlson, E.D., Gan, R., Hodgman, C.E., and Jewett, M.C. (2012). Cell-free protein synthesis: applications come of age. *Biotechnol. Adv.* 30 (5): 1185–1194.

33 Bharti, C., Nagaich, U., Pal, A.K., and Gulati, N. (2015). Mesoporous silica nanoparticles in target drug delivery system: a review. *Int. J. Pharm. Investig.* 5 (3): 124–133.

34 Cen, Y.K., Liu, Y.X., Xue, Y.P., and Zheng, Y.G. (2019). Immobilization of enzymes in/on membranes and their applications. *Adv. Synth. Catal.* 361 (24): 5500–5515.

35 Luo, J., Meyer, A.S., Mateiu, R.V., and Pinelo, M. (2015). Cascade catalysis in membranes with enzyme immobilization for multi-enzymatic conversion of CO_2 to methanol. *N. Biotechnol.* 32 (3): 319–327.

36 Zhu, D., Ao, S., Deng, H. et al. (2019). Ordered coimmobilization of a multienzyme cascade system with a metal organic framework in a membrane: reduction of CO_2 to methanol. *ACS Appl. Mater. Interfaces* 11 (37): 33581–33588.

37 Bolivar, J.M., Eisl, I., and Nidetzky, B. (2016). Advanced characterization of immobilized enzymes as heterogeneous biocatalysts. *Catal. Today* 259: 66–80.

38 Grazú, V., Abian, O., Mateo, C. et al. (2005). Stabilization of enzymes by multipoint immobilization of thiolated proteins on new epoxy-thiol supports. *Biotechnol. Bioeng.* 90 (5): 597–605.

39 Jesionowski, T., Zdarta, J., and Krajewska, B. (2014). Enzyme immobilization by adsorption: a review. *Adsorption* 20 (5): 801–821.

40 Nelson, J.M. and Griffin, E.G. (1916). Adsorption of invertase. *J. Am. Chem. Soc.* 38 (5): 1109–1115.

41 Mateo, C., Abian, O., Fernández-Lafuente, R., and Guisan, J.M. (2000). Reversible enzyme immobilization via a very strong and nondistorting ionic adsorption on support–polyethylenimine composites. *Biotechnol. Bioeng.* 68 (1): 98–105.

42 Mateo, C., Pessela, B.C.C., Fuentes, M. et al. (2006). Very strong but reversible immobilization of enzymes on supports coated with ionic polymers. In: *Immobilization of Enzymes and Cells* (ed. J.M. Guisan), 205–216. Totowa, NJ: Humana Press.

43 Roy, I. and Gupta, M.N. (2006). Bioaffinity immobilization. In: *Immobilization of Enzymes and Cells* (ed. J.M. Guisan), 107–116. Totowa, NJ: Humana Press.

44 Andreescu, S., Bucur, B., and Marty, J.-L. (2006). Affinity immobilization of tagged enzymes. In: *Immobilization of Enzymes and Cells* (ed. J.M. Guisan), 97–106. Totowa, NJ: Humana Press.

45 Ovsejevi, K., Manta, C., and Batista-Viera, F. (2013). Reversible covalent immobilization of enzymes via disulfide bonds. In: *Immobilization of Enzymes and Cells*, 3 ed. (ed. J.M. Guisan), 89–116. Totowa, NJ: Humana Press.

46 López-Gallego, F., Fernández-Lorente, G., Rocha-Martin, J. et al. (2013). Stabilization of enzymes by multipoint covalent immobilization on supports activated with glyoxyl groups. In: *Immobilization of Enzymes and Cells*, 3 ed. (ed. J.M. Guisan), 59–71. Totowa, NJ: Humana Press.

47 López-Gallego, F., Guisán, J.M., and Betancor, L. (2013). Glutaraldehyde-mediated protein immobilization. In: *Immobilization of Enzymes and Cells*, 3 ed. (ed. J.M. Guisan), 33–41. Totowa, NJ: Humana Press.

48 Mateo, C., Grazu, V., Palomo, J.M. et al. (2007). Immobilization of enzymes on heterofunctional epoxy supports. *Nat. Protoc.* 2: 1022.

49 Reetz, M.T. (2013). Practical protocols for lipase immobilization via sol–gel techniques. In: *Immobilization of Enzymes and Cells*, 3 ed. (ed. J.M. Guisan), 241–254. Totowa, NJ: Humana Press.

50 Sassolas, A., Hayat, A., and Marty, J.-L. (2013). Enzyme immobilization by entrapment within a gel network. In: *Immobilization of Enzymes and Cells*, 3 ed. (ed. J.M. Guisan), 229–239. Totowa, NJ: Humana Press.

51 Ji, X., Wang, P., Su, Z. et al. (2014). Enabling multi-enzyme biocatalysis using coaxial-electrospun hollow nanofibers: redesign of artificial cells. *J. Mater. Chem. B* 2 (2): 181–190.

52 Flickinger, M.C., Rother, C., and Nidetzky, B. (2014). Enzyme immobilization by microencapsulation: methods, materials, and technological applications. In: *Encyclopedia of Industrial Biotechnology*, M.C. Flickinger (Ed.). John Wiley & Sons. https://doi.org/10.1002/9780470054581.eib275.

53 Cao, L., Lv, L., and Sheldon, R.A. (2003). Immobilised enzymes: carrier-bound or carrier-free? *Curr. Opin. Biotechnol.* 14 (4): 387–394.

54 Sheldon, R.A. (2011). Characteristic features and biotechnological applications of cross-linked enzyme aggregates (CLEAs). *Appl. Microbiol. Biotechnol.* 92 (3): 467–477.

55 Velasco-Lozano, S., López-Gallego, F., Mateos-Díaz Juan, C., and Favela-Torres, E. (2016). Cross-linked enzyme aggregates (CLEA) in enzyme improvement – a review. *Biocatalysis* 1: 166–177. doi: https://doi.org/10.1515/boca-2015-0012.

56 Gloria, F.-L., Fernando, L.-G., Juan, M.B. et al. (2015). Immobilization of proteins on highly activated glyoxyl supports: dramatic increase of the enzyme stability via multipoint immobilization on pre-existing carriers. *Curr. Org. Chem.* 19 (17): 1719–1731.

57 Palomo, J.M., Fuentes, M., Fernández-Lorente, G. et al. (2003). General trend of lipase to self-assemble giving bimolecular aggregates greatly modifies the enzyme functionality. *Biomacromolecules* 4 (1): 1–6.

58 Rocha-Martin, J., Acosta, A., Berenguer, J. et al. (2014). Selective oxidation of glycerol to 1,3-dihydroxyacetone by covalently immobilized glycerol dehydrogenases with higher stability and lower product inhibition. *Bioresour. Technol.* 170: 445–453.

59 Back, J.F., Oakenfull, D., and Smith, M.B. (1979). Increased thermal stability of proteins in the presence of sugars and polyols. *Biochemistry* 18 (23): 5191–5196.

60 Kaushik, J.K. and Bhat, R. (2003). Why is trehalose an exceptional protein stabilizer?: an analysis of the thermal stability of proteins in the presence of the compatible osmolyte trehalose. *J. Biol. Chem.* 278 (29): 26458–26465.

61 Boudrant, J., Woodley, J.M., and Fernández-Lafuente, R. (2020). Parameters necessary to define an immobilized enzyme preparation. *Process Biochem.* 90: 66–80.

62 Fernández-Lafuente, R., Rosell, C.M., Rodríguez, V. et al. (1993). Preparation of activated supports containing low pK amino groups. a new tool for protein immobilization via the carboxyl coupling method. *Enzyme Microb. Technol.* 15 (7): 546–550.

63 Illanes, A. (2008). *Enzyme Biocatalysis: Principles and Applications*, 1–391. The Netherlands: Springer.

64 Liese, A. and Hilterhaus, L. (2013). Evaluation of immobilized enzymes for industrial applications. *Chem. Soc. Rev.* 42 (15): 6236–6249.

65 Secundo, F. (2013). Conformational changes of enzymes upon immobilisation. *Chem. Soc. Rev.* 42 (15): 6250–6261.

66 Cantone, S., Ferrario, V., Corici, L. et al. (2013). Efficient immobilisation of industrial biocatalysts: criteria and constraints for the selection of organic polymeric carriers and immobilisation methods. *Chem. Soc. Rev.* 42 (15): 6262–6276.

67 García-García, P., Rocha-Martin, J., Fernández-Lorente, G., and Guisán, J.M. (2018). Co-localization of oxidase and catalase inside a porous support to improve the elimination of hydrogen peroxide: oxidation of biogenic amines by amino oxidase from *Pisum sativum*. *Enzyme Microb. Technol.* 115: 73–80.

68 Rocha-Martín, J., Rivas, B.d.l., Muñoz, R. et al. (2012). Rational co-immobilization of bi-enzyme cascades on porous supports and their applications in bio-redox reactions with in situ recycling of soluble cofactors. *ChemCatChem.* 4 (9): 1279–1288.

69 Zaak, H., El-Hocine, Siar., Kornecki, J.F. et al. (2017). Effect of immobilization rate and enzyme crowding on enzyme stability under different conditions. The case of lipase from Thermomyces lanuginosus immobilized on octyl agarose beads. *Process Biochem.* 56: 117–123.

70 Fernandez-Lopez, L., Pedrero, S.G., Lopez-Carrobles, N. (2017). Effect of protein load on stability of immobilized enzymes. *Enzyme Microb. Technol.* 98: 18–25.

71 Horvath, C. and Engasser, J.-M. (1974). External and internal diffusion in heterogeneous enzymes systems. *Biotechnol. Bioeng.* 16 (7): 909–923.

72 Henley, J.P. and Sadana, A. (1986). Deactivation theory. *Biotechnol. Bioeng.* 28 (8): 1277–1285.
73 Dias Gomes, M. and Woodley, J.M. (2019). Considerations when measuring biocatalyst performance. *Molecules* 24 (19): 3573.
74 Schüth, F., Ward, M.D., and Buriak, J.M. (2018). Common pitfalls of catalysis manuscripts submitted to chemistry of materials. *Chem. Mater.* 30 (11): 3599–3600.
75 Zdarta, J., Meyer, A.S., Jesionowski, T., and Pinelo, M. (2018). A general overview of support materials for enzyme immobilization: characteristics, properties, practical utility. *Catalysts* 8 (2): 92.
76 Lykourinou, V., Chen, Y., Wang, X.-S. et al. (2011). Immobilization of MP-11 into a mesoporous metal–organic framework, MP-11@mesoMOF: a new platform for enzymatic catalysis. *J. Am. Chem. Soc.* 133 (27): 10382–10385.
77 Zare, A., Bordbar, A.K., Razmjou, A., and Jafarian, F. (2019). The immobilization of Candida rugosa lipase on the modified polyethersulfone with MOF nanoparticles as an excellent performance bioreactor membrane. *J. Biotechnol.* 289: 55–63.
78 Liu, Q., Chapman, J., Huang, A. et al. (2018). User-tailored metal–organic frameworks as supports for carbonic anhydrase. *ACS Appl. Mater. Interfaces* 10 (48): 41326–41337.
79 Mehta, J., Bhardwaj, N., Bhardwaj, S.K. et al. (2016). Recent advances in enzyme immobilization techniques: metal-organic frameworks as novel substrates. *Coord. Chem. Rev.* 322: 30–40.
80 Lian, X., Fang, Y., Joseph, E. et al. (2017). Enzyme–MOF (metal–organic framework) composites. *Chem. Soc. Rev.* 46 (11): 3386–3401.
81 Li, P., Chen, Q., Wang, T.C. et al. (2018). Hierarchically engineered mesoporous metal-organic frameworks toward cell-free immobilized enzyme systems. *Chem* 4 (5): 1022–1034.
82 Correa, S., Puertas, S., Gutiérrez, L. et al. (2019). Design of stable magnetic hybrid nanoparticles of Si-entrapped HRP. *PLOS One* 14 (4): e0214004.
83 Je, H.H., Noh, S., Hong, S.-G. et al. (2017). Cellulose nanofibers for magnetically-separable and highly loaded enzyme immobilization. *Chem. Eng. J.* 323: 425–433.
84 Esmaeilnejad-Ahranjani, P., Kazemeini, M., Singh, G., and Arpanaei, A. (2016). Study of molecular conformation and activity-related properties of lipase immobilized onto core–shell structured polyacrylic acid-coated magnetic silica nanocomposite particles. *Langmuir* 32 (13): 3242–3252.
85 Zhang, X.-J., Wang, W.-Z., Zhou, R. et al. (2019). Asymmetric synthesis of *tert*-butyl (3*R*,5*S*)-6-chloro-3,5-dihydroxyhexanoate using a self-sufficient biocatalyst based on carbonyl reductase and cofactor co-immobilization. *Bioprocess Biosyst. Eng.* 43: 21–31.
86 Velasco-Lozano, S., Benítez-Mateos, A.I., and López-Gallego, F. (2017). Co-immobilized phosphorylated cofactors and enzymes as self-sufficient heterogeneous biocatalysts for chemical processes. *Angew. Chem. Int. Ed.* 56 (3): 771–775.
87 Huffman, M.A., Fryszkowska, A., Alvizo, O. et al. (2019). Design of an in vitro biocatalytic cascade for the manufacture of islatravir. *Science* 366 (6470): 1255–1259.
88 Betancor, L. and Luckarift, H.R. (2010). Co-immobilized coupled enzyme systems in biotechnology. *Biotechnol. Genet. Eng. Rev.* 27 (1): 95–114.
89 Tsitkov, S. and Hess, H. (2019). Design principles for a compartmentalized enzyme cascade reaction. *ACS Catal.* 9 (3): 2432–2439.

90 Zhang, G., Quin, M.B., and Schmidt-Dannert, C. (2018). Self-assembling protein scaffold system for easy in vitro coimmobilization of biocatalytic cascade enzymes. *ACS Catal.* 8 (6): 5611–5620.

91 Pan, Y., Li, H., Farmakes, J. et al. (2018). How do enzymes orient when trapped on metal–organic framework (mof) surfaces? *J. Am. Chem. Soc.* 140 (47): 16032–16036.

92 Armenia, I., Grazu Bonavia, M.V., De Matteis, L. et al. (2019). Enzyme activation by alternating magnetic field: importance of the bioconjugation methodology. *J. Colloid Interface Sci.* 537: 615–628.

93 Schöffer, J.N., Matte, C.R., Charqueiro, D.S. et al. (2017). Directed immobilization of CGTase: the effect of the enzyme orientation on the enzyme activity and its use in packed-bed reactor for continuous production of cyclodextrins. *Process Biochem.* 58: 120–127.

94 Mendez, M.B., Rivero, C.W., López-Gallego, F. et al. (2018). Development of a high efficient biocatalyst by oriented covalent immobilization of a novel recombinant 2'-N-deoxyribosyltransferase from *Lactobacillus animalis*. *J. Biotechnol.* 270: 39–43.

95 Benitez-Mateos, A.I., Llarena, I., Sanchez-Iglesias, A., and López-Gallego, F. (2018). Expanding one-pot cell-free protein synthesis and immobilization for on-demand manufacturing of biomaterials. *ACS Synth. Biol.* 7 (3): 875–884.

96 Benítez-Mateos, A.I., Nidetzky, B., Bolivar, J.M., and López-Gallego, F. (2018). Single-particle studies to advance the characterization of heterogeneous biocatalysts. *ChemCatChem* 10: 654–665.

4

Compartmentalization in Biocatalysis

Robert Kourist[1] and Javier González-Sabín[2]

[1] *Graz University of Technology, Institute for Molecular Biotechnology, NAWI Graz, Petersgasse14, 8010, Graz, Austria*
[2] *EntreChem SL, Vivero Ciencias de la Salud, 33011, Oviedo, Spain*

4.1 Introduction

The concept of compartmentalization has its origin in cells, the building blocks of life. The metabolism of both bacterial cells and eukaryotic cells with complex systems of specialized cell organelles is regulated by well-defined compartments, each specializing in a particular function. Thus, the cell is divided into different regions displaying specific microenvironments where each organelle can exhibit optimized performance. With regard to biocatalysis, whole cells containing a particular enzyme can serve as individual compartments. In fact, biological metabolism and biocatalytic processes conceived by chemists face the same problems when multi-catalytic systems get to work. The past years have witnessed the impressive development of cascade reactions (chemo- and enzymatic processes) as a powerful tool in organic synthesis (see Chapters 13 and 14) [1–4]. Conducting consecutive reactions in *one pot* saves unit operations for the isolation and purification of reaction intermediates. Moreover, the immediate conversion of unstable intermediates [5, 6] avoids losses due to side reactions, which is an important aspect regarding downstream processing. In particular, concurrent cascades, also referred to as tandem reactions, can overcome synthetic hurdles for reactions that would not be possible in isolated steps. Classical examples are the regeneration [7, 8] or *in situ generation* of redox cofactors for enzymatic transformations (see also Chapter 10) and dynamic kinetic resolutions (DKRs), which combine a stereoselective reaction with a simultaneous isomerization (chemical catalyst [9] or biocatalyst [10]) in order to overcome the yield limitation of 50% of kinetic resolutions [11]. Additionally, unfavorable equilibria can be shifted by coupling a further reaction to remove the product. A typical example is the amination of ketones by amine transaminases (ATAs) (see also Chapter 6), where the reaction equilibrium lies on the side of the ketone [12]. On the other hand, the

Biocatalysis for Practitioners: Techniques, Reactions and Applications, First Edition.
Edited by Gonzalo de Gonzalo and Iván Lavandera.

combination of chemical and biological catalysts opens new perspectives by exploiting the advantages of both "catalytic worlds" [13], which (i) present important differences in terms of productivity, reactivity, and selectivity and (ii) constitute two of the most important pillars in the toolbox of synthetic organic chemistry.

The development of multienzyme cascade reactions profits from both the tremendous progress in molecular biology techniques and the fact that many enzymes work under rather similar reaction conditions. Yet, the complexity of cascade reactions is often underestimated. In particular, the assembly of enzymes from different organisms and of enzymes with chemical catalysts has to cope with side reactivity and compatibility issues. In a cell, different rates of *in situ* enzyme production provide stable ratios of biocatalysts for the desired metabolic pathway. As *in situ* production in cell-free biocatalytic reactions is not possible, the different stability of enzymes represents an additional challenge. Moreover, there are no standard procedures for the development and optimization of cascades, and the intensification of such processes requires close collaboration between molecular and process engineering. Usually, the advantages of a cascade approach must be balanced against the complexity. The case of the aforementioned chemoenzymatic cascades is particularly significant, and despite the wider pool of water-compatible metal catalysts, the number of concurrent processes (concomitant action of catalysts) reported to date does not exceed 10 examples (Scheme 4.1, compare also Chapter 14). As a result, chemists and engineers have developed ingenious, if not convoluted, strategies such as biphasic systems, supramolecular hosts, artificial metalloenzymes, and spatial separation by flow chemistry to tear down the barriers between both catalytic toolboxes (reciprocal poisoning of catalysts; degradation due to additives, cofactors, or cosolvents; and incompatibility of reaction conditions).

Here, it is also important to note that cascades can be conducted in different stages of interaction and exposure between the different reactions and their components. Temporal and spatial separation of reaction steps often reduces complexity. The perhaps easiest way is a simple "telescoping", in which the crude product of a step is used for the next one. Often, avoiding the purification of an intermediate product offers considerable cost advantage, whereas an extraction step is simpler than a cascade approach. A sequential cascade requires compatibility of the catalyst of the second step with the reaction conditions of the first step. Immobilization allows removal of the first catalyst, but even traces of a protein can inactivate a metal catalyst such as the palladium catalyst for a Suzuki coupling [20]. Thus, a sequential cascade setup allows to change temperature and pH value and to add cosolvents [21, 22]. In contrast, a concurrent cascade is only successful if the inhibitory effect between catalysts and other components is avoided, the side reactivity is suppressed, and full compatibility of the catalysts and their reaction conditions is ensured. Despite this, methodological progress has greatly facilitated cascade development. Enzyme engineering (see also Chapter 2) is a powerful tool to improve parameters such as activity [10] or stability [23] and can thus both increase the general productivity in the conversion of a substrate and serve to adapt an enzyme to the reaction conditions. A striking example (see also Chapter 15) is the adaptation of an ATA for high dimethyl sulfoxide concentrations by Codexis [24]. Yet, enzyme engineering offers mostly gradual improvements and requires usually a starting point, i.e. some residual activity at least under similar conditions.

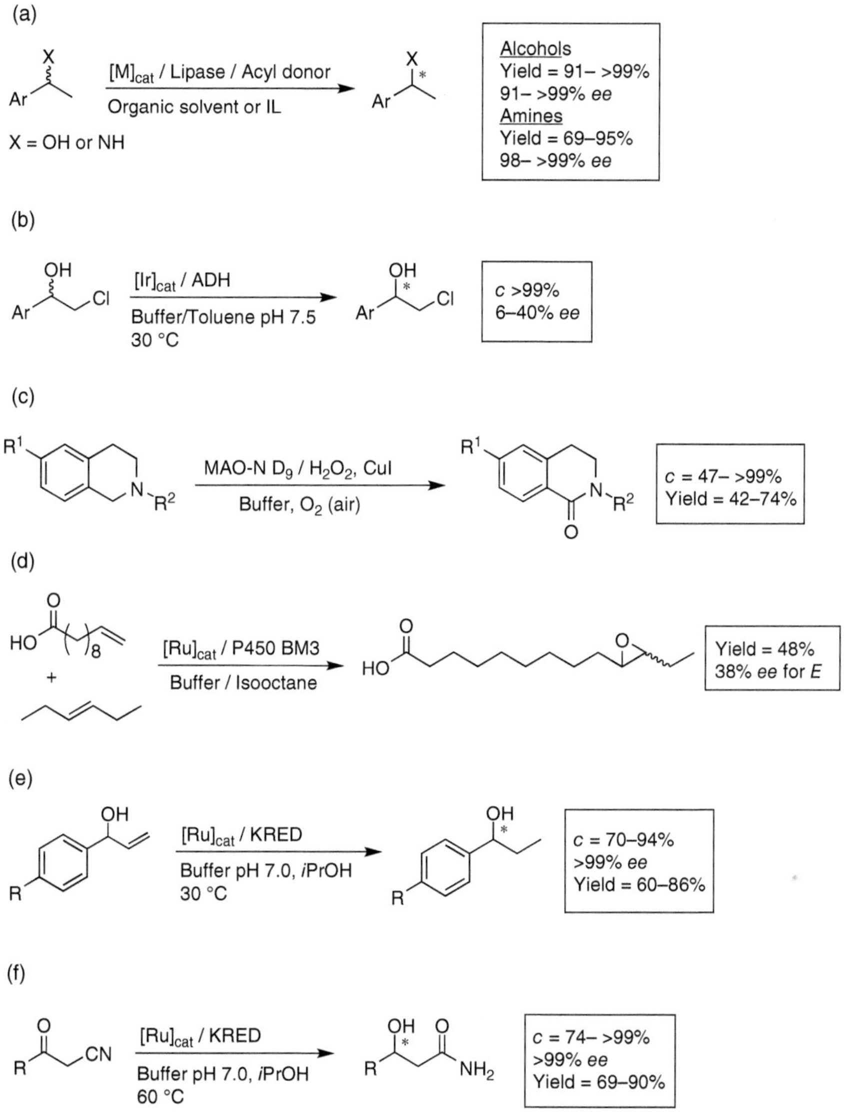

(a)

X

Ar

$[M]_{cat}$ / Lipase / Acyl donor

Organic solvent or IL

X = OH or NH

Alcohols
Yield = 91– >99%
91– >99% *ee*
Amines
Yield = 69–95%
98– >99% *ee*

(b)

OH

Ar Cl

$[Ir]_{cat}$ / ADH

Buffer/Toluene pH 7.5
30 °C

c >99%
6–40% *ee*

(c)

R^1 N R^2

MAO-N D_9 / H_2O_2, CuI

Buffer, O_2 (air)

c = 47– >99%
Yield = 42–74%

(d)

HO 8

+

$[Ru]_{cat}$ / P450 BM3

Buffer / Isooctane

Yield = 48%
38% *ee* for *E*

(e)

OH

R

$[Ru]_{cat}$ / KRED

Buffer pH 7.0, *i*PrOH
30 °C

c = 70–94%
>99% *ee*
Yield = 60–86%

(f)

R CN

$[Ru]_{cat}$ / KRED

Buffer pH 7.0, *i*PrOH
60 °C

R NH_2

c = 74– >99%
>99% *ee*
Yield = 69–90%

Scheme 4.1 Examples of concurrent cascade metal-biocatalytic processes. (a) DKR of alcohols and amines. Source: Modified from Verho and Bäckvall [14]. (b) Metal-catalyzed oxidation-asymmetric bioreduction. Source: Modified from Mutti et al. [15]. (c) Combo enzymatic-chemical oxidation. Source: Modified from Bechi et al. [16]. (d) Metal-catalyzed cross-metathesis-enzymatic epoxidation. Source: Modified from Denard et al. [17]. (e) Metal-catalyzed allylic alcohol isomerization-asymmetric bioreduction. Source: Modified from Ríos-Lombardía et al. [18]. (f) Metal-catalyzed nitrile hydration-asymmetric bioreduction. Source: Modified from Liardo et al. [19].

Solvent engineering can also solve some difficulties. On the one hand, the solvent is crucial for the activity and stability of the catalysts. Although lipases and many proteases show outstanding activity in organic solvents, the majority of enzymes require aqueous media. On the other hand, the choice of the solvent is decisive for the solubility of the substrate and thus on the possible space–time yields. Moreover, the solvent is crucial for the separation of the product. In an ideal case, a soluble substrate is converted to an insoluble product that can be isolated simply by purification [25]. Biphasic systems are a pragmatic approach to increase the solubility of hydrophobic compounds, and many enzymes that do not operate in organic media at low water activity can be successfully applied in them. These systems are, however, complex and difficult to use in continuous flow. Chemoenzymatic cascade reactions are often affected by a solvent dilemma that lies in the preference of many enzymes for water and of many chemical catalysts for water-free reaction conditions. Recently, nonconventional solvents (see also Chapter 9) have offered a third option as they often provide suitable reaction media for catalysts that otherwise require very different solvents [20].

Compartmentalization and continuous flow allow separating some components of cascade reactions. In contrast to compartmentalized *one-pot* reactions, flow chemistry allows to operate at different temperatures and to add additional components to the solvents (see also Chapter 11). The most applied form of compartmentalization is the classical enzyme immobilization by attaching the enzyme on a carrier, enzyme encapsulation, or carrier-free methods such as cross-linking. Although Chapter 3 deals with this topic in detail, a few points should be mentioned here. Immobilization of a single enzyme is routine and a method par excellence to stabilize a biocatalyst improves the productivity and facilitates its separation from the reaction mixture. With a large number of successful applications, the development of new carriers and coupling methods is still a thriving field. Yet, application of immobilization for cascade reactions is quite troublesome. One important aspect is the stability loss of the biocatalyst during repeated reaction cycles. With single-enzyme immobilization, this can be addressed quite easily by either prolonging the duration of batches after several repetitions [26] or by performing the reaction in flow with a series of columns of different lifetimes [27]. A depleted column can then easily be exchanged *en route*. Co-immobilization of several enzymes on the same carrier, however, has to deal with the different half-life times of the cascade enzymes. Therefore, the duration of the formulated catalyst is determined by the weakest enzyme. Moreover, enzymes usually differ in their preferences for carriers, pore sizes, and attachment methods, which lead to difficulty in finding a common immobilization approach to all enzymes of a cascade. Immobilization on different carriers would avoid this, although mixing of different formulated enzymes add mass transfer limitations of the reaction intermediates to those of the substrate and product and the carrier mixture cannot be separated easily. Similar to the co-immobilization case, the half-life time of the mixture is determined by the least-stable biocatalyst.

Throughout the present chapter, different compartmentalization techniques will be discussed, aimed at separating reactions and their catalysts from each other, and also useful to simplify catalyst separation. As deduced from this introduction section, one-pot cascade processes will be a major topic in the chapter, so it seems convenient to highlight the two types of processes that we will show, namely, sequential processes where a second catalyst or a certain key reagent is added only after the completion of the first step and concurrent processes in which the reaction conditions do not change among the consecutive steps and no new reagents are added since the beginning of the process. Compartmentalized cascades can be performed either with or without spatial separation.

As of 2020, the majority of published compartmentalized chemoenzymatic cascades are performed in sequential mode rather than concurrently [4]. Compartmentalization therefore can be considered as a means to support sequential cascades as well as a method to enable new concurrent ones.

4.2 Cell as a Compartment

Compartmentalization of multi-catalytic reactions is mostly done in academia and has not reached industrial implementation yet. In nature, however, as stated in the Introduction section, compartmentalization represents one of the fundamental tools to control complexity. The universal compartment of nature is the cell. It should be noted that this does not only divide the world into "inside" and "outside" but allows for a very precise control of the transfer of molecules. Transporters are highly substrate specific: Nerve cells build up inverse gradients of sodium and potassium cations by using selective transporters. The transporters can be rapidly inhibited or activated and their activity can be regulated on a genetic level. Several strategies use the cell directly as a compartment. Co-expression of enzymes within the same cell limits the transport phenomena between cells (Figure 4.1a). The adjustment of the concentrations of both catalysts, often a crucial factor for cascade reactions, can proceed on the level of the copy number of different plasmids, on the level of different transcription of genes under the control of different promoters, or on the level of translation by using different ribosome binding sites for genes assembled in one operon and thus under the control of one promoter. The amount of transcript from a promoter under certain cultivation conditions can be relatively well estimated. However, an accurate prediction of the resulting correctly folded protein resulting from a certain promoter and ribosome binding site is not possible, particularly if several heterologous genes are expressed at the same time. Therefore, randomization of promoter sequences is often the method of choice for optimization [28]. A much easier approach for the adjustment of biocatalyst ratios is to use mixtures of cells expressing different enzymes (Figure 4.1b) [29]. Use of cell mixtures poses an additional barrier for the transport of reaction intermediates. A recent example is the coupling of an ene-reductase to regeneration of the nicotinamide adenine dinucleotide reduced (NADH) cofactor by an intracellular hydrogenase in the cells of *Cupriavidus necator* [30]. Although the cells showed activity to several known substrates of the enzyme, a series of maleimides were not converted. After cell disruption, however, product formation was observed. As the biotransformation of these maleimides is possible in *Escherichia coli* [31] and cyanobacteria [32], the missing production of specific transporter molecules prohibited the use of *C. necator* for this biotransformation. Nevertheless, transport across cell membranes is often not the limiting factor, and it is much easier to determine the amounts of separate cells than to adjust gene co-expression. Moreover, it has been known that in many cases, fermentations using the so-called "mixed cultures" show superior performance compared to the cultivation of pure strains [33].

Using mixtures of cells of different metabolic capacities expands this concept and allows combining different species with complementary metabolic strengths and weaknesses (Figure 4.1c). In particular, it allows to couple autotrophic and heterotrophic strains. For

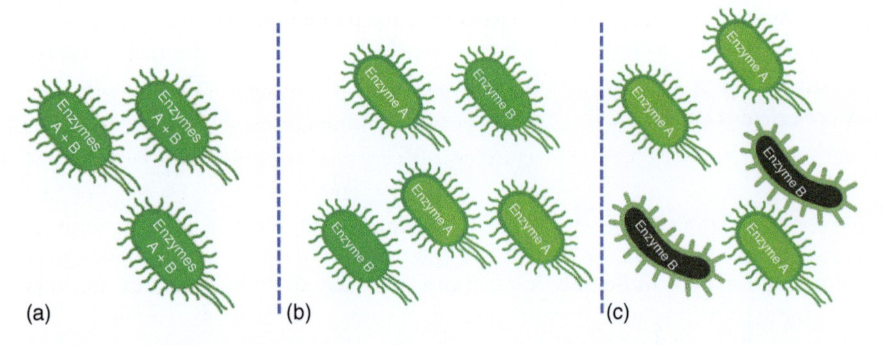

Figure 4.1 Strategies for using cells for compartmentalization of a two-enzyme cascade. (a) Co-expression of the genes within the same cell; (b) mixing cells from the same strain producing different enzymes; and (c) mixed cultures of different species.

example, coupling the saccharose-producing mutant of cyanobacterium *Synechococcus elongatus* cscB to a polyhydroxyalkanoate producing *Pseudomonas putida* strain enabled the synthesis of biopolymers from CO_2 with a maximal rate of 23.8 mg $(l\,day)^{-1}$ and a titer of 156 mg l^{-1} [34]. In a similar approach, Fedeson et al. reported the use of a mixed culture of barium-alginate-encapsulated cells of *S. elongatus* PCC 7942 with recombinant cells of *P. putida* for the coupling of photosynthetic saccharose production in the cyanobacterium with the reductive biotransformation of dinitrotoluene in *P. putida* [35].

In addition to the cell, nature offers a number of additional compartments. In prokaryotes, these are the periplasm and spores, and in eukarya, these are the different organelles. Although these compartments are used by nature for the compartmentalization of the reactions of the metabolism, use in biotechnology is still far from being routine. On top of these structures limited by membranes, nature uses assembly of multi-catalytic cascades. Scaffolding allows the rapid conversion of intermediates. This includes multistep enzymes such as the fatty acid biosynthesis, polyketide synthases, and nonribosomal peptide synthases. Sometimes, the same pathway might be done by dissolved enzymes in one kingdom of life and scaffolded enzymes in another. This is the case with the fatty acid biosynthesis in bacteria and eukarya. A prominent example is the cellulosome that brings together different enzymes for the depolymerization of cellulose. Even a simple bacterial cell has a clear orientation, and biocatalysts can be selectively allocated at the cell center, the periphery, or close to the sites of specific metabolic activities. Membranes are prominent sites for enzyme assembly. Classic examples are here the photosynthetic machinery on the thylakoid membrane and the membrane-bound denitrification apparatus in the bacterium *Pseudomonas aeruginosa* [36] (which connects enzymes from the tricarboxylic acid cycle with electron transport chains in the membrane with nitrate as an electron acceptor). In biocatalysis, a large number of highly interesting works report the potential of compartmentalization to solve synthetic problems. Despite these encouraging findings, however, the principle is far from the universal role that it assumes in metabolism. The following sections will outline different principles of compartmentalization and scaffolding. This includes artificial, bio-inspired, and biological compartmentalization systems. Then, possible challenges for implementation and strategies to overcome them will be discussed.

4.3 Compartmentalization Using Protein Assemblies

Next to cells, self-assembling protein containers can be used to provide a confined reaction environment for reaction steps of a cascade. Protein containers have the advantage that they can be easily produced in cellular systems. Hilvert et al. exploited the fact that the container as a protein itself is amenable to directed enzyme evolution [37]. The loading capacity of the container was coupled to growth of *E. coli* by sequestering human immunodeficiency virus (HIV) protease, which is toxic for the cell, within a lumazine synthase capsid. Four rounds of mutagenesis achieved variants with a 5–10-fold higher loading capacity.

Ferritin is a naturally occurring protein cage with an approximate size of 12 nm. A wide diversity of organisms including bacteria, archaea, fungi, plants, and several animals employ ferritin for iron storage within the cell. The hyperthermophilic bacterium *Pyrococcus furiosus* (Pf) proved a highly thermostable ferritin (PfFerritin), which makes it particularly suitable for applications in chemical catalysis. Incorporation of nanometallic palladium into the core of the ferritin cage creates a nanostructured catalyst. One application is the oxidation of alcohols at higher temperatures in water as reaction medium (Figure 4.2) [38]. Abe et al. used the ferritin cage to immobilize norbornadiene ruthenium complexes (Rh[nbd]). The limited space within the apo-Ft was used to create a defined molecular weight distribution in the polymerization of phenylacetylene. Using the encapsulated catalyst, they achieved an average distribution of 130 monomers per molecule [39].

Virus capsids have usually a size of >20 nm and complex mechanical properties. Their natural role is the encapsulation of the viral genome in one host and its protection during

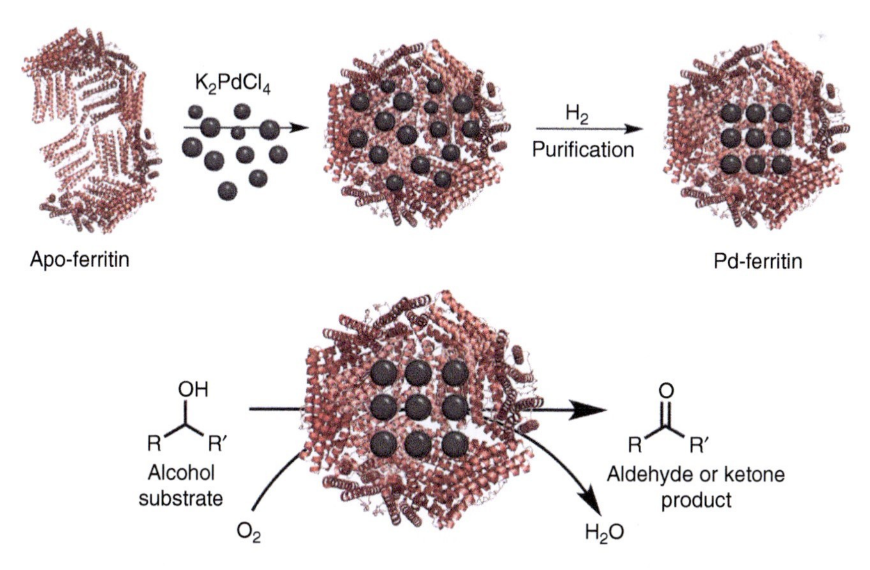

Figure 4.2 The incorporation of nano-metallic palladium (Pd) into the core of the ferritin cage (from *Pyrococcus furiosus*) resulted in a nanostructured hybrid catalyst (Pd-ferritin) catalyzing highly specifically aerobic alcohol oxidations in water. Source: Schmidt et al. [4].

transfer and finally the release inside another host cell. The capacity for self-assembly and their regular and well-defined structure makes virus capsids highly promising systems for encapsulation. As virus capsids are larger than ferritin, they are used for the incorporation of larger molecules, including enzymes [4]. A typical example is the incorporation of horseradish peroxidase (HRP) by Comellas-Aragonès in the particles of cowpea chlorotic mottle virus with an outer diameter of 28 nm [40]. An interesting property of these particles is that the gating behavior is pH dependent, which provides a mechanism for the controlled incorporation and release of entrapped molecules. Since this pioneering work, the encapsulation of enzymes in virus particles as bio-based nanoreactors has developed as a very dynamic field.

Cells are able to form dynamic membraneless compartments in response to changes in the environment. These organelles without membrane boundaries result from liquid–liquid-phase separation of proteins and nucleic acids and participate in several biological functions. Proteins in these compartments contain low complexity domains (LCDs), that is, disordered domains enriched in specific amino acids. Arosio et al. achieved to functionalize inorganic nanoparticles with LCDs resulting in hybrid biological–inorganic materials with promising applications [41]. As a result, the properties of the LCDs were conferred to the nanoparticles, which showed controlled self-assembly and stimulus responsiveness to pH and ionic strength. Then, the LCDs were conjugated with the globular enzyme adenylate kinase (AK) and the resulting material acted as a multifunctional reactor exploiting dual enzymatic and chemical catalytic activity. On the one hand, the interconversion of adenine nucleotides catalyzed by the enzyme AK was monitored. On the other hand, the peroxidase-like catalytic activity of the iron nanoparticles was tested through the reduction of hydrogen peroxide as a model reaction. This contribution represents a novel platform for many research fields including the catalysis one throughout the chemoenzymatic cascade processes.

4.4 Compartmentalization Using Emulsion and Micellar Systems

In the past years, the traditional detergents derived from petroleum feedstock have given way to rationally designed surfactants for a particular use. These amphiphilic molecules in a certain concentration can self-aggregate into micelles in water, resulting in smart reaction media able to set up catalytic processes. Initially, the benefits of micellar catalysis as a sustainable alternative to traditional organic solvents were demonstrated by examples on well-established metal-catalyzed processes: Heck, Suzuki, Stille, Negishi, or Sonogashira C–C couplings, among others [42]. Likewise, Park and Kim developed a highly active biocatalyst by co-lyophilization of a lipase and a designed ionic tensoactive [43]. The resulting ionic surfactant-coated enzyme promoted, in combination with a ruthenium racemization catalyst, the DKR of secondary alcohols in organic solvent. The remarkable enhancement of enzymatic activity was attributed to a dual effect of coating, namely, the lower deactivation during lyophilization and a better dispersion of the enzyme in the organic solvent.

With this background, Lipshutz et al. envisaged the micellar systems as plausible nanoreactors for setting up chemoenzymatic cascades in aqueous media [44]. Ideally, in such hybrid catalytic systems, the enzymes would remain in the aqueous solution, while the micelles act as a solvent and a reservoir for substrates, products, and metallic or organocatalysts. As a result, the separation of both reactions would allow bypassing the incompatibility issues between catalysts and their reaction conditions. Upon this idea, the micellar medium resulting from TPGS-750-M (**X**, Scheme 4.2a), a vitamin E-based surfactant able to form micelles of 50 nm size, was essayed as the medium for combining a palladium-mediated Sonogashira cross-coupling reaction with a further bioreduction on the triggered cetoaryl intermediate in a one-pot, two-step procedure. The coupling step occurred inside the micelle; meanwhile, the enzyme exerted its activity on the external aqueous medium. Interestingly, the alcohol dehydrogenases (ADHs) showed excellent tolerance toward the detergent, which did not display any effect on the enzyme structure. Similarly, the micelle remained unaltered throughout the overall catalytic process. After this proof of concept, several sequential chemoenzymatic cascades were assessed in such a reaction medium; Sonogashira and Heck couplings (Pd catalysts), alkyne hydrations (Au and Ag catalysts), or 1,4-additions (Rh catalysts) connected with subsequent bioreductions gave access to optically active secondary alcohols in high yield and enantiomeric excess (Scheme 4.2a). Later, other research groups exploited the concept of compartmentalization based on micellar media to obtain stilbene derivatives by assembling an enzymatic decarboxylation of

(a)

TPGS-750-M (X)

(1) $[Pd]_{cat}$, CuI_{cat}
H_2O (2 wt% **X**)

(2) ADH, *i*-PrOH
buffer (2 wt% **X**)

(1) $[Pd]_{cat}$
H_2O (2 wt% **X**)

(2) ADH, *i*-PrOH
buffer (2 wt% **X**)

(b)

Cremophor EL (Y)

(1) PAD
H_2O (2 wt% **Y**)

(2) $[Pd]_{cat}$, PhI
H_2O/EtOH
(1 wt% **Y**)

(c)

(1) Laccase/TEMPO
H_2O (1 wt% Y)

(2) ATA
buffer (1 wt% Y)

(1) Laccase/TEMPO
H_2O (1 wt% Y)

(2) ADH
buffer (1 wt% Y)

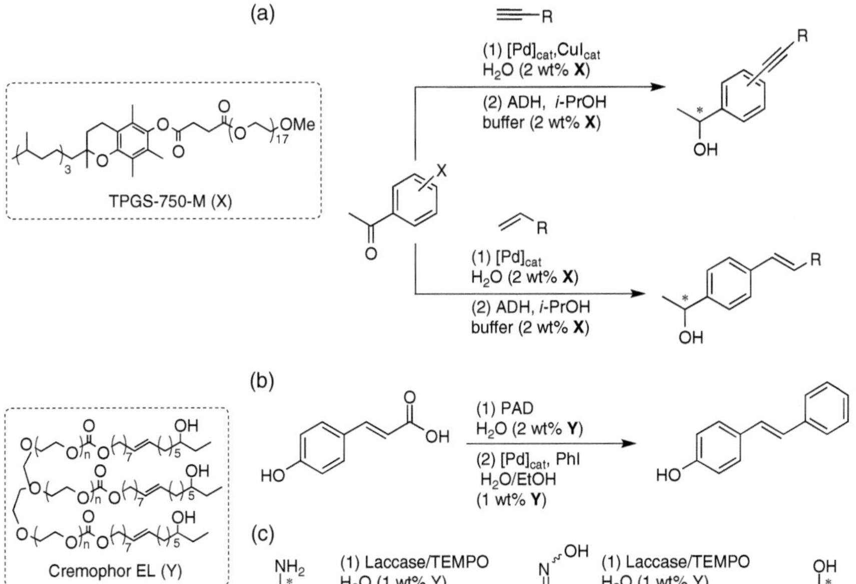

Scheme 4.2 Examples of sequential chemoenzymatic and multienzyme cascades in micellar media. (a) Pd-catalyzed Sonogashira and Heck couplings followed by bioreduction. (b) Enzymatic decarboxylation of *p*-hydroxycinnamic acids followed by metal-catalyzed Heck coupling. (c) Chemoenzymatic deoximation coupled with a bioreduction/bioamination process.

p-hydroxycinnamic acids with a Heck C–C coupling reaction in a one-pot, two-step process (Scheme 4.2b) [45]. In this case, the role of the micelle based on Cremophor EL® (**Y**) was critical to avoid the strong inhibition produced by the phenolic acid decarboxylase (PAD) on the Pd catalyst responsible for the second step. Additionally, the solubilizing properties of the surfactant enabled to attain high substrate concentrations.

As an example of the surfactant technology on multienzyme cascades, a one-pot, two-step process consisting of a laccase/2,2,6,6-tetramethylpiperidin-1-oxyl (TEMPO)-catalyzed deoximation of ketoximes followed by bioreduction or bioamination of the intermediate ketone was established (Scheme 4.2c) [46]. The addition to the aqueous medium of Cremophor EL® (**Y**, 1 wt%), a castor oil typically used as a formulation vehicle for hydrophobic drugs, was innocuous for both enzymes (laccases, ketoreductases [KREDs], and ATAs) and chemical mediator (TEMPO) and enabled high substrate concentrations, namely, 200 mM for the starting deoximation and 100 mM in the following ketoreduction. As a result, both antipodes of chiral alcohols and amines were isolated in very high overall yield and enantiomeric excess. The main feature of this nonconventional medium, free of volatile organic compounds (VOCs) as cosolvents, was the possibility to assess enzymatic cascades at reasonable high substrate concentration, a traditional pitfall in aqueous media. Presumably, micelles in the buffer prevent phenomena such as enzyme saturation and inhibition of substrate/product on the enzyme by housing the organic molecules and enabling a controlled supply toward the active site of biocatalysts.

In addition to micelles, media based on conventional emulsions also offer the possibility of compartmentalizing enzymes. Some years ago, ATAs were employed in organic solvents with moderate success. Mutti and Kroutil co-lyophilized ATAs and the cofactor pyridoxal-5′-phosphate (PLP) for further use in *tert*-butyl methyl ether (MTBE) at a water activity (a_w) of 0.6. The resulting lyophilized crude cell-free extract showed good activity toward some selected ketones [47]. In a further approach, the ATA from *Chromobacterium violaceum* (Cv-ATA) was treated with a surfactant before the lyophilization and then used as a suspension of stabilized lyophilized powder for the bioamination of ketones in dry isooctane [48]. Although successful in some examples, such strategies remained ineffective for hydrophobic substrates. Very recently, Tessaro et al. devised a disperse system composed of a hydrophobic solvent, a nonionic surfactant, and traces of water [49]. The authors revealed how the ternary mixtures led to reverse micelles ("water in oil" micelles) if the water content is lower than 0.2% v/v or macroscopically homogeneous "water in oil" emulsions at 1% v/v of aqueous solution. Then, they focused on the latter and a medium based on isooctane or MTBE, Brij® C10 as a surfactant and 1% v/v of aqueous solution of ATA (*Cv*-ATA or *Vf*-ATA) efficiently enabled the bioamination of a library of 12 substituted cyclohexanones (25 mM) working at 37 °C and using (*S*)-phenylethylamine [(S)-PEA] as an amine donor (Scheme 4.3). On the one hand, the challenging solubility of hydrophobic substrates was circumvented and, in addition, the enzyme stability was enhanced significantly because of the confined aqueous phase encapsulating the ATA.

Pickering emulsions are emulsions stabilized by solids locating at the oil–water interface. The benefits of these media have been demonstrated through many examples of interfacial enzymes displaying higher robustness upon hard reaction conditions, being an innovative platform to develop improved biocatalysts, which do not need stirring and immobilization. Such biocatalysts act as both emulsifiers and catalysts and enable easy recycling and

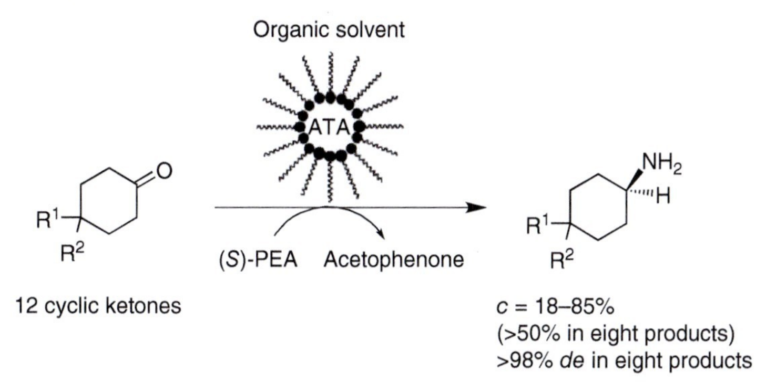

Scheme 4.3 Bioamination of hydrophobic substrates by a compartmentalized ATA in a disperse system (organic solvent/nonionic surfactant/traces of water).

reduced diffusional limitations. As a practical example, the addition of a small amount of solid particle emulsifier produced compartmentalization of lipase B form *Candida antarctica* (CAL-B) within millions of micron-sized water droplets, while organic substrates remained in the oil phase outside the droplets [50]. Then, the kinetic resolution of several racemic esters was accomplished without stirring with enhanced efficiency and enantioselectivity regarding the biphasic stirred system (Scheme 4.4a). The reaction conditions involved 90 mM substrate concentration in a mixture of *n*-hexane/buffer 100 mM pH 7.0 (1 : 2.5), 35 °C, and a reaction time ranging from 6 to 10 h depending on the substrate. In all cases, the resulting (*R*)-alcohols and the remaining (*S*)-esters exhibited >99% and >96% *ee*, respectively, for a measured conversion higher than 49%. The unique properties of the emulsion allow easy separation of the product and catalyst by centrifugation, the enzyme featuring excellent activity after 27 cycles because of the absence of erosion produced by stirring. On a different application, a Pickering emulsion based on silica nanoflowers

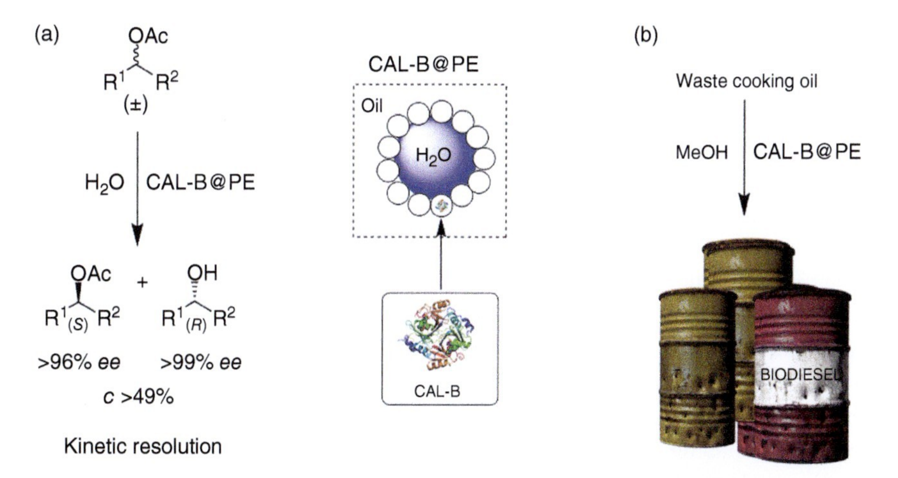

Scheme 4.4 Practical applications of CAL-B Pickering emulsions (CAL-B@PE): (a) enantioselective kinetic resolution of racemic esters. (b) Production of biodiesel via methanolysis of waste cooking oil. Source: (b) Robert Kourist, Javier González-Sabín.

(SNFs) acting as the carrier for CAL-B (CAL-B@SNFs-PE) proved to be effective for biodiesel production via methanolysis of waste cooking oil (Scheme 4.4b) [51]. Under the optimized conditions, namely, a methanol/oil ratio of 2.63 : 1, 46 °C, and eight hours of reaction, the experimental yield was >98%. The catalytic system maintained over 76% of initial biodiesel yield after 15 cycles, rendering a better outcome than free CAL-B or the conventional formulations of CAL-B (Novozym 435).

Pickering emulsions have also enabled to compartmentalize incompatible reagents for cascade reactions such as deacetalization–Knoevenagel, deacetalization–Henry, and diazotization–iodization processes [52]. With regard to enzymatic cascades, a one-pot biphasic tandem reaction was recently demonstrated in a Pickering emulsion. Specifically, hemoglobin attached on the surface of silica nanoparticles (SiO_2@Hb) acted as an interfacial biocatalyst and emulsifier and was coupled to glucose oxidase (GOx) remaining in the aqueous phase to set up a biphasic cascade process [53]. First, glucose was oxidized to gluconic acid and H_2O_2 by the action of GOx and oxygen in solution. Then, SiO_2@Hb oxidized pyrogallol at an expense of the formed H_2O_2. As a consequence of the large interfacial area and higher stability of the biocatalyst, the Pickering emulsion led to higher enzymatic performances than the neat biphasic system.

The examples covered in this section highlight medium engineering as a means to compartmentalize enzymes either through emulsions or micelles. It is worth noting the ease and accessibility of some of these methodologies as well as the major improvements on the enzymatic performance.

4.5 Compartmentalization Using Encapsulation

Molecular encapsulation consists of the confinement of an individual molecule within a larger molecule. In chemistry, self-assembling molecular capsules can serve as containers, separation tools, and even as reaction vessels and catalysts. Likewise, encapsulation allows to offer different reaction conditions for the different catalysts of a cascade. Although it permits to use the enzyme in the free form, it shares with immobilization techniques the benefit of a facilitated separation of the catalyst. We used polyvinyl alcohol (PVA) cryogels [5] to encapsulate a bacterial decarboxylase for the combination with a ruthenium catalyst in a chemoenzymatic tandem reaction (Scheme 4.5). Bacterial PAD catalyzes the conversion of bio-based *p*-hydroxycinnamic acid derivatives to *p*-hydroxystyrenes and is a highly promising catalyst for the synthesis of bio-based olefins. We reasoned that a coupling of two *p*-hydroxystyrene molecules would give access to symmetrical 4,4′-dihydroxystilbenes. These molecules have a similar structure to the well-known antioxidant resveratrol and were shown to possess antioxidant activity. Their natural synthesis requires coenzyme A derivatives, which makes this reaction too expensive in cell-free systems. The conversion of hydroxystyrenes to the 4,4′-dihydroxystilbenes could be achieved by combining the PAD-catalyzed enzymatic decarboxylation of *p*-coumaric acid with a cross-metathesis of two molecules of the intermediate *p*-hydroxystyrene. A stepwise fashion of this route suffered from the tendency of *p*-hydroxystyrenes to undergo spontaneous polymerization during the work-up. A chemoenzymatic cascade reaction was envisioned as a practical approach to solve this problem. Although styrene is a model compound for

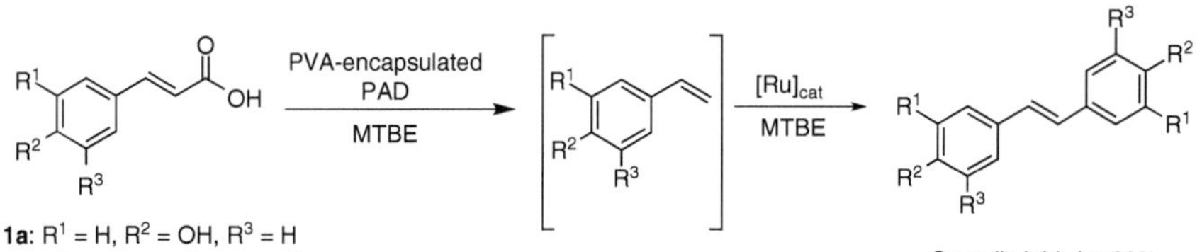

PVA-encapsulated PAD, MTBE → [Ru]$_{cat}$, MTBE

1a: R^1 = H, R^2 = OH, R^3 = H
2a: R^1 = OCH_3, R^2 = OH, R^3 = H
3a: R^1 = OH, R^2 = OH, R^3 = H

Overall yield: **1c**: 90%, **2c**: 36%, **3c**: 37%

Scheme 4.5 Combination of encapsulated phenolic acid decarboxylase (PAD) with an olefin metathesis catalyst for the synthesis of symmetric 4,4′-dihydroxystilbenes.

olefin metathesis catalysts, finding a suitable ruthenium catalyst for the conversion of *p*-hydroxystyrenes proved to be hard. Although there have been reported metathesis catalysts with high stability in water, screening of commercially available kits led to the identification of several catalysts that were active but required hydrophobic solvents. These results were confirmed in a later study, employing the same cascade in deep eutectic solvents, leading to limited conversions [45]. Besides, this cascade is thus a typical example for the dilemma that the choice of an appropriate solvent poses to the researcher: As water remains the preferred medium for many enzymes, the development of chemoenzymatic cascade reactions is supported by the ongoing development of chemical catalysts active and stable in aqueous media [54, 55]. Yet, it should be noted that the majority of homogeneous catalysts still requires organic solvents, which create a demand for methods facilitating the use of enzymes in these solvents. Nonconventional solvents such as deep eutectic solvents could alleviate this situation [20]. PAD from *Bacillus subtilis* (*Bs*PAD) is in fact highly active in aqueous media and in deep eutectic solvents [21, 56, 57] but in our work did not show any activity (as free enzyme and in immobilized form) in organic solvents. The limited mutual compatibility of enzyme and chemical catalyst made a compartmentalization strategy the method of choice. Encapsulation of PAD in PVA cryogels [58] allows to apply the enzyme in a biphasic system where the encapsulation protects the enzyme from denaturing and facilitates the separation of the biocatalyst for reuse after the reaction. In a one-pot, two-step sequential approach, the catalyst was added after removal of the biocatalyst, resulting in complete conversions and isolated yields up to 90%.

Polymersomes are nanometer-sized vesicles formed by amphiphilic block copolymers. They represent the macromolecular counterparts to the well-established lipid vesicles or liposomes and have been used for biomedical applications as therapeutic drug carriers. More recently, and seeing as their ability to encapsulate biomolecules, polymersomes have received attention as nanoreactors for enzymatic cascade reactions [59–62]. Compared to natural phospholipid membranes, they have higher mechanical stability and much lower permeability. With a thickness of up to 40 nm, they act as a considerable diffusional barrier, which can be possibly exploited for compartments with selective transport properties by insertion of natural or engineered channel proteins and porous membranes or increase of permeability by adding external stimuli. Castiglione et al. showed the feasibility of the addition of a selective transporter by assembling a three-step enzyme cascade for the synthesis of cytidine-5′-monophosphate-*N*-acetylneuraminic acid (CMP-NeuAc, Scheme 4.6) [62]. The incompatibilities between the reaction steps require

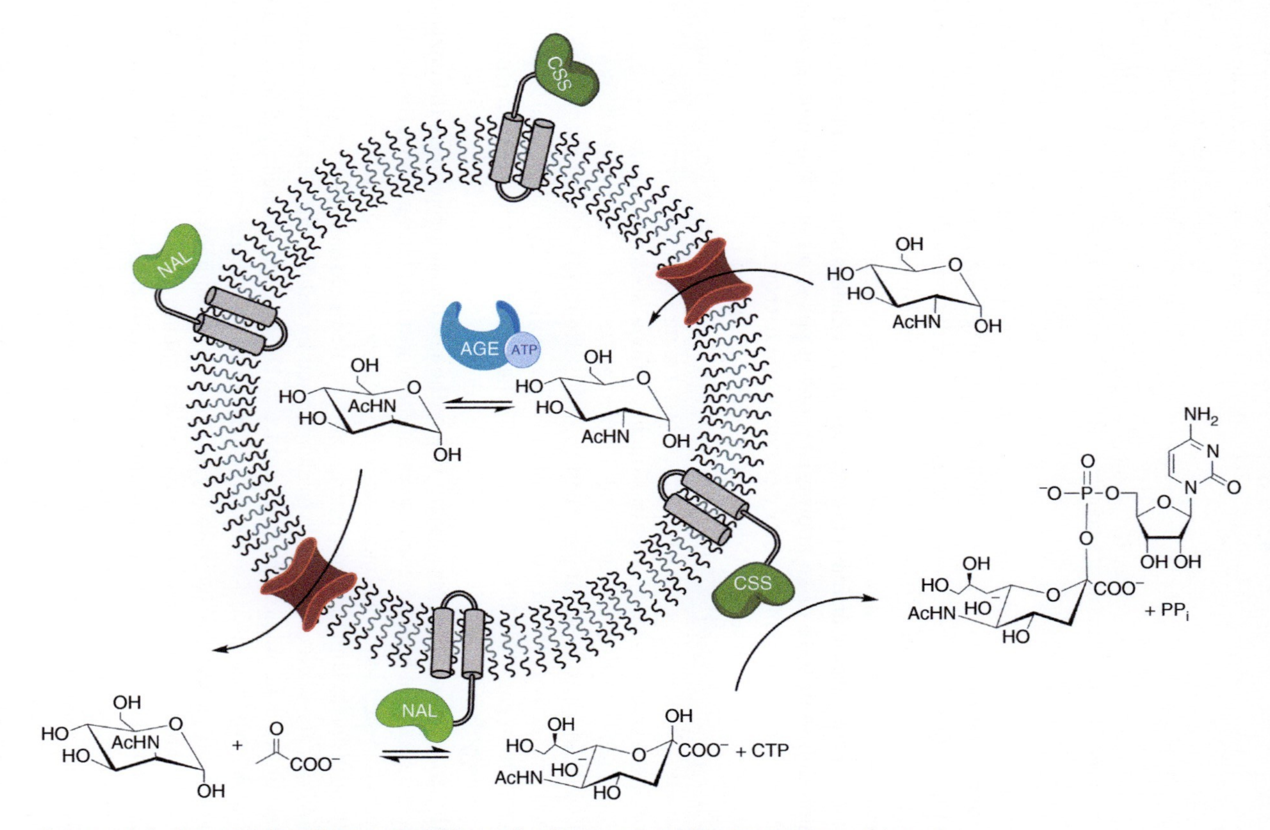

Scheme 4.6 Use of polymersomes to achieve selective transport as a strategy to avoid cross-inhibition within multienzyme cascades. Source: Schmidt et al. [4].

successful compartmentalization and selective transport. In this cascade, the first enzyme, *N*-acyl-D-glucosamine 2-epimerase (AGE), is inhibited by cytidine triphosphate (CTP), the substrate of the third enzyme CMP-sialic acid synthetase (CSS). Moreover, CSS showed much lower stability than other enzymes, which could be solved by immobilization of CSS on the polymersome surface. Using an engineered membrane channel allowed to achieve spatial separation of the first and third reaction step and thus to avoid the cross-inhibition. This visionary approach shows the potential of nano- and microcompartments for the establishment of selective transport. The ultimate goal here is the construction of an artificial cell that allows to exploit its strengths (such as controlled transport) without the complexity and the side reactions of metabolism and genetic regulation.

The possibility to modify the membranes of polymersomes has offered a wide field of different potential applications. Essential processes of the energy metabolism utilize biomembranes for assembly of the involved components and for the formation of proton gradients. Using the membranes of polymersomes has thus emerged as a strategy to generate artificial photosynthetic systems in biomimetic approaches. In a highly interesting example, Li and coworkers reported the attempt to assemble the photosystem II and the F_0/F_1 ATPase to form artificial chloroplasts [63], perhaps has the considerable difficulty to prepare them in a precise way in the laboratory, preventing so far practical applications of these systems in concrete biotechnological processes; nevertheless, considerable progress in this field is expected within the next years.

4.6 Compartmentalization Using Tea Bags and Thimbles

In the quest to conveniently isolate enzymes by imitating the cells in nature, chemists have designed ingenious practical solutions. Thus, from the basic concepts of everyday life have emerged the so-called "tea bag synthetic technology" or the smart polydimethylsiloxane (PDMS) thimbles.

"Tea bags" for containing solid catalysts were introduced by Houghthen in the combinatorial synthesis of peptides [64]. In biocatalysis, they were initially conceived as a way to tackle the recalcitrant work-up demanded by some enzymatic processes, especially those with whole cells and living organisms. Ideally, confining the biocatalyst would avoid steps such as filtration and centrifugation and allow an easy catalyst recycling. More recently, the emergence of chemo- and multienzyme cascade reactions as a primary topic on current catalysis has revalued and expanded the tea bag applications. Rother et al. set out an enzymatic cascade toward 1,2-diols in excellent diastereo- and enantioselectivity (>99% *de* and >99% *ee*) by melting a carboligation reaction of aldehydes with a further reduction on the transiently formed hydroxy ketone (Scheme 4.7) [65]. Different operating modes were explored employing the enzymes, namely, the *Pseudomonas fluorescens,* benzaldehyde lyase (BAL) and the *Ralstonia* sp. alcohol dehydrogenase (RADH) in sequential and concurrent fashion. In the most ambitious one, both biocatalysts were conveniently compartmentalized in tea bags of polyvinylidene fluoride (PVDF) and worked from the beginning of the process simultaneously. Tea bags showed easy catalyst recyclability and enabled to run the process at high substrate concentration (up to 339 mM). Later, Rother et al. combined the tea bags approach with the SpinChem rotating bed reactor (Nordic ChemQuest

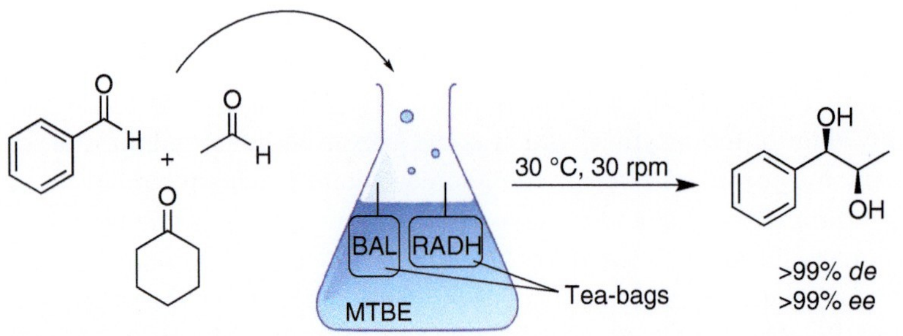

Scheme 4.7 Concurrent enzymatic cascade toward enantiopure diols employing tea bags for compartmentalization.

AB) [66]. Originally conceived for heterogeneous chemocatalysis, such reactor is compatible with other solid phases such as immobilized enzymes, encapsulated cells, and ion exchangers and minimizes mass transfer limitations and catalyst degradation. To reach this, the corresponding solid phase is isolated as a packed bed inside the rotating cylinder. The spinning of the reactor triggers a circulating flow in which the reaction solution is aspirated from the bottom of the vessel, percolated through the solid phase and returned to the vessel. On melting both tea bag and SpinChem technologies, the previous cascade toward 1,2-diols was established on a preparative 140 ml reactor with a remarkable product concentration and space–time yield (32.9 $g\,l^{-1}$ and 8.2 $g\,l^{-1}\,day^{-1}$).

With regard to polydimethylsiloxane (PDMS) thimbles, the groundbreaking studies of Bowden's group led to a profound paradigm shift in the way s-block organometallic elements can be used in organic synthesis [66]. Thus, the employment of PDMS enabled to site-isolate water from $LiAlH_4$, Grignard, or cuprate reagents and also to establish cascade processes involving these reagents. The success of the approach relies on the high hydrophobicity of PDMS, which allows the diffusion of small organic molecules. After some years, Gröger et al. implemented this concept as a way to compartmentalize enzymes and metal catalysts from each other through chemoenzymatic one-pot, two-step cascades in water [67]. In preliminary studies, the aim of connecting a Wacker oxidation of styrene with a further bioreduction collided with the strong inhibition exerted by Cu and Pd catalysts on ADHs. Accordingly, the authors compartmentalized the metal catalyst inside a PDMS thimble with an outer aqueous phase housing the enzyme (Scheme 4.8a). Further optimization unveiled a sequential process as the best operating mode. First, the metal-catalyzed step was accomplished in the internal aqueous phase. Then, a second buffer containing the enzyme (ADH from *Lactobacillus kefir*), cofactor ($NADP^+$), and propan-2-ol for cofactor recycling were added as an external phase. In this way, the resulting acetophenone from the first step went across the PDMS membrane into the outer phase and the ADH led to (*R*)-1-phenylethanol in excellent overall conversion (85%) and *ee* (99%). The study covered a library of nine substituted styrenes, eight of them being efficiently converted into the corresponding (*R*)-alcohols (70–94% overall yield, 98–99% *ee*). Interestingly, the Pd-catalyst could be recycled for further use more than 15 cycles. A further extension of this methodology delivered a highly regioselective arylation of unactivated aromatic compounds, a challenging task under conventional methodologies [68]. The chemoenzymatic hybrid system

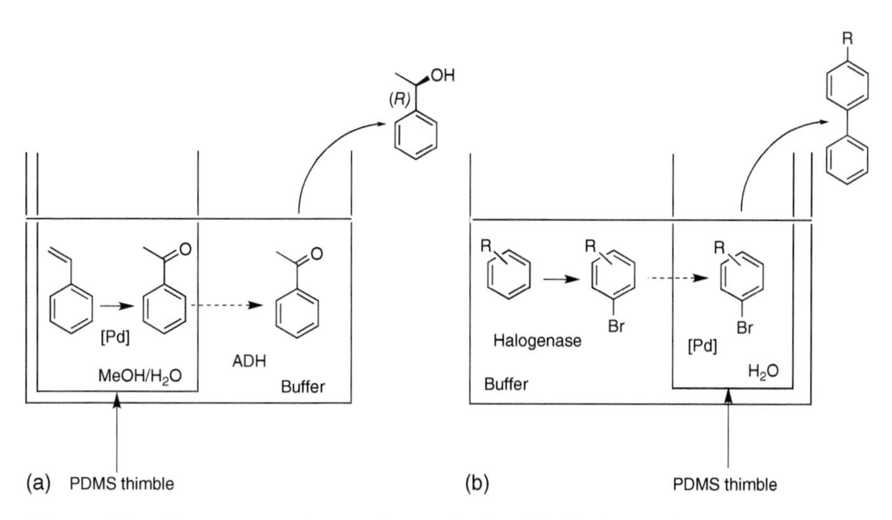

Scheme 4.8 Chemoenzymatic cascades employing PDMS thimbles for compartmentalization.

was based on an enzymatic halogenation, followed by a Pd-catalyzed Suzuki–Miyaura cross-coupling, with all reagents placed in the pot from the outset in an aqueous buffer (Scheme 4.8b). First, the enzymatic halogenation (2 mM substrate concentration) was catalyzed by the tryptophan 5-halogenase PyrH in the external phase of the PDMS membrane at room temperature overnight. Then, the only adjustment was an increase on temperature to 80 °C to trigger the Pd-catalyzed coupling inside the thimble (10% mol $Pd(OAc)_2$). The resulting biaryl compounds were isolated in moderate yields (57–74%).

4.7 Separation of Reaction Steps Using Continuous Flow

Although compartmentalization facilitates the combination of catalysts in the same reaction vessel, separation of the reactions by immobilization in different columns in a continuous system allows a greater variation of reaction parameters. Continuous flow synthesis yields a number of benefits such as consistent product quality and reduced reaction time as well as the suppression of the catalyst removal step from the reaction solution (see also Chapter 11). For chemoenzymatic and multienzyme cascades, an additional benefit lies in the possibility to change the reaction conditions of the individual reaction steps and thus to alleviate incompatibility issues. In particular, the possibility to offer an optimal reaction temperature for all catalysts is a crucial advantage over the concurrent cascade mode. Sieber et al. exploited this principle to synthesize 2-keto-3-deoxy sugar acids (KDS) by combining the gold-catalyzed chemical oxidation of a carbohydrate, which demanded a low temperature to suppress side product formation with a thermostable dihydroxyacid dehydratase, namely, DHAD from *Sulfolobus solfataricus* (SsDHAD), that required reaction temperatures exceeding 50 °C (Scheme 4.9) [2, 22]. The compartmented continuous process operated at substrate concentrations close to 40 mM and flow rate 0.3–1.0 ml min^{-1}, and the promiscuity of the DHAD enabled to obtain several KDS (58–86% isolated yield).

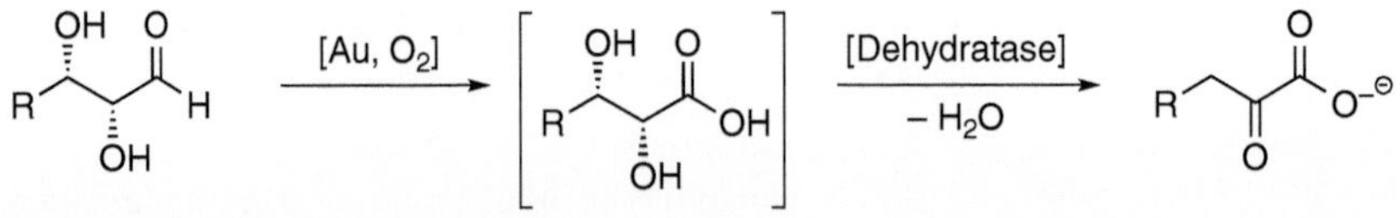

Scheme 4.9 Flow separation of a gold-catalyzed oxidation coupled to an enzymatic dehydration for the synthesis of 2-keto-3-deoxy sugar acids.

Although the coupling of hydroxystyrols by olefin metathesis leads necessarily to symmetrical stilbenes (Scheme 4.5), Gruber-Wölfler et al. expanded this approach to asymmetrically substituted stilbene derivatives taking advantage of a Pd-catalyzed Heck cross-coupling reaction in a continuous flow system (Scheme 4.10) [21]. In a cascade reaction in combination with phenolic acid decarboxylation (*Bs*PAD), different solvent requirements of both reactions posed a problem. The low solubility of coumaric acid in buffer limited the substrate concentration for the enzymatic step to a few millimolar (5 mM). Likewise, the need to add ethanol for the Heck-coupling reaction required a further dilution of the solution, leading to very low volumetric yields. Gratefully, the use of a 1 : 1 mixture of the deep eutectic solvent composed by choline chloride : glycerol (1 : 2 mol mol^{-1}) with aqueous buffer allowed to increase the substrate solubility to 20 mM and thus to improve the volumetric yield. Moreover, the implementation of a continuous flow system allowed to operate the enzymatic and the chemical steps at 30 and 80 °C, respectively. It should be noted that even if the synthesis of asymmetric stilbenes was achieved in full conversion, the overall yield was reduced by the rather similar reactivity of the two alkene C-atoms of the hydroxystyrol in the Heck reaction, leading to the formation of side products.

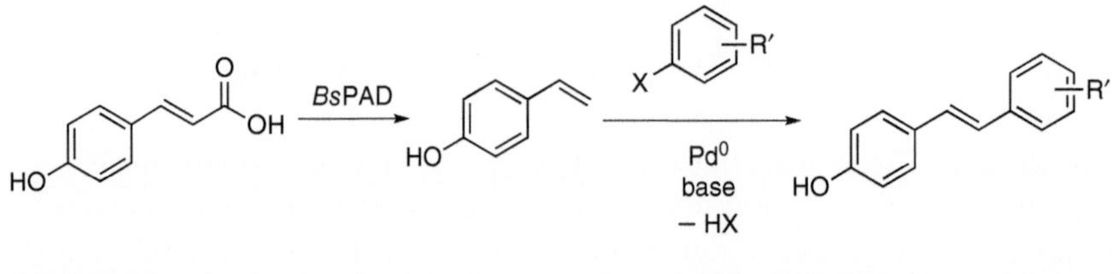

Scheme 4.10 Continuous flow coupling of an enzymatic decarboxylation mediated by phenolic acid decarboxylase from *Bacillus subtilis* (*Bs*PAD) with a Pd-catalyzed Heck coupling in a mixture of deep eutectic solvent (ChCl/glycerol 1 : 2 mol/mol) with water (1 : 1 v/v).

Corma and Iborra developed a two-bed continuous system to deracemize valuable chiral alcohols through hybrid chemoenzymatic catalysis (Scheme 4.11) [69]. In a first reactor, racemic alcohols are oxidized to the corresponding ketones via Oppenauer oxidation catalyzed by a Lewis acid zeolite (Zr-Beta) and with the assistance of acetone as a hydrogen acceptor (191 mM substrate concentration, 50 °C, and 0.5 ml min^{-1}). Then, after partial removal of acetone and further dilution, an ADH supported on a 2D zeolite promotes the stereoselective reduction of the transiently formed ketone through the second reactor (30 mM substrate concentration, 25 °C, and 0.55 ml min^{-1}). The overall conversion for the two-step process exceeded 95%, and depending on the stereopreference of the ADH, both enantiomers of the target alcohols exhibited perfect optical purity (>99% *ee*). In particular, commercially available (Evocatal) crude cell extract of Prelog ADH (ADH030, NADH dependent) and anti-Prelog ADH (ADH270, nicotinamide adenine dinucleotide phosphate (NADPH) dependent) was used. The rational design of the system leads to reach the principle of 100% atom economy because the isopropanol co-product coming from the first step (acetone reduction) feds up the bioreduction via cofactor reduction. Finally, the acetone accompanying the target chiral alcohol is reincorporated to the outset of the process. Interestingly, the catalytic system could operate up to 16 days.

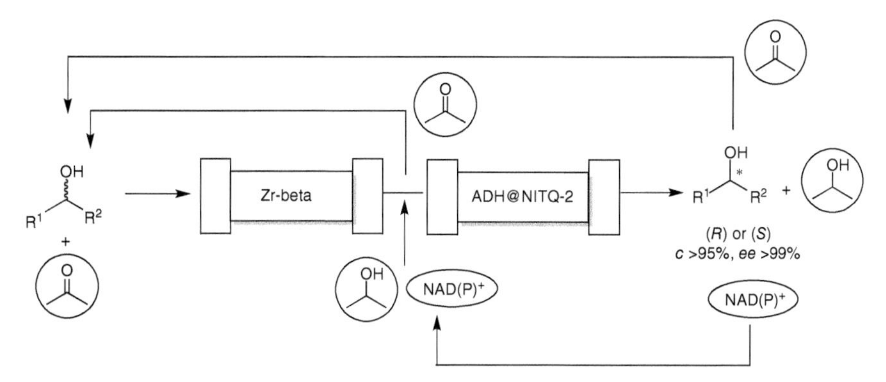

Scheme 4.11 Chemoenzymatic deracemization of chiral alcohols employing a continuous flow system combining a zeolite-mediated oxidation and an enzymatic bioreduction.

4.8 Conclusions and Prospects

The development of systems for the compartmentalization of chemoenzymatic reactions has led to a tremendous diversity of systems, ranging from nanoencapsulation to continuous flow chemistry. Mixed-culture cultivations and flow chemistry are perhaps the most mature approaches for spatial separation of reaction steps. Similar to temporal separation in subsequent reaction steps, continuous flow with an immobilized chemical and biological catalyst can rely on know-how from existing industrial applications. Other approaches are attractive because of their potential for the combination of elements that would be otherwise incompatible, such as the coupling of an enzyme that is inactive in organic solvent with a chemocatalyst that requires exactly this solvent. The rise of nonconventional solvents such as ionic liquids and deep eutectic solvents are expected to further increase the options for the combination of chemical and biological catalysts.

Many systems that are currently under investigation are of visionary character. This means that they have attracted much attention in academic works, but that the concrete application in synthesis lacks behind, and that the scalability remains to be shown. Perhaps this can be explained with the intrinsic complexity of most compartmentalization approaches. This is a difference to immobilization of single enzymes that has arrived at the industrial application but still offers a lot potential for further development. In contrast, compartmentalized cascades have not yet fulfilled their promise in the development of concrete processes, which can be attributed to some extent by their complexity and the lack of strategies for intensification and up-scale of these systems. Nevertheless, the dynamics of the field and the increasing number of interesting approaches raise the expectation that the intensive research will lead to technologies that facilitate the development of cost-efficient, sustainable multienzyme, and multi-catalyst reaction sequences.

References

1 Schrittwieser, J.H., Velikogne, S., Hall, M. et al. (2018). Artificial biocatalytic linear cascades for preparation of organic molecules. *Chem. Rev.* 118 (1): 270–348.

2 Sperl, J.M. and Sieber, V. (2018). Multienzyme cascade reactions - status and recent advances. *ACS Catal.* 8 (3): 2385–2396.

3 Schmidt-Dannert, C. and López-Gallego, F. (2016). A roadmap for biocatalysis – functional and spatial orchestration of enzyme cascades. *Microb. Biotechnol.* 9 (5): 601–609.

4 Schmidt, S., Castiglione, K., and Kourist, R. (2018). Overcoming the incompatibility challenge in chemoenzymatic and multi-catalytic cascade reactions. *Chem. Eur. J.* 24 (8): 1755–1768.

5 Gómez-Baraibar, Á., Reichert, D., Mügge, C. et al. (2016). A sequential one-pot cascade reaction combining an encapsulated decarboxylase with metathesis for the synthesis of bio-based antioxidants. *Angew. Chem. Int. Ed.* 55 (47): 14823–14827.

6 Enoki, J., Linhorst, M., Busch, F. et al. (2019). Preparation of optically pure flurbiprofen via an integrated chemo-enzymatic synthesis pathway. *Mol. Catal.* 467: 135–142.

7 Wichmann, R. and Vasic-Racki, D. (2005). Cofactor regeneration at the lab scale. In: *Technology Transfer in Biotechnology* (ed. U. Kragl), 225–260. Berlin, Heidelberg: Springer.

8 Strohmeier, G.A., Eiteljörg, I.C., Schwarz, A. et al. (2019). Enzymatic one-step reduction of carboxylates to aldehydes with cell-free regeneration of ATP and NADPH. *Chem. Eur. J.* 25 (24): 6119–6123.

9 Engström, K., Johnston, E.V., Verho, O. et al. (2013). Co-immobilization of an enzyme and a metal into the compartments of mesoporous silica for cooperative tandem catalysis: an artificial metalloenzyme. *Angew. Chem. Int. Ed.* 125 (52): 14256–14260.

10 Enoki, J., Meisborn, J., Müller, A.C. et al. (2016). A multi-enzymatic cascade reaction for the stereoselective production of γ-oxyfunctionalyzed amino acids. *Front. Microbiol.* 7: 425.

11 Larsson, A.L.E., Persson, B.A., and Bäckvall, J.E. (1997). Enzymatic resolution of alcohols coupled with ruthenium-catalyzed racemization of the substrate alcohol. *Angew. Chem., Int. Ed.* 36 (11): 1211–1212.

12 Mutti, F.G., Knaus, T., Scrutton, N.S. et al. (2015). Conversion of alcohols to enantiopure amines through dual-enzyme hydrogen-borrowing cascades. *Science* 349 (6255): 1525–1529.

13 Rudroff, F., Mihovilovic, M.D., Gröger, H. et al. (2018). Opportunities and challenges for combining chemo-and biocatalysis. *Nat. Catal.* 1 (1): 12–22.

14 Verho, O. and Bäckvall, J.E. (2015). Chemoenzymatic dynamic kinetic resolution: a powerful tool for the preparation of enantiomerically pure alcohols and amines. *J. Am. Chem. Soc.* 137 (12): 3996–4009.

15 Mutti, F.G., Orthaber, A., and Schrittwieser, J.H. (2010). Simultaneous iridium catalysed oxidation and enzymatic reduction employing orthogonal reagents. *Chem. Commun.* 46 (42): 8046–8048.

16 Bechi, B., Herter, S., McKenna, S. et al. (2014). Catalytic bio–chemo and bio–bio tandem oxidation reactions for amide and carboxylic acid synthesis. *Green Chem.* 16 (10): 4524–4529.

17 Denard, C.A., Huang, H., Bartlett, M.J. et al. (2014). Cooperative tandem catalysis by an organometallic complex and a metalloenzyme. *Angew. Chem., Int. Ed.* 53 (2): 465–469.

18 Ríos-Lombardía, N., Vidal, C., Liardo, E. et al. (2016). From a sequential to a concurrent reaction in aqueous medium: ruthenium-catalyzed allylic alcohol isomerization and asymmetric bioreduction. *Angew. Chem., Int. Ed.* 55 (30): 8691–8695.

19 Liardo, E., González-Fernández, R., Ríos-Lombardía, N. et al. (2018). Strengthening the combination between enzymes and metals in aqueous medium: concurrent ruthenium-catalyzed nitrile hydration - asymmetric ketone bioreduction. *ChemCatChem* 10 (20): 4676–4682.

20 González-Sabín, J. and Kourist, R. (2020). Non-conventional media as strategy to overcome the solvent dilemma in chemoenzymatic tandem catalysis. *ChemCatChem* 12 (7): 1903–1920.

21 Grabner, B., Schweiger, A.K., Gavric, K. et al. (2020). A chemo-enzymatic tandem reaction in a mixture of deep eutectic solvent and water in continuous flow. *React. Chem. Eng.* 5 (2): 263–269.

22 Sperl, J.M., Carsten, J.M., Guterl, J.K. et al. (2016). Reaction design for the compartmented combination of heterogeneous and enzyme catalysis. *ACS Catal.* 6 (10): 6329–6334.

23 Telzerow, A., Paris, J., Håkansson, M. et al. (2019). Amine transaminase from *Exophiala Xenobiotica* - crystal structure and engineering of a fold IV transaminase that naturally converts biaryl ketones. *ACS Catal.* 9 (2): 1140–1148.

24 Savile, C.K., Janey, J.M., Mundorff, E.C. et al. Biocatalytic asymmetric synthesis of chiral amines from ketones applied to sitagliptin manufacture. *Science* 329 (5989): 305–309.

25 Aßmann, M., Stöbener, A., Mügge, C. et al. (2017). Reaction engineering of biocatalytic (*S*)-naproxen synthesis integrating in-line process monitoring by Raman spectroscopy. *React. Chem. Eng.* 2 (4): 531–540.

26 Aßmann, M., Mügge, C., Gaßmeyer, S.K. et al. (2017). Improvement of the process stability of arylmalonate decarboxylase by immobilization for biocatalytic profen synthesis. *Front. Microbiol.* 8: 448.

27 DiCosimo, R., McAuliffe, J., Poulose, A.J. et al. (2013). Industrial use of immobilized enzymes. *Chem. Soc. Rev.* 42 (15): 6437–6474.

28 Ajikumar, P.K., Xiao, W.H., Tyo, K.E.J. et al. (2010). Isoprenoid pathway optimization for taxol precursor overproduction in *Escherichia coli*. *Science* 330 (6000): 70–74.

29 Schmidt, S., Scherkus, C., Muschiol, J. et al. (2015). An enzyme cascade synthesis of ε-caprolactone and its oligomers. *Angew. Chem. Int. Ed.* 54 (9): 2784–2787.

30 Assil-Companioni, L., Schmidt, S., Heidinger, P. et al. (2019). Hydrogen-driven cofactor regeneration for stereoselective whole-cell C=C bond reduction in *Cupriavidus necator*. *ChemSusChem* 12 (11): 2361–2365.

31 Winkler, C.K., Tasnádi, G., Clay, D. et al. (2012). Asymmetric bioreduction of activated alkenes to industrially relevant optically active compounds. *J. Biotechnol.* 162 (4): 381–389.

32 Köninger, K., Gómez-Baraibar, A., Mügge, C. et al. (2016). Recombinant cyanobacteria as tools for asymmetric C=C bond reduction fueled by biocatalytic water oxidation. *Angew. Chem., Int. Ed.* 55 (18): 5582–5585.

33 Driessen, F.M. (1981). Protocooperation of yogurt bacteria in continuous culture. In: *Mixed Culture Fermentations* (eds. E. Bushell and J.H. Slater), 99–120. London: Academic Press.

34 Löwe, H., Hobmeier, K., Moos, M. et al. (2017). Photoautotrophic production of polyhydroxyalkanoates in a synthetic mixed culture of *Synechococcus elongatus* cscB and *Pseudomonas putida* cscAB. *Biotechnol. Biofuels* 10 (1): 190.

35 Fedeson, D.T., Saake, P., Calero, P. et al. (2020). Biotransformation of 2, 4-dinitrotoluene in a phototrophic co-culture of engineered *Synechococcus elongatus* and *Pseudomonas putida*. *Microb. Biotechnol.* 13 (4): 997–1011. https://doi.org/10.1111/1751-7915.13544.

36 Borrero-de Acuña, J.M., Rohde, M., Wissing, J. et al. (2016). Protein network of the *Pseudomonas aeruginosa* denitrification apparatus. *J. Bacteriol.* 198 (9): 1401–1413.

37 Wörsdörfer, B., Woycechowsky, K.J., and Hilvert, D. (2011). Directed evolution of a protein container. *Science* 331 (6017): 589–592.

38 Kanbak-Aksu, S., Hasan, M.N., Hagen, W.R. et al. (2012). Ferritin-supported palladium nanoclusters: selective catalysts for aerobic oxidations in water. *Chem. Commun.* 48 (46): 5745–5747.

39 Abe, S., Hirata, K., Ueno, T. et al. (2009). Polymerization of phenylacetylene by rhodium complexes within a discrete space of apo-ferritin. *J. Am. Chem. Soc.* 131 (20): 6958–6960.

40 Comellas-Aragonès, M., Engelkamp, H., Claessen, V.I. et al. A virus-based single-enzyme nanoreactor. *Nat. Nanotechnol.* 2 (10): 635–639.

41 Palmiero, U.C., Küffner, A.M., Krumeich, F. et al. (2020). Adaptive chemo-enzymatic microreactors composed of inorganic nanoparticles and bio-inspired intrinsically disordered proteins. *Angew. Chem. Int. Ed.* 59 (7970): 8138–8142. https://doi.org/10.1002/anie.202000835.

42 Lipshutz, B.H., Ghorai, S., and Cortes-Clerget, M. (2018). The hydrophobic effect applied to organic synthesis: recent synthetic chemistry "in water". *Chem. Eur. J.* 24 (26): 6672–6695.

43 Kim, H., Choi, Y.K., Lee, J. et al. (2011). Ionic-surfactant-coated *Burkholderia cepacia* lipase as a highly active and enantioselective catalyst for the dynamic kinetic resolution of secondary alcohols. *Angew. Chem. Int. Ed.* 50 (46): 10944–10948.

44 Cortes-Clerget, M., Akporji, N., Zhou, J. et al. (2019). Bridging the gap between transition metal-and bio-catalysis via aqueous micellar catalysis. *Nat. Commun.* 10 (1): 1–10.

45 Ríos-Lombardía, N., Rodríguez-Álvarez, M.J., and Morís, F. (2020). DESign of sustainable one-pot chemoenzymatic organic transformations in deep eutectic solvents for the synthesis of 1,2-disubstituted aromatic olefins. *Front. Chem.* 8: 139.

46 Correia Cordeiro, R.S., Ríos-Lombardía, N., and Morís, F. (2019). One-pot transformation of ketoximes into optically active alcohols and amines by sequential action of laccases and ketoreductases or ω-transaminases. *ChemCatChem* 11 (4): 1272–1277.

47 Mutti, F.G. and Kroutil, W. (2012). Asymmetric bio-amination of ketones in organic solvents. *Adv. Synth. Catal.* 354 (18): 3409–3413.

48 Chen, S., Land, H., Berglund, P. et al. (2016). Stabilization of an amine transaminase for biocatalysis. *J. Mol. Catal. B Enzym.* 124: 20–28.

49 Fiorati, A., Berglund, P., Humble, S.M. et al. (2020). Application of transaminases in a disperse system for the bioamination of hydrophobic substrates. *Adv. Synth. Catal.* 362 (5): 1156–1166.

50 Wei, L., Zhang, M., Zhang, X. et al. (2016). Pickering emulsion as an efficient platform for enzymatic reactions without stirring. *ACS Sustain. Chem. Eng.* 4 (12): 6838–6843.

51 Wang, L., Liu, X., Jiang, Y. et al. (2019). Silica nanoflowers-stabilized Pickering emulsion as a robust biocatalysis platform for enzymatic production of biodiesel. *Catalysts* 9 (12): 1026.

52 Yang, H.Q., Fu, L.M., Wei, L.J. et al. (2015). Compartmentalization of incompatible reagents within Pickering emulsion droplets for one-pot cascade reactions. *J. Am. Chem. Soc.* 137 (3): 1362–1371.

53 Pan, Y., Qiu, W., Li, Q. et al. (2019). Assembling two-phase enzymatic cascade pathways in Pickering emulsion. *ChemCatChem* 11 (7): 1878–1883.

54 Dumeignil, F., Guehl, M., Gimbernat, A. et al. (2018). From sequential chemoenzymatic synthesis to integrated hybrid catalysis: taking the best of both worlds to open up the scope of possibilities for a sustainable future. *Catal. Sci. Technol.* 8 (22): 5708–5734.

55 Runge, M.B., Mwangi, M.T., Miller, A.L. et al. (2008). Cascade reactions using $LiAlH_4$ and grignard reagents in the presence of water. *Angew. Chem. Int. Ed.* 47 (5): 935–939.

56 Kourist, R., Schweiger, A., and Büchsenschütz, H. (2018). Enzymatic decarboxylation as a tool for the enzymatic defunctionalization of hydrophobic bio-based organic acids. In: *Lipid Modification by Enzymes and Engineered Microbes* (ed. U.T. Bornscheuer), 89–118. AOCS Press.

57 Schweiger, A., Ríos Lombardía, N., Winkler, C. et al. (2019). Using deep eutectic solvents to overcome limited substrate solubility in the enzymatic decarboxylation of bio-based phenolic acids. *ACS Sustain. Chem. Eng.* 7 (19): 16364–16370.

58 Hischer, T., Steinsiek, S., and Ansorge-Schumacher, M.B. (2006). Use of polyvinyl alcohol cryogels for the compartmentation of biocatalyzed reactions in non-aqueous media. *Biocatal. Biotransform.* 24 (6): 437–442.

59 Discher, B.M. (1999). Polymersomes: tough vesicles made from diblock copolymers. *Science* 284 (5417): 1143–1146.

60 Ahmed, F., Photos, P.J., and Discher, D.E. (2006). Polymersomes as viral capsid mimics. *Drug Dev. Res.* 67 (1): 4–14.

61 Klermund, L., Poschenrieder, S.T., and Castiglione, K. (2017). Biocatalysis in polymersomes: improving multienzyme cascades with incompatible reaction steps by compartmentalization. *ACS Catal.* 7 (6): 3900–3904.

62 Klermund, L., Poschenrieder, S.T., and Castiglione, K. (2016). Simple surface functionalization of polymersomes using non-antibacterial peptide anchors. *J. Nanobiotechnol.* 14 (1): 48.

63 Feng, X., Jia, Y., Cai, P. et al. (2016). Coassembly of photosystem II and ATPase as artificial chloroplast for light-driven ATP synthesis. *ACS Nano* 10 (1): 556–561.

64 Houghten, R.A. (1985). General method for the rapid solid-phase synthesis of large numbers of peptides: specificity of antigen-antibody interaction at the level of individual amino acids. *Proc. Natl. Acad. Sci.* 82 (15): 5131–5135.

65 Wachtmeister, J., Jakoblinnert, A., Kulig, J. et al. (2014). Whole-cell teabag catalysis for the modularisation of synthetic enzyme cascades in micro-aqueous systems. *ChemCatChem* 6 (4): 1051–1058.

66 Wachtmeister, J., Mennicken, P., Hunold, A. et al. (2016). Modularized biocatalysis: immobilization of whole cells for preparative applications in microaqueous organic solvents. *ChemCatChem* 8 (3): 607–614.

67 Sato, H., Hummel, W., and Gröger, H. (2015). Cooperative catalysis of noncompatible catalysts through compartmentalization: Wacker oxidation and enzymatic reduction in a one-pot process in aqueous media. *Angew. Chem. Int. Ed.* 54 (15): 4488–4492.

68 Latham, J., Henry, J.M., Sharif, H.H. et al. (2016). Integrated catalysis opens new arylation pathways via regiodivergent enzymatic CH activation. *Nat. Commun.* 7: 11873.

69 Carceller, J.M., Mifsud, M., Climent, M.J. et al. (2020). Production of chiral alcohols from racemic mixtures by integrated heterogeneous chemoenzymatic catalysis in fixed bed continuous operation. *Green Chem.* 22 (9): 2767–2777.

Part II

Enzymes Handling and Applications

5

Promiscuous Activity of Hydrolases

Erika V. M. Orozco and André L. M. Porto

University of São Paulo, São Carlos Institute of Chemistry, Av. João Dagnone, 1100, Santa Angelina, 13563-120, São Carlos, São Paulo, Brazil

5.1 Introduction

The importance of promiscuity in biocatalysis is remarkable because the vast majority of biocatalytic reactions applied industrially and studied in the academic field make use of non-natural substrates under reaction conditions that are different from those found naturally. Thus, this promiscuous behavior of enzymes provides an opportunity for the directed evolution of artificial enzymes with tailored properties through molecular biology techniques. These artificial enzymes may be useful for applications in organic synthesis and as tools for a wide range of technological applications [1, 2].

Within biocatalysis, the use of hydrolases has made considerable contributions not only in the hydrolysis of the carbon–heteroatom bond of esters and amides or of carbon–carbon bonds [3, 4] but also in synthetic reactions of these types of bonds, which are usually performed in organic solvents. The high versatility of hydrolases toward a broad range of substrates allows them to catalyze reactions that involve the formation of C—C bonds, such as aldol reactions [5] and Michael additions [6, 7] and three-component reactions such as the Mannich reaction and the Ugi reaction, and the synthesis of different tetrahydro chromene derivatives. Other reactions include the formation of peroxycarboxylic acids [8], Baeyer–Villiger oxidations [9], and epoxidations [10–12].

This chapter will introduce the unnatural activity of hydrolases, which can achieve a substantial number of practical applications, and other promiscuous reactions that are a source of new publications that to date have not found industrial applications. However, industrial applications are expected to develop because of progress in areas linked to biocatalysis, such as molecular biology, directed evolution, and high-throughput detection [13], it is possible to obtain high-performance mutant enzymes from wild-type enzymes at affordable prices for most research laboratories and industrial applications (see also Chapter 2). Finally, the aldol reaction catalyzed by lipases is presented as an example of

Biocatalysis for Practitioners: Techniques, Reactions and Applications, First Edition.
Edited by Gonzalo de Gonzalo and Iván Lavandera.

this promiscuous catalysis by pancreatic porcine lipase type II (PPL-II) and *Rhizopus niveus* lipase (RNL).

5.2 Catalytic Promiscuity

Catalytic promiscuity is defined by evolutionary theories as the ability of an enzyme to catalyze secondary reactions that are irrelevant or inefficient or that do not occur under physiological conditions due to the absence of natural substrates [14–17].

Because enzymes diverge from a common ancestor, enzymatic promiscuity is a mechanism associated with evolution by gene duplication [18] and aided by cooperativity and conformational flexibility [19–21]. Under selective pressure, enzymatic promiscuity has led to a specific function that has become their primary function, turning enzymes into specialists (specific) and causing other functions to be lost or silenced [22–25].

A high percentage of enzymes (75%) in *Archaea* microorganisms stands out for having a wide acceptance of substrates, while 25% catalyzes secondary reactions related to different transition states [26, 27].

The promiscuous behavior of enzymes has become increasingly useful because it allows the consideration of other types of reactions that previously have not been explored in the presence of biocatalysts. Currently, promiscuous enzymes provide the opportunity for new reaction applications through molecular evolution in the laboratory. In order to favor new catalytic activities or improve the existing ones, much effort is being made to rationally modify enzymes by protein engineering [23, 26, 28, 29] through point mutagenesis or directed evolution [22], random mutagenesis and DNA shuffling, and reaction engineering [30].

Promiscuous enzymes have been explored for the incorporation of unnatural amino acid residues into proteins [31] and for the synthesis of DNA labeled with a fluorophore or with chemically reactive groups for cross-linking to other molecules [32]. The promiscuity of avidin to bind 8-oxodeoxyguanosine was explored in commercial ELISA kits for the detection of oxidatively damaged DNA [33].

The approach of exploring catalytic promiscuity as a starting point for creating tailor-made biocatalysts may support these areas and may be the key to further application of biocatalysis in the industry [1]. Indeed, because of their broad substrate specificity, enantioselectivity, and regioselectivity, hydrolases have found industrial applications in biotransformations during the past two decades [34].

Among the hydrolases, lipases (EC 3.1.1.3, triacylglycerol (TG) lipases) stand out for their wide use in organic synthesis via kinetic resolution (KR) of racemic compounds. This is due to the high number of commercial preparations, their wide tolerance to substrates, and their good stability in media containing organic solvents when compared to other enzymes [27].

Lipases also catalyze unusual reactions in aqueous media such as esterification, transesterification, interesterification, polyesterification [35], and even the transfer of acyl groups to nucleophiles as amines (see also Chapter 6), oximes, and thiols [36].

The spectra of reactions catalyzed by some promiscuous lipases also include peroxycarboxylic acid formation, epoxidation of unsaturated compounds under mild conditions, and formation of C—C bonds through aldolic reactions, Michael additions, and Mannich reactions [36].

One of the most widely studied and industrially applied lipases is *Candida antarctica* B (CAL-B) [37], which has good stability at high temperatures (70–90 °C) and in nonpolar organic solvents. Also, it has a broad scope of substrates for acylation, transacylation [38], aminolysis, and ammonolysis reactions [39] when water is replaced by an appropriate nucleophile.

Another example of a promiscuous enzyme is 4-oxalocrotonate tautomerase (4-OT) and its mutant Phe50Ala, which catalyze aldol condensations. The presence of proline at the active site has been shown to be essential for catalysis in these enzymes [40]. The mechanism proposed in this study shows the formation of an iminium ion between the ketone and the proline residue **I**, which, by eliminating a proton, forms the enamine **II** that carries out the benzaldehyde attack to generate the β-hydroxyl-iminium **III**. This intermediate **III** eliminates water and produces intermediate **IV**, which finally forms cinnamaldehyde by hydrolysis (Scheme 5.1). The reaction was carried out between acetaldehyde (20 mM) and benzaldehyde (20 mM) with 4-OT Phe50Ala (0.5 mg, 114 μM, 0.6 mol%) in NaH_2PO_4 buffer (20 mM, pH 7.3) over 24 hours to produce cinnamaldehyde in 16% yield.

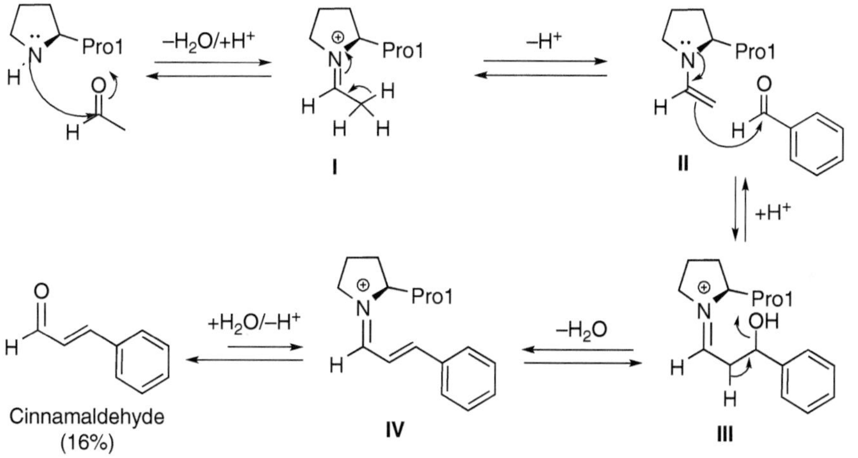

Scheme 5.1 Mechanism of the aldol condensation reaction between acetaldehyde and benzaldehyde catalyzed by 4-OT to produce cinnamaldehyde. Source: Adapted from Baas et al. [40].

5.3 Hydrolases

Hydrolysis of esters and amides is the natural reaction of hydrolases. Lipases, esterases, aminoacylases, amidases, and proteases integrate the large family of hydrolases that are so important for maintaining the metabolism of living organisms. Also, they provide a catalyst tool for organic synthesis, bioremediation, and fuel production [4, 41, 42].

Hydrolases carry out hydrolytic reactions using water as the nucleophile with no cofactor requirements. They differ mainly in the kind of substrate they hydrolyze under physiological conditions. Thus, esterases catalyze the hydrolysis of water-soluble esters, short and/or branched side-chain esters, while lipases hydrolyze triglycerides and are particularly tolerant of hydrophobic environments [43]. Aminoacylases, amidases, and proteases cleave the amide bond in *N*-acyl-L-amino acid, monocarboxylic acid amide, and proteins or peptides, respectively [44]. However, this criterion is not decisive when hydrolases are used in organic synthesis. Proteases catalyze the hydrolysis of both amides and esters, or lipases catalyze reactions on water-soluble esters, but the reaction rates will be different from those of natural substrates.

Other differences lie in the catalytic machinery. Even though it is formed by the triad of serine-histidine-aspartate residues (Ser-His-Asp), it turns out that the active site of proteases has the mirror image orientation of lipases and esterases. As a consequence, enantioselectivity is opposite between proteases and lipases or esterases. Proteases usually react with (*S*)-secondary alcohols, while lipases and esterases usually transform (*R*)-secondary alcohols [41]. The absolute *R* or *S* configuration depends on the Cahn-Ingold-Prelog priority rule given by the respective substitution groups.

Another big difference between proteases and lipases or esterases is the depth of the active site. The active site in lipases and esterases is distant from the surface; so the substrate interacts in a folded conformation. This is unlike the interaction of the substrates with the shallow active site in proteases, where they approximate an extended conformation [41]. This difference in their active sites allows lipases to accommodate two substituents of a secondary alcohol in their hydrophobic pockets, while proteases can only accommodate one of the two substituents; the other is faced toward the solvent. In this way, sterically hindered substituents are better accepted by proteases. Because of this characteristic shallow active site, the solvent greatly influences enantioselectivity [41, 45].

There is less information available in the literature for carbon–heteroatom hydrolases. Because their genes have been poorly identified, they have been less investigated; so their reaction mechanisms are not well understood [4]. Some of them require cofactors, such as pyridoxal 5′-phosphate in kynureninase or a thiamine diphosphate in α-diketone hydrolase. A review article by Kroutil and coworkers [3] provides important details on these enzymes and their generic reaction mechanisms.

5.3.1 Applications of Hydrolases to Organic Synthesis

Hydrolases are useful for a wide range of applications. They are used as additives in detergents to remove grease or protein-based stains, in the production of biofuels (making use of their potential in transesterification reactions), and in the synthesis of synthetic intermediaries for the manufacture of active pharmaceutical ingredients (APIs).

The reactions where hydrolases are most widely used in the industry are the asymmetric synthesis of optically active esters and amides through KRs, dynamic kinetic resolutions (DKRs), and desymmetrizations of prochiral or *meso* compounds. These are all based on hydrolytic reactions or synthetic reactions, and the latter are carried out in organic solvents [41, 46].

KRs facilitate the separation of enantiomers. Thus, in the case of racemic mixtures, the highest yield can only reach 50% for one of the enantiomers. This makes it possible to obtain pure enantiomers of many compounds, which in turn can be employed in the synthesis of more complex molecules. For example, PPL (lipase from porcine pancreas – PPL) catalyzes the hydrolysis of *rac*-glycidyl butyrate to (*R*)-glycidol in 89% yield and 92% *ee*. (*R*)-glycidol can be used as an intermediate in the synthesis of β-adrenoblockers such as propanolol (Scheme 5.2) [36].

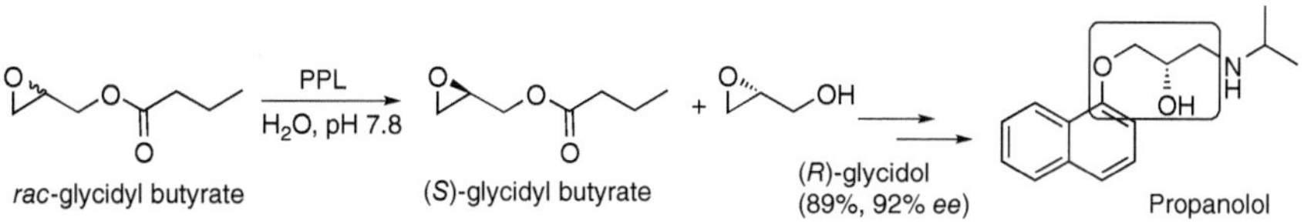

rac-glycidyl butyrate —PPL, H_2O, pH 7.8→ (S)-glycidyl butyrate + (R)-glycidol (89%, 92% ee) → Propanolol

Scheme 5.2 Enantioselective hydrolysis of glycidyl butyrate by PPL. Source: Adapted from Eremeev and Zaitsev [36].

DKRs are based on the combination of KRs and simultaneous racemization reactions, leading to theoretical yields of 100% for one of the enantiomers. Racemizations are performed by racemases or metal catalysts such as Ru or Ni, as the preparation of an enantiopure amine intermediate in the synthesis of Rasagilin (Azilect®), an irreversible monoamine oxidase-B inhibitor used in the treatment of Parkinson's disease, in the presence of *Candida rugosa* lipase (Scheme 5.3).

rac-1-Aminoindane is transformed into the (*R*)-enantiomer through DKR using the Ni catalyst KT-02, which acts as a racemizing agent of the remaining (*S*)-enantiomer of 1-aminoindane (Scheme 5.3). The lipase from *C. rugosa* catalyzes the acetylation of the (*R*)-enantiomer of 1-aminoindane using (*R*)-*O*-acetyl mandelic acid. A high yield and high enantioselectivity of the acetylated product were obtained (96% and 99% *ee*). Sequentially, an acid hydrolysis reaction of the acetylated (*R*)-1-aminoindane was necessary to obtain the (*R*)-1-aminoindane salt. The free amine was regenerated by increasing the pH with a base, and it was recovered from the organic phase in high yield (94%) and enantioselectivity (99.7%).

The specificity and enantioselectivity of a hydrolase usually follow empirical rules such as the lower reactivity with tertiary alcohols, in some cases favored by the large active site of esterases and lipases, and the low reactivity of aryl acid esters and α-substituted esters in carboxylic acids. Also, there is enantiopreference based on the relative size of the substituents of secondary alcohols. In general, alcohols of (*R*)-configuration react faster than those of (*S*)-configuration [48]. However, this assumption is not fulfilled for proteases which, as mentioned previously, have shallow sites. Also, polar solvents such as water influence the enantioselectivity of reactions with secondary alcohols and primary amines because hydrophobic substituents tend to target the hydrophobic pockets of the active site, while the other part of the substrate is exposed to the solvent [41, 45].

The model presented in Scheme 5.4 was proposed by Kazlauskas to explain the different enantiopreference of subtilisins for 1-(*p*-tolyl)ethanol in water and organic solvents. The reaction in water will favor the aryl substituent in the hydrophobic nonpolar pocket S_1', while the methyl group is oriented toward the solvent, opposite to the prediction based only on the size. In this way, subtilisin favors the (*R*)-enantiomer of 1-(*p*-tolyl)ethanol in water. In organic solvents, where the polarity differences of the solvent and the hydrophobic pocket S1′ is smaller, the (*S*)-enantiomer of 1-(*p*-tolyl)ethanol is favored. On the other

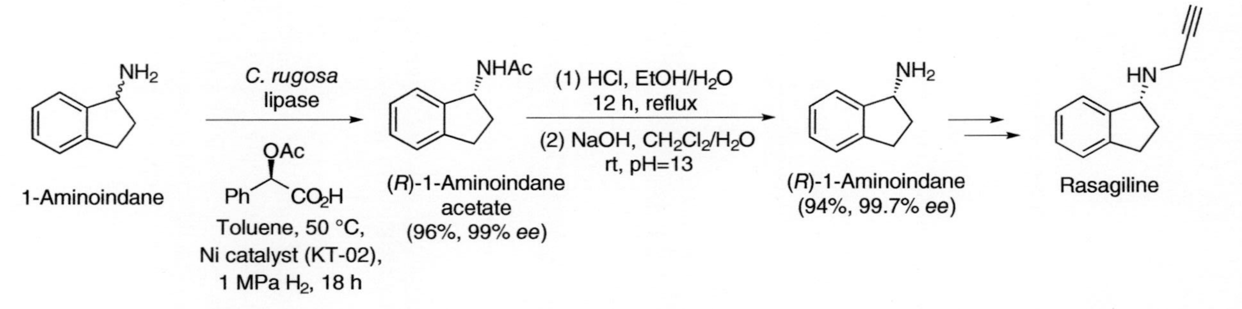

Scheme 5.3 Dynamic kinetic resolution in the preparation of the enantiopure intermediate amine for the synthesis of Rasagiline catalyzed by *C. rugosa* lipase. Source: Adapted from Domínguez de María et al. [47].

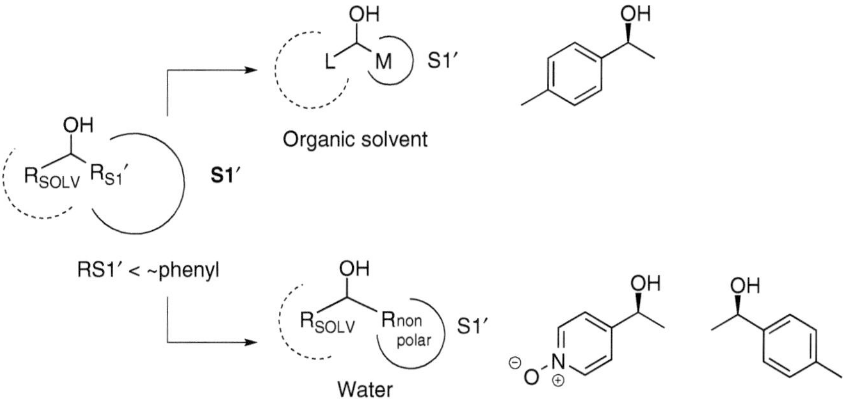

Enantioselectivity, *E* (preferred enantiomer)

	1-(4-pyridyl N-oxide)ethanol	1-(4-methylphenyl)ethanol
Subtilisin Carlsberg	3.1 (*S*)	1.1 (*S*)
Subtilisin E	4.5 (*S*)	16 (*S*)
Subtilisin BPN'	2.5 (*S*)	37 (*S*)

Scheme 5.4 Kazlauskas model to explain the enantiopreference of subtilisin enzymes toward secondary alcohols in water and in organic solvent. The numerical values correspond to the enantioselectivity (*E*) of the relative rate of the fast vs slow enantiomer. The preferred enantiomer is between parentheses. Source: Adapted from Kazlauskas [41].

hand, when the aryl substituent is more polar, as in the case of 4-pyridine *N*-oxide, the (*S*)-enantiomer is favored both, in water where the solvation of the pyridine *N*-oxide is favorable, and in organic solvent where the pyridine *N*-oxide is placing towards solvent prevents steric interactions in the S_1' pocket. Thus, this model predicts that increasing the polarity difference between substituents will increase the enantioselectivity of subtilisins in hydrolytic reactions as well as increase the enantioselectivity in alcohols with hydrophobic substituents by changing the organic medium polarity to a nonpolar one because this improves the solvation of the substituent exposed to the solvent [41, 45].

According to a recent literature survey by Domínguez de María et al. [47], most of the patents granted between 2014 and 2019 in applied biocatalysis correspond to hydrolases (68%). A large number of these hydrolases correspond to lipases applied to the synthesis of precursors of APIs based on the resolution of enantiomers by hydrolysis or acetylation reactions.

A recent patent has described the synthesis of the precursor to Ticagrelor, an inhibitor of platelet aggregation, originally produced by AstraZeneca. The precursor was obtained from the hydrolysis of the racemic mixture of N-protected *cis*-4-aminocyclopent-2-en-1-ol ($20\,g\,l^{-1}$, Scheme 5.5) catalyzed by Novozyme 435 at pH 5.0, 60 °C and 4 hours. The reaction provided the enantiopure (1*S*,4*R*)-alcohol in 45% conversion and 99% *ee*.

Generally, the use of biocatalysts is presented in the initial stages of a synthetic route that leads to the desired API; although some less frequent applications of biocatalysts in the final stage of synthesis have been patented [49].

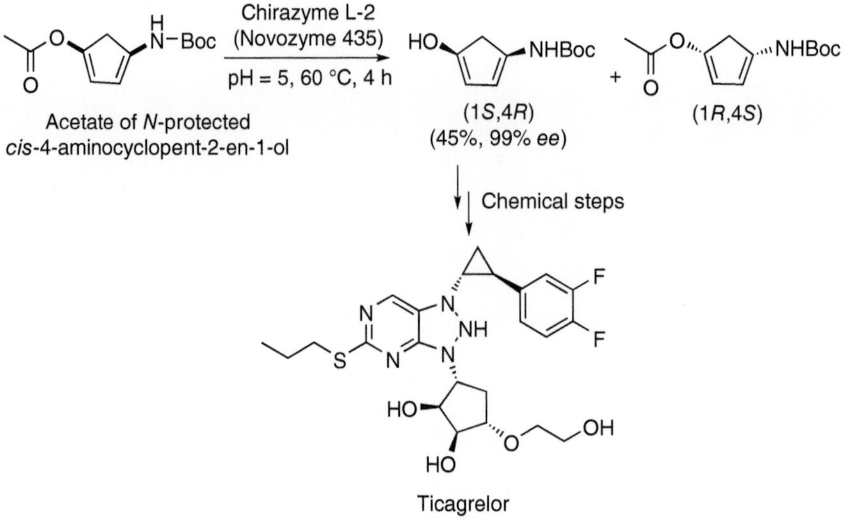

Chirazyme L-2 (Novozyme 435)
pH = 5, 60 °C, 4 h
Acetate of *N*-protected *cis*-4-aminocyclopent-2-en-1-ol
(1*S*,4*R*) (45%, 99% *ee*)
(1*R*,4*S*)
Chemical steps
Ticagrelor

Scheme 5.5 Synthesis of enantiopure Ticagrelor precursor catalyzed by lipase B from *Candida antarctica*. Source: Adapted from Domínguez de María et al. [47].

5.3.2 Lipases and Their Hydrolysis Mechanism

Lipases are TG acylhydrolases (E.C. 3.1.1.3) that in aqueous emulsions hydrolyze TGs releasing fatty acids, diacylglycerols (DG), monoglycerides (MG), and glycerol esters with different structures of acyl groups and alkyl chains [50] (Scheme 5.6).

Lipases have a hydrophobic β central leaf composition that consists of eight different β chains linked to six α-helices. This arrangement of its tertiary structure is called α/β-hydrolase folding, showing the α/β-hydrolase folds of CAL-B (Figure 5.1) [50].

The active site of lipases is formed by a catalytic triad of residues of L-serine (Ser), L-aspartate (Asp) (or glutamate), and L-histidine (His), an oxyanion hole, and in most cases a hydrophobic "cap" formed by an α-helix that covers the active enzyme site [53].

The widely accepted reaction mechanism of hydrolytic enzymes involves the activation of the serine assisted by His and Asp to form an alkoxide anion through the transfer of the proton from the hydroxyl group of serine to the nitrogen of histidine and from nitrogen to the carboxylate of aspartate. Activated serine makes a nucleophilic attack on the carbonyl of the ester or amide forming an acyl-enzyme intermediate (Step 1, Scheme 5.7). A second nucleophilic attack is carried out by a water molecule forming a tetrahedral intermediate (not shown), returning to the C=O bond, and detaching from the enzyme as carboxylic acid. The catalytic triad returns to its original protonation state (Step 2, Scheme 5.7) [54, 55].

5.3.3 Catalytic Promiscuity of Hydrolases

Hydrolases can accept different ester and amide substrates to be hydrolyzed through an acyl-enzyme intermediate if the reaction is performed in the presence of water. However, other products can be obtained when the reaction is carried out in an organic solvent and in the presence of other nucleophilic species. These products can be derived from

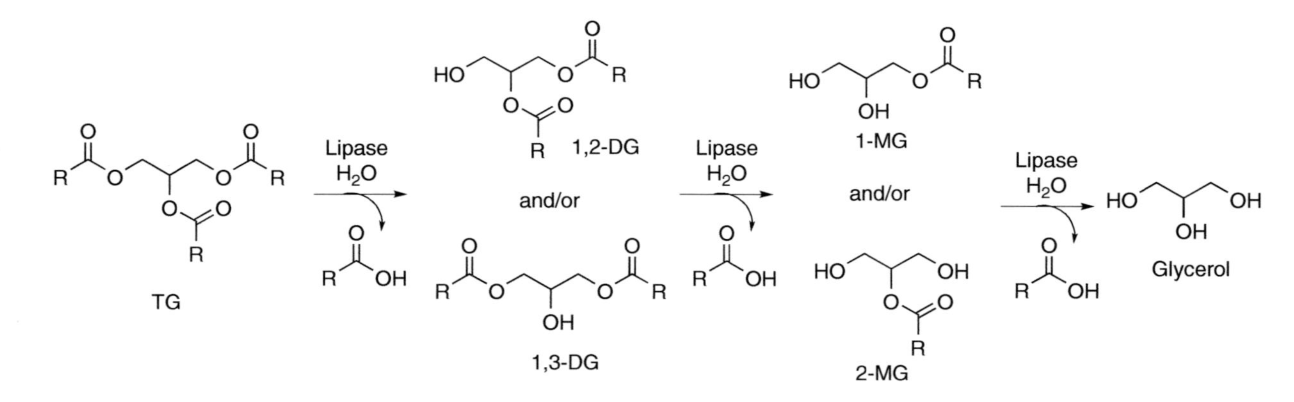

Scheme 5.6 Hydrolysis reaction of triacylglycerols catalyzed by lipases. Source: Adapted from Benito-Gallo et al. [51].

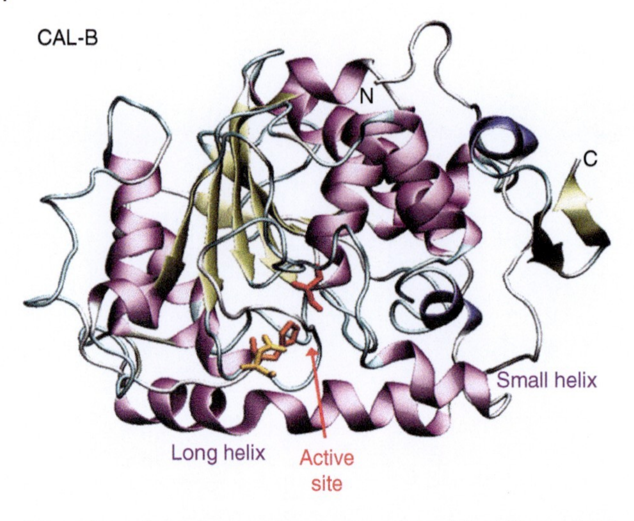

Figure 5.1 Crystallographic structure of the CAL-B lipase (PDB code: 1TCA). The catalytic triad at the active site is displayed: Ser105 (red), His224 (orange) and Asp187 (yellow). Source: Klahn et al. [52].

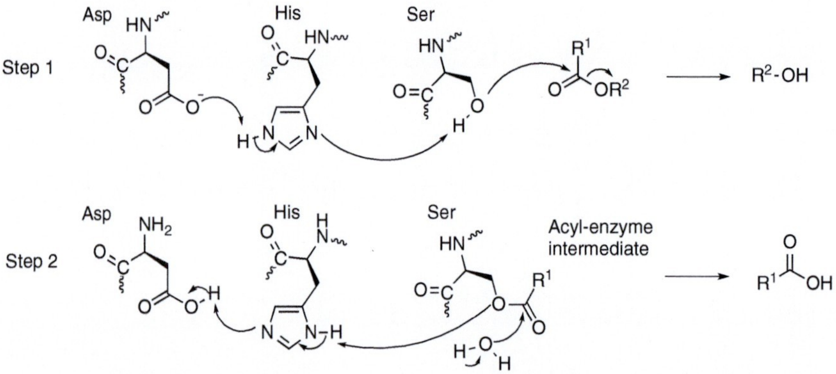

Scheme 5.7 General simplified mechanism of lipase-catalyzed ester hydrolysis. Source: Adapted from Faber [54].

transesterification reactions (esters) or from aminolysis or ammonolysis reactions (amides) [44].

The intermediate amide for the synthesis of Saxagliptin was prepared through ammonolysis with ammonium carbamate mediated by CAL-B in a continuous flow process (Scheme 5.8). The transformation was performed on a preparative scale using $220\,g\,l^{-1}$ of the *N*-Boc-(*S*)-2,3-dihydro-1*H*-pyrrole-2-carboxylic acid ethyl ester, $90\,g\,l^{-1}$ of ammonium carbamate, $33\,g\,l^{-1}$ of CALB, $110\,g\,l^{-1}$ calcium chloride, and $216\,g\,l^{-1}$ of ascarite (desiccant agent in the headspace). After three days, there was complete conversion of 96% ($182\,g\,l^{-1}$) of the amide in >99% *ee* (81% isolated yield) [56].

The use of a desiccant agent to displace the reaction equilibrium and avoid hydrolysis is a strategy that allows an increase in the yield as exemplified in Scheme 5.8. Also, examples

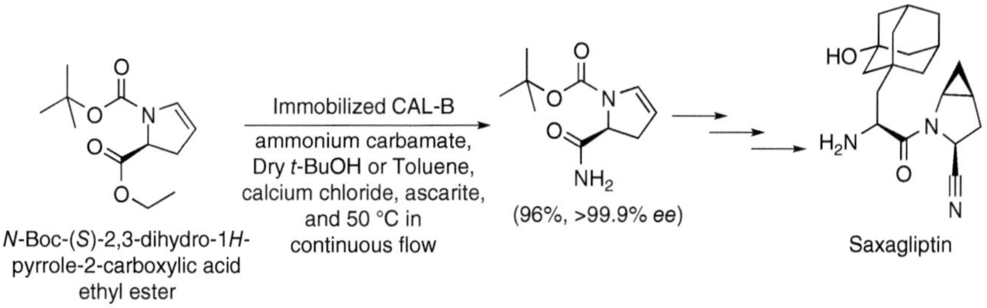

Immobilized CAL-B
ammonium carbamate,
Dry *t*-BuOH or Toluene,
calcium chloride, ascarite,
and 50 °C in
continuous flow

N-Boc-(*S*)-2,3-dihydro-1*H*-pyrrole-2-carboxylic acid ethyl ester

(96%, >99.9% *ee*)

Saxagliptin

Scheme 5.8 Preparation of the amide intermediate for the synthesis of Saxagliptin through CAL-B catalysis. Source: Adapted from Patel [56].

showing non-hydrolytic reactions in water have been reported. *Mycobacterium smegmatis* acyltransferase (MsAcT) catalyzed the synthesis of flavor esters by transesterification in water. This study does not constitute an example of promiscuity of condition or substrate, so it is an intrinsic property of this enzyme, which has a hydrophobic tunnel to the active site that hampers water access [57]. However, this activity opens up the range of catalyst options for transesterifications and even more, it could be a promiscuous catalyst for the formation of amides, with the advantage of the reaction being performed in water.

There are few examples in which hydrolases catalyze C—C bond formation reactions through catalytic promiscuity. A recent example has shown the use of several lipases as catalysts of the Knoevenagel reaction between indolin-2-one and different aromatic aldehydes to obtain 3-arylidenes. These compounds find extensive applications as therapeutic substances (antibacterial, antitumor, or anti-inflammatory). Lipase from pig pancreas (PPL) was the outstanding catalyst for this application in a water/dimethyl sulfoxide (DMSO) mixture (1/4 v/v) at 45 °C (Scheme 5.9) [47].

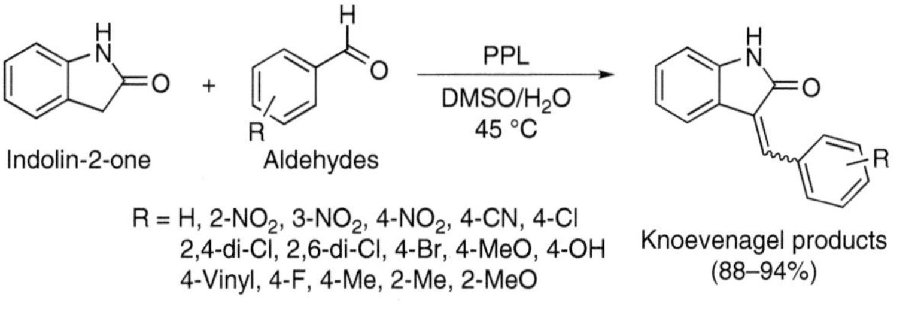

Indolin-2-one + Aldehydes

PPL
DMSO/H_2O
45 °C

R = H, 2-NO_2, 3-NO_2, 4-NO_2, 4-CN, 4-Cl
2,4-di-Cl, 2,6-di-Cl, 4-Br, 4-MeO, 4-OH
4-Vinyl, 4-F, 4-Me, 2-Me, 2-MeO

Knoevenagel products
(88–94%)

Scheme 5.9 Synthesis of 3-arylidenes through a Knoevenagel condensation catalyzed by lipase from pig pancreas (PPL). Source: Adapted from Domínguez de María et al. [47].

The recognition of other types of organic substrates and their good performance in organic solvents are examples of substrate promiscuity and condition promiscuity, even if it is not another type of reaction (catalytic promiscuity). An interesting example is the transesterification of Lobucavir with the protected amino acid Cbz-L-valine for the synthesis of the prodrug Lobucavir L-valine, an antiviral agent for the treatment of herpes virus and hepatitis B. The reaction catalyzed by lipase from *Pseudomonas cepacia* was selective and yielded the Cbz-protected intermediate of Lobucavir L-valine at 87% yield (Scheme 5.10) [49].

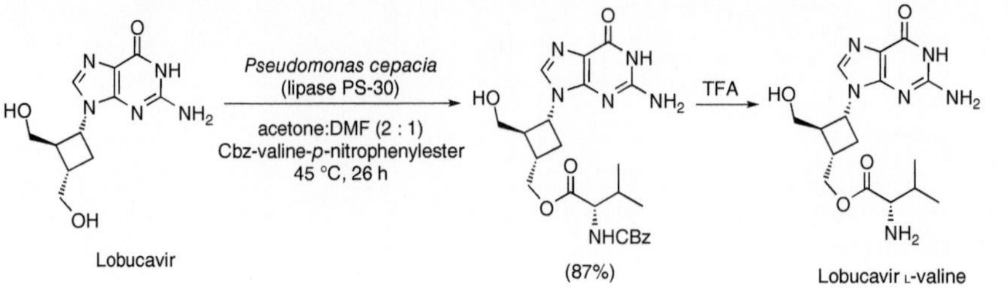

Scheme 5.10 Regioselective enzymatic transesterification of Lobucavir catalyzed by *Pseudomonas cepacia* lipase. Source: Adapted from Patel [49].

Hydrolases can catalyze other types of reactions that proceed through mechanisms and transition-state geometries other than those normally established where new C—C bonds are formed, such as aldol reactions [5, 58–62], Michael additions [6, 7], and three-component reactions such as the Mannich reaction [63–65], Ugi reaction, and the synthesis of different tetrahydrochromene derivatives [66]. Other reactions include the formation of peroxycarboxylic acids [8, 67], Baeyer–Villiger oxidation [9], and epoxidations [10–12]. A review addressing the promiscuity of lipases as catalysts for organic reactions has been described by Dwivedee et al. [68].

The Mannich reaction was initially reported by Li et al. where the product (Mannich base or β-amino-carbonyl compound) was obtained in yields between 43% and 87% using lipase from *Mucor miehei* (Scheme 5.11a) [63]. Other lipases catalyzed the Mannich asymmetric addition of ketimines with excellent yields of optically active β-amino ketone derivatives, such as the lipase from *C. rugosa* (Scheme 5.11b) [64] and the lipase from wheat germ (Scheme 5.11c) [65]. For reactions shown in Scheme 5.11a,b, enantiomeric excesses and absolute configurations of the stereogenic centers were not reported.

α-Chymotrypsin catalyzed the synthesis of two-substituted benzimidazole from *o*-phenylenediamines and several β-keto esters via retro-Claisen reaction in good to excellent yields by using ethanol, 50 °C, and 0.25 mmol of the *o*-phenylenediamine derivative and the β-keto ester (Scheme 5.12) [69].

Some examples of lipase-mediated promiscuous reactions on benzaldehyde derivatives are presented in Scheme 5.13. In Scheme 5.13a, a decarboxylative Knoevenagel reaction catalyzed by CAL-B used primary amines as additives such as aniline, *p*-toluidine, or benzylamine. Good to excellent yields were obtained for others β-keto ester substrates and different substituted aromatic aldehydes [70].

The second example in Scheme 5.13b is a Knoevenagel reaction, followed by a transesterification. These sequential reactions catalyzed by pancreatic porcine lipase (PPL) gave the 85% yield to the final product. In the same study, other lipases were tested for this reaction, highlighting lipase A from *Aspergillus niger* (35%) and lipase M from *Mucor javanicus* (40%) [71]. The third example is a three-component reaction in a

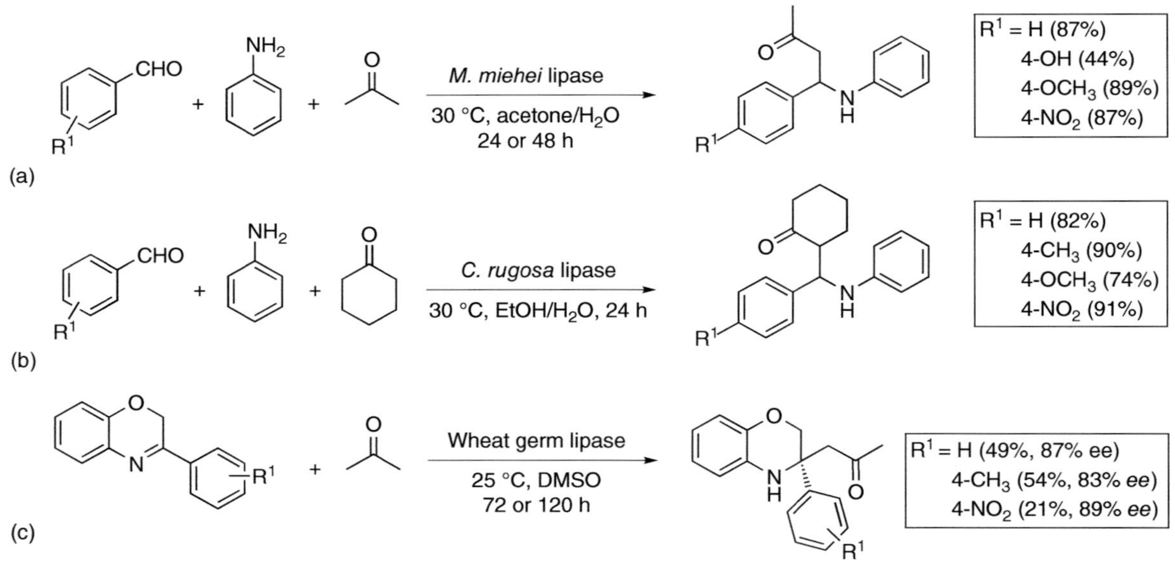

(a) R^1 CHO + NH_2 + O → *M. miehei* lipase, 30 °C, acetone/H_2O, 24 or 48 h

R^1 = H (87%)
4-OH (44%)
4-OCH_3 (89%)
4-NO_2 (87%)

(b) R^1 CHO + NH_2 + O → *C. rugosa* lipase, 30 °C, EtOH/H_2O, 24 h

R^1 = H (82%)
4-CH_3 (90%)
4-OCH_3 (74%)
4-NO_2 (91%)

(c) O N R^1 + O → Wheat germ lipase, 25 °C, DMSO, 72 or 120 h

R^1 = H (49%, 87% ee)
4-CH_3 (54%, 83% *ee*)
4-NO_2 (21%, 89% *ee*)

Scheme 5.11 Mannich reaction catalyzed by lipases. Source: (a) Adapted from Li et al. [63], (b) adapted from He et al. [64], and (c) adapted from Wu et al. [65].

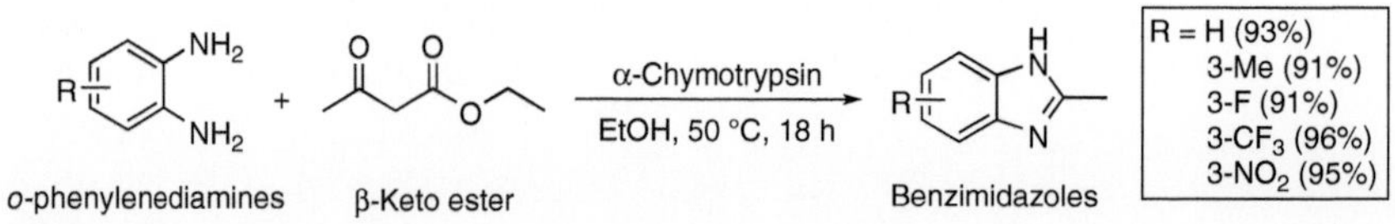

Scheme 5.12 Synthesis of two-substituted benzimidazoles catalyzed by α-chymotrypsin. Source: Adapted from Liu et al. [69].

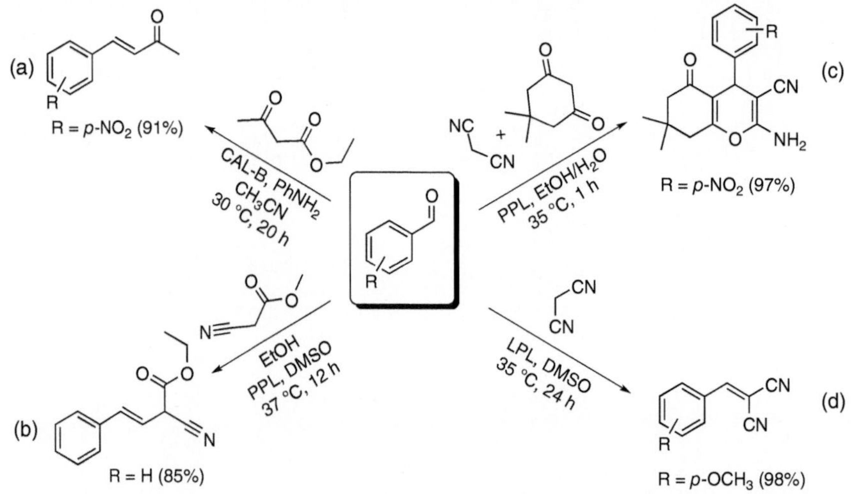

Scheme 5.13 Examples of promiscuous reactions on benzaldehyde derivatives catalyzed by lipases: (a) decarboxylative Knoevenagel reaction. Source: Adapted from Feng et al. [70]. (b) Sequential Knoevenagel and transesterification reactions catalyzed by PPL. Source: Adapted from Lai et al. [71]. (c) Three-component reaction in a one-pot biocatalytic approach for tetrahydrochromene synthesis catalyzed by PPL. Source: Adapted from Xu et al. [66]. (d) Knoevenagel condensation between an aromatic aldehyde and malononitrile catalyzed by LPL from *Aspergillus niger*. Source: Adapted from Ding et al. [72].

one-pot biocatalytic approach for tetrahydrochromene synthesis catalyzed by PPL (Scheme 5.13c). In general, high yields (91–97%) were obtained for different aromatic aldehydes [66].

Another example in Scheme 5.13d shows a Knoevenagel condensation between an aromatic aldehyde and malononitrile catalyzed by lipase lipoprotein (LPL) from *A. niger* with excellent yield (98%). In the same work, other aromatic aldehydes and activated methylene groups, such as cyanoacetate, β-keto esters, or 1,3-diketones, were tested and, in general good to excellent yields were obtained (44–98%) [72].

Scheme 5.14 presents some reactions such as epoxidation, Michael addition, and aldol condensation, which have an unsaturated carbonyl compound as one of the substrates. The epoxidation reaction using hydrogen peroxide and catalyzed by CAL-B Ser105Ala (Scheme 5.14a) was studied experimentally and theoretically by Svedendahl et al. [73]. A comparison between the mutant CAL-B Ser105Ala and the wild-type enzyme in the reaction with but-2-enal showed a better activity toward epoxidation by the mutant.

Examples of Michael additions are presented in Scheme 5.14b–f. The reaction of acetylacetone with acrolein (Scheme 5.14b) proceeded extremely fast with the mutant CAL-B Ser105Ala, obtaining 100% of conversion in less than 10 minutes. In a study of the kinetics of this biocatalyzed reaction, the specific rate ($4000\,s^{-1}$) was found to be 36 times higher than the specific rate for the wild-type lipase [74].

In Scheme 5.14d, a Michael addition reaction followed by a lactonization process to form a 2-hydroxy-2*H*-chromenone product is shown. 75% yield was obtained when the reaction was catalyzed by the lipase of *Pseudomonas fluorescens* (PFL) in CH_2Cl_2. However, when the reaction was catalyzed by the bovine pancreatic lipase (BPL) in acetonitrile, the preferred product (Scheme 5.14c) was 2*H*-chromenone (91%), which was obtained through the sequence of the aldol condensation reaction (biocatalyzed) and electrocyclization of 6π electrons (spontaneous reaction) [6].

PPL catalyzed a Michael addition (Scheme 5.14e) to produce warfarin in 87% yield and 22% *ee* from 4-hydroxycoumarin (0.5 mmol) and an α,β-unsaturated enal (2.5 mmol) in DMSO (0.9 ml) and H_2O (0.1 ml) at 25 °C [7]. The regioselective aza-Michael addition of

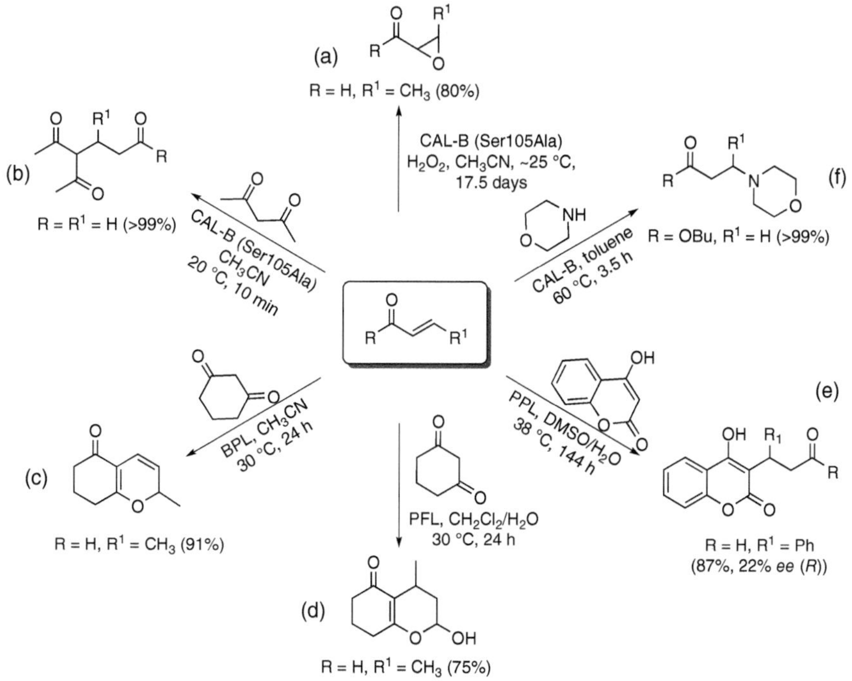

Scheme 5.14 Examples of promiscuous lipase-mediated reactions on α,β-unsaturated carbonyl substrates: (a) Epoxidation reaction catalyzed by CALB Ser105Ala. Source: Adapted from Svedendahl et al. [73]. (b) Michael addition reaction catalyzed by CALB Ser105Ala. Source: Adapted from Svedendahl et al. [74]. (c) 2*H*-Chromenone synthesis catalyzed by bovine pancreatic lipase (BPL). Source: Adapted from Yang et al. [6]. (d) 2-Hydroxy-2*H*-chromenone synthesis catalyzed by *Pseudomonas fluorescens* lipase (PFL). Source: Adapted from Yang et al. [6]. (e) Synthesis of warfarin through the reaction catalyzed by porcine pancreatic lipase (PPL). Source: Adapted from Xie et al. [7]. (f) Aza-Michael addition of morpholine to butyl acrylate catalyzed by CAL-B. Source: Adapted from Dhake et al. [75].

amines (primary and secondary) to acrylates using CAL-B as a biocatalyst at 60 °C is exemplified in Scheme 5.14f. The β-amino ester products were obtained in good to excellent yields (31–96%) when the reaction was carried out in toluene at 60 °C. Once again, this demonstrates the versatility of CAL-B under drastic reaction conditions [75].

5.3.4 Promiscuous Aldol Reaction Catalyzed by Hydrolases

The aldol reaction is one of the most common methods employed for the formation of carbon–carbon bonds in organic synthesis [76]. It is a reaction where carbonyl compounds that act as an anion enolate (donor) and electrophilic (aldehydes or ketones as acceptors) nucleophiles are used to form the β-hydroxy aldehydes or β-hydroxy ketones as products. The aldol adducts may undergo dehydration to form α,β-unsaturated carbonyl compounds.

Some aldolases, such as 2-deoxyribose 5-phosphate aldolase (DERA), are industrially employed in the synthesis of key intermediates of bioactive molecules such as statins or some antivirals such as Zanamavir [77, 78]. However, most are not in common use because of high substrate specificity and cofactor requirements, which are expensive and unstable under most reaction conditions. In this sense, it is convenient to explore other types of enzymes that, besides promoting the synthesis of organic compounds, have advantages in in terms of their thermal stability and exposure to organic solvents, as well as, do not require the use of cofactors to carry out their function.

CAL-B was the first lipase reported with promiscuous aldolase activity by Berglund and coworkers [60]. This enzyme catalyzed the reaction of simple aldehydes such as hexenal or propanal with acetone (see also later), but the reaction times were long (50 days).

However, a stereoselective aldol reaction was firstly described in 2008 (Scheme 5.15). The reaction between 4-nitrobenzaldehyde and acetone catalyzed by PPL was moderately enantioselective [61], obtaining the (*S*)-aldol product with 44% enantiomeric excess but in low yield (12%). An increase in water content to 20% v/v improved the yield to 56% although it affected the enantioselectivity, which decreased to 16% *ee*.

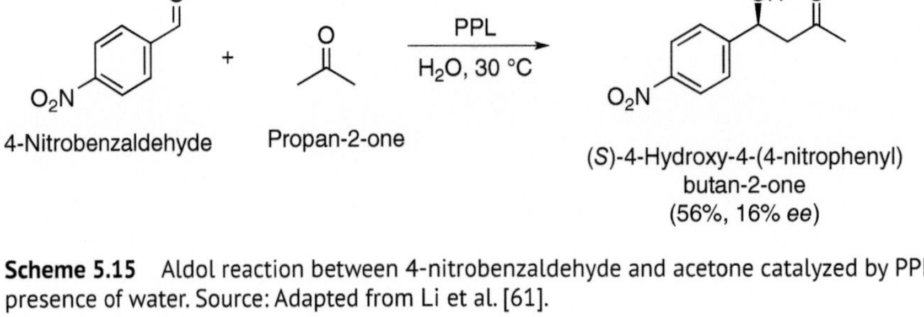

4-Nitrobenzaldehyde + Propan-2-one $\xrightarrow[H_2O,\ 30\ °C]{PPL}$ (*S*)-4-Hydroxy-4-(4-nitrophenyl)butan-2-one (56%, 16% *ee*)

Scheme 5.15 Aldol reaction between 4-nitrobenzaldehyde and acetone catalyzed by PPL in the presence of water. Source: Adapted from Li et al. [61].

The scope of substrates studied in the biocatalyzed aldol reaction includes cyclic ketones and heterocyclic and alicyclic ketones as donor substrates. However, the substrate scope for acceptors is limited to aromatic aldehydes.

In Table 5.1, some examples of the promiscuous aldol reactions of hydrolases are presented. Entries 1 and 2 show the conversions and selectivities of two heterocyclic ketones when the process was catalyzed by PPL-II, using acetonitrile and water as a solvent (9 : 1) [5]. Hence, *N*-Boc-piperidin-4-one reacted with 4-nitrobenzaldehyde (entry 1, Table 5.1), yielding the aldol product in 49%, with a diastereoselectivity of 62 : 38 favoring the *anti*-isomer and 62% *ee* for the *anti*-aldol pair. Tetrahydro-4*H*-pyran-4-one reacted with 4-nitrobenzaldehyde (entry 2, Table 5.1) and yielded the aldol product in 56%, favoring the *syn* diastereoisomer (38 : 62), opposite to the previous reaction, and the enantiomeric excess was lower for the *anti*-aldol pair (46%).

Lipase of pancreas bovine catalyzed the reaction between 3-nitrobenzaldehyde and cyclohexanone providing a 91% conversion when the medium was composed of cyclohexanone and water as a solvent mixture (9 : 1). The diastereoselectivity was 72 : 28 favoring the *anti*-aldol product with a 66% of *ee* of the major isomer (entry 3, Table 5.1) [79].

Halogenated aromatic aldehydes are substrates well tolerated by the active site of lipases, as shown by the high conversions and selectivities in aldol reactions catalyzed by PPL in cyclohexane and water (9 : 1). When the substituent in the aldehyde aromatic ring is electron-withdrawing, such as in 4-fluorobenzaldehyde, its reaction with cyclohexanone yielded 75% of the aldol product with diastereoselectivity 88 : 12 (*anti:syn*) and 90% enantiomeric excess in the reaction catalyzed by PPL (entry 4, Table 5.1). When the substrate was 4-bromobenzaldehyde, 99% yield, 88 : 12 (*anti:syn*) diastereoselectivity and 87% *ee* (*anti*-aldol product) was obtained (entry 5, Table 5.1) [80].

Other hydrolases, such as the alkaline protease from *Bacillus licheniformis* [81] and trypsin [82], catalyzed the aldol reaction between substituted aromatic aldehydes and cyclic ketones with conversions for aldol products from 28% to 99% with low dr but high to excellent *ee* (entries 6–7, Table 5.1). Cysteine proteases (entries 8–9, Table 5.1) such as chymopapain [83] and ficin [84] provided similar aromatic aldol products with conversions of 23% (4-Me) and 39% (3-NO_2), respectively, with low to moderate diastereoselectivity but high *ee*. Aspartarte proteases from *Aspergillus usamii* [85] and *Aspergillus melleus* [86] showed conversions for aldol products of 29–63%, but with high levels of diastereomeric ratios and enantiomeric excess (entries 10–12, Table 5.1). The zinc-dependent endonuclease P1 from *Penicillium citrinum* [87] catalyzed the aldol reaction with 25% conversion for the respective aldol product with 60% *de* and >99% *ee* favoring the *anti*-diastereoisomer (entry 13, Table 5.1).

In general, promiscuous enzyme catalyzed aldol reactions showed preferentially *anti*-diastereoselectivity. In the cases where good conversions were attained, they were accompanied by low enantiomeric excess. In contrast, when there were low conversions, high enantiomeric excess was usually observed. For example, the aldol reaction promoted by the *Bacillus licheniformis protease* (BLAP) [82] and endonuclease P1 from *P. citrinum* [87] provided enantiomeric excess greater than 99% when cyclohexanone and 4-methylbenzaldehyde were used. However, conversions were 28% and 25%, respectively (entries 6 and 13, Table 5.1).

The reaction between *N*-Boc-piperidin-4-one and 4-nitrobenzaldehyde in CH_3CN/H_2O was also catalyzed by the lipase from *A. niger* (Amano), providing 26% yield, 40 : 60 (*anti:syn*) diastereoselectivity, and 13% *ee* for the *anti*-diastereoisomer. However, with *C. rugosa* and RNLs, no aldolase activity was observed. This reaction was also catalyzed by a non-catalytic protein, bovine serum albumin (BSA), attaining 43% yield, 37 : 63 (*anti:syn*) diastereoselectivity, and 5% *ee* for the *anti*-diastereoisomer [5].

Table 5.1 Catalytic promiscuity of some enzymes in the aldol reaction between substituted benzaldehydes and cyclohexanone derivatives.

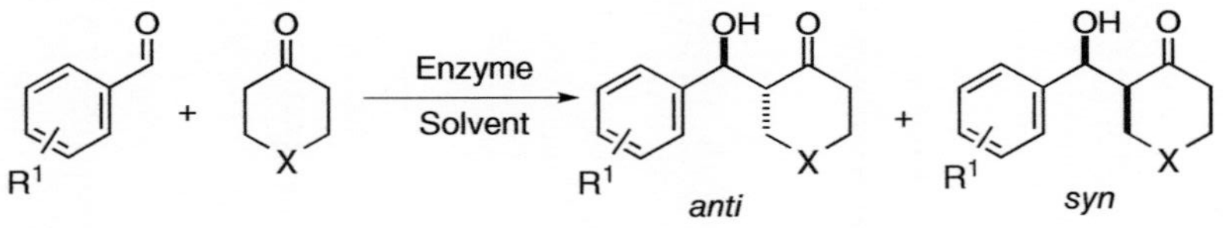

Entry	R^1	X	Enzyme	Solvent	Conv. (%)	dr (*anti:syn*)	ee_{anti} (%)	Reference
1	4-NO_2	*N*-Boc	PPL-II	CH_3CN/H_2O	49	62 : 38	62	[5]
2	4-NO_2	O	PPL-II	$MeCN/H_2O$	56	38 : 62	46	[5]
3	3-NO_2	CH_2	BPL	Cyhex[a)]/H_2O	91	72 : 28	66	[79]
4	4-F	CH_2	PPL	Cyhex[a)]/H_2O	75	88 : 12	90	[80]
5	4-Br	CH_2	PPL	Cyhex[a)]/H_2O	99	88 : 12	87	[80]
6	4-Me	CH_2	BLAP[b)]	$DMSO/H_2O$	28	70 : 30	>99	[81]
7	4-CF_3	CH_2	Trypsin	H_2O	34	59 : 41	65	[82]
8	4-Me	CH_2	Chymopapain	$MeCN/H_2O$	23	63 : 37	96	[83]
9	3-NO_2	CH_2	Ficin	$MeCN/H_2O$	39	86 : 14	81	[84]
10	3-Cl	CH_2	AUAP[c)]	$MeCN/H_2O$	29	92 : 8	88	[85]
11	4-NO_2	CH_2	AUAP[c)]	$MeCN/H_2O$	63	83 : 17	82	[85]
12	2-NO_2	CH_2	AMP[d)]	$MeCN/H_2O$	52	92 : 8	91	[86]
13	4-Me	CH_2	Nuclease p1	H_2O	25	80 : 20	>99	[87]

a) Cyhex, cyclohexanone.
b) BLAP, protease alkaline from *Bacillus licheniformis*.
c) AUAP, acidic protease from *Aspergillus usamii*.
d) AMP, protease from *Aspergillus melleus*. dr, diastereomeric ratio.
Source: Adapted from Miao et al. [88].

Promiscuous aldol condensation catalyzed by lipases has been rarely reported in the literature because most of the time, the reaction stops in the aldol addition product. One of the few examples is the one-pot process catalyzed by the lipase of *M. miehei* catalyzed two sequential reactions in a one-pot process [89], which involved a hydrolysis reaction of vinyl acetate to yield acetaldehyde. *In situ* generated acetaldehyde reacted with aromatic aldehydes, which led to the corresponding condensation products. In the case of 4-nitrobenzaldehyde 78% yield was obtained (Scheme 5.16).

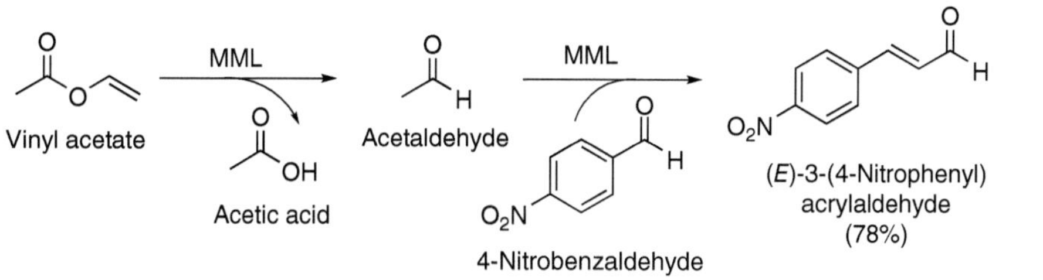

Vinyl acetate → (MML; Acetic acid) → Acetaldehyde → (MML; 4-Nitrobenzaldehyde) → (*E*)-3-(4-Nitrophenyl) acrylaldehyde (78%)

Scheme 5.16 Aldol condensation between *p*-nitrobenzaldehyde with *in situ* generated acetaldehyde from vinyl acetate hydrolysis. The reaction was performed in *one-pot* and catalyzed by *Mucor miehei* lipase. Source: Adapted from Wang et al. [89].

Catalytic promiscuity in lipases toward aldol reaction is favored because of the oxyanion hole that promotes the polarization of the carbonyl C=O double bond, leaving the carbon susceptible to a nucleophilic attack by the catalytic serine (Ser) for the subsequent formation of a hemiacetal (Scheme 5.17) [90].

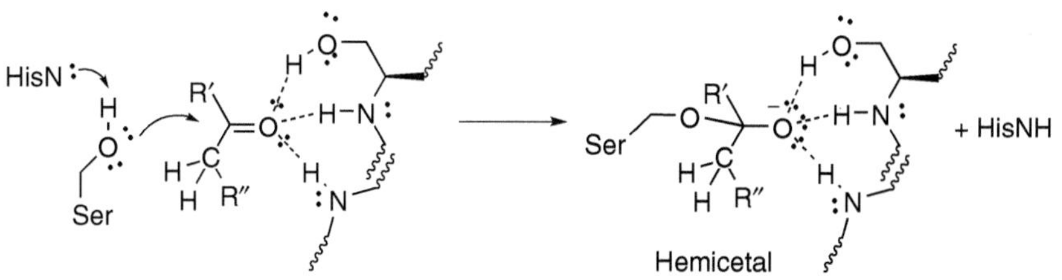

HisN, Ser → Hemicetal + HisNH

Scheme 5.17 Hemiacetal formation and its stabilization through interaction with amino acid residues of the active site in lipases in the presence of carbonyl substrates. Source: Adapted from Hult and Berglund [90].

This type of catalysis is assisted by enzyme–substrate interactions such as hydrogen bonds and electrostatic and hydrophobic interactions. These interactions possibly have a relevant contribution to promiscuous reactions, are less dependent on substrate specificity and may contribute to substrate orientation in the catalytic site [15].

To improve the interactions that promote catalysis, mutant enzymes are designed or tailored in the laboratory by use of different accelerated molecular evolution techniques. One of the first rational designs applied to lipases was in Ser105Ala mutant CAL-B. It increased aldolase activity in CAL-B mutant compared to native CAL-B for

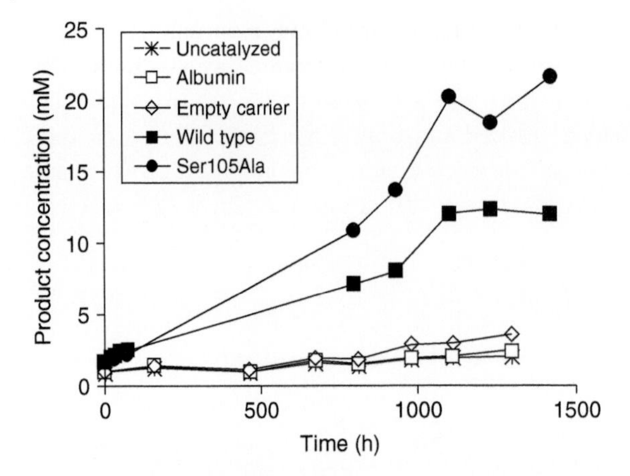

Figure 5.2 Aldol reaction progress between hexenal and acetone catalyzed by Ser105Ala mutant CAL-B, wild-type CAL-B, and control reactions: empty carrier, albumin, and uncatalyzed reaction. Source: Branneby et al. [60].

the reaction between hexenal and acetone when the enzymes were immobilized on the polypropylene carrier EP100. A comparison of reaction progress curves in Figure 5.2 for the mutant enzyme, the native enzyme and control reactions (empty carrier, albumin and uncatalyzed reaction) stand out a higher reaction rate for the mutated enzyme [60].

The subtraction of the nucleophilic functionality in the Ser105Ala mutant CAL-B promoted an increase in the reaction rate that is probably due to the absence of the hemiacetal formation. However, as mentioned previously, the non-enzymatic protein BSA was able to catalyze the aldol reaction. An explanation for this may be a nonspecific protein catalysis, which would lead to the same products through different modes of interaction. In this case, a set of amino acid residues such as some residues of lysine, which are not part of the active site, can catalyze the reaction through enamine formation.

As shown in the study by Birolli et al. [59], the aldolic reaction between 4-nitrobenzaldehyde and cyclohexanone by RNL was nonspecific protein catalysis. Thus, the enzyme subjected to denaturing conditions continued to exert a catalytic effect. In the case where the enzyme was denatured in water at 100 °C or an aqueous urea solution at 100 °C, the product was obtained in 61% yield with 42 : 58 diastereoselectivity (*anti:syn*) and 23% *ee*. Similar results were found when the enzyme was treated with urea (95%, dr 51 : 49, *anti:syn*, 11% ee_{anti}) or inhibited with the irreversible hydrolase inhibitor phenylmethylsulfonylfluoride, PMSF, (30%, dr 53 : 47, *anti:syn*, 46% ee_{anti}).

Although the catalytic mechanism of lipases is well defined, in many cases, there is no proven mode of substrate interaction at the active site for promiscuous aldol reactions, which invites an investigation into understanding which amino acids can promote the specific and/or nonspecific catalysis.

Despite the progress in biocatalysis, the number of biocatalytic methodologies in asymmetric aldol reactions are still limited, especially in the formation of carbon–carbon bonds.

We carried out a study of the catalytic promiscuity in lipases using PPL-II and RNL presented in the next section.

5.3.5 Aldol Reaction Between 4-Cyanobenzaldehyde and Cyclohexanone Catalyzed by Porcine Pancreatic Lipase (PPL-II) and *Rhizopus niveus* Lipase (RNL)[1]

Commercial extracts of PPL-II and RNL were used in this study. PPL is a 50 kDa protein with 448 amino acid residues. PPL is made up of two domains: α/β-fold N-terminal domain and β-sandwich C-terminal domain called domain 1 and domain 2, respectively. The domains are linked through a hinge residue Ala333. The active site is covered by a region extended between amino acids Cys238 and Cys262, which forms a lid that regulates the entry and exit of the substrates. The catalytic triad is composed of the amino acid residues Ser153, Asp177, and His264 [92–96]. RNL is made up of two polypeptides (chain A is a small peptide consisting of 52 amino acids linked to a sugar molecule and chain B is a large peptide with an active center of 34 000 molecular weight consisting of 297 acidic amino acids) that are non-covalently associated [97].

The biocatalytic aldol reactions of the PPL-II and RNL lipases were studied by us through the reaction between 4-cyanobenzaldehyde (4-CNB) and cyclohexanone (CHX) (Scheme 5.18).

This reaction provided the main products, the *anti-* and *syn*-aldols and the *E*-enone and *Z*-enone condensation products in minor proportions (Scheme 5.18).

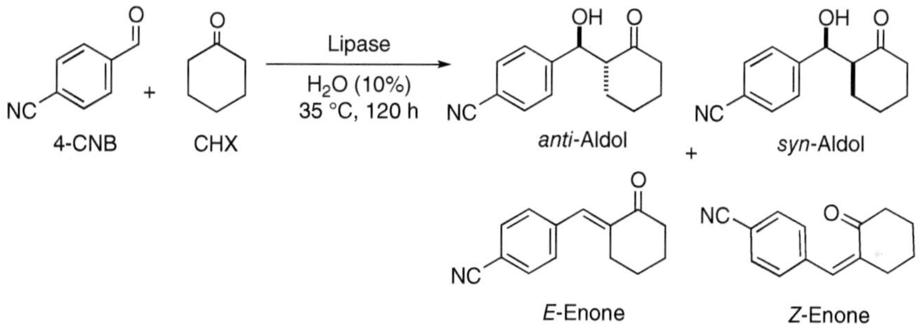

Scheme 5.18 Aldol reaction between 4-cyanobenzaldehyde and cyclohexanone catalyzed by lipases PPL-II and RNL. Source: Adapted from Meñaca [91].

Table 5.2 summarizes the conversion and diastereoselectivities obtained along 96 hours when the aldol addition reaction was catalyzed by PPL-II. In 24 hours of reaction, the *anti*-aldol diastereoisomer was slightly favored. After 48 hours of reaction, the *anti*-diastereoselectivity increased and remained at 81 : 19 until 96 hours. The aldol conversion was maximum at 72 hours and showed a slight decrease between 72 and 96 hours.

In Table 5.2, the conversion and diastereoselectivity of the aldol reaction between 4-CNB and cyclohexanone catalyzed by RNL are also presented. In this case, besides *anti-* and *syn*-aldols, the condensation products were formed. Their conversion increased over time, opposite to the reaction catalyzed by PPL-II under identical reaction conditions.

[1]The results presented here are part of the Doctoral Thesis of M. V. Erika Orozco entitled in Portuguese: "*Promiscuidade enzimática de lipases na síntese de aldóis e 2H-cromenonas*" [91].

Table 5.2 Conversions and diastereoselectivities obtained for the aldol addition reaction between 4-cyanobenzaldehyde and cyclohexanone catalyzed by PPL-II and RNL.

4-CNB + CHX → (Lipase, H_2O (10%), 35 °C, 120 h) → *anti*-Aldol + *syn*-Aldol

Time (h)	Conversion (%)	Aldol diastereoselectivity (*anti:syn*)
Aldol reaction mediated by PPL-II		
24	50	56 : 44
48	64	82 : 18
72	68	81 : 19
96	61	81 : 19
Aldol reaction mediated by lipase from *Rhizopus niveus*		
24	5	70 : 30
48	18	73 : 27
72	26	73 : 27
96	28	73 : 27

PPL-II, Porcine pancreatic lipase type II and RNL, *Rhizopus niveus* lipase.
Source: Adapted from Meñaca [91].

The oscillations of the concentrations of *syn*- and *anti*-aldols during the reaction time showed that the catalytic behavior of lipase for the addition and condensation reactions (*E*- and *Z*-enone) need to be investigated in more detail, such as a study of the retro-aldol reaction in the presence of the biocatalyst (results unpublished).

A comparative study of both biocatalysts allowed us to observe a higher reaction rate for the PPL-II than the RNL, yielding aldols in conversions of 61% and 28%, respectively, after 96 hours of reaction. The diastereoselectivities were slightly different, mainly 81 : 19 and 73 : 27, in favor of the *anti*-diastereoisomers. The enantioselectivities were higher for the *anti*-enantiomeric pair in both lipases, 74% and 52% *ee*, for PPL-II and RNL, respectively. Recently the aldolic reaction was carried out with 4-NO_2-benzaldehyde in a miniemulsion with the lipase of *R. niveus*. The reaction developed in miniemulsion allowed a decrease in the catalyst concentration from 20 mg mL^{-1} to 6 mg mL^{-1} compared to that carried out in organic solvents. Besides, the stability of the enzyme improved [98].

5.4 Conclusions

The role that biocatalysis has been playing in industrial applications has its greatest representativeness in hydrolases. Within this large class of enzymes, lipases are the group of enzymes most widely used in organic synthesis. This is due to their remarkable characteristics as promiscuous catalysts. However, other types of reactions different from hydrolysis or synthesis of esters and amides have been barely applied at an industrial level. This panorama is different from an academic point of view, where the catalytic promiscuity of hydrolases is increasingly demonstrated, even though the mechanisms of promiscuous reactions remain unveiled in many cases.

Notwithstanding the low catalytic activities in some cases, these enzymes can constitute a starting point for directed evolution creating mutant enzymes with higher performance for promiscuous reactions such as C—C bond formation.

On the other hand, there are other hydrolases, such as those that cleave C—C bonds, that have been less explored and this constitutes a fertile ground for future biocatalytic applications.

References

1 Leveson-Gower, R.B., Mayer, C., and Roelfes, G. (2019). The importance of catalytic promiscuity for enzyme design and evolution. *Nat. Rev. Chem.* 3 (12): 687–705.

2 Sheldon, R.A., Brady, D., and Bode, M.L. (2020). The Hitchhiker's guide to biocatalysis: recent advances in the use of enzymes in organic synthesis. *Chem. Sci.* 11 (10): 2587–2605.

3 Siirola, E., Frank, A., Grogan, G., and Kroutil, W. (2013). C-C hydrolases for biocatalysis. *Adv. Synth. Catal.* 355 (9): 1677–1691.

4 Siirola, E. and Kroutil, W. (2014). Organic solvent tolerance of retro-Friedel-Crafts hydrolases. *Top. Catal.* 57 (5): 392–400.

5 Guan, Z., Fu, J., and He, Y. (2012). Biocatalytic promiscuity: lipase-catalyzed asymmetric aldol reaction of heterocyclic ketones with aldehydes. *Tetrahedron Lett.* 53 (37): 4959–4961.

6 Yang, Q., Zhou, L., Wu, W. et al. (2015). Lipase-catalyzed regioselective domino reaction for the synthesis of chromenone derivatives. *RSC Adv.* 5 (96): 78927–78932.

7 Xie, B., Guan, Z., and He, Y. (2012). Promiscuous enzyme-catalyzed Michael addition: synthesis of warfarin and derivatives. *J. Chem. Technol.* 87 (12): 1709–1714.

8 Carboni-Oerlemans, C., Domínguez de María, P., Tuin, B. et al. (2006). Hydrolase-catalysed synthesis of peroxycarboxylic acids: biocatalytic promiscuity for practical applications. *J. Biotechnol.* 126 (2): 140–151.

9 Lemoult, S.C., Richardson, P.F., and Roberts, S.M. (1995). Lipase-catalysed Baeyer-Villiger reactions. *J. Chem. Soc., Perkin Trans.* 1 (1): 89–91.

10 Björkling, F., Godtfredsen, S.E., and Kirk, O. (1990). Lipase-mediated formation of peroxycarboxylic acids used in catalytic epoxidation of alkenes. *J. Chem. Soc., Chem. Commun.* (19): 1301–1303.

11 Rusch Gen Klaas, M. and Warwel, S. (1999). Chemoenzymatic epoxidation of alkenes by dimethyl carbonate and hydrogen peroxide. *Org. Lett.* 1 (7): 1025–1026.

12 Sarma, K., Bhati, N., Borthakur, N., and Goswami, A. (2007). A novel method for the synthesis of chiral epoxides from styrene derivatives using chiral acids in presence of *Pseudomonas* lipase G6 [PSL G6] and hydrogen peroxide. *Tetrahedron* 63 (36): 8735–8741.

13 Bunzel, H.A., Garrabou, X., Pott, M., and Hilvert, D. (2018). Speeding up enzyme discovery and engineering with ultrahigh-throughput methods. *Curr. Opin. Struct. Biol.* 48: 149–156.

14 Copley, S.D. (2015). An evolutionary biochemist's perspective on promiscuity. *Trends Biochem. Sci.* 40 (2): 72–78.

15 Babtie, A., Tokuriki, N., and Hollfelder, F. (2010). What makes an enzyme promiscuous ? *Curr. Opin. Chem. Biol.* 14 (2): 200–207.

16 Gupta, R.D. (2016). Recent advances in enzyme promiscuity. *Sustain Chem. Process* 4 (1): 1–7.

17 Singla, P. and Bhardwaj, R.D. (2020). Enzyme promiscuity–A light on the "darker" side of enzyme specificity. *Biocatal. Biotransfor.* 38 (2): 81–92.

18 Soskine, M. and Tawfik, D.S. (2010). Mutational effects and the evolution of new protein functions. *Nat. Rev. Genet.* 11 (8): 572–582.

19 Pabis, A., Risso, V.A., Sanchez-Ruiz, J.M., and Kamerlin, S.C. (2018). Cooperativity and flexibility in enzyme evolution. *Curr. Opin. Struct. Biol.* 48: 83–92.

20 Devamani, T., Rauwerdink, A.M., Lunzer, M. et al. (2016). Catalytic promiscuity of ancestral esterases and hydroxynitrile lyases. *J. Am. Chem. Soc.* 138 (3): 1046–1056.

21 Gatti-Lafranconi, P. and Hollfelder, F. (2013). Flexibility and reactivity in promiscuous enzymes. *ChemBioChem* 14 (3): 285–292.

22 Bolt, A., Berry, A., and Nelson, A. (2008). Directed evolution of aldolases for exploitation in synthetic organic chemistry. *Arch. Biochem. Biophys.* 474 (2): 318–330.

23 Gijsen, H. and Wong, C. (1994). Unprecedented asymmetric aldol reactions with three aldehyde substrates catalyzed by 2-deoxyribose-5-phosphate aldolase. *J. Am. Chem. Soc.* 116 (18): 8422–8423.

24 Mahmoudian, M., Noble, D., Drake, C.S. et al. (1997). An efficient process for production of N-acetylneuraminic acid using N-acetylneuraminic acid aldolase. *Enzyme Microb. Technol.* 20 (5): 393–400.

25 Nam, H., Lewis, N.E., Lerman, J.A. et al. (2012). Network context and selection in the evolution to enzyme specificity. *Science* 337 (6098): 1101–1104.

26 Martínez, N.M., Rodríguez, E.Z., Rodríguez, V.K., and Pérez, R.E. (2017). Tracing the repertoire of promiscuous enzymes along the metabolic pathways in archaeal organisms. *Life* 7 (3): 1–14.

27 Kapoor, M. and Gupta, M.N. (2012). Lipase promiscuity and its biochemical applications. *Process Biochem.* 47 (4): 555–569.

28 Arnold, F.H. (1990). Engineering enzymes for non-aqueous solvents. *Trends Biotechnol.* 8: 244–249.

29 Sheldon, R.A. and Woodley, J.M. (2018). Role of biocatalysis in sustainable chemistry. *Chem. Rev.* 118 (2): 801–838.

30 Dean, S.M., Greenberg, W.A., and Wong, C. (2007). Recent advances in aldolase-catalyzed asymmetric synthesis. *Adv. Synth. Catal.* 349 (8–9): 1308–1320.

31 Liu, C.C. and Schultz, P.G. (2010). Adding new chemistries to the genetic code. *Annu. Rev. Biochem.* 79 (1): 413–444.

32 Hocek, M. (2014). Synthesis of base-modified 2′-deoxyribonucleoside triphosphates and their use in enzymatic synthesis of modified DNA for applications in bioanalysis and chemical biology. *J. Org. Chem.* 79 (21): 9914–9921.

33 Struthers, L., Patel, R., Clark, J., and Thomas, S. (1998). Direct detection of 8-oxodeoxyguanosine and 8-oxoguanine by avidin and its analogues. *Anal. Biochem.* 255 (1): 20–31.

34 López-Iglesias, M. and Gotor-Fernández, V. (2015). Recent advances in biocatalytic promiscuity: hydrolase-catalyzed reactions for nonconventional transformations. *Chem. Rec.* 15 (4): 743–759.

35 Klibanov, A.M. (2001). Improving enzymes by using them in organic solvents. *Nature* 409: 241–246.

36 Eremeev, N.L. and Zaitsev, S.Y. (2016). Porcine pancreatic lipase as a catalyst in organic synthesis. *Mini Rev. Org. Chem.* 13 (1): 78–85.

37 Widersten, M. (2014). Protein engineering for development of new hydrolytic biocatalysts. *Curr. Opin. Chem. Biol.* 21: 42–47.

38 Khersonsky, O., Roodveldt, C., and Tawfik, D.S. (2006). Enzyme promiscuity: evolutionary and mechanistic aspects. *Curr. Opin. Chem. Biol.* 10: 498–508.

39 Lima, R.N., dos Anjos, C.S., Orozco, E.V.M., and Porto, A.L.M. (2019). Versatility of *Candida antarctica* lipase in the amide bond formation applied in organic synthesis and biotechnological processes. *Mol. Catal.* 466: 75–105.

40 Baas, B., Zandvoort, E., Geertsema, E.M., and Poelarends, G.J. (2013). Recent advances in the study of enzyme promiscuity in the tautomerase superfamily. *ChemBioChem* 14 (8): 917–926.

41 Kazlauskas, R. (2016). Hydrolysis and formation of carboxylic acid and alcohol derivatives. In: *Organic Synthesis Using Biocatalysis*, 127–148. Elsevier.

42 Nikolaivits, E., Kanelli, M., Dimarogona, M., and Topakas, E. (2018). A middle-aged enzyme still in its prime: recent advances in the field of cutinases. *Catalysts* 8 (12): 612.

43 Mehta, A., Bodh, U., and Gupta, R. (2017). Fungal lipases: a review. *J. Biotech. Res.* 8 (1): 58–77.

44 Gotor, V., Gotor-Fernández, V., and Busto, E. (2012). Hydrolysis and reverse hydrolysis: hydrolysis and formation of amides. In: *Comprehensive Chirality*, 101–121. Elsevier.

45 Savile, C.K. and Kazlauskas, R.J. (2005). How substrate solvation contributes to the enantioselectivity of subtilisin toward secondary alcohols. *J. Am. Chem. Soc.* 127 (35): 12228–12229.

46 García-Urdiales, E., Alfonso, I., and Gotor, V. (2011). Enantioselective enzymatic desymmetrizations in organic synthesis. Update 1. *Chem. Rev.* 105 (1): 313–354.

47 Domínguez de María, P., de Gonzalo, G., and Alcántara, A.R. (2019). Biocatalysis as useful tool in asymmetric synthesis: an assessment of recently granted patents (2014-2019). *Catalysts* 9 (10): 802.

48 Kazlauskas, R.J., Weissfloch, A.N.E., Rappaport, A.T., and Cuccia, L.A. (1991). A rule to predict which enantiomer of a secondary alcohol reacts faster in reactions catalyzed by cholesterol esterase, lipase from *Pseudomonas cepacia*, and lipase from *Candida rugosa*. *J. Org. Chem.* 56 (8): 2656–2665.

49 Patel, R.N. (2016). Applications of biocatalysis for pharmaceuticals and chemicals. In: *Organic Synthesis Using Biocatalysis*, 339–411. Elsevier Inc.

50 Jaeger, K.-E., Ransac, S., Dijkstra, B.W. et al. (1994). Bacterial lipases. *FEMS Microbiol. Rev.* 15 (1): 29–63.

51 Benito-Gallo, P., Franceschetto, A., Wong, J.C.M. et al. (2015). Chain length affects pancreatic lipase activity and the extent and pH – time profile of triglyceride lipolysis. *Eur. J. Pharm. Biopharm.* 93: 353–362.

52 Klähn, M., Lim, G.S., Seduraman, A., and Wu, P. (2011). On the different roles of anions and cations in the solvation of enzymes in ionic liquids. *Phys. Chem. Chem. Phys.* 13 (4): 1649–1662.

53 De Jesus, C.M.C. (2010). *Resolução cinética enzimática de álcoois secundários em água por tecnologia de miniemulsões*. Universidade de Lisboa. Faculdade de ciências e tecnologia.

54 Faber, K. (2011). *Biotransformations in Organic Chemistry*, 423. Berlin: Springer-Verlag Berlin Heidelberg.

55 Chen, H., Meng, X., Xu, X. et al. (2018). The molecular basis for lipase stereoselectivity. *Appl. Microbiol. Biotechnol.* 102 (8): 3487–3495.

56 Patel, R.N. (2018). Biocatalysis for synthesis of pharmaceuticals. *Bioorg. Med. Chem.* 26 (7): 1252–1274.

57 Perdomo, I.C., Gianolio, S., Pinto, A. et al. (2019). Efficient enzymatic preparation of flavor esters in water. *J. Agric. Food Chem.* 67 (23): 6517–6522.

58 Li, C., Zhou, Y.J., Wang, N. et al. (2010). Promiscuous protease-catalyzed aldol reactions: a facile biocatalytic protocol for carbon-carbon bond formation in aqueous media. *J. Biotechnol.* 150 (4): 539–545.

59 Birolli, W.G., Fonseca, L.P., and Porto, A.L.M. (2017). Aldol reactions by lipase from *Rhizopus niveus*, an example of unspecific protein catalysis. *Catal. Lett.* 147 (8): 1977–1987.

60 Branneby, C., Carlqvist, P., Magnusson, A. et al. (2002). Carbon—Carbon bonds by hydrolytic enzymes. *J. Am. Chem. Soc.* 125 (4): 874–875.

61 Li, C., Feng, X., Wang, N. et al. (2008). Biocatalytic promiscuity: the first lipase-catalysed asymmetric aldol reaction. *Green Chem.* 10 (6): 616–618.

62 Mitrev, Y.N., Mehandzhiyski, A.Y., Batovska, D.I. et al. (2016). Original enzyme-catalyzed synthesis of chalcones: utilization of hydrolase promiscuity. *J. Serbian Chem. Soc.* 81 (11): 1231–1237.

63 Li, K., He, T., Li, C. et al. (2009). Lipase-catalysed direct Mannich reaction in water: utilization of biocatalytic promiscuity for C-C bond formation in a "one-pot" synthesis. *Green Chem.* 11 (6): 777–779.

64 He, T., Li, K., Wu, M.Y. et al. (2010). Utilization of biocatalytic promiscuity for direct Mannich reaction. *J. Mol. Catal. B Enzym.* 67 (3–4): 189–194.

65 Wu, L.L., Xiang, Y., Yang, D.C. et al. (2016). Biocatalytic asymmetric Mannich reaction of ketimines using wheat germ lipase. *Catal. Sci. Technol.* 6 (11): 3963–3970.

66 Xu, J.C., Li, W.M., Zheng, H. et al. (2011). One-pot synthesis of tetrahydrochromene derivatives catalyzed by lipase. *Tetrahedron* 67 (49): 9582–9587.

67 Yin, D. and Kazlauskas, R.J. (2012). Revised molecular basis of the promiscuous carboxylic acid perhydrolase activity in serine hydrolases. *Chem. Eur. J.* 18 (26): 8130–8139.

68 Dwivedee, B.P., Soni, S., Sharma, M. et al. (2018). Promiscuity of lipase-catalyzed reactions for organic synthesis: a recent update. *ChemistrySelect* 3 (9): 2441–2466.

69 Liu, L.S., Xie, Z.B., Zhang, C. et al. (2018). α-Chymotrypsin-catalyzed synthesis of 2-substituted benzimidazole through retro-Claisen reaction. *Green Chem. Lett. Rev.* 11 (4): 503–507.

70 Feng, X.W., Li, C., Wang, N. et al. (2009). Lipase-catalysed decarboxylative aldol reaction and decarboxylative Knoevenagel reaction. *Green Chem.* 11 (12): 1933–1936.

71 Lai, Y.F., Zheng, H., Chai, S.J. et al. (2010). Lipase-catalysed tandem Knoevenagel condensation and esterification with alcohol cosolvents. *Green Chem.* 12 (11): 1917–1918.

72 Ding, Y., Ni, X., Gu, M. et al. (2015). Knoevenagel condensation of aromatic aldehydes with active methylene compounds catalyzed by lipoprotein lipase. *Catal. Commun.* 64: 101–104.

73 Svedendahl, M., Carlqvist, P., Branneby, C. et al. (2008). Direct epoxidation in *Candida antarctica* lipase B studied by experiment and theory. *ChemBioChem* 9 (15): 2443–2451.

74 Svedendahl, M., Hult, K., and Berglund, P. (2005). Fast carbon-carbon bond formation by a promiscuous lipase. *J. Am. Chem. Soc.* 127 (51): 17988–17989.

75 Dhake, K.P., Tambade, P.J., Singhal, R.S., and Bhanage, B.M. (2010). Promiscuous *Candida antarctica* lipase B-catalyzed synthesis of β-amino esters via aza-Michael addition of amines to acrylates. *Tetrahedron Lett.* 51 (33): 4455–4458.

76 Li, C. (2005). Organic reactions in aqueous media with a focus on carbon—carbon bond formations: a decade update. *Chem. Rev.* 105 (8): 3095–3165.

77 Patel, R.N. (2015). Biocatalytic key steps in semisynthesis and total synthesis. In: *Biocatalysis in Organic Synthesis*, 403–459. https://doi.org/10.1055/sos-SD-216-00250

78 Bezborodov, A.M. and Zagustina, N.A. (2016). Enzymatic biocatalysis in chemical synthesis of pharmaceuticals. *Appl. Biochem. Microbiol.* 52 (3): 237–249.

79 Xie, Z., Wang, N., Jiang, G., and Yu, X. (2013). Biocatalytic asymmetric aldol reaction in buffer solution. *Tetrahedron Lett.* 54 (8): 945–948.

80 Xie, Z., Wang, N., Zhou, L. et al. (2013). Lipase-catalyzed stereoselective cross-aldol reaction promoted by water. *ChemCatChem* 5 (7): 1935–1940.

81 Li, H., He, Y., and Guan, Z. (2011). Protease-catalyzed direct aldol reaction. *Catal. Commun.* 12 (7): 580–582.

82 Chen, Y., Li, W., Liu, Y. et al. (2013). Trypsin-catalyzed direct asymmetric aldol reaction. *J. Mol. Catal. B Enzym.* 87: 83–87.

83 He, Y., Li, H., Chen, Y. et al. (2012). Chymopapain-catalyzed direct asymmetric aldol reaction. *Adv. Synth. Catal.* 354 (4): 712–719.

84 Fu, J., Gao, N., Yang, Y. et al. (2013). Ficin-catalyzed asymmetric aldol reactions of heterocyclic ketones with aldehydes. *J. Mol. Catal. B Enzym.* 97: 1–4.

85 Xie, B., Li, W., Liu, Y. et al. (2012). The enzymatic asymmetric aldol reaction using acidic protease from *Aspergillus usamii*. *Tetrahedron* 68 (15): 3160–3164.

86 Yuan, Y., Guan, Z., and He, Y. (2013). Biocatalytic direct asymmetric aldol reaction using proteinase from *Aspergillus melleus*. *Sci. China Chem.* 56 (7): 939–944.

87 Li, H.-H., He, Y.-H., Yuan, Y., and Guan, Z. (2011). Nuclease p1: a new biocatalyst for direct asymmetric aldol reaction under solvent-free conditions. *Green Chem.* 13 (1): 185–189.

88 Miao, Y., Rahimi, M., Geertsema, E.M., and Poelarends, G.J. (2015). Recent developments in enzyme promiscuity for carbon – carbon bond-forming reactions. *Curr. Opin. Chem. Biol.* 25: 115–123.

89 Wang, N., Zhang, W., Zhou, L.H. et al. (2013). One-pot lipase-catalyzed aldol reaction combination of in situ formed acetaldehyde. *Appl. Biochem. Biotechnol.* 171 (7): 1559–1567.

90 Hult, K. and Berglund, P. (2007). Enzyme promiscuity: mechanism and applications. *Trends Biotechnol.* 25 (5): 231–238.

91 Meñaca, O.E.V. (2019). Promiscuidade enzimática de lipases na síntese de aldóis e 2*H*-cromenonas. PhD Thesis. University of São Paulo (Brazil).

92 Haque, N. and Prabhu, N.P. (2016). Lid closure dynamics of porcine pancreatic lipase in aqueous solution. *Biochim. Biophys. Acta Gen. Subj.* 1860 (10): 2313–2325.

93 Haque, N. and Prabhu, N.P. (2016). Lid dynamics of porcine pancreatic lipase in non-aqueous solvents. *Biochim. Biophys. Acta Gen. Subj.* 1860 (10): 2326–2334.

94 Segura, R.L., Palomo, J.M., Cortes, A. et al. (2004). Different properties of the lipases contained in porcine pancreatic lipase extracts as enantioselective biocatalysts. *Biotechnol. Prog.* 20 (3): 825–829.

95 Segura, R.L., Betancor, L., Palomo, J.M. et al. (2006). Purification and identification of different lipases contained in PPL commercial extracts: a minor contaminant is the main responsible of most esterasic activity. *Enzym. Microb. Technol.* 39 (4): 817–823.

96 Chaitanya, P.K. and Prabhu, N.P. (2014). Stability and activity of porcine lipase against temperature and chemical denaturants. *Appl. Biochem. Biotechnol.* 174 (8): 2711–2724.

97 Kohno, M., Enatsu, M., Takee, R., and Kugimiya, W. (2000). Thermal stability of *Rhizopus niveus* lipase expressed in a kex2 mutant yeast. *J. Biotechnol.* 81 (2–3): 141–150.

98 Birolli, W.G., Porto, A.L.M. and Fonseca, L.P. (2019). Miniemulsion in biocatalysis, a new approach employing a solid reagent and an easy protocol for product isolation applied to the aldol reaction by *Rhizopus niveus* lipase. Bioresource Technology. 297 1–43.

6

Enzymes Applied to the Synthesis of Amines

Francesco G. Mutti and Tanja Knaus

University of Amsterdam, Van't Hoff Institute for Molecular Sciences, HIMS-Biocat, Science Park 904, 1098 XH, Amsterdam, The Netherlands

6.1 Introduction

Amines are the most widely used chemical intermediates for the production of active pharmaceutical ingredients (APIs), fine chemicals, agrochemicals, polymers, dyestuffs, pigments, emulsifiers, and plasticizing agents [1, 2]. In particular, α-chiral amines constitute approximately 40% of the optically active drugs that are currently commercialized mainly as single enantiomers [3], and the chiral amine market size is expected to reach USD 14 billion in 2020. Significant progress in organometallic catalysis and organocatalysis for the asymmetric reductive amination of carbonyl-containing compounds and hydroamination of alkenes has been achieved during the past decade [4–6]. However, the application of enzyme catalysis for α-chiral amine synthesis is gaining prominence because of the diversity of available methods that often yield products with excellent selectivities (i.e. chemo-, regio-, and stereoselectivity) and atom efficiency [7].

Figure 6.1 depicts the enzyme families that are commonly applied for the synthesis of amines, in particular, α-chiral amines. The kinetic resolution (KR) – either hydrolytic or oxidative – of a racemic mixture is the classical biocatalytic methodology for obtaining enantiopure α-chiral amines, and it is frequently applied in industrial-scale processes. Hydrolytic KR is generally performed by stereoselective acylation catalyzed by hydrolases [8], whereas oxidative KR is performed by stereoselective oxidation catalyzed by monoamine oxidases [9, 10]. The drawback of KR processes is that the theoretical maximum attainable yield of the enantiopure amine product cannot exceed 50%. Dynamic kinetic resolution (DKR) and deracemization can resolve this limitation, thus in principle giving access to an enantiopure amine product in quantitative yield. Classical DKR entails the combination of a hydrolase with either a metal or an organometallic catalyst that performs the *in situ* racemization of the unreacted amine enantiomer [11–14]. Cyclic deracemization commonly combines a monoamine oxidase with a reducing reagent (or another enzymatic reaction) that performs the reduction of the imine intermediate – obtained in the first enzymatic

Biocatalysis for Practitioners: Techniques, Reactions and Applications, First Edition.
Edited by Gonzalo de Gonzalo and Iván Lavandera.

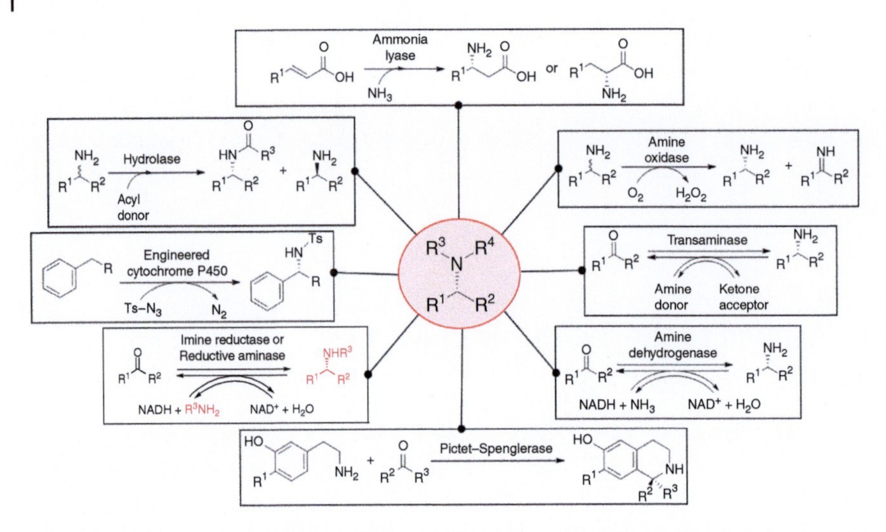

Figure 6.1 The most commonly applied enzymes for the synthesis of α-chiral amines and terminal amines. Hydrolases and amine oxidases are applied for the resolution or deracemization of racemic mixtures of amines. In contrast, the inherent reversibility of biocatalytic transamination and reductive amination reactions enables the application of ω-transaminases, amine dehydrogenases, imine reductases, and reductive aminases for either the asymmetric synthesis of α-chiral amines from prochiral ketones or the kinetic resolution/deracemization of racemic amines. Finally, ammonia lyases, engineered cytochrome P450s, and Pictet–Spenglerases catalyze the synthesis of optically active amines from non-chiral substrates.

step – back to the racemic amine [9, 14, 15]; the iteration of this process results in an enantiopure product in theoretical quantitative yield.

Transaminases (also called aminotransferases) catalyze the formal reversible transfer of an amine moiety from an amine donor to a ketone acceptor, and this property can be exploited for either the asymmetric reduction of a prochiral ketone or the KR of a racemic mixture of the amine [16–23]. The oxidoreductase enzymes, amine dehydrogenases, and imine reductases (including the subclass of the reductive aminases) perform a "truly" reductive amination of carbonyl compounds at the expense of an amine donor (e.g. ammonia, primary or secondary amine) and the reduced form of a nicotinamide adenine dinucleotide (NADH or NADPH) as the hydride source [24–29]. Similar to the enzymatic transamination, the reversibility of the reductive amination enables the asymmetric synthesis of an amine from a prochiral ketone as well as the KR (or deracemization) of a racemic amine mixture [24, 30–33].

Ammonia lyases (ALs) perform the biocatalytic hydroamination, which is restricted to α,β unsaturated carboxylic acids as substrate acceptors, although a wide variety of amine nucleophiles are accepted [34, 35]. Engineered cytochrome P450s have recently been shown to catalyze the direct intermolecular and enantioselective amination of sp^3-hybridized C—H bonds by consuming tosyl-azide as the nitrogen source [36]. Finally, Pictet–Spenglerases can generate enantiopure cyclic aromatic secondary amines through the coupling of an aldehyde or a ketone with a substituted phenethylamine or structural analogs [37, 38].

Nature has developed an arsenal of enzymes that are suitable for the synthesis of (chiral) amines. This chapter aims to orient nonspecialists to the available biocatalytic methodologies with a focus on the enzymes' substrate scope and selectivity as well as the structural diversity of the obtained products. Notably, some enzyme classes are highly specific for the synthesis of primary amines (e.g. ω-transaminases [ωTAs] and amine dehydrogenases), whereas other classes can produce secondary and tertiary amines (e.g. monoamine oxidases and imine reductases). Finally, this chapter describes the synthesis of important amine products and related practical procedures.

6.2 Hydrolases

KR catalyzed by hydrolases is the classical method for obtaining α-chiral amines in optically pure form. Hydrolases (see also Chapter 5) from the serine protease family (EC 3.4.21.X) naturally catalyze the stereoselective hydrolysis of amide bonds, and they possess a typical "aspartate-histidine-serine" (Asp–His–Ser) catalytic triad in their active site (Scheme 6.1) that generates a deprotonated serine residue. The catalytic cycle is initiated when this serine performs the nucleophilic attack on the carbon atom of the carbonyl moiety. In the presence of a racemic mixture of acylated amines, the enzyme's serine preferentially attacks one enantiomer, thereby leading to the release of the enantiopure α-chiral amine. Upon formation of the covalent enzyme–substrate intermediate, the subsequent nucleophilic attack of a water molecule produces the carboxylic acid coproduct and restores the enzyme's resting state.

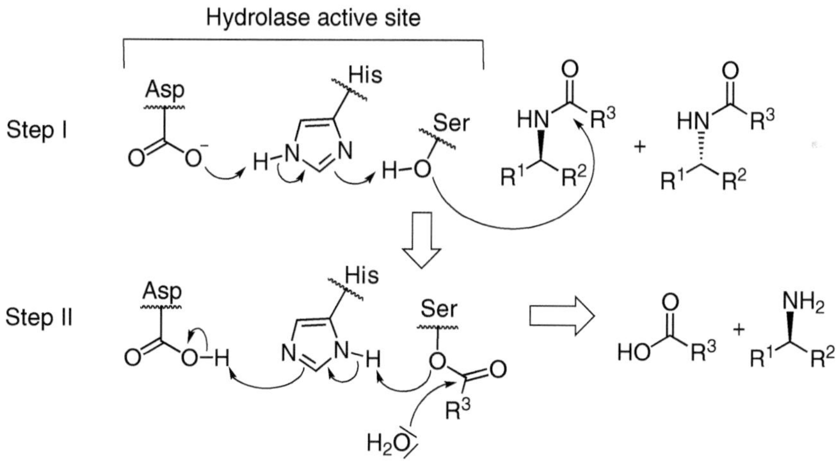

Scheme 6.1 Simplified catalytic mechanism of serine proteases and lipases.

6.2.1 Practical Approaches with Hydrolases

6.2.1.1 Kinetic Resolution

Although peptide hydrolysis is the natural activity of hydrolases such as proteases, the most practical approach is the reverse reaction (i.e. KR via amide formation; Scheme 6.2) [39, 40]; this enantioselective amide bond formation starting from racemic α-chiral amines

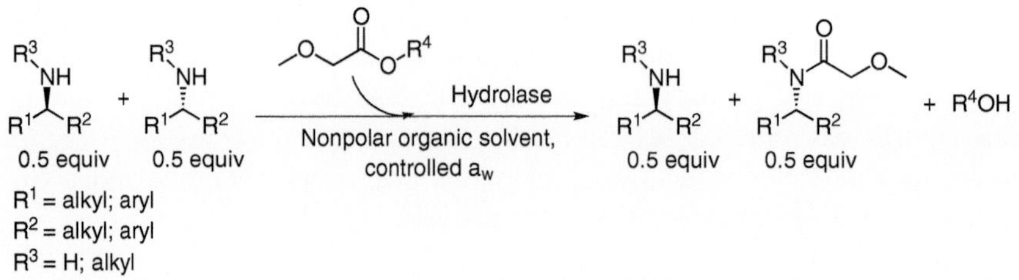

Hydrolase
Nonpolar organic solvent, controlled a_w
0.5 equiv 0.5 equiv → 0.5 equiv 0.5 equiv + R^4OH
R^1 = alkyl; aryl
R^2 = alkyl; aryl
R^3 = H; alkyl

Scheme 6.2 Depiction of a classical kinetic resolution of racemic primary or secondary amines catalyzed by hydrolases in a nonaqueous medium. Sources: Gutman et al. [39], Balkenhohl et al. [40].

is often conducted with a lipase in a nonpolar organic solvent at a defined water content (i.e. water activity) [41–43]. In fact, lipases are classified as carboxylic ester hydrolases (EC 3.1.1.X) but possess the same catalytic triad of serine proteases and can promiscuously catalyze amide bond formation in an organic solvent. Notably, peptide hydrolysis in aqueous environments proceeds at a low rate or does not occur at all; therefore, lipases offer superior applicability than proteases for KR [44]. The acyl group is transferred from an acyl donor molecule, which must be added in at least equimolar amount to the amine enantiomer. The most widely applied acyl donors are alkyl methoxyacetates, which enable rapid enzymatic acylation while also precluding nonenzymatic side reactions [40]. The substrate scope comprises aromatic and aliphatic amines (e.g. α-methylbenzylamine, amphetamine, 1-(1-naphthyl)ethylamine, aminoindane, 1-(heteroaryl)ethanamines, and derivatives thereof, arylalkylamines, alkylamines (including cycloalkylamines), phenoxyalkylamines, and amino alcohols.

6.2.1.2 Dynamic Kinetic Resolution

DKR is the most efficient hydrolytic method to obtain an enantiopure amine from a racemate because a theoretical quantitative yield is attainable (Scheme 6.3). DKR combines the *in situ* racemization of the amine starting material, which is catalyzed by a metal or a metal–organic catalyst, with the enzymatic enantioselective acylation. The reaction is conducted in a nonpolar organic solvent that must be compatible with the racemization metal catalyst and enable the preservation of enzyme activity. In the context of the KR depicted in Scheme 6.2, the use of a nonaqueous medium is mandatory in DKR to enable the shift of the reaction equilibrium toward substrate acylation rather than product hydrolysis. Toluene is the most frequently applied solvent with the addition of molecular sieves that enable capture of the water by-product that is generated during the enzymatic acylation. To enable full irreversibility of the enzymatic acylation, highly activated esters or enol esters such as vinyl esters, *para*-chlorophenylacetate, and dibenzylcarbonate are preferred, albeit 2-propylacetate and 2-propylmethoxyacetate are also frequently applied. Ethyl acetate can also be used, thus acting as both a neat organic solvent and an acyl donor. The most well-known racemization catalyst is the dimeric ruthenium-Shvo catalyst (Scheme 6.3, structure f), and improved generations thereof have been created during the past two decades. However, DKR is more difficult for racemic amines than alcohols because the Ru-catalyzed racemization of the former normally requires significantly higher

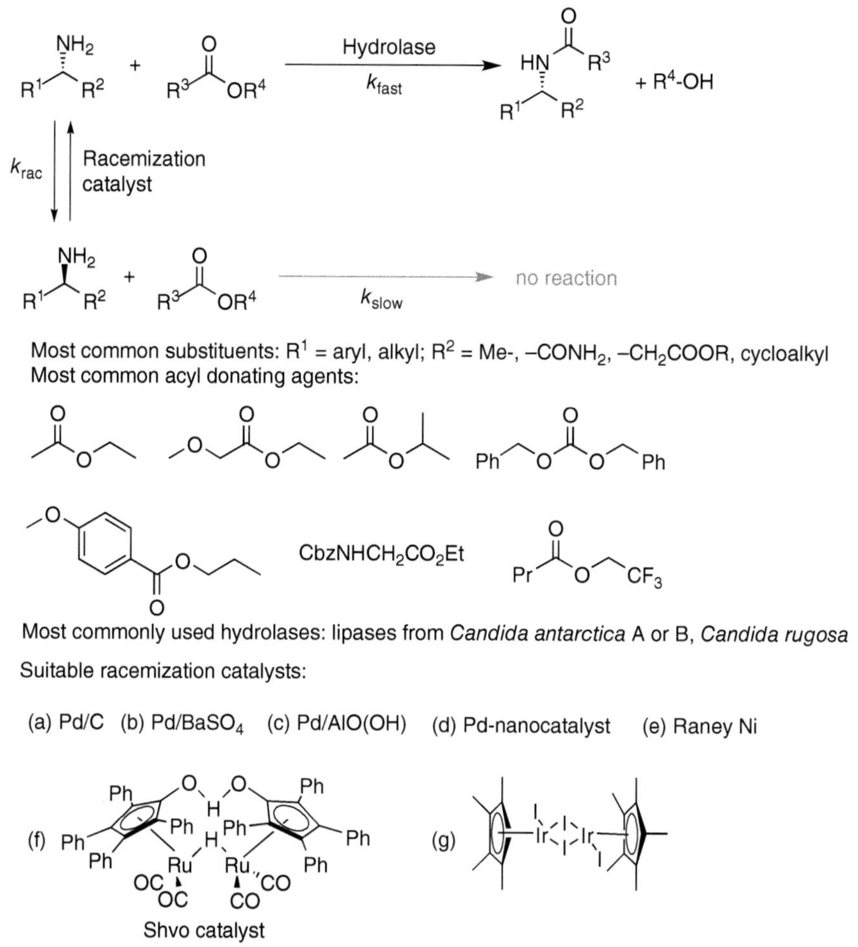

Hydrolase k_{fast}

$+ R^4$-OH

k_{rac} Racemization catalyst

no reaction

k_{slow}

Most common substituents: R^1 = aryl, alkyl; R^2 = Me-, $-CONH_2$, $-CH_2COOR$, cycloalkyl
Most common acyl donating agents:

$CbzNHCH_2CO_2Et$

Most commonly used hydrolases: lipases from *Candida antarctica* A or B, *Candida rugosa*

Suitable racemization catalysts:

(a) Pd/C (b) $Pd/BaSO_4$ (c) Pd/AlO(OH) (d) Pd-nanocatalyst (e) Raney Ni

(f) Shvo catalyst (g)

Scheme 6.3 Dynamic kinetic resolution through the combination of a chemocatalytic racemization reaction with an enantioselective enzymatic acylation catalyzed by a hydrolase.

temperatures (i.e. above 100 °C), which can be detrimental for the stability of the hydrolase involved in the concomitant acylation step [11–14, 45]. Furthermore, the amine starting material becomes more reactive around 100 °C, thus acting as a potential ligand toward the Ru-center of the racemization catalyst and thereby leading to a rate reduction or even interruption of the DKR process. A number of catalysts operating racemization at 50 °C or below that are suitable for the DKR of racemic amines comprise iridium organometallic catalysts, nickel(Ni)-Raney, and palladium (Pd) nanoparticles as well as Pd combined with either $BaSO_4$, $CaCO_3$, or AlO(OH) [13]. Typically reported temperatures for the DKR of amines are between 40 °C and 100 °C, although a temperature of 60 °C was most frequently applied. Depending on the type of the substrate, acyl donor, metal catalyst, and hydrolase, the reaction time for complete DKR can take from few hours to several days [13].

The selection criteria for a practical DKR can be summarized as follows:

1) The organometallic or metal catalyst must be compatible with the hydrolase in order to assure operational stability under the same reaction conditions and precluding mutual deactivation;
2) The hydrolase must be sufficiently enantioselective so that the reaction rate for the acylation of the desired amine enantiomer will be at least 20-fold greater than that for the other enantiomer ($k_{fast}/k_{slow} \geq 20$);
3) The racemization's reaction rate must be at least 10-fold higher than the enzymatic acylation's reaction rate of the slow reacting amine enantiomer ($k_{rac}/k_{slow} > 10$);
4) The racemization catalyst must react with neither the racemic amine substrate nor the acylated product.

The substrate scope of DKR appears to be slightly narrower than that of KR, which is probably due to the limitation of finding compatible substrates and conditions for chemical racemization and enzymatic acylation. In general, DKR is applicable on aromatic amines (e.g. α-methylbenzylamines, tetrahydroisoquinolines, aminotetralines, α- and β-amino esters, and amino amides) and arylalkylamines (e.g. amphetamines and 4-phenylbutan-2-amine).

6.2.2 Practical Examples with Hydrolases

6.2.2.1 Kinetic Resolution of Racemic α-Methylbenzylamine Through the Methoxyacetylation Catalyzed by a Lipase

Researchers from BASF have studied the KR of aromatic amines such as α-methylbenzylamine (*rac*-**1**, Scheme 6.4) [40]. A crude enzyme preparation of the lipase from *Burkholderia plantarii* (activity: 1 kU $mg_{prep.}^{-1}$; enzyme content: 30% w/w) was applied for this process, and the optimal reaction conditions were obtained using ethyl methoxyacetate as the acyl donor in *tert*-butyl methyl ether (MTBE) as the solvent. Quantitative KR of *rac*-**1** was accomplished using 10 mass% of crude lipase (equal to c. 3.3 mass% of pure enzyme compared to *rac*-**1**) with a one-day reaction time, and (*S*)-**1** was obtained in >45% isolated yield and >99% ee. Free (*R*)-**1** could also be obtained by hydrolysis of the corresponding amide (*R*)-**2** (ee 93%) upon separation. A higher ee for (*R*)-**1** could be achieved by running the KR below 50% conversion. The practical protocol is described in Section 6.9.1.1. A similar process was scaled-up in a flow reactor (450 ml) in which 5.7 g

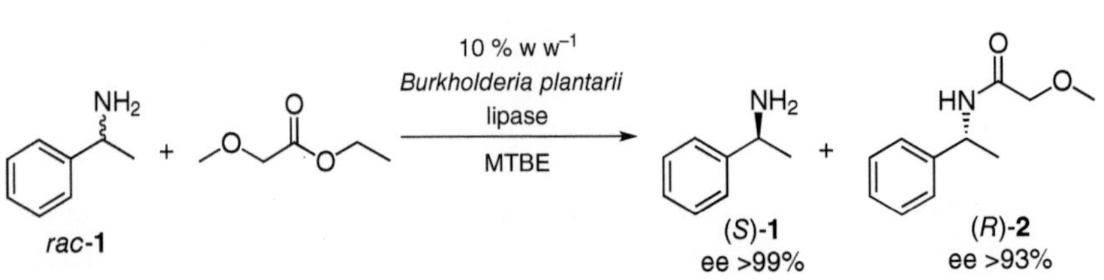

10 % w w^{-1} *Burkholderia plantarii* lipase, MTBE

rac-**1** + ethyl methoxyacetate → (*S*)-**1** (ee >99%) + (*R*)-**2** (ee >93%)

Scheme 6.4 BASF industrial-scale KR of α-methylbenzylamine (*rac*-**1**) catalyzed by a lipase in MTBE. Source: Modified from Balkenhohl et al. [40].

of immobilized subtilisin over glass beads (1% enzyme loading, w/w) could resolve 1.6 kg of 1-(1-naphthyl)ethylamine in 320 hours of continuous operation [39].

6.2.2.2 Dynamic Kinetic Resolution for the Synthesis of Norsertraline

Bäckvall's group developed a synthetic route to enantiopure norsertraline ((1*R*,4*S*)-**5**), an antidepressant pharmaceutical (Scheme 6.5) [45]. Racemic 1-aminotetralin (*rac*-**3**, 0.50 mmol) underwent DKR to yield the corresponding (*R*)-amide (**4**) in 70% isolated yield and 99% ee; the DKR was performed using Ru-Shvo (0.02 mmol) as a racemization catalyst, *Candida antarctica* lipase B (CALB, 20 mg) as the biocatalyst, and isopropyl acetate (3.50 mmol) as the acyl donor in anhydrous toluene as the solvent. The practical protocol for the DKR is described in Section 6.9.1.2. The obtained (*R*)-**4** intermediate was converted into ((1*R*,4*S*)-**5**) over five additional chemical steps, which ultimately yielded 28% overall isolated yield, 98% de, and 99% ee. A more recently developed strategy entails the use of a Pd-CALB CLEA biohybrid catalyst to perform DKR of primary benzylic amines [46].

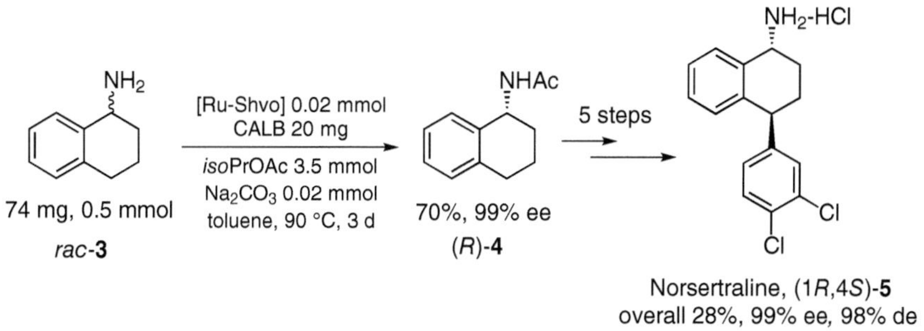

Scheme 6.5 DKR of 1-aminotetralin (*rac*-**3**) toward the synthesis of norsertraline (1*R*,4*S*)-**5**. Source: Based on Thalen et al. [45].

6.3 Amine Oxidases

Amine oxidases of the EC 1.4.3.X category catalyze the enantioselective oxidation of C—N bonds to generate an imine product at the expense of dioxygen as the ultimate electron acceptor [9]. This category is further divided into subcategories that mainly comprise flavin-dependent and copper-dependent amine oxidases. The most studied and applied amine oxidases in biocatalysis are flavin-dependent (flavin adenine dinucleotide [FAD] or flavin mononucleotide [FMN]) [47]. Some of them specifically act on either L- or D-configured α-amino acids (EC 1.4.3.1-3), but these subcategories are not discussed any further; the reader can refer to a number of reviews and book chapters on this topic [10, 48]. In contrast, the FAD-dependent monoamine oxidases (MAO, EC 1.4.3.4) catalyze the enantioselective oxidation of α-chiral amines lacking an α-carboxylic moiety, as depicted in Scheme 6.6. In the first step, the hydride is transferred from the amine substrate to the oxidized flavin (FAD) to generate the imine product and reduced flavin ($FADH_2$). $FADH_2$ then reacts with dioxygen from air to produce a flavin-hydroperoxide intermediate, which evolves in hydrogen peroxide formation and regeneration of FAD.

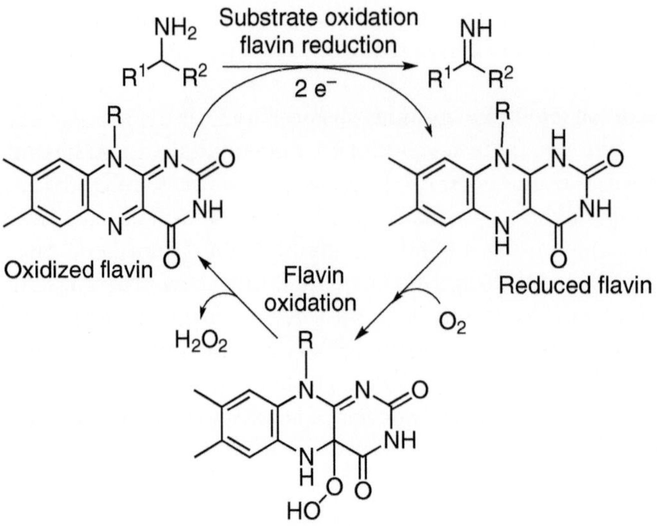

Scheme 6.6 Simplified catalytic cycle of FAD-dependent amine oxidases.

6.3.1 Practical Approaches with Amine Oxidases

6.3.1.1 Kinetic Resolution and Deracemization

The monoamine oxidase N from *Aspergillus niger* (MAO-N) and engineered variants thereof are the most applied biocatalysts for the oxidative KR and deracemization of α-chiral amines [9, 49]. Directed evolution of MAO-N during the past two decades has generated libraries of variants that can oxidize a wide range of structurally diverse primary [50], secondary, and tertiary α-chiral amines with high selectivity [50–53]. Notably, the substrate scope has been extended to include pharmaceutically relevant alkaloids such as (*R*)-configured nicotine, eleagnine, harmicine, crispine A and intermediates for the synthesis of levocetirizine, and solifenacin [52, 54, 55]. The most efficient approach to harness the catalytic power of MAO-N is deracemization, in which the enantioselective oxidation reaction is combined with a concurrent nonselective chemical reduction (Scheme 6.7) [14, 15]. The process runs iteratively, thereby leading to a progressive enantio-enrichment of the amine substrate mixture after each oxidation–reduction cycle; thus, a theoretical

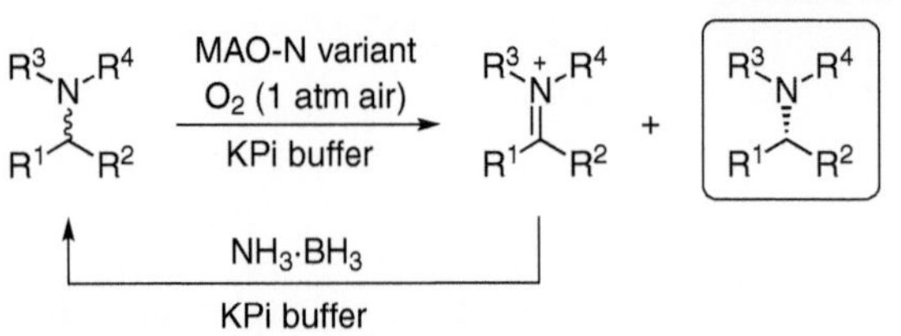

Scheme 6.7 Classical deracemization of primary, secondary, and tertiary (mainly heterocyclic) α-chiral amines through the simultaneous combination of a MAO-N enantioselective oxidation with a chemical nonselective reduction. Sources: Musa et al. [14], Turner [15].

quantitative yield can be obtained, as also illustrated for DKR processes. Assuming that the MAO-N enzyme is highly enantioselective (ee of oxidation ≥99%), seven cycles will be required to increase the amine's optical purity from racemic to >99% ee. Nevertheless, the advantage of deracemization is that a perfectly enantioselective enzyme is not necessary, as the desired final ee value is in theory always achievable by running a certain number of cycles. Therefore, the number of equivalents of the reducing agent must be calculated depending on the inherent enantioselectivity of the enzyme that determines the minimum number of required cycles. In this context, ammonia borane ($NH_3{\cdot}BH_3$) is the most versatile and biocompatible reducing agent [50–55]; however, more atom-efficient deracemization cycles combine the MAO-N enantioselective oxidation with a concurrent enantioselective reduction catalyzed by an imine reductase (see Section 6.5) or an artificial oxidoreductase enzyme. Thus, the consumption of the reducing reagent can be limited to a theoretical minimum of half equivalent compared with the racemic amine substrate [56, 57].

6.3.2 Practical Examples with Amine Oxidases

6.3.2.1 One-pot, One-enzyme Oxidative Pictet–Spengler Approach Combined with Deracemization

Turner's group has created subsequent generations of MAO-N enzyme libraries through the combination of directed evolution and structurally guided engineering. One of these variants, namely MAO-N D9, was successfully employed for the deracemization of 530 mg of racemic harmicine (**8**) to yield (*R*)-**8** in 95% yield and >99% ee [54]. The practical protocol for the deracemization is described in Section 6.9.2.1. This strategy was then implemented in a one-pot "oxidative Pictet–Spengler-type" cyclization of 3-(2-(pyrrolidin-1-yl) ethyl)-1*H*-indole (**7**, Scheme 6.8). This domino reaction entails the MAO-N-catalyzed oxidation of achiral **7**, Pictet–Spengler cyclization to afford *rac*-**8**, and deracemization at the expense of NH_3BH_3 as the reducing agent. (*R*)-**8** was obtained from **7** in 83% conversion and >99% ee. Considering that **7** can be synthesized by abundant tryptamine (**6**) and

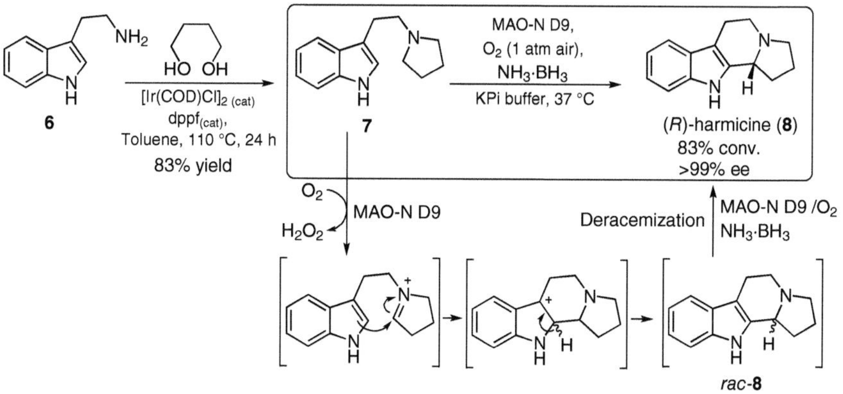

Scheme 6.8 Chemoenzymatic synthetic route from tryptamine (**6**) to (*R*)-harmicine (**8**) involving MAO-N/$NH_3{\cdot}BH_3$ deracemization of the racemic intermediate.

butane-1,4-diol in 83% yield, Scheme 6.8 depicts the shortest and more efficient route to the natural alkaloid (*R*)-harmicine, which is used for the treatment of leishmania disease.

6.3.2.2 Desymmetrization of *meso*-compounds

Researchers from Merck/Codexis developed a chemoenzymatic asymmetric synthesis of the anti-hepatitis C drug boceprevir [58]. The key step of the synthesis is the desymmetrization of the *meso* starting material 6,6-dimethyl-3-azabicyclo[3.1.0]hexane, catalyzed by a MAO variant. This industrial process is discussed in detail in Chapter 15 (Section 15.2.6).

6.4 Transaminases (or Aminotransferases)

Transaminases (EC 2.6.1.X) are pyridoxal-5′-phosphate (PLP)-dependent transferase enzymes. The transamination reaction follows a ping-pong bi-bi mechanism in which the PLP cofactor shuttles an amine group from an amine donor (most commonly L- or D-alanine) to a carbonyl compound acceptor. Therefore, the cofactor changes between two forms in the catalytic cycle, namely the amine-form pyridoxamine (PMP) and the aldehyde-form PLP, and Scheme 6.9 shows the reversible nature of the biotransformation. Transaminases reversibly acting on α-keto/α-amino or β-keto/β-amino carboxylic acids are called α- or β-transaminases, respectively. Synthetic applications of α- and β-transaminases are reviewed elsewhere [17, 19, 59] and will not be discussed here any further. ωTAs (also known as ATAs) are a family of enzymes that can aminate prochiral ketone and/or aldehyde substrates, which lack any further carboxylic acid functionality (or this functionality is in ω-position). They have attracted great interest in synthetic chemistry during the past decade because of their robustness under process conditions, broad substrate scope, exquisite regio- and stereoselectivity, and excellent evolvability along with the availability of stereocomplementary enzymes [19–22]. Therefore, the industrial synthesis of a great number

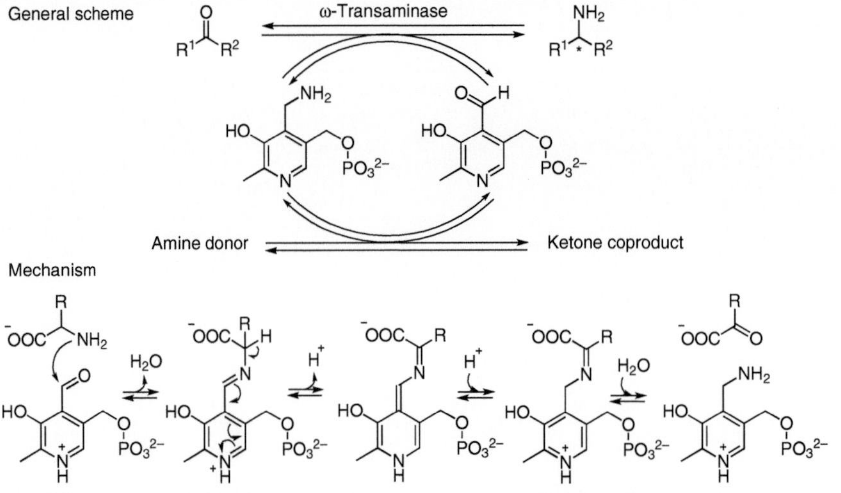

Scheme 6.9 Reversible catalytic cycle of the enzymatic transamination reaction.

of pharmaceuticals nowadays involves a pivotal enzymatic transamination step to install the α-chiral amine moiety with high stereoselectivity [18, 23, 60].

6.4.1 Practical Approaches with Transaminases

The reversibility of the transamination reaction enables the performance of either KR of a racemic mixture of α-chiral amines or asymmetric synthesis from a prochiral ketone (Scheme 6.10). Considering the most frequently accepted pyruvate/L- or D-alanine couple, KR is the thermodynamically favored process in aqueous reaction media (this also applies to the other α-keto carboxylic acid/α-amino acid couples); therefore, half equivalent of pyruvate is sufficient to resolve one equivalent of the racemic amine (Scheme 6.10a). Conversely, asymmetric synthesis is more challenging to accomplish with high yields (Scheme 6.10b). A number of strategies have been developed to shift the reaction thermodynamic equilibrium by either recycling (b1) or removing (b2) the by-product (e.g. pyruvate if the amine donor is alanine) using a second enzyme such as an alanine dehydrogenase (AADH), or a lactate dehydrogenase (LDH), or a pyruvate decarboxylase (PDC), or an acetolactate synthase (ALS) [19, 20]. An excess of L-/D-alanine (typically, a minimum of three equivalents) is required to achieve quantitative conversion in a reasonable amount of time. Alternative protocols combine this equilibrium shift with *in situ* sequestration of the amine product using selective ion exchange resins; the additional advantage in this case is the avoidance of enzyme inhibition by the chiral amine product [61]. A number of "smart" amine donors have also been designed and tested; these all rely on a sacrificial amine donor that produces a carbonyl compound by-product, which then undergoes a further spontaneous and irreversible chemical reaction (b3). Thus, one equivalent of the amine donor is generally sufficient to enable quantitative conversion [62–66]. The applicability of these latter methods is currently limited by either the high cost of the amine donor [62] or the polymerization of the by-product that complicates the reaction work-up [63, 64]. In summary, all of the methods depicted in Scheme 6.10 are useful and often applied in lab-scale synthesis or during biocatalyst development and the initial phases of industrial process development. Another option is the application of ωTAs in a neat organic solvent at defined water activity, which has the advantage of providing remarkably more favorable thermodynamic equilibrium for the asymmetric amination of prochiral ketones using simple amine donors (e.g. 2-propylamine), while also preserving the catalytic activity [67–69]. However, industrial-scale enzymatic transamination is nowadays most commonly run in an aqueous buffer in the presence of dimethyl sulfoxide (DMSO) as a cosolvent (i.e. for improved substrate solubility), using inexpensive 2-propylamine as an amine donor and performing selective removal of the acetone by-product by applying vacuum pressure or dinitrogen sweep [60, 70].

6.4.2 Practical Examples with Transaminases

6.4.2.1 Kinetic Resolution and Deracemization

Kroutil's group deracemized mexiletine (*rac*-**9**), a drug used for the treatment of cardiac disease, general chronic pain, and muscle stiffness [71]. The one-pot protocol is based on two steps that are separated in time (Scheme 6.11) and is described in Section 6.9.3.1. The

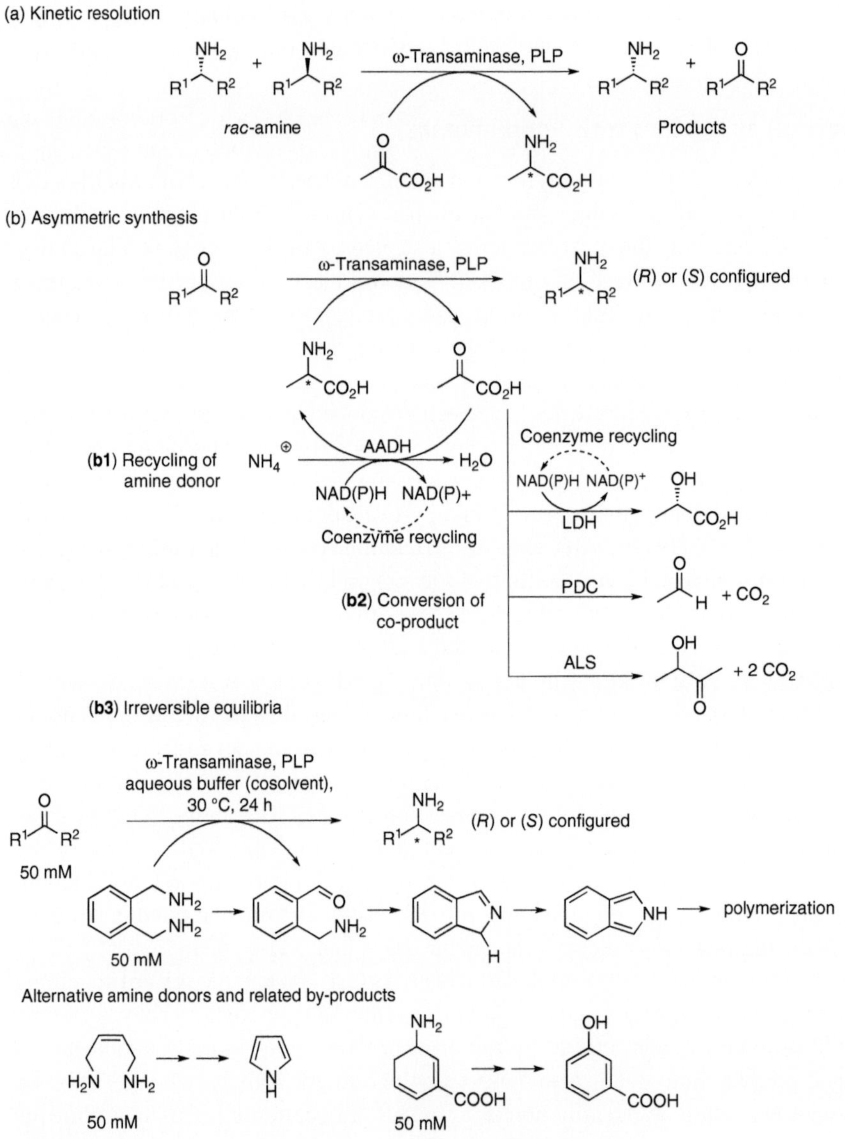

(a) Kinetic resolution

NH_2 + NH_2 ω-Transaminase, PLP NH_2 + O
R^1 R^2 R^1 R^2 R^1 R^2 R^1 R^2
rac-amine Products
O NH_2
CO_2H * CO_2H

(b) Asymmetric synthesis

O ω-Transaminase, PLP NH_2 (*R*) or (*S*) configured
R^1 R^2 R^1 * R^2
NH_2 O
* CO_2H CO_2H

(**b1**) Recycling of amine donor $NH_4^{\oplus}$ AADH H_2O
NAD(P)H NAD(P)+
Coenzyme recycling

Coenzyme recycling
NAD(P)H NAD(P)$^+$ OH
LDH CO_2H
O
PDC H + CO_2
(**b2**) Conversion of co-product
OH
ALS + 2 CO_2
O

(**b3**) Irreversible equilibria

ω-Transaminase, PLP
aqueous buffer (cosolvent),
30 °C, 24 h
O NH_2 (*R*) or (*S*) configured
R^1 R^2 R^1 * R^2
50 mM
NH_2 O N NH polymerization
NH_2 NH_2 H
50 mM

Alternative amine donors and related by-products

NH_2 OH
H_2N NH_2 N H COOH COOH
50 mM 50 mM

Scheme 6.10 Practical approaches to biocatalytic transamination.

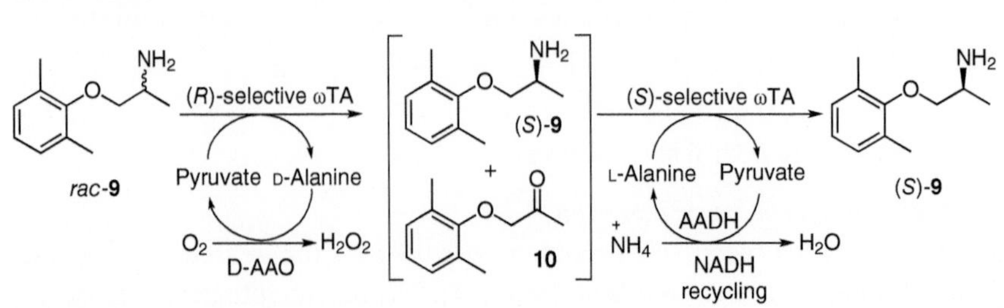

NH_2 NH_2 NH_2
O (*R*)-selective ωTA O (*S*)-selective ωTA O
(*S*)-**9**
rac-**9** Pyruvate D-Alanine + O L-Alanine Pyruvate (*S*)-**9**
O AADH
O_2 H_2O_2 + NH_4 H_2O
D-AAO **10** NADH recycling

Scheme 6.11 One-pot, two-step (separated in time) deracemization of mexiletine (*rac*-**9**).

first step is a KR in which one enantiomer ((*R*)-configured in the example) is deaminated to the ketone intermediate **10**. Although the deamination can be quantitatively run with a stoichiometric amount of pyruvate, a recycling system was implemented by using an amino acid oxidase that converts alanine (D-configured in the example) back to pyruvate at the expense of dioxygen. Notably, the use of catalytic pyruvate avoids ωTA-induced co-substrate inhibition and precludes the generation of a stoichiometric amount of D-alanine as a coproduct, which would impede the second step. Upon completion of the KR step, the first ωTA is deactivated by heat treatment and the second stereocomplementary ωTA is added along with the complementary alanine amine donor (L-configured in the example). Thus, both (*S*) and (*R*)-mexiletine can be obtained from racemic starting material in 97 to >99% isolated yield and >99% ee, by reversing the order of the applied ωTAs in the one-pot process.

6.4.2.2 Asymmetric Synthesis from Prochiral Ketone

Through several rounds of directed evolution and structurally guided protein engineering, researchers from Merck/Codexis have created a variant of the (*R*)-selective ωTA from *Arthrobacter* sp. (i.e. ArRmut11-ωTA), which was applied for the asymmetric amination of prositagliptin to yield sitagliptin – a medication for the treatment of diabetes mellitus type 2 [60]. This industrial process is discussed in detail in Chapter 15 (see also Section 15.2.3).

Bornscheuer's group has engineered ωTAs from fold class I to perform the asymmetric transamination of bulky–bulky ketones [72]. Starting from an ωTA from *Ruegeria* sp., they identified a characteristic motif and related mutations (Y59W, Y87F, Y152F, and T213A) whose introduction could extend the enzyme's substrate scope toward the desired sterically demanding substrates. Furthermore, the motif and its mutations proved to be applicable to another six wild-type (WT) ωTAs that have c. 70% sequence identity with *Ruegeria sp.* ωTA. The obtained variants were applied for the 100 mg scale amination of aromatic ketones, including a tricyclic one (Scheme 6.12). Notably, these variants possess a stereocomplementary selectivity compared with ArRmut11-ωTA for the amination of bulky–bulky ketones.

6.5 Amine Dehydrogenases, Imine Reductases, and Reductive Aminases

This section describes oxidoreductases that catalyze the NAD(P)H-dependent reduction of imine substrates or intermediates to give α-chiral amines [24, 25]. The imine can be either a stable cyclic imine or formed *in situ* by the reaction between a carbonyl compound and ammonia or another amine. Imine reductases (IReds, EC 1.5.1.X) have been known for many decades, and one of the first characterized members was purified in 1975 [73]. It was long thought that IReds could only catalyze the asymmetric reduction of cyclic (non-hydrolyzable) imines; however, recent studies have shown that they can catalyze reductive amination between prochiral ketones and an amine, albeit a considerable excess of the latter was required [74–78]. An IRed subfamily was recently found to catalyze reductive amination between selected ketones and primary or secondary amines with a 1 : 1 molar ratio in certain cases. The members of this subfamily were named reductive aminases (RedAms) [79–81]. The first report on a naturally occurring amine dehydrogenase (AmDHs, EC

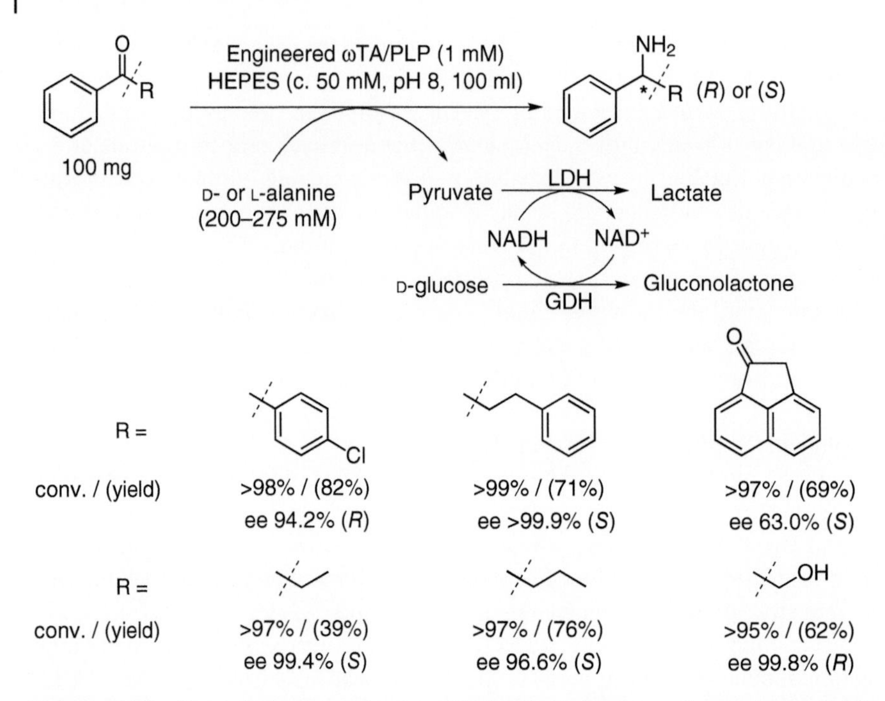

Scheme 6.12 Biocatalytic transamination of bulky–bulky ketones using engineered ωTAs from fold class I.

1.4.1.X) was published in 2000; however, its enzyme sequence is still unknown [82]. In 2013, Bommarius' group reported the first synthetically applicable AmDH, which was obtained by protein engineering of an L-α-amino acid dehydrogenase (L-AADH) [83]. Although IReds, RedAms, and AmDHs are commonly referred to as different enzyme families, they essentially share a similar reaction mechanism, which is hydride transfer from the NAD(P)H coenzyme to the prochiral carbon of an iminium intermediate in the active site. The main difference lies in the substrate scope, as AmDHs almost exclusively accept ammonia as a substrate to generate primary amines, whereas IReds/RedAms can also form secondary and tertiary amines. Therefore, the advantage of oxidoreductases compared with transferases for amine synthesis is that the latter class can only give access to primary amines.

Structural and computational analysis of AmDHs compared with the catalytic cycle of naturally occurring L-AADHs (Scheme 6.13a) showed that a catalytic lysine is essential to effect the protonation of the carbonyl moiety of the ketone substrate (intermediate I), whereas an aspartate residue stabilizes the iminium intermediate (III) [84]. The elucidated catalytic mechanism of an IRed (Scheme 6.13b) showed that a catalytic tyrosine effects the same type of protonation of the ketone substrate (intermediate I), whereas an aspartate is again involved in the interaction with the amine moiety of the other substrate [85].

6.5.1 Practical Approaches with Amine Dehydrogenases, Imine Reductases, and Reductive Aminases

IReds preferentially catalyze the reduction of cyclic imines because they are chemically inert to hydrolysis in an aqueous medium. A group of IReds can reduce cyclic imino acids

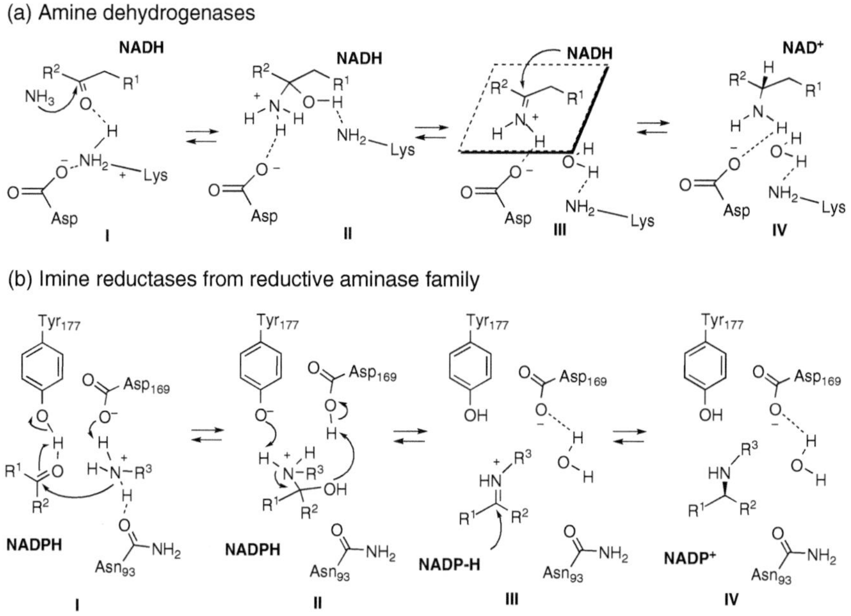

(a) Amine dehydrogenases

(b) Imine reductases from reductive aminase family

Scheme 6.13 Comparison between the catalytic mechanisms of (a) amine dehydrogenases and (b) imine reductases from the reductive aminase family.

(e.g. substituted Δ1-pyrroline and Δ1-piperideine) to produce L-configured α-amino acids, as described in an excellent review [25]. Conversely, most IReds acting on imine substrates that are devoid of any additional carboxylic acid moiety have been found in various *Streptomyces* species; however, it is currently unclear whether their "imine reductase" catalytic activity is a natural or promiscuous reaction [86]. IReds have been reported to convert heterocyclic compounds, preferentially containing aryl groups (Scheme 6.14) [74, 86–95]. The ee and absolute configuration of the amine product for any IRed-catalyzed reaction is typically elevated. However, both vary greatly for the same enzyme from one substrate to another [92, 93], thereby complicating the classification depending on the stereoselective outcome of the reaction. IRed-catalyzed reactions are commonly carried out using *Escherichia coli* whole cells overexpressing the recombinant enzyme [90].

IReds have begun to attract greater interest since the publication of an initial report on the reductive amination of prochiral ketones with methylamine and aniline [74, 75]. During the past five years, new family members have been discovered and characterized in terms of their catalytic activity, substrate scope, and structure [76–78, 85]. This work culminated with the discovery of IReds (also called RedAms) that can catalyze reductive amination between certain ketone acceptors and amine donors with a 1 : 1 (or nearly 1 : 1) stoichiometry [79–81]. The first RedAm was identified from the genome of *Aspergillus oryzae* through bioinformatics analysis based on sequence similarity with previously known IReds, which supports the close relation between the two families [79]. The amine products were obtained with up to >98% conversion, up to >98% ee, and displaying a maximum turnover of 32 000. Another IRed/RedAm was engineered to perform reductive amination between an aldehyde and a racemic amine (1 : 2 molar ratio) with concomitant KR of the amine (for details, see Section 6.5.2) [81]. Scheme 6.15a depicts the substrate scope of the

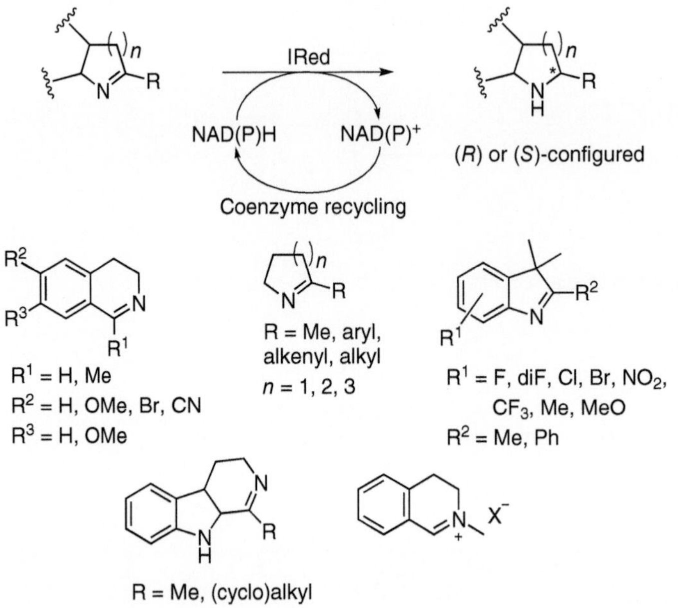

Scheme 6.14 Characterized substrate scope for the reduction of cyclic imines catalyzed by IReds. Sources: Huber et al. [74], Roth et al. [86], Mitsukura et al. [87–89], Leipold et al. [90], Gand et al. [91], Man et al. [92], Hussain et al. [93], Li et al. [94], Wetzl et al. [95].

available IReds/RedAms for reductive amination of ketones, which comprises arylaliphatic (including bicyclic), alicyclic (including heterocyclic) ketone substrates, albeit the R^2 substituent currently appears to be limited to the small methyl group. Acceptance of the amine donor is thus far limited to C1–C4 alkyl amines; reactions with longer-chain alkylamines have not yet been reported. Other amine donors are aromatic ones such as aniline, benzylamine, and amines containing pyridine, thiophene, and methylene-thiophene groups.

AmDHs differ from IReds and RedAms in that they appear to almost exclusively accept ammonia (or ammonium species) as the donor [24], although the acceptance of methylamine, ethylamine, cyclopropylamine, *n*-propylamine, and prop-2-yn-1-amine was reported when applied in high concentrations [84]. Another important feature of AmDHs is their extremely high stereoselectivity, as in the large majority of the cases, the undesired enantiomer is below detection limits (i.e., ee >99.9 has been reported). This virtually perfect stereoselectivity is not such a general feature with IReds or even with ωTAs. The first AmDHs were obtained by protein engineering starting from L-AADHs such as L-leucine-AADH, L-phenylalanine-AADH, or L-valine-AADH and by generating their chimeras [83, 96–100]. All these variants were obtained by mutating two highly conserved amino acid residues in the active site – namely, a Lys and an Asp, which are involved in binding the carboxylic moiety of the natural substrate. Because of that, they eventually showed the same stereoselectivity, thereby producing (*R*)-configured α-chiral amines [99–101]; the characterized substrate scope (Scheme 6.15b) comprises arylaliphatic ketones such as phenylpropan-2-one, 4-phenylbutan-2-one, phenoxypropan-2-one, and derivatives thereof, as well as aliphatic substrates such as linear short-/medium-chain methyl ketones,

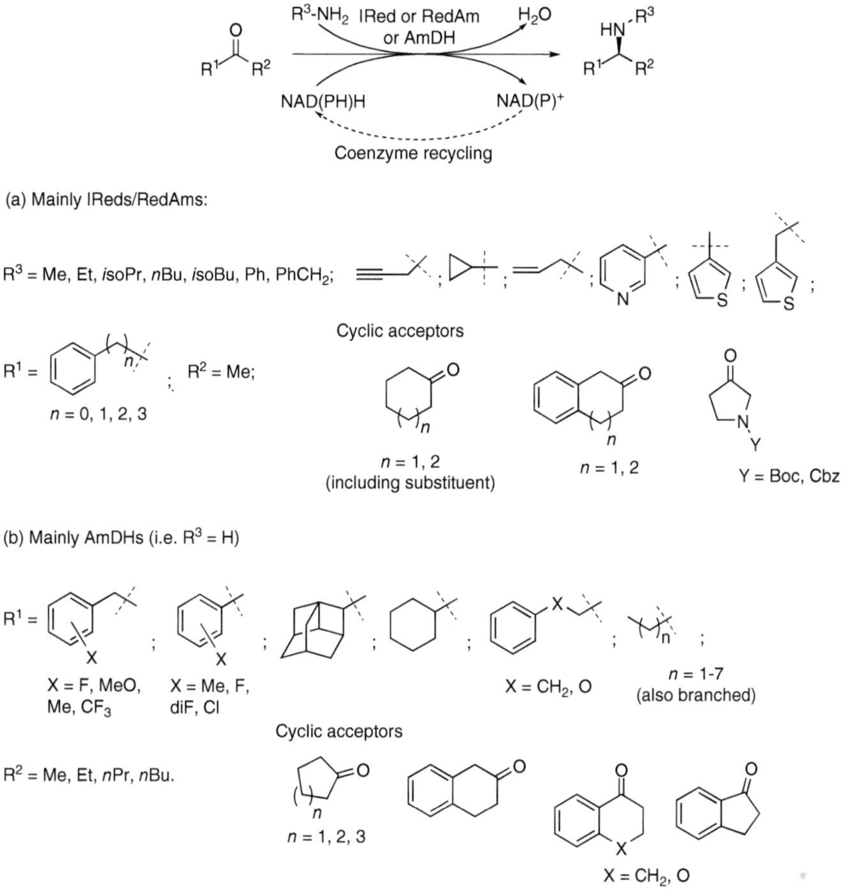

Scheme 6.15 Substrate scope of the currently available IReds/RedAms (a) and AmDHs (b).

1-adamantylmethyl ketone, and substituted cyclohexanones [99–103]. Modest activity toward acetophenone and few derivatives thereof was detected [97, 101, 103]. Further structurally guided protein engineering enabled the extension of the substrate scope to longer aliphatic chain methyl ketones [104]. The substrate scope and synthetic applicability of AmDHs were enhanced by the engineering of a new family of variants starting from the scaffold of the (ε-deaminating) L-lysine amino acid dehydrogenase (LysEDH). Substitution of Phe173 with an alanine in the active site of LysEDH created a variant that was capable of catalyzing the reductive amination of acetophenone and its derivatives with remarkably improved catalytic efficiency. Moreover, the substrate scope was extended to include bulky–bulky ketones and bicyclic aromatic ketones, the latter of which are important for pharmaceutical manufacturing. Because the WT LysEDH is a highly thermostable enzyme, the variant displayed high temperature and pH robustness under process conditions [105]. Most LysEDH variants produce the (*R*)-configured amines with excellent optical purity, although (*S*)-configured products are also accessible. However, the breakthrough in the synthesis of (*S*)-configured amines with AmDHs came through the discovery of natural enzymes by bioinformatics analysis [106, 107]. The substrate scope of the four characterized WT AmDHs included methyl and dimethyl substituted cyclohexanones and cyclopentanones, 1-indanone, and a number of aliphatic ketones possessing up to six carbon atoms. AmDHs are commonly utilized as isolated and purified enzymes; however, protocols using immobilized enzymes [108, 109] or whole cells have been reported – especially in cascade reactions [110, 111].

IReds, RedAms, and AmDHs are preferentially used for the reductive amination of prochiral ketones. However, application in KR and deracemization is also possible by combining the aminating enzyme with a system for the *in situ* regeneration of the $NAD(P)^+$ coenzyme [32, 33]. This is conveniently achieved by the use of a water-forming NAD(P)H oxidase (NOx), which consumes dioxygen as an oxidant. In practice, the RedAm/NOx or AmDH/NOx combination emulates the catalytic activity of the MAO-N enzymes described in Section 6.3. Scheme 6.16 illustrates the current substrate scope of this methodology. The bi-enzymatic system was also run in deracemization mode via the addition of NH_3BH_3 as a reductant.

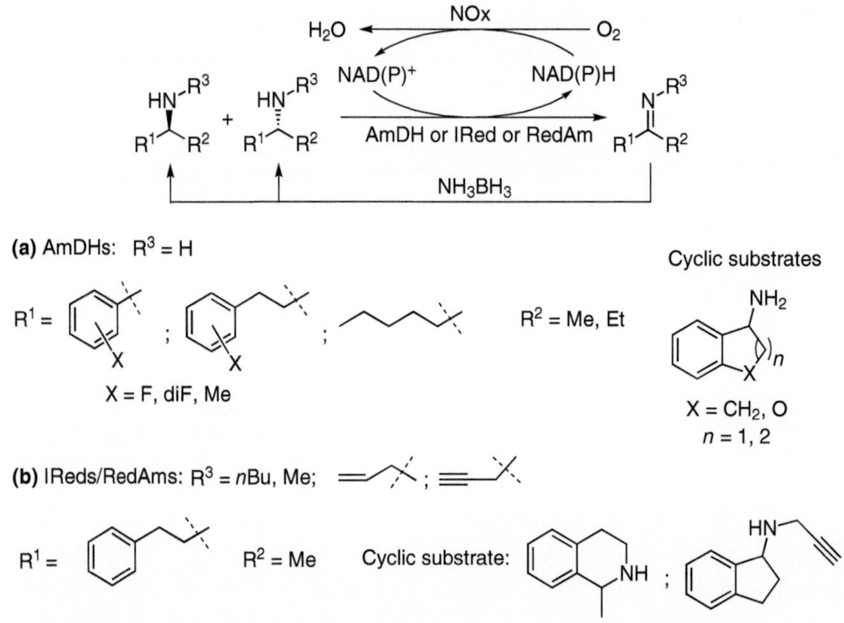

Scheme 6.16 KR and deracemization of α-chiral amines: (a) combination of an amine dehydrogenase with a NAD(P)H oxidase (NOx) [33]; (b) combination of an imine reductase with NOx [32].

6.5.2 Practical Examples with Amine Dehydrogenases, Imine Reductases, and Reductive Aminases

6.5.2.1 IRed-Catalyzed Reductive Amination of an Aldehyde Combined with KR of a Racemic Amine

Researchers from GlaxoSmithKline (GSK) engineered a WT IRed to perform reductive amination of an aldehyde with concomitant KR of a racemic amine to yield the lysine-specific demethylase-1 (LSD1) inhibitor GSK2879552 ((1*R*,2*S*)-**13**), which is under investigation for the treatment of lung cancer and leukemia (Scheme 6.17) [81]. The screening of a panel of IReds showed that several enzymes could catalyze the reductive amination between *t*-butyl 4-((4-formylpiperidin-1-yl)methyl)benzoate (**11**, 1 equiv) and *rac-trans*-tranylcypromine sulfate (*rac*-**12**, 1.1 equiv of the desired enantiomer), the best of which (IR-46) gave 86% conversion and >99.9% ee. However, IR-46 did not fulfill the requirement for an industrial process because of high biocatalyst loading (>450% w/w compared to substrate), low substrate concentration ($10\,g\,l^{-1}$) related to narrow pH tolerance (pH >6), and moderate conversion and isolated yield (43%). Three rounds of directed evolution were performed aiming at subsequently improving (i) tolerance to acid pH and increased substrate concentration; (ii) decrease of biocatalyst loading; and (iii) all parameters to meet desired operating space (<5% w/w biocatalyst loading, $20\,g\,l^{-1}$ substrate concentration, pH <5, >95% conversion, and >80% yield). Compared with the WT enzyme, the best variant (IRed-M3) contained 13 mutations (Y142S, L37Q, A187V, L201F, V215I, Q231F, S258N, G44R, V92K, F97V, L198M, T260C, and A303D) and resulted in >38 000-fold improvement of TON. In a 20-l scale reaction, IRed-M3 converted the aldehyde **11** (329 g, 54 mM, 1 equiv) with *rac-trans*-amine sulfate (*rac*-**12**, 476 g, 1.2 equiv of each enantiomer) in the presence of $NADP^+$ coenzyme (1.3 mM) to give the final product in 84% isolated yield, 99.9% purity, and >99.7% ee. Notably, the relatively high loading of $NADP^+$ (2.5 mol%) – which is also c. fourfold more expensive than NAD^+ – did not prevent the economic profitability of the process. The practical protocol for this reaction is described in Section 6.9.4.1.

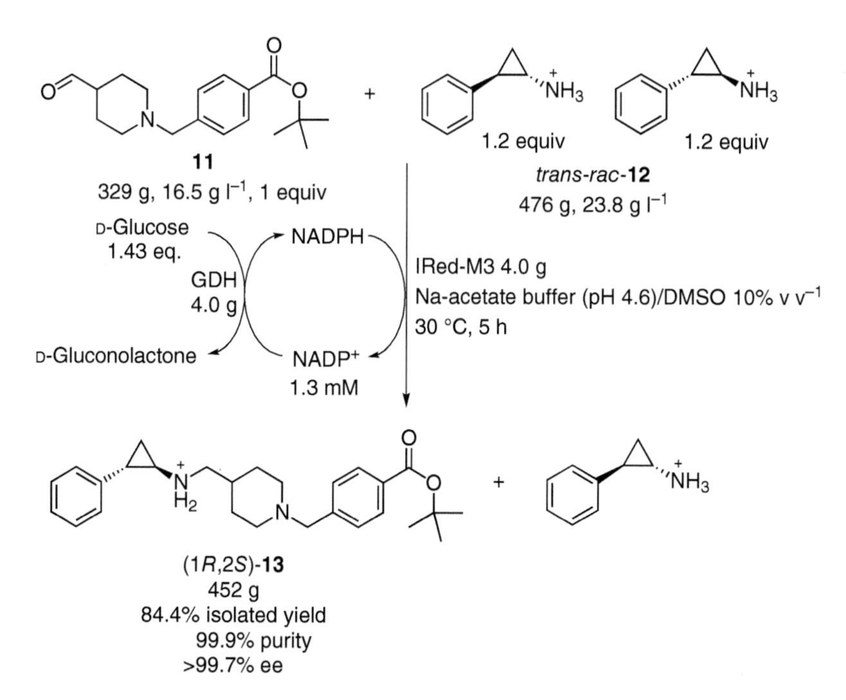

Scheme 6.17 Biocatalytic synthesis of lysine-specific demethylase-1 (LSD1) inhibitor GSK2879552 ((1*R*,2*S*)-**13**) through reductive amination of an aldehyde combined with the KR of the amine donor. Source: Based on Schober et al. [81].

6.5.2.2 Asymmetric Reductive Amination Catalyzed by AmDH

Mutti's group has created a family of AmDHs by structurally guided protein engineering of the WT ε-(deaminating) L-lysine dehydrogenase (LysEDH) [105]. Notably, LysEDH does not operate any apparent asymmetric transformation on its natural substrate, L-lysine. The best variant (LE-AmDH-v1) contained the F173A mutation and could catalyze the reductive amination of pharmaceutically relevant aromatic ketones (e.g. 1-tetralone, 4-chromanone, and indanone), as well as a number of additional aromatic bulky–bulky ketones (e.g. 1-phenylpentan-1-one and 1-phenylbutan-1-one) with enantiomeric excess up to >99.9%. LE-AmDH-v1 (and the other variants) generally retained the high thermostability of the parent LysEDH; for instance, LE-AmDH-v1 exhibited a melting temperature (T_m) of 69 °C and maintained >99% and 80% of its catalytic activity after incubation for seven days at room temperature and 50 °C, respectively. Asymmetric reductive aminations were conducted at a 600 mg scale or higher amounts in an aqueous solution of ammonium formate, which simultaneously acts as the reaction buffer, amine donor, and ultimate hydride source. In fact, a formate dehydrogenase (Cb-FDH) recycled the reduced form of the cofactor (NADH) through the consumption of a stoichiometric amount of formate, thereby generating bicarbonate as a side product. For instance, acetophenone (**14**) was converted into (*R*)-α-methylbenzylamine ((*R*)-**15**) with >99% conversion and >99.7% enantiomeric excess (Scheme 6.18). The practical protocol for this reaction is described in Section 6.9.4.2.

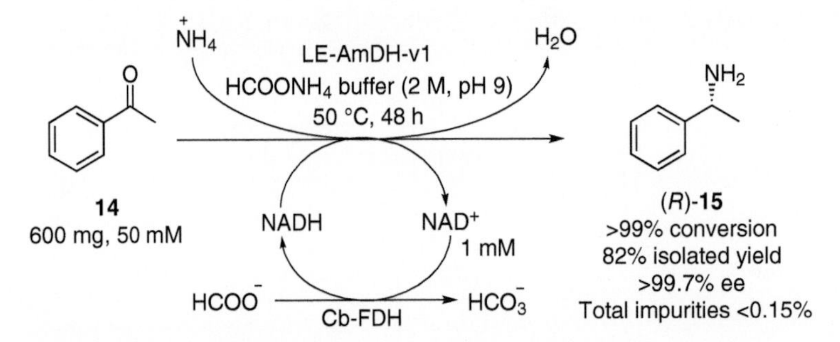

Scheme 6.18 Biocatalytic asymmetric reductive amination of acetophenone (**14**) at the expense of aqueous ammonium formate catalyzed by LE-AmDH-v1.

6.6 Ammonia Lyases

ALs (EC 4.3.1.X) consist of a heterogeneous enzyme category that catalyzes the addition of ammonia to α,β-unsaturated carboxylic acids to yield α- and β-amino acids [34, 35]. This reaction has the advantage of being atom-neutral, thus displaying a theoretically perfect atom efficiency for chiral amine synthesis; however, the ability to obtain only α- and β-amino acids limits its synthetic applicability. Although we have not discussed the synthesis of amino acids for the other enzyme classes, an exception is made in this section, as amino acids are the only attainable product with ALs. Aspartate ALs and aromatic amino acid ALs and aminomutase (e.g. phenylalanine ammonia lyase [PAL] or phenylalanine aminomutase [PAM]) are among the most studied family members. They perform the addition of ammonia to the C=C double bonds following two completely different mechanisms.

Aspartate AL (also named as aspartase) and 3-methylaspartate AL catalyze the reversible transfer of ammonia onto fumarate by the action of a catalytic deprotonated serine residue (Scheme 6.19a) or a lysine residue [112, 113]. In contrast, PAL and PAM utilize 5-methylene-3,5-dihydro-4*H*-imidazol-4-one (MIO) as a cofactor (Scheme 6.19b), which results from the autocatalytic cyclization of the highly conserved Thr/Ala-Ser-Gly motif [114]. In the catalytic amination cycle, the ammonia molecule is first added to the MIO's alkene moiety and then transferred to the α,β-unsaturated carboxylic acid substrate, thereby regenerating the enzyme's resting state. Interestingly, a competitive non-stereoselective MIO-independent catalytic mechanism has been proposed [115].

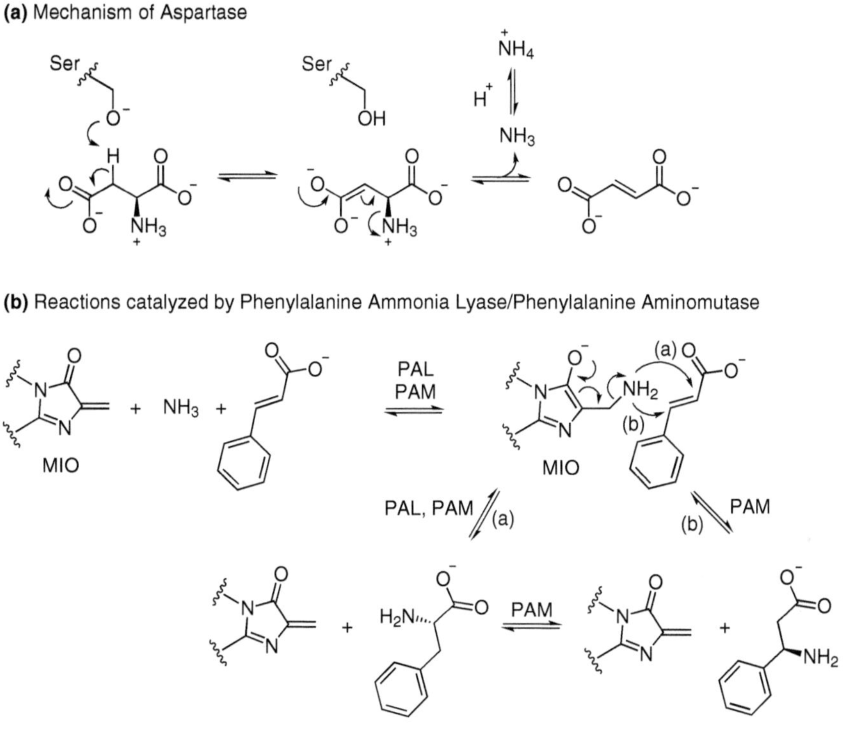

(a) Mechanism of Aspartase

(b) Reactions catalyzed by Phenylalanine Ammonia Lyase/Phenylalanine Aminomutase

Scheme 6.19 Simplified depiction of two typical reaction mechanisms of ammonia lyases: (a) aspartase and (b) aromatic amino acid ammonia lyases and aminomutases (e.g. PAL or PAM). Sources: (a) Puthan Veetil et al. [112], Asuncion et al. [113], (b) Based on Turner [114].

6.6.1 Practical Approaches with Ammonia Lyases

6.6.1.1 Aspartase, 3-Methylaspartate Ammonia Lyase, and Related Enzymes

The reactivity of aspartase is practically restricted to the reversible addition of ammonia to fumarate to yield aspartate. In contrast, the natural substrate scope of 3-methylaspartate AL already comprises different α,β-unsaturated dicarboxylic acids as acceptors and different amines and other nitrogen-containing nucleophiles as donors [116, 117]. The substrate scope of 3-methylaspartate AL was further extended by protein engineering [118, 119]. Common acceptors of natural and engineered 3-methylaspartate ALs are fumarate and their two-substituted derivatives such as 2-halo (i.e. chloro and bromo), 2-alkyl, 2-benzyl, 2-alkoxy, 2-thioalkoxy, and a number of more complex ether-containing derivatives (Scheme 6.20a-1) [116–120]. Apart from natural ammonia, nitrogen-containing donors include C1-C6 alkyl-/cycloalkyl-amines (i.e. from methylamine to hexylamines), hydrazine, hydroxylamine, ω-methoxy- and ω-hydroxy-substituted alkylamines, benzylamines, and α,ω-aliphatic diamines. The catalytic activity of ethylenediamine-*N,N′*-disuccinic acid lyase (EDDS lyase) – which naturally catalyzes the double addition of one molecule of ethylenediamine to two molecules of fumarate – was recently harnessed for the asymmetric synthesis of aspergillomarasmine A and related amino carboxylic compounds that possess antibiotic activity or applicability as radiotracers [123]. EDDS lyase was also used for the amination of fumarate to obtain a number of N-cycloalkyl-substituted L-aspartic acids [124]. In another recent work, the active site of the aspartase YM55-1 from *Bacillus* sp. was redesigned to accomplish the asymmetric addition of ammonia onto acrylic acids such as crotonic acid and *trans*-2-pentenoic acid (Scheme 6.20a-2) [121].

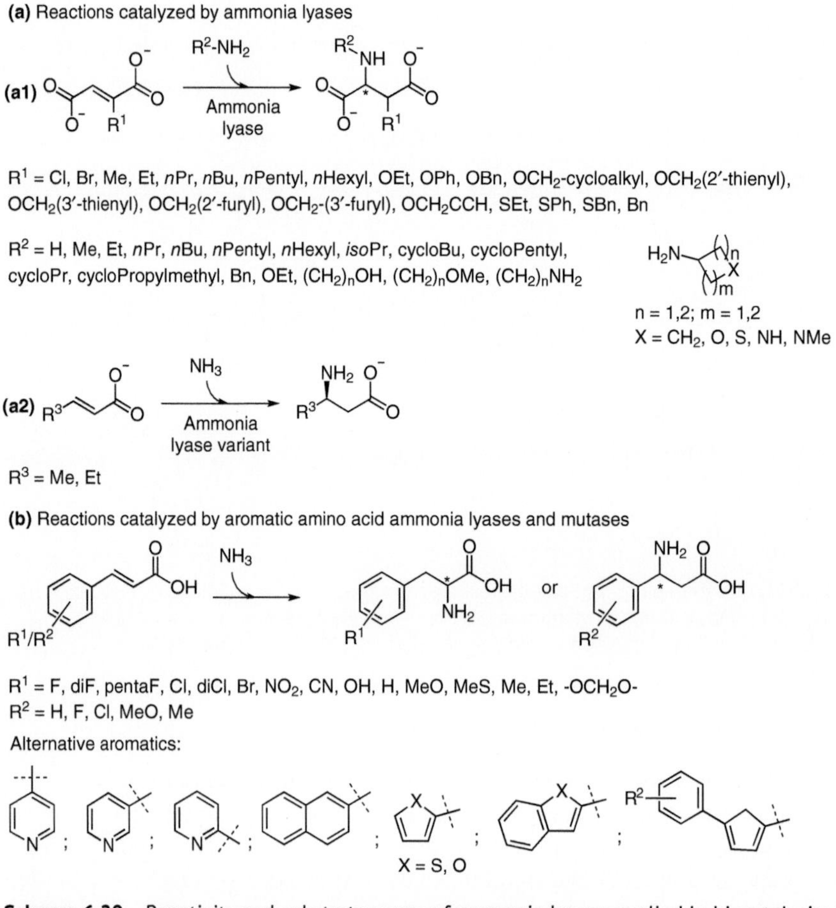

(a) Reactions catalyzed by ammonia lyases

(a1) R^2-NH_2, Ammonia lyase

R^1 = Cl, Br, Me, Et, *n*Pr, *n*Bu, *n*Pentyl, *n*Hexyl, OEt, OPh, OBn, OCH_2-cycloalkyl, OCH_2(2′-thienyl), OCH_2(3′-thienyl), OCH_2(2′-furyl), OCH_2-(3′-furyl), OCH_2CCH, SEt, SPh, SBn, Bn

R^2 = H, Me, Et, *n*Pr, *n*Bu, *n*Pentyl, *n*Hexyl, *iso*Pr, cycloBu, cycloPentyl, cycloPr, cycloPropylmethyl, Bn, OEt, $(CH_2)_nOH$, $(CH_2)_nOMe$, $(CH_2)_nNH_2$

n = 1,2; m = 1,2
X = CH_2, O, S, NH, NMe

(a2) NH_3, Ammonia lyase variant

R^3 = Me, Et

(b) Reactions catalyzed by aromatic amino acid ammonia lyases and mutases

NH_3 ... or ...

R^1 = F, diF, pentaF, Cl, diCl, Br, NO_2, CN, OH, H, MeO, MeS, Me, Et, -OCH_2O-
R^2 = H, F, Cl, MeO, Me

Alternative aromatics:

X = S, O

Scheme 6.20 Reactivity and substrate scope of ammonia lyases applied in biocatalysis: (a) aspartase, 3-methylaspartate ammonia lyase and related enzymes; (b) aromatic amino acid ammonia lyases and mutases. Sources: (a1) Gulzar et al. [116], Akhtar et al. [117], Raj et al. [118], de Villiers et al. [119], Fu et al. [120], (a2) Based on Li et al. [121]; (b) Gloge et al. [122].

6.6.1.2 Aromatic Amino Acid Ammonia Lyases and Mutases

PAL, tyrosine ammonia lyase (TAL), and histidine ammonia lyase (HAL) are the most studied aromatic amino acid ALs. Their natural reaction is the deamination of α-amino acids, but the reaction can be reversed by applying an excess of ammonia (>4 M) [125]. PAL is the most applied family member in chemical synthesis because it accepts a considerable number of substituted cinnamic acids bearing different substituents (including multiple substitutions) on the phenyl ring, including halogens, nitro, cyano, hydroxy, methyl, ethyl, methoxy, methylthio, and cyclic di-ethers (Scheme 6.20b) [122, 126, 127]. PAL can also accept other aromatic 3-acrylic acids possessing 2-, 3-, and 4-pyridyl groups; 2-thienyl and 2-benzothienyl groups; 1-naphthyl group; 2-furyl and 2-benzofuryl groups; and 5-arylfuran-2-yl group [126, 128–130]. Chemoenzymatic synthesis enables access to a wider structural diversity; notable examples are the synthesis of 2-indolinecarboxylic acid ((*S*)-**19** in Scheme 6.21) or biarylalanines by AL-catalyzed reactions combined with either a copper-catalyzed ring closure or a Suzuki coupling, respectively [128, 131]. Both (*R*)- and (*S*)-configured α-amino acids can be obtained with high enantiomeric excess that, if required, can be even further enhanced by applying multienzymatic cascades [132].

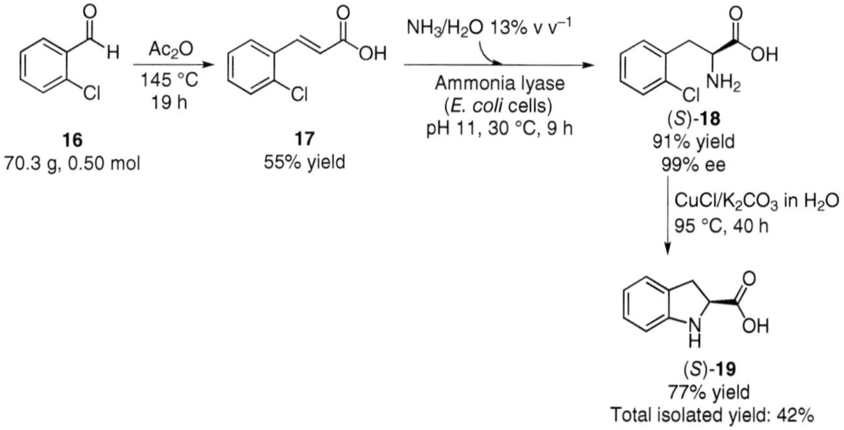

16
70.3 g, 0.50 mol
Ac_2O
145 °C
19 h
17
55% yield
NH_3/H_2O 13% v v^{-1}
Ammonia lyase
(*E. coli* cells)
pH 11, 30 °C, 9 h
(*S*)-**18**
91% yield
99% ee
$CuCl/K_2CO_3$ in H_2O
95 °C, 40 h
(*S*)-**19**
77% yield
Total isolated yield: 42%

Scheme 6.21 Chemoenzymatic synthesis of (*S*)-2-indolinecarboxylic acid ((*S*)-**19**) using PAL enzyme.

PAM and tyrosine aminomutase (TAM) naturally catalyze the interconversion between α- and β-phenylalanine derivatives; therefore, these enzymes can be exploited for asymmetric amination of cinnamic acids to yield β-phenylalanine derivatives. These aminomutases normally produce a mixture of α- and β-amino acids, albeit with high enantiomeric excess (e.g. (*S*)-configured α- along with (*R*)-configured β-phenylalanines) [133–137]. In other cases, better regioselectivity toward the formation of optically pure β-amino acids is observed in nature or is attainable by enzyme engineering [133, 138–140]. However, the regioselectivity of the PAL- and PAM-catalyzed aminations always remains – at least partially – influenced by the electron-withdrawing or electron-donating properties of the groups on the phenyl ring [133, 136, 137, 140].

6.6.2 Practical Examples with Ammonia Lyases

6.6.2.1 Chemoenzymatic Synthesis of (*S*)-2-Indolinecarboxylic Acid

Researchers at DSM developed a chemoenzymatic route for the ton-scale manufacture of (*S*)-2-indolinecarboxylic acid ((*S*)-**19**, Scheme 6.21), which is an intermediate for the synthesis of antihypertension drugs [128]. The route starts from the Perkin condensation of 2′-chlorobenzaldehyde (**16**) to yield the cinnamic acid derivative (**17**), which is then fed in small aliquots to an aqueous ammonia solution (pH 11, 13% v/v) containing a suspension of *E. coli* cells expressing a PAL enzyme. After approximately nine hours of reaction time, (*S*)-2′-chlorophenylalanine ((*S*)-**18**) was obtained in 91% yield and 99% ee. The final step is a copper-catalyzed ring closure, in which the intermediate itself serves as a ligand for the metal ion. Thus, (*S*)-**19** is obtained in 42% total isolated yield and 99% ee. The practical protocol for this reaction is described in Section 6.9.5.1.

6.6.2.2 Synthesis of L-Aspartate from Fumarate

Researchers from Genex Co. improved the previously reported biocatalytic conversion of fumarate (**20**) and ammonia to yield L-aspartate (**21**), 10^5 tonnes of which is produced per year as intermediates for food and pharmaceuticals production. *E. coli* whole cells expressing the aspartase enzyme were immobilized by entrapment into a flow reactor, following which the reactor was loaded with a 1.8 M solution of ammonium fumarate at a flow rate of 90 ml h^{-1}. Quantitative conversion and productivity of 140 g $l^{-1} h^{-1}$ were achieved. Notably, the system proved very robust, with a half-life of two years (Scheme 6.22).

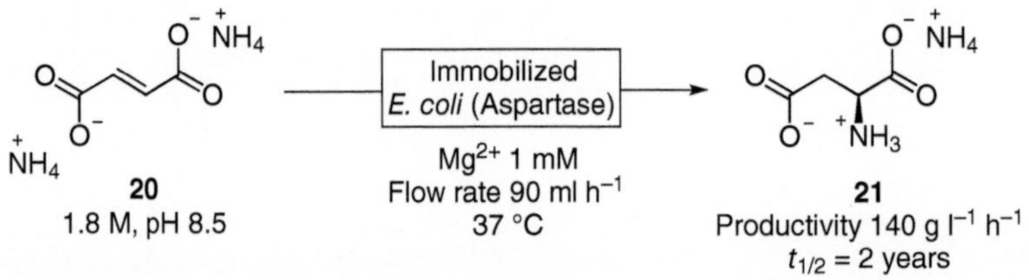

20
1.8 M, pH 8.5

Immobilized
E. coli (Aspartase)
Mg^{2+} 1 mM
Flow rate 90 ml h^{-1}
37 °C

21
Productivity 140 g l^{-1} h^{-1}
$t_{1/2}$ = 2 years

Scheme 6.22 *E. coli*/aspartase-immobilized cells catalyze the multiton-scale synthesis of L-aspartate (**21**) in a flow reactor.

6.6.2.3 Enzymatic and Chemoenzymatic Synthesis of Toxin A and Aspergillomarasmine A and B

Poelarends' group reported a biocatalytic methodology for the asymmetric enzymatic and chemoenzymatic synthesis of natural products toxin A (**24**), aspergillomarasmine A and B (AMA and AMB, **25**), and their homologs [123]. The research team harnessed the catalytic promiscuity of WT EDDS lyase to catalyze the C—N bond formation between fumarate and diverse diamines as nucleophiles. Scheme 6.23 depicts one example in which EDDS performs the C–N addition of (*S*)-2,3-diaminopropanoic acid ((*S*)-**23**) to fumarate (**22**). Applying an excess of fumarate (3 equiv) enables the complete consumption of the amine donor. Full conversion of (*S*)-**23** is necessary to avoid cross-reactivity in the subsequent chemical nucleophilic substitution with bromoacetic acid. The enzymatic step furnished (2*S*,2'*S*)-toxin A in 98% conversion (related to (*S*)-**23**), >99% chemoselectivity, and de >98%.

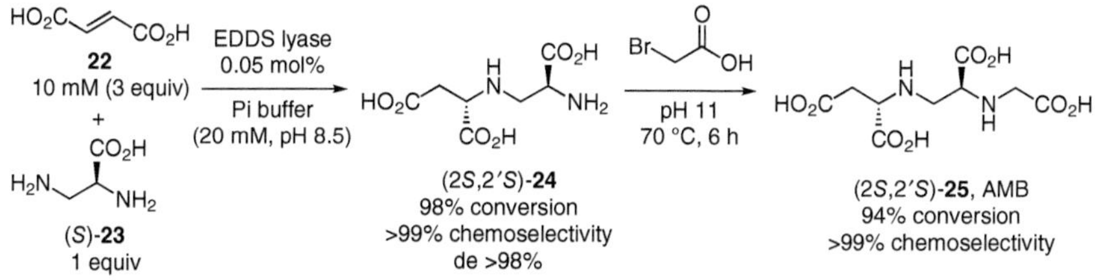

Scheme 6.23 Enzymatic and chemoenzymatic synthesis of toxin A (**24**) and aspergillomarasmine A and B (**25**) catalyzed by EDDS lyase.

6.7 Pictet–Spenglerases

Pictet–Spenglerases catalyze asymmetric Mannich-type cyclization between an electron-rich aromatic carbon and the prochiral carbon of an imine/iminium moiety [38, 141]. The first step of the reaction is the condensation between an aldehyde (**26**, or even a ketone for engineered Pictet–Spenglerases) and a primary amine (**27**) to generate the imine intermediate (**28**, Scheme 6.24).

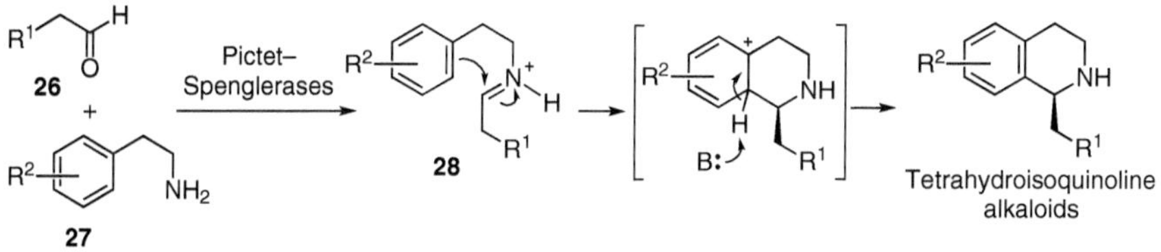

Scheme 6.24 Catalytic mechanism of the enzymatic Pictet–Spengler reaction.

6.7.1 Practical Approaches with Pictet–Spenglerases

Norcoclaurine synthase (NCS; EC 4.2.1.78) and strictosidine synthase (STR; EC.4.3.3.2) are the most applied Pictet–Spenglerase family members for synthetic purposes. The substrate scopes of NCS from *Thalictrum flavum* (Tf) and *Coptis japonica* (Cj) have been elucidated [142–144]. The WT enzymes display a certain level of promiscuity for aromatic aldehyde acceptors that can be a substituted phenylacetaldehyde, 2-(thiophen-3-yl)acetaldehyde, 2-(1*H*-indol-3-yl)acetaldehyde, and 3-phenylpropanal (Scheme 6.25). Interestingly, TfNCS accepted aliphatic aldehydes possessing a ω-carboxyl-methyl ester moiety and were exploited for the synthesis of (*S*)-trollines [145]. In contrast, the substrate scope is more restricted for the amine donor, which comprises dopamine and hydroxy-substituted phenethylamines or phenylpropanolamines. An important extension of the substrate scope was reported in a recent work by Hailes' and Ward's groups, in which engineered TfNCS variants could accept ketones such as methoxy- and hydroxy-substituted phenylacetones and cyclohexanones besides the aldehydes [146]. The best TfNCS variants (A79I and A79F) were applied in the synthesis of six spiro-tetrahydroisoquinolines with 33–87% isolated yields (c. 20–30 mg scale). Furthermore, NCS has recently been applied in a number of multienzymatic processes for the conversion of inexpensive aldehydes into valuable alkaloids [147–149]. These cascades normally entail a first ωTA-catalyzed step to generate the amine donor, which

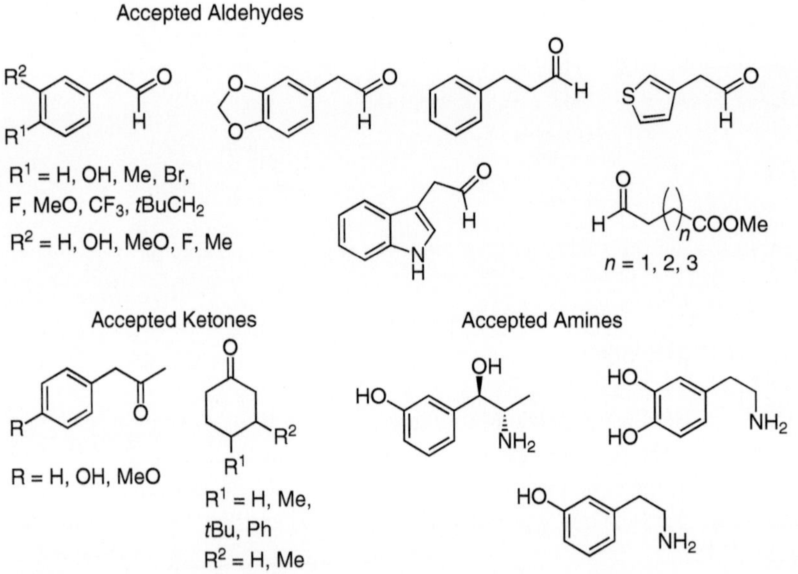

Accepted Aldehydes

R^1 = H, OH, Me, Br, F, MeO, CF_3, *t*BuCH_2
R^2 = H, OH, MeO, F, Me

n = 1, 2, 3

Accepted Ketones

R = H, OH, MeO

R^1 = H, Me, *t*Bu, Ph
R^2 = H, Me

Accepted Amines

Scheme 6.25 Typical substrate scope of the available norcoclaurine synthases (NCS) and related variants.

subsequently reacts with a phenylacetaldehyde (or analog) in the Pictet–Spengler enzymatic reaction. STR catalyzes the analogous reaction of NCS by using tryptamine (**7**) as a preferred amine donor with secologanin as a natural acceptor [141, 150]. STRs were shown to accept aliphatic C2-C6 aldehydes and methyl 4-oxobutanoate; this activity was exploited for the synthesis of (*R*)-1,2,3,4-tetrahydro-β-carbolines ((*R*)-**30** in Scheme 6.26) [151]. The full synthetic potential of this enzyme family is still untapped, as it was demonstrated that alternative alkaloid scaffolds are accessible [152]. Finally, STRs were also applied in cascade reactions with ωTAs for the synthesis of 3-methylated strictosidine [153].

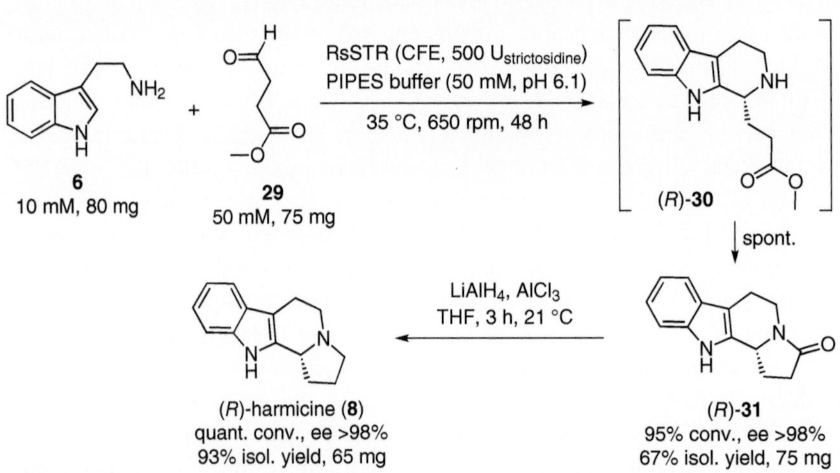

6 10 mM, 80 mg + **29** 50 mM, 75 mg

RsSTR (CFE, 500 $U_{strictosidine}$)
PIPES buffer (50 mM, pH 6.1)
35 °C, 650 rpm, 48 h

(*R*)-**30**

spont.

(*R*)-**31**
95% conv., ee >98%
67% isol. yield, 75 mg

$LiAlH_4$, $AlCl_3$
THF, 3 h, 21 °C

(*R*)-harmicine (**8**)
quant. conv., ee >98%
93% isol. yield, 65 mg

Scheme 6.26 Chemoenzymatic synthesis of (*R*)-harmicine ((*R*)-**8**) using *Rauvolfia serpentina* STR. Source: Modified from Pressnitz et al. [151].

6.7.2 Practical Examples with Pictet–Spenglerases

6.7.2.1 Biocatalytic Synthesis of (*R*)-Harmicine

Kroutil's group investigated an asymmetric Pictet–Spengler reaction between tryptamine (**6**) and a number of aliphatic aldehydes and obtained (*R*)-configured-1,2,3,4-tetrahydro-β-carbolines [151]. Notably, the natural reaction with secologanin provides the opposite enantiomer. The most active strictosidine synthase from *Rauvolfia serpentina* (RsSTR) was then applied for the chemoenzymatic synthesis of the anti-leishmania natural product (*R*)-harmicine ((*R*)-**8**, Scheme 6.26). The enzymatic Pictet–Spengler reaction between **6** (10 mM) and methyl 4-oxobutanoate (**29**, 50 mM) provides the tricyclic intermediate (*R*)-**30** that spontaneously cyclizes to obtain tetracyclic alkaloid (*R*)-**31** in 67% isolated yield and >98% ee. The enzymatic product is then reduced with $LiAlH_4$ in a separated step to yield (*R*)-**8** in 93% isolated yield and full retention of the ee from the previous step. This approach represents an alternative route to (*R*)-**8** from the MAO-N route described in Section 6.3.2, and the practical protocol is described in Section 6.9.6.1.

6.7.2.2 Biocatalytic Synthesis of (*S*)-Trolline and Analogs

Hailes' group developed a one-pot synthesis of the antiviral agent (*S*)-trolline ((*S*)-**33**) and analogs through the combination of an asymmetric Pictet–Spengler reaction catalyzed by NCS from *T. flavum* with a chemical lactam-ring formation (Scheme 6.27) [145]. The first enzymatic step occurred at pH 7.5 in HEPES buffer and provided only the *para*-substituted product (**32**). Next, the cyclization to yield (*S*)-**33** was accomplished following the addition of Na_2CO_3 in a four hour reaction at 60 °C under argon.

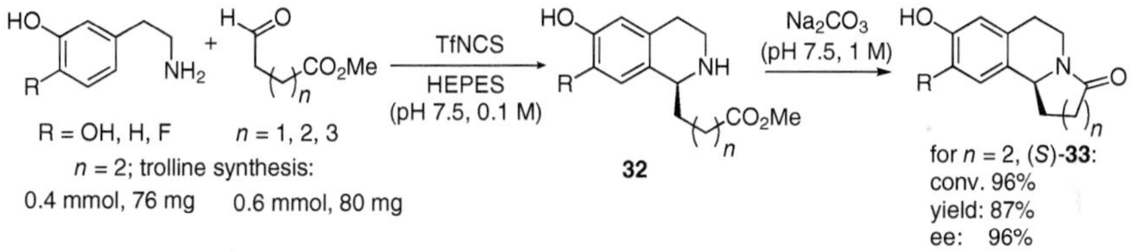

HO, R, NH_2 + H, O, CO_2Me, *n*
R = OH, H, F   *n* = 1, 2, 3
n = 2; trolline synthesis:
0.4 mmol, 76 mg   0.6 mmol, 80 mg
TfNCS, HEPES (pH 7.5, 0.1 M)
HO, R, NH, CO_2Me, *n*
32
Na_2CO_3 (pH 7.5, 1 M)
HO, R, N, O, *n*
for *n* = 2, (*S*)-**33**:
conv. 96%
yield: 87%
ee: 96%

Scheme 6.27 Chemoenzymatic synthesis of (*S*)-trolline ((*S*)-**33**) and analogs. Source: Based on Zhao et al. [145].

6.8 Engineered Cytochrome P450s (Cytochrome "P411")

Cytochrome P450s (EC 1.6.2.X) are heme-dependent enzymes that naturally catalyze the hydroxylation of sp^3-hybridized C—H bonds as well as the hydroxylation and/or epoxidation of C=C double bonds [154, 155]. The classical (simplified) view of the catalytic cycle of P450s entails (i) the one-electron reduction of the Fe(III) state of the heme cofactor, (ii) the addition of a dioxygen molecule to generate a Fe(III)-superoxo intermediate, (iii) further one-electron reduction to generate the Fe(III)-peroxo intermediate, (iv) heterolytic O–O cleavage to obtain the formal heme Fe(V)-oxo intermediate, and v) oxygen atom transfer from heme Fe(V)-oxo to the substrate (i.e. C–H hydroxylation), and regeneration of the ferric enzyme's resting state [156]. The redox equivalents are ultimately coming from one molecule of NADH. In 1985, Dawson's group demonstrated that a cytochrome P450 can catalyze intramolecular nitrene transfer into the sp^3-hybridized C—H bond of a substrate [157]. Arnold's and Fasan's groups revived this research by showing that engineered cytochrome P450s from *Bacillus megaterium* can catalyze this intramolecular C–H amination with higher efficiency [158–161]. Arnold's group further proposed that nitrene activity is enhanced if the WT heme's axial cysteine is substituted with a serine, and this family of variants was named as cytochrome P411. Finally, further evolved P411 variants exhibited activity toward the intermolecular C–H amination, although reactivity is currently limited to the C–H group in α-position to an aromatic ring [36]. A catalytic cycle was proposed in which the ferric resting state of the P411 is reduced *in situ* to the active ferrous state and then reacts with tosyl azide (TsN_3) to give the iron nitrenoid species. At this stage, the nitrene can be transferred to the substrate, thereby regenerating the ferrous state of the heme cofactor. However, unproductive nitrene reduction can also occur, thereby regenerating the ferric enzyme's resting state (Scheme 6.28).

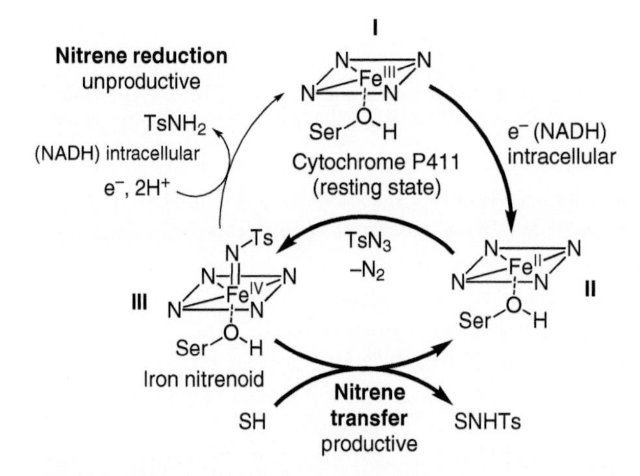

Scheme 6.28 Proposed catalytic cycle for nitrene transfer catalyzed by engineered cytochrome P450s. S = substrate.

6.8.1 Practical Approaches with Engineered Cytochrome P450s

$P411_{BM3}$ P-4, which contains 17 mutations compared with the WT P450 from *B. megaterium*, showed moderate activity for the amination of 4-ethylanisole [36, 162]. The enzyme was further evolved to obtain the final $P411_{CHA}$ variant, which was capable of aminating a number of aromatic substrates at the α-position to the phenyl ring (**34**, Scheme 6.29). Typical substrates were substituted: ethylbenzenes, 2,3-dihydro-1*H*-indene, 2,3-dihydrobenzofuran, 1,3-dihydroisobenzofuran, isochromane, and methylnaphthalene. On a 0.25 mmol scale reaction, 4-ethylanisole (**34**, 60 mg) was aminated to give (*R*)-*N*-tosyl-1-(*p*-methoxyphenyl) ethylamine ((*R*)-**35**) in 78% isolated yield (610 TON) and >99% ee. Further treatment with samarium diiodide could remove the tosyl group without erosion of the ee. In general, P411s exhibit significantly higher activity and stability when used as *E. coli* whole cell biocatalysts. The reaction proceeds better under anoxic conditions because air exposure reduces the activity seven-fold. The practical protocol for this reaction is described in Section 6.9.7.1.

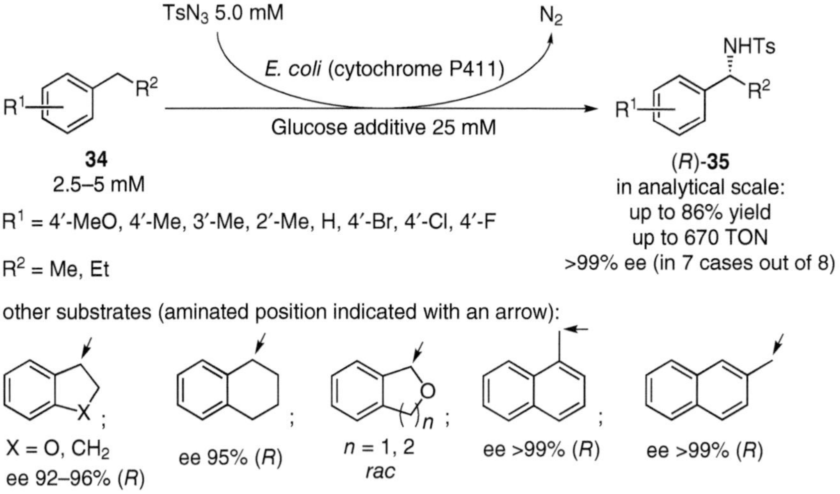

TsN_3 5.0 mM
N_2
E. coli (cytochrome P411)
Glucose additive 25 mM
NHTs
R^1 R^2
34
2.5–5 mM
(*R*)-**35**
in analytical scale:
up to 86% yield
up to 670 TON
>99% ee (in 7 cases out of 8)
R^1 = 4′-MeO, 4′-Me, 3′-Me, 2′-Me, H, 4′-Br, 4′-Cl, 4′-F
R^2 = Me, Et
other substrates (aminated position indicated with an arrow):
X ; X = O, CH_2 ee 92–96% (*R*)
ee 95% (*R*)
O n ; *n* = 1, 2 *rac*
ee >99% (*R*)
ee >99% (*R*)

Scheme 6.29 Reaction conditions and substrate scope of cytochrome $P411_{CHA}$.

6.9 Protocols for Selected Reactions

6.9.1 Hydrolases

6.9.1.1 Kinetic Resolution *rac*-Methylbenzylamine (*rac*-1)

rac-**1** (20 g, 165 mmol), methoxyacetic acid ethyl ester (19.5 g, 165 mmol), and *B. plantarii* lipase (2 $g_{crude\ enzyme}$, activity 1000 U mg^{-1} measured with tributyrin) were suspended in *tert*-butyl methyl ether (MTBE) (200 ml) and stirred for 24 hours until 52% conversion was obtained (determined by GC). Following removal by filtration of the lipase, the organic reaction mixture was treated with aqueous HCl (1 N); the unreacted (*S*)-configured amine ((*S*)-**1**) was extracted into the aqueous phase as hydrochloride, whereas the (*R*)-configured amide product ((*R*)-**2**) remained in the MTBE. After separation, the MTBE phase was dried with $MgSO_4$ and evaporated to yield 15.5 g of white crystals of the (*R*)-**2** (48% yield, 93% ee; T_M = 63 °C, $[\alpha]_D^{20}$ = +86.5°(c = 5 in 1,4-dioxane)). Finally, the aqueous phase was treated with NaOH until a pH 10 was obtained, and it was then extracted with MTBE. The organic phase was dried and evaporated to yield 9.2 g of the (*S*)-**1** as colorless liquid (46% yield, >99% ee; $[\alpha]_D^{20}$ = −39.5° [pure compound]) (Scheme 6.4) [40].

6.9.1.2 Dynamic Kinetic Resolution of Norsertraline Intermediate (*rac*-3)

C. antarctica lipase B (Novozyme 435, 20 mg), Na_2CO_3 (0.20 mmol, 20 mg), and a *p*-MeO-C_6H_4-substituted Ru-Shvo catalyst (26.5 mg, 0.02 mmol) were charged in a flame dried 20 ml flask. The flask was evacuated and filled with argon three times. Toluene (8 ml), *rac*-**3** (0.50 mmol, 76 μl), isopropyl acetate (3.50 mmol, 400 μl), and pentadecane (30 μl) were added, and the reaction mixture was stirred at 90 °C for three days. The reaction was cooled to room temperature, and the solids were removed by filtration and washed with DCM (3 × 20 ml). The solvent was evaporated and the crude product (*R*)-**4** was purified by column chromatography (SiO_2, acetone/DCM, 1 : 5 v/v). (*R*)-**4** was obtained as a white solid (66 mg, 0.35 mmol, 70% yield, 99% ee; $[\alpha]^{27}{}_D$ = +106.0 (c = 1.0, $CHCl_3$)) (Scheme 6.5) [45].

6.9.2 Monoamine Oxidases

6.9.2.1 Chemoenzymatic Deracemization of Harmicine (*rac*-8)

In a 500 ml flask, *rac*-**8** (530 mg, 2.5 mmol) and BH_3-NH_3 (330 mg, 10 mmol) were dissolved in a KPi buffer (165 ml, 1 M, pH 7.8). The pH of the solution was adjusted to 7.8 by addition of aqueous HCl. *E. coli*/MAO-N D9 cell pellet (5 g) was added, and the flask was shaken in an incubator at 37 °C, 250 rpm for 48 hours. Quantitative conversion was detected by HPLC analysis. The reaction was stopped by the addition of aqueous NaOH (2 ml, 10 M) and DCM (200 ml). The mixture was transferred in Falcon tubes, and the two phases were separated by centrifugation (4000 rpm, 5 minutes). The aqueous phase was extracted with DCM (2 × 100 ml), and the combined organic phases were dried over $MgSO_4$ and evaporated. As a result, 504 mg of (*R*)-**8** was obtained (95% yield, 99% ee) (Scheme 6.8) [54].

6.9.3 ω-Transaminases

6.9.3.1 Deracemization of Mexiletine (*rac*-9, Kinetic Resolution, Followed by Formal Reductive Amination)

rac-**9** (100 mg, 0.46 mmol, 23 mM), sodium pyruvate (0.9 mM), D-amino acid oxidase from porcine kidney (30 mg, 45 U), and ωTA ATA-117 (Codexis, 30 mg) were added in KPi (20 ml, 100 mM, pH 7.0, 1 mM PLP) in a 50 ml flask. The reaction was shaken at 30 °C, 300 rpm for 24 hours. Next, the mixture was heated to 75 °C for 30 minutes to deactivate ωTA ATA-117 and then cooled to RT. L-Alanine amino acid dehydrogenase from *Bacillus subtilis* (100 μl, 40 U), ammonium formate (70 mM), NAD^+ (1 mM), and FDH (200 μl, 40 U) were added, and the mixture was shaken for one hour at 30 °C. Next, L-alanine (100 mM) and lyophilized cells of *E. coli*/ωTA from *Chromobacterium violaceum* (100 mg) were added, and the mixture was shaken at 30 °C for 24 hours. Aqueous HCl (5 M) was added to the solution until pH 1 was obtained, and the remaining ketone was extracted with DCM (5 × 10 ml). The pH was then adjusted to pH 12 with KOH (10 N), and (*S*)-**9** was extracted with DCM (4 × 10 ml). The combined organic phase was evaporated and 95.2 mg of the (*S*)-**9** were obtained (95% yield, >99% ee). $[\alpha]_D^{20} = +3.2$ (c 2.0, $CHCl_3$) (Scheme 6.11) [71].

6.9.4 Imine Reductases and Amine Dehydrogenases

6.9.4.1 Reductive Amination of Aldehyde (11) with Kinetic Resolution of Amine Nucleophile (*rac-trans*-12)

In a 400 ml EasyMax reactor conditioned at 30 °C and equipped with an overhead stirrer, *rac-trans*-**12** (6.2 g, 2.04 equiv), D-glucose (5.8 g, 1.94 equiv), and $NADP^+$ (0.66 g, 0.05 equiv) were added to a sodium acetate buffer (100 mM, pH 5.0, 216 ml) and DMSO (54 ml). The mixture was stirred at 300 rpm, at 30 °C for 30 minutes. Next, IRED M2 lyophilized cell extract (0.6 g) and GDH-CDX-901 (0.2 g) were added and the mixture was stirred for additional 10 minutes. Aldehyde **11** was added dropwise as a stock solution in DMSO (5.0 g in 30 ml final volume) over 30 minutes. The reaction was run for additional 2.5 hours and then quenched by adding glacial acetic acid (30 ml). Conversion to the product (with respect to the limiting reagent **11**) was 87.6%. The work-up was performed by adding Celite 545 (1.7 g, 0.34 w/w), stirring for one hour, and filtering the resulting mixture through a glass fiber filter paper using a Buchner funnel. The filtrate was charged in the EasyMax reactor, cooled to 5 °C, and stirred at 300 rpm. Next, solid NaCl (7.8 g) was added in portions and stirring was continued overnight. The obtained mixture was cold-filtered using a Buchner funnel, and the product dihydrochloride salt precipitate was washed with cold water (2×15 ml) and acetone (15 ml). The solid was then dried overnight in a vacuum oven at 40 °C to afford product (1*R*,2*S*)-**13** as a white solid (5.6 g, 68.5% yield, 99.3% purity, 99.4% ee) (Scheme 6.17) [81].

6.9.4.2 Asymmetric Reductive Amination of Acetophenone (14) Using Amine Dehydrogenase

In a 250 ml flask, **14** (50 mM, 600 mg, 5 mmol), LE-AmDH-v1 (90 μM, 4 mg ml^{-1}), FDH (16 μM, 0.68 mg ml^{-1}), and NAD^+ (1 mM) were added in a $HCOONH_4/NH_3$ buffer (100 ml, 2 M, pH 9.0). The reaction was stirred at 50 °C and the conversion reached >99% within 48 hours. The reaction was acidified with concentrated HCl (6 ml) and the unreacted **14** was extracted with MTBE (2×80 ml). After the addition of KOH (10 M, 12 ml), (*R*)-methylbenzylamine ((*R*)-**15**)) was extracted with MTBE (2×120 ml). The combined organic layers were dried over $MgSO_4$, the solvent was evaporated, and (*R*)-**15** was obtained as a colorless liquid in 82% isolated yield (495 mg), >99.9% ee. Any possible side product was below detection limits (analyzed by GC and ^{1}H-NMR) (Scheme 6.18) [105].

6.9.5 Ammonia Lyases

6.9.5.1 Asymmetric Ammonia Addition to 2′-Chlorocinnamic Acid (17)

17 (1.8 g, 10 mmol) was dissolved in aqueous NH_3 (0.5 l, 13 v/v) and the pH was adjusted to 11 with aqueous H_2SO_4 (25 w/w). *E. coli* cell pellets containing the expressed *Rhodotorula glutinis* PAL (130 g) were suspended in aqueous NH_3 (0.2 l, 13 v/v) previously adjusted to pH 11 with aqueous H_2SO_4 (25 w/w). The cell suspension was added to the substrate solution, and aqueous NH_3 (13 v/v) was added until reaching a total volume of 1 l. The reaction was stirred at 200 rpm and 30 °C. Aliquots of **17** (total of 0.91 g, 5.0 mmol) were added every five minutes to the reaction medium during the first one hour and then every 25 minutes during an additional seven hours. After 8.5 hours, the reaction mixture contained approximately 18.1 g (91 mmol) of (*S*)-2′-chlorophenylalanine ((*S*)-**18**, yield 91% measured by HPLC). The aqueous phase (850 ml, pH 10.8) was concentrated under reduced pressure (150 to 1 mPa; at 60 °C), and after the removal of about 65% of the water, a large quantity of precipitate was formed (solution pH 7.5). The precipitate was removed by filtration and stirred in water to remove inorganic salts. The resulting solid material was dried to give 14.9 g of crude product, which contained 57% weight of (*S*)-**18** (8.5 g, 47% yield, 99% ee), 24% weight of water (as co-crystallized molecules), and 1.7% weight of **17**. The mother liquor still contained the remaining 9.5 g of (*S*)-**18** (Scheme 6.21) [128].

6.9.6 Pictet–Spenglerases

6.9.6.1 Asymmetric Pictet–Spengler Reaction with Strictosidine Synthase

In a 150 ml flask, lyophilized cell extract of recombinant RsSTR was added in a PIPES–tryptamine (**6**)·HCl buffer system (50 ml, 50 mm PIPES, 10 mm **6**·HCl, pH 6.1). Methyl

4-oxobutanoate (**29**, 75 mg, 0.33 mmol, final concentration: 50 mM) was added and the mixture was stirred at 470 rpm and 35 °C for 48 hours. The reaction was quenched by the addition of a NaOH aqueous solution (5 ml, 10 N) and extracted with EtOAc (3 × 100 ml). The combined organic phase was dried over Na_2SO_4 and the solvent was evaporated. (*R*)-**31** (precursor of (*R*)-harmicine, **8**) was obtained as a white solid (75.4 mg, 67% yield) (Scheme 6.26) [151].

6.9.7 Engineered Cytochrome P450s

6.9.7.1 Intermolecular Alkane C–H Amination Using Cytochrome P411

In a 250 ml flask, a suspension of wet cells of *E. coli*/cytochrome $P411_{CHA}$ (OD_{600} = 30, 80 ml) and a solution of D-glucose (10 ml, 250 mM, final concentration in the reaction mixture 25 mM) in M9-N growth media were flushed with argon for 40 minutes. The reaction flask was transferred into an anaerobic chamber and a dioxygen depletion system consisting of a KPi solution (0.1 M, pH 8.0, 5 ml), catalase (14 000 U ml^{-1}), and glucose oxidase (1000 U ml^{-1}) was added. 4-Ethylanisole (**34**, 2.5 ml of a 100 mM stock solution in DMSO, final concentration 2.5 mM) and tosyl azide (2.5 ml of a 200 mM stock solution in DMSO, final concentration 5.0 mM) were added; thus, the final reaction volume was 100 ml. The flask was sealed with parafilm, removed from the anaerobic chamber, and shaken at 130 rpm and RT for 20 hours. The reaction was quenched by adding acetonitrile (50 ml) and then centrifuged (4000 × g, 10 minutes). The supernatant was concentrated and extracted with EtOAc (3 × 25 ml). The organic layers were washed with brine (20 ml), dried over $MgSO_4$, filtered, concentrated, and purified by chromatography (SiO_2, EtOAc/hexane, 5–30%, v/v). *N*-Tosyl-1-(*p*-methoxyphenyl)ethylamine ((*R*)-**35**) was isolated in 59.5 mg (78% yield, 610 TON, >99% ee) (Scheme 6.29) [36].

6.10 Conclusions

Industrial-scale production of α-chiral amines greatly relies on the asymmetric hydrogenation of activated intermediates such as enamides, enamines, or preformed N-substituted imines [4–6]. These procedures often require precious and toxic transition metals – which are coordinated to costly and sophisticated organic ligands – and usually operate at high hydrogen gas pressure. Another common disadvantage is the requirement for protection and deprotection steps that reduce the atom efficiency and increase the number of chemical steps. Recrystallization of the chiral amine product is often required to upgrade the enantiomeric excess. Finally, traces of heavy metals also have to be removed from the final product to comply with legislative requirements (e.g. API manufacture).

In this context, the biocatalytic synthesis of α-chiral amines (and amino acids) offers an environmentally friendly and economically profitable alternative to more traditional chemical and chemocatalytic methods. Hydrolases remain industry's first choice for α-chiral amine synthesis because of the large availability of low-cost enzymes, the lack of any cofactor requirement, and the existence of optimized processes. Nevertheless, MAOs and ωTAs are increasingly attracting attention, particularly from the pharmaceutical and fine chemical industry. Engineered variants have been developed to substitute chemical catalysts into existing synthetic routes in order to increase atom efficiency and stereo- and chemoselectivity as well as reduce waste generation. However, the full synthetic potential of enzyme catalysis for chiral amine synthesis is still untapped because, in most cases, the enzyme was engineered and adapted to an existing synthetic route [60]. Nonetheless, considering the available biocatalyst toolbox in the early conception of a synthesis route would result in better integration between biocatalysis and other types of available synthetic methodologies toward more efficient chemical synthesis [163]. Another aspect of biocatalysis is the possibility to construct one-pot multienzyme processes for the synthesis of α-chiral amines starting from inexpensive and abundant material [164–167]. As the concept of "biocatalytic retrosynthesis" is becoming more popular in organic synthesis, it is expected that in the near future, both established biocatalysts and novel enzymes such as AmDHs, IReds, and Pictet–Spenglerases will be more widely employed for α-chiral amine synthesis in academic and industrial settings.

Acknowledgments

This project has received funding from the European Research Council (ERC) under the European Union's Horizon 2020 research and innovation programme (ERC-StG, grant agreement No 638271, BioSusAmin). Dutch funding from the NWO Sector Plan for Physics and Chemistry is also acknowledged.

References

1 Wittcoff, H.A., Rueben, B.G., and Plotkin, J.S. (2004). *Industrial Organic Chemicals*, 2 ed. New York, NY: Wiley-Interscience.
2 Constable, D.J.C., Dunn, P.J., Hayler, J.D. et al. (2007). *Green Chem.* 9: 411–420.
3 Ghislieri, D. and Turner, N.J. (2013). *Top. Catal.* 57: 284–300.
4 Nugent, T.C. (2010). *Chiral Amine Synthesis: Methods, Developments and Applications.* Wiley-VCH.
5 Wang, C. and Xiao, J. (2014). Asymmetric reductive amination. In: *Stereoselective Formation of Amines*, vol. 343 (eds. W. Li and X. Zhang), 261–282. Berlin Heidelberg: Springer-Verlag.
6 Bauer, H., Alonso, M., Färber, C. et al. (2018). *Nat. Catal.* 1: 40–47.
7 Kohls, H., Steffen-Munsberg, F., and Hohne, M. (2014). *Curr. Opin. Chem. Biol.* 19: 180–192.
8 Bornscheuer, U. and Kazlauskas, R.J. (2006). *Hydrolases in Organic Synthesis*, 2 ed. Weinheim: Wiley-VCH.
9 Turner, N.J. (2011). *Chem. Rev.* 111: 4073–4087.
10 Pollegioni, L. and Molla, G. (2015). *C-N Oxidation with amine oxidases and amino acid oxidases.* In: *Science of Synthesis, Biocatalysis in Organic Synthesis*, vol. 3 (eds. K. Faber, W.-D. Fessner and N.J. Turner), 235–284. Stuttgart (Germany): Georg Thieme Verlag KG.
11 Verho, O. and Backvall, J.E. (2015). *J. Am. Chem. Soc.* 137: 3996–4009.
12 Pellissier, H. (2016). *Tetrahedron* 72: 3133–3150.
13 Kim, Y., Park, J., and Kim, M.-J. (2011). *ChemCatChem* 3: 271–277.
14 Musa, M.M., Hollmann, F., and Mutti, F.G. (2019). *Catal. Sci. Technol.* 9: 5487–5503.

15 Turner, N.J. (2010). *Curr. Opin. Chem. Biol.* 14: 115–121.

16 Koszelewski, D., Tauber, K., Faber, K., and Kroutil, W. (2010). *Trends Biotechnol.* 28: 324–332.

17 Mathew, S. and Yun, H. (2012). *ACS Catal.* 2: 993–1001.

18 Fuchs, M., Farnberger, J.E., and Kroutil, W. (2015). *Eur. J. Org. Chem.* 2015: 6965–6982.

19 Guo, F. and Berglund, P. (2017). *Green Chem.* 19: 333–360.

20 Slabu, I., Galman, J.L., Lloyd, R.C., and Turner, N.J. (2017). *ACS Catal.* 7: 8263–8284.

21 Patil, M.D., Grogan, G., Bommarius, A., and Yun, H. (2018). *Catalysts* 8: 254.

22 Gomm, A. and O'Reilly, E. (2018). *Curr. Opin. Chem. Biol.* 43: 106–112.

23 Kelly, S.A., Pohle, S., Wharry, S. et al. (2018). *Chem. Rev.* 118: 349–367.

24 Sharma, M., Mangas-Sanchez, J., Turner, N.J., and Grogan, G. (2017). *Adv. Synth. Catal.* 359: 2011–2025.

25 Schrittwieser, J.H., Velikogne, S., and Kroutil, W. (2015). *Adv. Synth. Catal.* 357: 1655–1685.

26 Mangas-Sanchez, J., France, S.P., Montgomery, S.L. et al. (2017). *Curr. Opin. Chem. Biol.* 37: 19–25.

27 Grogan, G. and Turner, N.J. (2016). *Chem. Eur. J.* 22: 1900–1907.

28 Patil, M.D., Grogan, G., Bommarius, A., and Yun, H. (2018). *ACS Catal.* 8: 10985–11015.

29 Dong, J., Fernandez-Fueyo, E., Hollmann, F. et al. (2018). *Angew. Chem. Int. Ed.* 57: 9238–9261.

30 Jeon, H., Yoon, S., Ahsan, M. et al. (2017). *Catalysts* 7: 251.

31 Patil, M.D., Yoon, S., Jeon, H. et al. (2019). *Catalysts* 9: 600.

32 Aleku, G.A., Mangas-Sanchez, J., Citoler, J. et al. (2018). *ChemCatChem* 10: 515–519.

33 Tseliou, V., Knaus, T., Vilím, J. et al. (2020). *ChemCatChem* 12: 2184–2188.

34 Parmeggiani, F., Weise, N.J., Ahmed, S.T., and Turner, N.J. (2018). *Chem. Rev.* 118: 73–118.

35 Bartsch, S. and Vogel, A. (2015). Addition of ammonia and amines to C=C bonds. In: *Science of Synthesis, Biocatalysis in Organic Synthesis*, vol. 2 (eds. K. Faber, W.-D. Fessner and N.J. Turner), 291–311. Stuttgart (Germany): Georg Thieme Verlag KG.

36 Prier, C.K., Zhang, R.K., Buller, A.R. et al. (2017). *Nat. Chem.* 9: 629–634.

37 Ilari, A., Bonamore, A., and Boffi, A. (2015). Addition to C=N bonds. In: *Science of Synthesis, Biocatalysis in Organic Synthesis*, vol. 2 (eds. K. Faber, W.-D. Fessner and N.J. Turner), 159–175. Stuttgart (Germany): Georg Thieme Verlag KG.

38 Schmidt, N.G., Eger, E., and Kroutil, W. (2016). *ACS Catal.* 6: 4286–4311.

39 Gutman, A.L., Meyer, E., Kalerin, E. et al. (1992). *Biotechnol. Bioeng.* 40: 760–767.

40 Balkenhohl, F., Ditrich, K., Hauer, B., and Ladner, W. (1997). *J. Prakt. Chem./Chem. Ztg.* 339: 381–384.

41 Rebolledo, F., González-Sabín, J., and Gotor, V. (2002). *Tetrahedron: Asymmetry* 13: 1315–1320.

42 González-Sabín, J., Gotor, V., and Rebolledo, F. (2004). *Tetrahedron: Asymmetry* 15: 481–488.

43 Alatorre-Santamaría, S., Gotor-Fernández, V., and Gotor, V. (2009). *Eur. J. Org. Chem.* 2009: 2533–2538.

44 Klibanov, A.M. (2001). *Nature* 409: 241–246.

45 Thalen, L.K., Zhao, D., Sortais, J.B. et al. (2009). *Chem. Eur. J.* 15: 3403–3410.

46 Gustafson, K.P.J., Gorbe, T., de Gonzalo-Calvo, G. et al. (2019). *Chem. Eur. J.* 25: 9174–9179.

47 Dijkman, W.P., De Gonzalo, G., Mattevi, A., and Fraaije, M.W. (2013). *Appl. Microbiol. Biotechnol.* 97: 5177–5188.

48 Asano, Y. and Yasukawa, K. (2019). *Curr. Opin. Chem. Biol.* 49: 76–83.
49 Alexeeva, M., Enright, A., Dawson, M.J. et al. (2002). *Angew. Chem.* 114: 3309–3312.
50 Carr, R., Alexeeva, M., Enright, A. et al. (2003). *Angew. Chem. Int. Ed.* 42: 4807–4810.
51 Carr, R., Alexeeva, M., Dawson, M.J. et al. (2005). *ChemBioChem* 6: 637–639.
52 Dunsmore, C.J., Carr, R., Fleming, T., and Turner, N.J. (2006). *J. Am. Chem. Soc.* 128: 2224–2225.
53 Herter, S., Medina, F., Wagschal, S. et al. (2018). *Bioorg. Med. Chem.* 26: 1338–1346.
54 Ghislieri, D., Green, A.P., Pontini, M. et al. (2013). *J. Am. Chem. Soc.* 135: 10863–10869.
55 Rowles, I., Malone, K.J., Etchells, L.L. et al. (2012). *ChemCatChem* 4: 1259–1261.
56 Heath, R.S., Pontini, M., Hussain, S., and Turner, N.J. (2016). *ChemCatChem* 8: 117–120.
57 Kohler, V., Wilson, Y.M., Durrenberger, M. et al. (2013). *Nat. Chem.* 5: 93–99.
58 Li, T., Liang, J., Ambrogelly, A. et al. (2012). *J. Am. Chem. Soc.* 134: 6467–6472.
59 Hwang, B.-Y., Cho, B.-K., Yun, H. et al. (2005). *J. Mol. Catal. B: Enzym.* 37: 47–55.
60 Savile, C.K., Janey, J.M., Mundorff, E.C. et al. (2010). *Science* 329: 305–309.
61 Truppo, M.D., Rozzell, J.D., and Turner, N.J. (2010). *Org. Process Res. Dev.* 14: 234–237.
62 Wang, B., Land, H., and Berglund, P. (2013). *Chem. Commun.* 49: 161–163.
63 Green, A.P., Turner, N.J., and O'Reilly, E. (2014). *Angew. Chem. Int. Ed.* 53: 10714–10717.
64 Martínez-Montero, L., Gotor, V., Gotor-Fernández, V., and Lavandera, I. (2016). *Adv. Synth. Catal.* 358: 1618–1624.
65 Gomm, A., Lewis, W., Green, A.P., and O'Reilly, E. (2016). *Chem. Eur. J.* 22: 12692–12695.
66 Payer, S.E., Schrittwieser, J.H., and Kroutil, W. (2017). *Eur. J. Org. Chem.* 2017: 2553–2559.
67 Mutti, F.G. and Kroutil, W. (2012). *Adv. Synth. Catal.* 354: 3409–3413.
68 Truppo, M.D., Strotman, H., and Hughes, G. (2012). *ChemCatChem* 4: 1071–1074.
69 Böhmer, W., Volkov, A., Engelmark Cassimjee, K., and Mutti, F.G. (2020). *Adv. Synth. Catal.* 362: 1858–1867.
70 Mutti, F.G., Sattler, J., Tauber, K., and Kroutil, W. (2011). *ChemCatChem* 3: 109–111.
71 Koszelewski, D., Pressnitz, D., Clay, D., and Kroutil, W. (2009). *Org. Lett.* 11: 4810–4812.
72 Pavlidis, I.V., Weiß, M.S., Genz, M. et al. (2016). *Nat. Chem.* 8: 1076–1082.
73 Roberts, M.F. (1975). *Phytochemistry* 14: 2393–2397.
74 Huber, T., Schneider, L., Präg, A. et al. (2014). *ChemCatChem* 6: 2248–2252.
75 Scheller, P.N., Lenz, M., Hammer, S.C. et al. (2015). *ChemCatChem* 7: 3239–3242.
76 Wetzl, D., Gand, M., Ross, A. et al. (2016). *ChemCatChem* 8: 2023–2026.
77 Roiban, G.-D., Kern, M., Liu, Z. et al. (2017). *ChemCatChem* 9: 4475–4479.
78 Matzel, P., Gand, M., and Höhne, M. (2017). *Green Chem.* 19: 385–389.
79 Aleku, G.A., France, S.P., Man, H. et al. (2017). *Nat. Chem.* 9: 961–969.
80 France, S.P., Howard, R.M., Steflik, J. et al. (2018). *ChemCatChem* 10: 510–514.
81 Schober, M., MacDermaid, C., Ollis, A.A. et al. (2019). *Nat. Catal.* 2: 909–915.
82 Itoh, N., Yachi, C., and Kudome, T. (2000). *J. Mol. Catal. B: Enz.* 10: 281–290.
83 Abrahamson, M.J., Vazquez-Figueroa, E., Woodall, N.B. et al. (2012). *Angew. Chem. Int. Ed.* 51: 3969–3972.
84 Tseliou, V., Masman, M.F., Böhmer, W. et al. (2019). *ChemBioChem* 20: 800–812.
85 Sharma, M., Mangas-Sanchez, J., France, S.P. et al. (2018). *ACS Catal.* 8: 11534–11541.
86 Roth, S., Prag, A., Wechsler, C. et al. (2017). *ChemBioChem* 18: 1703–1706.
87 Mitsukura, K., Suzuki, M., Tada, K. et al. (2010). *Org. Biomol. Chem.* 8: 4533–4535.
88 Mitsukura, K., Suzuki, M., Shinoda, S. et al. (2011). *Biosci. Biotechnol. Biochem.* 75: 1778–1782.

89 Mitsukura, K., Kuramoto, T., Yoshida, T. et al. (2013). *Appl. Microbiol. Biotechnol.* 97: 8079–8086.

90 Leipold, F., Hussain, S., Ghislieri, D., and Turner, N.J. (2013). *ChemCatChem* 5: 3505–3508.

91 Gand, M., Müller, H., Wardenga, R., and Höhne, M. (2014). *J. Mol. Catal. B: Enzym.* 110: 126–132.

92 Man, H., Wells, E., Hussain, S. et al. (2015). *ChemBioChem* 16: 1052–1059.

93 Hussain, S., Leipold, F., Man, H. et al. (2015). *ChemCatChem* 7: 579–583.

94 Li, H., Luan, Z.-J., Zheng, G.-W., and Xu, J.-H. (2015). *Adv. Synth. Catal.* 357: 1692–1696.

95 Wetzl, D., Berrera, M., Sandon, N. et al. (2015). *ChemBioChem* 16: 1749–1756.

96 Abrahamson, M.J., Wong, J.W., and Bommarius, A.S. (2013). *Adv. Synth. Catal.* 355: 1780–1786.

97 Bommarius, B.R., Schurmann, M., and Bommarius, A.S. (2014). *Chem. Commun.* 50: 14953–14955.

98 Ye, L.J., Toh, H.H., Yang, Y. et al. (2015). *ACS Catal.* 5: 1119–1122.

99 Chen, F.-F., Liu, Y.-Y., Zheng, G.-W., and Xu, J.-H. (2015). *ChemCatChem* 7: 3838–3841.

100 Pushpanath, A., Siirola, E., Bornadel, A. et al. (2017). *ACS Catal.* 7: 3204–3209.

101 Knaus, T., Böhmer, W., and Mutti, F.G. (2017). *Green Chem.* 19: 453–463.

102 Au, S.K., Bommarius, B.R., and Bommarius, A.S. (2014). *ACS Catal.* 4: 4021–4026.

103 Lowe, J., Ingram, A.A., and Groger, H. (2018). *Bioorg. Med. Chem.* 26: 1387–1392.

104 Chen, F.-F., Zheng, G.-W., Liu, L. et al. (2018). *ACS Catal.* 8: 2622–2628.

105 Tseliou, V., Knaus, T., Masman, M.F. et al. (2019). *Nat. Commun.* 10: 3717.

106 Mayol, O., David, S., Darii, E. et al. (2016). *Catal. Sci. Technol.* 6: 7421–7428.

107 Mayol, O., Bastard, K., Beloti, L. et al. (2019). *Nat. Catal.* 2: 324–333.

108 Liu, J., Pang, B.Q.W., Adams, J.P. et al. (2017). *ChemCatChem* 9: 425–431.

109 Böhmer, W., Knaus, T., and Mutti, F.G. (2018). *ChemCatChem* 10: 731–735.

110 Houwman, J.A., Knaus, T., Costa, M., and Mutti, F.G. (2019). *Green Chem.* 21: 3846–3857.

111 Yu, H.L., Li, T., Chen, F.F. et al. (2018). *Metab. Eng.* 47: 184–189.

112 Puthan Veetil, V., Fibriansah, G., Raj, H. et al. (2012). *Biochemistry* 51: 4237–4243.

113 Asuncion, M., Blankenfeldt, W., Barlow, J.N. et al. (2002). *J. Biol. Chem.* 277: 8306–8311.

114 Turner, N.J. (2011). *Curr. Opin. Chem. Biol.* 15: 234–240.

115 Lovelock, S.L., Lloyd, R.C., and Turner, N.J. (2014). *Angew. Chem. Int. Ed.* 53: 4652–4656.

116 Gulzar, M.S., Akhtar, M., and Gani, D. (1997). *J. Chem. Soc., Perkin Trans.* 1: 649–656.

117 Akhtar, M., Cohen, M.A., and Gani, D. (1987). *Tetrahedron Lett.* 28: 2413–2416.

118 Raj, H., Szymanski, W., de Villiers, J. et al. (2012). *Nat. Chem.* 4: 478–484.

119 de Villiers, J., de Villiers, M., Geertsema, E.M. et al. (2015). *ChemCatChem* 7: 1931–1934.

120 Fu, H., Zhang, J., Tepper, P.G. et al. (2018). *J. Med. Chem.* 61: 7741–7753.

121 Li, R., Wijma, H.J., Song, L. et al. (2018). *Nat. Chem. Biol.* 14: 664–670.

122 Gloge, A., Zoń, J., Kövári, Á. et al. (2000). *Chem. Eur. J.* 6: 3386–3390.

123 Fu, H., Zhang, J., Saifuddin, M. et al. (2018). *Nat. Catal.* 1: 186–191.

124 Zhang, J., Fu, H., Tepper, P.G., and Poelarends, G.J. (2019). *Adv. Synth. Catal.* 361: 2433–2437.

125 Poppe, L. and Retey, J. (2005). *Angew. Chem. Int. Ed.* 44: 3668–3688.

126 Paizs, C., Katona, A., and Retey, J. (2006). *Chem. Eur. J.* 12: 2739–2744.

127 Ahmed, S.T., Parmeggiani, F., Weise, N.J. et al. (2018). *ACS Catal.* 8: 3129–3132.

128 de Lange, B., Hyett, D.J., Maas, P.J.D. et al. (2011). *ChemCatChem* 3: 289–292.
129 Rétey, J., Paizs, C., Ioana Toşa, M. et al. (2010). *Heterocycles* 82: 1217.
130 Gloge, A., Langer, B., Poppe, L., and Retey, J. (1998). *Arch. Biochem. Biophys.* 359: 1–7.
131 Ahmed, S.T., Parmeggiani, F., Weise, N.J. et al. (2015). *ACS Catal.* 5: 5410–5413.
132 Parmeggiani, F., Lovelock, S.L., Weise, N.J. et al. (2015). *Angew. Chem. Int. Ed.* 54: 4608–4611.
133 Wu, B., Szymanski, W., Wybenga, G.G. et al. (2012). *Angew. Chem. Int. Ed.* 51: 482–486.
134 Verkuijl, B.J., Szymanski, W., Wu, B. et al. (2010). *Chem. Commun.* 46: 901–903.
135 Bartsch, S., Wybenga, G.G., Jansen, M. et al. (2013). *ChemCatChem* 5: 1797–1802.
136 Wu, B., Szymanski, W., Wietzes, P. et al. (2009). *ChemBioChem* 10: 338–344.
137 Szymanski, W., Wu, B., Weiner, B. et al. (2009). *J. Org. Chem.* 74: 9152–9157.
138 Ratnayake, N.D., Wanninayake, U., Geiger, J.H., and Walker, K.D. (2011). *J. Am. Chem. Soc.* 133: 8531–8533.
139 Krug, D. and Muller, R. (2009). *ChemBioChem* 10: 741–750.
140 Weise, N.J., Parmeggiani, F., Ahmed, S.T., and Turner, N.J. (2015). *J. Am. Chem. Soc.* 137: 12977–12983.
141 Maresh, J.J., Giddings, L.A., Friedrich, A. et al. (2008). *J. Am. Chem. Soc.* 130: 710–723.
142 Bonamore, A., Rovardi, I., Gasparrini, F. et al. (2010). *Green Chem.* 12: 1623–1627.
143 Ruff, B.M., Brase, S., and O'Connor, S.E. (2012). *Tetrahedron Lett.* 53: 1071–1074.
144 Pesnot, T., Gershater, M.C., Ward, J.M., and Hailes, H.C. (2012). *Adv. Synth. Catal.* 354: 2997–3008.
145 Zhao, J., Lichman, B.R., Ward, J.M., and Hailes, H.C. (2018). *Chem. Commun.* 54: 1323–1326.
146 Lichman, B.R., Zhao, J.X., Hailes, H.C., and Ward, J.M. (2017). *Nat. Commun.* 8: 14883.
147 Erdmann, V., Lichman, B.R., Zhao, J. et al. (2017). *Angew. Chem. Int. Ed.* 56: 12503–12507.
148 Wang, Y., Tappertzhofen, N., Méndez-Sánchez, D. et al. (2019). *Angew. Chem. Int. Ed.* 58: 10120–10125.
149 Lichman, B.R., Lamming, E.D., Pesnot, T. et al. (2015). *Green Chem.* 17: 852–855.
150 Fischereder, E., Pressnitz, D., Kroutil, W., and Lutz, S. (2014). *Bioorg. Med. Chem.* 22: 5633–5637.
151 Pressnitz, D., Fischereder, E.M., Pletz, J. et al. (2018). *Angew. Chem. Int. Ed.* 57: 10683–10687.
152 Wu, F., Zhu, H., Sun, L. et al. (2012). *J. Am. Chem. Soc.* 134: 1498–1500.
153 Fischereder, E.-M., Pressnitz, D., and Kroutil, W. (2016). *ACS Catal.* 6: 23–30.
154 Urlacher, V.B. and Girhard, M. (2012). *Trends Biotechnol.* 30: 26–36.
155 Lundemo, M.T. and Woodley, J.M. (2015). *Appl. Microbiol. Biotechnol.* 99: 2465–2483.
156 Guengerich, F.P. (2018). *ACS Catal.* 8: 10964–10976.
157 Svastits, E.W., Dawson, J.H., Breslow, R., and Gellman, S.H. (1985). *J. Am. Chem. Soc.* 107: 6427–6428.
158 McIntosh, J.A., Coelho, P.S., Farwell, C.C. et al. (2013). *Angew. Chem. Int. Ed.* 52: 9309–9312.
159 Singh, R., Bordeaux, M., and Fasan, R. (2014). *ACS Catal.* 4: 546–552.
160 Bordeaux, M., Singh, R., and Fasan, R. (2014). *Bioorg. Med. Chem.* 22: 5697–5704.
161 Hyster, T.K., Farwell, C.C., Buller, A.R. et al. (2014). *J. Am. Chem. Soc.* 136: 15505–15508.
162 Prier, C.K., Hyster, T.K., Farwell, C.C. et al. (2016). *Angew. Chem. Int. Ed.* 55: 4711–4715.

163 Hönig, M., Sondermann, P., Turner, N.J., and Carreira, E.M. (2017). *Angew. Chem. Int. Ed.* 56: 8942–8973.

164 Mutti, F.G., Knaus, T., Scrutton, N.S. et al. (2015). *Science* 349: 1525–1529.

165 Sehl, T., Hailes, H.C., Ward, J.M. et al. (2013). *Angew. Chem. Int. Ed.* 52: 6772–6775.

166 Corrado, M.L., Knaus, T., and Mutti, F.G. (2019). *Green Chem.* 21: 6246–6251.

167 Martínez-Montero, L., Gotor, V., Gotor-Fernández, V., and Lavandera, I. (2017). *Green Chem.* 19: 474–480.

7

Applications of Oxidoreductases in Synthesis: A Roadmap to Access Value-Added Products

Mélanie Hall

University of Graz, Department of Chemistry, Heinrichstrasse 28, 8010 Graz, Austria;
University of Graz, Field of Excellence BioHealth, Austria

7.1 Introduction

The use of oxidoreductases in organic synthesis is now recognized as a suitable alternative to chemical methods, especially in those cases where the most attractive features of enzymes (chemo-, regio-, and stereoselectivity) are recruited [1]. Stereoselectivity in particular is of great relevance for the production of optically active molecules, which are becoming prevalent in the pharmaceutical industry, and several industrial processes, currently running on multi-kilogram scale, rely on one or more enzymes for the key reaction(s) of a synthetic sequence toward formation of enantiopure molecules (see also Chapter 15) [2–6].

Oxidoreductases (EC 1) encompass a large diversity of enzymes (total of 22 subclasses) capable of electron transfer between two molecules. For simultaneous transfer of two electrons in the form of a hydride, the cofactor nicotinamide adenine dinucleotide (phosphate) (NAD(P)H) is typically used as a reductant and must be added externally. In some cases, additional coenzymes are required as redox active mediators (e.g. flavin, heme-bound iron, and iron–sulfur cluster) and are usually produced by the cells during enzyme expression, so that they are already present in the biocatalyst preparation. In most cases, coenzymes are tightly bound to the enzyme active site and the experimenter does not need to take care of these molecules directly. Special attention may be however required in the case of coenzyme loss during preparation of the biocatalyst (such as with flavin) or sensitivity to the environment (such as oxygen lability of iron–sulfur cluster).

Several biocatalytic strategies are available for the synthesis of enantiopure molecules using oxidoreductases, which often imply the creation of a new stereogenic center

Biocatalysis for Practitioners: Techniques, Reactions and Applications, First Edition.
Edited by Gonzalo de Gonzalo and Iván Lavandera.

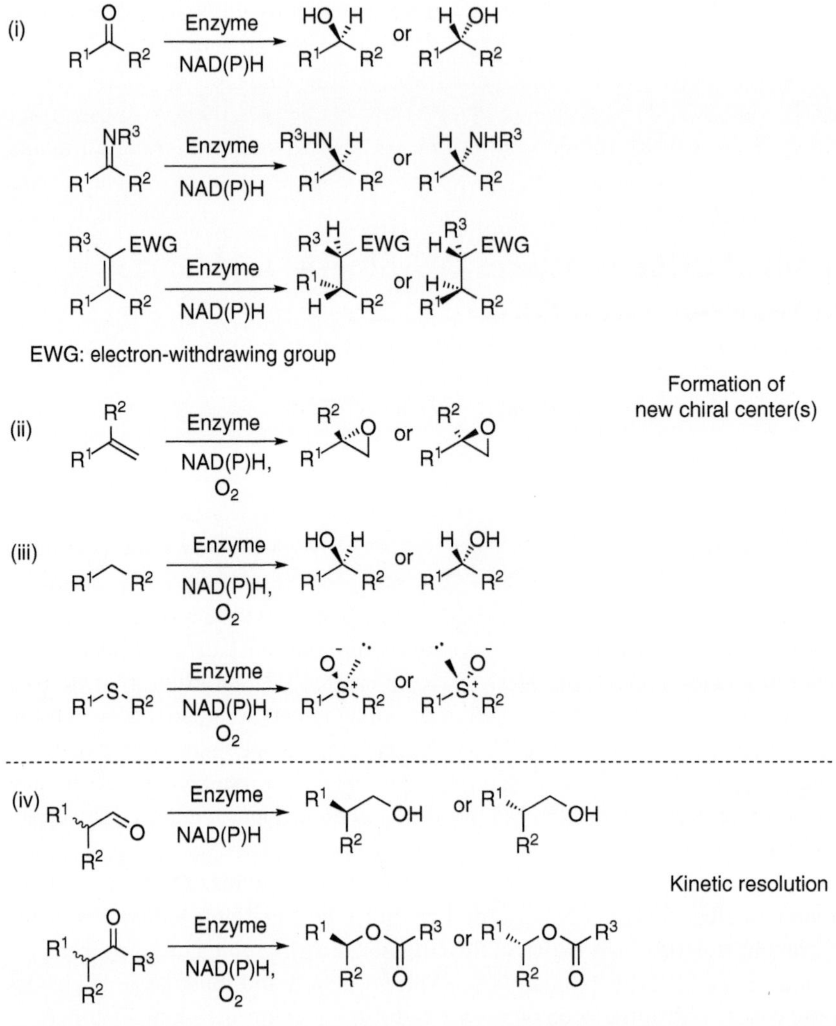

Scheme 7.1 Strategies to obtain enantiopure molecules using oxidoreductases: (i) reduction of prochiral molecules (e.g. ketone, imine, and alkene); (ii) oxygenation of prochiral substrates (e.g. alkene); (iii) enantioselective hydroxylation and heteroatom oxidation (e.g. alkane and sulfide, respectively); and (iv) (dynamic) kinetic resolution of racemic substituted carbonyl compounds (e.g. through carbonyl reduction or Baeyer–Villiger oxidation). The enzymes required for each strategy are detailed in this chapter.

(see Scheme 7.1): (i) reduction of prochiral molecules (e.g. of C=O, C=C, or C=N bond); (ii) stereoselective oxygenation of prochiral substrates (e.g. alkene epoxidation); and (iii) enantioselective formation of chiral compounds (e.g. hydroxylation or heteroatom oxidation). In some cases, the chiral center may be already present in the starting material and the reaction involves (iv) stereorecognition in the form of a (dynamic) kinetic resolution of the racemic substrate, such as in the reduction of racemic α-substituted aldehydes or the Baeyer–Villiger oxidation of substituted carbonyl compounds (*vide infra*).

The aim of this chapter is to provide a set of practical guidelines for the selection and use of oxidoreductases in synthesis, with a strong focus on the generation of enantiopure molecules. Well-established enzymes in industry as well as emerging biocatalysts will be discussed and include alcohol dehydrogenases (ADH), ene-reductases, monooxygenases, and peroxygenases. The growing importance of imine synthesis and correspondingly available biocatalytic strategies are highlighted in a companion chapter (see Chapter 6). The number of commercially available enzymes keeps increasing, and for many reactions displayed in this chapter, screening kits can be purchased, which grandly facilitate the identification of a suitable biocatalyst for the targeted reaction (see also Chapter 16). Thus, less emphasis will be put on the techniques necessary to obtain biocatalyst preparations. Instead, where relevant, practical "tricks" will be shared in order to facilitate the experience of first-time users in implementing biocatalysis in organic synthesis.

Most enzymes treated in this chapter share in common their dependence on a nicotinamide cofactor. Despite being highly unattractive for cost reasons when used in stoichiometric amounts, the use of NAD(P)H/NAD(P)$^+$ could be established on large scale owing to the availability of cofactor regeneration strategies that rely on a sacrificial and inexpensive reductant (see also Chapter 10). Through coupled-enzyme and coupled-substrate approaches [1, 7–9], NAD(P)H can be employed in catalytic amounts, and this has greatly favored the incorporation of nicotinamide-dependent biocatalytic schemes in industrial processes (see Scheme 7.2). Well-established coupled-enzyme systems include notably glucose (6-phosphate)/glucose (6-phosphate) dehydrogenase, formate/formate dehydrogenase, isopropanol/alcohol dehydrogenase, and phosphite/phosphite dehydrogenase [8]. To alleviate the stoichiometric requirement for the natural nicotinamide cofactor, other strategies are emerging, such as the design of self-sufficient redox processes ([11] and references therein), the replacement of NAD(P)H with artificial cofactors (NAD(P)H-mimics) [12], or the *in situ* production of NADPH by light-dependent biological processes (e.g. photosynthetic cyanobacteria) [13]. Although less developed, strategies are also available for the recycling of NAD(P)$^+$ [2]. Elegant methods involve the reduction of oxygen by the

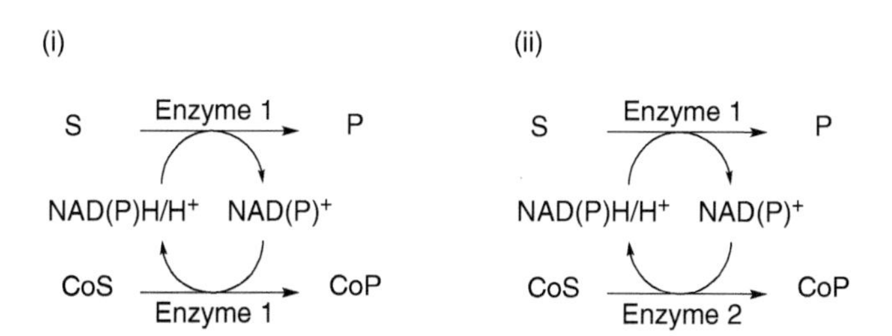

Scheme 7.2 NAD(P)H regeneration strategies via (i) coupled-substrate approach: the cosubstrate is employed as a hydride source and delivers one equivalent of coproduct through conversion by the same enzyme in charge of the targeted reaction (exemplary is the use of alcohol dehydrogenase for reduction of a carbonyl substrate through oxidation of isopropanol to acetone [10]), and (ii) coupled-enzyme approach: a second enzyme is required for the transformation of the cosubstrate (see text for examples); S, substrate; P, product; CoS, cosubstrate; CoP, coproduct. In both cases, the cofactor is used in catalytic amounts as a hydride shuttle.

water-forming NAD(P)H oxidase (NOX) [2, 14] or a glutathione-based recycling system, which uses for instance 2-hydroxyethyl disulfide as a final oxidizing agent [15].

A diversity of enzyme preparations can be employed to catalyze the biotransformations reported in this chapter. Although the use of wild-type strains was dominating at the onset of biocatalysis, advances in molecular biology have contributed to generalize the use of recombinant enzymes obtained from heterologous overexpression, typically performed in *Escherichia coli*. These recombinant enzymes can be used as isolated purified proteins, as cell-free extracts or as whole cells, either in resting, growing, or lyophilized form. Provided cloning techniques are available, the use of lyophilized recombinant whole cells appears the most straightforward because no additional step beyond cloning and cell growth is necessary [1]. However, the choice of the preparation will mostly depend on the targeted reaction and parameters such as enzyme stability, substrate concentration, process conditions, recycling of cofactor, or competing reactions should be taken into consideration before designing a biocatalyzed reaction [16, 17].

7.2 Reductive Processes

7.2.1 Reduction of C=O Bonds

Carbonyl compounds (aldehydes or ketones) can be reduced to their corresponding alcohols (primary or secondary) by ADH (EC: 1.1.1.X), also named carbonyl reductases or ketoreductases (KRED). In the case of a prochiral carbonyl substrate, a stereoselective reduction may take place and the degree of enantiopurity of the final secondary alcohol product depends on the stereoselectivity of the enzyme (*vide infra*).

ADHs are nicotinamide-dependent enzymes that rely on external nicotinamide adenine dinucleotide (phosphate) cofactor NAD(P)H as the reductant. Notably, the reaction is reversible (oxidation of alcohols by $NAD(P)^+$). This feature has important practical implications, especially when strategies are implemented for the recycling of the expensive cofactor, and overall, ADHs are well suited for the coupled-substrate approach (see Scheme 7.2). In such cases, isopropanol is usually used as an ultimate hydride source in reduction reaction; conversely, acetone may be employed for oxidation reactions.

ADHs are found mostly within four protein superfamilies [18–20]:

(i) short-chain dehydrogenases/reductases (SDR) show great variety in substrate acceptance and are, in general, nonmetal-dependent enzymes. However, in the case of ADHs from *Lactobacillus brevis* and *Lactobacillus kefir*, magnesium was found necessary to retain activity because of structural reasons (i.e. no role in catalysis) [21, 22]. (ii) Medium-chain dehydrogenases/reductases (MDR) are usually zinc-dependent enzymes and employ the metal as a Lewis acid for activation of the carbonyl group; for these enzymes, the necessary divalent metal can be added to the reaction mixture in the form of $ZnCl_2$ salt to ensure saturation and maximum catalytic activity of the enzymes, as loss or chelation of the metal may happen during biocatalyst preparation and/or the reaction. (iii) Long-chain dehydrogenases/reductases (LDR) constitute a relatively heterogeneous protein family and accept substrates such as polyols, including many kinds of sugars; metal dependency is not conserved throughout this family [23]. (iv) Aldo-keto reductases (AKR) are often involved in metabolic reactions of endogenous compounds and xenobiotics and, as a consequence, tolerate a wide range of substrates [24].

7.2.1.1 Selection of Alcohol Dehydrogenase (ADH) for Stereoselective Reduction Reactions

The selection of a suitable ADH for the targeted reaction must take into account several factors that are connected to the structure of the carbonyl compound to be reduced and/or the reaction conditions [25].

7.2.1.1.1 Absolute Configuration of the Product

The binding of the substrate in the enzyme active site will determine the absolute configuration of the product because the relative positioning of the carbonyl group *vis-à-vis* NAD(P)H will dictate the side of attack by the hydride (*Si*- or *Re*-face of the carbonyl). For a large number of ADHs, the use of the "Prelog's rule" [26] allows the successful prediction of the stereochemical outcome of the reaction (see Scheme 7.3). Although most ADHs are known to deliver "Prelog"-type alcohols (attack on the *Re*-face of the carbonyl to deliver the (*S*)-alcohol[1]), a significant number of ADHs with opposite stereopreference have emerged over the past decade [27], mostly because of the advances in genome mining [28] and access to new sources of enzymes. This has opened up the opportunity for stereocomplementary reductive processes. ADHs from *L. brevis* [29] and *L. kefir* [30] are two examples of "anti-Prelog" ADHs. A selection of common ADHs displaying "Prelog" and "anti-Prelog" selectivity is displayed in Table 7.1.

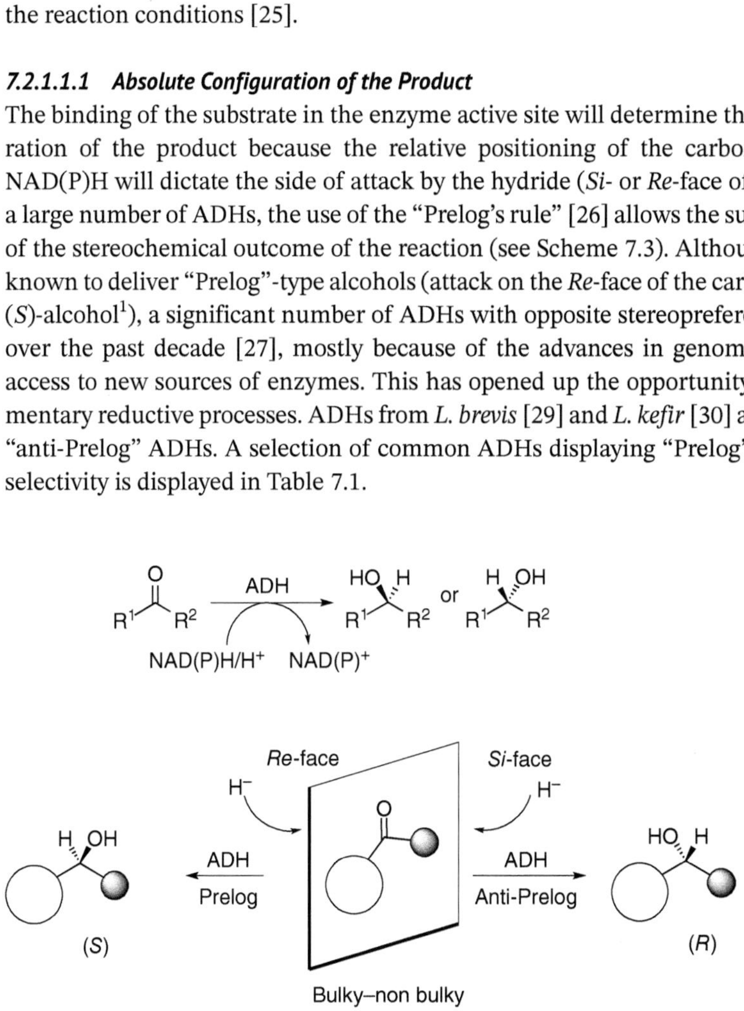

Scheme 7.3 Stereoselective reduction of carbonyl compounds with an alcohol dehydrogenase (ADH). ADHs can follow "Prelog" or "anti-Prelog" rule, thereby producing opposite enantiomers, according to their stereoselectivity. H^- denotes attack of a hydride from NAD(P)H.

[1]The (*S*)-alcohol is obtained under the Prelog rule provided the smaller substituent has the lower priority according to the Cahn–Ingold–Prelog (CIP) sequence rule, which is used to name stereoisomers with fixed absolute configuration.

Table 7.1 Selection of common ADHs with "Prelog" or "anti-Prelog" selectivity. Corresponding Uniprot accession number and PDB number (in the case where a crystal structure is available) are indicated, as well as the cofactor preference.

"Prelog"-ADH			"anti-Prelog" ADH		
Source (organism)	**Uniprot (PDB [a]) accession number**	**Cofactor**	**Source (organism)**	**Uniprot (PDB) accession number**	**Cofactor**
Horse liver [31] [b]	E chain: P00327 (PDB 2JHF) S chain: P00328 (PDB 1EE2)	NADH	*Candida parapsilosis* [32]	B2KJ46 (PDB 3CTM)	NADPH
Rhodococcus ruber (ADH-A) [33]	Q8KLT9 (PDB 3JV7)	NADH	*Candida magnoliae* [34]	Q9C4B3	NADPH
Rhodococcus erythropolis [35]	Q6YBW1	NADH	*Lactobacillus brevis* [29]	Q84EX5 (PDB 1ZK4)	NADPH
Saccharomyces cerevisiae [36]	Q12068 (PDB 4PVC)	NADPH	*Lactobacillus kefir* [30]	Q6WVP7 (PDB 4RF5)	NADPH
Thermoanaerobacter brockii [37]	P14941 (PDB 3FPC)	NADPH	*Leifsonia* sp. [38]	Q4R1E0	NADH

a) PDB, Protein Data Bank (freely accessible under: https://www.rcsb.org).
b) HLADH can exist as a dimer of chain E (isoenzyme EE, active on alcohols), chain S (isoenzyme SS, active on steroids and alcohols), or a combination of both (isoenzyme ES) [29].

7.2.1.1.2 Substrate Type

(i) Bulky non-bulky substrates

Most ADHs naturally display stereopreference on prochiral carbonyl compounds that present major differences in size between the two substituents (R^1 and R^2, see Scheme 7.3) [39].

(ii) Bulky–bulky substrates

Because of the lower discrepancy between the size of the two substituents of the prochiral molecule (R^1 and R^2, see Scheme 7.3), bulky–bulky compounds, such as diaryl ketones, are typically more challenging to reduce with high stereoselectivity [40]. A few ADHs have gained notoriety in the reduction of bulky–bulky ketones and include ADH from *Ralstonia* sp. DSM 6428 [41], ADH from *Sphingobium yanoikuyae* DSM 6900 [42, 43], or ADH from *Sporobolomyces salmonicolor* AKU442 [44].

With the combined advances in molecular biology, genetic engineering, and *in silico* study of proteins, enzymes, which do not present the desired substrate specificity or stereopreference, can now be engineered "on demand," i.e. the biocatalyst is tuned to fit a desired application/substrate. Some prominent examples from the industry are available [45] (see Section 7.2.1.2), while other proof-of-concepts at the laboratory scale have also demonstrated the excellent tunability of ADHs via protein engineering [46–49]. ADH from *Kluyveromyces polysporus*, for instance, naturally accepts bulky–bulky ketones, but the enantiopurity of some products thereby obtained was not satisfactory [50–52]. Applying (iterative) combinatorial saturation mutagenesis, the stereoselectivity of the enzyme could be significantly improved on a range of diaryl ketones, and on some substrates, a switch of stereopreference was also attained. ADH from *Ralstonia* sp. was also successfully engineered to catalyze the reductive desymmetrization of prochiral 2,2-disubstituted-1,3-cyclodiketones and allowed the production of a single isomer from all four stereoisomers possible through reduction of one single carbonyl group in a stereoselective fashion [49].

7.2.1.1.3 Thermostability

In some cases, higher reaction temperatures may be required, such as with poorly soluble substrates or to prevent bacterial contamination during the process. A number of ADHs have been isolated from (hyper)thermophilic organisms [53] and were shown to display remarkable thermostability (characterized by a high melting temperature $-T_m$ - up to 90–95 °C). This feature allows the use of such biocatalysts at elevated temperatures, which can be of major advantage [54] because of higher reaction rates. ADH from *Sulfolobus solfataricus* could be employed at 80 °C [55], while ADH from *Thermoanaerobacter brockii* can sustain operational temperatures up to 85 °C. Although the thermotolerance of ADH-A from *Rhodococcus ruber* is not as high, the enzyme shows a half-life of 35 h at 50 °C [10], and conveniently, the protein can be partially purified by heat precipitation at 65 °C [56].

7.2.1.1.4 Cofactor Preference

A key information necessary before setting up a bioreduction using ADH is the cofactor preference of the enzyme for NADPH or NADH (see Table 7.1). This property can be altered and preference for the other cofactor may be engineered. This may be relevant when a particular cofactor recycling strategy is envisaged, which strongly favors one form of the cofactor. The cofactor specificity of NADPH-dependent ADH from *Ralstonia* sp. was successfully altered by rational design and the resulting variant, displaying four to six mutations, eventually preferred NAD^+ over $NADP^+$ (reaction only tested in the oxidative direction) [57].

7.2.1.1.5 Kits

Several commercial suppliers of enzymes exist and provide screening kits containing diverse and complementary ADHs. These kits are straightforward to implement [58, 59] as the enzymes are typically provided as lyophilized cell-free extracts (i.e. powders that can be stored between −20 and +4 °C). They should be preferred by the nonspecialists to identify the suitable biocatalyst for a particular reaction.

7.2.1.2 Practical Approach

The successful implementation of ADHs in a number of industrial processes is mostly connected to their robustness in organic synthesis [25, 60]. Several ADHs are known to be tolerant to rather high concentration of organic solvents [61]. This feature becomes particularly practical for cofactor regeneration in the coupled-substrate approach, in which ADH concurrently reduces the substrate of choice while it oxidizes a cheap alcohol such as isopropanol, which serves not only as the ultimate hydride source but also as an organic cosolvent. Concentrations up to 20–50 vol% are frequently encountered, such as with

ADH-A, the naturally occurring solvent-tolerant ADH from *R. ruber* [10, 62]. The oxidation of this co-substrate delivers acetone, which can be stripped out of the reaction vessel and thus does not accumulate. ADH-A was shown to also operate under micro-aqueous conditions (up to 99 vol% hexane) [56]. Going in hand with this property, high substrate concentrations may also be employed, and values up to the molar range are not rare [51, 63]. In most applications, however, protein engineering is implemented in order to allow the enzyme to operate under rather extreme conditions (elevated temperature and high substrate concentration).

In this section, selected examples applied in industry are presented as well as systems that employ ADHs beyond C=O reduction reactions. For a larger sampling of bioreduction reactions developed in synthesis, the reader is referred to some comprehensive contributions [5, 39, 64, 65].

7.2.1.2.1 Montelukast

Merck has developed a chemoenzymatic process for the synthesis of montelukast, an anti-asthmatic agent, to supplant the previously existing fully chemical route, which relied on the asymmetric reduction of a ketone precursor by an asymmetric reagent ((−)-*B*-chlorodiisopinocampheylborane, (−)-Ipc_2BCl). The targeted substrate for bioreduction was a bulky–bulky ketone presenting several other functionalities ((*E*)-methyl 2-(3-(3-(2-(7-chloroquinolin-2-yl)vinyl)phenyl)-3-oxopropyl)benzoate), and from the initial panel of Codexis KRED enzymes screened, several candidates showed high stereoselectivity for the ketone but low levels of activity. After a few rounds of directed evolution, the best variant displayed not only a 3000-fold improved activity but was also stable under process conditions, presenting tolerance to high concentrations of substrate and organic solvent (up to 60 vol%) and robustness at elevated temperatures. Isopropanol was used as the organic cosolvent (together with toluene) and as the hydride source (coupled-substrate approach), allowing the regeneration *in situ* of NADPH from $NADP^+$. Finally, the process could be scaled up (230 kg). The precipitation of the product during the reaction drove the reaction to completion and a yield of 97.2% along with *ee* > 99.9% for the (*S*)-alcohol were obtained (see Scheme 7.4) [66].

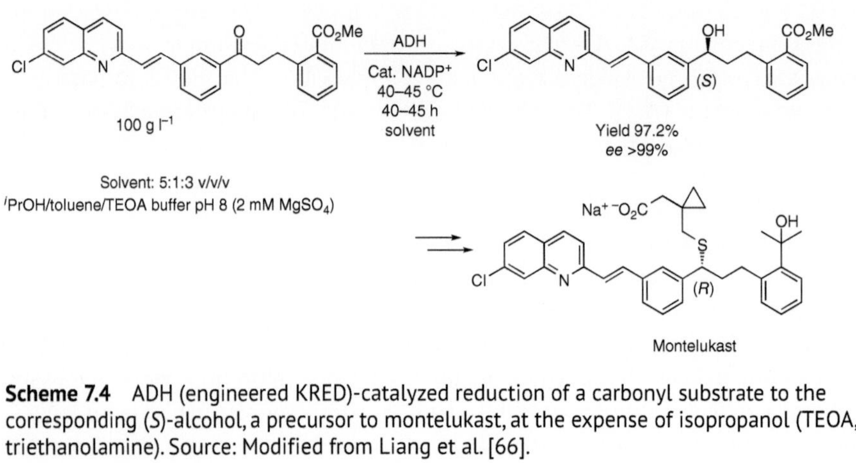

Cl N O CO_2Me
100 g l^{-1}
ADH
Cat. $NADP^+$
40–45 °C
40–45 h
solvent
Cl N OH CO_2Me (*S*)
Yield 97.2%
ee >99%
Solvent: 5:1:3 v/v/v
^{i}PrOH/toluene/TEOA buffer pH 8 (2 mM $MgSO_4$)
$Na^+ {}^-O_2C$ S OH Cl N (*R*)
Montelukast

Scheme 7.4 ADH (engineered KRED)-catalyzed reduction of a carbonyl substrate to the corresponding (*S*)-alcohol, a precursor to montelukast, at the expense of isopropanol (TEOA, triethanolamine). Source: Modified from Liang et al. [66].

7.2.1.2.2 Atorvastatin

Several biocatalytic strategies have been developed for the synthesis of atorvastatin, a blockbuster drug used to treat high level of cholesterol. The key part of the molecule is the statin chain, which contains the chiral information required for biological activity in the form of two secondary alcohol groups. In one route developed by Codexis, a bi-enzymatic cascade was applied to the transformation of 4-chloroacetoacetate into (*R*)-4-cyano-3-hydroxybutyrate (see Scheme 7.5). The reaction involved the action of an ADH (engineered KRED) for the reduction to (*S*)-4-chloro-3-hydroxybutyrate and a halohydrin dehydrogenase for the subsequent step (enzymatic dehalogenation–epoxidation/chemical cyanolysis). The reaction was scaled up to 240 g and led over two steps to the formation of (*R*)-4-cyano-3-hydroxybutyrate in >99.5% *ee* after 50 h reaction time [67, 68].

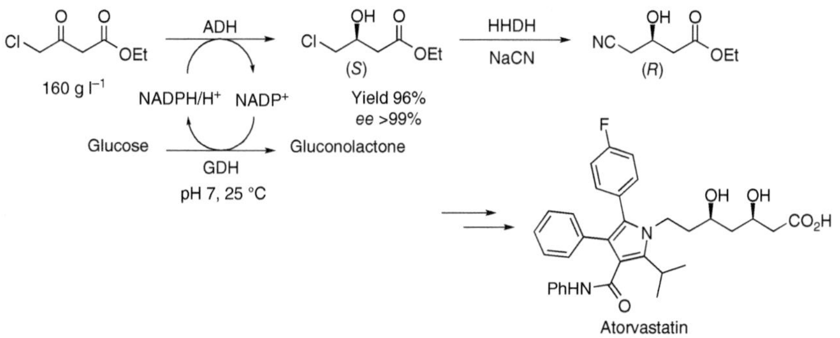

Scheme 7.5 Bi-enzymatic sequence toward the formation of a chiral precursor to atorvastatin. ADH (engineered KRED)-catalyzed reduction of the β-keto ester relies on glucose as a reductant (coupled-enzyme approach with glucose/glucose dehydrogenase (GDH)). Second step not detailed for clarity (HHDH, halohydrin dehydrogenase). Source: Modified from Fox et al. [67].

In a complementary route, ADH from *L. brevis* was used for the stereoselective reduction of the precursor *tert*-butyl-6-chloro-3,5-dioxohexanoate to (*S*)-6-chloro-5-hydroxy-3-oxohexanoate. Isopropanol was employed as the final reductant and the reaction was finally run on a 100 g scale [69].

7.2.1.2.3 Dynamic Kinetic Resolutions

The conversion of α-substituted chiral carbonyl compounds by ADHs may proceed with high stereoselectivity and enantioselectivity and thereby generate products with two chiral centers with high *de* and *ee* values (four stereoisomers possible) [70]. Because the α-proton is labile under aqueous conditions at close to neutral pH, *in situ* racemization takes place (see Scheme 7.6), and a dynamic kinetic resolution (DKR) leads to high conversion levels (up to >99% attainable with highly stereoselective enzymes). A number of examples have been reported in the literature, such as for the generation of one stereoisomer of α-alkyl-β-hydroxy esters [71], α-alkyl-β-hydroxy ketones, α-amino-β-hydroxy esters, or variously substituted halohydrins [65, 70, 72].

In the case of α-substituted aldehydes, DKR under reductive conditions delivers enantiopure α-substituted primary alcohols. This strategy was elegantly applied by the group of Berkowitz to access (*S*)-profenols [55]. A thermostable ADH from *S. solfataricus* was

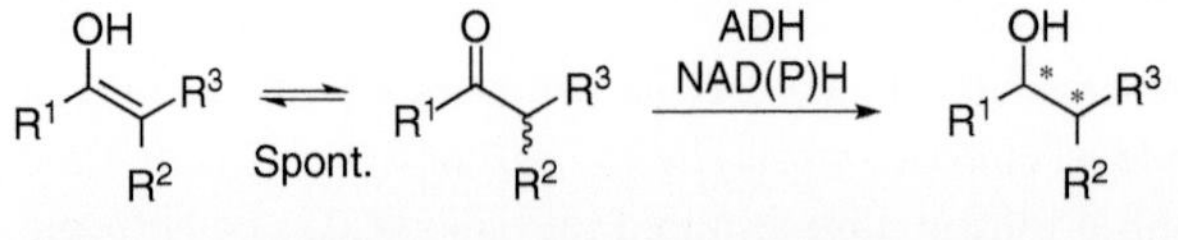

Scheme 7.6 General concept for the dynamic kinetic resolution of α-substituted chiral carbonyl compounds ($R^2 \neq R^3$) by a stereoselective ADH. Spontaneous racemization occurs under the selected reaction conditions. Stereorecognition of the chiral center ensures high *de* ($R^1 \neq H$) and *ee* values.

employed for the stereoselective reduction of a series of 2-arylpropionaldehydes under DKR conditions (basic pH of 9 was selected). Ethanol was used as the hydride source at low concentration (5 vol%) and the reaction was run at 80 °C, which contributed to dissolution of the substrates. After completion of the reaction, and upon cooling down, the products precipitated and could be recovered by filtration, allowing recycling of the enzyme kept in solution. High yields were obtained (up to 99%) along with excellent *ee* values (up to 99%). The reaction was run on the gram scale toward formation of the (*S*)-alcohol precursor of naproxen (see Scheme 7.7).

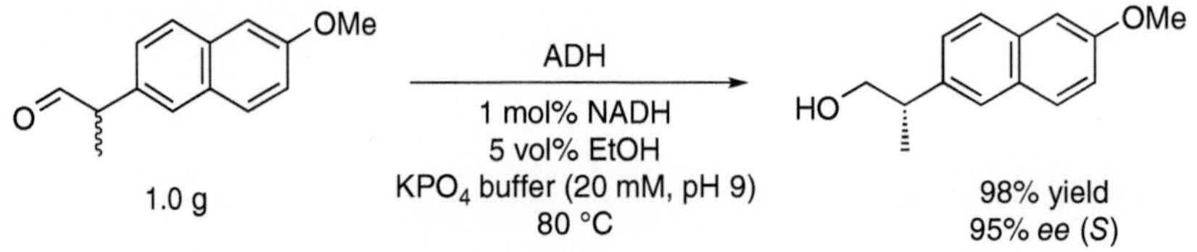

Scheme 7.7 Dynamic reductive kinetic resolution of *rac*-2-(6-methoxynaphthalen-2-yl)propanal by ADH from *Sulfolobus solfataricus* toward the alcohol precursor of naproxen with ethanol as the final reductant. Source: Modified from Friest et al. [55].

7.2.1.2.4 Disproportionation

Building on the spontaneous racemization of α-substituted aldehydes (see Scheme 7.6, $R^1 = H$), a novel system was designed for the parallel interconnected dynamic asymmetric transformation of a series of racemic 2-arylpropanals. ADH from horse liver (HLADH) catalyzed the disproportionation of aldehydes, thereby delivering a 1 : 1 mixture of corresponding alcohols and carboxylic acids [73]. (*S*)-Profens and profenols were thereby obtained with high enantiopurity (up to 99% *ee*) and high conversion levels. The major advantage of this process is the absence of co-substrate and the generation of no by-product. NAD^+ was used in catalytic amounts and functioned as a hydride shuttle between two molecules of aldehyde, which were concurrently oxidized and reduced (see Scheme 7.8). The reaction was scaled up to 100 mg with *rac*-2-phenylpropanal and yielded, after purification, 31 mg of (*S*)-2-phenylpropanol in 97% *ee* and 35 mg of (*S*)-2-phenylpropanoic acid in 86% *ee* in overall 62% isolated yield (see Section 7.4.1 for experimental details).

7.2.1.2.5 Redox Isomerization

Recently, an elegant redox isomerization protocol has been developed for Achmatowicz pyranones, which were found good substrates for ADHs. The sequence started by the

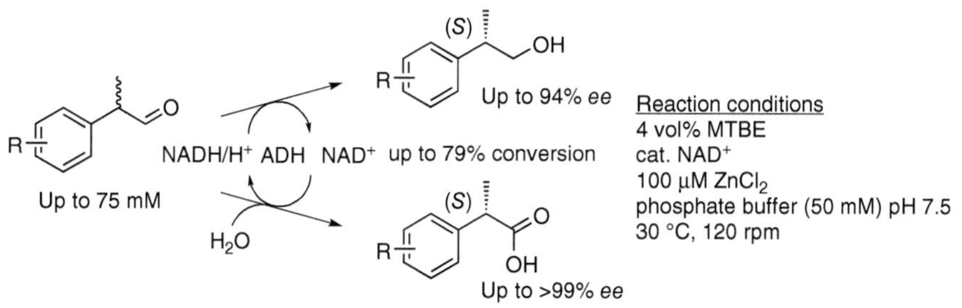

(S) OH

Up to 94% ee

NADH/H+ ADH NAD+ up to 79% conversion

Up to 75 mM

H_2O

(S) O OH

Up to >99% ee

Reaction conditions
4 vol% MTBE
cat. NAD^+
100 μM $ZnCl_2$
phosphate buffer (50 mM) pH 7.5
30 °C, 120 rpm

Scheme 7.8 Disproportionation of *rac*-2-arylpropanals catalyzed by horse liver ADH (HLADH) for the generation of (*S*)-profens and profenols. Racemization of the substrate occurs spontaneously under the reaction conditions (MTBE: Methyl *tert*-butyl ether). Source: Modified from Tassano et al. [73].

oxidation of the free hydroxy group, generating the corresponding γ-oxo-lactones, which were reduced in the subsequent step with high stereoselectivity to yield enantiopure γ-hydroxylated δ-lactones as final products. The oxidation – stereoselective reduction sequence relied on an intramolecular hydride shift and was overall redox-neutral, allowing the use of catalytic amounts of the nicotinamide cofactor in the form of $NADP^+$ (see Scheme 7.9) [74]. Although commercial screening kits were used initially and led to the formation of either (*R*)- or (*S*)-products, ADH from *L. kefir* was also investigated in recombinant form as preparation of *E. coli* resting cells yielding the (*S*)-products.

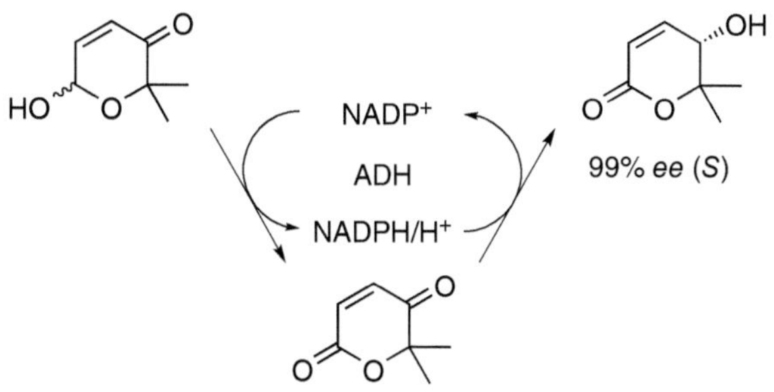

HO O O

$NADP^+$

ADH

NADPH/H^+

OH

99% ee (S)

Scheme 7.9 Redox isomerization of Achmatowicz pyranones to enantiopure γ-hydroxylated δ-lactones via oxidation – stereoselective reduction sequence catalyzed by ADH. Catalytic amounts of $NADP^+$ are employed in this overall redox-neutral cascade. Source: Modified from Liu et al. [74].

7.2.2 Reduction of C=C Bonds

7.2.2.1 Mechanism

Ene-reductases catalyze the asymmetric reduction of activated alkenes at the expense of NAD(P)H in a *trans*-fashion [75–78], thereby offering attractive synthetic strategies for the generation of chiral molecules complementary to established *cis*-specific homogeneous hydrogenation protocols [79, 80]. Most ene-reductases belong to the family of old yellow enzymes (OYEs), which are flavin mononucleotide (FMN)-dependent proteins found in bacteria, plants, fungi, and protozoa. The reaction proceeds according to a ping-pong bi-bi mechanism: (i) NAD(P)H first reduces the coenzyme FMN, which is noncovalently (usually tightly) bound in the enzyme active site and exists under aerobic conditions in the oxidized form (reductive half-reaction); (ii) NAD(P)$^+$ exits the active site and is replaced by the substrate, which is an alkene activated by an electron-withdrawing group (*vide infra*); (iii) the reduced FMN stereoselectively transfers a hydride onto Cβ of the α,β-unsaturated compound in a Michael-type manner (oxidative half-reaction); (iv) protonation at Cα is performed in an anti-fashion, in most cases by a conserved tyrosine; (v) the reduced product finally leaves the active site, leaving behind FMN back in its oxidized state and ready for the next catalytic cycle. The tyrosine eventually gets reprotonated from the solvent (see Scheme 7.10). The occurrence of a large number of diverse OYE homologs active in C=C bond reduction reactions has contributed to the development of stereoselective and stereocomplementary strategies for a broad range of substrates (*vide infra*); other enzymes known to catalyze the reduction of activated alkenes (e.g. flavin-dependent iron–sulfur containing enoate reductases, SDR, and MDR) will not be discussed here [81].

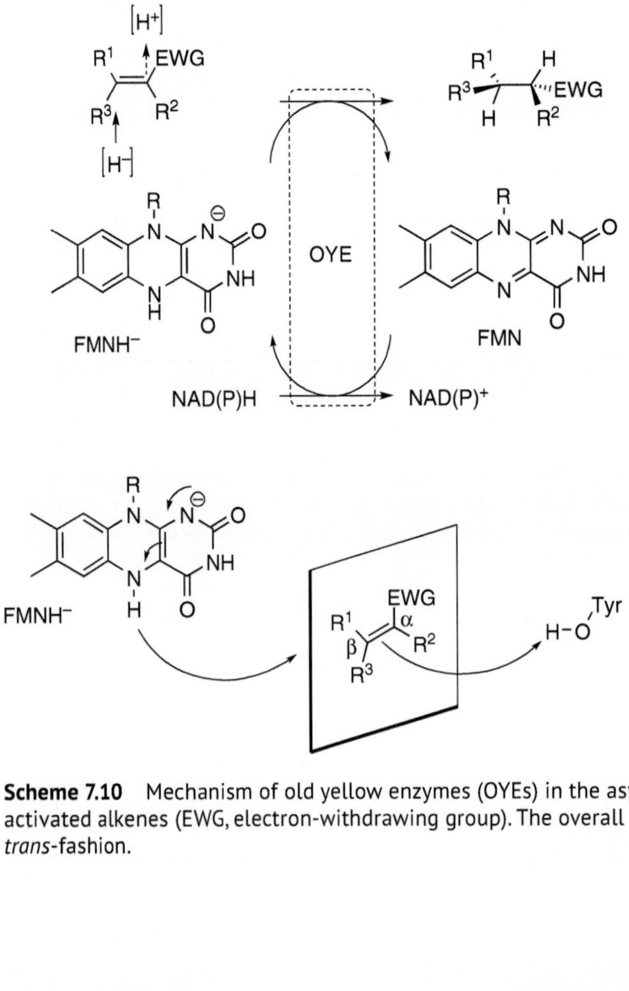

Scheme 7.10 Mechanism of old yellow enzymes (OYEs) in the asymmetric reduction of activated alkenes (EWG, electron-withdrawing group). The overall addition of [2H] proceeds in a *trans*-fashion.

7.2.2.2 Enzymes and Substrates

7.2.2.2.1 Enzymes

The number of identified OYE homologs has increased dramatically over the past decade, and the family now includes highly diverse proteins from various kingdoms. Yeast OYEs, historically the first identified members of this family [82, 83], are now complemented with bacterial and plant homologs, and even parasites were shown to express OYEs [84]. For many homologs, crystal structures are available (see Table 7.2). Common to all members of the family is the FMN dependency as well as a few conserved residues responsible for binding of the substrate (usually in the form of a His/Asn or His/His pair) and for protonation (typically a tyrosine – only morphinone reductase is known to have a cysteine in place of the proton donor [86]). The cofactor preference is enzyme dependent and most OYEs, although displaying some differences in kinetic data (k_{cat} and K_M) with NADPH and NADH, typically accept these two cofactors equally well in biotransformation reactions [76, 85, 87]. Exception is again morphinone reductase, which is strictly specific for NADH [88].

7.2.2.2.2 Substrates

Activating Group A conjugated electron-withdrawing group is required for activity. Its presence reduces the electron density of the C=C double bond, thereby rendering the alkene more susceptible to hydride attack. The electron-withdrawing group is also

Table 7.2 OYE homologs from various sources most frequently applied in the bioreduction of activated alkenes. Corresponding Uniprot accession number and PDB number (in the case where a crystal structure is available) are indicated (see [85] for recent survey of most members of the family).

Name	Source (organism)	Uniprot (PDB) accession number
OYE1	*Saccharomyces pastorianus*	Q02899 (PDB 1OYA)
OYE2	*Saccharomyces cerevisiae*	Q03558
OYE3	*Saccharomyces cerevisiae*	P41816 (PDB 5V4P)
OYE	*Candida macedoniensis*	Q6I7B7 (PDB 4TMB)
OYE 2.6	*Pichia stipitis* CB5064	A3LT82 (PDB 3UPW)
YqjM	*Bacillus subtilis*	P54550 (PDB 1Z41)
NCR	*Zymomonas mobilis*	Q5NLA1 (PDB 4A3U)
Morphinone reductase	*Pseudomonas putida* M10	Q51990 (PDB 2R14)
PETN reductase	*Enterobacter cloacae* PB2	P71278 (PDB 1GVO)
TOYE	*Thermoanaerobacter pseudoethanolicus*	B0KAH1 (PDB 3KRU)
XenA	*Pseudomonas putida* II-B	Q3ZDM6 (PDB 3N19)
OPR1	*Solanum lycopersicum*	Q9XG54 (PDB 1ICP)
OPR3	*Solanum lycopersicum*	Q9FEW9 (PDB 2HSA)

responsible for anchoring the substrate in the enzyme active site by hydrogen bonding with the His/Asn or His/His pair. A range of electronically activated alkenes has been investigated, and from the large body of data obtained in bioreduction reactions, four distinct substrate groups appear (see Scheme 7.11):

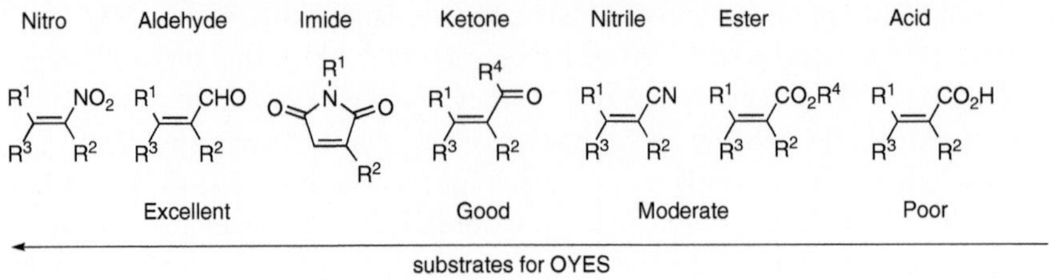

Nitro | Aldehyde | Imide | Ketone | Nitrile | Ester | Acid

Excellent | Good | Moderate | Poor

substrates for OYES

Scheme 7.11 Overview of activated α,β-unsaturated compounds classified according to substrate acceptance by old yellow enzyme (OYE) homologs (substitution arbitrary).

- α,β-Unsaturated nitro-compounds and aldehydes, as well as maleimides, are highly activated and thus excellent substrates for most OYEs. With α-substituted nitro-compounds, strong racemization of the resulting products may be avoided through reaction engineering (e.g. reaction at low temperature and slightly acidic pH) [89].
- α,β-Unsaturated ketones are good substrates; however, for these compounds, the substitution pattern has a major influence on the activity level (see the section on *Substitution Pattern and Alkene Configuration*).
- α,β-Unsaturated nitriles and esters are only moderately activated and thus substrate acceptance is highly enzyme dependent; further activation may be provided by an adjacent halogen atom [90].
- Finally, carboxylic acids are poorly activated and overall poor substrates and can only be converted in very specific cases [91].

Activated alkynes can be converted and generate (*E*)-alkenes because of the *trans*-specificity of the reaction. Examples are rare and specific to alkynones, such as 4-phenylbut-3-yn-2-one [92]. The reaction typically proceeds to the fully saturated product, provided enough reducing equivalents are available.

Isolated C═C bonds remain inert to OYEs, and this feature allows the regioselective reduction of polyunsaturated compounds, such as citronellal [93].

Substitution Pattern and Alkene Configuration In theory, the reduction of activated alkenes can generate up to two chiral centers; however, such highly substituted compounds have been rarely investigated. A few published examples have indicated that the corresponding products can be obtained with high diastereoselectivity [94].

The stereoselective reduction of alkenes with disubstitution at Cα or Cβ generates enantiomeric products, whose absolute configuration depends on the binding of the substrate in relation to the reduced flavin. In addition, the substitution pattern determines the configuration of the alkene (*E* vs *Z*), which also influences the absolute configuration of the product at a given substrate binding mode.

All possible binding poses are summarized in Scheme 7.12. Poses A/C may be described as "up" binding mode, poses B/D as "down" binding mode, and a switch from "up" to

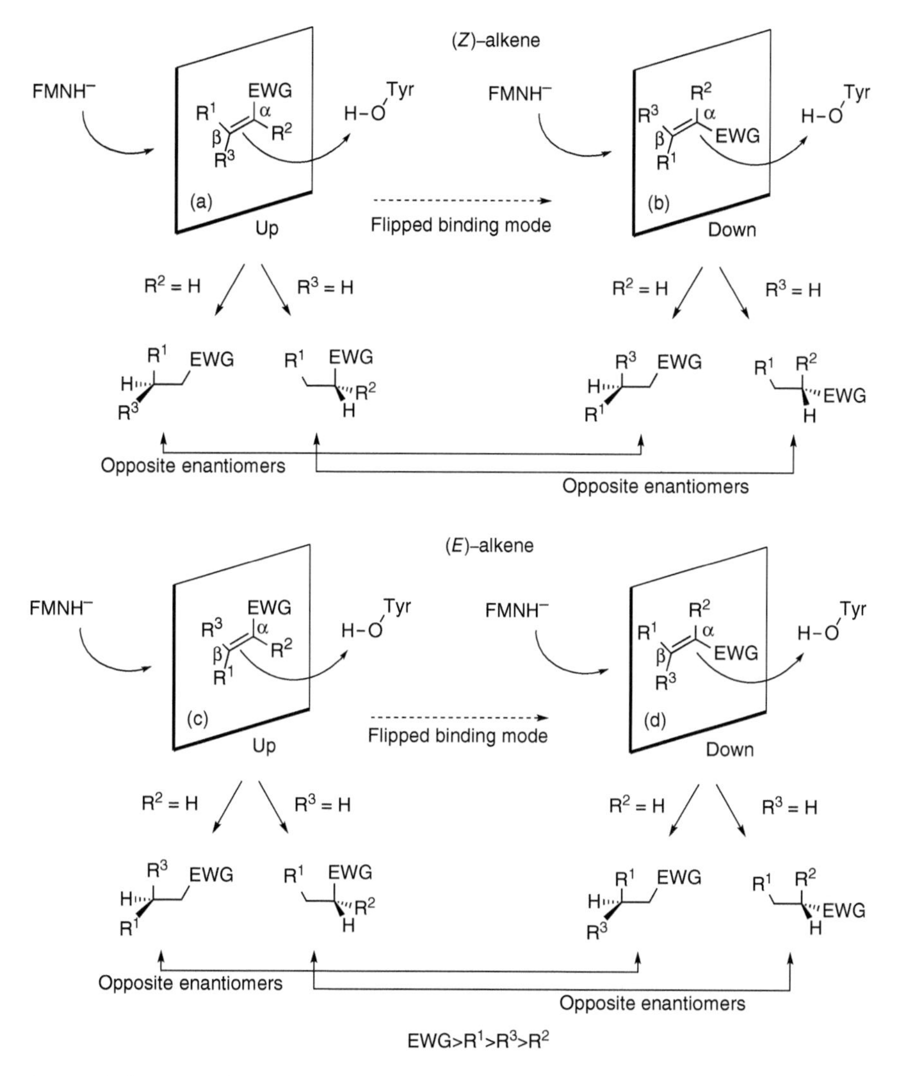

Scheme 7.12 Overview of all possible binding poses for trisubstituted alkenes depending on alkene configuration (*E* and *Z*) and positioning of the anchoring group (electron-withdrawing group, EWG) in the enzyme active site ("up"/"down" and flipped binding modes).

"down" implies a 180° flip of the substrate around the C=C bond axis. This is both substitution and enzyme dependent. For a given alkene, the two isomers may bind in a similar or flipped fashion (A/C or B/D). Thus, the following relationships between enantiomeric products will be obtained:

- α-substituted product (R^3 = H):
  Poses A and B yield opposite enantiomer with (*Z*)-alkene (flip)
  Poses C and D yield opposite enantiomer with (*E*)-alkene (flip)
  Poses A and C yield same enantiomer in "up"-mode (*Z* and *E*)
  Poses B and D yield same enantiomer in "down"-mode (*Z* and *E*)

- β-substituted product ($R^2 = H$):
  Poses A and C yield opposite enantiomer with "up"-mode (*Z* and *E*)
  Poses B and D yield opposite enantiomer with "down"-mode (*Z* and *E*)
  Poses A and D yield same enantiomer (*Z* and *E* and "up" to "down")
  Poses B and C yield same enantiomer (*Z* and *E* and "down" to "up")

With carbonyl compounds, rotamers may exist, which increases the number of possible binding modes [95].

Although a large part of the substrates investigated so far present only one functional group (the electron-withdrawing group), a trend toward bioreduction of chemically richer molecules is appearing [78]. A number of difunctionalized products have been obtained and, in some cases, the corresponding substrates displayed double activation, which overall favors the reduction [78, 96–98]. The presence of a halogen atom at either Cα or Cβ is more frequent; however, cases of dehydrohalogenation of the obtained (unstable) products have been monitored with halogen substitution at Cβ [99].

Both cyclic and acyclic substrates are accepted by OYEs, and cyclic enones and lactones in particular have been well studied [85, 87]. Larger substituents (e.g. aryl moiety) are well tolerated, in particular on Cβ [76].

7.2.2.3 Practical Approach

7.2.2.3.1 Stereocontrol

A major advantage of employing OYEs for the bioreduction of C=C double bonds is the possibility to design stereocomplementary strategies to provide access to both enantiomers of a given product. To that end, two methodologies can be employed: substrate- and enzyme-based stereocontrol [91, 95].

Substrate-Based Approach (See Scheme 7.13) As discussed above (see Scheme 7.12), the configuration of the substrate (*E* or *Z*) plays a major role and, combined with the two possible binding modes ("up" and "down"), may provide access to both enantiomers of α- and β-substituted products [87, 95]. For instance, this has been observed in the reduction of methyl (*E*) and (*Z*)-2-chloropentenoate by YqjM, which yielded the corresponding (*R*)- and (*S*)-products in 93% and 98% *ee*, respectively [90]. In that case, additional rotation of the

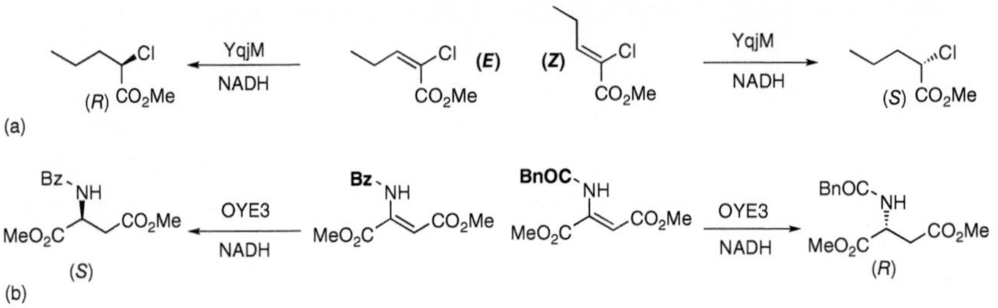

(a) (*R*) CO_2Me ← YqjM, NADH ← (*E*) CO_2Me Cl | (*Z*) CO_2Me Cl → YqjM, NADH → (*S*) CO_2Me Cl

(b) Bz-NH, MeO_2C, CO_2Me (*S*) ← OYE3, NADH ← Bz-NH MeO_2C CO_2Me | BnOC-NH MeO_2C CO_2Me → OYE3, NADH → BnOC-NH MeO_2C CO_2Me (*R*)

Scheme 7.13 Substrate-based stereocontrol via (a) alkene configuration (bioreduction of methyl (*E*) and (*Z*)-2-chloropentenoate by YqjM). Source: Modified from Tasnádi et al. [90] and (b) substituent effect (bioreduction of (*Z*)-dimethyl-2-benzamidofumarate and (*Z*)-dimethyl 2-(2-phenylacetamido)fumarate by OYE3). Source: Modified from Stueckler et al. [98].

alkene moiety around the Cα−C = O σ-bond (involving s-*cis*/*trans*-isomers) was highlighted through docking studies and explained the outcome of the reaction [95].

In addition, the size of the substituent may induce steric clashes with the neighboring amino acids in the enzyme active site and induce a flip of binding, thereby providing access to the opposite enantiomer. This has been mostly observed by changing the substitution at Cα [100]. With protected α,β-dehydroamino acid esters, the insertion of an additional CH_2 in the capping of the amino group (from benzamido to 2-phenylacetamido) was responsible for a change in the binding group (two ester groups are available), which led to opposite enantiomeric products after reduction with OYE3 ((*S*)-product in >99% *ee* and (*R*)-product in 92% *ee*, respectively) [98].

Enzyme-Based Approach (See Scheme 7.14) The broad protein diversity of the OYE family has, in several cases, contributed to the identification of stereocomplementary biocatalysts. OPR1 and OPR3 from *Solanum lycopersicum* are isoenzymes, which share 52% sequence identity and surprisingly similar crystal structures (RSMD 0.86 Å). Both enzymes showed high level of activity on (*E*)-1-nitro-2-phenylpropene with opposite stereopreference, yielding (*R*)- and (*S*)-1-nitro-2-phenylpropane, respectively, with high enantiopurity (up to 98% *ee*) [101].

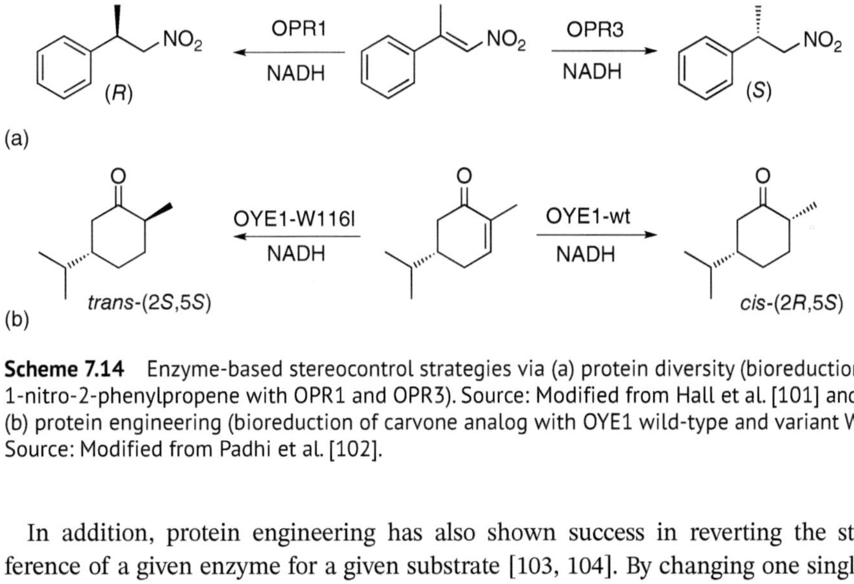

Scheme 7.14 Enzyme-based stereocontrol strategies via (a) protein diversity (bioreduction of 1-nitro-2-phenylpropene with OPR1 and OPR3). Source: Modified from Hall et al. [101] and (b) protein engineering (bioreduction of carvone analog with OYE1 wild-type and variant W116I). Source: Modified from Padhi et al. [102].

In addition, protein engineering has also shown success in reverting the stereopreference of a given enzyme for a given substrate [103, 104]. By changing one single amino acid in OYE1, the stereopreference on (*S*)-5-isopropyl-2-methyl-cyclohex-2-enone, a carvone analog, could be dramatically inverted, and conversion level even improved. While the wild-type enzyme delivered the *cis*-(2*R*,5*S*)-product in >98% *de* and 59% conversion, Trp116Ile variant generated the *trans*-(2*S*,5*S*)-product in >98% *de* and >98% conversion [102].

7.2.2.3.2 (Dynamic) Kinetic Resolution

Stereorecognition of a distant γ-chiral center allowed the reductive kinetic resolution of racemic substituted lactones. The degree of enantioselectivity was found dependent on the presence of a cosolvent, and highest enantioselectivities (E-values) were obtained with OYE3 in the reduction of *rac*-5-oxo-2,5-dihydrofuran-2-yl acetate in *t*-butyl methyl ether at 30 vol% (E-value up to 49) [105]. In the case of *rac*-3-methyl-5-phenylfuran-2(5*H*)-one, spontaneous racemization of the substrate under the reaction conditions, combined with high enantio- and stereoselectivity of NCR from *Zymomonas mobilis*, accounted for a case of DKR, yielding the final reduced product (3*R*,5*S*)-3-methyl-5-phenyldihydrofuran-2(3*H*)-one in 73% conversion with >99% *ee* and 96% *de* (see Scheme 7.15 and Section 7.4.2 for experimental details).

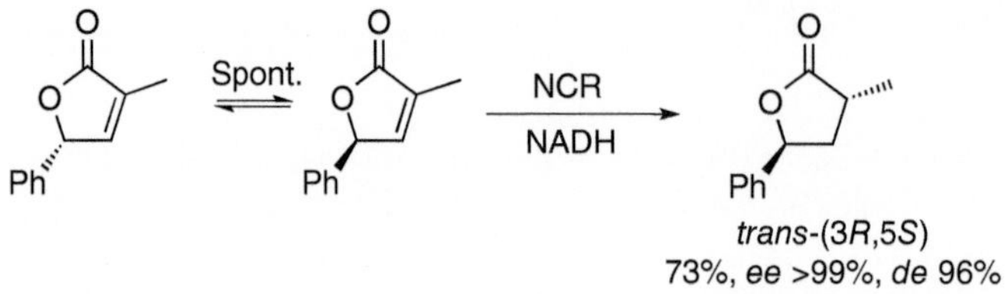

Spont.
NCR
NADH
trans-(3*R*,5*S*)
73%, *ee* >99%, *de* 96%

Scheme 7.15 Dynamic kinetic resolution by reduction of *rac*-3-methyl-5-phenylfuran-2(5*H*)-one by NCR toward the formation of *trans*-(3*R*,5*S*)-3-methyl-5-phenyldihydrofuran-2(3*H*)-one. Source: Modified from Turrini et al. [105].

Although ene-reductases from the OYE family appear well suited for the laboratory, only a few processes have been applied on large scale [106] because of the typically low substrate concentration tolerance of OYEs (10–20 mM). In addition, as opposed to the case with ADH (see Section 7.2.1.1), the absence of predictability in the substrate acceptance and stereopreference still implies that screening efforts are necessary to identify the biocatalyst suitable for a targeted reaction; this is rendered more straightforward with the availability of commercial screening kits. In the near future, protein engineering and dedicated efforts from various research groups may provide more robust enzymes suitable for industrial applications.

7.3 Oxidative Processes

7.3.1 Oxygenations

7.3.1.1 Baeyer–Villiger Oxidations

The enzymatic Baeyer–Villiger oxidation of carbonyl compounds is catalyzed by flavin-dependent monooxygenases, also denoted as Baeyer–Villiger monooxygenases (BVMOs). Although the reaction is an oxidation of a substrate, the biocatalyst requires a reductant in the form of nicotinamide cofactor. Indeed, only the reduced flavin can activate oxygen for the reaction, generating the reactive peroxyflavin intermediate, which is responsible for the nucleophilic attack of the carbonyl group (see Scheme 7.16a,b). In the course of the reaction, one oxygen atom is incorporated into the final product while the second oxygen atom is reduced to water. In the case of a poorly reactive substrate, the peroxyflavin may decompose to form hydrogen peroxide – this nonproductive decay is called uncoupling – leading to protein deactivation and/or side reactions. For this reason, in several cases, a catalase

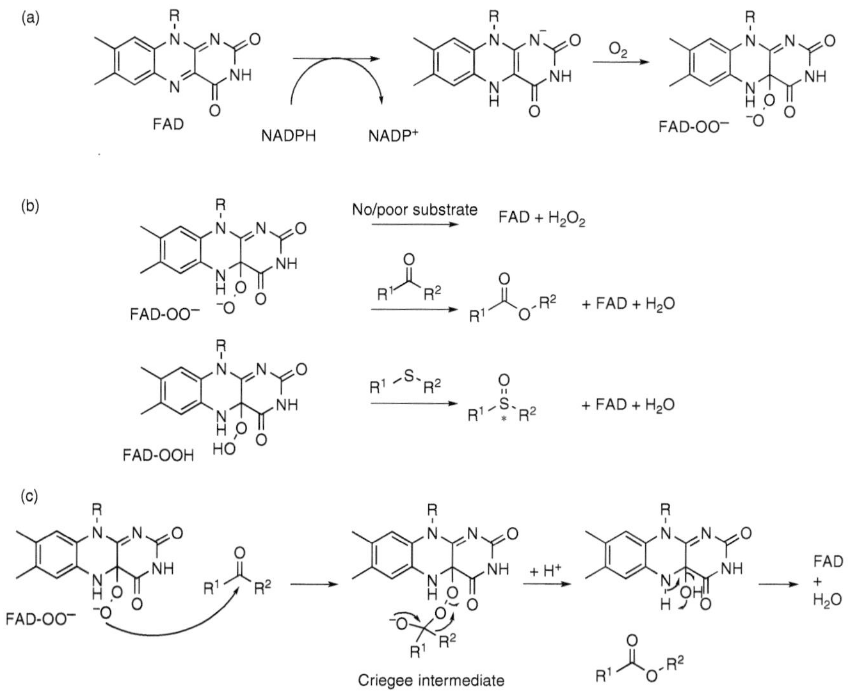

Scheme 7.16 Overview of the reactions of flavin-dependent monooxygenases. (a) Reduction of the flavin FAD by NADPH and activation of oxygen through attack by $FADH^-$ to form the reactive peroxyflavin ($FAD\text{-}OO^-$); (b) reactions possible with the (hydro)peroxyflavin (nonproductive decay with release of hydrogen peroxide, Baeyer–Villiger oxidation, and sulfoxidation); and (c) mechanism of the Baeyer–Villiger oxidation of carbonyl compounds via the peroxyflavin and the Criegee intermediate.

may be added to the reaction mixture in order to rapidly consume hydrogen peroxide via a dismutation reaction to water and oxygen.

Similar to the chemical Baeyer–Villiger reaction using alkylhydroperoxids or peracids, the mechanism of the biocatalyzed reaction involves formation of a Criegee intermediate, which collapses upon rearrangement, but in contrary to most chemical methods, two products can be obtained using enzymes: "normal" and "abnormal" esters or lactones (see Scheme 7.16c). The regiopreference of the enzyme determines the product ratio on a given substrate (*vide infra*) [107].

Two types of BVMOs are known: type I includes NADPH-specific flavin adenine dinucleotide (FAD)-dependent monooxygenases, while type II BVMOs are FMN-dependent and specific for NADH. Biocatalytic applications rely on type I BVMOs. Most BVMOs are categorized in the following types: cyclohexanone monooxygenase (CHMO), cyclopentanone monooxygenase (CPMO), phenylacetone monooxygenase (PAMO), cyclopentadecanone monooxygenase (CPDMO), 4-hydroxyacetophenone monooxygenase (HAPMO), and steroid monooxygenase (SMO), with prototypical enzymes listed in Table 7.3. Many more sources of BVMOs have been identified and each type includes several homologs [107, 115].

Table 7.3 Baeyer–Villiger monooxygenases (BVMO)[a)] frequently used for biocatalytic applications and their prototypical substrate.

Name	Source (organism)	Prototypical substrate
CHMO [108, 109]	*Acinetobacter calcoaceticus* NCIMB 9871 *Xanthobacter* sp. ZL5	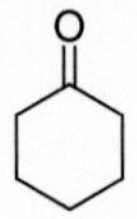
CPMO [110]	*Comamonas* sp. NCIMB 9872	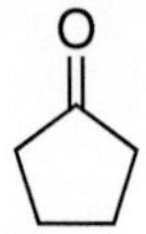
PAMO [111]	*Thermobifida fusca*	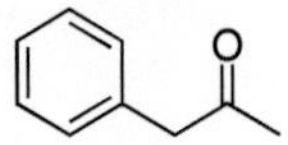
CPDMO [112]	*Pseudomonas* sp. HI-70	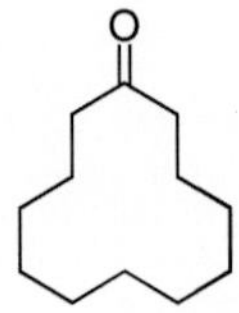
HAPMO [113]	*Pseudomonas putida* JD1	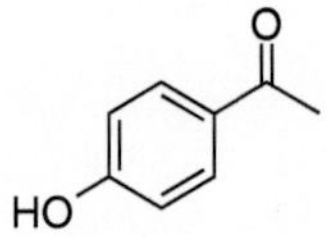 HO
SMO [114]	*Rhodococcus rhodochrous*	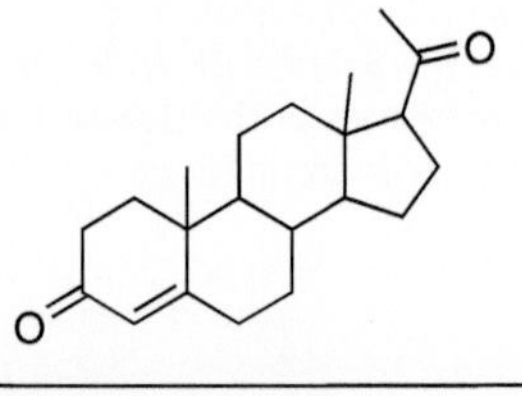

a) CHMO, cyclohexanone monooxygenase; CPMO, cyclopentanone monooxygenase; PAMO, phenylacetone monooxygenase; CPDMO, cyclopentadecanone monooxygenase; HAPMO, 4-hydroxyacetophenone monooxygenase; SMO, steroid monooxygenase.

Linear, alicyclic, and aromatic ketones are accepted [107, 116] and substrate preference is somewhat relaxed and dependent on the type of BVMO.

7.3.1.1.1 Regiopreference

The conversion of nonsymmetrical ketones by BVMOs leads to the formation of two possible regioisomers. The regiopreference is determined at the stage of the collapse of the Criegee intermediate (see Scheme 7.16c), and "normal" as well as "abnormal" esters or lactones may be obtained. As opposed to the preference for "normal" products in the chemical Baeyer–Villiger oxidation (i.e. rearrangement of R^2 over R^1 according to *tert*-alkyl > *sec*-alkyl ~ phenyl > *prim*-alkyl > methyl), both regiopreferences may be achieved with BVMOs depending on the substrate and the biocatalyst, but on an average, the normal lactones are still most commonly obtained. This feature can be modulated upon protein engineering: for instance, directed evolution applied to BVMO has allowed a complete switch of regiopreference in the Baeyer–Villiger oxidation of 4-phenyl-2-butanone with a change of ratio of abnormal/normal products from 1 : 99 to 98 : 2 [117].

7.3.1.1.2 Stereoselectivity

Most BVMOs display stereoselectivity in the oxygenation of carbonyl compounds and several strategies have been developed to obtain enantioenriched products through monooxygenation reactions (see Scheme 7.17).

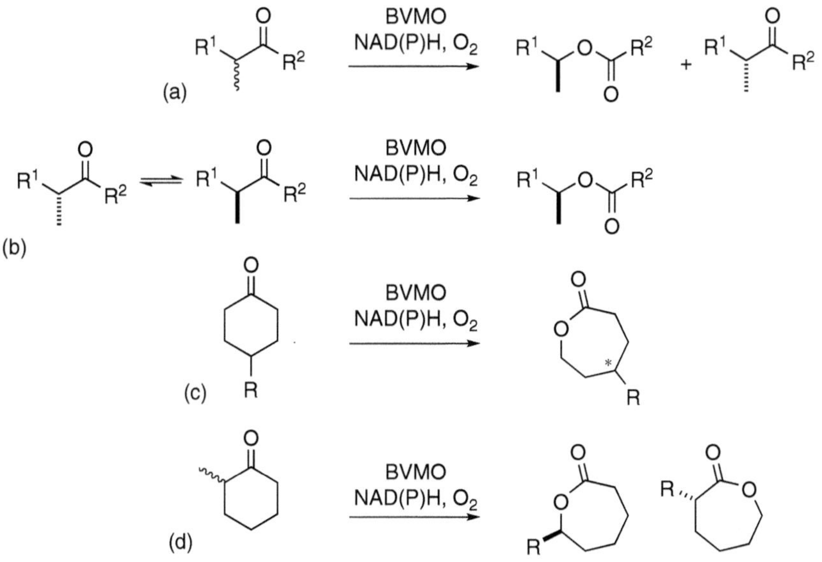

Scheme 7.17 Strategies employed on carbonyl compounds with Baeyer–Villiger monooxygenases (BVMO) to access enantiopure products (regiopreference arbitrary). (a) Kinetic resolution of racemic substrates; (b) dynamic kinetic resolution of racemic substrates; (c) desymmetrization of prochiral cyclic ketones; and (d) enantiocomplementary regiodivergent conversion of racemic substrates.

In a kinetic resolution, the enzyme shows enantiopreference on racemic chiral compounds, typically in the case where the chiral center is adjacent to the carbonyl [118]. Under particular reaction conditions, those substrates may racemize and the resulting DKR enables the generation of the products in high yields and optical purity. Next, prochiral carbonyl compounds can undergo desymmetrization in the case of a strong stereopreference by the biocatalyst. Finally, the two enantiomers of a chiral compound may be converted with regiodivergent preference [119], yielding an enantiocomplementary process with high conversion levels.

Under perfect kinetic resolution, several 2-substituted cyclopentanone and cyclohexanone derivatives [112, 120, 121], as well as (aryl) aliphatic ketones [122, 123], were converted with E-value >200 (see Scheme 7.18a). By working at basic pH (9–10) or in

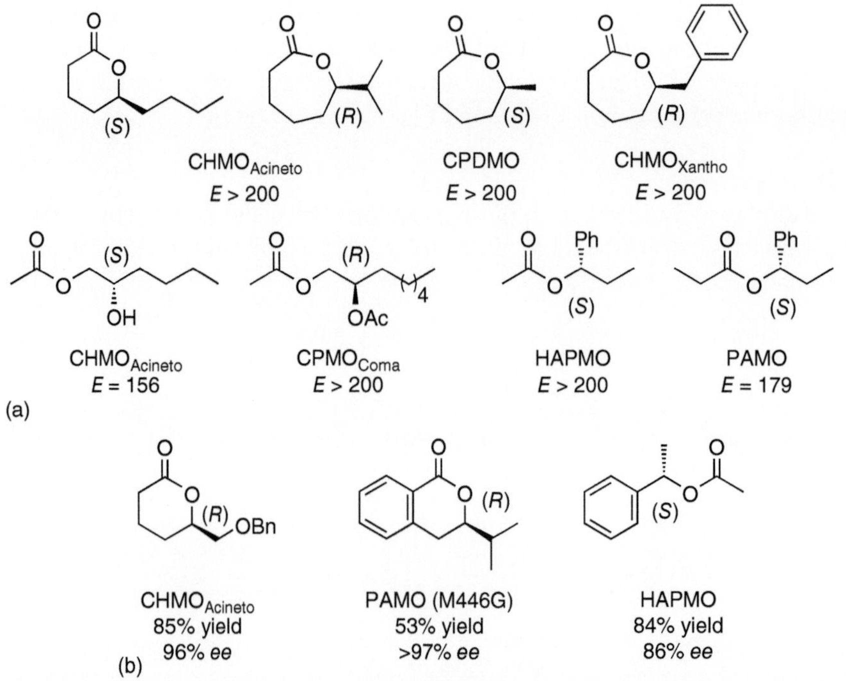

(S) (R) (S) (R)

$CHMO_{Acineto}$ $E > 200$ CPDMO $E > 200$ $CHMO_{Xantho}$ $E > 200$

(S) OH (R) OAc 4 Ph (S) Ph (S)

$CHMO_{Acineto}$ $E = 156$ $CPMO_{Coma}$ $E > 200$ HAPMO $E > 200$ PAMO $E = 179$

(a)

(R) OBn (R) (S)

$CHMO_{Acineto}$ 85% yield 96% *ee* PAMO (M446G) 53% yield >97% *ee* HAPMO 84% yield 86% *ee*

(b)

Scheme 7.18 Overview of products obtained by (a) kinetic resolution. Source: Refs. [112, 120–123] and (b) dynamic kinetic resolution of the corresponding ketones with BVMOs ($CHMO_{Acineto}$, cyclohexanone monooxygenase from *Acinetobacter calcoaceticus*; $CHMO_{Xantho}$, cyclohexanone monooxygenase from *Xanthobacter* sp. ZL5; $CPMO_{Coma}$, cyclopentanone monooxygenase from *Comamonas* sp.; and PAMO (M446G), variant of phenylacetone monooxygenase with mutation Met446Gly). Source: Refs. [124–126].

combination with a weekly basic anionic exchange resin [124–126], racemization of α-alkylated ketones can be promoted and leads to DKR. Such strategy was applied for instance to 2-substituted pentanones and benzyl ketones and to access substituted dihydroisocoumarins (see Scheme 7.18b and Section 7.4.3).

Prochiral carbonyl compounds, usually substituted cyclic ketones, can undergo desymmetrization with stereoselective BVMOs, thereby providing access in high enantiopurity to chiral lactones that are otherwise difficult to obtain because of the remote location of the chiral center compared to the reactive functional group. That way, several 4-substituted [109, 127–129], 3,5-dimethyl-substituted [130, 131], and 4,4-disubstituted cyclohexanones [109], as well as 3-substituted cyclobutanones [132], and many more prochiral ketones [107] were successfully converted to enantioenriched products, and in several cases, stereocomplementary enzymes were identified (see Scheme 7.19a).

Regiodivergent and enantiocomplementary conversion of racemic chiral cyclic ketones may lead to the generation of two regioisomeric products with high enantiopurity (see Scheme 7.19b) [133, 134]. If the strategy is applied separately to the two enantiomers in pure form, the two optically pure regioisomers may be easily obtained without the need for separation and purification [135].

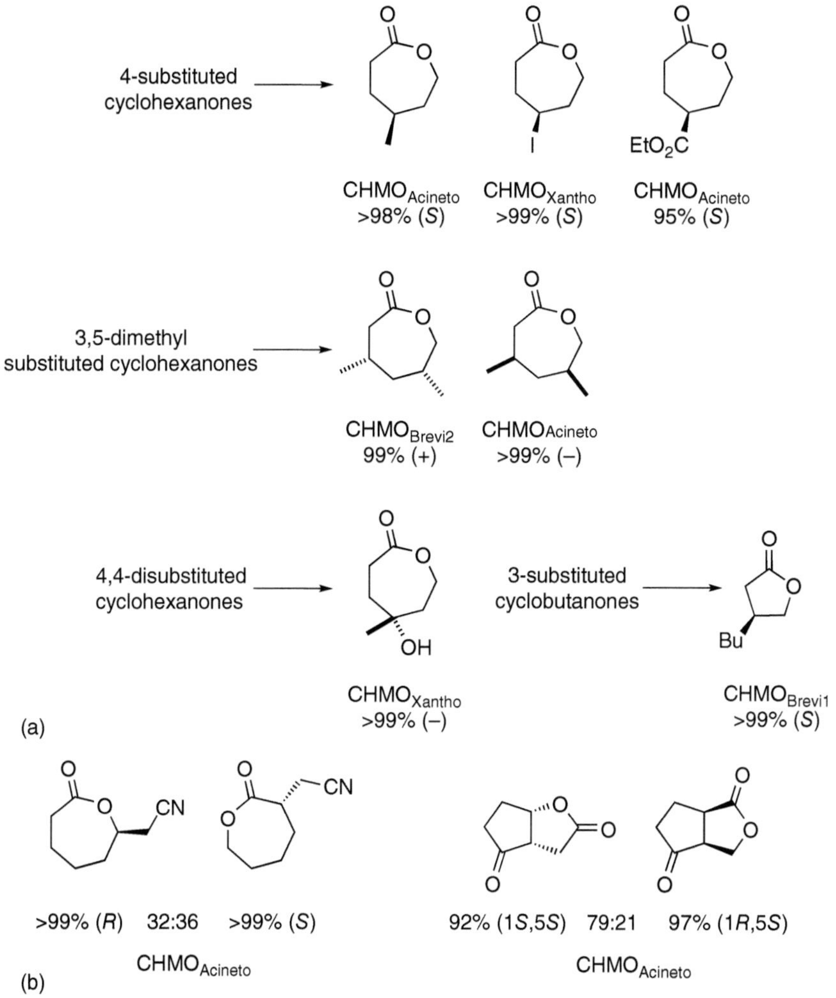

4-substituted cyclohexanones
EtO_2C
$CHMO_{Acineto}$ >98% (*S*)
$CHMO_{Xantho}$ >99% (*S*)
$CHMO_{Acineto}$ 95% (*S*)

3,5-dimethyl substituted cyclohexanones
$CHMO_{Brevi2}$ 99% (+)
CHMOAcineto >99% (−)

4,4-disubstituted cyclohexanones
OH
$CHMO_{Xantho}$ >99% (−)

3-substituted cyclobutanones
Bu
$CHMO_{Brevi1}$ >99% (*S*)

(a)

CN
CN
>99% (*R*) 32:36 >99% (*S*)
$CHMO_{Acineto}$

92% (1*S*,5*S*) 79:21 97% (1*R*,5*S*)
$CHMO_{Acineto}$

(b)

Scheme 7.19 Overview of products obtained by (a) desymmetrization of prochiral cyclic ketones. Source: Refs. [109, 127–132] and (b) enantiocomplementary regiodivergent reactions of racemic substrates using BVMOs ($CHMO_{Acineto}$, cyclohexanone monooxygenase from *Acinetobacter calcoaceticus*; $CHMO_{Xantho}$, cyclohexanone monooxygenase from *Xanthobacter* sp. ZL5; and $CHMO_{Brevi1/2}$, cyclohexanone monooxygenase from *Brevibacterium epidermis* HCU). Source: Refs. [133, 134].

7.3.1.1.3 Practical Approach

As highlighted throughout this section, only few BVMOs have gathered notoriety because of their broad substrate acceptance and excellent stereoselectivity and are frequently employed in a diversity of biocatalytic processes (e.g. $CHMO_{Acineto}$, $CHMO_{Xantho}$, HAPMO, and PAMO). Beyond the activity pattern of a given enzyme, other factors are important to correctly select and employ the biocatalyst for a given application. For instance, BVMOs tend to display an optimum pH in the alkaline range [136]. Performing reactions at a pH not basic enough may therefore favor formation of the flavin hydroperoxide, which is inactive in Baeyer–Villiger monooxygenation reactions. Influence of the pH on the enantioselectivity has also been observed [107, 137, 138]. Identifying the optimum pH window is particularly crucial when considering the use of such enzymes because most of them are also active in sulfoxidation reactions (see Section 7.3.2), in which the hydroperoxyflavin is the central catalytic species. The protonation process of the peroxyflavin intermediate is however not globally understood [139, 140].

7.3.1.2 Epoxidation of Alkenes

The asymmetric enzymatic *cis*-epoxidation of alkenes is catalyzed by a diversity of enzymes [141]. Prominent biocatalysts include flavin-dependent monooxygenases, which display more complex biochemistry compared to other monooxygenases, such as BVMOs. The prototype monooxygenase for epoxidation reactions is styrene monooxygenase from *Pseudomonas* sp., which is composed of two components, the FAD-dependent epoxidase StyA and the NADH-dependent flavin reductase StyB. The reaction proceeds through the activation of the flavin via reduction with NADH by the reductase. The reduced FAD then transfers to the monooxygenase component, where it reacts with molecular oxygen – similar to the reduced FAD in the case of BVMO (see Scheme 7.16a) – allowing subsequent reaction with the alkene substrate. Styrene monooxygenase-type enzymes have been extensively studied and employed on a broad variety of substrates, mostly poorly functionalized small alkenes. The stereoselectivity is high and favors almost in all cases the formation of (*S*)-epoxides, regardless of the source of the monooxygenase.

Other non-flavin-dependent monooxygenases are active in epoxidation reactions [141], such as xylene monooxygenase, or alkane and alkene monooxygenases, which are iron-dependent nonheme monooxygenases, known to catalyze hydroxylation reactions. Finally, cytochrome P450 monooxygenases can also catalyze epoxidation reactions, however, with usually lower enantioselectivity compared to the other available biocatalysts. All these enzymes rely on external nicotinamide for activation, which is obtained through reduction of the iron and further reaction with oxygen, and show varied substrate and stereopreferences. Often, these enzymes are employed in the form of their native organisms and as whole cells because of the complexity of their catalytic machinery, which renders heterologous expression challenging.

To guide the user, an overview of the most suited substrate types for each class of epoxidases is provided.

- Styrene monooxygenases appear particularly well suited for α- and β-substituted styrenes, as well as terminal and cyclic alkenes (see Scheme 7.20). A kinetic resolution strategy based on the enantioselective epoxidation of racemic allylic alcohols was developed with styrene monooxygenase from *Pseudomonas* sp. LQ26. The reaction proceeded well on a series of substrates and the corresponding products were obtained with excellent *ee* values, while the *de* values were more dependent on the substitution pattern (see Scheme 7.20c) [145].
- Xylene monooxygenase is a membrane-bound protein and, as a result, is less frequently used because of the low practicality and difficulty to obtain the protein in high yields. It behaves on average similarly to – or is less performing than – styrene monooxygenases [146, 147].

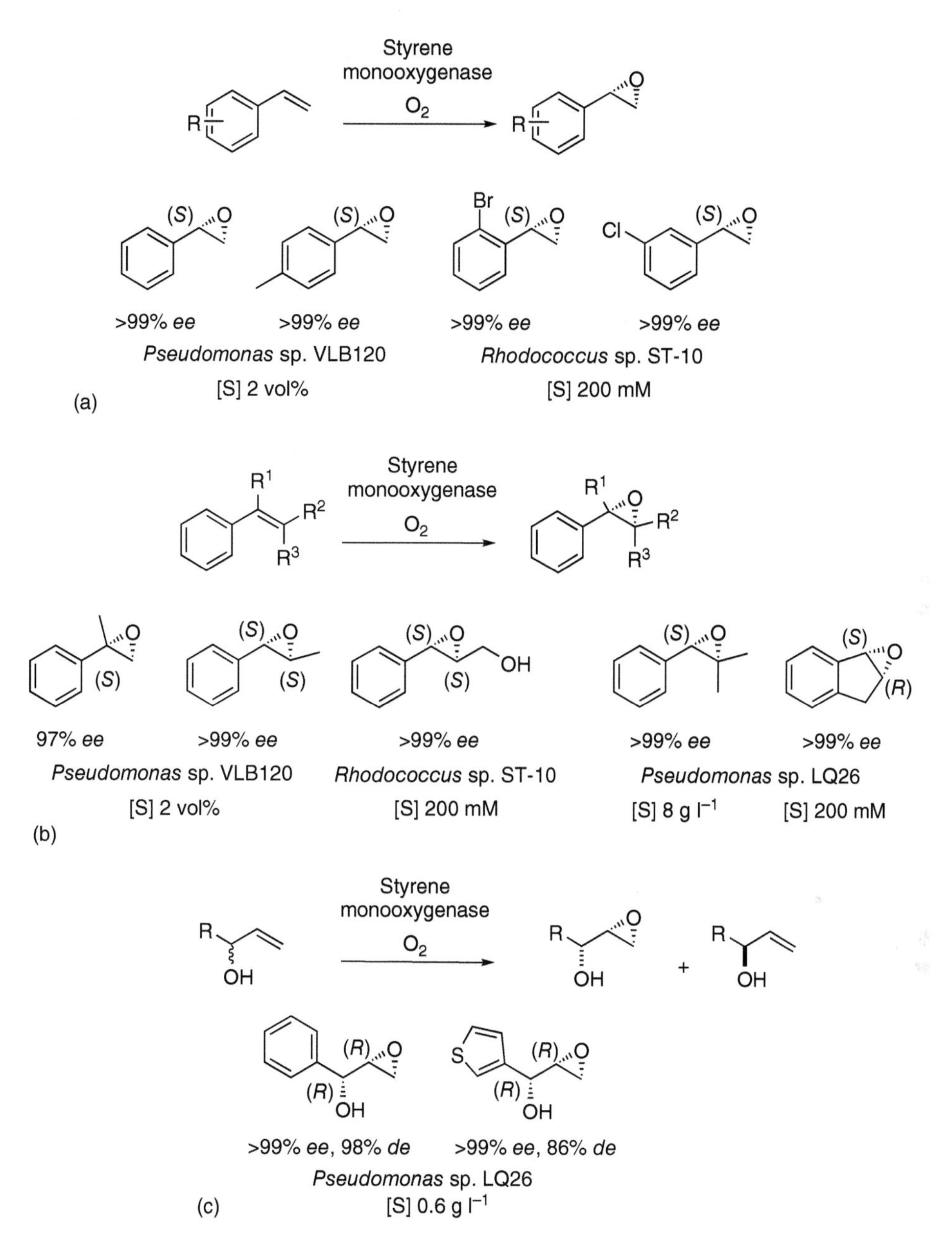

Scheme 7.20 Selection of well reacting substrates with styrene monooxygenases. Reaction with (a) substituted styrenes. Source: Refs. [142, 143]; (b) substituted 1-phenylethenes. Source: Refs. [142–144]; and (c) racemic secondary allylic alcohols proceeding via kinetic resolution. Source: Modified from Lin et al. [145]. The source of enzyme (organism) as well as the substrate concentration ([S]) used to achieve the reported *ee* and *de* values are indicated.

- *Pseudomonas oleovorans* expresses an alkane monooxygenase that presents dual activity: it catalyzes the asymmetric epoxidation of alkenes and shows activity on alkanes as ω-hydroxylase. The strain is usually employed as a whole-cell biocatalyst, either in the growing or resting state [148], and preferentially converts linear terminal alkenes with (*S*)- or (*R*)-selectivity (see Scheme 7.21a) [149–151]. Overall, yields are low and the enzyme is inactive on styrene.
- Alkene monooxygenases are multicomponent systems specific for short-chain alkenes and are used typically as whole-cell native biocatalysts. A few strains showing both stereopreferences are particularly active (see Scheme 7.21b) [152, 153].
- Cytochrome P450 monooxygenases, while accepting a wide range of substrates (terminal alkenes, styrene derivatives, and unsaturated fatty acids) [141], typically work at low substrate concentrations (2–10 mM) and are not ideal for preparative synthesis. In addition, they are challenging enzymes to manipulate by the nonspecialist and will not be covered here. Notably, P450s can provide access to (*R*)-styrene epoxides [154].

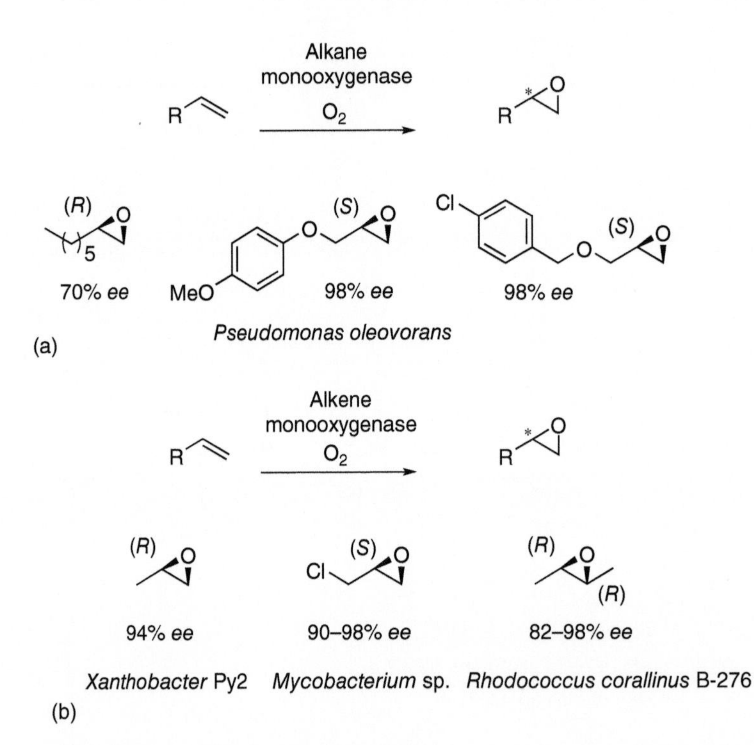

Scheme 7.21 Substrate acceptance of (a) alkane monooxygenase. Source: Refs. [149–151] and (b) alkene monooxygenases with the corresponding products and source of enzymes. Source: Refs. [152, 153]. The biocatalyst is employed as a whole-cell preparation of the native strain.

7.3.2 Heteroatom Oxidation

7.3.2.1 Reaction

In addition to Baeyer–Villiger oxidation reactions (see Section 7.3.1.1), flavin-dependent monooxygenases can catalyze the oxidation of heteroatoms (typically sulfur and nitrogen atoms). The reaction is initiated by the reduction of FAD by NADPH. Upon subsequent attack of molecular oxygen, a peroxyflavin is generated and gets protonated to the hydroperoxyflavin intermediate. This electrophilic species is directly attacked by soft nucleophiles to produce (see Scheme 7.16a,b), in the case of sulfides, the corresponding sulfoxides (cases of over-oxidation to the sulfones are known, but of limited synthetic value), and in the case of amines, the corresponding *N*-oxides.

Although the generation of *N*-oxides finds highest relevance in the study of metabolic pathways of xenobiotics, including drugs [155, 156], the oxidation of sulfides is of particular interest in organic chemistry because of the possibility to generate chiral sulfoxides [1, 157]. Several monooxygenases display stereopreference and, in consequence, optically pure sulfoxides can be produced via asymmetric oxidation of thioethers. BVMOs and flavin-containing monooxygenases (FMOs) have turned very useful for this reaction and typically achieve high yield and high *ee* values, but other enzymes, such as peroxidases, cytochrome P450 monooxygenases, and dioxygenases, can also catalyze the reaction [158]. Although whole-cell biocatalysts have been used for a long time in this area, the trend is toward the selection and use of recombinant enzymes in pure form or as whole-cell preparation of the heterologous host, which ensures more consistent and reproducible results.

7.3.2.2 Substrates

A variety of thioethers can be converted with BVMOs, ranging from alkyl aryl sulfides [159–161], to dialkyl sulfides [162–165], to dithioacetals [166, 167], and even disulfides [168] (see Scheme 7.22 and Section 7.4.4). FMOs are usually applied to the sulfoxidation of larger functionalized molecules, such as drug precursors [157, 158]. The stereopreference depends strongly on the biocatalyst source and the substitution pattern of the sulfide substrate.

Notably, an engineered BVMO has been used for the synthesis of esomeprazole by asymmetric oxidation of the precursor pyrmetazole on 30 g scale (see Scheme 7.23). NADPH was regenerated via the coupled-enzyme approach using isopropanol as the hydride source and an alcohol dehydrogenase (KRED from Codexis), and the final (*S*)-enantiomer was obtained in 87% yield with exquisite purity and >99% *ee* [169, 170].

7.3.3 Peroxygenases: One Catalyst – Many Reactions

Several of the oxidoreductases mentioned throughout this chapter display chemical versatility and can be employed for more than just one type of reaction. BVMOs and FMOs for instance can catalyze both Baeyer–Villiger monooxygenation and heteroatom oxidation reactions, while cytochrome P450 monooxygenases – although not presented in detail here – are notorious for their broad spectrum of chemical activities. Related to P450 monooxygenases, heme-dependent peroxygenases have lately emerged as promising biocatalysts for two main reasons: (i) contrary to P450s, which rely on a complex electron transport chain and several cofactors for activity [171], peroxygenases require only hydrogen peroxide to initiate a catalytic cycle and (ii) the spectrum of catalyzed oxidation reactions is vast and includes epoxidation, hydroxylation, and heteroatom oxidation [172, 173].

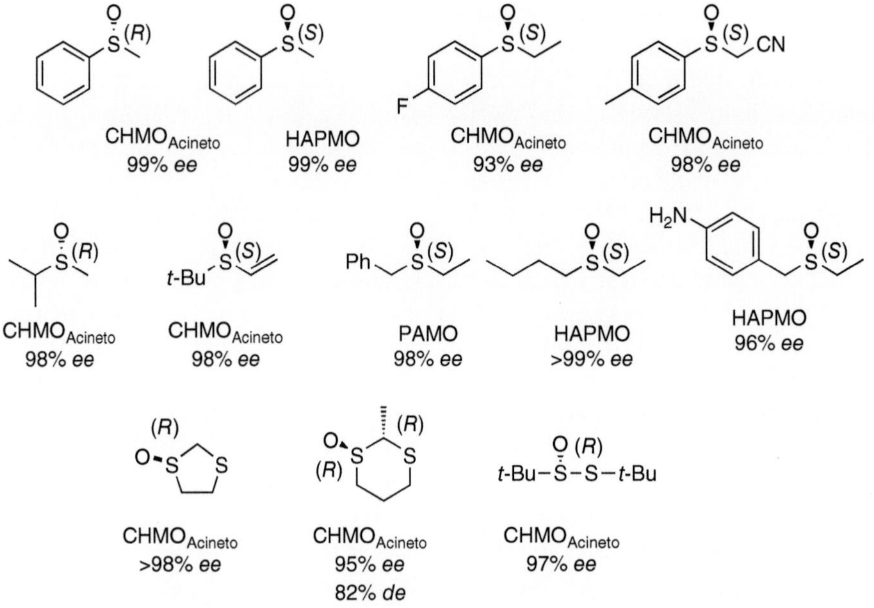

CHMO$_{Acineto}$ 99% *ee* | HAPMO 99% *ee* | CHMO$_{Acineto}$ 93% *ee* | CHMO$_{Acineto}$ 98% *ee*

CHMO$_{Acineto}$ 98% *ee* | CHMO$_{Acineto}$ 98% *ee* | PAMO 98% *ee* | HAPMO >99% *ee* | HAPMO 96% *ee*

CHMO$_{Acineto}$ >98% *ee* | CHMO$_{Acineto}$ 95% *ee* 82% *de* | CHMO$_{Acineto}$ 97% *ee*

Scheme 7.22 Range of chiral sulfoxides obtained in optically pure form through asymmetric oxidation reactions using BVMOs (CHMO$_{Acineto}$, cyclohexanone monooxygenase from *Acinetobacter calcoaceticus*; HAPMO, 4-hydroxyacetophenone monooxygenase; and PAMO, phenylacetone monooxygenase). Source: Refs. [159–168].

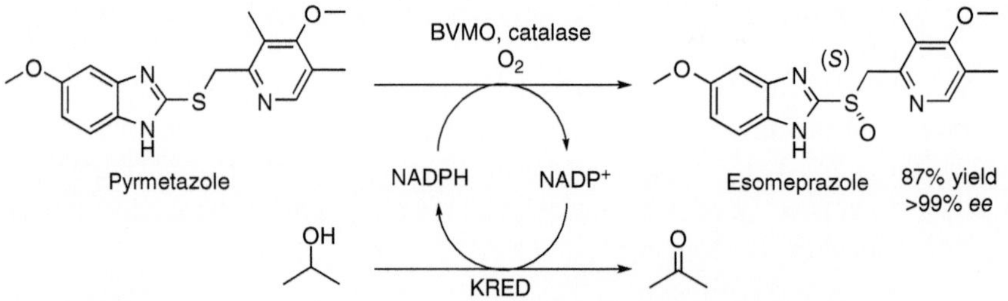

Pyrmetazole → Esomeprazole (BVMO, catalase, O_2; NADPH/NADP$^+$; KRED) 87% yield >99% *ee*

Scheme 7.23 Asymmetric sulfoxidation of pyrmetazole to esomeprazole using an engineered Baeyer–Villiger monooxygenase developed by Codexis. Source: Bong et al. [169].

Peroxygenases, also termed unspecific peroxygenases (UPOs), are vastly found in fungi [174]; however, only a few UPOs have been so far well characterized: UPOs from *Agrocybe aegerita* (*Aae*UPO) [175], *Coprinellus radians* (*Cra*UPO) [176], and *Marasmius rotula* (*Mro*UPO) [177].

In addition to high catalytic efficiency (up to 110 000 TTN [178]), UPOs often display excellent enantioselectivities [173] and are attractive catalysts for selective oxyfunctionalization reactions, which are otherwise difficult to achieve. Such reactions include benzylic hydroxylation, *cis*-epoxidation of styrene derivatives [178], and regioselective aromatic hydroxylation, as shown with diclofenac [179] (Scheme 7.24).

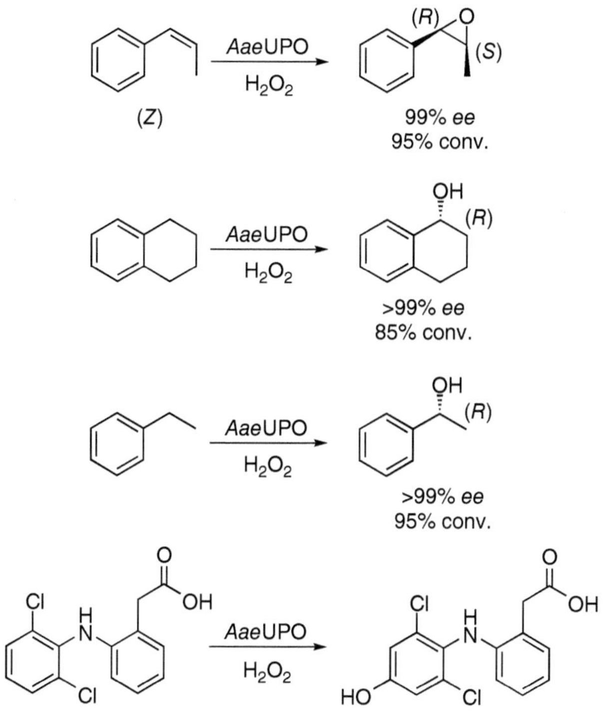

Scheme 7.24 Selective oxyfunctionalization reactions catalyzed by unspecific peroxygenase from *Agrocybe aegerita* using hydrogen peroxide as the oxidant. Source: Refs. [178, 179].

Despite all these promises, peroxygenases are still facing major issues that prevent these enzymes to enter the portfolio of broadly applicable biocatalytic routes. The low tolerance to synthetically relevant substrate concentration currently prevents the development of reactions on large scale. In addition, the lability of peroxygenases to hydrogen peroxide, connected to the heme center, remains a major issue; however, promising protocols are being developed, such as (enzymatic) *in situ* production of H_2O_2 [172, 180–182].

7.4 Protocols for Selected Reactions Employing Oxidoreductases

7.4.1 Alcohol Dehydrogenase (ADH): Disproportionation of *rac*-2-Phenylpropanal

The disproportionation of *rac*-2-phenylpropanal under parallel interconnected dynamic asymmetric transformation was performed on a 100 mg scale by using purified HLADH according to Scheme 7.8 [73].

7.4.1.1 Biotransformation

The substrate *rac*-2-phenylpropanal (100 mg, 0.75 mmol) was added as a *t*-butyl methyl ether solution (0.4 ml) to a solution of phosphate buffer (50 mM, pH 7.5, final volume

10 ml) supplemented with 100 μM $ZnCl_2$ (essential metal for enzymatic activity). 5 mg of purified HLADH was added along with NAD^+ (3 mg, 5 μmol, 0.0067 equivalent) and the reaction mixture was incubated at 30 °C and 120 rpm for 24 h. A second aliquot of both enzyme (5 mg) and cofactor (3 mg) was added and the incubation was pursued for another 24 h.

7.4.1.2 Product Recovery and Purification

The reaction mixture was extracted with ethyl acetate and the solvent was removed under reduced pressure. The crude was purified by flash chromatography (cyclohexane/ethyl acetate 4 : 1) and furnished 31 mg (0.23 mmol) of (*S*)-2-phenylpropanol in 97% *ee*, as determined by chiral gas chromatography (GC) on a Macherey–Nagel Hydrodex β-TBDAc column (25 m × 0.25 mm × 0.25 μm).

The aqueous phase was acidified with 3 N HCl and extracted with ethyl acetate. Removal of the solvent under reduced pressure yielded 35 mg (0.23 mmol) of (*S*)-2-phenylpropanoic acid in 86% *ee*, as determined by chiral GC on an Agilent CP-Chirasil-DEX CB column (25 m × 0.32 mm × 0.25 μm).

The overall isolated yield was 62% with a ratio carboxylic acid/alcohol of 1.03.

7.4.2 Ene-reductase/Old Yellow Enzyme (OYE): Dynamic Kinetic Resolution of a γ-substituted Lactone

The DKR of *rac*-3-methyl-5-phenylfuran-2(5*H*)-one was performed on a 16 ml scale according to Scheme 7.15 [105].

7.4.2.1 Biotransformation

To a solution of Tris–HCl buffer (50 mM, pH 7.5, final volume: 16 ml) was added the substrate (final concentration: 10 mM) as a 0.5 M stock solution in dimethyl sulfoxide (DMSO) (final: 1 vol%). NADH (20 mM) and the purified enzyme NCR (100 μM) were added and the reaction mixture was incubated at 30 °C and 120 rpm for 72 h. Additional batches of NCR and NADH were added after 24 and 48 h.

7.4.2.2 Product Recovery and Purification

After extraction with dichloromethane (4 × 5 ml), the combined organic layers were dried over Na_2SO_4, filtered, and the solvent was evaporated under reduced pressure. The product was purified by preparative TLC (hexane/diethyl ether 2 : 1) and afforded 13.8 mg (0.078 mmol, 49% isolated yield) of (3*R*,5*S*)-3-methyl-5-phenyldihydrofuran-2(3*H*)-one in 99% *ee* and 70% *de*, slightly below the *de* value of 96% obtained under screening reaction conditions ($[\alpha]_D = -6.8$, $c = 0.69$, $CHCl_3$). *de* and *ee* values were measured by GC on a Restek Rt-βDEXse column (30 m × 0.32 mm, 0.25 μm film).

7.4.3 Baeyer–Villiger Monooxygenase (BVMO): Kinetic Resolution of a Racemic Ketone

The kinetic resolution of *rac*-4-phenylhexan-3-one was performed on a 20 ml scale according to Scheme 7.18a [123]. NADPH was used in catalytic amounts and regenerated with the glucose 6-phosphate/glucose 6-phosphate dehydrogenase recycling system.

7.4.3.1 Biotransformation

To a solution of the ketone (0.5 mmol) in Tris–HCl buffer (50 mM, pH 8.0, final volume: 20 ml) was added NADPH (0.02 mM), glucose 6-phosphate (0.5 mmol, 1 eq.), glucose 6-phosphate dehydrogenase (10 U), and HAPMO from *Pseudomonas fluorescens* as a purified enzyme (1 U). The reaction was incubated at 20 °C and 250 rpm for 72 h.

7.4.3.2 Product Recovery and Purification

After extraction with EtAOc (4 × 5 ml), the combined organic layers were dried over Na_2SO_4 and concentrated. The products were separated and purified by flash chromatography on silica gel (hexane/diethyl ether 4 : 1), yielding (*R*)-4-phenylhexan-3-one (45.8 mg, 52%) with 98% *ee* ($[\alpha]_D = -61.2$, $c = 0.75$, $CHCl_3$) and (*S*)-1-phenylpropyl propionate (37.2 mg, 39%) in 90% *ee* ($[\alpha]_D = -41.7$, $c = 0.83$, $CHCl_3$). *ee* values were measured by GC on a Restek Rt-βDEXse column (30 m × 0.25 mm, 0.25 μm film).

7.4.4 Baeyer–Villiger Monooxygenase (BVMO): Asymmetric Sulfoxidation

(*S*)-*n*-Butyl methyl sulfoxide was obtained from *n*-butyl methyl sulfide according to Scheme 7.22 [164].

7.4.4.1 Biotransformation

To a solution of Tris–HCl buffer (50 mM, pH 9.0, 40 ml) containing the substrate *n*-butyl methyl sulfide (250 mg, 2.4 mmol), DMSO (1 vol%) and NADPH (0.2 mM), glucose 6-phosphate (2 eq.), and glucose 6-phosphate dehydrogenase (1250 U) were added as a cofactor recycling system, and finally recombinant HAPMO (HAPMO as a cell-free extract, 46 ml). The reaction mixture was incubated at 250 rpm and 20 °C for 24 h.

7.4.4.2 Product Recovery and Purification

After extraction with ethyl acetate (5 × 30 ml), the combined organic layers were dried over Na_2SO_4 and the solvent was evaporated under reduced pressure. The product was purified by flash chromatography (dichloromethane/methanol 9 : 1) and yielded the pure (*S*)-enantiomer (>99% *ee*, $[\alpha]_D = +32.5$, $c = 0.5$, CH_2Cl_2) in 88% yield. *ee* values were measured by GC on a Restek Rt-βDEXse column (30 m × 0.25 mm, 0.25 μm film).

7.5 Conclusions

The examples treated in this chapter demonstrate how oxidoreductases can be valuable in organic synthesis, in particular, for the generation of products with optical activity. Synthetic chemists more and more recognize the advantages of biocatalysis as synthetic methodology; however, further progresses are still required to establish biocatalysis as a general tool in the laboratory, especially in the area of activity and stereoselectivity predictions and access to stereocomplementary biocatalysts for any given substrate. Given the current pace of development in the areas of computational study of enzymes, artificial intelligence, and genome mining [183], one can anticipate that such expected challenges will be met in the near future. In addition, the booming of biotechnology companies and enzyme producers offers highly practical benefits to chemists, who are encouraged to explore biosynthetic technologies to complement already existing chemical processes. The availability of a growing number of screening kits and single enzymes as lyophilized powders that can be easily stored and used should contribute to demystify a field long considered out of reach for the nonspecialists.

The field of asymmetric reduction of ketones using ADHs is a testimony of what collective efforts in the field can achieve. Nowadays, ADHs are becoming the first choice in the industry for the synthesis of enantiopure chiral alcohols and many processes running on multi-kilogram scale have found their way into the chemical manufacturing and pharmaceutical industry.

Acknowledgments

The Austrian Science Fund is gratefully acknowledged for financial support (P30519-N36 and P32815-N).

References

1 Faber, K. (2018). *Biotransformations in Organic Chemistry*, 7 ed. Heidelberg: Springer.
2 Hall, M. and Bommarius, A.S. (2011). Enantioenriched compounds via enzyme-catalyzed redox reactions. *Chem. Rev.* 111: 4088–4110.
3 Patel, R.N. (2008). Synthesis of chiral pharmaceutical intermediates by biocatalysis. *Coord. Chem. Rev.* 252: 659–701.
4 Truppo, M.D. (2017). Biocatalysis in the pharmaceutical industry: the need for speed. *ACS Med. Chem. Lett.* 8: 476–480.
5 Patel, R.N. (2016). Green processes for the synthesis of chiral intermediates for the development of drugs. In: *Green Biocatalysis* (ed. R.N. Patel), 71–114. Hoboken, NJ: Wiley.
6 Hughes, D.L. (2018). Biocatalysis in drug development-highlights of the recent patent literature. *Org. Process. Res. Dev.* 22: 1063–1080.
7 Chenault, H.K., Simon, E.S., and Whitesides, G.M. (1988). Cofactor regeneration for enzyme-catalysed synthesis. *Biotechnol. Genet. Eng. Rev.* 6: 221–270.
8 van der Donk, W.A. and Zhao, H.M. (2003). Recent developments in pyridine nucleotide regeneration. *Curr. Opin. Biotechnol.* 14: 421–426.
9 Wu, H., Tian, C.Y., Song, X.K. et al. (2013). Methods for the regeneration of nicotinamide coenzymes. *Green Chem.* 15: 1773–1789.
10 Kosjek, B., Stampfer, W., Pogorevc, M. et al. (2004). Purification and characterization of a chemotolerant alcohol dehydrogenase applicable to coupled redox reactions. *Biotechnol. Bioeng.* 86: 55–62.
11 Tassano, E. and Hall, M. (2019). Enzymatic self-sufficient hydride processes. *Chem. Soc. Rev.* 48 (23): 5596–5615.
12 Zachos, I., Nowak, C., and Sieber, V. (2019). Biomimetic cofactors and methods for their recycling. *Curr. Opin. Chem. Biol.* 49: 59–66.

13 Koninger, K., Baraibar, A.G., Mugge, C. et al. (2016). Recombinant *Cyanobacteria* for the asymmetric reduction of C=C bonds fueled by the biocatalytic oxidation of water. *Angew. Chem. Int. Ed.* 55: 5582–5585.

14 Riebel, B.R., Gibbs, P.R., Wellborn, W.B., and Bommarius, A.S. (2003). Cofactor regeneration of both NAD(+) from NADH and NADP(+) from NADPH: NADH oxidase from *Lactobacillus sanfranciscensis*. *Adv. Synth. Catal.* 345: 707–712.

15 Angelastro, A., Dawson, W.M., Luk, L.Y.P., and Allemann, R.K. (2017). A versatile disulfide-driven recycling system for NADP(+) with high cofactor turnover number. *ACS Catal.* 7: 1025–1029.

16 Illanes, A., Wilsona, L., and Vera, C. (2018). Technical biocatalysis. In: *Modern Biocatalysis: Advances towards Synthetic Biological Systems* (eds. G. Williams and M. Hall), 475–515. Croydon: The Royal Society of Chemistry.

17 Woodley, J.M. (2018). Biocatalytic process engineering. In: *Modern Biocatalysis: Advances towards Synthetic Biological Systems* (eds. G. Williams and M. Hall), 516–538. Croydon: The Royal Society of Chemistry.

18 Jornvall, H., Hoog, J.O., and Persson, B. (1999). SDR and MDR: completed genome sequences show these protein families to be large, of old origin, and of complex nature. *FEBS Lett.* 445: 261–264.

19 An, J.H., Nie, Y., and Xu, Y. (2019). Structural insights into alcohol dehydrogenases catalyzing asymmetric reductions. *Crit. Rev. Biotechnol.* 39: 366–379.

20 Gröger, H., Hummel, W., Borchert, S., and Kraußer, M. (2010). Reduction of ketones and aldehydes to alcohols. In: *Enzyme Catalysis in Organic Synthesis*, 3 ed. (eds. K. Drauz, H. Gröger and O. May), 1037–1110. Weinheim: Wiley-VCH.

21 Niefind, K., Muller, J., Riebel, B. et al. (2003). The crystal structure of *R*-specific alcohol dehydrogenase from *Lactobacillus brevis* suggests the structural basis of its metal dependency. *J. Mol. Biol.* 327: 317–328.

22 Weckbecker, A. and Hummel, W. (2006). Cloning, expression, and characterization of an (*R*)-specific alcohol dehydrogenase from *Lactobacillus kefir*. *Biocatal Biotransformation* 24: 380–389.

23 Klimacek, M., Kavanagh, K.L., Wilson, D.K., and Nidetzky, B. (2003). *Pseudomonas fluorescens* mannitol 2-dehydrogenase and the family of polyol-specific long-chain dehydrogenases/reductases: sequence-based classification and analysis of structure-function relationships. *Chem. Biol. Interact.* 143: 559–582.

24 Penning, T.M. (2015). The aldo-keto reductases (AKRs): overview. *Chem. Biol. Interact.* 234: 236–246.

25 Matsuda, T., Yamanaka, R., and Nakamura, K. (2009). Recent progress in biocatalysis for asymmetric oxidation and reduction. *Tetrahedron Asymmetry* 20: 513–557.

26 Prelog, V. (1964). Specification of the stereospecificity of some oxido-reductases by diamond lattice sections. *Pure Appl. Chem.* 9: 119–130.

27 Itoh, N. (2014). Use of the anti-Prelog stereospecific alcohol dehydrogenase from *Leifsonia* and *Pseudomonas* for producing chiral alcohols. *Appl. Microbiol. Biotechnol.* 98: 3889–3904.

28 Itoh, N., Isotani, K., Makino, Y. et al. (2014). PCR-based amplification and heterologous expression of *Pseudomonas* alcohol dehydrogenase genes from the soil metagenome for biocatalysis. *Enzym. Microb. Technol.* 55: 140–150.

29 Schlieben, N.H., Niefind, K., Muller, J. et al. (2005). Atomic resolution structures of *R*-specific alcohol dehydrogenase from *Lactobacillus brevis* provide the structural bases of its substrate and cosubstrate specificity. *J. Mol. Biol.* 349: 801–813.
30 Bradshaw, C.W., Hummel, W., and Wong, C.H. (1992). *Lactobacillus kefir* alcohol dehydrogenase: a useful catalyst for synthesis. *J. Org. Chem.* 57: 1532–1536.
31 Park, D.H. and Plapp, B.V. (1991). Isoenzymes of horse liver alcohol-dehydrogenase active on ethanol and steroids. *J. Biol. Chem.* 266: 13296–13302.
32 Nie, Y., Xu, Y., Mu, X.Q. et al. (2007). Purification, characterization, gene cloning, and expression of a novel alcohol dehydrogenase with anti-Prelog stereospecificity from *Candida parapsilosis*. *Appl. Environ. Microbiol.* 73: 3759–3764.
33 Karabec, M., Lyskowski, A., Tauber, K.C. et al. (2010). Structural insights into substrate specificity and solvent tolerance in alcohol dehydrogenase ADH-'A' from *Rhodococcus ruber* DSM 44541. *Chem. Commun.* 46: 6314–6316.
34 Yasohara, Y., Kizaki, N., Hasegawa, J. et al. (2000). Molecular cloning and overexpression of the gene encoding an NADPH-dependent carbonyl reductase from *Candida magnoliae*, involved in stereoselective reduction of ethyl 4-chloro-3-oxobutanoate. *Biosci. Biotechnol. Biochem.* 64: 1430–1436.
35 Abokitse, K. and Hummel, W. (2003). Cloning, sequence analysis, and heterologous expression of the gene encoding a (*S*)-specific alcohol dehydrogenase from *Rhodococcus erythropolis* DSM 43297. *Appl. Microbiol. Biotechnol.* 62: 380–386.
36 Choi, Y.H., Choi, H.J., Kim, D. et al. (2010). Asymmetric synthesis of (*S*)-3-chloro-1-phenyl-1-propanol using *Saccharomyces cerevisiae* reductase with high enantioselectivity. *Appl. Microbiol. Biotechnol.* 87: 185–193.
37 Korkhin, Y., Kalb, A.J., Peretz, M. et al. (1998). NADP-dependent bacterial alcohol dehydrogenases: crystal structure, cofactor-binding and cofactor specificity of the ADHs of *Clostridium beijerinckii* and *Thermoanaerobacter brockii*. *J. Mol. Biol.* 278: 967–981.
38 Inoue, K., Makino, Y., Dairi, T., and Itoh, N. (2006). Gene cloning and expression of *Leifsonia alcohol dehydrogenase* (LSADH) involved in asymmetric hydrogen-transfer bioreduction to produce (*R*)-form chiral alcohols. *Biosci. Biotechnol. Biochem.* 70: 418–426.
39 Gröger, H., Hummel, W., Borchert, S., and Kraußer, M. (2012). Reduction of ketones and aldehydes to alcohols. In: *Enzyme Catalysis in Organic Synthesis*, 3 ed. (eds. K. Drauz, H. Gröger and O. May), 1037–1110. Weinheim: Wiley-VCH.
40 Truppo, M.D., Pollard, D., and Devine, P. (2007). Enzyme-catalyzed enantioselective diaryl ketone reductions. *Org. Lett.* 9: 335–338.
41 Lavandera, I., Kern, A., Ferreira-Silva, B. et al. (2008). Stereoselective bioreduction of bulky-bulky ketones by a novel ADH from *Ralstonia* sp. *J. Org. Chem.* 73: 6003–6005.
42 Lavandera, I., Kern, A., Resch, V. et al. (2008). One-way biohydrogen transfer for oxidation of *sec*-alcohols. *Org. Lett.* 10: 2155–2158.
43 Man, H., Kedziora, K., Kulig, J. et al. (2014). Structures of alcohol dehydrogenases from *Ralstonia* and *Sphingobium* spp. reveal the molecular basis for their recognition of 'bulky–bulky' ketones. *Top. Catal.* 57: 356–365.
44 Li, H.M., Zhu, D.M., Hua, L., and Biehl, E.R. (2009). Enantioselective reduction of diaryl ketones catalyzed by a carbonyl reductase from *Sporobolomyces salmonicolor* and its mutant enzymes. *Adv. Synth. Catal.* 351: 583–588.

45 Lalonde, J. (2016). Highly engineered biocatalysts for efficient small molecule pharmaceutical synthesis. *Curr. Opin. Biotechnol.* 42: 152–158.

46 Nealon, C.M., Musa, M.M., Patel, J.M., and Phillips, R.S. (2015). Controlling substrate specificity and stereospecificity of alcohol dehydrogenases. *ACS Catal.* 5: 2100–2114.

47 Noey, E.L., Tibrewal, N., Jimenez-Oses, G. et al. (2015). Origins of stereoselectivity in evolved ketoreductases. *Proc. Natl. Acad. Sci. U. S. A.* 112: E7065–E7072.

48 Maria-Solano, M.A., Romero-Rivera, A., and Osuna, S. (2017). Exploring the reversal of enantioselectivity on a zinc-dependent alcohol dehydrogenase. *Org. Biomol. Chem.* 15: 4122–4129.

49 Chen, X., Zhang, H.L., Maria-Solano, M.A. et al. (2019). Efficient reductive desymmetrization of bulky 1,3-cyclodiketones enabled by structure-guided directed evolution of a carbonyl reductase. *Nat. Catal.* 2: 931–941.

50 Zhou, J.Y., Xu, G.C., Han, R.Z. et al. (2016). Carbonyl group-dependent high-throughput screening and enzymatic characterization of diaromatic ketone reductase. *Catal. Sci. Technol.* 6: 6320–6327.

51 Xu, G.C., Wang, Y., Tang, M.H. et al. (2018). Hydroclassified combinatorial saturation mutagenesis: reshaping substrate binding pockets of KpADH for enantioselective reduction of bulky-bulky ketones. *ACS Catal.* 8: 8336–8345.

52 Zhou, J.Y., Wang, Y., Xu, G.C. et al. (2018). Structural insight into enantioselective inversion of an alcohol dehydrogenase reveals a "polar gate" in stereorecognition of diaryl ketones. *J. Am. Chem. Soc.* 140: 12645–12654.

53 Atomi, H. (2005). Recent progress towards the application of hyperthermophiles and their enzymes. *Curr. Opin. Chem. Biol.* 9: 166–173.

54 Rigoldi, F., Donini, S., Redaelli, A. et al. (2018). Review: engineering of thermostable enzymes for industrial applications. *APL Bioeng.* 2: 011501.

55 Friest, J.A., Maezato, Y., Broussy, S. et al. (2010). Use of a robust dehydrogenase from an archael hyperthermophile in asymmetric catalysis-dynamic reductive kinetic resolution entry into (*S*)-profens. *J. Am. Chem. Soc.* 132: 5930–5931.

56 de Gonzalo, G., Lavandera, I., Faber, K., and Kroutil, W. (2007). Enzymatic reduction of ketones in "micro-aqueous" media catalyzed by ADH-A from *Rhodococcus ruber*. *Org. Lett.* 9: 2163–2166.

57 Lerchner, A., Jarasch, A., Meining, W. et al. (2013). Crystallographic analysis and structure-guided engineering of NADPH-dependent *Ralstonia* sp. alcohol dehydrogenase toward NADH cosubstrate specificity. *Biotechnol. Bioeng.* 110: 2803–2814.

58 Liz, R., Liardo, E., and Rebolledo, F. (2019). Highly efficient asymmetric bioreduction of 1-aryl-2-(azaaryl)ethanones. Chemoenzymatic synthesis of lanicemine. *Org. Biomol. Chem.* 17: 8214–8220.

59 Moore, J.C., Pollard, D.J., Kosjek, B., and Devine, P.N. (2007). Advances in the enzymatic reduction of ketones. *Acc. Chem. Res.* 40: 1412–1419.

60 Huisman, G.W., Liang, J., and Krebber, A. (2010). Practical chiral alcohol manufacture using ketoreductases. *Curr. Opin. Chem. Biol.* 14: 122–129.

61 Grunwald, J., Wirz, B., Scollar, M.P., and Klibanov, A.M. (1986). Asymmetric oxidoreductions catalyzed by alcohol-dehydrogenase in organic-solvents. *J. Am. Chem. Soc.* 108: 6732–6734.

62 Stampfer, W., Kosjek, B., Moitzi, C. et al. (2002). Biocatalytic asymmetric hydrogen transfer. *Angew. Chem. Int. Ed.* 41: 1014–1017.

63 Daussmann, T., Hennemann, H.G., Rosen, T.C., and Dunkelmann, P. (2006). Enzymatic technologies for the synthesis of chiral alcohol derivatives. *Chem. Ing. Technol.* 78: 249–255.

64 Tasnádi, G. and Hall, M. (2014). Relevant practical applications of bioreduction processes in the synthesis of active pharmaceutical ingredients. In: *Synthetic Methods for Biologically Active Molecules* (ed. E. Brenna), 329–374. Weinheim: Wiley-VCH.

65 Moody, T., Mix, S., Brown, G., and Beecher, D. (2015). Ketone and aldehyde reductions. In: *Science of Synthesis - Biocatalysis in Organic Synthesis* (eds. K. Faber, W.-D. Fessner and N.J. Turner), 421–458. Stuttgart: Thieme.

66 Liang, J., Lalonde, J., Borup, B. et al. (2010). Development of a biocatalytic process as an alternative to the (−)-DIP-Cl-mediated asymmetric reduction of a key intermediate of montelukast. *Org. Process. Res. Dev.* 14: 193–198.

67 Fox, R.J., Davis, S.C., Mundorff, E.C. et al. (2007). Improving catalytic function by ProSAR-driven enzyme evolution. *Nat. Biotechnol.* 25: 338–344.

68 Ma, S.K., Gruber, J., Davis, C. et al. (2010). A green-by-design biocatalytic process for atorvastatin intermediate. *Green Chem.* 12: 81–86.

69 Wolberg, M., Villela, M., Bode, S. et al. (2008). Chemoenzymatic synthesis of the chiral side-chain of statins: application of an alcohol dehydrogenase catalysed ketone reduction on a large scale. *Bioprocess Biosyst. Eng.* 31: 183–191.

70 Kalaitzakis, D. and Smonou, I. (2014). Dynamic kinetic resolutions based on reduction processes. In: *Synthetic Methods for Biologically Active Molecules* (ed. E. Brenna), 307–327. Weinheim: Wiley-VCH.

71 Cuetos, A., Rioz-Martinez, A., Bisogno, F.R. et al. (2012). Access to enantiopure alpha-alkyl-beta-hydroxy esters through dynamic kinetic resolutions employing purified/overexpressed alcohol dehydrogenases. *Adv. Synth. Catal.* 354: 1743–1749.

72 Cuetos, A., Díaz-Rodríguez, A., and Lavandera, I. (2014). Synthetic strategies based on C=O bioreductions for the preparation of biologically active molecules. In: *Synthetic Methods for Biologically Active Molecules* (ed. E. Brenna), 85–111. Weinheim: Wiley-VCH.

73 Tassano, E., Faber, K., and Hall, M. (2018). Biocatalytic parallel interconnected dynamic asymmetric disproportionation of alpha-substituted aldehydes: atom-efficient access to enantiopure (*S*)-profens and profenols. *Adv. Synth. Catal.* 360: 2742–2751.

74 Liu, Y.C., Merten, C., and Deska, J. (2018). Enantioconvergent biocatalytic redox isomerization. *Angew. Chem. Int. Ed.* 57: 12151–12156.

75 Stuermer, R., Hauer, B., Hall, M., and Faber, K. (2007). Asymmetric bioreduction of activated C=C bonds using enoate reductases from the old yellow enzyme family. *Curr. Opin. Chem. Biol.* 11: 203–213.

76 Toogood, H.S., Gardiner, J.M., and Scrutton, N.S. (2010). Biocatalytic reductions and chemical versatility of the old yellow enzyme family of flavoprotein oxidoreductases. *ChemCatChem* 2: 892–914.

77 Williams, R.E. and Bruce, N.C. (2002). 'New uses for an old enzyme' - the old yellow enzyme family of flavoenzymes. *Microbiology* 148: 1607–1614.

78 Winkler, C.K., Faber, K., and Hall, M. (2018). Biocatalytic reduction of activated C=C-bonds and beyond: emerging trends. *Curr. Opin. Chem. Biol.* 43: 97–105.

79 Knowles, W.S. (2002). Asymmetric hydrogenations (Nobel lecture). *Angew. Chem. Int. Ed.* 41: 1999–2007.

80 Noyori, R. (2002). Asymmetric catalysis: science and opportunities (Nobel lecture). *Angew. Chem. Int. Ed.* 41: 2008–2022.

81 Bougioukou, D.J. and Stewart, J.D. (2012). Reduction of C=C double bonds. In: *Enzyme Catalysis in Organic Synthesis*, 3 ed. (eds. K. Drauz, H. Gröger and O. May), 1111–1163. Weinheim: Wiley-VCH.

82 Warburg, O. and Christian, W. (1932). On a new oxidation enzyme and its absorption spectrum. *Biochem. Z.* 254: 438–458.

83 Weygand, F. and Birkofer, L. (1939). On the preparation in a pure state from the "old" yellow enzyme from yeast and a new method for reversible fission. *Hoppe Seylers Z. Physiol. Chem.* 261: 172–182.

84 Kubata, B.K., Kabututu, Z., Nozaki, T. et al. (2002). A key role for old yellow enzyme in the metabolism of drugs by *Trypanosoma cruzi*. *J. Exp. Med.* 196: 1241–1251.

85 Scholtissek, A., Tischler, D., Westphal, A.H. et al. (2017). Old yellow enzyme-catalysed asymmetric hydrogenation: linking family roots with improved catalysis. *Catalysts* 7: 130.

86 French, C.E. and Bruce, N.C. (1995). Bacterial morphinone reductase is related to old yellow enzyme. *Biochem. J.* 312: 671–678.

87 Faber, K. and Hall, M. (2015). Addition of hydrogen to C=C bonds: alkene reductions. In: *Science of Synthesis - Biocatalysis in Organic Synthesis* (eds. K. Faber, W.-D. Fessner and N.J. Turner), 213–260. Stuttgart: Thieme.

88 French, C.E. and Bruce, N.C. (1994). Purification and characterization of morphinone reductase from *Pseudomonas putida* M10. *Biochem. J.* 301: 97–103.

89 Burda, E., Ress, T., Winkler, T. et al. (2013). Highly enantioselective reduction of alpha-methylated nitroalkenes. *Angew. Chem. Int. Ed.* 52: 9323–9326.

90 Tasnádi, G., Winkler, C.K., Clay, D. et al. (2012). A substrate-driven approach to determine reactivities of alpha,beta-unsaturated carboxylic esters towards asymmetric bioreduction. *Chem. Eur. J.* 18: 10362–10367.

91 Stueckler, C., Hall, M., Ehammer, H. et al. (2007). Stereocomplementary bioreduction of alpha,beta-unsaturated dicarboxylic acids and dimethyl esters using enoate reductases: enzyme- and substrate-based stereocontrol. *Org. Lett.* 9: 5409–5411.

92 Müller, A., Sturmer, R., Hauer, B., and Rosche, B. (2007). Stereospecific alkyne reduction: novel activity of old yellow enzymes. *Angew. Chem. Int. Ed.* 46: 3316–3318.

93 Müller, A., Hauer, B., and Rosche, B. (2007). Asymmetric alkene reduction by yeast old yellow enzymes and by a novel *Zymomonas mobilis* reductase. *Biotechnol. Bioeng.* 98: 22–29.

94 Classen, T., Korpak, M., Schölzel, M., and Pietruszka, J. (2014). Stereoselective enzyme cascades: an efficient synthesis of chiral γ-butyrolactones. *ACS Catal.* 4: 1321–1331.

95 Oberdorfer, G., Gruber, K., Faber, K., and Hall, M. (2012). Stereocontrol strategies in the asymmetric bioreduction of alkenes. *Synlett*: 1857–1864.

96 Turrini, N.G., Cioc, R.C., van der Niet, D.J.H. et al. (2017). Biocatalytic access to nonracemic gamma-oxo esters via stereoselective reduction using ene-reductases. *Green Chem.* 19: 511–518.

97 Swiderska, M.A. and Stewart, J.D. (2006). Asymmetric bioreductions of beta-nitro acrylates as a route to chiral beta(2)-amino acids. *Org. Lett.* 8: 6131–6133.

98 Stueckler, C., Winkler, C.K., Hall, M. et al. (2011). Stereo-controlled asymmetric bioreduction of alpha,beta-dehydroamino acid derivatives. *Adv. Synth. Catal.* 353: 1169–1173.

99 Tasnádi, G., Winkler, C.K., Clay, D. et al. (2012). Reductive dehalogenation of beta-haloacrylic ester derivatives mediated by ene-reductases. *Catal. Sci. Technol.* 2: 1548–1552.

100 Winkler, C.K., Stueckler, C., Mueller, N.J. et al. (2010). Asymmetric synthesis of O-protected acyloins using enoate reductases: stereochemical control through protecting group modification. *Eur. J. Org. Chem.*: 6354–6358.

101 Hall, M., Stueckler, C., Kroutil, W. et al. (2007). Asymmetric bioreduction of activated alkenes using cloned 12-oxophytodienoate reductase isoenzymes OPR-1 and OPR-3 from *Lycopersicon esculentum* (tomato): a striking change of stereoselectivity. *Angew. Chem. Int. Ed.* 46: 3934–3937.

102 Padhi, S.K., Bougioukou, D.J., and Stewart, J.D. (2009). Site-saturation mutagenesis of tryptophan 116 of *Saccharomyces pastorianus* old yellow enzyme uncovers stereocomplementary variants. *J. Am. Chem. Soc.* 131: 3271–3280.

103 Amato, E.D. and Stewart, J.D. (2015). Applications of protein engineering to members of the old yellow enzyme family. *Biotechnol. Adv.* 33: 624–631.

104 Toogood, H.S. and Scrutton, N.S. (2018). Discovery, characterization, engineering, and applications of ene-reductases for industrial biocatalysis. *ACS Catal.* 8: 3532–3549.

105 Turrini, N.G., Hall, M., and Faber, K. (2015). Enzymatic synthesis of optically active lactones via asymmetric bioreduction using ene-reductases from the old yellow enzyme family. *Adv. Synth. Catal.* 357: 1861–1871.

106 Mangan, D., Miskelly, I., and Moody, T.S. (2012). A three-enzyme system involving an ene-reductase for generating valuable chiral building blocks. *Adv. Synth. Catal.* 354: 2185–2190.

107 Leisch, H., Morley, K., and Lau, P.C.K. (2011). Baeyer–Villiger monooxygenases: more than just green chemistry. *Chem. Rev.* 111: 4165–4222.

108 Donoghue, N.A., Norris, D.B., and Trudgill, P.W. (1976). Purification and properties of cyclohexanone oxygenase from *Nocardia globerula* Cl1 and *Acinetobacter* NCIB-9871. *Eur. J. Biochem.* 63: 175–192.

109 Rial, D.V., Bianchi, D.A., Kapitanova, P. et al. (2008). Stereoselective desymmetrizations by recombinant whole cells expressing the Baeyer–Villiger monooxygenase from *Xanthobacter* sp. ZL5: a new biocatalyst accepting structurally demanding substrates. *Eur. J. Org. Chem.*: 1203–1213.

110 Iwaki, H., Hasegawa, Y., Wang, S.Z. et al. (2002). Cloning and characterization of a gene cluster involved in cyclopentanol metabolism in *Comamonas* sp. strain NCIMB 9872 and biotransformations effected by *Escherichia coli*-expressed cyclopentanone 1,2-monooxygenase. *Appl. Environ. Microbiol.* 68: 5671–5684.

111 Fraaije, M.W., Wu, J., Heuts, D.P.H.M. et al. (2005). Discovery of a thermostable Baeyer–Villiger monooxygenase by genome mining. *Appl. Microbiol. Biotechnol.* 66: 393–400.

112 Iwaki, H., Wang, S.Z., Grosse, S. et al. (2006). Pseudomonad cyclopentadecanone monooxygenase displaying an uncommon spectrum of Baeyer–Villiger oxidations of cyclic ketones. *Appl. Environ. Microbiol.* 72: 2707–2720.

113 Kamerbeek, N.M., Moonen, M.J.H., van der Ven, J.G.M. et al. (2001). 4-Hydroxyacetophenone monooxygenase from *Pseudomonas fluorescens* ACB - a novel flavoprotein catalyzing Baeyer–Villiger oxidation of aromatic compounds. *Eur. J. Biochem.* 268: 2547–2557.

114 Morii, S., Sawamoto, S., Yamauchi, Y. et al. (1999). Steroid monooxygenase of *Rhodococcus rhodochrous*: sequencing of the genomic DNA, and hyperexpression, purification, and characterization of the recombinant enzyme. *J. Biochem.* 126: 624–631.

115 Balke, K., Kadow, M., Mallin, H. et al. (2012). Discovery, application and protein engineering of Baeyer–Villiger monooxygenases for organic synthesis. *Org. Biomol. Chem.* 10: 6249–6265.

116 de Gonzalo, G., van Berkel, W.J.H., and Fraaije, M.W. (2015). Baeyer–Villiger oxidation. In: *Science of Synthesis - Biocatalysis in Organic Synthesis* (eds. K. Faber, W.-D. Fessner and N.J. Turner), 187–233. Stuttgart: Thieme.

117 Li, G.Y., Garcia-Borras, M., Furst, M.J.L.J. et al. (2018). Overriding traditional electronic effects in biocatalytic Baeyer–Villiger reactions by directed evolution. *J. Am. Chem. Soc.* 140: 10464–10472.

118 Kirschner, A. and Bornscheuer, U.T. (2006). Kinetic resolution of 4-hydroxy-2-ketones catalyzed by a Baeyer–Villiger monooxygenase. *Angew. Chem. Int. Ed.* 45: 7004–7006.

119 Cernuchova, P. and Mihovilovic, M.D. (2007). Microbial Baeyer–Villiger oxidation of terpenones by recombinant whole-cell biocatalysts - formation of enantiocomplementary regioisomeric lactones. *Org. Biomol. Chem.* 5: 1715–1719.

120 Wang, S.Z., Chen, G., Kayser, M.M. et al. (2002). Baeyer–Villiger oxidations catalyzed by engineered microorganisms: enantioselective synthesis of delta-valerolactones with functionalized chains. *Can. J. Chem./Rev. Can. Chim.* 80: 613–621.

121 Stewart, J.D., Reed, K.W., Zhu, J. et al. (1996). A "designer yeast" that catalyzes the kinetic resolutions of 2-alkyl-substituted cyclohexanones by enantioselective Baeyer–Villiger oxidations. *J. Org. Chem.* 61: 7652–7653.

122 Rehdorf, J., Lengar, A., Bornscheuer, U.T., and Mihovilovic, M.D. (2009). Kinetic resolution of aliphatic acyclic beta-hydroxyketones by recombinant whole-cell Baeyer–Villiger monooxygenases-formation of enantiocomplementary regioisomeric esters. *Bioorg. Med. Chem. Lett.* 19: 3739–3743.

123 Rodriguez, C., de Gonzalo, G., Fraaije, M.W., and Gotor, V. (2007). Enzymatic kinetic resolution of racemic ketones catalyzed by Baeyer–Villiger monooxygenases. *Tetrahedron Asymmetry* 18: 1338–1344.

124 Berezina, N., Alphand, V., and Furstoss, R. (2002). Microbiological transformations. Part 51: the first example of a dynamic kinetic resolution process applied to a microbiological Baeyer–Villiger oxidation. *Tetrahedron Asymmetry* 13: 1953–1955.

125 Rioz-Martinez, A., de Gonzalo, G., Pazmino, D.E.T. et al. (2010). Synthesis of chiral 3-Alkyl-3,4-dihydroisocoumarins by dynamic kinetic resolutions catalyzed by a Baeyer–Villiger monooxygenase. *J. Org. Chem.* 75: 2073–2076.

126 Rodriguez, C., de Gonzalo, G., Rioz-Martinez, A. et al. (2010). BVMO-catalysed dynamic kinetic resolution of racemic benzyl ketones in the presence of anion exchange resins. *Org. Biomol. Chem.* 8: 1121–1125.

127 Taschner, M.J. and Black, D.J. (1988). The enzymatic Baeyer–Villiger oxidation - enantioselective synthesis of lactones from mesomeric cyclohexanones. *J. Am. Chem. Soc.* 110: 6892–6893.

128 Wang, S.Z., Kayser, M.M., Iwaki, H., and Lau, P.C.K. (2003). Monooxygenase-catalyzed Baeyer–Villiger oxidations: CHMO versus CPMO. *J. Mol. Catal. B Enzym.* 22: 211–218.

129 Rudroff, F., Rydz, J., Ogink, F.H. et al. (2007). Comparing the stereoselective biooxidation of cyclobutanones by recombinant strains expressing bacterial Baeyer–Villiger monooxygenases. *Adv. Synth. Catal.* 349: 1436–1444.

130 Mihovilovic, M.D., Rudroff, F., Grotzl, B., and Stanetty, P. (2005). Microbial Baeyer–Villiger oxidation of prochiral polysubstituted cyclohexanones by recombinant whole-cells expressing two bacterial monooxygenases. *Eur. J. Org. Chem.*: 809–816.

131 Mihovilovic, M.D., Rudroff, F., Muller, B., and Stanetty, P. (2003). First enantiodivergent Baeyer–Villiger oxidation by recombinant whole-cells expressing two monooxygenases from *Brevibacterium*. *Bioorg. Med. Chem. Lett.* 13: 1479–1482.

132 Gagnon, R., Grogan, G., Groussain, E. et al. (1995). Oxidation of some prochiral 3-substituted cyclobutanones using monooxygenase enzymes - a single-step method for the synthesis of optically enriched 3-substituted gamma-lactones. *J. Chem. Soc., Perkin Trans. 1*: 2527–2528.

133 Berezina, N., Kozma, E., Furstoss, R., and Alphand, V. (2007). Asymmetric Baeyer–Villiger biooxidation of alpha-substituted cyanocyclohexanones: influence of the substituent length on regio- and enantioselectivity. *Adv. Synth. Catal.* 349: 2049–2053.

134 Konigsberger, K. and Griengl, H. (1994). Microbial Baeyer–Villiger reaction of bicyclo 3.2.0 heptan-6-ones--a novel approach to sarkomycin A. *Bioorg. Med. Chem.* 2: 595–604.

135 Wang, S.Z., Kayser, M.M., and Jurkauskas, V. (2003). Access to optically pure 4- and 5-substituted lactones: a case of chemical-biocatalytical cooperation. *J. Org. Chem.* 68: 6222–6228.

136 Zhang, Y., Liu, F., Xu, N. et al. (2018). Discovery of two native Baeyer–Villiger monooxygenases for asymmetric synthesis of bulky chiral sulfoxides. *Appl. Environ. Microbiol.* 84: e00638-18.

137 Sheng, D.W., Ballou, D.P., and Massey, V. (2001). Mechanistic studies of cyclohexanone monooxygenase: chemical properties of intermediates involved in catalysis. *Biochemistry (Mosc)* 40: 11156–11167.

138 Zambianchi, F., Fraaije, M.W., Carrea, G. et al. (2007). Titration and assignment of residues that regulate the enantioselectivity of phenylacetone monooxygenase. *Adv. Synth. Catal.* 349: 1327–1331.

139 Furst, M.J.L.J., Gran-Scheuch, A., Aalbers, F.S., and Fraaije, M.W. (2019). Baeyer–Villiger monooxygenases: tunable oxidative biocatalysts. *ACS Catal.* 9: 11207–11241.

140 Chenprakhon, P., Wongnate, T., and Chaiyen, P. (2019). Monooxygenation of aromatic compounds by flavin-dependent monooxygenases. *Protein Sci.* 28: 8–29.

141 Li, A.T. and Li, Z. (2015). Asymmetric synthesis of enantiopure epoxides using monooxygenases. In: *Science of Synthesis - Biocatalysis in Organic Synthesis* (eds. K. Faber, W.-D. Fessner and N.J. Turner), 479–505. Stuttgart: Thieme.

142 Schmid, A., Hofstetter, K., Feiten, H.J. et al. (2001). Integrated biocatalytic synthesis on gram scale: the highly enantio selective preparation of chiral oxiranes with styrene monooxygenase. *Adv. Synth. Catal.* 343: 732–737.

143 Toda, H., Imae, R., and Itoh, N. (2012). Efficient biocatalysis for the production of enantiopure (*S*)-epoxides using a styrene monooxygenase (SMO) and *Leifsonia* alcohol dehydrogenase (LSADH) system. *Tetrahedron Asymmetry* 23: 1542–1549.

144 Lin, H., Liu, Y., and Wu, Z.L. (2011). Asymmetric epoxidation of styrene derivatives by styrene monooxygenase from *Pseudomonas* sp. LQ26: effects of alpha- and beta-substituents. *Tetrahedron Asymmetry* 22: 134–137.

145 Lin, H., Liu, Y., and Wu, Z.L. (2011). Highly diastereo- and enantio-selective epoxidation of secondary allylic alcohols catalyzed by styrene monooxygenase. *Chem. Commun.* 47: 2610–2612.

146 Panke, S., Meyer, A., Huber, C.M. et al. (1999). An alkane-responsive expression system for the production of fine chemicals. *Appl. Environ. Microbiol.* 65: 2324–2332.

147 Wubbolts, M.G., Hoven, J., Melgert, B., and Witholt, B. (1994). Efficient production of optically-active styrene epoxides in 2-liquid phase cultures. *Enzym. Microb. Technol.* 16: 887–894.

148 Desmet, M.J., Wynberg, H., and Witholt, B. (1981). Synthesis of 1,2-epoxyoctane by *Pseudomonas oleovorans* during growth in a 2-phase system containing high-concentrations of 1-octene. *Appl. Environ. Microbiol.* 42: 811–816.

149 Desmet, M.J., Witholt, B., and Wynberg, H. (1981). Practical approach to high-yield enzymatic stereospecific organic-synthesis in multiphase systems. *J. Org. Chem.* 46: 3128–3131.

150 Fu, H., Shen, G.J., and Wong, C.H. (1991). Asymmetric epoxidation of allyl alcohol derivatives by omega-hydroxylase from *Pseudomonas oleovorans*. *Recl. Trav. Chim. Pays-Bas* 110: 167–170.

151 Fu, H., Newcomb, M., and Wong, C.H. (1991). *Pseudomonas oleovorans* monooxygenase catalyzed asymmetric epoxidation of allyl alcohol derivatives and hydroxylation of a hypersensitive radical probe with the radical ring-opening rate exceeding the oxygen rebound rate. *J. Am. Chem. Soc.* 113: 5878–5880.

152 Habetscrutzen, A.Q.H., Carlier, S.J.N., Debont, J.A.M. et al. (1985). Stereospecific formation of 1,2-epoxypropane, 1,2-epoxybutane and 1-chloro-2,3-epoxypropane by alkene-utilizing bacteria. *Enzym. Microb. Technol.* 7: 17–21.

153 Weijers, C.A.G.M., Vanginkel, C.G., and Debont, J.A.M. (1988). Enantiomeric composition of lower epoxyalkanes produced by methane-utilizing, alkane-utilizing, and alkene-utilizing bacteria. *Enzym. Microb. Technol.* 10: 214–218.

154 Li, A.T., Wu, S.K., Adams, J.P. et al. (2014). Asymmetric epoxidation of alkenes and benzylic hydroxylation with P450tol monooxygenase from *Rhodococcus coprophilus* TC-2. *Chem. Commun.* 50: 8771–8774.

155 Geier, M., Bachler, T., Hanlon, S.P. et al. (2015). Human FMO_2-based microbial whole-cell catalysts for drug metabolite synthesis. *Microb. Cell Factories* 14: 82.

156 Hanlon, S.P., Camattari, A., Abad, S. et al. (2012). Expression of recombinant human flavin monooxygenase and moclobemide-*N*-oxide synthesis on multi-mg scale. *Chem. Commun.* 48: 6001–6003.

157 Brondani, P.B., Fraaije, M.W., and de Gonzalo, G. (2016). Recent developments in flavin-based catalysis: enzymatic sulfoxidation. In: *Green Biocatalysis* (ed. R.N. Patel), 149–164. Hoboken, NJ: Wiley.

158 Grogan, G. (2015). Oxidation at sulfur. In: *Science of Synthesis - Biocatalysis in Organic Synthesis* (eds. K. Faber, W.-D. Fessner and N.J. Turner), 285–312. Stuttgart: Thieme.

159 Carrea, G., Redigolo, B., Riva, S. et al. (1992). Effects of substrate structure on the enantioselectivity and stereochemical course of sulfoxidation catalyzed by cyclohexanone monooxygenase. *Tetrahedron Asymmetry* 3: 1063–1068.

160 Secundo, F., Carrea, G., Dallavalle, S., and Franzosi, G. (1993). Asymmetric oxidation of sulfides by cyclohexanone monooxygenase. *Tetrahedron Asymmetry* 4: 1981–1982.

161 Kamerbeek, N.M., Olsthoorn, A.J.J., Fraaije, M.W., and Janssen, D.B. (2003). Substrate specificity and enantioselectivity of 4-hydroxyacetophenone monooxygenase. *Appl. Environ. Microbiol.* 69: 419–426.

162 Colonna, S., Gaggero, N., Carrea, G., and Pasta, P. (1997). A new enzymatic enantioselective synthesis of dialkyl sulfoxides catalysed by monooxygenases. *Chem. Commun.*: 439–440.

163 de Gonzalo, G., Pazmino, D.E.T., Ottolina, G. et al. (2005). Oxidations catalyzed by phenylacetone monooxygenase from *Thermobifida fusca*. *Tetrahedron Asymmetry* 16: 3077–3083.

164 Rioz-Martinez, A., de Gonzalo, G., Pazmino, D.E.T. et al. (2010). Enzymatic synthesis of novel chiral sulfoxides employing Baeyer–Villiger monooxygenases. *Eur. J. Org. Chem.*: 6409–6416.

165 de Gonzalo, G., Pazmino, D.E.T., Ottolina, G. et al. (2006). 4-Hydroxyacetophenone monooxygenase from *Pseudomonas fluorescens* ACB as an oxidative biocatalyst in the synthesis of optically active sulfoxides. *Tetrahedron Asymmetry* 17: 130–135.

166 Colonna, S., Gaggero, N., Bertinotti, A. et al. (1995). Enantioselective oxidation of 1,3-dithioacetals catalyzed by cyclohexanone monooxygenase. *J. Chem. Soc. Chem. Commun.*: 1123–1124.

167 Colonna, S., Gaggero, N., Carrea, G., and Pasta, P. (1996). Enantio and diastereoselectivity of cyclohexanone monooxygenase catalyzed oxidation of 1,3-dithioacetals. *Tetrahedron Asymmetry* 7: 565–570.

168 Colonna, S., Gaggero, N., Carrea, G. et al. (2001). Enantioselective synthesis of *tert*-butyl *tert*-butanethiosulfinate catalyzed by cyclohexanone monooxygenase. *Chirality* 13: 40–42.

169 Bong, Y.K., Song, S.W., Nazor, J. et al. (2018). Baeyer–Villiger monooxygenase-mediated synthesis of esomeprazole as an alternative for kagan sulfoxidation. *J. Org. Chem.* 83: 7453–7458.

170 Wilson, R. (2015). Evolving BVMOs for increasing the yield and purity of enzyme catalyzed oxidation reactions. *Chim. Oggi – Chem. Today* 33: 50–52.

171 de Montellano, P.R.O. (2010). Hydrocarbon hydroxylation by cytochrome P450 enzymes. *Chem. Rev.* 110: 932–948.

172 Bormann, S., Baraibar, A.G., Ni, Y. et al. (2015). Specific oxyfunctionalisations catalysed by peroxygenases: opportunities, challenges and solutions. *Catal. Sci. Technol.* 5: 2038–2052.

173 Hofrichter, M. and Ullrich, R. (2014). Oxidations catalyzed by fungal peroxygenases. *Curr. Opin. Chem. Biol.* 19: 116–125.

174 Faiza, M., Huang, S.F., Lan, D.M., and Wang, Y.H. (2019). New insights on unspecific peroxygenases: superfamily reclassification and evolution. *BMC Evol. Biol.* 19: 76.

175 Ulrich, R., Nuske, J., Scheibner, K. et al. (2004). Novel haloperoxidase from the agaric basidiomycete *Agrocybe aegerita* oxidizes aryl alcohols and aldehydes. *Appl. Environ. Microbiol.* 70: 4575–4581.

176 Anh, D.H., Ullrich, R., Benndorf, D. et al. (2007). The coprophilous mushroom *Coprinus radians* secretes a haloperoxidase that catalyzes aromatic peroxygenation. *Appl. Environ. Microbiol.* 73: 5477–5485.

177 Grobe, G., Ullrich, R., Pecyna, M.J. et al. (2011). High-yield production of aromatic peroxygenase by the agaric fungus *Marasmius rotula*. *AMB Express* 1: 31.

178 Kluge, M., Ullrich, R., Scheibner, K., and Hofrichter, M. (2012). Stereoselective benzylic hydroxylation of alkylbenzenes and epoxidation of styrene derivatives catalyzed by the peroxygenase of *Agrocybe aegerita*. *Green Chem.* 14: 440–446.

179 Kinne, M., Poraj-Kobielska, M., Aranda, E. et al. (2009). Regioselective preparation of 5-hydroxypropranolol and 4′-hydroxydiclofenac with a fungal peroxygenase. *Bioorg. Med. Chem. Lett.* 19: 3085–3087.

180 Tieves, F., Willot, S.J.P., van Schie, M.M.C.H. et al. (2019). Formate oxidase (FOx) from *Aspergillus oryzae*: one catalyst enables diverse H_2O_2-dependent biocatalytic oxidation reactions. *Angew. Chem. Int. Ed.* 58: 7873–7877.

181 Wang, Y.H., Lan, D.M., Durrani, R., and Hollmann, F. (2017). Peroxygenases en route to becoming dream catalysts. What are the opportunities and challenges? *Curr. Opin. Chem. Biol.* 37: 1–9.

182 Sheldon, R.A.A., Brady, D., and Bode, M.L.L. (2020). The Hitchhiker's guide to biocatalysis: recent advances in the use of enzymes in organic synthesis. *Chem. Sci.* 11: 2587–2605.

183 Williams, G. and Hall, M. (eds.) (2018). *Modern Biocatalysis: Advances towards Synthetic Biological Systems*. Croydon: Royal Society of Chemistry.

8

Glycosyltransferase Cascades Made Fit For the Biocatalytic Production of Natural Product Glycosides

Bernd Nidetzky[1,2]

[1] *Graz University of Technology, Institute of Biotechnology and Biochemical Engineering, NAWI Graz, Petersgasse 12, A-8010 Graz, Austria*
[2] *Austrian Centre of Industrial Biotechnology (acib), Krenngasse 37, A-8010 Graz, Austria*

8.1 Introduction: Glycosylated Natural Products and Leloir Glycosyltransferases

Natural products often have one or more sugars attached to their core structure (e.g. flavonoids, polyketides, and terpenoids) [1, 2]. The sugar component can be essential for their physiological activity, selectivity, and pharmacological properties [3]. Small molecules become better soluble and often more stable through glycosylation. Variation in the extent and type of glycosyl moieties attached may affect profoundly their function (e.g. as an antibiotic agent) and govern the suitability for different applications (e.g. through modulation of their bioavailability). Glycosylation thus becomes critical for efficacy in the uses considered. Glycosylated natural products have important therapeutic uses but are also applied as functional food additives and cosmetic ingredients [4–9]. Figure 8.1 shows selected examples of glycosylated products having potential for industrial applications. The development of robust methods enabling glycosylation to be carried out in a scalable manner is therefore a centrally important objective in the field. A large toolbox of chemical methods of glycosylation exists [10–13]. Glycosylation by enzyme catalysis is a powerful alternative [9, 14]. It can be used in a complementary fashion with chemical glycosylation [15, 16]. Engineering of the glycosylation is a promising approach to achieve functional diversification of natural products [1, 2]. New bioactive substances and drug leads might be developed in that way.

In nature, the selective modification of target compounds with sugars is catalyzed by glycosyltransferases (EC 2.4) [9, 17–20]. These enzymes use an activated donor substrate, typically a nucleoside diphosphate (NDP) sugar, for transfer of the glycosyl residue onto a specific position of an acceptor molecule (Figure 8.2). In terms of reaction stereochemical course, glycosyltransferases can be of the inverting or retaining type [18].

Biocatalysis for Practitioners: Techniques, Reactions and Applications, First Edition.
Edited by Gonzalo de Gonzalo and Iván Lavandera.

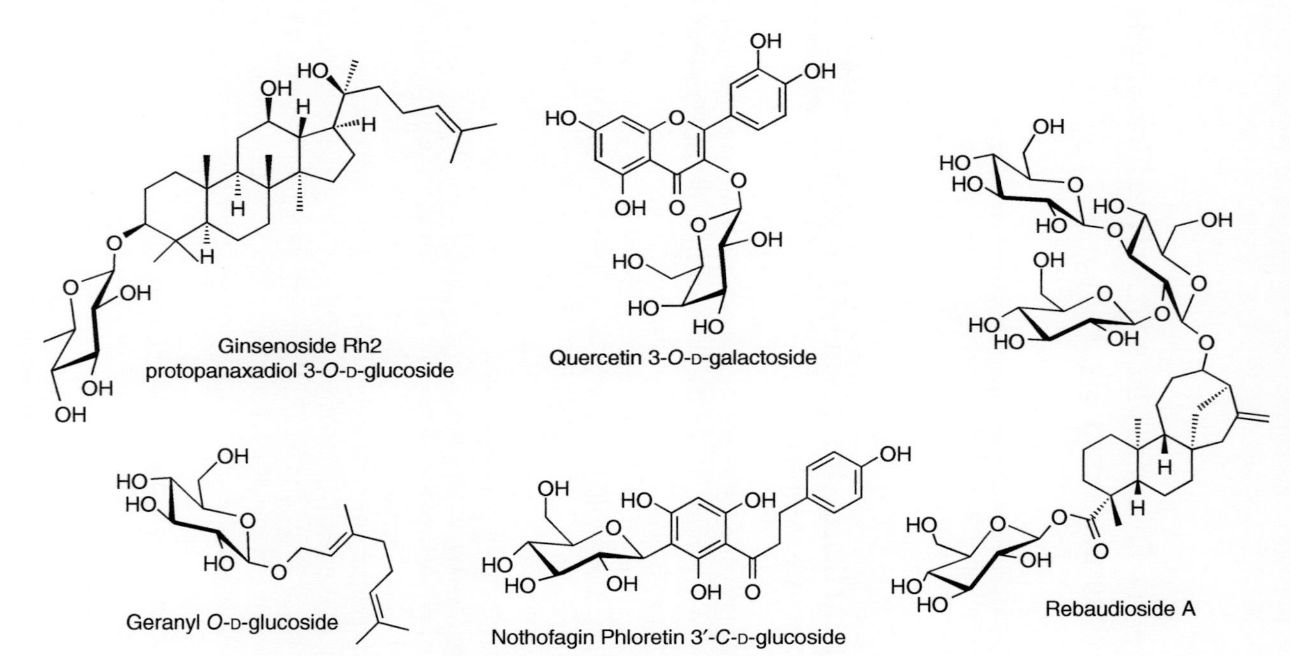

Figure 8.1 Structures of natural product glycosides that are of interest for industrial production. Source: Nidetzky et al. [9]. © 2018, American chemical society.

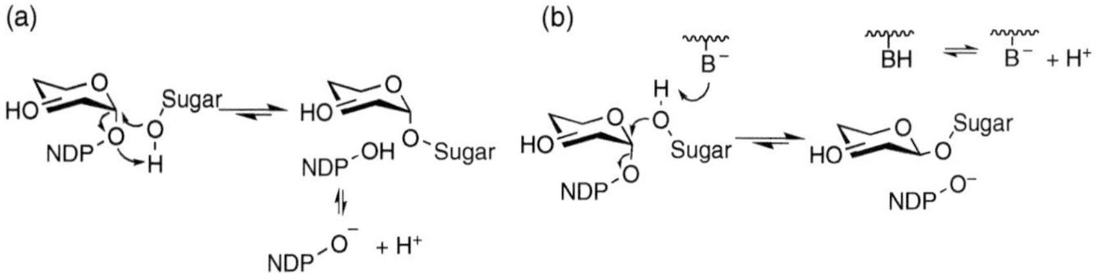

NDP: Nucleoside diphosphate

Figure 8.2 Retaining (a) and inverting (b) glycosyltransferase reaction. A sugar nucleotide is the donor substrate for the glycosylation of an acceptor molecule. The proton release occurs at pH conditions (pH ≥ 7.0) in which the nucleoside diphosphate is unprotonated. Source: Nidetzky et al. [9]. © 2018, American chemical society.

Glycosyltransferases usually show high regioselectivity and stereochemical control in the transformations catalyzed, and they are therefore recognized as highly valuable glycosylation catalysts [9, 17, 19, 21]. However, synthetic applications of glycosyltransferases have so far been quite restricted because of the complexities of these enzymes (e.g. low specific enzyme activity and stability) and the supply of donor and acceptor substrates for the enzymatic reactions [9].

In this chapter, the example of nothofagin (3′-*C*-β-D-glucoside of phloretin; Figure 8.1) is used to highlight issues encountered, and advances made, in the biocatalytic application of Leloir glycosyltransferases for natural product glycosylation. For other interesting examples, the reader is referred to [9] and references cited therein. Generally, a biocatalytic process is composed of three main steps: (i) the catalyst preparation, (ii) the biotransformation, and (iii) the product isolation. When using glycosyltransferases for nothofagin synthesis, we noticed that each step requires careful consideration; and an integrative development is important for a successful procedure [9, 22, 23]. Besides standard parameters of process efficiency (e.g. yield, atom economy, productivity, and catalyst usage), sustainability parameters such as waste formation in terms of the E-factor or related metrics become progressively more important as the production increases in scale [24]. Product quality in terms of a uniform glycostructure is also important in the synthesis. Development of the nothofagin synthesis has taken into consideration each of these aspects of the biocatalytic process.

8.2 Glycosylated Flavonoids and Nothofagin

Flavonoids are plant-derived polyphenols widely known for their important roles in the human diet related to health [25, 26]. Many flavonoids occur naturally as glycosides. The glycosylation impacts the solubility, stability, and bioactivity of flavonoids, thus making their glycosides attractive as food additives, therapeutics, and nutraceuticals [27, 28]. Nothofagin is a prominent example of a unique class of flavonoid glycosides featuring a rare-in-nature *C*-glycosidic structure [29]. Unlike the naturally widespread flavonoid

O-glycosides, the corresponding *C*-glycosides cannot be degraded via canonical hydrolysis of the glycoside linkage [30, 31]. They therefore show superior biological stability, resulting in improved long-term efficacy during application [32, 33]. Nothofagin (Figure 8.1) is almost exclusively found in rooibos plant (*Aspalathus linearis*) and linked to various health-promoting effects of its herbal tea [34, 35]. Nothofagin has attracted considerable interest because of its various biological activities [36–40].

Despite recent progress in isolating nothofagin from rooibos [41], natural scarcity restricts the application of the *C*-glucoside in food and pharmaceutical products. In addition, the chemical synthesis of nothofagin is difficult and apparently not suitable for large-scale production [42]. These problems are not just relevant for nothofagin but concern the flavonoid glycoside production in general [43–45]. The biocatalytic synthesis, via enzymatic glycosylation of the flavonoid core, was often found useful in preparing the corresponding flavonoid *O*-glycosides [46–50]. The specific problem of nothofagin is that the glycosylation strategies established for *O*-glycosides are inappropriate for *C*-glycosides. Installing the *C*-glycosidic linkage necessitates special enzymes from the class of Leloir glycosyltransferases, distinct from the enzymes able and used previously to form flavonoid *O*-glycosides [51–53]. These Leloir glycosyltransferases are not well established for application in biocatalytic synthesis. A *C*-glucosyltransferase from rice (*Oryza sativa*) catalyzes the site- and chemo-selective 3′-*C*-β-D-glucosylation of phloretin from uridine 5′-diphospho (UDP)-glucose [51]. Promising design of a biocatalytic process for nothofagin involves a parallel reaction cascade of *C*-glucosyltransferase and sucrose synthase, as depicted in Scheme 8.1 [54]. The one-pot glucosylation of phloretin thus occurs from sucrose via an UDP/UDP-glucose shuttle [55]. Sucrose is an expedient donor substrate for (enzymatic) glucosylation processes [56, 57].

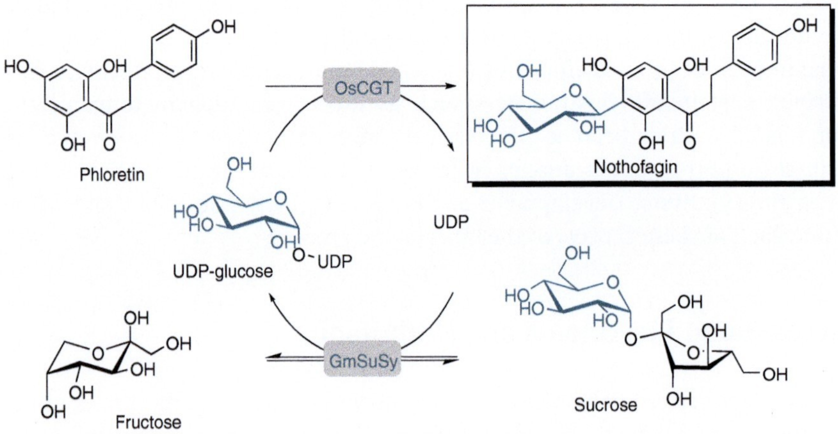

Scheme 8.1 Parallel cascade reaction for synthesis of nothofagin from phloretin and sucrose in the presence of UDP. *Os*CGT, *C*-glycosyltransferase from rice (*Oryza sativa*); *Gm*SuSy, sucrose synthase from soybean (*Glycine max*). The CGT reaction is largely irreversible. The SuSy reaction is freely reversible. Source: Schmölzer et al. [23]. © 2018 John Wiley & Sons.

8.3 Glycosyltransferase Cascades for Biocatalytic Synthesis of Nothofagin

For efficient nothofagin synthesis, the *C*-glucosyltransferase reaction was coupled to enzymatic *in situ* supply of the glucosyl donor substrate (Scheme 8.1): UDP-glucose is produced from sucrose and UDP using recombinant sucrose synthase from soybean (*Glycine max*; *Gm*SuSy) [55, 58]. Although the applied internal UDP-glucose regeneration is known in principle and has been applied to enzymatic reactions involving different glycosyl donor substrates [57], critical test of its performance capability in the synthesis of natural product glucosides such as nothofagin was outstanding. Thermodynamic, kinetic, and stability parameters of the coupled enzymatic reaction were evaluated systematically.

Enzymes are obtained from *Escherichia coli* expression cultures and purified to apparent homogeneity by Strep-tag affinity chromatography. Their activities are determined using enzymatic or HPLC-based assays. Reaction of *Os*CGT is monitored with an HPLC assay capable of distinguishing between nothofagin and potential alternative products resulting from *O*-glucosyl transfer at the 2′ or 4′ position of the acceptor (see Scheme 8.1) [59]. *Os*CGT displays absolute selectivity (within error limit of ≤0.5%) for 3′-*C*-glycosylation of phloretin. We also determined pH-activity dependencies for *Gm*SuSy (sucrose cleavage and synthesis) and *Os*CGT (nothofagin synthesis) at 30 °C. The resulting pH profiles show suitable overlap of the enzyme activities in the pH range 6.5–8.0. *C*-Glucosylation of phloretin at pH 7.5 resulted in high conversion of the substrates (≥95%).

The *C*-glycosylation of phloretin is not detectably reversible (see Figure 8.3). It therefore does not require thermodynamic push from a coupled reaction, which the conversion of sucrose could provide in principle. Only to note, the *O*-glycosylation of phloretin represents a more readily reversible reaction. Its coupling to the sucrose synthase reaction was therefore accompanied by a clear thermodynamic benefit. Inhibition of the *O. sativa C*-glycosyltransferase by UDP is not pronounced. Continuous removal of UDP from the *C*-glycosyltransferase reaction was not as critical as it appeared to be in other glycosyltransferase reactions. Important function of the sucrose synthase reaction in the overall nothofagin synthesis was therefore mainly in supplying UDP-glucose for the *C*-glucosylation. The equilibrium for sucrose conversion is pH-dependent, and a low pH of 6.0 or smaller favors the formation of UDP-glucose [58]. At pH 7.5, the equilibrium constant (K_{eq}) for conversion of sucrose and UDP has a value of approximately 0.49 (Figure 8.3). However, thermodynamic constraints on the supply of UDP-glucose at elevated pH can be eliminated effectively using sucrose in excess. Conversion studies were carried out at pH 7.5 and 30 °C where both glycosyltransferases showed useful activity and stability, and quantitative transformation of sucrose into nothofagin was feasible.

Kinetic characterization of different sucrose synthases revealed that a low UDP K_m was very important for enzyme efficiency in UDP-glucose regeneration. The *G. max* sucrose synthase [54, 55] was superior to the *Acidithiobacillus caldus* sucrose synthase [60] according to this kinetic criterion. The *A. caldus* enzyme preferred adenosine 5′-diphosphate

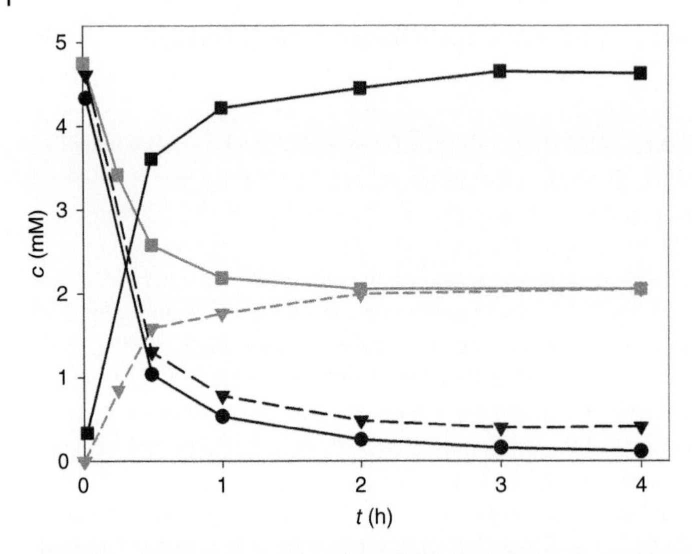

Figure 8.3 Time course analysis for individual enzymatic reactions catalyzed by *Os*CGT and GmSuSy at pH 7.5 and 30 °C. Nothofagin synthesis by *Os*CGT (black symbols): 80 mU ml^{-1}, 5 mM phloretin (triangle, dashed line), 4.75 mM UDP-glucose (circle, solid line), and nothofagin (square, solid line). Reactions of *Gm*SuSy (gray symbols): 50 mU ml^{-1}, 5 mM of each substrate, UDP-glucose in sucrose synthesis (squares, solid line), and cleavage (triangles, dashed line). Source: Bungaruang et al. [54]. © 2013 John Wiley & Sons.

(ADP) over UDP. Structure-guided engineering of the sucrose synthase was used to enhance the enzyme's affinity for UDP [61]. Engineered variants of the *A. caldus* sucrose synthase exhibited a K_m for UDP about 60-fold lowered compared to the wild-type enzyme. These variants showed improved performance in UDP-glucose recycling. However, the *G. max* sucrose synthase that naturally exhibits a high affinity for UDP ($K_m = 0.13$ mM) was the best enzyme candidate for coupling with the *C*-glycosyltransferase [62].

Basic parameters of the cascade reaction requiring optimization are the enzyme activity ratio (fivefold excess of *C*-glycosyltransferase) as well as the sucrose (100 mM) and the UDP concentration (0.5–1.0 mM). The final nothofagin concentration, the nothofagin yield on phloretin utilized, the productivity, and the number of cycles of the UDP/UDP-glucose shuttle (referred to as RC_{max}) are important parameters of the process efficiency. The total turnover number of the enzymes, limited by their operational stability (*Os*CGT: ~14 hours; *Gm*Suy: ~19 hours), is another important parameter [54].

8.4 Enzyme Expression

Limitation on the applicability of (plant) Leloir glycosyltransferases arises from their difficult expression in standard microbial hosts [9, 56, 63]. *Escherichia coli* is mostly used. Expression is low ($\leq$10 mg l^{-1}) in general ([55, 64]) or poor ($\leq$1 mg l^{-1}) in various instances

([65, 66]). With notable exceptions [64, 67–69], the enzyme production has received relatively little attention for systematic process development. The main bottlenecks on production efficiency thus remain largely unknown. Besides specific requirements an individual glycosyltransferase may have, it seems probable that there are also important factors of a more general if not universal relevance. Discovery of such factors, and process optimization along the lines thus suggested, would present important advance for the biocatalytic application of glycosyltransferases.

In a study of a bacterial Leloir glycosyltransferase (sucrose synthase from *A. caldus*; [60]), we showed that constitutive expression in *E. coli* BL21 shifted the production of recombinant protein mainly to the stationary growth phase [22]. Once the glucose carbon source had been depleted, active enzyme was accumulated gradually to a substantial titer of ~350 mg l^{-1} of culture. This result gave rise to the working hypothesis that expression in growth-arrested *E. coli* might constitute a general strategy for efficiency-enhanced production of (plant) glycosyltransferases.

We used a synthetic biology-based approach to decouple *E. coli* BL21(DE3) cell growth from target gene overexpression [70, 71]. The underlying concept is built upon the Gp2 protein from the bacteriophage T7. Gp2 inhibits the *E. coli* endogenous RNA polymerase [72] while it leaves T7 RNA polymerase unaffected [73]. An inducible, T7 RNA polymerase-based expression of the Gp2 gene thus allows for a constant supply of the inhibitor. This enables the *E. coli* transcription to be shut off, hence the cell growth to be arrested, in a controllable fashion. Under conditions managed by Gp2, therefore, the *E. coli* protein synthesis machinery is taken over for recombinant production of the target protein(s). The practical design embodied in enGenes technology involves genome integration of the Gp2 coding gene under control of the ara*B* promoter inducible by L-arabinose. Within this strain background, a pET plasmid vector is used that contains the gene(s) of interest inducible by isopropyl-β-D-galactoside. This design provides flexibility and temporal control for cell proliferation to be switched off and protein production to be induced [71, 73]. It presents a new approach toward quiescent *E. coli* cells applied to recombinant protein production (for alternative approaches, see [74–76]).

Application of the outlined approach to the production of *Gm*SuSy was demonstrated (Figure 8.4) [77]. Enzyme production in growth-arrested *E. coli* gave significant improvements in the amount and quality of the recombinant glycosyltransferase as compared to the exactly comparable production in the growing *E. coli* reference. *Gm*SuSy was obtained at 115 U g^{-1} cell dry weight, corresponding to ~5% of total intracellular protein. Transferability of the production strategy to a high-cell density fed-batch culture of *E. coli* at 20 l operating scale was shown and up to 830 mg glycosyltransferase protein (equivalent to 2300 activity units) l^{-1} of culture were obtained in that way. Analyzing the isolated glycosyltransferase, we showed that improvement in the enzyme production was due to enhancement of both amount (2.5-fold) and quality (2.4-fold) of the soluble sucrose synthase. Enzyme preparation from decoupled production comprised an increased portion (62% compared with 26%) of the active sucrose synthase homo-tetramer. In summary, therefore, expression in growth-arrested *E. coli* is promising for recombinant production of plant Leloir glycosyltransferases.

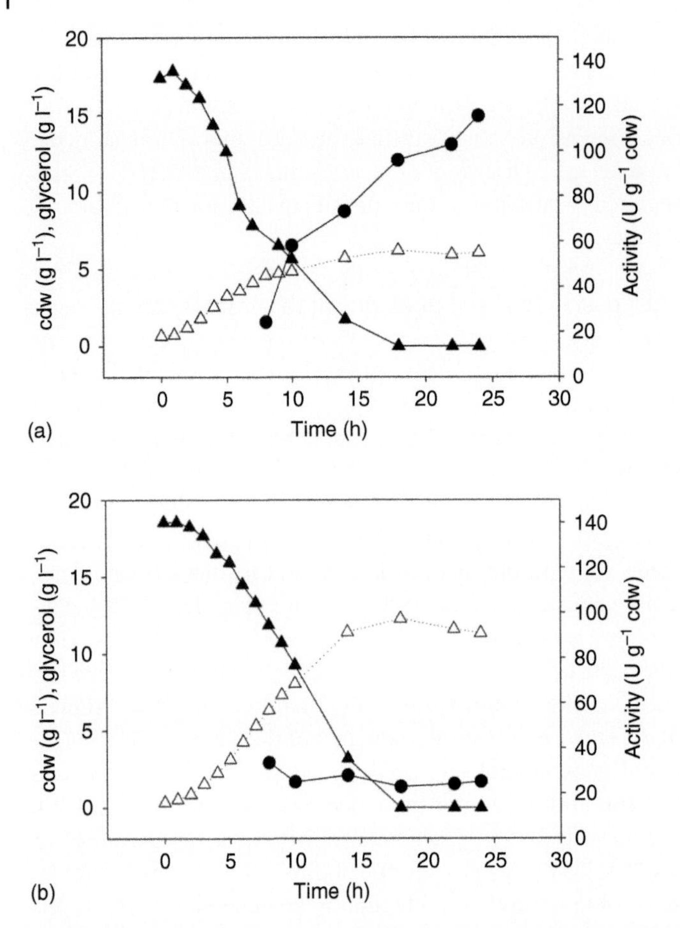

Figure 8.4 Time courses of growth, glycerol consumption, and enzyme formation in batch bioreactor cultivations of enGenes-X-press (a) and *E. coli* BL21(DE3) (b) producing GmSuSy. The induction temperature was 30 °C. Isopropyl-thio-β-galactoside (0.1 mM) was used for induction in both experiments. The cultivation in panel (a) used additionally 100 mM L-arabinose. The symbols show cell dry mass concentration, open triangles; glycerol concentration, full triangles; and enzyme activity/cell dry weight (cdw), full circles. Source: Lemmerer et al. [77]. © 2019 John Wiley & Sons.

8.5 Solvent Engineering for Substrate Solubilization

The low water solubility of phloretin (≤0.5 mM) was identified as the main bottleneck of an efficient nothofagin synthesis. The problem is important generally because natural product glycosylation very often involves a highly hydrophobic acceptor substrate. When using dimethyl sulfoxide (DMSO, 20% by volume) as a cosolvent, about 5 mM phloretin was dissolved and the same concentration of nothofagin was produced. To enhance the final nothofagin concentration, phloretin was added in 5 mM portions during the course of the reaction and about 45 mM product could so be formed (Figure 8.5) [54].

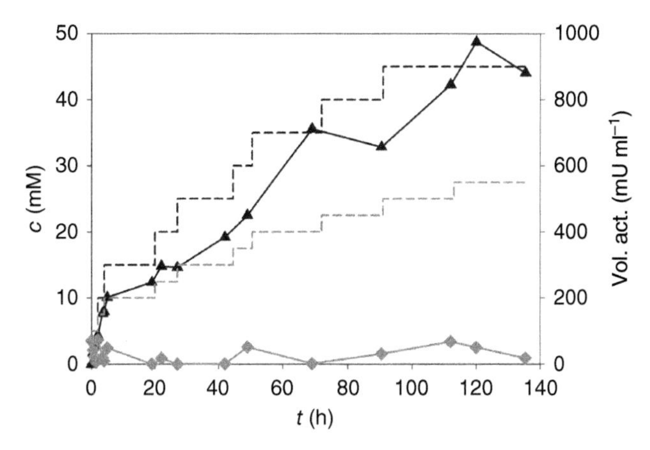

Figure 8.5 Synthesis of nothofagin in fed-batch reaction. Reaction conditions: 100 mU ml^{-1} *Os*CGT/*Gm*SuSy, 5 mM phloretin, 300 mM sucrose, and 1 mM UDP; 30 °C; 50 mM HEPES buffer, pH 7.5; and 20% (by vol.) DMSO. After acceptor substrate depletion, 5 mM phloretin and 50 mU ml^{-1} *Os*CGT/*Gm*SuSy were added. Symbols: phloretin added (black dashed), *Os*CGT/*Gm*SuSy added (gray dashed), nothofagin (black), and phloretin (gray). Source: Bungaruang et al. [54]. © 2013 John Wiley & Sons.

Among various strategies further considered to enhance phloretin solubility, inclusion complexation by β-cyclodextrins was identified as the option best compatible with enzyme stability and activity (Figures 8.6 and 8.7) [78].

The expedient and well-soluble 2-hydroxypropyl-β-cyclodextrin enabled solubilization of phloretin to a point (~150 mM) where viscous fluid mixing became a physical boundary on the process operation. The nothofagin synthesis thus pushed to upper practical limits gave a final product concentration of ~120 mM (50 g l^{-1}) (Figure 8.8) [23]. The UDP was regenerated 220 times. A limiting mass-based total turnover number of 100 was obtained for the *C*-glycosyltransferase in a single-batch conversion lasting 24 hours. Stability of the enzymes, especially that of the sucrose synthase, remains an issue and currently does not support recycling of the enzymes. Immobilization of the sucrose synthase, or of both enzymes, could present a useful solution for stabilization and reuse.

8.6 Nothofagin Production at 100 g Scale

Figure 8.7 shows the synthesis of ~50 g of nothofagin in a single batch (1 l) under optimized conditions. The initial nothofagin production rate was 20 g l^{-1} h^{-1}. Nothofagin was obtained in excellent yield (97%; based on the phloretin used) and concentration (50 g l^{-1}, 110 mM) within 16 hours of reaction. The space–time yield of the biotransformation overall was 3 g l^{-1} h^{-1}. The high product concentration was vital for efficient downstream processing, as is shown later and is generally recognized as important [79]. Compared to the experiments described above or reported previously [54, 78], nothofagin synthesis was up-scaled by 500-fold. Synthesis of different quantities of nothofagin gave almost identical yields and space–time yields, indicating a highly successful scale-up. In contrast to the enzymatic reaction

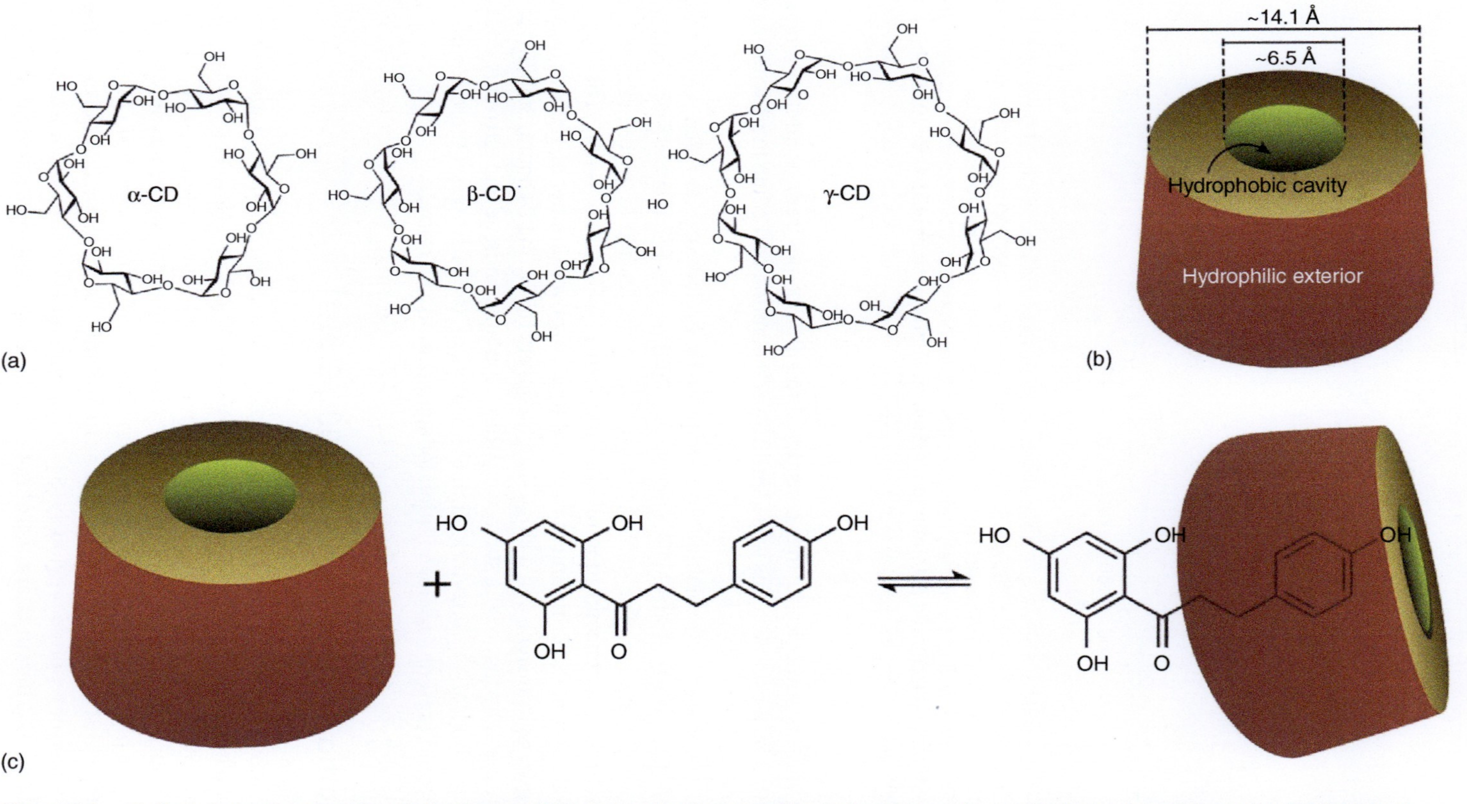

Figure 8.6 (a) Chemical structures of cyclodextrins and (b) molecular shape of β-cyclodextrin as relevant for host–guest interaction. (c) Proposed complexation of free phloretin through inclusion by β-cyclodextrin. Source: Bungaruang et al. [78]. © 2016 John Wiley & Sons.

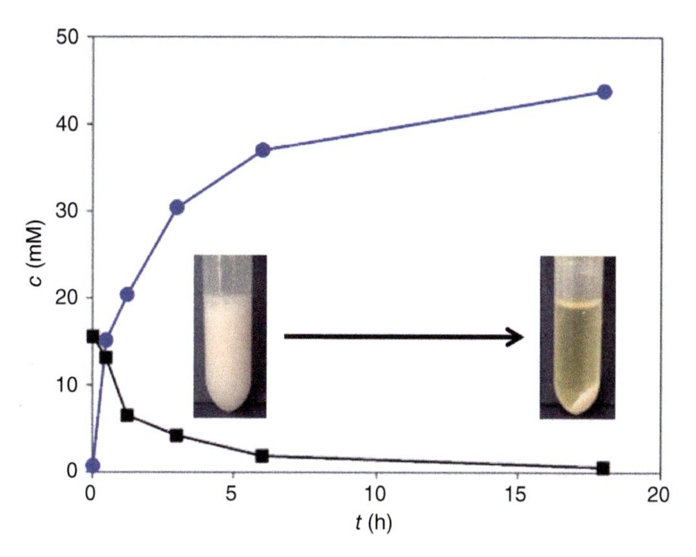

Figure 8.7 Nothofagin synthesis from phloretin suspension that contained phloretin/β-cyclodextrin inclusion complex formed *ex situ* by coprecipitation. Phloretin and β-cyclodextrin were used equimolar. Soluble phloretin (black) and nothofagin (blue) were analyzed. The time course of nothofagin production from 50 mM phloretin using 5 U ml^{-1} of *Os*CGT and 1 U ml of *Gm*SuSy is shown. Other conditions: 300 mM sucrose, 0.5 mM UDP, 50 mM HEPES buffer, pH 7.5, and 30 °C. The inset shows the appearance of the reaction mixture before and after conversion. The precipitate at reaction end is mostly unreacted phloretin. Source: Bungaruang et al. [78]. © 2016 John Wiley & Sons.

using the native β-cyclodextrin [78], phloretin precipitation was prevented effectively in the process and none of the nothofagin formed was lost as insoluble. The UDP-glucose was regenerated up to ~220 times (RC_{max}), representing a 2.5-fold improvement in recycling efficiency compared to the reference reaction using β-cyclodextrin. With efficiency parameters of the batch conversion largely pushed to their respective limits, further intensification of nothofagin synthesis could only be realized by changing to a continuous or semicontinuous process mode. This would however necessitate improvement of enzyme activities and stabilities, especially the stability of the *Gm*SuSy. However, even as it stands, the current nothofagin synthesis is performance-wise without precedent in preparative glycosyltransferase-catalyzed glycosylations of flavonoids and other polyphenols ([47, 50, 53, 54, 67, 78, 80–84]).

The downstream processing had to remove sugars (390 mM sucrose; 110 mM fructose) and ~140 mM 2-hydroxypropyl-β-cyclodextrin from the nothofagin (110 mM). We considered anion exchange chromatography (AEC) to be promising for capture and initial purification of the product. Preparative AEC on a SuperQ-650M column (~500 ml bed volume) enabled isolation of about 11 g of nothofagin in a single AEC run with a run time of 11 hours. The total solvent used was only ~0.61 $g^{-1}{}_{nothofagin}$. A low solvent consumption is vital for the scalability of AEC. About 80% of the 2-hydroxypropyl-β-cyclodextrin was removed with the flow through. Recycling of the 2-hydroxypropyl-β-cyclodextrin would thus be possible at this process step by extraction [85] or chromatography [86]. The

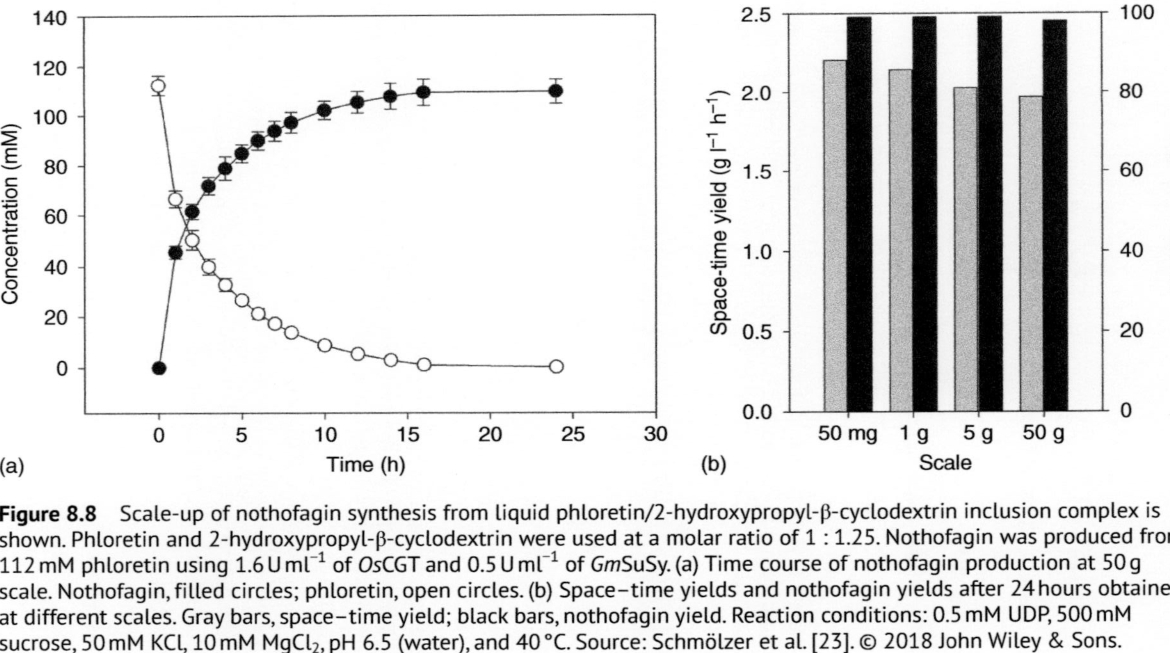

Figure 8.8 Scale-up of nothofagin synthesis from liquid phloretin/2-hydroxypropyl-β-cyclodextrin inclusion complex is shown. Phloretin and 2-hydroxypropyl-β-cyclodextrin were used at a molar ratio of 1 : 1.25. Nothofagin was produced from 112 mM phloretin using 1.6 U ml^{-1} of *Os*CGT and 0.5 U ml^{-1} of *Gm*SuSy. (a) Time course of nothofagin production at 50 g scale. Nothofagin, filled circles; phloretin, open circles. (b) Space–time yields and nothofagin yields after 24 hours obtained at different scales. Gray bars, space–time yield; black bars, nothofagin yield. Reaction conditions: 0.5 mM UDP, 500 mM sucrose, 50 mM KCl, 10 mM $MgCl_2$, pH 6.5 (water), and 40 °C. Source: Schmölzer et al. [23]. © 2018 John Wiley & Sons.

SuperQ-650M was also described as the most effective material for AEC of glucoses nucleotides [87, 88].

Then, precipitation of nothofagin was used as an alternative to size exclusion chromatography (SEC) for desalting. SEC is difficult to scale up because its throughput is low and highly diluted product is obtained [88, 89]. The AEC eluate was concentrated (~3-fold), paying attention to not exceed the solubility limit of sodium chloride (~6 M in water). A pH of 6 was used to precipitate the nonionic nothofagin from solution [90]. The precipitate had its residual water removed by freeze-drying and nothofagin was then extracted with acetone, which we found to be an excellent solvent for the *C*-glucoside (~100 $g\,l^{-1}$ at 20 °C). The 2-hydroxypropyl-β-cyclodextrin still present in the solid sample (~20% of total) was removed at this step because of its low solubility in dry acetone (<0.1 $g\,l^{-1}$ at 20 °C). Note: because 2-hydroxypropyl-β-cyclodextrin is well soluble in water, drying of the solid sample was important for selective acetone extraction of the nothofagin. Isopropanol was not a suitable solvent because of a 2-hydroxypropyl-β-cyclodextrin solubility of ~15 $g\,l^{-1}$ (at 20 °C). About 9 g of nothofagin was obtained in a single batch. We were able to isolate 102 g of nothofagin in 12 batches with constant quality. Nothofagin was obtained as a yellowish powder in ≥65% yield and ≥95% purity.

8.7 Concluding Remarks

The successful nothofagin process involves upstream and downstream processing optimized for efficient interconnected performance. It strongly supports the idea of a holistic process development for biocatalytic production of flavonoid glycosides by coupled Leloir glycosyltransferases in general. Numerous further applications of these enzymes in natural product glycosylation are envisaged [9]. Besides flavonoids (e.g. quercetin), a large class of terpenes is of interest, for their glycosides find increased uses as sweeteners (e.g. rebaudioside A) and as flavor and fragrance ingredients (e.g. geranyl β-D-glucoside). By including additional enzymatic steps into the reaction cascade (e.g. epimerization, oxidation at C′6), the scope of glycosylation from sucrose can be expanded to glycosyl residues (e.g. D-galactosyl and D-glucuronyl) other than D-glucosyl. In the production of flavonoid glycosides, building up the product by glycosylation presents a conceptually important alternative to trimming down a natural di-glycoside substrate ([91]).

References

1 Thibodeaux, C.J., Melançon, C.E., and Liu, H.W. (2008). Natural-product sugar biosynthesis and enzymatic glycodiversification. *Angew. Chem. Int. Ed.* 47: 9814–9859.
2 Elshahawi, S.I., Shaaban, K.A., Kharel, M.K., and Thorson, J.S. (2015). A comprehensive review of glycosylated bacterial natural products. *Chem. Soc. Rev.* 44: 7591–7697.
3 Gantt, R.W., Peltier-Pain, P., and Thorson, J.S. (2011). Enzymatic methods for glyco(diversification/randomization) of drugs and small molecules. *Nat. Prod. Rep.* 28: 1811–1853.

4 Xiao, J., Muzashvili, T.S., and Georgiev, M.I. (2014). Advances in the biotechnological glycosylation of valuable flavonoids. *Biotechnol. Adv.* 32: 1145–1156.

5 Schwab, W., Fischer, T., Giri, A., and Wüst, M. (2015). Potential applications of glucosyltransferases in terpene glucoside production: impacts on the use of aroma and fragrance. *Appl. Microbiol. Biotechnol.* 99: 165–174.

6 Schwab, W., Fischer, T., and Wüst, M. (2015). Terpene glucoside production: Improved biocatalytic processes using glycosyltransferases. *Eng. Life Sci.* 15: 376–386.

7 Kim, B.G., Yang, S.M., Kim, S.Y. et al. (2015). Biosynthesis and production of glycosylated flavonoids in *Escherichia coli*: current state and perspectives. *Appl. Microbiol. Biotechnol.* 99: 2979–2988.

8 Antonopoulou, I., Varriale, S., Topakas, E. et al. (2016). Enzymatic synthesis of bioactive compounds with high potential for cosmeceutical application. *Appl. Microbiol. Biotechnol.* 100: 6519–6543.

9 Nidetzky, B., Gutmann, A., and Zhong, C. (2018). Leloir glycosyltransferases as biocatalysts for chemical production. *ACS Catal.* 8: 6283–6300.

10 Yu, B. and Yang, Y. (2017). Recent advances in the chemical synthesis of *C*-glycosides. *Chem. Rev.* 117 (19): 12281–12356.

11 Kitamura, K., Ando, Y., Matsumoto, T., and Suzuki, K. (2018). Total synthesis of aryl *C*-glycoside natural products: strategies and tactics. *Chem. Rev.* 118: 1495–1598.

12 Kulkarni, S.S., Wang, C.-C., Sabbavarapu, N.M. et al. (2018). "One-pot" protection, glycosylation, and protection–glycosylation strategies of carbohydrates. *Chem. Rev.* 118: 8025–8104.

13 Nielsen, M.M. and Pedersen, C.M. (2018). Catalytic glycosylations in oligosaccharide synthesis. *Chem. Rev.* 118: 8285–8358.

14 Danby, P.M. and Withers, S.G. (2016). Advances in enzymatic glycoside synthesis. *ACS Chem. Biol.* 11: 1784–1794.

15 Li, T.H., Liu, L., Wei, N. et al. (2019). An automated platform for the enzyme-mediated assembly of complex oligosaccharides. *Nat. Chem.* 11: 229–236.

16 Wen, L., Edmunds, G., Gibbons, C. et al. (2018). Toward automated enzymatic synthesis of oligosaccharides. *Chem. Rev.* 118: 8151–8187.

17 Bowles, D., Lim, E.-K., Poppenberger, B., and Vaistij, F.E. (2006). Glycosyltransferases of lipophilic small molecules. *Annu. Rev. Plant Biol.* 57: 567–597.

18 Lairson, L.L., Henrissat, B., Davies, G.J., and Withers, S.G. (2008). Glycosyltransferases: structures, functions, and mechanisms. *Annu. Rev. Biochem.* 77: 521–555.

19 Palcic, M.M. (2011). Glycosyltransferases as biocatalysts. *Curr. Opin. Chem. Biol.* 15: 226–233.

20 Liang, D.-M., Liu, J.-H., Wu, H. et al. (2015). Glycosyltransferases: mechanisms and applications in natural product development. *Chem. Soc. Rev.* 44: 8350–8374.

21 Hofer, B. (2016). Recent developments in the enzymatic *O*-glycosylation of flavonoids. *Appl. Microbiol. Biotechnol.* 100: 4269–4281.

22 Schmölzer, K., Lemmerer, M., Gutmann, A., and Nidetzky, B. (2017). Integrated process design for biocatalytic synthesis by a Leloir glycosyltransferase: UDP-glucose production with sucrose synthase. *Biotechnol. Bioeng.* 114: 924–928.

23 Schmölzer, K., Lemmerer, M., and Nidetzky, B. (2018). Glycosyltransferase cascades made fit for chemical production: integrated biocatalytic process for the natural polyphenol *C*-glucoside nothofagin. *Biotechnol. Bioeng.* 115: 545–556.

24 Sheldon, R.A. and Woodley, J.M. (2018). Role of biocatalysis in sustainable chemistry. *Chem. Rev.* 118: 801–838.

25 Andersen, O.M. and Markham, K.R. (2005). *Flavonoids: Chemistry, Biochemistry and Applications*. Boca Raton, FL, USA: CRC Press.

26 Rodriguez-Mateos, A., Vauzour, D., Krueger, C.G. et al. (2014). Bioavailability, bioactivity and impact on health of dietary flavonoids and related compounds: an update. *Arch. für Toxikol.* 88: 1803–1853.

27 De Bruyn, F., Maertens, J., Beauprez, J. et al. (2015). Biotechnological advances in UDP-sugar based glycosylation of small molecules. *Biotechnol. Adv.* 33: 288–302.

28 Kren, V. and Martinkova, L. (2001). Glycosides in medicine: the role of glycosidic residue in biological activity. *Curr. Med. Chem.* 8: 1303–1328.

29 Veitch, N.C. and Grayer, R.J. (2011). Flavonoids and their glycosides, including anthocyanins. *Nat. Prod. Rep.* 28: 1626–1695.

30 Bililign, T., Hyun, C.-G., Williams, J.S. et al. (2004). The hedamycin locus implicates a novel aromatic PKS priming mechanism. *Cell Chem. Biol.* 11: 959–969.

31 Härle, J., Günther, S., Lauinger, B. et al. (2011). Rational design of an aryl-*C*-glycoside catalyst from a natural product *O*-glycosyltransferase. *Chem. Biol.* 18: 520–530.

32 Lee, J., Lee, S.-H., Seo, H.J. et al. (2010). Novel *C*-aryl glucoside SGLT2 inhibitors as potential antidiabetic agents: 1,3,4-Thiadiazolylmethylphenyl glucoside congeners. *Bioorg. Med. Chem.* 18: 2178–2194.

33 Meng, W., Ellsworth, B.A., Nirschl, A.A. et al. (2008). Discovery of dapagliflozin: a potent, selective renal sodium-dependent glucose cotransporter 2 (SGLT2) inhibitor for the treatment of type 2 diabetes. *J. Med. Chem.* 51: 1145–1149.

34 Joubert, E., Beelders, T., de Beer, D. et al. (2012). Variation in phenolic content and antioxidant activity of fermented rooibos herbal tea infusions: role of production season and quality grade. *J. Agric. Food Chem.* 60: 9171–9179.

35 McKay, D.L. and Blumberg, J.B. (2007). A review of the bioactivity of south African herbal teas: rooibos (*Aspalathus linearis*) and honeybush (*Cyclopia intermedia*). *Phytother. Res.* 21: 1–16.

36 Ku, S.K., Lee, W., Kang, M., and Bae, J.S. (2014). Antithrombotic activities of aspalathin and nothofagin via inhibiting platelet aggregation and FIIa/FXa. *Arch. Pharm. Res.* 38: 1080–1089.

37 Ku, S.K., Kwak, S., Kim, Y., and Bae, J.S. (2015). Aspalathin and nothofagin from rooibos (*Aspalathus linearis*) inhibits high glucose-induced inflammation in vitro and in vivo. *Inflammation* 38: 445–455.

38 Schloms, L., Storbeck, K.-H., Swart, P. et al. (2012). The influence of *Aspalathus linearis* (rooibos) and dihydrochalcones on adrenal steroidogenesis: quantification of steroid intermediates and end products in H295R cells. *J. Steroid Biochem. Mol. Biol.* 128: 128–138.

39 Snijman, P.W., Swanevelder, S., Joubert, E. et al. (2007). The antimutagenic activity of the major flavonoids of rooibos (*Aspalathus linearis*): some dose-response effects on mutagen activation-flavonoid interactions. *Mutat. Res.* 631: 111–123.

40 Snijman, P.W., Joubert, E., Ferreira, D. et al. (2009). Antioxidant activity of the dihydrochalcones aspalathin and nothofagin and their corresponding flavones in relation to other rooibos (*Aspalathus linearis*) flavonoids, epigallocatechin gallate, and Zrolox. *J. Agric. Food Chem.* 57: 6678–6684.

41 de Beer, D., Malherbe, C.J., Beelders, T. et al. (2015). Isolation of aspalathin and nothofagin from rooibos (*Aspalathus linearis*) using high-performance countercurrent chromatography: sample loading and compound stability considerations. *J. Chromatogr. A* 1381: 29–36.

42 Yepremyan, A., Salehani, B., and Minehan, T.G. (2010). Concise total syntheses of aspalathin and nothofagin. *Org. Lett.* 12: 1580–1583.

43 Chen, R. (2018). Enzyme and microbial technology for synthesis of bioactive oligosaccharides: an update. *Appl. Microbiol. Biotechnol.* 102: 3017–3026.

44 Lim, E.K., Ashford, D.A., Hou, B. et al. (2004). *Arabidopsis* glycosyltransferases as biocatalysts in fermentation for regioselective synthesis of diverse quercetin glucosides. *Biotechnol. Bioeng.* 87: 623–631.

45 Zhang, T.-J., Liang, J.-Q., Wei, X.-Y. et al. (2017). Development of an enzymatic synthesis approach to produce phloridzin using *Malus* x *domestica* glycosyltransferase in engineered *Pichia pastoris* GS115. *Process Biochem.* 59: 187–193.

46 Bertrand, A., Morel, S., Lefoulon, F. et al. (2006). *Leuconostoc mesenteroides* glucansucrase synthesis of flavonoid glucosides by acceptor reactions in aqueous-organic solvents. *Carbohydr. Res.* 341: 855–863.

47 De Bruyn, F., Van Brempt, M., Maertens, J. et al. (2015). Metabolic engineering of *Escherichia coli* into a versatile glycosylation platform: production of bio-active quercetin glycosides. *Microb. Cell Factories* 14: 138.

48 Gao, C., Mayon, P., MacManus, D.A., and Vulfson, E.N. (2000). Novel enzymatic approach to the synthesis of flavonoid glycosides and their esters. *Biotechnol. Bioeng.* 71: 235–243.

49 Ono, Y., Tomimori, N., Tateishi, N. et al. (2005). Quercetin glycoside composition and method of preparing the same. Patent WO 2006/070883 A1.

50 Xia, T. and Eiteman, M.A. (2017). Quercetin glucoside production by engineered *Escherichia coli*. *Appl. Biochem. Biotechnol.* 182: 1358–1370.

51 Brazier-Hicks, M., Evans, K.M., Gershater, M.C. et al. (2009). The *C*-glycosylation of flavonoids in cereals. *J. Biol. Chem.* 284: 17926–17934.

52 Chen, D., Chen, R., Wang, R. et al. (2015). Probing the catalytic promiscuity of a regio- and stereospecific *C*-glycosyltransferase from *Mangifera indica*. *Angew. Chem. Int. Ed.* 54: 12678–12682.

53 Ito, T., Fujimoto, S., Shimosaka, M., and Taguchi, G. (2014). Production of *C*-glucosides of flavonoids and related compounds by *Escherichia coli* expressing buckwheat *C*-glucosyltransferase. *Plant Biotechnol.* 31: 519–524.

54 Bungaruang, L., Gutmann, A., and Nidetzky, B. (2013). Leloir glycosyltransferases and natural product glycosylation: biocatalytic synthesis of the *C*-glucoside nothofagin, a major antioxidant of redbush herbal tea. *Adv. Synth. Catal.* 355: 2757–2763.

55 Schmölzer, K., Gutmann, A., Diricks, M. et al. (2016). Sucrose synthase: a unique glycosyltransferase for biocatalytic glycosylation process development. *Biotechnol. Adv.* 34: 88–111.

56 Desmet, T., Soetaert, W., Bojarová, P. et al. (2012). Enzymatic glycosylation of small molecules: challenging substrates require tailored catalysts. *Chem. Eur. J.* 18: 10786–10801.

57 Rupprath, C., Kopp, M., Hirtz, D. et al. (2007). An enzyme module system for in situ regeneration of deoxythymidine 5′-diphosphate (dTDP)-activated deoxy sugars. *Adv. Synth. Catal.* 349: 1489–1496.

58 Gutmann, A. and Nidetzky, B. (2016). Unlocking the potential of Leloir glycosyltransferases for applied biocatalysis: efficient synthesis of uridine 5'-diphosphate-glucose by sucrose synthase. *Adv. Synth. Catal.* 358: 3600–3609.

59 Gutmann, A. and Nidetzky, B. (2012). Switching between *O*- and *C*-glycosyltransferase through exchange of active-site motifs. *Angew. Chem. Int. Ed.* 51: 12879–12883.

60 Diricks, M., de Bruyn, F., van Deale, P. et al. (2015). Identification of sucrose synthase in non-photosynthetic bacteria and characterization of the recombinant enzymes. *Appl. Microbiol. Biotechnol.* 99: 8465–8474.

61 Diricks, M., Gutmann, A., Debacker, S. et al. (2017). Sequence determinants of nucleotide binding in sucrose synthase: improving the affinity of a bacterial sucrose synthase for UDP by introducing plant residues. *Protein Eng. Des. Sel.* 30: 141–148.

62 Gutmann, A., Lepak, A., Diricks, M. et al. (2017). Glycosyltransferase cascades for natural product glycosylation: use of plant instead of bacterial sucrose synthases improves the UDP-glucose recycling from sucrose and UDP. *Biotechnol. J.* 12: 1600557.

63 Lim, E.-K. (2005). Plant glycosyltransferases: their potential as novel biocatalysts. *Chem. Eur. J.* 11: 5486–5494.

64 Arend, J., Warzecha, H., Hefner, T., and Stöckigt, J. (2001). Utilizing genetically engineered bacteria to produce plant-specific glucosides. *Biotechnol. Bioeng.* 76: 126–131.

65 Cai, R., Chen, C., Li, Y. et al. (2017). Improved soluble bacterial expression and properties of the recombinant flavonoid glucosyltransferase UGT73G1 from *Allium cepa*. *J. Biotechnol.* 255: 9–15.

66 Weiner, D.H., Shin, D., Tomaleri, G.P. et al. (2017). Plant cell wall glycosyltransferases: high-throughput recombinant expression screening and general requirements for these challenging enzymes. *PLoS One* 12: e0177591.

67 Dewitte, G., Walmagh, M., Diricks, M. et al. (2016). Screening of recombinant glycosyltransferases reveals the broad acceptor specificity of stevia UGT-76G1. *J. Biotechnol.* 233: 49–55.

68 Schmideder, A., Priebe, X., Rubenbauer, M. et al. (2016). Non-water miscible ionic liquid improves biocatalytic production of geranyl glucoside with *Escherichia coli* overexpressing a glucosyltransferase. *Bioprocess Biosyst. Eng.* 39: 1409–1414.

69 Priebe, X., Daschner, M., Schwab, W., and Weuster-Botz, D. (2018). Rational selection of biphasic reaction systems for geranyl glucoside production by *Escherichia coli* whole-cell biocatalysts. *Enzym. Microb. Technol.* 112: 79–87.

70 Mairhofer, J., Wittwer, A., Cserjan-Puschmann, M., and Striedner, G. (2015). Preventing T7 RNA polymerase read-through transcription: a synthetic termination signal capable of improving bioprocess stability. *ACS Synth. Biol.* 4: 265–273.

71 Mairhofer, J., Stargardt, P., Feuchtenhofer, L. et al. (2016). Innovation without growth: non-growth associated recombinant protein production in *Escherichia coli*. *New Biotechnol.* 33: S26.

72 Mekler, V., Minakhin, L., Sheppard, C. et al. (2011). Molecular mechanism of transcription inhibition by phage T7 gp2 protein. *J. Mol. Biol.* 413: 1016–1027.

73 Mairhofer, J., Striedner, G., Grabherr, R., and Wilde, M. (2016). Uncoupling growth and protein production. WO2016/174195 A1.

74 Chen, C.-C., Walia, R., Mukherjee, K.J. et al. (2015). Indole generates quiescent and metabolically active *Escherichia coli* cultures. *Biotechnol. J.* 10: 636–646.

75 Gosh, C., Gupta, R., and Mukherjee, K.J. (2012). An inverse metabolic engineering approach for the design of an improved host platform for over-expression of recombinant proteins in *Escherichia coli*. *Microb. Cell Factories* 11: 93.

76 Mahalik, S., Sharma, A.K., and Mukherjee, K.J. (2014). Genome engineering for improved recombinant protein expression in *Escherichia coli*. *Microb. Cell Factories* 13: 177.

77 Lemmerer, M., Mairhofer, J., Lepak, A. et al. (2019). Decoupling of recombinant protein production from *Escherichia coli* cell growth enhances functional expression of plant Leloir glycosyltransferases. *Biotechnol. Bioeng.* 116: 1259–1268.

78 Bungaruang, L., Gutmann, A., and Nidetzky, B. (2016). β-Cyclodextrin improves solubility and enzymatic *C*-glucosylation of the flavonoid phloretin. *Adv. Synth. Catal.* 358: 486–493.

79 Straathof, A.J., Panke, S., and Schmid, A. (2002). The production of fine chemicals by biotransformations. *Curr. Opin. Biotechnol.* 13: 548–556.

80 Dai, L., Li, J., Yao, P. et al. (2017). Exploiting the aglycon promiscuity of glycosyltransferase Bs-YjiC from *Bacillus subtilis* and its application in synthesis of glycosides. *J. Biotechnol.* 248: 69–76.

81 De Bruyn, F., De Paepe, B., Maertens, J. et al. (2015). Development of an in vivo glucosylation platform by coupling production to growth: production of phenolic glucosides by a glycosyltransferase of *Vitis vinifera*. *Biotechnol. Bioeng.* 112: 1594–1603.

82 Lepak, A., Gutmann, A., Kulmer, S.T., and Nidetzky, B. (2015). Creating a water-soluble resveratrol-based antioxidant by site-selective enzymatic glucosylation. *ChemBioChem* 16: 1870–1874.

83 Michlmayr, H., Malachová, A., Varga, E. et al. (2015). Biochemical characterization of a recombinant UDP-glucosyltransferase from rice and enzymatic production of deoxynivalenol-3-*O*-β-D-glucoside. *Toxins* 7: 2685.

84 Shen, X., Wang, J., Wang, J. et al. (2017). High-level de novo biosynthesis of arbutin in engineered *Escherichia coli*. *Metab. Eng.* 42: 52–58.

85 Pitha, J., Szabo, L., and Fales, H.M. (1987). Reaction of cyclodextrins with propylene oxide or with glycidol: analysis of product distribution. *Carbohydr. Res.* 168: 191–198.

86 Szathmary, S.C. (1989). Determination of hydroxypropyl-β-cyclodextrin in plasma and urine by size-exclusion chromatography with post-column complexation. *J. Chromatogr. Biomed. Sci. Appl.* 487: 99–105.

87 Kulmer, S.T., Gutmann, A., Lemmerer, M., and Nidetzky, B. (2017). Biocatalytic cascade of polyphosphate kinase and sucrose synthase for synthesis of nucleotide-activated derivatives of glucose. *Adv. Synth. Catal.* 359: 292–301.

88 Lemmerer, M., Schmölzer, K., Gutmann, A., and Nidetzky, B. (2016). Downstream processing of nucleotide-diphospho-sugars from sucrose synthase reaction mixtures at decreased solvent consumption. *Adv. Synth. Catal.* 358: 3113–3122.

89 Ó'Fágáin, C., Cummins, P.M., and O'Connor, B.F. (2011). Gel-filtration chromatography. In: *Protein Chromatography: Methods and Protocols* (eds. D. Walls and T.S. Loughran), 25–33. New York: Humana Press.

90 Gutmann, A., Krump, C., Bungaruang, L., and Nidetzky, B. (2014). A two-step *O*- to *C*-glycosidic bond rearrangement using complementary glycosyltransferase activities. *Chem. Commun.* 50: 5465–5468.

91 Weignerová, L., Marhol, P., Gerstorferová, D., and Křen, V. (2012). Preparatory production of quercetin-3-β-D-glucopyranoside using alkali-tolerant thermostable α-L-rhamnosidase from *Aspergillus terreus*. *Bioresour. Technol.* 115: 222–227.

Part III

Ways to Improve Enzymatic Transformations

9

Application of Nonaqueous Media in Biocatalysis

Afifa A. Koesoema[1] *and Tomoko Matsuda*[2]

[1] *Okayama University, Research Institute for Interdisciplinary Science (RIIS), Division of Photosynthesis and Structural Biology, 3-1-1 Tsushimanaka, Kita-ku, 700-8530, Okayama-shi, Japan*
[2] *Tokyo Institute of Technology, Department of Life Science and Technology, 4259, Nagatsuta-cho Midori-ku, Yokohama, 226-8501, Japan*

9.1 Introduction

In nature, enzymes catalyze reactions inside the living cells or organisms in aqueous media [1]. However, aqueous media are not always ideal for organic synthesis of active pharmaceutical intermediates and fine chemicals, given the lack of solubility and undesirable side reactions of hydrophobic substrates in water [2, 3]. As an alternative, nonaqueous reaction media are being utilized for biocatalysis.

Organic solvents are the first nonaqueous media used for biocatalysis. The activities of enzymes in organic solvents have been reported as early as 1936 by observing the activity of esterase in various organic solvents [4]. Sym prepared various active esterase preparations and placed in contact with various water-soluble or water-insoluble organic solvents for esterification. It was found that water-insoluble solvents, such as benzene and carbon tetrachloride, gave significantly higher ester yields (>95%) than the aqueous system. In the presence of a water-insoluble organic solvent, the ester will be removed from the aqueous phase to the water-insoluble organic solvent phase, thus shifting the reaction equilibrium to near completion. Although the finding by Sym was significant, biocatalysis in organic solvents was only rediscovered again in 1977 by Berezin and coworkers [5]. They reported the esterification of *N*-acetyl-L-tryptophan in a water–chloroform biphasic system with 100% yield, whereas the reaction yield was only 0.01% in the aqueous system. This finding then led to more applications of organic solvents as well as other nonaqueous reaction media for biocatalysis.

In recent decades, enzymes have been utilized in various nonaqueous reaction media from ionic liquids (ILs) [6] to pressurized gas such as CO_2-expanded liquids (CXL) [7]. Concurrently, efforts are being put to improve the stability of enzymes in nonaqueous media by isolation of robust enzymes, immobilization (see also Chapter 3), modification, and protein engineering (see also Chapter 2) [8]. This chapter focuses on describing novel

Biocatalysis for Practitioners: Techniques, Reactions and Applications, First Edition.
Edited by Gonzalo de Gonzalo and Iván Lavandera.

nonaqueous media with more environmentally friendly properties such as bio-based liquids [9], liquid CO_2 [10], and deep eutectic solvents (DESs) [11]. The recent practical approach to stabilize enzymes in nonaqueous media is also discussed. We mainly discuss the recent examples in two classes of enzymes mostly used for organic synthesis, lipases (see also Chapter 5), and alcohol dehydrogenases (ADHs, see also Chapter 7).

9.2 Advantages and Disadvantages of Reactions in Nonaqueous Media

Various nonaqueous media have been employed for biocatalysis owing to their advantages. However, some disadvantages still hamper the utilization of biocatalysts in nonaqueous media. The advantages and disadvantages of nonaqueous media compiled from the literature [3, 12, 13] are described as follows:

Advantages

- Facilitate the reactions infeasible in the aqueous media due to the unfavorable thermodynamic equilibrium, such as lipase-catalyzed transesterification that can only be done in nonaqueous solvents.
- Suppression of water-dependent side reactions and microbial contamination.
- Increase the solubility of hydrophobic substrates, thus improve the reaction yield.
- Elimination of substrate or product inhibition by regulating the partition of substrate and product between aqueous and nonaqueous solvent in a biphasic system.
- Enhance the enzymes' stability by structural rigidification in nonaqueous media.
- Tune the substrate specificity and chemo-, regio-, and stereoselectivities of enzymes.
- Facilitate the recovery and reusability of enzymes.
- Simplified work-up procedures.

Disadvantages:

- Enzyme inactivation in nonaqueous media.
- Cost-intensive preparation of immobilized or engineered enzymes.
- Mass transfer limitation in viscous media or a biphasic system.
- Necessity of water activity control.

9.3 Nonaqueous Media Used for Biocatalysis

Currently, there are four kinds of media used for biocatalysis as shown in Figure 9.1: aqueous media, and three kinds of nonaqueous media consisting of organic solvents, pressurized gas, and ILs. It can be monophasic, micro-aqueous, and biphasic system employing both water-miscible and water-immiscible solvents. Upon choosing the best media for biocatalysis, we have to consider the reaction rates as well as the work-up procedure. Because of the raising awareness of the detrimental effect caused by the utilization of organic solvents, more environmentally friendly alternatives are also explored such as bio-based liquid, liquid CO_2, CXL, DESs, and natural deep eutectic solvents (NADES). Recently, some

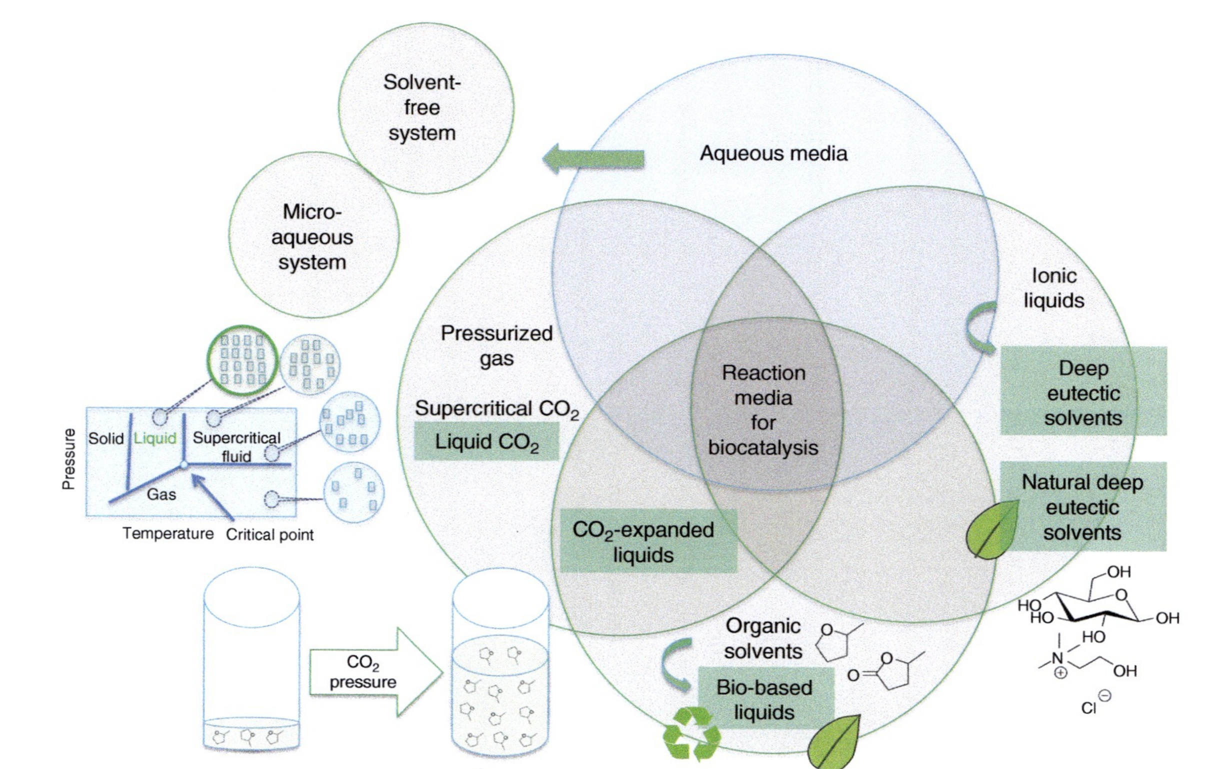

Figure 9.1 Nonaqueous media for biocatalysis with their more environmentally friendly alternatives highlighted in green boxes.

studies also reported the utilization of a solvent-free system or a neat substrate reaction, enabling the biocatalytic reaction without any addition of solvents.

With respect to sustainable chemistry, the best reaction is by using no solvent [14]. However, solvent-free reactions still face many challenges, such as the mass transfer limitation, viscosity, and the stability of the biocatalyst itself. To reduce the detrimental effect of conventional organic solvents, bio-based liquids are currently being utilized as green solvents [10]. Bio-based liquids are derived from renewable raw materials such as lignocellulosic biomass, vegetable oils, and biomass waste. Their advantages are sustainable production, lower toxicity, and biodegradability compared to the conventional organic solvents. Several bio-based liquids such as 2-methyltetrahydrofuran (MeTHF), γ-valerolactone (GVL), and glycerol derivative have been utilized in biocatalysis with promising results [10].

Pressurized gas such as supercritical CO_2 ($scCO_2$) achieved above the critical point of CO_2 (7.4 MPa at 31 °C) is used for biocatalysis because of its environmentally friendly properties, enzyme compatibility, and tenability [15, 16]. Below its critical temperature, CO_2 exists as liquid and gas, and the liquid can be used as a solvent for biocatalysis. It has lower operating pressure and temperature (i.e. 4.5 MPa at 10 °C) than $scCO_2$. This milder operation condition of liquid CO_2 is preferable for the operation cost of the instrument and might as well induce higher reaction performance [7].

Another recently emerging pressurized gas used for reaction media is CXL. CXL is a condensed phase consisting of compressed CO_2 with liquid, mainly organic solvents dissolving a substantial amount of CO_2. CXL has the merit of lower reaction pressure with wide reaction temperature and better solvent power than $scCO_2$ and liquid CO_2 [17]. Recently, as a greener alternative of organic solvents, bio-based liquids are also being employed as CXL and utilized in enantioselective biotransformations [18].

The last nonaqueous media discussed in this chapter is IL and DESs. ILs are salts with poorly coordinated ions causing them to be liquid below 100 °C. The cations of ILs are usually large, asymmetric, and organic (e.g. imidazolium and pyridinium ions), and the anions are inorganic (e.g. BF_4^-, PF_6^-, and SbF_6^- ions) [6, 14]. Room temperature ILs have been applied for biocatalysis since the 2000s, and it is considered to be promising media [19–22]. ILs are considered to be green solvents because of its low or no vapor pressure and low flammability. Moreover, ILs are thermally robust and have enzyme-stabilizing as well as substrate-solubilizing properties. The properties of ILs are also easily tunable by changing the cation and anion.

Although initially regarded as a green solvent, ILs presented several problems such as environmental effect and toxicity [23]. Since 2004, a more environmentally friendly alternative solvent, DES, has been proposed to replace ILs [11, 23]. DES is a mixture of two compounds with one compound acting as a hydrogen bond acceptor (HBA) and another as a hydrogen bond donor (HBD). DES often employs a quaternary ammonium salt such as choline chloride (ChCl) as the HBA and acids or urea as the HBD. The hydrogen-bonding formation in DES will lower the overall melting point and form a eutectic mixture. By adjusting the mole ratio of the mixture, then we can adjust the solid–liquid phase of DES [24]. DES is considered to be a greener solvent than ILs because of its waste-free synthesis process and low toxicity. Moreover, the tunable properties of DES (e.g. polarity, viscosity, density, and surface tension) made its potential application to be diverse. The application of DES is reported not only for biocatalysis but also for extractions [25], organometallic chemistry reactions [26], and material production [27]. In 2013, Choi and coworkers reported the emergence of natural DES, in which natural compounds such as sugars, organic acids, and amino acids are acting as the HBD [28]. A large combination of NADES could be explored and utilized as a green media for biocatalysis.

9.4 Enzymatic Activity and Inactivation in Nonaqueous Media

9.4.1 Enzymatic Activity in Nonaqueous Media

Enzymatic activity in nonaqueous media depends highly on the kind of enzyme and media used. Lipase and other lipolytic enzymes naturally existing in the environment with a high content of hydrophobic molecules are usually more active in nonaqueous media. However, the activity of other kinds of enzymes decreases in these nonaqueous media. Thus, lyophilization or immobilization is often necessary before their utilization [2]. We can preserve the enzymatic activity by controlling the pH of the solution before the lyophilization. The catalytic activity of an enzyme in nonaqueous media reflects the pH of the last aqueous solvent they are exposed to before lyophilization. This phenomenon is called "pH memory" [29].

In nonaqueous media, enzymatic activity is preserved as long as an essential amount of water is kept in the reaction medium [30]. Understanding on how enzyme remains active in nonaqueous media with high pressure can be done by molecular dynamics simulation. For instance, Martínez and coworkers conducted molecular dynamics simulation of *Candida antarctica* lipase B (CALB) in a $scCO_2$–water biphasic system [31]. It is shown that water and CO_2 molecules distribute heterogeneously in the enzyme surface (Figure 9.2). Water molecules cover specific site hindering CO_2 molecules from penetrating to the active

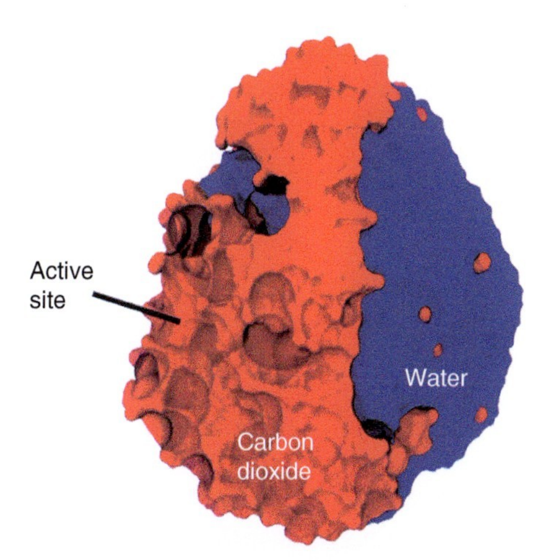

Figure 9.2 Distribution of water and CO_2 molecules on the surface of *Candida antarctica* lipase B in the $scCO_2$–water biphasic system by molecular dynamics simulation. Source: Silveira et al. [31]. © 2012 American Chemical Society.

site of the enzyme, thus preventing the disruption of the active site conformation. Meanwhile, CO_2 molecules mainly bind at the entrance of the catalytic tunnel, which may increase the accessibility of hydrophobic substrates.

Overall, the enzyme conformation in nonaqueous media is more rigid because of the lack of lubrication by water [2, 12], which may also improve enzymes' stability [29, 32]. Ligand binding causes a conformational change before the lyophilization, and it may "rigidify" the enzyme with different conformations than the enzyme prepared in the absence of the ligand. The phenomenon is called "molecular memory" [33, 34]. As a result, catalytic activity, stability, and selectivity may also be affected. Several studies have reported the enhancement of activity [35, 36] and change of selectivity [37] in nonaqueous media because of this "molecular memory" phenomenon.

9.4.2 Factors Causing Inactivation of Enzymes in Nonaqueous Media

In order for an enzyme to be functional, its structure has to be catalytically active. The enzyme structure is maintained by intramolecular interaction such as hydrophobic and electrostatic interactions, van der Waals forces, and hydrogen bonds. Putting an enzyme in nonaqueous media risks the disruption of these interactions by the change of hydrophobicity of solvents. Conformational changes that can cause protein unfolding may happen especially by the polar solvents with higher ability to protrude to the enzyme's interior [12, 38].

The loss of essential water is one of the contributing factors for enzyme inactivation in nonaqueous media. As mentioned earlier, the loss of water will rigidify the enzyme, thus limiting the conformational changes needed for catalysis. Depending on the enzyme, water may also participate directly in the reaction mechanism; i.e. in ADH, the proton relay system transfers a proton to water as a bulk solvent from the alcohol bound to the catalytic zinc [39]. The kind of nonaqueous media also affects the essential water content for the enzymes. For example, the hydrophilic solvent is able to strip off the essential water from the enzyme, rendering it inactive, whereas the hydrophobic solvent has the less capability to do so [40]. At last, employing enzymes in nonaqueous media, especially the biphasic system, requires vigorous stirring to enhance the mass transfer rate across the interface, possibly causing inactivation of enzymes [23, 41].

9.5 Practical Approaches to Stabilize Enzymes in Nonaqueous Media

9.5.1 Utilization of Nonaqueous Media-Tolerant Enzymes or Host Cells

The most practical way to perform biocatalysis in nonaqueous media is by utilizing enzymes or host cells that are tolerant toward nonaqueous media. Organic solvent-tolerant organisms are one of the sources of organic solvent-tolerant enzymes, mostly lipolytic and proteolytic enzymes [12, 42]. One notable example is lipase isolated from organic solvent-tolerant *Pseudomonas aeruginosa* LST-03 [43]. This lipase is shown to have higher stability in up to 80% of *n*-decane [44], or cyclohexane [45], than in aqueous media.

Although not as many as lipases, organic solvent-tolerant ADHs are also being reported. ADH from *Paracoccus pantotrophus* DSM 11072 was described to be highly tolerant toward dimethyl sulfoxide (DMSO) [46], while ADH from *Rhodococcus ruber* DSM 44541 was highly stable in the presence of nonpolar solvents, such as *n*-hexane, toluene, and diisopropylether [47, 48]. ADH from *Geotrichum candidum* NBRC 4597 was reported to be active in the presence of 59% 2-propanol [49]. Halophilic and alkaliphilic organisms may be the source of organic solvent-tolerant enzymes [8, 50, 51].

Lipophilic organisms have also been used as a host expression system to conduct reaction using enzymes in nonaqueous media. Kawabata and coworkers described the expression of organic solvent-tolerant mutant of carbonyl reductase from *Ogataea minuta* (OCR) [52] in *Rhodococcus opacus* strain B-4. This microorganism has a lipophilic cell surface structure and can form small aggregates in anhydrous solvents. The utilization of this host expressing carbonyl reductase, coupled with the cofactor regeneration enzyme, was proven to be effective in reducing 2,2,2-trifluoroacetophenone in various organic solvents [53].

Recently, Thompson and coworkers reported the engineering of a $scCO_2$-tolerant *Bacillus megaterium* strain isolated from the McElmo Dome CO_2 field [54]. This strain was engineered to overexpress two enzymes employed in the isobutanol production pathway, α-ketoisovalerate decarboxylase, and ADH by using exogenous plasmids. Isobutanol production was accomplished with more than 40% yield and the coproduction of isopentanol under $scCO_2$ condition. Remarkably, the whole process was incorporated with *in situ* extraction methods by using $scCO_2$, yielding a high-purity biofuel with very low toxicity. This study opens a new possibility to isolate pressurized gas-tolerant organisms. Despite these promising results, finding a suitable organism for enzyme isolation or a host remains a difficult challenge. Thus, more approaches are made to stabilize mesophilic enzymes in nonaqueous solvents.

9.5.2 Enzyme Immobilization

Enzyme immobilization is an effective technique to improve the enzyme's stability in nonaqueous media. By immobilization, an enzyme is less prone to the unfolding as well as malformation on its active site [55]. Besides, the enzyme can be easily recovered and be suitable for continuous operation. The immobilization methods commonly used comprise adsorption, covalent binding, entrapment and encapsulation, and carrier-free immobilization by cross-linking. One of the most successful examples is the immobilization of CALB by adsorption on Lewatit ion exchange resin (Novozym 435), which shows excellent activity in neat organic solvents [56] and pressurized gas [15]. For ADH, immobilization is often done together with its cofactor NAD(P)H. ADHs from *Lactobacillus kefir* and *Thermoanaerobacter ethanolicus* were successfully immobilized by entrapment in polyvinyl alcohol gel beads [57] and xerogel [58], respectively.

Computational tools can also be used as a guideline to do immobilization, which is known as structure-based immobilization [56]. Grid analysis can map regions predicted to be suitable for interaction with immobilization support after establishing the hydrophobic and hydrophilic area of the enzyme surface [59]. A functional group such as amino group could be easily located, and this strategy was implemented on the immobilization of a lipase from *Thermomyces lanuginosus*. Lysine residues on the surface of the protein was removed except for the one located opposite to the active site. The active site of the enzyme is then oriented toward the solvent, and higher activity was observed [60].

Recently, nanoparticles immobilization has also drawn considerable interest because of its high surface-to-volume ratio, which allows a considerable decrease of the carrier size, thus increasing the exposure of enzyme to the reaction media. For example, immobilization by using organic–inorganic hybrid nanoflower [61, 62] was found as an effective and simple method. Several lipases [63–65], and a few ADHs [66, 67], have been immobilized with this method showing a remarkable stability as well as recyclability. At last, biological scaffold such as porous protein crystals can be utilized as a robust template for enzyme immobilization [68]. Ueno and coworkers immobilized protein kinase C by using a polyhedral crystal with retention of activity after storage under harsh conditions [69]. *Bacillus subtilis* lipase A was also immobilized into protein crystal Cry3Aa for biodiesel production [70]. Most recently, Snow and coworkers immobilized horseradish peroxidase and glucose oxidase to *Campylobacter jejuni* protein crystal to perform cascade reaction with increased thermal tolerance [71].

9.5.3 Modification of the Enzyme Preparation

Another approach to improve the enzyme stability and activity in nonaqueous media is by modifying the enzyme preparation. To prevent activity decrease induced by lyophilization, polyols, sugars, or inorganic salts are often added before this process. These compounds may rigidify the enzymes through hydrophobic interaction among nonpolar amino acid residues [72]. Chemical modification by forming covalent bond with polyethylene glycol (PEG) or PEGylation can increase the solvent ordering within medium. PEGylation has been shown to increase the activity of α-chymotrypsin in nonpolar organic solvents [73]. Enzymes may also be prepared by repeatedly rinsing the enzyme in *n*-propanol solution or known as propanol-rinsed enzyme preparation (PREP). In acetonitrile or tetrahydrofuran, silica-immobilized subtilisin Carlsberg prepared with PREP exhibited 1000-fold higher activities than its lyophilized powders. The removal of water associated with PREP is believed to be nondestructive toward the protein structure [74].

Recently, a novel chemical modification of glucosidase by surface cationization followed by conjugation with surfactant was reported. The modification successfully increased the stability of the glucosidase in ILs as well as its thermostability up to 137 °C. The glucosidase was then employed for cellobiose hydrolysis in various ILs. Unexpectedly, not only the enzyme exhibited higher activity toward cellobiose but also the enzyme exhibited activity toward cellulose. This solvent-induced enzyme promiscuity is facilitated by the solubilization of cellulose induced by the presence of ILs. It is hypothesized that glucosidase will indiscriminately attack glycosidic linkage in cellulose to form oligosaccharides, which will be hydrolyzed to form glucose [75].

Coating an enzyme with ILs can also enhance its stability in nonaqueous media. Enhanced activity with maintaining excellent enantioselectivity was achieved by coating *Burkholderia cepacia* lipase with various imidazolium poly(oxyethylene) alkyl sulfate ILs [76]. It is shown that ILs as a coating agent can bind to the enzyme, providing an ideal microenvironment for a reaction. Enzymes can also be prepared as reverse micelles or water-in-oil microemulsions by introducing amphiphilic surfactants [8]. ILs can also be utilized for water-in-IL microemulsion with a superior activity compared to the water-in-oil microemulsion system for *Pseudomonas* sp. lipase [77]. Although some enzymes can be stabilized by modifying the preparation, this method is proven to be empirical and cannot be generalized.

9.5.4 Protein Engineering

The approach commonly done to enhance the stability of an enzyme in nonaqueous media is protein engineering by employing directed evolution (DE), rational design, and semi-rational design (Figure 9.3). DE was first employed by Arnold and coworkers to engineer subtilisin E with better hydrolysis efficiency in 60% v/v *N,N*-dimethylformamide (DMF) [78].

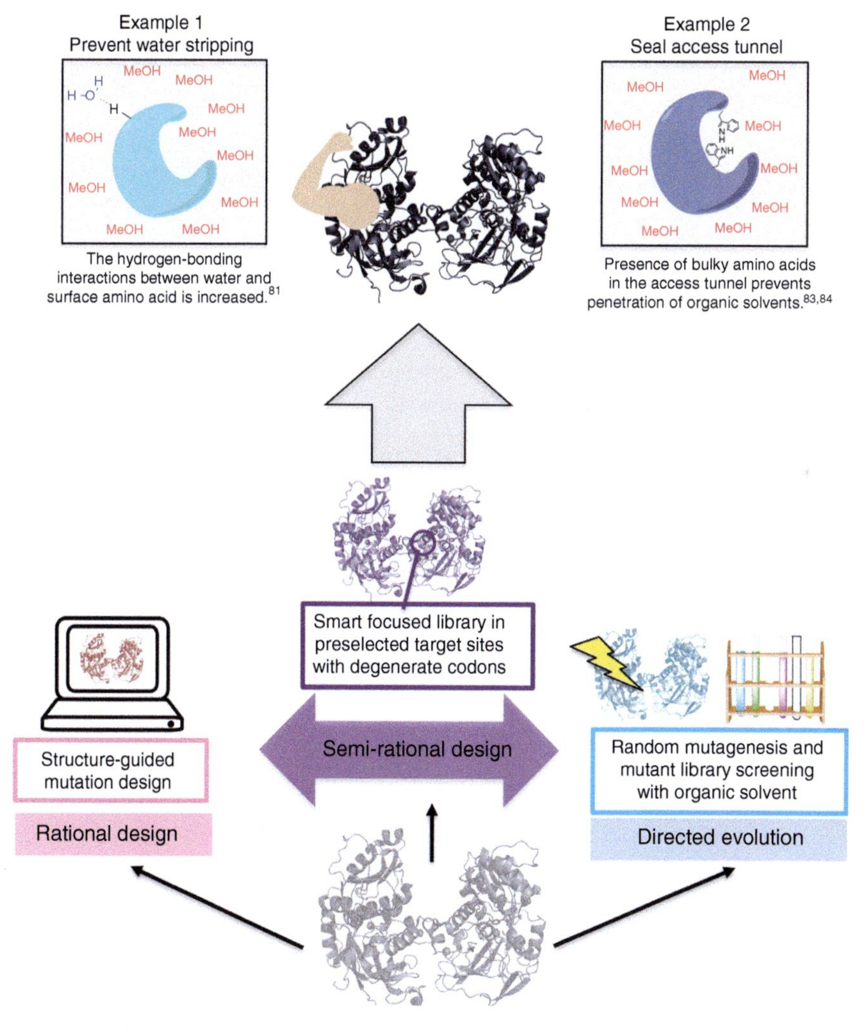

Figure 9.3 Protein engineering methods employed to enhance enzyme stability in nonaqueous media.

DE was also employed for enhancing the organic solvent stability of the *P. aeruginosa* LST-03 lipase. Analysis of the mutants' structure revealed that the substituted amino acid residues, located on the surface of the protein, may prevent the penetration of organic solvents that can disrupt the core of the protein [79].

One of the drawbacks of DE was the need to develop an effective screening procedure for the vast mutant library generated. Thus, rational design or structure-based engineering is employed as an alternative. Most of the rational design methods focus on introducing interaction such as disulfide bonds [80], hydrogen bonds [81], and ionic interaction in enzymes. Most of the stabilization occurs by mutating the surface amino acid residue. A notable example includes the stabilization of CALB in hydrophilic organic solvents conducted by Park and coworkers. By increasing the hydrogen-bonding interactions between water molecules and surface amino acid, the stability of CALB is increased by 1.5-fold in the presence of 80% v/v methanol [81]. Two surface residues of *B. megaterium* penicillin G acylase were also mutated to enhance its stability in 40% v/v DMF [82].

Other than surface residue, mutating the access tunnel residue has also proven to enhance the stability and resistance of enzyme toward organic solvents. By employing random and site-directed mutagenesis, the half-life of a haloalkane dehalogenase from *Rhodococcus rhodochrous* NCIMB 13064 was improved by 4000-fold in the presence of 40% v/v DMSO [83]. Aromatic interactions were introduced to the access tunnel of *Geobacillus stearothermophilus* T6 lipase, thus increasing its stability in methanol [84]. In both studies, the presence of bulkier amino acid substitution is believed to "seal" the access tunnel and preventing organic solvent from getting inside and disrupting the active site. However, the lack of understanding about the interaction between enzyme and nonaqueous media hampers a more extensive application of rational design for enzyme stabilization. Thus, a semi-rational design, a combination of DE and rational design, is employed. By using a semi-rational design, we can create a focused library in preselected target sites by using a degenerate codon library.

Computational methods also played an important role to help designing enzyme with improved stability. By the methods, we can calculate the free energy [85], optimize the protein surface charge [86], analyze the B-factor [87, 88], and do homologous comparison [89] to design the best mutant variants. Recently, Janssen and coworkers develop the Framework for Rapid Enzyme Stabilization by Computational libraries (FRESCO), enabling us to design a library with chemically diverse stabilizing mutations by *in silico* screening. As a result, the amount of variants to be screened is far less than that in the DE method [90]. Moreover, Fasan and coworkers reported the development of Rosetta-guided protein stapling (R-GPS) to improve the thermostability of Mb-derived cyclopropanation biocatalyst as well as its stability in the presence of 45% v/v DMSO. Rosetta was used to determine sites where the covalent staples should be placed [91]. It is expected that more study about the enzyme stabilization in the nonaqueous media with the aid of computational method will be reported in the near future.

9.6 Examples of Biocatalyzed Reactions in Solvent-Free Systems

In a solvent-free system, the substrate and product are acting as the reaction media. This system can prevent the degradation of unstable substrate or product in the reaction media, as well as simplify the work-up procedure by lowering the formation of the water-induced emulsion [92]. Enzymes are active in the solvent-free system as long as the essential amount of water is present. Lipases are active in the medium with a water activity (a_w) as low as 0.0001 [93].

The lipase solvent-free system has been used to synthesize bio-based materials. A versatile monomer, (meth)acrylate ester synthesis by lipase in a solvent-free, was reported [94]. The esterification of tetrahydrofurfuryl alcohol, citronellol, and other alcohols with aliphatic hydroxyl groups resulted in near full conversion. Hoffmann-Jacobsen and coworkers reported the solvent-free synthesis of bio-based biscyclocarbonate by using CALB *Immo* Plus, the yield and purity of purified product reached 65% and >95%, respectively (Scheme 9.1) [95].

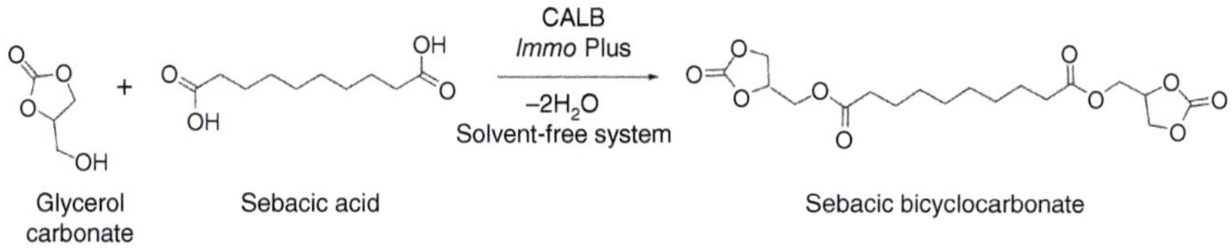

Glycerol carbonate + Sebacic acid → (CALB *Immo* Plus, $-2H_2O$, Solvent-free system) → Sebacic bicyclocarbonate

Scheme 9.1 Synthesis of sebacic bicyclocarbonate by CALB *Immo* Plus in a solvent-free system. Source: Modified from Wunschik et al. [95].

Whole cells expressing one or two ADHs are used to catalyze the asymmetric reduction of ketones in a solvent-free system (Figure 9.4). The lyophilized *Escherichia coli* cells expressing a *Candida parapsilosis* carbonyl reductase (CPCR) utilizing 2-propanol as the cofactor regeneration system catalyzed the reduction of acetophenone with excellent enantioselectivity (>99% *ee*). This system enabled the production of up to 500 g l^{-1} of (*S*)-1-phenylethanol. This system was also applied in the reduction of 3-butyn-2-one, which was unstable in aqueous media, achieving excellent enantioselectivity [92]. Similarly, ADH evo 1.1.200 expressed in *E. coli* cells was also utilized to reduce 2-butanone in the solvent-free system [96]. Enantiomerically pure 2-butanol is a valuable building block for the preparation of liquid crystals. Although the turnover number of ADH evo-1.1.200 was significant

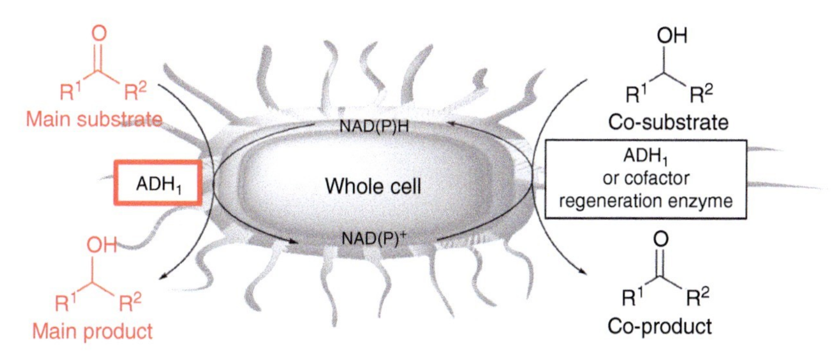

Figure 9.4 Whole cell expressing ADH and cofactor regenerating enzyme utilized for the asymmetric reduction of a ketone in a solvent-free system.

(~7800), the conversion was limited to <1%. The low conversion was probably caused by the highly hydrophilic co-substrate used, stripping the essential water from ADH. The further study is introduced in the "micro-aqueous system" section. Jakoblinnert and coworkers performed a continuous solvent-free reaction system to reduce 2-butanone. In order to produce both (*S*)- and (*R*)-2-butanol, two stereo-complementary ADHs were employed. High enantioselectivity of both alcohols (up to >98% ee) was achieved by optimizing the residence time [97].

9.7 Examples of Reactions in Micro-aqueous Systems

Micro-aqueous system has been used for the asymmetric reduction of ketones catalyzed by ADHs. The xerogel-encapsulated *Thermoanaerobacter ethanolicus* ADH was reported to be active in the micro-aqueous system containing only the essential water remained in the immobilization material [58]. It was found that the kind of organic solvents used can affect the enantioselectivity of the reduction. For example, 1-phenyl-2-propanone reduction enantioselectivities in diisopropyl ether, hexane, and toluene were 73% (*S*), 69% (*S*), and 55% (*S*), respectively. Nevertheless, high conversion (up to >99%) and enantioselectivity [>99% ee (*S*)] could be observed in the reduction of other substrates such as 4-phenyl-2-butanone and 1-phenoxy-2-propanone in hexane. However, the reaction concentration employed in this study was still relatively low (~ 140 mM).

Reactions in the micro-aqueous system were also conducted in a higher substrate concentration. Kroutil and coworkers reported the reduction of ketones in a micro-aqueous system (99% v/v organic solvent) using the lyophilized *E. coli* expressing ADH-A from *R. ruber* DSM 44541 [98]. Hydrophobic solvents were better than the hydrophilic one. High space–time yield at $1.42\,g\,l^{-1}\,h^{-1}$ and excellent enantioselectivity were observed in the reduction of 2-octanone. The reduction of other ketones proceeded with >99% ee (*S*) with moderate to high conversion depending on the substrates' steric requirement.

As a continuation of their previous study, Kara and coworkers reported the asymmetric reduction of ethyl 4,4,4-trifluoroacetoacetate by using lyophilized whole-cell harboring ADH evo-1.1.200 in methyl *tert*-butyl ether (MTBE) micro-aqueous system using a "smart co-substrate," 1,4-butanediol (1,4-BD) [99]. Lactone-forming diol co-substrate will lead to a thermodynamically favorable lactone coproduct (γ-butyrolactone [GBL]), thus shifting the overall reaction equilibrium toward the alcohol product. The reaction was performed at a 2 l scale with a substrate concentration of 527 mM and a minimal amount of water (2.5% v/v). The substrate-to-enzyme ratio of the reaction was also optimized to >300 by employing only $\sim 0.3\,g\,l^{-1}$ of the ADH evo-1.1.200 per $100\,g\,l^{-1}$ reaction. Following the reaction, the crude product was purified by distillation, and in total, 150 g (70% yield) of enantiomerically pure product could be isolated with excellent ee (>99%) and high purity (94%). The only drawback reported is the need for distillation to separate the desired alcohol and the lactone coproduct (GBL). To overcome the drawback, the author suggested performing the hydrolytic cleavage of the GBL to its corresponding water-soluble γ-hydroxy acids, which will ease the separation process by aqueous extraction.

Several studies also reported the possibility of cascade reaction in the micro-aqueous system. The lyophilized cell-free extract of monooxygenase-ADH fusion was applied in a micro-aqueous system with only 5% v/v buffer to produce lactone from cyclic ketone (Figure 9.5a) [100]. Fusion enzyme was created to minimize the "cofactor travel distance," thus avoiding the degradation of the labile cofactor. MTBE and cyclopentyl methyl ether (CPME) were found to be the best reaction media. In this cascade, when the monooxygenase performs Baeyer–Villiger oxidation of the cyclic ketone to the corresponding lactone, the cofactor is regenerated by the ADH while the ADH simultaneously produces the lactone from the precursor diol co-substrate. A two-step biocatalytic cascade to produce pharmaceutically important vicinal diol was reported by Rother and Jakoblinnert (Figure 9.5b) [101]. At first, the carboligation of benzaldehyde and acetaldehyde to produce (*R*)-2-hydroxy-1-phenylpropan-1-one ((2*R*)-HPP) was done by *E. coli* whole cells express benzaldehyde lyase (BAL) from *Pseudomonas fluorescens*. The α-hydroxy ketone was then reduced by whole cells expressing ADH from *Ralstonia* sp. (*Ras*ADH) to produce (1*R*,2*R*)-1-phenylpropane-1,2-diol, (1*R*,2*R*)-PPD. MTBE and 10% v/v of 1.0 M triethanolamine (TEA) buffer was the optimum solvent and aqueous media, respectively. Under optimized conditions, up to 327 $g\,l^{-1}\,day^{-1}$ of vicinal diol could be obtained with a low environmental factor (*E*-factor).

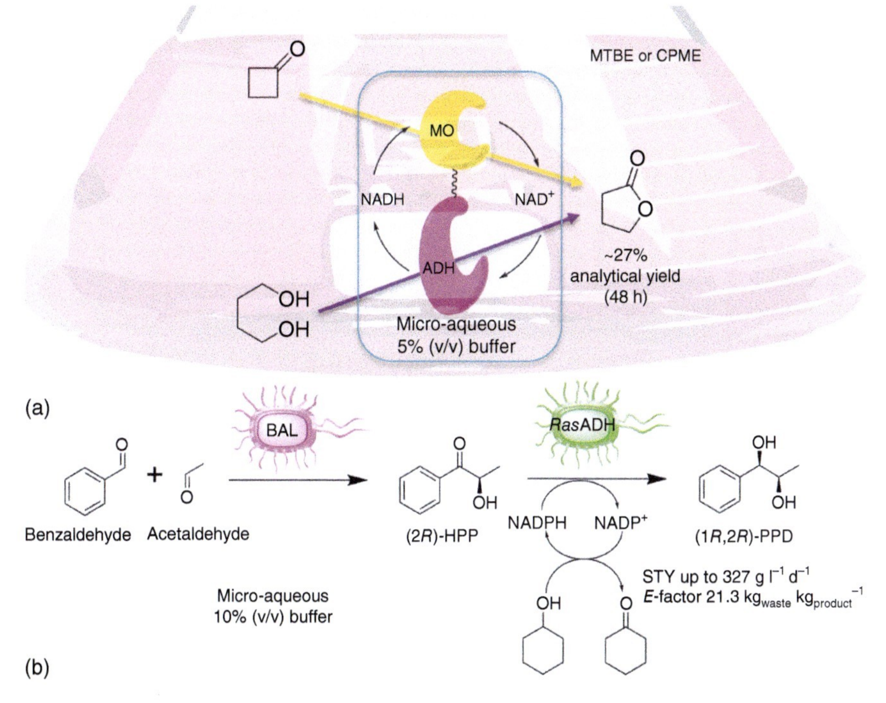

Figure 9.5 Example of cascade reactions conducted in micro-aqueous system: (a) lactone synthesis by fusion of a monooxygenase (MO) and ADH. Source: Based on Huang et al. [100] and (b) pharmaceutically important vicinal diol synthesis by two whole cells expressing benzaldehyde lyase (BAL) and *Ras*ADH, respectively. Source: Modified from Jakoblinnert and Rother [101].

9.8 Examples of Reactions in Bio-Based Liquids

9.8.1 2-Methyltetrahydrofuran (MeTHF)

MeTHF, derived from lignocellulosic biomass, is one of the most widely used bio-based liquids. MeTHF has low volatility, water miscibility, and low toxicity. It was proven to be stable and biocompatible. Thus, it was extensively used for biocatalysis in the past decade [102]. Many examples of lipase-catalyzed reactions in MeTHF are reported. Ascorbyl undecylenate was synthesized by the acylation of L-ascorbic acid in MeTHF : *t*-butanol (4 : 1) with high yield and >99% regioselectivity [103]. The fatty acid ester of ascorbic acid was reported to have higher lipo-solubility as well as improved antioxidative properties. MeTHF acted to lower the activation energy of the reaction while *t*-butanol solubilizes the substrate. MeTHF was also employed in the *Candida rugosa* lipase (CRL)-catalyzed hydrolysis of 3,4-DHP-2-one phenolic ester to prepare optically pure 3,4-DHP-2-one, an essential structural unit in medicinal and synthetic chemistry [104]. By using MeTHF:water (99 : 1) as the reaction media, the highest *E* value could be achieved when the methyl ester at the 5-position of the pyridine ring was benzyl ester. Immobilized lipase from *Pseudomonas cepacia* (PS 30) catalyzed the kinetic resolution of *rac*-1-(2-fluoro-4-iodophenyl)-3-hydroxypyrrolidin-2-one in MeTHF, yielding the (*S*)-non-acylated derivative with high yield (~40%) and >99% enantioselectivity (Scheme 9.2) [105]. The non-acylated derivative could be used as an intermediate for a potential G-protein-coupled receptor agonist for antidiabetic drugs.

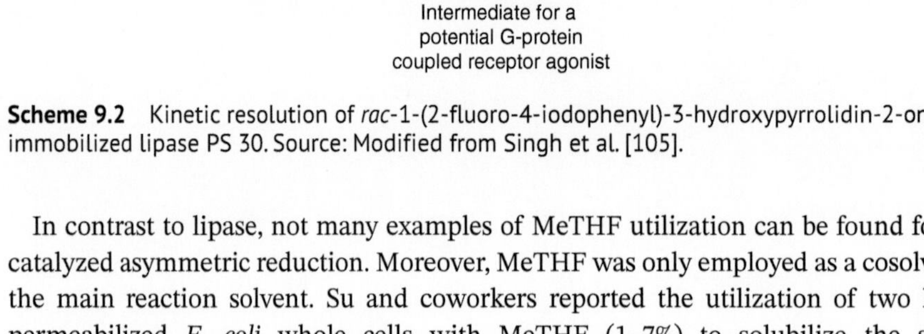

Lipase PS 30
MeTHF, 4 °C
(*S*)
Intermediate for a potential G-protein coupled receptor agonist
(*R*)

Scheme 9.2 Kinetic resolution of *rac*-1-(2-fluoro-4-iodophenyl)-3-hydroxypyrrolidin-2-one by immobilized lipase PS 30. Source: Modified from Singh et al. [105].

In contrast to lipase, not many examples of MeTHF utilization can be found for ADH-catalyzed asymmetric reduction. Moreover, MeTHF was only employed as a cosolvent, not the main reaction solvent. Su and coworkers reported the utilization of two kinds of permeabilized *E. coli* whole cells with MeTHF (1–7%) to solubilize the substrate (Scheme 9.3) [106]. The permeabilized cells containing YOL151W reductase was employed to synthesize (*S*)-3-chloro-1-phenylpropanol, an intermediate of antidepressant drug, achieving excellent enantioselectivity (>99% ee (*S*)) and yield (98%). Several ADH from Codexis KRED (ketoreductase) screening kits were also employed to synthesize β-hydroxydioxinones, an intermediate for natural products [107]. MeTHF was employed as a cosolvent (10% v/v) to afford 74% yield and 98% ee (*S*).

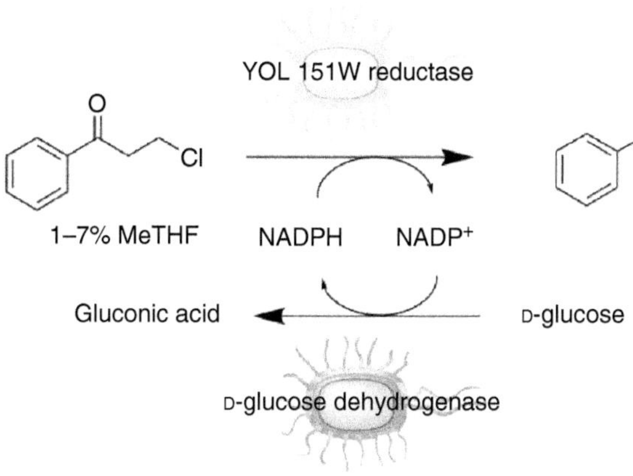

YOL 151W reductase

1–7% MeTHF NADPH NADP+

Gluconic acid D-glucose

D-glucose dehydrogenase

Scheme 9.3 The synthesis of (*S*)-3-chloro-1-phenylpropanol by asymmetric reduction of the corresponding ketone in permeabilized whole cells. Source: Modified from Tian et al. [106].

9.8.2 Cyclopentyl Methyl Ether (CPME)

Because the chemical precursors that may be used for CPME synthesis might be derived from biomass (lignin and sugar), there is a potential for a biomass-derived route for the production of CPME in the future [108]. CPME has been considered as an environmentally friendly solvent because of its low toxicity. CPME has been employed in *Pseudomonas cepacia* lipase (PSL)-catalyzed transesterification of racemic 2,2-dimethyl-1,3-dioxolane-4-methanol (solketal) [109]. CPME was found to be superior compared to diisopropyl ether (DIPE). Transesterification catalyzed by Novozym 435 was also performed in CPME to synthesize 5-formylfuran-2-yl-methyl 4-oxopentanonate (HMF levulinate), a fuel additive, from 5-hydroxymethylfurfural (5-HMF) and levulinic acid [110]. After 24 hours, HMF levulinate was obtained in >90% conversion. Dynamic kinetic resolution to obtain (*S*)-benzoin butyrate was performed by using *Pseudomonas stutzeri* (lipase TL) and a mesoporous silicate-containing zirconium (Zr-TUD-1) as a chemical racemization catalyst (Scheme 9.4) [111]. By using dry CPME, up to 20 g l^{-1} of *rac*-benzoin was solubilized at 50 °C. Remarkably, (*S*)-benzoin butyrate was obtained in >99% ee (*S*).

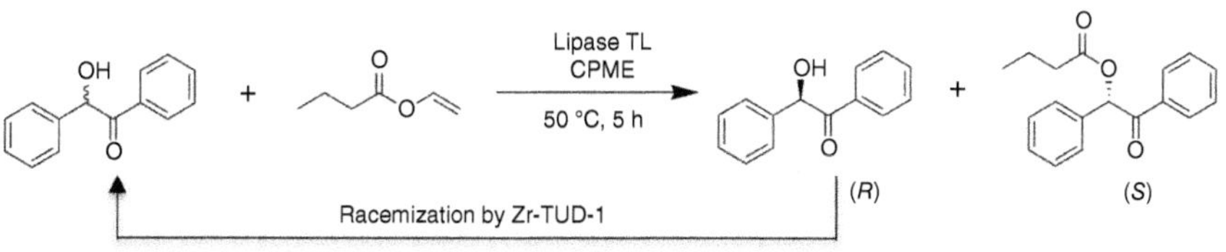

Lipase TL CPME 50 °C, 5 h

(*R*) (*S*)

Racemization by Zr-TUD-1

Scheme 9.4 Dynamic kinetic resolution to obtain (*S*)-benzoin butyrate. Source: Modified from Petrenz et al. [111].

CPME has also been employed as a cosolvent in ADH-catalyzed reduction of ketones. One notable example is the reduction of an α-chloroketone by using the ADH in Codex KRED Screening Kit in the presence of 5% v/v CPME (Scheme 9.5) [112]. CPME helps to dissolve the substrate; thus, excellent conversion and enantioselectivity (99% *de*) was achieved. The enantiomerically pure chlorohydrin obtained is a building block of human immunodeficiency virus (HIV) protease inhibitor, Nelfanivir.

Codex KRED
Buffer/CPME (5% (v/v))
30 °C, 24 h
NADPH, 2-propanol
Buiding block of
HIV protease inhibitor, Nelfanivir

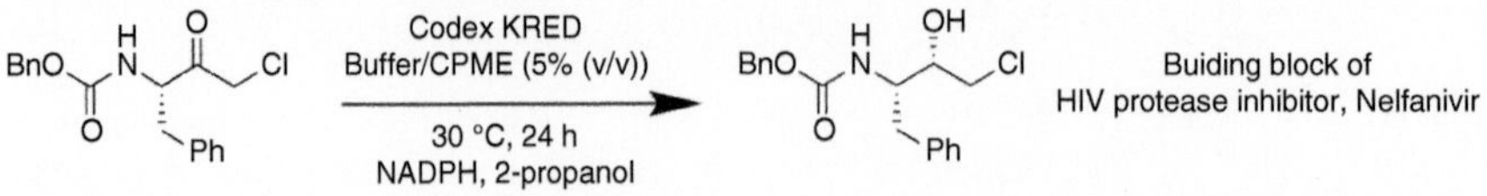

Scheme 9.5 Reduction of an α-chloroketone by using the Codex KRED screening kit. Source: Modified from Castoldi et al. [112].

9.8.3 Potential Application of other Bio-based Liquids

Several other bio-based liquids also have potential applications for biocatalysis. Glycerol is an ideal candidate for reaction media because of its renewable, biocompatible, and non-flammable properties. Glycerol carbonate has been employed for the CALB- and CRL-catalyzed transesterification of ethyl butyrate with *n*-butanol [113]. Glycerol and its derivatives have also been employed for baker's yeast-catalyzed asymmetric reduction of ethyl acetoacetate with a slight decrease in *E*-value compared to the reaction conducted in water [114]. More hydrophobic glycerol derivatives generally showed better cell viability after incubation. Glycerol is also widely utilized as the HBD in NADES [24]. GVL was also a good candidate for biocatalysis because of its low toxicity, low melting point, and stability. Duan and Hu reported the phospholipase D from *Streptomyces chromofuscus*-mediated transphosphatidylation of phosphatidylcholine with L-serine in GVL with up to 95% yield. The yield and selectivity were better than employing ethyl acetate, a conventional organic solvent [115].

9.9 Examples of Reactions in Liquid CO_2

Liquid CO_2 as a reaction media for biocatalysis was first reported by Hoang and Matsuda in 2015 [10]. Novozym 435-catalyzed transesterification of *rac*-1-phenylethanol proceeded with higher activity in liquid CO_2 than in commonly used hydrophobic solvents such as hexane and toluene. The reactions of other alcohols in liquid CO_2 proceeded with a comparable or even higher rate and an enantioselectivity value (*E*-value) than that in hexane. The potential applicability of liquid CO_2 as a biocatalysis reaction media in industrial scale was assessed through large-scale kinetic resolution in a packed-column reactor with a continuous flow of liquid CO_2 and substrate. Three operation cycles (24 hours/cycle) were performed to stably afford the corresponding ester with excellent conversion (49.9%). Both ester and alcohols were obtained with excellent enantioselectivity ($ee_S > 99\%$ (*S*), $ee_P > 99\%$ (*R*), for an *E*-value >200). By using no organic solvent and an adequate amount of the acyl donor, the reaction achieved a very low *E*-factor (<0.3), indicating that a very minimum waste product was generated [10]. Convinced by the rate acceleration induced by liquid CO_2, further investigation was done for the transesterification of challenging bulky 1-phenylakanols known to have almost no reactivity in the conventional organic solvents, such as 1-phenyl-1-dodecanol, α-cyclopropylbenzyl alcohol, and 2-methyl-1-phenyl-1-propanol [116]. Lipase showed considerable activity toward these challenging substrates. High enantioselectivity ($E > 200$) was also observed for the transesterification of α-cyclopropylbenzyl alcohol (Scheme 9.6).

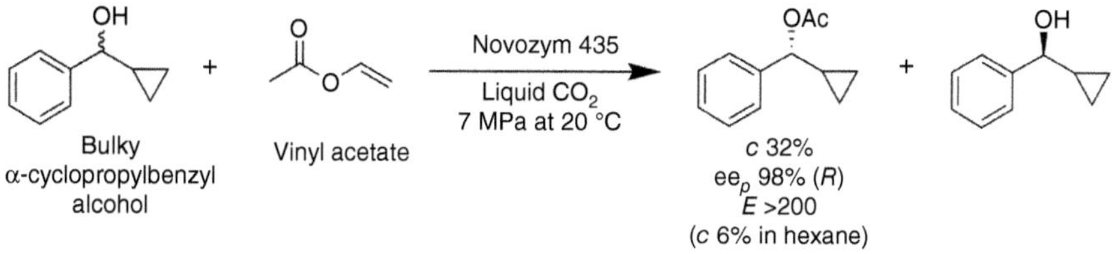

OH + Novozym 435 / Liquid CO_2 7 MPa at 20 °C → OAc + OH

Bulky α-cyclopropylbenzyl alcohol; Vinyl acetate; c 32%, ee_p 98% (R), E >200 (c 6% in hexane)

Scheme 9.6 Novozym 435-catalyzed transesterification of a bulky α-cyclopropylbenzyl alcohol. Source: Modified from Hoang and Matsuda [116].

The activity enhancement promoted by liquid CO_2 was also reported by Wang and coworkers in the *Pseudomonas fluorescens* (PFL)-catalyzed transesterification of 3-hydroxy-3-(2-thienyl)propanenitrile, yielding a beneficial intermediate of the antidepressant drug, duloxetine [117]. Although no exact mechanism of the activity enhancement is known yet, Hoang and Matsuda proposed that the lipase became more flexible in liquid CO_2, particularly in the stereoselective pocket. This proposal is supported by the fact that high conversion could be achieved for the series of bulky 1-phenylalkan-1-ols tested with elongating alkyl chain without a change in selectivity.

Only one study reported the activity of ADH in liquid CO_2, the reduction of butyraldehyde in liquid CO_2 by using horse liver alcohol dehydrogenase (HLADH) and fluorinated nicotinamide cofactor (FNADH), which is soluble in liquid CO_2 (Scheme 9.7) [118]. 3–5 mM of FNADH was soluble in liquid CO_2 with a pressure of 8.3–9.7 MPa. The reduction/oxidation reaction with FNADH was performed in liquid CO_2 (18 MPa, room temperature) and afforded a better yield than the reaction performed with the insoluble cofactor.

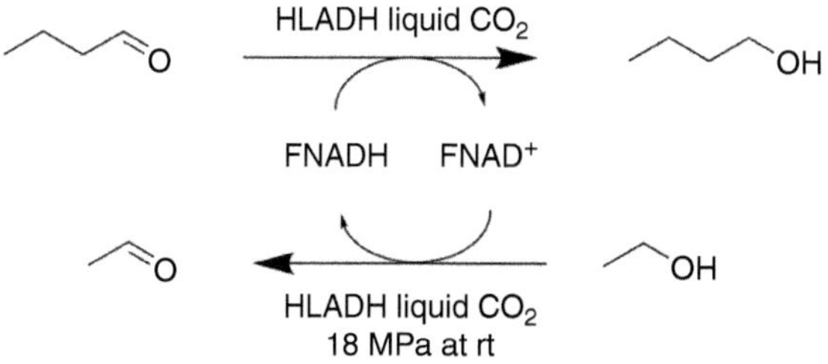

HLADH liquid CO_2 → OH

FNADH FNAD$^+$

O ← OH

HLADH liquid CO_2 18 MPa at rt

Scheme 9.7 Butyraldehyde reduction catalyzed by HLADH in liquid CO_2 with soluble FNADH. Source: Modified from Panza et al. [118].

9.10 Examples of Reactions in CO_2-Expanded Bio-based Liquids

CO_2-expanded bio-based liquids are used because of the combined benefit obtained from the properties of pressurized gas and the environmental friendliness of bio-based liquids. The utilization of CO_2-expanded bio-based liquids was first reported by Matsuda and coworkers in 2017 [18]. Most bio-based liquids exhibiting low hydrophobicity such as MeTHF are generally not a favorable solvent for lipase-catalyzed transesterifications. However, by expanding MeTHF with CO_2, lipase exhibited high activity. The activity toward alcohol with a bulky quaternary carbon substituent, *rac*-1-adamantylethanol, was greatly improved from <1% in neat MeTHF to 29% in CO_2-expanded MeTHF. The kinetic resolution of this bulky substrate was also performed in a gram scale by employing Novozym 435 (0.50 g) in CO_2-expanded MeTHF (6.0 MPa, 20 °C) for five days. Enantiomerically pure alcohol [0.94 g, ee > 99% (*S*)] and ester [1.21 g, ee > 99% (*R*)] were isolated (Scheme 9.8). Several plausible reasons for the improvement by the solvent expansion by CO_2 are (i) conformational change of the lipase induced by the carbamate formation and free amino groups at the surface of the lipase, (ii) higher flexibility of the lipase [119], and (iii) enhanced transport and physicochemical properties [18, 120]. The activity of Novozym 435-catalyzed transesterification in CO_2-expanded MeTHF was tested for other kinds of bulky substrates. The reaction could only happen in the presence of CO_2-expanded MeTHF for substrates with a bulky substituent at the α-position, such as anthracene, 1-naphthyl, and icosahedral boron cluster.

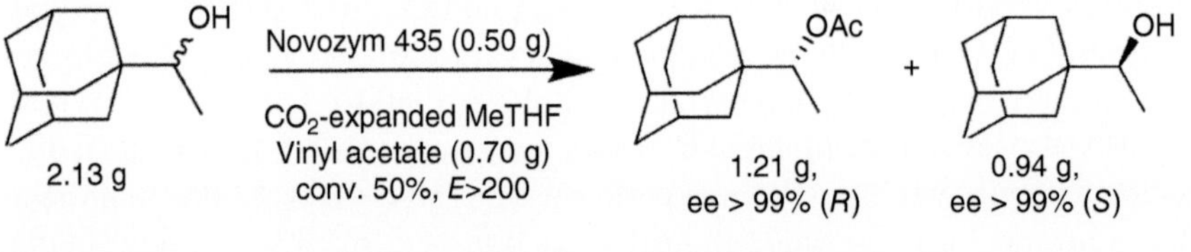

OH
2.13 g
Novozym 435 (0.50 g)
CO_2-expanded MeTHF
Vinyl acetate (0.70 g)
conv. 50%, *E*>200
OAc
1.21 g,
ee > 99% (*R*)
+
OH
0.94 g,
ee > 99% (*S*)

Scheme 9.8 Gram-scale Novozym 435-catalyzed transesterification of *rac*-1-adamantylethanol in CO_2-expanded MeTHF. Source: Modified from Hoang et al. [18].

The solvent properties of CO_2-expanded MeTHF were studied experimentally and by molecular dynamics simulation [120]. The transport properties of CO_2-expanded MeTHF were improved by the decrease in its density and viscosity. The higher reaction conversion was attributed to the decreased dipolarity/polarizability, π^*, with an increasing concentration of CO_2. The activity of Novozym 435 in different temperatures correlate well with its π^* or ability to stabilize a charge or a hydrogen bond (Figure 9.6). Thus, the activity of the lipase in this solvent system can then be easily modulated by tuning the CO_2 mole fraction, which in turn will tune the π^*.

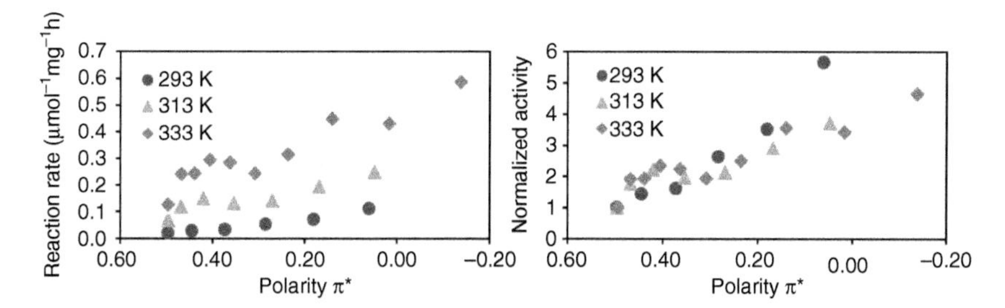

Figure 9.6 Relationship between π^* and the activity of Novozym 435 in various temperatures. Source: Hoang et al. [120]. © 2017 American Chemical Society.

9.11 Examples of Reactions in Natural Deep Eutectic Solvents

Various applications of lipases and ADHs in NADES have been reported since 2014 [11]. In 2017, several groups reported the utilization of NADES for the chemoenzymatic epoxidation mediated by lipase-catalyzed acid perhydrolysis. At first, it was found that the lipase G from *Penicillium camemberti* exhibited no activity toward triglycerides, which opened the possibility of its application for epoxidation reaction [121]. The low stability of this lipase was overcome by employing ChCl:xylitol (1 : 1 mol/mol) as the reaction media. As a result, the chemoenzymatic epoxidation of glyceryl triolate, beneficial substitutes for phthalates, reached a full conversion. The author reported that the utilization of NADES conserved the secondary structure of lipase; thus, stabilization could be achieved despite the addition of H_2O_2.

Moreover, CALB Ser105Ala mutant with eliminated hydrolysis activity was found to be more selective toward the formation of the lactone product by employing NADES ChCl:sorbitol (1 : 1 mol/mol) as shown in Scheme 9.9 [122]. The highest reaction selectivity (100%) was observed in the perhydrolysis reaction of octanoic acid to peroctanoic acid, which spontaneously transforms cyclobutanone to γ-butyrolactone. Sieber and coworkers reported the epoxidation of monoterpenes in ChCl-based DES. The epoxidation of 3-carene, camphene, limonene, and α-pinene could proceeds until completion by using NADES ChCl:glycerol (1 : 2 mol/mol) and ChCl :sorbitol (1 : 1 mol/mol) [123]. In this study, they

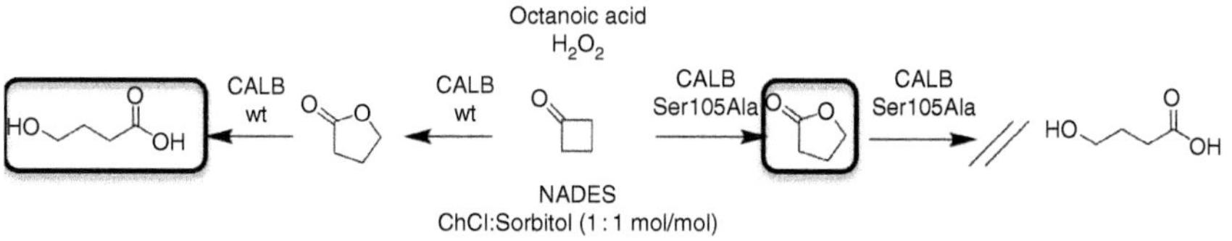

Octanoic acid
H_2O_2
CALB wt
CALB wt
CALB Ser105Ala
CALB Ser105Ala
NADES
ChCl:Sorbitol (1 : 1 mol/mol)

Scheme 9.9 The comparison between wild-type CALB and CALB Ser105Ala mutant in the epoxidation of cyclobutanone. The main product formed by each variant is shown inside a box. Source: Modified from Wang et al. [122].

also reported the utilization of a "minimal DES," ChCl :U·H_2O_2 (1 : 2 mol/mol), which not only act as a solvent but also as the source of H_2O_2 for the perhydrolysis reaction. High conversions could be achieved for all monoterpenes tested by using the "minimal DES" system.

Wild-type whole cells and recombinant whole cells expressing ADHs have been utilized for asymmetric reduction of ketones in NADES. For example, baker's yeast cells were employed to perform the reduction of phenylacetone to 1-phenylpropan-2-ol, an important intermediate of drug used in Alzheimer's treatment [124]. By employing 90% v/v of NADES ChCl:glycerol (2 : 1 mol/mol), the enantioselectivity of the reaction inverted completely from 98% (*S*) in water to 98% (*R*) in NADES. It is hypothesized that the presence of NADES may inhibit the activity of (*S*)-ADHs in the baker's yeast cells. De María and coworkers reported the first utilization of recombinant whole cells expressing ADH [125]. *Ralstonia* sp. ADH (*Ras*ADH) was found to be active with up to 95% v/v of ChCl : glycerol (1 : 2 mol/mol). NADES was found to improve the enantioselectivity of bulky–bulky ketones. The enantioselectivity improvement might be attributed to the interactions between NADES composition and the active site.

Purified ADH from a Codex® Screening Kit exhibited high activity by using 50% or 80% w/w of ChCl : glycerol (1 : 2 mol/mol) [126]. The enantioselectivities of all propiophenone derivatives tested were improved by using 80% w/w of ChCl : glycerol (1 : 2 mol/mol). This study is the first example of the chemoenzymatic cascade in NADES–buffer mixtures. The ruthenium-catalyzed isomerization of racemic allylic alcohols was performed both in concurrent and in sequential ways with the reduction by using an ADH, achieving the enantiomerically pure alcohol products. A designer NADES that can also work as a co-substrate for cofactor regeneration system in ADH-catalyzed reduction was reported recently (Scheme 9.10) [127]. Glucose in a ChCl:Glucose (1.5 : 1 mol/mol) mixture was utilized as a co-substrate to regenerate NAD(P)H. Excellent enantioselectivity and conversion values were reported by using 30–50% v/v of the NADES. Semi-preparative biotransformations yielding 78–89% of the corresponding alcohols demonstrate the applicability of this process.

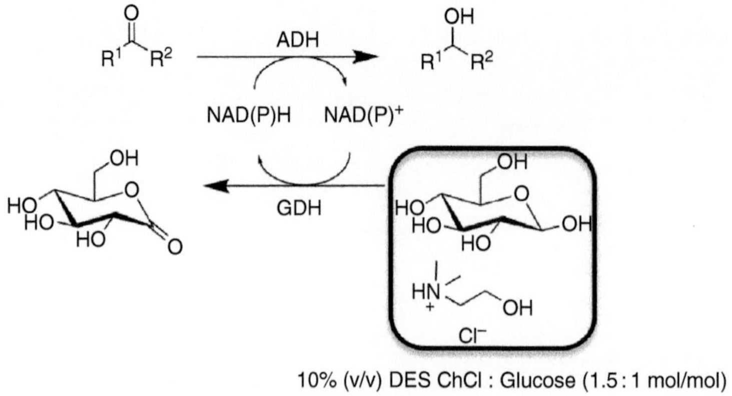

10% (v/v) DES ChCl : Glucose (1.5 : 1 mol/mol)

Scheme 9.10 Design of a designer NADES that can be utilized as a cosolvent and a co-substrate. Source: Modified from Mourelle-Insua et al. [127].

9.12 Conclusions and Future Perspectives

The reaction in nonaqueous media has many benefits that can increase the productivity of biocatalysis for the synthesis of beneficial intermediates. However, with the labile nature of most enzymes, exploration of biocatalysis in nonaqueous media has to be done together with enzyme stabilization. It is expected that by the advances of metagenomic, more nonaqueous media-tolerant enzymes will be isolated. Nanoparticle immobilization and the utilization of biological scaffolds are expected to develop more because of the versatility and activity enhancement displayed by these methods. Moreover, the aid of computational methods will lead to a better and more efficient discovery in protein engineering for enzyme stabilization.

Discovery of more environmentally friendly alternatives to organic solvents for nonaqueous biocatalysis is expected to happen in the near future. The preferable reaction media is the solvent-free system, however, if not possible, bio-based liquids are preferable. More bio-based liquids will evolve by derivatization of the biomass by biocatalytic processes. Pressurized gas (liquid CO_2 and CO_2-expanded bio-based liquids) also displayed a promising potential, especially because of its tunability. Novel NADES with high biocompatibility should be explored because of the limitless availability of their source, not only botanical resources commonly used nowadays. Future works should focus on developing highly effective biocatalysis in nonaqueous media with simple as well as "green" downstream processing.

References

1. Huang, L., de María, P.D., and Kara, S. (2018). The 'water challenge' opportunities and challenges of using oxidoreductases in non-conventional media. *Chem. Today* 36: 48–56.
2. Klibanov, A.M. (2001). Improving enzymes by using them in organic solvents. *Nature* 409: 241–246.
3. Cao, C. and Matsuda, T. (2016). *Organic Synthesis Using Biocatalysis*, 67–97. Elsevier.
4. Sym, E.A. (1936). Action of esterase in the presence of organic solvents. *Biochem. J.* 30: 609–617.
5. Klibanov, A.M., Samokhin, G.P., Martinek, K., and Berezin, I.V. (1977). A new approach to preparative enzymatic synthesis. *Biotechnol. Bioeng.* XIX: 1351–1361.
6. Itoh, T. (2017). Ionic liquids as tool to improve enzymatic organic synthesis. *Chem. Rev.* 117: 10567–10607.
7. Hoang, H.N., Are, K.R.A., and Matsuda, T. (2018). *Supercritical and Other High-pressure Solvent Systems: For Extraction, Reaction and Material Processing*, 191–220. The Royal Society of Chemistry.
8. Stepankova, V., Bidmanova, S., Koudelakova, T. et al. (2013). Strategies for stabilization of enzymes in organic solvents. *ACS Catal.* 3: 2823–2836.
9. Gu, Y. and Jérôme, F. (2013). Bio-based solvents: an emerging generation of fluids for the design of eco-efficient processes in catalysis and organic chemistry. *Chem. Soc. Rev.* 42: 9550–9570.

10 Hoang, H.N. and Matsuda, T. (2015). Liquid carbon dioxide as an effective solvent for immobilized *Candida antarctica* lipase B catalyzed transesterification. *Tetrahedron Lett.* 56: 639–641.

11 Gotor-Fernández, V. and Paul, C.E. (2019). Deep eutectic solvents for redox biocatalysis. *J. Biotechnol.* 293: 24–35.

12 Doukyu, N. and Ogino, H. (2010). Organic solvent-tolerant enzymes. *Biochem. Eng. J.* 48: 270–282.

13 Bommarius, A.S. and Riebel, B.R. (2005). *Biocatalysis*, 339–372. Wiley-VCH Verlag GmbH & Co. KGaA.

14 Sheldon, R.A. (2005). Green solvents for sustainable organic synthesis: state of the art. *Green Chem.* 7: 267–278.

15 Matsuda, T., Harada, T., and Nakamura, K. (2004). Organic synthesis using enzymes in supercritical carbon dioxide. *Green Chem.* 6: 440–444.

16 Matsuda, T., Kanamaru, R., Watanabe, K. et al. (2003). Asymmetric synthesis using hydrolytic enzymes in supercritical carbon dioxide. *Tetrahedron Asymmetry* 14: 2087–2091.

17 Subramaniam, B. (2010). Gas-expanded liquids for sustainable catalysis and novel materials: recent advances. *Coord. Chem. Rev.* 254: 1843–1853.

18 Hoang, H.N., Nagashima, Y., Mori, S. et al. (2017). CO_2-expanded bio-based liquids as novel solvents for enantioselective biocatalysis. *Tetrahedron* 73: 2984–2989.

19 Madeira Lau, R., Van Rantwijk, F., Seddon, K.R., and Sheldon, R.A. (2000). Lipase-catalyzed reactions in ionic liquids. *Org. Lett.* 2: 4189–4191.

20 Erbeldinger, M., Mesiano, A.J., and Russell, A.J. (2000). Enzymatic catalysis of formation of Z-aspartame in ionic liquid – an alternative to enzymatic catalysis in organic solvents. *Biotechnol. Prog.* 16: 1129–1131.

21 Cull, S.G., Holbrey, J.D., Vargas-Mora, V. et al. (2000). Room-temperature ionic liquids as replacements for organic solvents in multiphase bioprocess operations. *Biotechnol. Bioeng.* 69: 227–233.

22 Itoh, T., Akasaki, E., Kudo, K., and Shirakami, S. (2001). Lipase-catalyzed enantioselective acylation in the ionic liquid solvent system: reaction of enzyme anchored to the solvent. *Chem. Lett.*: 262–263.

23 de María, P.D. and Hollmann, F. (2015). On the (Un)greenness of biocatalysis: some challenging figures and some promising options. *Front. Microbiol.* 6: 6–10.

24 Paiva, A., Craveiro, R., Aroso, I. et al. (2014). Natural deep eutectic solvents–solvents for the 21st century. *ACS Sustain. Chem. Eng.* 2: 1063–1071.

25 Pena-Pereira, F. and Namieśnik, J. (2014). Ionic liquids and deep eutectic mixtures: sustainable solvents for extraction processes. *ChemSusChem* 7: 1784–1800.

26 García-Álvarez, J., Hevia, E., and Capriati, V. (2018). The future of polar organometallic chemistry written in bio-based solvents and water. *Chem. Eur. J.* 24: 14854–14863.

27 Wagle, D.V., Zhao, H., and Baker, G.A. (2014). Deep eutectic solvents: sustainable media for nanoscale and functional materials. *Acc. Chem. Res.* 47: 2299–2308.

28 Dai, Y., van Spronsen, J., Witkamp, G.J. et al. (2013). Natural deep eutectic solvents as new potential media for green technology. *Anal. Chim. Acta* 766: 61–68.

29 Zaks, A. and Klibanov, A.M. (1988). Enzymatic catalysis in nonaqueous solvents. *J. Biol. Chem.* 263: 3194–3201.

30 Zaks, A. and Klibanov, A.M. (1985). Enzyme-catalyzed processes in organic solvents. *Proc. Natl. Acad. Sci. U. S. A.* 82: 3192–3196.
31 Silveira, R.L., Martínez, J., Skaf, M.S., and Martínez, L. (2012). Enzyme microheterogeneous hydration and stabilization in supercritical carbon dioxide. *J. Phys. Chem. B* 116: 5671–5678.
32 Zaks, A. and Klibanov, A.M. (1984). Enzymatic catalysis in organic media at 100 degrees C. *Science* 224: 1249–1251.
33 Klibanov, A.M. (2003). Asymmetric enzymatic oxidoreductions in organic solvents. *Curr. Opin. Biotechnol.* 14: 427–431.
34 Klibanov, A.M. (1995). Enzyme memory. What is remembered and why? *Nature* 374: 596.
35 Dai, L. and Klibanov, A.M. (2000). Peroxidase-catalyzed asymmetric sulfoxidation in organic solvents versus in water. *Biotechnol. Bioeng.* 70: 353–357.
36 Wu, M., Lin, M., Lee, C. et al. (2019). Enhancement of laccase activity by pre-incubation with organic solvents. *Sci. Rep.* 9: 1–11.
37 Schumacher, J., Eckstein, M., and Kragl, U. (2006). Influence of water-miscible organic solvents on kinetics and enantioselectivity of the (*R*)-specific alcohol dehydrogenase from *Lactobacillus brevis*. *Biotechnol. J.* 1: 574–581.
38 Serdakowski, A.L. and Dordick, J.S. (2008). Enzyme activation for organic solvents made easy. *Trends Biotechnol.* 26: 48–54.
39 Plapp, B.V. (2010). Conformational changes and catalysis by alcohol dehydrogenase. *Arch. Biochem. Biophys.* 493: 3–12.
40 Klibanov, A.M. (1997). Why are enzymes less active in organic solvents than in water? *Trends Biotechnol.* 15: 97–101.
41 Ghatorae, A.S., Bell, G., and Halling, P.J. (1994). Inactivation of enzymes by organic solvents: new technique with well-defined interfacial area. *Biotechnol. Bioeng.* 43: 331–336.
42 Kumar, A., Dhar, K., Kanwar, S.S., and Arora, P.K. (2016). Lipase catalysis in organic solvents: advantages and applications. *Biol. Proced. Online* 18: 1–11.
43 Ogino, H., Yasui, K., Shiotani, T. et al. (1995). Organic solvent-tolerant bacterium which secretes an organic solvent-stable proteolytic enzyme. *Appl. Environ. Microbiol.* 61: 4258–4262.
44 Ogino, H., Nakagawa, S., Shinya, K. et al. (2000). Purification and characterization of organic solvent-stable lipase from organic solvent-tolerant *Pseudomonas aeruginosa* LST-03. *J. Biosci. Bioeng.* 89: 451–457.
45 Ogino, H., Miyamoto, K., Yasuda, M. et al. (1999). Growth of organic solvent-tolerant *Pseudomonas aeruginosa* LST-03 in the presence of various organic solvents and production of lipolytic enzyme in the presence of cyclohexane. *Biochem. Eng. J.* 4: 1–6.
46 Lavandera, I., Kern, A., Schaffenberger, M. et al. (2008). An exceptionally DMSO-tolerant alcohol dehydrogenase for the stereoselective reduction of ketones. *ChemSusChem* 1: 431–436.
47 Stampfer, W., Kosjek, B., Kroutil, W., and Faber, K. (2003). On the organic solvent and thermostability of the biocatalytic redox system of *Rhodococcus ruber* DSM 44541. *Biotechnol. Bioeng.* 81: 865–869.
48 Kosjek, B., Stampfer, W., Pogorevc, M. et al. (2004). Purification and characterization of a chemotolerant alcohol dehydrogenase applicable to coupled redox reactions. *Biotechnol. Bioeng.* 86: 55–62.

49 Yamamoto, T., Nakata, Y., Cao, C. et al. (2013). Acetophenone reductase with extreme stability against a high concentration of organic compounds or an elevated temperature. *Appl. Microbiol. Biotechnol.* 97: 10413–10421.

50 Alsafadi, D., Alsalman, S., and Paradisi, F. (2017). Extreme halophilic alcohol dehydrogenase mediated highly efficient syntheses of enantiopure aromatic alcohols. *Org. Biomol. Chem.* 15: 9169–9175.

51 Silva, C., Martins, M., Jing, S. et al. (2018). Practical insights on enzyme stabilization. *Crit. Rev. Biotechnol.* 38: 335–350.

52 Honda, K., Inoue, M., Ono, T. et al. (2017). Improvement of operational stability of *Ogataea minuta* carbonyl reductase for chiral alcohol production. *J. Biosci. Bioeng.* 123: 673–678.

53 Honda, K., Ono, T., Okano, K. et al. (2019). Expression of engineered carbonyl reductase from *Ogataea minuta* in *Rhodococcus opacus* and its application to whole-cell bioconversion in anydrous solvents. *J. Biosci. Bioeng* 127: 145–149.

54 Boock, J.T., Freedman, A.J.E., Tompsett, G.A. et al. (2019). Engineered microbial biofuel production and recovery under supercritical carbon dioxide. *Nat. Commun.* 10: 587.

55 Azevedo, A.M., Prazeres, D.M.F., Cabral, J.M.S., and Fonseca, L.P. (2001). Stability of free and immobilised peroxidase in aqueous–organic solvents mixtures. *J. Mol. Catal. B Enzym.* 15: 147–153.

56 Hanefeld, U., Gardossi, L., and Magner, E. (2009). Understanding enzyme immobilisation. *Chem. Soc. Rev.* 38: 453–468.

57 De Temino, D.M.R., Hartmeier, W., and Ansorge-Schumacher, M.B. (2005). Entrapment of the alcohol dehydrogenase from *Lactobacillus kefir* in polyvinyl alcohol for the synthesis of chiral hydrophobic alcohols in organic solvents. *Enzym. Microb. Technol.* 36: 3–9.

58 Musa, M.M., Ziegelmann-Fjeld, K.I., Vieille, C. et al. (2007). Xerogel-encapsulated W110A secondary alcohol dehydrogenase from *Thermoanaerobacter ethanolicus* performs asymmetric reduction of hydrophobic ketones in organic solvents. *Angew. Chem. Int. Ed.* 46: 3091–3094.

59 Goodford, P.J. (1985). A computational procedure for determining energetically favorable binding sites on biologically important macromolecules. *J. Med. Chem.* 28: 849–857.

60 Svendsen, A., Skjot, M., Brask, J. et al. (2007). *World Intellect. Prop. Organ.* 080197: A2.

61 Ge, J., Lei, J., and Zare, R.N. (2012). Protein-inorganic hybrid nanoflowers. *Nat. Nanotechnol.* 7: 428–432.

62 Cui, J. and Jia, S. (2017). Organic–inorganic hybrid nanoflowers: a novel host platform for immobilizing biomolecules. *Coord. Chem. Rev.* 352: 249–263.

63 Fotiadou, R., Patila, M., Hammami, M.A. et al. (2019). Development of effective lipase-hybrid nanoflowers enriched with carbon and magnetic nanomaterials for biocatalytic transformations. *Nanomaterials* 9: 808–825.

64 Ke, C., Fan, Y., Chen, Y. et al. (2016). A new lipase–inorganic hybrid nanoflower with enhanced enzyme activity. *RSC Adv.* 6: 19413–19416.

65 Hua, X., Xing, Y., and Zhang, X. (2016). Enhanced promiscuity of lipase-inorganic nanocrystal composites in the epoxidation of fatty acids in organic media. *ACS Appl. Mater. Interfaces* 8: 16257–16261.

66 Chen, X., Xu, L., Wang, A. et al. (2019). Efficient synthesis of the key chiral alcohol intermediate of Crizotinib using dual-enzyme@CaHPO4 hybrid nanoflowers assembled by mimetic biomineralization. *J. Chem. Technol. Biotechnol.* 94: 236–243.

67 López-Gallego, F. and Yate, L. (2015). Selective biomineralization of $Co_3(PO_4)_2$-sponges triggered by His-tagged proteins: efficient heterogeneous biocatalysts for redox processes. *Chem. Commun.* 51: 8753–8756.

68 Ueno, T. (2013). Porous protein crystals as reaction vessels. *Chem. Eur. J.* 19: 9096–9102.

69 Abe, S., Ijiri, H., Negishi, H. et al. (2015). Design of enzyme-encapsulated protein containers by in vivo crystal engineering. *Adv. Mater.* 27: 7951–7956.

70 Heater, B.S., Lee, M.M., and Chan, M.K. (2018). Direct production of a genetically-encoded immobilized biodiesel catalyst. *Sci. Rep.* 8: 12783.

71 Kowalski, A.E., Johnson, L.B., Dierl, H.K. et al. (2019). Porous protein crystals as scaffolds for enzyme immobilization. *Biomater Sci.* 7: 1898–1904.

72 Iyer, P.V. and Ananthanarayan, L. (2008). Enzyme stability and stabilization-aqueous and non-aqueous environment. *Process Biochem.* 43: 1019–1032.

73 Castillo, B., Mendez, J., Al-Azzam, W. et al. (2006). On the relationship between the activity and structure of PEG-α-chymotrypsin conjugates in organic solvents. *Biotechnol. Bioeng.* 94: 565–574.

74 Partridge, J., Hailing, P.J., and Moore, B.D. (1998). Practical route to high activity enzyme preparations for synthesis in organic media. *Chem. Commun.*: 841–842.

75 Brogan, A.P.S., Bui-Le, L., and Hallett, J.P. (2018). Non-aqueous homogenous biocatalytic conversion of polysaccharides in ionic liquids using chemically modified glucosidase. *Nat. Chem.* 10: 859–865.

76 Itoh, T., Matsushita, Y., Abe, Y. et al. (2006). Increased enantioselectivity and remarkable acceleration of lipase-catalyzed transesterification by using an imidazolium PEG-alkyl sulfate ionic liquid. *Chem. Eur. J.* 12: 9228–9237.

77 Moniruzzaman, M., Kamiya, N., Nakashima, K., and Goto, M. (2008). Water-in-ionic liquid microemulsions as a new medium for enzymatic reactions. *Green Chem.* 10: 497–500.

78 Chen, K. and Arnold, F.H. (1991). Enzyme engineering for nonaqueous solvents: random mutagenesis to enhance activity of subtilisin E in polar organic media. *Nat. Biotechnol.* 9: 1073–1077.

79 Kawata, T. and Ogino, H. (2009). Enhancement of the organic solvent-stability of the LST-03 lipase by directed evolution. *Biotechnol. Prog.* 25: 1605–1611.

80 Yin, X., Hu, D., Li, J.F. et al. (2015). Contribution of disulphide bridges to the thermostability of a type A feruloyl esterase from *Aspergillus usamii*. *PLoS One* 10: e0126864.

81 Park, H.J., Joo, J.C., Park, K., and Yoo, Y.J. (2012). Stabilization of *Candida antarctica* lipase B in hydrophilic organic solvent by rational design of hydrogen bond. *Biotechnol. Bioprocess Eng.* 728: 722–728.

82 Yang, S., Zhou, L., Tang, H. et al. (2002). Rational design of a more stable penicillin G acylase against organic cosolvent. *J. Mol. Catal. B Enzym.* 18: 285–290.

83 Koudelakova, T., Chaloupkova, R., Brezovsky, J. et al. (2013). Engineering enzyme stability and resistance to an organic cosolvent by modification of residues in the access tunnel. *Angew. Chem. Int. Ed.* 52: 1959–1963.

84 Gihaz, S., Kanteev, M., Pazy, Y., and Fishman, A. (2018). Filling the void: introducing aromatic interactions into solvent tunnels to enhance lipase stability in methanol. *Appl. Environ. Microbiol.* 84: e02143-18.

85 Kellogg, E.H., Leaver-Fay, A., and Baker, D. (2011). Role of conformational sampling in computing mutation-induced changes in protein structure and stability. *Proteins* 79: 830–838.

86 Gribenko, A.V., Patel, M.M., Liu, J. et al. (2009). Rational stabilization of enzymes by computational redesign of surface charge-charge interactions. *Proc. Natl. Acad. Sci. U. S. A.* 106: 2601–2606.

87 Reetz, M.T., Carballeira, J.D., and Vogel, A. (2006). Iterative saturation mutagenesis on the basis of B factors as a strategy for increasing protein thermostability. *Angew. Chem. Int. Ed.* 45: 7745–7751.

88 Reetz, M.T., Soni, P., Fernández, L. et al. (2010). Increasing the stability of an enzyme toward hostile organic solvents by directed evolution based on iterative saturation mutagenesis using the B-FIT method. *Chem. Commun.* 46: 8657–8658.

89 Lehmann, M., Pasamontes, L., Lassen, S.F., and Wyss, M. (2000). The consensus concept for thermostability engineering of proteins. *Biochim. Biophys. Acta* 1543: 408–415.

90 Wijma, H.J., Floor, R.J., Jekel, P.A. et al. (2014). Computationally designed libraries for rapid enzyme stabilization. *Protein Eng. Des. Sel.* 27: 49–58.

91 Moore, E.J., Zorine, D., Hansen, W.A. et al. (2017). Enzyme stabilization via computationally guided protein stapling. *Proc. Natl. Acad. Sci. U. S. A.* 114: 12472–12477.

92 Jakoblinnert, A., Mladenov, R., Paul, A. et al. (2011). Asymmetric reduction of ketones with recombinant *E. coli* whole cells in neat substrate. *Chem. Commun.* 47: 12230–12232.

93 Valivety, R.H., Halling, P.J., and Macrae, A.R. (1992). *Rhizomucor miehei* lipase remains highly active at water activity below 0.0001. *FEBS Lett.* 301: 258–260.

94 Heeres, A., Vanbroekhoven, K., and Van Hecke, W. (2019). Solvent-free lipase-catalyzed production of (meth) acrylate monomers: experimental results and kinetic modeling. *Biochem. Eng. J.* 142: 162–169.

95 Wunschik, D.S., Ingenbosch, K.N., Zähres, M. et al. (2018). Biocatalytic and solvent-free synthesis of a bio-based biscyclocarbonate. *Green Chem.* 20: 4738–4745.

96 Kara, S., Spickermann, D., Weckbecker, A. et al. (2014). Bioreductions catalyzed by an alcohol dehydrogenase in non-aqueous media. *ChemCatChem* 6: 973–976.

97 Erdmann, V., Mackfeld, U., Rother, D., and Jakoblinnert, A. (2014). Enantioselective, continuous (*R*)- and (*S*)-2-butanol synthesis: achieving high space-time yields with recombinant *E. coli* cells in a micro-aqueous, solvent-free reaction system. *J. Biotechnol.* 191: 106–112.

98 de Gonzalo, G., Lavandera, I., Faber, K., and Kroutil, W. (2007). Enzymatic reduction of ketones in "micro-aqueous" media catalyzed by ADH-A from *Rhodococcus ruber*. *Org. Lett.* 9: 2163–2166.

99 Zuhse, R., Leggewie, C., Hollmann, F., and Kara, S. (2015). Scaling-up of "smart cosubstrate" 1, 4-butanediol promoted asymmetric reduction of ethyl-4, 4, 4-trifluoroacetoacetate in organic media. *Org. Process. Res. Dev.* 19: 369–372.

100 Huang, L., Aalbers, F.S., Tang, W. et al. (2019). Convergent cascade catalyzed by monooxygenase-alcohol dehydrogenase fusion applied in organic media. *ChemBioChem* 20: 1653–1658.

101 Jakoblinnert, A. and Rother, D. (2014). A two-step biocatalytic cascade in micro-aqueous medium: using whole cells to obtain high concentrations of a vicinal diol. *Green Chem.* 16: 3472–3482.

102 Alcántara, A.R. and de Maria, P.D. (2018). Recent advances on the use of 2-methyltetrahydrofuran (2-MeTHF) in biotransformations. *Curr. Green Chem.* 5: 86–103.

103 Hu, Y.D., Qin, Y.Z., Li, N., and Zong, M.H. (2014). Highly efficient enzymatic synthesis of an ascorbyl unstaturated fatty acid ester with ecofriendly biomass-derived 2-methyltetrahydrofuran as cosolvent. *Biotechnol. Prog.* 30: 1005–1011.

104 Torres, S.Y., Brieva, R., and Rebolledo, F. (2017). Chemoenzymatic synthesis of optically active phenolic 3,4-dihydropyridin-2-ones: a way to access enantioenriched 1,4-dihydropyridine and benzodiazepine derivatives. *Org. Biomol. Chem.* 15: 5171–5181.

105 Singh, A., Falabella, J., LaPorte, T.L., and Goswami, A. (2015). Enzymatic process for N-substituted (3*S*)- and (3*R*)-3-hydroxypyrrolidin-2-ones. *Org. Process. Res. Dev.* 19: 819–830.

106 Tian, Y., Ma, X., Yang, M. et al. (2017). Synthesis of (*S*)-3-chloro-1-phenylpropanol by permeabilized recombinant *Escherichia coli* harboring *Saccharomyces cerevisiae* YOL151W reductase in 2-methyltetrahydrofuran cosolvent system. *Catal. Commun.* 97: 56–59.

107 Betori, R.C., Miller, E.R., and Scheidt, K.A. (2017). A biocatalytic route to highly enantioenriched β-hydroxydioxinones. *Adv. Synth. Catal.* 359: 1131–1137.

108 de Gonzalo, G., Alcántara, A.R., and Domínguez de María, P. (2019). Cyclopentyl methyl ether (CPME): a versatile eco-friendly solvent for applications in biotechnology and biorefineries. *ChemSusChem* 12: 2083–2097.

109 Mine, Y., Zhang, L., Fukunaga, K., and Sugimura, Y. (2005). Enhancement of enzyme activity and enantioselectivity by cyclopentyl methyl ether in the transesterification catalyzed by *Pseudomonas cepacia* lipase co-lyophilized with cyclodextrins. *Biotechnol. Lett.* 27: 383–388.

110 Qin, Y.Z., Zong, M.H., Lou, W.Y., and Li, N. (2016). Biocatalytic upgrading of 5-hydroxymethylfurfural (HMF) with levulinic acid to HMF levulinate in biomass-derived solvents. *ACS Sustain. Chem. Eng.* 4: 4050–4054.

111 Petrenz, A., De, M.P.D., Ramanathan, A. et al. (2015). Medium and reaction engineering for the establishment of a chemo-enzymatic dynamic kinetic resolution of *rac*-benzoin in batch and continuous mode. *J. Mol. Catal. B Enzym.* 114: 42–49.

112 Castoldi, L., Ielo, L., Hoyos, P. et al. (2018). Merging lithium carbenoid homologation and enzymatic reduction: a combinative approach to the HIV-protease inhibitor Nelfinavir. *Tetrahedron* 74: 2211–2217.

113 Ou, G., He, B., and Yuan, Y. (2012). Design of biosolvents through hydroxyl functionalization of compounds with high dielectric constant. *Appl. Biochem. Biotechnol.* 166: 1472–1479.

114 Wolfson, A., Snezhko, A., Meyouhas, T., and Tavor, D. (2012). Glycerol derivatives as green reaction mediums. *Green Chem. Lett. Rev.* 5: 7–12.

115 Duan, Z.Q. and Hu, F. (2012). Highly efficient synthesis of phosphatidylserine in the eco-friendly solvent γ-valerolactone. *Green Chem.* 14: 1581–1583.

116 Hoang, H.N. and Matsuda, T. (2016). Expanding substrate scope of lipase-catalyzed transesterification by the utilization of liquid carbon dioxide. *Tetrahedron* 72: 7229–7234.

117 Zhang, J., Li, Y., Qian, W. et al. (2018). Lipase-catalyzed enantioselective transesterification of 3-hydroxy-3-(2-thienyl) propanenitrile in liquid carbon dioxide. *Green Chem. Lett. Rev.* 11: 224–229.

118 Panza, J.L., Russell, A.J., and Beckman, E.J. (2002). Synthesis of fluorinated NAD as a soluble coenzyme for enzymatic chemistry in fluorous solvents and carbon dioxide. *Tetrahedron* 58: 4091–4104.

119 Monhemi, H. and Housaindokht, M.R. (2012). How enzymes can remain active and stable in a compressed gas? New insights into the conformational stability of *Candida antarctica* lipase B in near-critical propane. *J. Supercrit. Fluids* 72: 161–167.

120 Hoang, H.N., Granero-Fernandez, E., Yamada, S. et al. (2017). Modulating biocatalytic activity toward sterically bulky substrates in CO_2-expanded biobased liquids by tuning the physicochemical properties. *ACS Sustain. Chem. Eng.* 5: 11051–11059.

121 Zhou, P., Wang, X., Zeng, C. et al. (2017). Deep eutectic solvents enable more robust chemoenzymatic epoxidation reactions. *ChemCatChem* 9: 934–936.

122 Wang, X.P., Zhou, P.F., Li, Z.G. et al. (2017). Engineering a lipase B from *Candida antactica* with efficient perhydrolysis performance by eliminating its hydrolase activity. *Sci. Rep.* 7: 44599.

123 Ranganathan, S., Zeitlhofer, S., and Sieber, V. (2017). Development of a lipase-mediated epoxidation process for monoterpenes in choline chloride-based deep eutectic solvents. *Green Chem.* 19: 2576–2586.

124 Vitale, P., Abbinante, V.M., Perna, F.M. et al. (2017). Unveiling the hidden performance of whole cells in the asymmetric bioreduction of aryl-containing ketones in aqueous deep eutectic solvents. *Adv. Synth. Catal.* 359: 1049–1057.

125 Müller, C.R., Lavandera, I., Gotor-Fernández, V., and Domínguez de María, P. (2015). Performance of recombinant whole-cell-catalyzed reductions in deep-eutectic solvent-aqueous-media mixtures. *ChemCatChem* 7: 2654–2659.

126 Cicco, L., Ríos-Lombardía, N., Rodríguez-Álvarez, M.J. et al. (2018). Programming cascade reactions interfacing biocatalysis with transition-metal catalysis in deep eutectic solvents as biorenewable reaction media. *Green Chem.* 20: 3468–3475.

127 Mourelle-Insua, Á., Lavandera, I., and Gotor-Fernández, V. (2019). A designer natural deep eutectic solvent to recycle the cofactor in alcohol dehydrogenase-catalysed processes. *Green Chem.* 21: 2946–2951.

10

Nonconventional Cofactor Regeneration Systems

Jiafu Shi[2,3,4,5,], Yizhou Wu[1,*], Zhongyi Jiang[1,2,3,4], Yiying Sun[1], Qian Huo[1], Weiran Li[5], Yang Zhao[5], and Yuqing Cheng[1,*]*

[1] *Tianjin University, Key Laboratory for Green Chemical Technology of Ministry of Education, Key Laboratory of Bioengineering of Ministry of Education, School of Chemical Engineering and Technology, 92 Weijin Road, Nankai District, Tianjin 300072, China*
[2] *East China University of Science and Technology, State Key Laboratory of Bioreactor Engineering, 130 Meilong Road, Shanghai 200237, P. R. China*
[3] *Collaborative Innovation Center of Chemical Science and Engineering (Tianjin), 92 Weijin Road, Nankai District, Tianjin 300072, China*
[4] *Joint School of National University of Singapore and Tianjin University, International Campus of Tianjin University, Binhai New City, Fuzhou, 350207, P. R. China*
[5] *Tianjin University, School of Environmental Science and Engineering, 92 Weijin Road, Nankai District, Tianjin 300072, China*

10.1 Introduction

The advances of biocatalysis, known as renewable catalysts, high activity/selectivity, and mild conditions, lead to broad applications in clean energy production, biomass conversion, and biomedicine manufacture [1]. The biocatalyst plays a critical role in the widening of green chemistry. The development of synthetic biology and directed enzyme evolution lowers the costs of the enzyme and improves the catalytic activity/stability, which could further broaden the industrial application of enzyme-involved catalytic processes [2–5]. Oxidoreductases (see also Chapters 6 and 7), as one of the largest categories, account for 25% of the total number of enzymes [6]. A cofactor is an organic molecule involved in an enzymatic reaction, acting as a shuttle to transport energy and materials among different enzymes. Typical cofactors, such as adenosine 5′-monophosphate (AMP)/adenosine 5′-diphosphate (ADP)/adenosine 5′-triphosphate (ATP), vitamins and their derivatives, ubiquinone (CoQ), flavin mononucleotide (FMN), and reduced nicotinamide adenine dinucleotide (NADH)/reduced nicotinamide adenine dinucleotide phosphate (NADPH), assist a series of important cellular metabolism processes, such as respiration and

*These authors contributed equally to this work.

Biocatalysis for Practitioners: Techniques, Reactions and Applications, First Edition.
Edited by Gonzalo de Gonzalo and Iván Lavandera.

photosynthesis [7]. Among them, NAD(P)H accounts for over 85% oxidoreductase transformations (Figure 10.1) [8]. The routes of NAD(P)H-dependent enzymes are also uncomplicated, which could be easily re-established for *in vitro* biosynthesis. However, the cycling nature of cofactor also means the rarity of NAD(P)H. The *in vitro* utilization of NAD(P)H requires efficient regeneration methods to achieve sustainability.

Conventional NAD(P)H regeneration mainly includes enzymatic, chemical, and electrochemical methods, and each has its own advantages and limitations. Enzymatic regeneration utilizes the specificity of an enzyme to achieve the interconversion of $NAD(P)^+$/NAD(P)H, which was first established by Rafter and coworkers [9]. At present, this regeneration method is most widely used in industrial applications, such as the regeneration of NADH by formate dehydrogenase and formic acid, with only a by-product of CO_2 [10]. Enzymatic regeneration could be mainly divided into two versions, substrate coupling and enzyme coupling (Figures 10.2a and 10.2b). Mangan and coworkers [11] synthesized a high-value prodrug (*S*)-2-bromo-2-cyclohexen-1-ol under the action of an alcohol dehydrogenase by adding isopropanol as a second substrate for NADH regeneration. The substrate coupling method is based on the fact that some enzymes have a wide range of substrate specificities. For example, horse liver alcohol dehydrogenase can catalyze various redox reactions to regenerate NADH using alcohols, ketones, and aldehydes as substrates, but it commonly reduces the catalytic efficiency of the enzyme and increases the complexity of the product separation. The method of enzyme coupling is a combination of multiple enzymes to achieve both substrate conversion and NAD(P)H regeneration. Glucose dehydrogenase and formate dehydrogenase are two typical enzymes used in the industry. Other more efficient systems, such as phosphite dehydrogenase and alcohol dehydrogenase, are still in the experimental stage because of the increased complexity of the system [6]. In general, enzymatic regeneration method is an environmental friendly process with high turnover number and high selectivity, but poor stability and high costs as well as the difficulty in product separation limit its further application.

Chemical regeneration method utilizes the redox reaction between chemical reagents (such as flavin derivatives, dithionite, or metal complexes) and an oxidized cofactor to achieve the regeneration (Figure 10.2c). Lehninger and coworkers [12] first demonstrated the concept of chemical regeneration by using a high-purity barium salt to reduce DPN to DPNH. This barium salt regenerated DPNH in the solid phase and was beneficial to the product separation, but the conversion efficiency and selectivity were low. Pentamethylcyclopentadienyl rhodium bipyridine complex $[Cp^*Rh(bpy)(H_2O)]^{2+}$ is a versatile chemical reagent for chemical regeneration of cofactor because of its high selectivity to enzymatically active 1,4-NAD(P)H regioisomer [13]. However, the metal complex was expensive and the product was difficult to separate for the contamination of chemical reagents. The special conditions required for the enzymatic reaction are difficult to satisfy, and the reaction medium seriously affected the stability of enzymes. Such method is rarely used for the regeneration of NADH nowadays.

Electrochemical regeneration method uses renewable electric energy as a driving force for NAD(P)H regeneration, including direct regeneration method and indirect regeneration method (Figure 10.2d). The direct regeneration method utilizes ultrahigh potential to drive direct electron transfer between NAD(P)H and the electrode, which cannot guarantee the selectivity of NAD(P)H and has an adverse impact on enzyme activity. Indirect

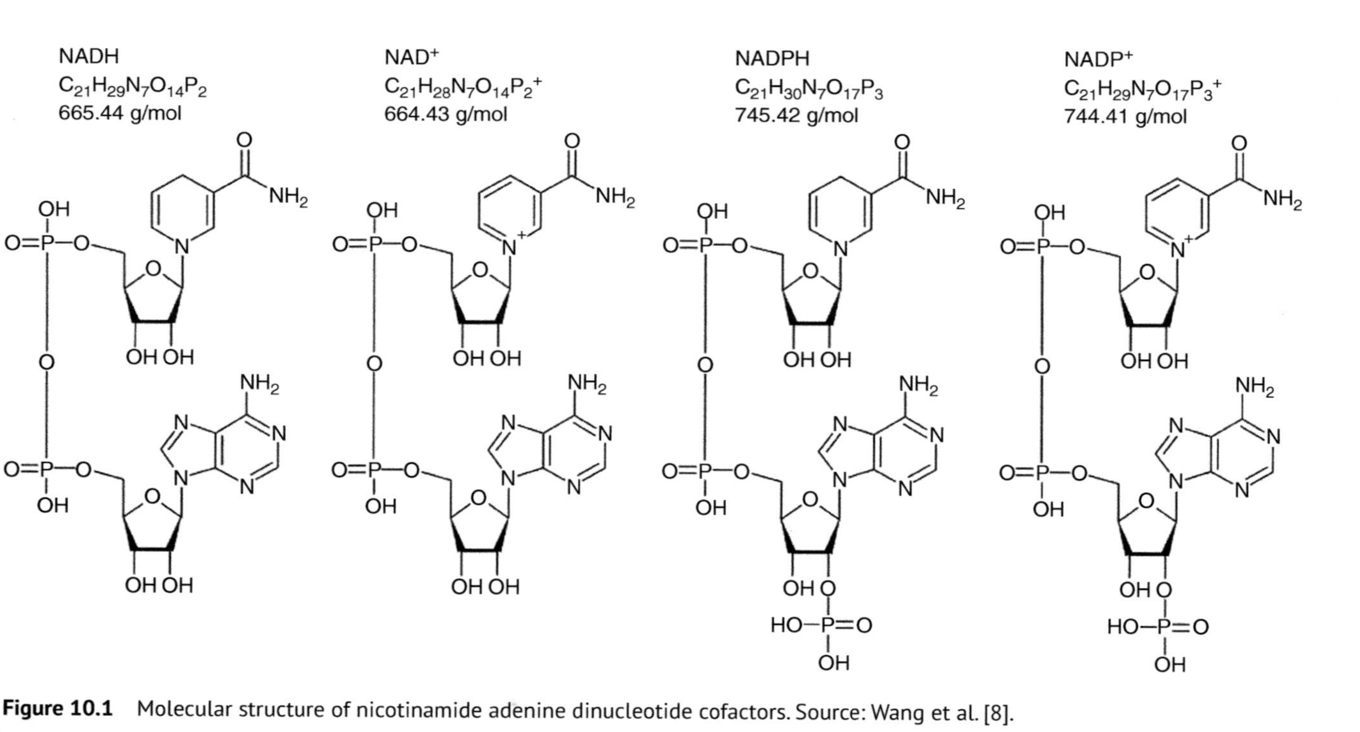

Figure 10.1 Molecular structure of nicotinamide adenine dinucleotide cofactors. Source: Wang et al. [8].

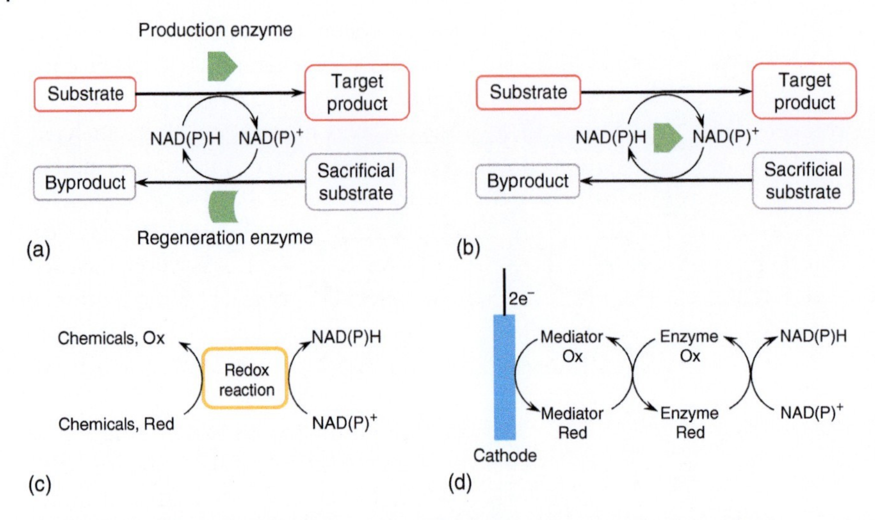

Figure 10.2 Schematic illustration of NAD(P)H enzymatic regeneration with (a) enzyme coupling and (b) substrate coupling; (c) schematic illustration of NAD(P)H chemical regeneration; and (d) schematic illustration of NAD(P)H electrochemical regeneration.

regeneration method is more frequently utilized by using an intermediate mediator (such as methyl viologen) to transfer electrons between NAD(P)H and the electrode. The method can achieve nearly 100% yield under sufficient overpotential, but too high overpotential will lead to lower NAD(P)H selectivity and higher separation cost. Limited by electrode materials, some enzymatic reactions are not compatible with them and the electrode is prone to fouling for a long time, which also limits the continuous regeneration process of NAD(P)H. These problems could be solved by the appropriate selection of the electron mediator or the surface modification of electrode. Miyawaki and Yano [14] found that phenazine methosulfate (PMS) has the highest rate constant as a mediator for the electrochemical oxidation of NADH, which could be firmly fixed to the surface of graphite electrode by adsorption. Schroder and coworkers [15] used ABTS 2,2′-azino-bis(3-ethylbenzothiazoline-6-sulfonic acid) as an electron mediator to regenerate NADH. The regenerated NADH was successfully used for an alcohol dehydrogenase to catalyze the oxidation of a *meso*-diol to a chiral lactone. Wagenknecht et al. [16] developed a system based on cerium(II) and cerium(III) organometallic compounds, catalyzing NADPH regeneration with H_2 as the hydrogen source.

Solar energy is considered as the next generation of renewable energy to replace fossil fuels. Photocatalysis directly converts solar energy to chemical energy, which holds great potential to achieve sustainability. Photocatalytic NADH regeneration was first reported by Krasnovsky and coworkers in 1992 [17]. The NADH regeneration was accomplished by a system that consists of the NAD-dependent hydrogenase from *Alcaligenes eutrophus* immobilized on cadmium sulfide (CdS) particles with formate as a sacrificial electron donor. CdS absorbed light and powered the hydrogenase to catalyze the reduction of NAD^+ by H_2. This was the first time that light was used as the driving force to trigger NADH regeneration. However, the use of enzyme and H_2 was not a sustainable approach and increased the complexity of the system, limiting its further coupling with a second enzymatic reaction. Jiang and coworkers developed a system consisted of carbon-doped TiO_2 and a metal catalyst, $[Cp^*Rh(bpy)(H_2O)]^{2+}$, for NADH regeneration. This metal complex could realize the selective hydrogenation of NAD^+ to 1,4-NADH and did not require H_2 as the hydrogen source, greatly decreasing the complexity of the system [18]. The yield of NADH reached 94.3% under visible light irradiation. Because of the superior performance of Rh complex, it was widely applied in subsequent investigations of photocatalytic NADH regeneration. As mentioned in the electrochemical regeneration method, photocatalysis could also regenerate NADH in the absence of the Rh complex. However, the catalytic process was less efficient and the selectivity of 1,4-NADH was low. The formation of 1,6-NADH and NAD-dimer would cause the irreversible consumption of NAD^+. In order to recycle NAD^+/NADH in a coupled reaction, the Rh complex or other catalyst is indispensable in photocatalytic NADH regeneration. In contrast to the conventional methods, the photocatalytic method only requires solar energy as the driving force to trigger the NADH regeneration. This nonconventional regeneration strategy may have the best chance to realize sustainable chemical production because of its inexhaustible power source, simple system, and zero emission process.

10.2 Basics of Photocatalytic NADH Regeneration

10.2.1 Processes and Mechanism Associated with Photocatalytic NADH Regeneration

Photocatalytic NADH regeneration consists of an electron donor, a photocatalyst, and a cocatalyst (also called as an electron mediator). As shown in Figure 10.3, the regeneration process could be divided into four steps: photoexcitation of the photocatalyst, photochemical reduction of the cocatalyst, selective hydrogenation of NAD^+, and recovery of the photocatalyst. This molecule harvests solar energy and converts it into active chemical energy as excited electrons. These excited electrons are utilized for the reduction of the cocatalyst. The reduced cocatalyst then catalyzes the selective hydrogenation of NADH. After extraction of the excited electrons, the left holes in the photocatalyst are quenched by the electron donor, accomplishing the whole process of photocatalytic NADH regeneration.

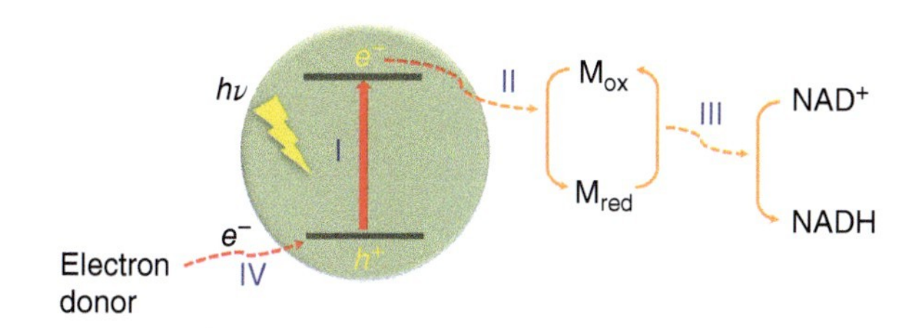

I. Photo-excitation of photocatalyst II. Photo-chemical reduction of mediator (M)
III. Selective hydrogenation of NAD^+ IV. Recovery of photcatalyst

Figure 10.3 Schematic illustration of photocatalytic NADH regeneration.

The first step in photocatalytic NADH regeneration is converting solar energy to excited electrons. These electrons are required to maintain sufficient energy and time for the subsequent chemical reaction. The intrinsic properties of the photocatalyst, such as light harvesting, charge transfer, and separation, normally determine the amount of electrons available for NADH regeneration. The light harvesting ability could be evaluated from two aspects: the absorption of solar spectrum and the utilization of incident light. On the one hand, the absorption spectrum of the photocatalyst should be expanded to the near-infrared region for full use of solar spectrum. On the other hand, the incident light could be absorbed, reflected, or scattered when reaching the surface of the photocatalyst. To utilize this reflected and scattered light, a hierarchical structure is designed to trap the light, improving light absorption. The absorbed light (photons) then excites the electrons on the valence band (VB) of the photocatalyst to its conduction band (CB), generating excited electron–hole pairs (as for the molecular photocatalyst, VB is called the highest occupied molecular orbital, highest occupied molecular orbital (HOMO), while CB is called the lowest unoccupied molecular orbital, LUMO). These electron–hole pairs are highly unstable, which tend to recombine with each other and return to the ground state. In order to maintain the separated state of electron–hole pairs, several types of heterostructures have been developed in designed photocatalysts, such as heterojunction, phase junction, and surface defects. These heterostructures could spatially separate electrons and holes, preventing their recombination. Moreover, conductive channels, normally called as charge transfer pathways, are also needed to reduce the loss of energy during charge transfer. With the coordination between light harvesting and charge transfer/separation, photocatalysts could efficiently convert solar energy into long-lived excited electrons, which could be further involved in NADH regeneration.

The regeneration of NADH requires a selective hydrogenation at N4 position of the pyridine ring. As mention above, the molecular catalyst, $[Cp^*Rh(bpy)H_2O]^{2+}$ (denoted as **Rh**), is normally utilized to guarantee the selectivity (Figure 10.4). The molecular catalyst needs to accept two electrons and one proton to its reduced form, $[Cp^*Rh(bpy)H]^+$ (denoted as **Rh-H**), which would then selectively catalyze the formation of 1,4-NADH [19]. It is possible that a photocatalyst can provide these two electrons for Rh via a one-step process. According to our recent investigation, the proton source should be water, which is different from the conventional chemical regeneration process. Using a chemical regeneration, **Rh** serves as a hydrogen transfer catalyst. The sacrificial substrate, such as formic acid, transfers its hydride to form **Rh-H**, which would then transfer this hydride to NAD^+. As in the case of photocatalysis, the electrons are already provided by the photocatalyst, only a proton is needed to form **Rh-H**. The ^{1}H NMR result represented in our study indicated that water, not the electron donor (triethanolamine, TEOA), provided the proton, indicating that photocatalytic NADH regeneration is potentially a green process.

The generated **Rh-H** then performs the selective hydrogenation of NADH. The metal center is bound to the amide carbonyl group at C3 of the pyridine ring in NAD^+. The hydrogen on the metal center then coordinates to C4, forming a hexatomic ring between **Rh-H** and NAD^+. The breaking of the bond between the Rh metal center and the hydrogen leads to the hydrogenation at the C4 position, thus forming 1,4-NADH. Some analogs of NADH could also be regenerated by **Rh-H**. The electron-withdrawing effect of the substituent on

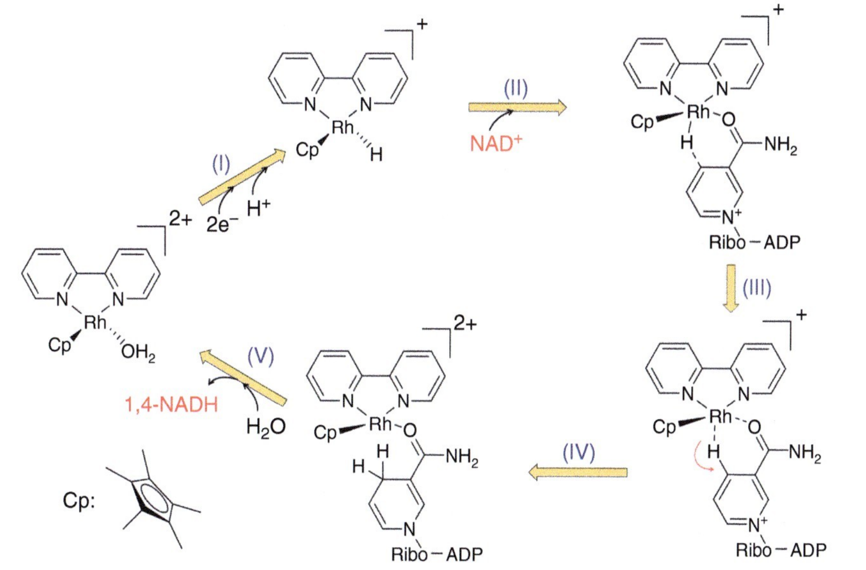

(I) Light induced electrons transfer (II) Formation of hydrogen bond
(III) Transfer hydrogenation (IV) Formation of 1,4-NADH (V) Recovery of catalyst

Figure 10.4 Mechanism of selective hydrogenation of NAD^+ to 1,4-NADH by **Rh** catalyst.

C3, as well as the substituent on C1, would influence the coordination at C4. Meanwhile, the steric hindrance of the group on C3 would also affect the formation of the hexatomic ring. These two aspects are critical for the selectivity regeneration of 1,4-NADH.

The extraction of electrons by **Rh** leaves electronic holes on the photocatalyst, which requires to be utilized or quenched for the sustainability of the reaction. The ideal option for the utilization of these holes is water oxidation. By using water as the electron donor, the generated gaseous oxygen and dissolved NADH could be easily separated and there is no by-product generation. However, using water as the electron donor cannot provide sufficient driving force to perform NADH regeneration. Sacrificial electron donors, such as TEOA and ascorbic acid, are often introduced to the system for acquiring an efficient NADH regeneration. Considering the high value of NADH and its enzyme-related products, the utilization of sacrificial electron donors is acceptable. Efficient photocatalytic NADH regeneration driven by water is still a challenge in this field.

10.2.2 Aspects of Measuring Photocatalytic NADH Regeneration

The activity of photocatalytic NADH regeneration is related to the light intensity, the photocatalyst, the concentration of electron donor, the concentration of **Rh**, and the pH value. The light intensity should be kept consistent when evaluating the efficiency of the system. We herein highly recommend to use $100\,mW\,cm^{-2}$ simulating sunlight to perform the reaction. The control and adjustment of the light source has been reported in a previous review [20]. The influence of the photocatalyst on the activity has been introduced in the previous section, where a high-efficient photocatalyst would benefit the activity. The concentration of the electron donor, to some extent, decides the driving force of the NADH regeneration. Low concentration should be applied in this reaction, which not only is a greener process to save energy but also motivates us to develop more efficient photocatalysts. According our experience, the concentration of **Rh**, which is normally one fifth (or more) of the concentration of NADH, is excessive to this reaction in most of the published research studies. The pH value would impact the activity of **Rh**, enzyme, and the effective concentration of TEOA, and all these aspects should be considered to choose the proper pH value.

Apart from the experiment conditions, the measurement of NADH concentration should also be carefully performed. The most facile and commonly used method to detect the NADH concentration is measuring the absorption at 340 nm by a UV–visible spectrophotometer. In particular, the background absorption before the photocatalysis should be utilized as the reference to correct the zero point. The yield of NADH is calculated based on the initial concentration of NAD^+ and the concentration of converted NADH. The measurement of NAD^+/NADH concentration should follow the Lambert–Beer Law with molar absorption coefficients of 16 900 cm^{-1} at 260 nm for NAD^+ and 6220 cm^{-1} at 340 nm for NADH.

10.3 Advancements in Photocatalytic NADH Regeneration

10.3.1 Nature Photosensitizers

In nature, plant photosynthesis converts solar energy into chemical energy through chloroplast, which is mainly composed by chloroplast membranes, thylakoids, and a chloroplast matrix. The photosensitizers in thylakoids absorb solar energy to produce photoexcited electrons, which are then stored in the form of NADPH and adenosine triphosphate (ATP) through an electron transfer chain called "Z-scheme" [21]. In detail, the photosensitizers P680, which are the reaction center of protein complex photosystem II (PS II) on the thylakoid membrane, are stimulated to form an excited state under visible light irradiation. The excited state P680 promotes water splitting to produce electrons, which are sequentially transferred to pheophytin, plastoquinone molecules Q_A and Q_B. The lipidic plastoquinone molecules move within the thylakoid membrane and introduce electrons into the cytochrome b_6f complex, simultaneously releasing protons over the thylakoid membrane. The plastocyanin (PC) obtains electrons from the cytochrome b_6f complex, which combines with protein complex photosystem I (PS I) to transfer electrons to it. The photosensitizers P700, which are the reaction center of PS I, are excited under the illumination. They can promote the subsequent transfer of electrons to ferredoxin (Fd) and flavoprotein (Fp) and finally transfer electrons to $NADP^+$ to form NADPH.

10.3.2 Organic Molecular Photosensitizers

Although these nature photosensitizers are highly efficient, most of them are small molecules. Structurally unstable, lowly recycling, and difficultly separation limit their application. According to the structure of natural photosensitizers, some organic molecular photosensitizers have been synthesized to solve these problems while ensuring efficient NAD(P)H regeneration.

Porphyrins, as the key light harvesting pigments in natural photosystem, have been applied for photocatalytic regeneration of NADH. Porphyrins are rigid, planar molecules composed of four pyrrole rings with multiple ligands that allow for the tuning of optical, physical, and electrochemical properties via functionalization. They have strong absorption in the visible light range, providing them with intrinsic advantages as photosensitizers [21]. For example, Park and coworkers investigated the effect of porphyrins with different active centers and pendent chain substituents for NADH regeneration [22]. The regeneration efficiency and turnover frequency (TOF) of zinc porphyrins (Zn-porphyrins) were superior to other active centers. The highest electronegativity of zinc compared with other inserted atoms benefited to the transfer of electrons from porphyrins to electron mediators. The axial coordination of the sacrificial agent TEOA and Zn-porphyrins also facilitated the transfer of electrons from TEOA to Zn-porphyrins (Figure 10.5a). Lee and coworkers prepared an organic molecular photosensitizer Tyr-ZnDPEG inspired by the structure of chloroplast, the composition of chlorophyll, and antenna protein [26]. The Tyr-ZnDPEG molecular photosensitizer used zinc porphyrins (ZnDPEG) as a photosensitizer and tyrosine (Tyr) as an antenna protein. They explored the effects of the photosensitizer, antenna protein, and mediator Ru(trpy)Cl_3 on the regeneration efficiency of NADH. The combination of the three components led to a highest activity, indicating that the presence of both antenna protein and mediator synergistically increased the energy transfer from solar energy to the photosensitizer.

These organic molecular photosensitizers also suffered from the poor stability and recyclability, which could be improved by immobilizing photosensitizers on supporting carriers. The immobilization methods could be divided into physical capture and chemical graft, according to the different interaction between photosensitizers and carriers.

Park and coworkers are pioneers in physical entrapment of organic molecular photosensitizers [27]. They have prepared a photosynthetic wood by mimicking the structure of thylakoid membrane with nature photosensitizers embedded [23]. The photosynthetic wood was obtained by *in situ* deposition of porphyrins in porous lignocellulose (Figure 10.5b). The NADH regeneration efficiency of photosynthetic wood was 3.3% for one-hour illumination. The same group further prepared a tubular molecular photosensitizer by embedding porphyrins in phenylalanine self-assembled peptide nanotubes [28]. Platinum nanoparticles were deposited on the surface to improve electron transfer efficiency. Tubular structure of phenylalanine self-assembled peptide nanotubes was favorable for electrons transfer. The NADH regeneration efficiency of the tubular molecular photosensitizer was 18% for one-hour illumination.

Because of the weak interaction between photosensitizers and carriers, the physical capture of photosensitizers has low NADH regeneration efficiency. The chemical grafted photosensitizers combine these compounds with carriers in stronger interactions. This can excite both groups and trigger the cross-linking reaction to obtain a higher NADH regeneration efficiency.

Baeg and coworkers prepared an organic molecular photosensitizer called CCGCMAQSP by covalently grafting multianthraquinone-substituted porphyrins

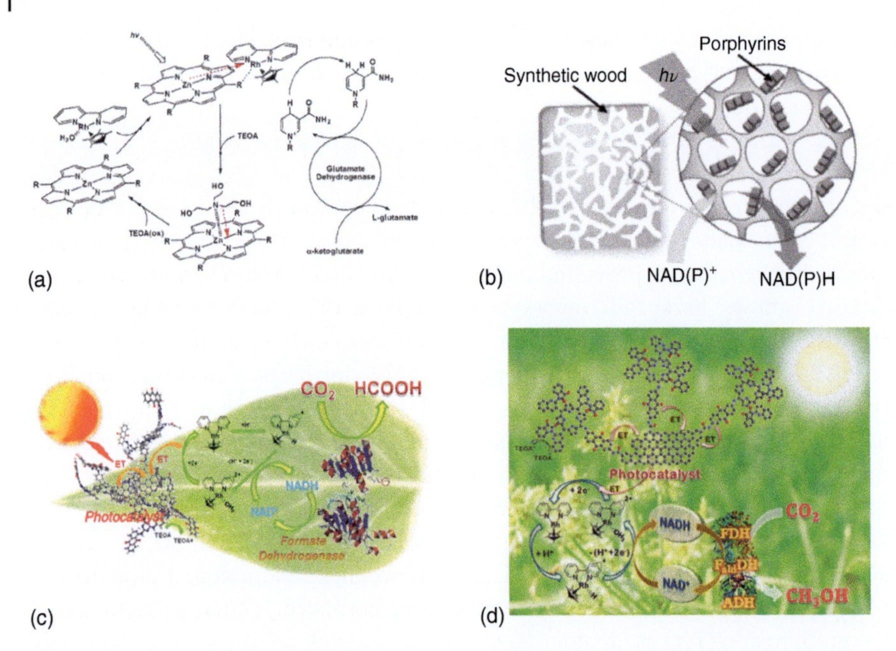

Figure 10.5 (a) Mechanism of photosensitized Zn-porphyrins in NADH photocatalytic regeneration. Source: Kim et al. [22]. © 2011 Royal society of chemistry.; (b) illustration of the light-harvesting synthetic wood (LSW) in NAD(P)H photocatalytic regeneration. Source: Lee et al. [23]. © 2011 John Wiley & Sons.; (c) schematic illustration of CCGCMAQSP in NADH photocatalytic regeneration. Source: Yadav et al. [24]. © 2012 American chemical society.; and (d) schematic illustration of CCG-IP in NADH photocatalytic regeneration. Source: Yadav et al. [25]. © 2014 American chemical society.

(MAQSP) onto chemically converted graphene (CCG) [24]. CCG could carry more catalytic active sites and promote electron transfer with good electrical conductivity and large specific surface area. The chemical interaction between MAQSP and CCG affected the π-conjugated structure of CCG. The opened band gap of CCG enabled electrons to pass from MAQSP to CCG and transfer to the mediator (Figure 10.5c). The NADH regeneration efficiency of CCGCMAQSP molecular photosensitizer could achieve 28.46% after one-hour illumination. Further, Baeg and coworkers prepared an organic molecular photosensitizer called CCG–IP by covalently grafting isatin and porphyrin (IP) together onto CCG (Figure 10.5d) [25]. The NADH regeneration efficiency of CCG-IP molecular photosensitizer reached 39.0% after one-hour illumination.

Electron transfer and product separation can be enhanced by photosensitizer immobilization, but mediators also confine the efficiency of photocatalytic cofactor regeneration. The free mediators need to diffuse to the surface of photosensitizers to obtain electrons for cofactor regeneration, which increase the complexity of the system and reduce its efficiency. Combining mediators with photosensitizers can shorten the distance of electrons transfer and improve the cofactor regeneration efficiency. Knör and coworkers immobilized 2,2′-bipyridyl-containing poly(arylene-ethynylene)-alt-poly(arylene-vinylene) polymer onto glass beads to perform the NADH regeneration process [29]. The bipyridyl group could be combined with **Rh** to integrate photosensitizers and mediators. The strategy can avoid the loss of mediator and shorten the distance between **Rh** and photosensitizers.

10.3.3 Inorganic Semiconductors

Inorganic semiconductors, such as metal oxide, metal sulfide, and metal nitride, are extensively investigated in photocatalysis. Among them, TiO_2 exhibits promising prospect in industrial applications because of its good hydrophilicity, low toxicity, low cost, and strong chemical stability. However, the wide band gap of 3.2 eV, which only absorbs high-energy ultraviolet light, limits the utilization of solar energy. A series of strategies have been carried out to narrow the band gap of TiO_2, such as construction of heterojunctions and doping.

Jiang et al. [18] explored a general strategy of doping nonmetal elements (including carbon, boron, nitrogen, phosphorus, etc.) to expand the band gap of TiO_2. Carbon-doped TiO_2 was prepared by a sol–gel process with ethanol and acetic acid as carbon sources. The carbon doping indeed narrowed the band gap of TiO_2, resulting in a red shift of the absorption edge, thus enhancing the absorption of visible light. With the assistance of a mediator, carbon-doped TiO_2 exhibited high activity and selectivity toward 1,4-NADH. The impact of several factors on NADH regeneration, including electron donor species, pH, and mediator concentration, was also explored to further understand the process. Phosphorus and nitrogen were also doped in TiO_2 and function similarly to carbon-doped TiO_2 (Figure 10.6a) [30, 34]. Compared with organic photosensitizers, TiO_2-based photosensitizers exhibit a much lower turnover frequency (TOF) in NADH regeneration. However, as the first generation of heterogeneous inorganic photosensitizers, modified TiO_2, paves the way for developing other types of inorganic semiconductors.

Quantum dot (QD) nanocrystals (including CdS, zinc sulfide [ZnS], and carbon nanodots [CDs]) are another type of inorganic semiconductors and are attractive visible light harvesting materials because of their suitable band gaps [27].

CDs are zero-dimensional semiconductors with UV/Vis activity, good stability, and good biocompatibility, which have been widely used in photocatalytic carbon dioxide reduction, water splitting, and NADH regeneration. Park and coworkers prepared N-doped carbon nanodots (N-CDs) by the one-step hydrothermal method for photocatalytic NADH regeneration and OYE-catalyzed asymmetric reduction (Figure 10.6b) [31]. Baeg et al. [35] prepared carbon nanodot-silica hybrid semiconductors (CNDSH) by a reverse microemulsion method to regenerate NADH, which was further coupled with formate dehydrogenase to convert CO_2 to formic acid. CNDSH has a wide visible light absorption (optical band gap of 2.25 eV), which may be ascribed to the formation of C—O—Si linkage between carbon QDs and silica. A NADH yield of 74.1% was obtained, 3.17-fold higher than the pure carbon nanodots. As a nanosize material (around 3 nm), carbon nanodots could be potentially integrated with molecular catalyst for high-efficient electron transfer between the photocatalyst and mediator in NADH regeneration.

CdS exhibits high catalytic activity under visible light irradiation because of the proper band gap of 2. 42 eV. Park and coworkers first applied CdS for NADH regeneration in the visible range ($\lambda > 420$ nm). The NADH regeneration efficiency of CdS was about 20% after 60 minutes light irradiation [36]. The low efficiency of photocatalytic NADH regeneration

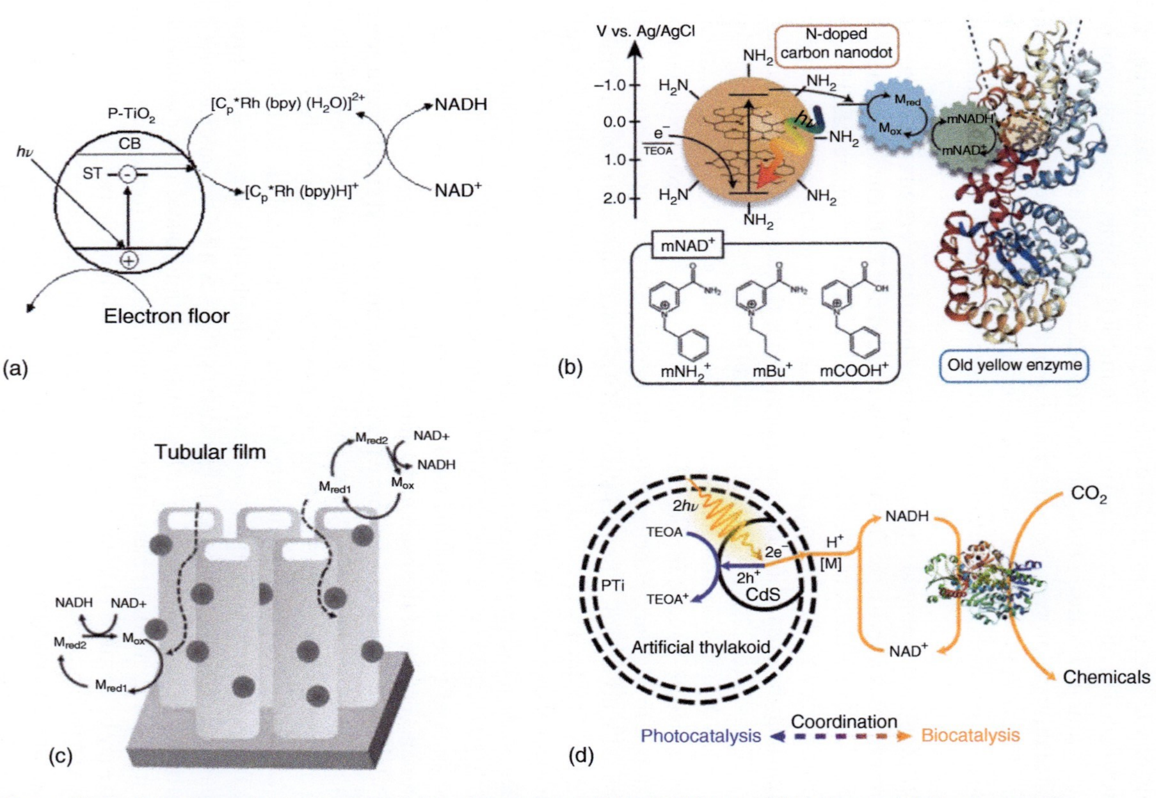

Figure 10.6 (a) Schematic illustration of phosphorus-doped TiO_2 (P-TiO_2) in NADH photocatalytic regeneration. Source: Shi et al. [30]. © 2006 Elsevier; (b) schematic illustration of N-CDs in NADH photocatalytic regeneration. Source: Kim et al. [31]. © 2018 John Wiley & Sons; (c) schematic illustration of TiO_2-CdS NT-film in NADH photocatalytic regeneration. Source: Ryu et al. [32]. © 2011 John Wiley & Sons; and (d) schematic illustration of CdS/PTi in NADH photocatalytic regeneration. Source: Zhang et al. [33]. © 2019 American chemical society.

could be attributed to fast charge recombination and photochemical corrosion, which could be improved by construction of heterostructures. Heterostructured photocatalysts are composed by two or more photocatalysts with proper interfaces. On the one hand, the surfaces of the two catalysts are both exposed to the external, so that both sides are involved in the possibility of interacting with the environment; on the other hand, the two photocatalysts independently form nanoparticles, so that the composite material has the characteristics of two photocatalysts and can realize more complicated functions. Because of the special band structure and carrier transport characteristics, the heterostructured photocatalysts can effectively suppress the recombination of photogenerated electron–hole pairs and improve the quantum efficiency. Park and coworkers loaded CdS nanoparticles on an anodized TiO_2 nanotube array to prepare a TiO_2–CdS nanotube film (NT-film) through successive ionic layer adsorption [32]. The NT film was used as a photocatalyst and TEOA was used as an electron donor to realize photocatalytic NAD(P)H regeneration (Figure 10.6c). The NAD(P)H regeneration efficiency of NT film was about 37% after one-hour light irradiation. Compared with TiO_2, the CB of CdS was more negative (~0.2 eV). Both materials matched the position of the energy band to each other and formed a heterojunction structure, which facilitated the electron transfer from CdS to TiO_2 and thus inhibited charge recombination.

Compared to molecule photosensitizers, inorganic semiconductors demonstrate higher activity of photocatalytic NADH regeneration, while the holes and radicals generated during photoexcitation may cause the inactivation of enzyme. It is necessary to construct a compatible interface between the photocatalysts and the enzyme to coordinate and optimize the photo-enzyme-coupled catalytic reaction. Inspired by the structure and function of chloroplasts, Jiang, Shi, and coworkers prepared TiO_2 microcapsule photocatalysts with CdS–QDs decorated on the inner wall (CdS/PTi) [33]. Calcium carbonate microspheres were used as a template to deposit CdS–QDs on the surface via an ion exchange adsorption–deposition process. Subsequently, TiO_2 coating was mineralized and deposited on the surface of CdS–QDs calcium carbonate. The template was then removed to obtain CdS/PTi. As shown in Figure 10.6d, although the CdS–QDs were deposited on the inner wall of the microcapsules, the larger band gap of TiO_2 allowed visible light to pass through, which did not affect the visible light absorption of the CdS–QDs. In addition, the microcapsule structure could achieve multiple scattering of light inside the microcapsule, which prolonged the optical path and increased the absorption of visible light by CdS–QDs. The electrons generated by CdS–QDs were transferred to the outer surface of the microcapsule through the wall TiO_2 and participated in NADH regeneration. After 10 minutes of illumination, the efficiency of NADH regeneration reached $70.4\% \pm 2.0\%$. The pores of the TiO_2 wall were distributed at around 3.9 nm and the size of enzymes was larger than the pore size of the microcapsule wall. Therefore, the wall of the capsule hindered the entrance of enzymes into the lumen and protected enzymes from inactivation by contacting with the photogenerated holes. The enzyme activity of internally modified CdS–QDs microcapsules (CdS/PTi) remained $96.1\% \pm 3.5\%$ after the photoreaction, higher than that of the externally modified CdS–QDs microcapsules (PTi/CdS; $71.9\% \pm 1.4\%$).

10.3.4 Organic Semiconductors

Organic semiconductors have been rapidly developed in recent years. Two kinds of typical organic semiconductors, graphitic carbon nitride (g-C_3N_4) and covalent–organic frameworks (COFs), exhibit promising applications in photocatalysis.

g-C_3N_4 has been widely investigated in the past decade because of its low cost and facile preparation [37]. The two-dimensional (2D) graphitic planes composed by a planar amino-linked tri-s-triazine endow g-C_3N_4 with similar band structures to inorganic semiconductors. COF is another kind of 2D organic semiconductor composed by periodical covalent-bonded organic monomers, which follows a donor–acceptor electron transfer mechanism [38]. The programmable organic monomers provide a flexible design of the COF photocatalyst. These two kinds of organic semiconductors have been applied in photocatalytic NADH regeneration and will be detailed in this section.

g-C_3N_4 was first applied in NADH regeneration in 2013 by Liu and Antonietti [39]. Diatomite was utilized as the backbone to construct DE-g-C_3N_4 with a NADH regeneration yield of nearly 100% after two hours of light irradiation (Figure 10.7a). DE-g-C_3N_4 could also regenerate NADH with a yield of up to 50% in a mediator-free system. However, the selectivity of NADH toward NADH-dependent enzymes was only 33.6% according to the subsequent enzymatic experiment. This work shows great potential of g-C_3N_4 performed as a metal-free photocatalyst for NADH regeneration. Liu and coworkers then constructed g-C_3N_4 mesoporous spheres (CNMS) with SiO_2 as the sacrificial template [43]. CNMS exhibited a BET surface area of $205\,m^2\,g^{-1}$, 17.1-fold higher than that of bulk g-C_3N_4. The increased surface and mesoporous structure area led to the enhanced light

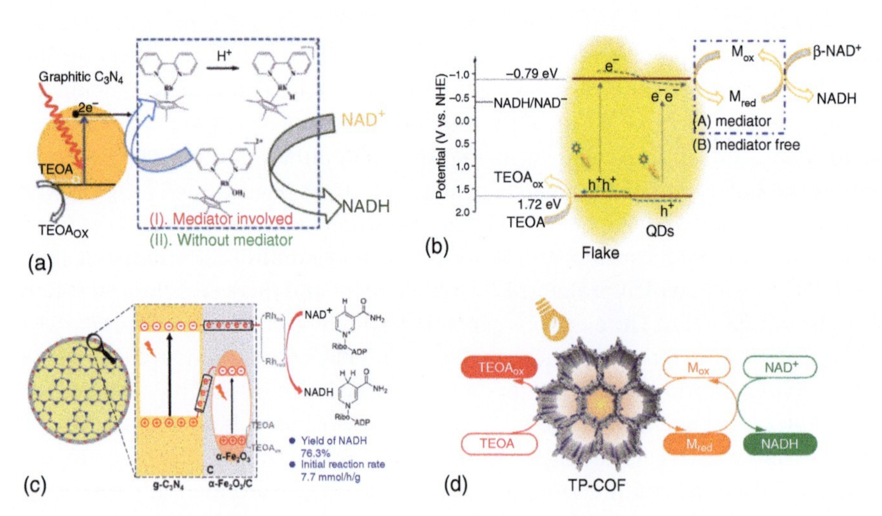

Figure 10.7 (a) Schematic illustration of graphitic C_3N_4 in NADH photocatalytic regeneration. Source: Liu and Antonietti [39]. © 2013 Royal society of chemistry; (b) schematic illustration of QDs@flake g-C_3N_4 in NADH photocatalytic regeneration. Source: Yang et al. [40]. © 2017 American chemical society; (c) schematic illustration of g-C_3N_4@α-Fe_2O_3/C in NADH photocatalytic regeneration. Source: Wu et al. [41]. © 2018 American chemical society; and (d) schematic illustration of TP-COF in NADH photocatalytic regeneration. Source: Zhao et al. [42]. © 2019 John Wiley & Sons.

absorption and charge separation. The yield of NADH regeneration reached 100% after 30 minutes of light irradiation. Photocatalytic NADH regeneration was then coupled with enzymatic lactate production with a conversion of 100% after seven hours of reaction, confirming the enzymatic reactivity of photogenerated NADH. The same group also prepared g-C_3N_4 nanorod and ordered a g-C_3N_4 array for photocatalytic NADH regeneration [44, 45]. The ordered hierarchical structure of g-C_3N_4 enhanced the light absorption and charge separation and thus improved the activity of NADH regeneration and corresponding enzymatic reactions.

Shi, Jiang, and coworkers focused on charge behaviors in photocatalytic NADH regeneration. g-C_3N_4-based materials were developed to discuss the coordination in different photocatalytic steps. The understanding of the photocatalytic mechanism could provide design principles of the photocatalyst for high-efficient solar-to-chemical energy conversion. Quantum dots@flake graphitic carbon nitride (QDs@flake g-C3N4) isotype heterojunctions were constructed to coordinate both charge generation and charge separation (Figure 10.7b) [40]. By immersing a cyanuric acid–melamine (CM) complex in a cyanamide solution, the hexagonal prism-like CM was exfoliated to a stacked structure by the absorbed cyanamide. The subsequent calcination led to the formation of a flake structure. Cyanamide on CM was *in situ* transformed into isotype QDs, forming quantum dots@flake g-C_3N_4 isotype heterojunctions. The 2D flake structure improved charge generation and planar charge transfer, while the isotype g-C_3N_4 QDs favored the charge separation between the heterojunctions. Quantum dots@flake g-C_3N_4 exhibited fivefold higher activity compared to g-C_3N_4 prepared by cyanamide. Furthermore, this concept was clearly pointed out and verified by their next investigation. The g-C_3N_4@α-Fe_2O_3/C core@shell photocatalyst was prepared by calcination of the Fe^{3+}/polyphenol-coated melamine precursor [41]. As shown in Figure 10.7c, the α-Fe_2O_3 moiety acts as an additional photosensitizer, offering more photogenerated charges, whereas the C moiety bridges a "highway" to facilitate the charge transfer between the α-Fe_2O_3 moiety and g-C_3N_4. This construction was confirmed by semiquantitative analysis of charge density. The coordination between charge generation and charge transfer could be adjusted by varying the component of the Fe^{3+}/polyphenol-coated melamine precursor. Under blue light emitting diode (LED) irradiation, an initial reaction rate of 7.7 $mmol\,h^{-1}\,g^{-1}$ among the highest rate for photocatalytic NADH regeneration was obtained. The utilization of an energy-intensive LED light source greatly benefited the activity of NADH regeneration. Song and coworkers shared a similar concept by the construction of a thiophene-modified double-shell hollow g-C_3N_4 nanosphere with synergistic enhancement of charge excitation and separation [46]. The double-shell hollow structure and the incorporation of thiophene ring enhanced the charge generation and separation while a yield of ~74% for NADH was obtained after 15-minutes of visible light (>420 nm) irradiation. Shi and coworkers incorporated phosphorus QDs on g-C_3N_4 hollow tubes [47]. The phosphorus QDs acted as an extra photogenerated electron pool, providing more electrons for g-C_3N_4 to perform NADH regeneration. The formed heterojunctions between phosphorus QDs and g-C_3N_4 hollow tubes could also facilitate the charge separation. The integrated photocatalyst exhibited a superior activity with a yield of 80.5% ± 4.1% after six minutes of light irradiation (LED, 405 nm).

Park and coworkers reported a construction of amorphous carbon nitride (ACN) via a simple re-calcination strategy [48]. ACN exhibited a much higher yield of NADH when compared with g-C_3N_4. The enhanced activity was originated from the unique microstructure of ACN that lacks hydrogen bonds that link polymeric melon units, leading to extended visible light absorption and less charge recombination. A construction of WS_2/g-C_3N_4 heterojunction was also applied for NADH regeneration with a moderate yield of 37.1% after six-hours of irradiation [49]. The low cost of g-C_3N_4 also exhibited potential applications in the fabrication of reactors. g-C_3N_4 and the mediator were integrated on a microfluidic chip to perform fast and continuous regeneration of NADH [50]. A g-C_3N_4-based drop reactor was also constructed, which exhibited enhanced activity when compared with the slurry systems [51].

Apart from g-C_3N_4, COFs and similar materials have also been applied in photocatalytic NADH regeneration. Baeg and coworkers prepared a 2D covalent triazine frameworks (CTFs) via the condensation polymerization between cyanuric chloride and perylene diimide [52]. These CTFs exhibited improved photoelectron conversion compared with its monomer, which was contributed to the highly ordered π electron channels. While performing NADH regeneration, CTFs exhibited a yield of 75.9% after two hours of reaction, about 3.7-fold higher than its monomer. A fully conjugated 2D sp^2-carbon COF (TP-COF) was prepared by Li and coworkers (Figure 10.7d) [42]. The formation of stable carbon–carbon double bonds greatly improved the conductivity of such kind of materials. The activity of NADH regeneration was thus promoted, leading to a yield of 90.3% after 10 minutes of light irradiation. The conductivity of organic semiconductors is normally lower than that of inorganic semiconductors. To promote the application of organic semiconductors, high crystallinity and ordered π stacking should be considered to improve the electron migration. Controllable construction of defects should also be considered to maintain charge separation while decreasing the energy loss during the charge transfer.

Amorphous organic monomer assembly also follows the donor–acceptor electron transfer mechanism in photocatalysis. Baeg and coworkers prepared a three-dimensional aromatic polymer (3DAP) by the polymerization of triptycene [53]. 3DAP exhibited a NADH yield of 91.1% while its monomer exhibited nearly no activity. The photocatalytic activity of 3DAP may be due to the efficient electron transfer through interconnecting 3D chain. Another 2D N-graphdiyne (NGDY) nanosheet containing different numbers of N atoms was prepared by Li and coworkers via a liquid/liquid interfacial synthesis strategy [54]. NGDY exhibited a highest NADH yield of 35%, which may be ascribed to the increased hydrophilicity related to the higher N contents (the number of doped N atoms). Carbon nanodots could also be considered as the self-polymerization of C6 rings. The edge of carbon nanodots could be functionalized to adjust its properties.

10.4 Expectations

Most of the present research studies about photocatalytic NADH regeneration focus on the development of the photocatalyst. Both catalyst and photocatalytic mechanism do play a critical role in NADH regeneration. However, after years of investigation, the efficiency of photocatalysis is high enough for NADH regeneration. Other challenges in this field should be settled in order to achieve a green, sustainable, and efficient regeneration of NADH. Herein, we would like to discuss some challenges that we think important for the development of this field.

First of all, useful oxidation reactions should be introduced to photocatalytic NADH regeneration. Most of the photocatalytic NADH regeneration systems require electron donors to sustain the reaction. The use of an electron donor not only causes the waste of oxidation energy but also generates oxidizing by-products. Moreover, the electron donor itself also increases the cost of separation because the concentration of the electron donor is much higher than that of NADH. Replacing typical electron donors with useful substrates, such as water and benzyl alcohol, in other words, performing NADH-related photocatalytic reactions, should develop a greener process. Water oxidation for oxygen evolution is the ideal option of oxidation reaction. However, the four-electron oxidation of water to oxygen requires efficient catalysts to overcome the energy barrier. The oxidation of water may not provide enough driving force for NADH regeneration. The intermediate generated during water oxidation may lead to the back reaction because NADH is also an active reductant. Compartmentalization strategy should be applied to spatially isolate water oxidation and NADH regeneration. Apart from water oxidation, other useful oxidation reactions could also be coupled with NADH regeneration. The oxidation of benzyl alcohol to benzaldehyde coupled to hydrogen evolution may offer some light to NADH regeneration [55]. Moreover, with a rational design, the oxidation product in the overall reaction may participate in another enzymatic reaction to produce high-value chemicals.

Secondly, the noble metal contented molecular catalyst, **Rh**, is a homogeneous component, which cannot be recycled after NADH regeneration. By loading **Rh** on an heterogeneous photocatalyst, **Rh** could be recycled for several times. The electron transfer between the photocatalyst and loaded **Rh**, as well as the hydrogen transfer between Rh and NAD^+, should be reconsidered because the heterogeneous reaction performs a different mechanism compared to the homogeneous reaction. However, the precursor of **Rh** used to construct the integrated photocatalyst is normally higher than the amount of loaded **Rh**, which would also lead to the waste of the material. The inherent problem is that rhodium is a noble metal. Therefore, a better solution is to replace it with noble-metal free or metal free materials. As far as we know, few research studies have been focused on the development of heterogeneous catalysts for the selective hydrogenation of NAD^+ to 1,4-NADH. The catalyst investigation in the transfer hydrogenation reaction may provide some inspiration in developing such kind of superior catalysts.

Additionally, the photostability of NADH needs to be improved for a more sustainable recycling of NAD^+/NADH in photo-enzyme-coupled systems. The active species generated during photocatalysis, such as hydroxyl radical, superoxide radical, and holes, may break the structure of NAD^+/NADH. Meanwhile, the active reductant NADH may also react with the active oxidant generated during photocatalysis. These two aspects lead to a lower yield of NADH. Strategies of protecting NAD^+/NADH and suppressing the back reaction must be investigated in photocatalytic NADH regeneration systems. NADH analogs could also be investigated to improve the photostability of cofactors.

10.5 Conclusions and Prospects

10.5.1 Conclusions

In summary, reduced NADH could be considered as "energy currency" in numerous oxidoreductases involved enzymatic reactions. Thanks to the clear route and simple process, such kind of reactions could be easily reconstructed *in vitro* for a diversity of chemical production (see also Chapters 6 and 7). Photocatalytic NADH regeneration provides a chance to directly store solar energy in chemicals more than just hydrogen or carbon fuels in a facile and flexible way. The recycling of NAD^+/NADH offers continuous reducing equivalents for the enzymatic reaction. As the pivot of these two kinds of energies, the behaviors of photocatalytic NADH regeneration would greatly influence the overall efficiency of the chemical production. 10 years of explorations in this field provide us deeper understanding of the mechanism and more flexible regulation of the process. The efficiency of photocatalysis is high enough for the supply of NADH to enzymatic reactions. The biocompatibility of enzyme toward photocatalysis also draws more attention for more efficient utilization of photogenerated NADH. Despite these achievements, there are still several challenges in this field. The terminal electron donors are normally high concentrated sacrificial agents such as triethanolamine and ascorbic acid, which should be replaced by greener electron donors such as water or useful chemicals. Cost-effective catalysts should be developed to replace the noble-metal-containing electron mediator ($[Cp^*Rh(bpy)(H_2O)]^{2+}$). The photodegradation mechanism of NADH should be investigated to improve the photostability of NADH for more sustainable enzymatic reactions.

10.5.2 Prospects

The energy conversion in natural photosynthesis inspires us to develop artificial systems for chemical production. By coupling photocatalytic NADH regeneration and enzymatic chemical production, we could manipulate chemical manufacture in a greener and simpler way. The directed evolution of enzymes and synthetic biology could provide more efficient catalysts, thus offering more efficiently the desired substrate (NADH) in return. With a rational design of the overall catalytic process, the "energy currency" of NADH will allow the solar-driven advanced biomanufacture.

List of Abbreviations

ACN	amorphous carbon nitride
ADP	adenosine 5′-diphosphate
AMP	adenosine 5′-monophosphate
ATP	adenosine 5′-triphosphate
bpy	2,2′-bipyridine
CB	conduction band
CDs	carbon nanodots

COFs	covalent organic frameworks
CoQ	ubiquinone
Cp	pentamethylcyclopentadienyl
CTFs	covalent triazine frameworks
DPNH	reduced diphosphopyridine nucleotide
Fd	ferredoxin
FMN	flavin mononucleotide
Fp	flavoprotein
g-C_3N_4	graphitic carbon nitride
HOMO	highest occupied molecular orbital
LUMO	lowest unoccupied molecular orbital
NADH	reduced nicotinamide adenine dinucleotide
NADPH	reduced nicotinamide adenine dinucleotide phosphate
PC	plastocyanin
PS I	photosystem I
PS II	photosystem II
Rh	$[Cp^*Rh(bpy)(H_2O)]^{2+}$
TEOA	triethanolamine
VB	valence band
QD	quantum dots

References

1 Shi, J., Jiang, Y., Jiang, Z. et al. (2015). Enzymatic conversion of carbon dioxide. *Chem. Soc. Rev.* 44 (17): 5981–6000.

2 Xie, Z.X., Li, B.Z., Mitchell, L.A. et al. (2017). "Perfect" designer chromosome V and behavior of a ring derivative. *Science* 355 (6329): eaaf4704.

3 Wu, Y., Li, B.Z., Zhao, M. et al. (2017). Bug mapping and fitness testing of chemically synthesized chromosome X. *Science* 355 (6329): eaaf4706.

4 Chen, K., Huang, X., Kan, S.B.J. et al. (2018). Enzymatic construction of highly strained carbocycles. *Science* 360 (6384): 71–75.

5 Kan, S.B., Lewis, R.D., Chen, K. et al. (2016). Directed evolution of cytochrome c for carbon-silicon bond formation: bringing silicon to life. *Science* 354 (6315): 1048–1051.

6 Hollmann, F., Arends, I.W.C.E., Holtmann, D. et al. (2011). Enzymatic reductions for the chemist. *Green Chem.* 13 (9): 2285–2313.

7 Wu, H., Tian, C., Song, X. et al. (2013). Methods for the regeneration of nicotinamide coenzymes. *Green Chem.* 15 (7): 1773–1789.

8 Wang, X., Saba, T., Yiu, H.H.P. et al. (2017). Cofactor NAD(P)H regeneration inspired by heterogeneous pathways. *Chem* 2 (5): 621–654.

9 Rafter, G.W. and Colowick, S.P. (1957). Enzymatic preparation of DPNH and TPNH. *Methods Enzymol.* 3: 887–890.

10 Wichmann, R., Wandrey, C., Andreas, F. et al. (2000). Continuous enzymatic transformation in an enzyme-membrane reactor with simultaneous NADH regeneration. *Biotechnol. Bioeng.* 67 (6): 791–804.

11 Calvin, S.J., Mangan, D., Miskelly, I. et al. (2011). Overcoming equilibrium issues with carbonyl reductase enzymes. *Org. Process Res. Dev.* 16 (1): 82–86.

12 Lehninger, A.L. (1957). Preparation of reduced DPN (chemical method). *Methods Enzymol.* 3: 885–887.

13 Rodríguez, C., Lavandera, I., and Gotor, V. (2012). Recent advances in cofactor regeneration systems applied to biocatalyzed oxidative processes. *Curr. Org. Chem.* 16 (21): 2525–2541.

14 Miyawaki, O. and Yano, T. (1993). Electrochemical bioreactor with immobilized glucose-6-phosphate dehydrogenase on the rotating graphite disc electrode modified with phenazine methosulfate. *Enzym. Microb. Technol.* 15 (6): 525–529.

15 Schröder, I., Steckhan, E., and Liese, A. (2003). In situ NAD^+ regeneration using 2,2'-azinobis(3-ethylbenzothiazoline-6-sulfonate) as an electron transfer mediator. *J. Electroanal. Chem.* 541: 109–115.

16 Wagenknecht, P.S., Penney, J.M., and Hembre, R.T. (2003). Transition-metal-catalyzed regeneration of nicotinamide coenzymes with hydrogen. *Organometallics* 22 (6): 1180–1182.

17 Shumilin, I.A., Nikandrov, V.V., Popov, V.O. et al. (1992). Photogeneration of NADH under coupled action of CdS semiconductor and hydrogenase from *Alcaligenes eutrophus* without exogenous mediators. *FEBS Lett.* 306 (2–3): 125–128.

18 Jiang, Z., Lü, C., and Wu, H. (2005). Photoregeneration of NADH using carbon-containing TiO_2. *Ind. Eng. Chem. Res.* 44 (12): 4165–4170.

19 Lo, H.C., Buriez, O., Kerr, J.B. et al. (2010). Regioselective reduction of NAD^+ models with $[Cp^*Rh(bpy)H]^+$: structure–activity relationships and mechanistic aspects in the formation of the 1,4-NADH derivatives. *Angew. Chem. Int. Ed.* 38 (10): 1429–1432.

20 Wang, Z., Li, C., Domen, K. et al. (2018). Recent developments in heterogeneous photocatalysts for solar-driven overall water splitting. *Chem. Soc. Rev.* 48 (7): 2109–2125.

21 Lee, S.H., Choi, D.S., Kuk, S.K. et al. (2018). Photobiocatalysis: activating redox enzymes by direct or indirect transfer of photoinduced electrons. *Angew. Chem. Int. Ed.* 57 (27): 7958–7985.

22 Kim, J.H., Lee, S.H., Lee, J.S. et al. (2011). Zn-containing porphyrin as a biomimetic light-harvesting molecule for biocatalyzed artificial photosynthesis. *Chem. Commun.* 47 (37): 10227–10229.

23 Lee, M., Kim, J.H., Lee, S.H. et al. (2011). Biomimetic artificial photosynthesis by light-harvesting synthetic wood. *ChemSusChem* 4 (5): 581–586.

24 Yadav, R.K., Baeg, J.O., Oh, G.H. et al. (2012). A photocatalyst-enzyme coupled artificial photosynthesis system for solar energy in production of formic acid from CO_2. *J. Am. Chem. Soc.* 134 (28): 11455–11461.

25 Yadav, R.K., Oh, G.H., Park, N.J. et al. (2014). Highly selective solar-driven methanol from CO_2 by a photocatalyst/biocatalyst integrated system. *J. Am. Chem. Soc.* 136 (48): 16728–16731.

26 Kwak, J., Kim, M.C., and Lee, S.Y. (2016). An enzyme-coupled artificial photosynthesis system prepared from antenna protein-mimetic tyrosyl bolaamphiphile self-assembly. *Nanoscale* 8 (32): 15064–15070.

27 Kim, J.H., Nam, D.H., and Park, C.B. (2014). Nanobiocatalytic assemblies for artificial photosynthesis. *Curr. Opin. Biotechnol.* 28: 1–9.

28 Kim, J.H., Lee, M., Lee, J.S. et al. (2012). Self-assembled light-harvesting peptide nanotubes for mimicking natural photosynthesis. *Angew. Chem. Int. Ed.* 51 (2): 517–520.

29 Oppelt, K.T., Gasiorowski, J., Egbe, D.A.M. et al. (2014). Rhodium-coordinated poly(arylene-ethynylene)-alt-poly(arylene-vinylene) copolymer acting as photocatalyst for visible-light-powered NAD^+/NADH reduction. *J. Am. Chem. Soc.* 136 (36): 12721–12729.

30 Shi, Q., Yang, D., Jiang, Z. et al. (2006). Visible-light photocatalytic regeneration of NADH using P-doped TiO_2 nanoparticles. *J. Mol. Catal. B Enzym.* 43 (1–4): 44–48.

31 Kim, J., Lee, S.H., Tieves, F. et al. (2018). Biocatalytic C=C bond reduction through carbon nanodot-sensitized regeneration of NADH analogues. *Angew. Chem. Int. Ed.* 57 (42): 13825–13828.

32 Ryu, J., Lee, S.H., Nam, D.H. et al. (2011). Rational design and engineering of quantum-dot-sensitized TiO_2 nanotube arrays for artificial photosynthesis. *Adv. Mater.* 23 (16): 1883–1888.

33 Zhang, S.H., Shi, J.F., Sun, Y.Y. et al. (2019). Artificial thylakoid for the coordinated photoenzymatic reduction of carbon dioxide. *ACS Catal.* 9 (5): 3913–3925.

34 Geng, J., Yang, D., Zhu, J. et al. (2009). Nitrogen-doped TiO_2 nanotubes with enhanced photocatalytic activity synthesized by a facile wet chemistry method. *Mater. Res. Bull.* 44 (1): 146–150.

35 Baeg, J.O., Yadav, R.K., Kumar, A. et al. (2017). New carbon nanodots-silica hybrid photocatalyst for highly selective solar fuel production from CO_2. *ChemCatChem* 9 (16): 3153–3159.

36 Nam, D.H., Lee, S.H., Park, C.B. et al. (2010). CdTe, CdSe, and CdS nanocrystals for highly efficient regeneration of nicotinamide cofactor under visible light. *Small* 6 (8): 922–926.

37 Wang, X., Maeda, K., Thomas, A. et al. (2009). A metal-free polymeric photocatalyst for hydrogen production from water under visible light. *Nat. Mater.* 8 (1): 76–80.

38 Rodríguez-San-Miguel, D. and Zamora, F. (2019). Processing of covalent organic frameworks: an ingredient for a material to succeed. *Chem. Soc. Rev.* 48 (16): 4375–4386.

39 Liu, J. and Antonietti, M. (2013). Bio-inspired NADH regeneration by carbon nitride photocatalysis using diatom templates. *Energy Environ. Sci.* 6 (5): 1486–1493.

40 Yang, D., Zou, H., Wu, Y. et al. (2017). Constructing quantum dots@flake graphitic carbon nitride isotype heterojunctions for enhanced visible-light-driven NADH regeneration and enzymatic hydrogenation. *Ind. Eng. Chem. Res.* 56 (21): 6247–6255.

41 Wu, Y., Ward-Bond, J., Li, D. et al. (2018). g-C_3N_4@α-Fe_2O_3/C photocatalysts: synergistically intensified charge generation and charge transfer for NADH regeneration. *ACS Catal.* 8 (7): 5664–5674.

42 Zhao, Y., Liu, H., Wu, C. et al. (2019). Fully conjugated two-dimensional sp^2-carbon covalent organic frameworks as artificial photosystem I with high efficiency. *Angew. Chem. Int. Ed.* 58 (16): 5376–5381.

43 Huang, J., Antonietti, M., and Liu, J. (2014). Bio-inspired carbon nitride mesoporous spheres for artificial photosynthesis: photocatalytic cofactor regeneration for sustainable enzymatic synthesis. *J. Mater. Chem. A* 2 (21): 7686–7693.

44 Liu, J., Huang, J., Zhou, H. et al. (2014). Uniform graphitic carbon nitride nanorod for efficient photocatalytic hydrogen evolution and sustained photoenzymatic catalysis. *ACS Appl. Mater. Interfaces* 6 (11): 8434–8440.

45 Liu, J., Cazelles, R., Chen, Z.P. et al. (2014). The bioinspired construction of an ordered carbon nitride array for photocatalytic mediated enzymatic reduction. *Phys. Chem. Chem. Phys.* 16 (28): 14699–14705.

46 Meng, J., Tian, Y., Li, C. et al. (2019). A thiophene-modified doubleshell hollow g-C_3N_4 nanosphere boosts NADH regeneration via synergistic enhancement of charge excitation and separation. *Catal. Sci. Technol.* 9 (8): 1911–1921.

47 Yang, D., Zhang, Y., Zou, H. et al. (2019). Phosphorus quantum dots-facilitated enrichment of electrons on g-C_3N_4 hollow tubes for visible-light-driven nicotinamide adenine dinucleotide regeneration. *ACS Sustain. Chem. Eng.* 7 (1): 285–295.

48 Son, E.J., Lee, Y.W., Ko, J.W. et al. (2018). Amorphous carbon nitride as a robust photocatalyst for biocatalytic solar-to-chemical conversion. *ACS Sustain. Chem. Eng.* 7 (2): 2545–2552.

49 Zeng, P., Ji, X., Su, Z. et al. (2018). WS_2/g-C_3N_4 composite as an efficient heterojunction photocatalyst for biocatalyzed artificial photosynthesis. *RSC Adv.* 8 (37): 20557–20567.

50 Huang, X., Liu, J., Yang, Q. et al. (2016). Microfluidic chip-based one-step fabrication of an artificial photosystem I for photocatalytic cofactor regeneration. *RSC Adv.* 6 (104): 101974–101980.

51 Huang, X., Hao, H., Liu, Y. et al. (2017). Rapid screening of graphitic carbon nitrides for photocatalytic cofactor regeneration using a drop reactor. *Micromachines* 8 (6): 175.

52 Yadav, R.K., Kumar, A., Park, N.J. et al. (2016). A highly efficient covalent organic framework film photocatalyst for selective solar fuel production from CO_2. *J. Mater. Chem. A* 4 (24): 9413–9418.

53 Yadav, R.K., Kumar, A., Yadav, D. et al. (2018). In situ prepared flexible 3D polymer film photocatalyst for highly selective solar fuel production from CO_2. *ChemCatChem* 10 (9): 2024–2029.

54 Pan, Q., Liu, H., Zhao, Y. et al. (2019). Preparation of N-graphdiyne nanosheets at liquid/liquid interface for photocatalytic nadh regeneration. *ACS Appl. Mater. Interfaces* 11 (3): 2740–2744.

55 Kasap, H., Caputo, C.A., Martindale, B.C.M. et al. (2016). Solar-driven reduction of aqueous protons coupled to selective alcohol oxidation with a carbon nitride-molecular Ni catalyst system. *J. Am. Chem. Soc.* 138 (29): 9183–9192.

11

Biocatalysis Under Continuous Flow Conditions

Bruna Goes Palma[1,2], Marcelo A. do Nascimento[2], Raquel A. C. Leão[2], Omar G. Pandoli[1], and Rodrigo O. M. A. de Souza[2]

[1] *Flow Chemistry and Nanochemistry Group, Chemistry Department, St. Marquês de São Vicente, 225, 22451-045, Rio de Janeiro, Brazil*
[2] *Biocatalysis and Organic Synthesis Group, Chemistry Institute, Federal University of Rio de Janeiro, Ave. Athos da Silveira Ramos, 149, 21941909, Rio de Janeiro, Brazil*

11.1 Introduction

In recent years, much attention has been paid to continuous flow protocols, and the advantages of flow chemistry are well documented by several reviews published in the recent literature [1–9]. Interestingly, although major advances in science in general transfer from academia to industry, the continuous flow technique made the inverse transition where the industry needed support from academia in order to develop better equipment and protocols. Active pharmaceutical ingredients (APIs) are synthesized in manufacturing plants and then shipped to other sites to be converted into a form that can be given to patients, such as tablets, drug solutions, or suspensions. This system offers less flexibility to respond to surges in demand and is susceptible to severe disruption if one of the plants has to shut down. Worldwide, a variety of companies such as Novartis, GSK, Lilly, Lonza, and others are investigating continuous manufacturing of new drug substances in order to reduce their manufacturing costs and to provide more robust ways of producing the desired molecules.

It is important to note that besides the fact that the fine chemical community knows continuous manufacturing as an emerging technology, the petrochemical industry has been using this technology for many years (since the 1960s) with several examples of success. The main idea behind the continuous process on petrochemical industry was to "fit the equipment to the process and not the process to the equipment.". One important difference between the two chemistry sectors, which must be pointed out, is the scale. Large-scale operations on petrochemical industry are far larger than the ones from the fine chemical or API industry [1–4].

This technology allows a continuous flow of reagents to be introduced at various points along a process stream, enabling interaction under highly controlled conditions. Flow

Biocatalysis for Practitioners: Techniques, Reactions and Applications, First Edition.
Edited by Gonzalo de Gonzalo and Iván Lavandera.

systems allow high-throughput chemistry to take place, often employing immobilized reagents or catalysts. When compared to batch processes, flow processes have minimal scale-up issues. Instead of scaling up for mass production by increasing reactor size as with batch processes, flow reactors can be scaled up by introducing more reactors in parallel (scaling out) while maintaining excellent mixing and heat transfer. The reduction in plant footprint while increasing plant production is known as process intensification. Carrying out a reaction in flow on a laboratory scale can have the advantage of faster reaction times, safer conditions, and faster optimization [3, 5, 8, 9]. Reaction times, temperature, reagents, pressure, and flow rates can all be rapidly varied to achieve the best conditions. Furthermore, some batch processes pose operational hazards, particularly with the use of highly reactive reagents. These hazards can be avoided under continuous flow conditions because of increased temperature control and short residence times [10].

Importantly, regulation of many parameters such as heat and mass transfer, mixing, and residence times is much improved over related batch processes. Mixing describes the way two phases come together and become intertwined. Batch and flow reactors exhibit different mixing mechanisms, where tube reactors inherently have much smaller diffusion times and achieve mixing much faster than in batch, which in combination with reaction kinetics will determine if flow conditions are beneficial, based most of the time on Reynolds numbers (*Re*)[1]. Reactions where mixing is not highly influential can still benefit from continuous flow conditions. Here, process intensification (high-temperature/high-pressure) can greatly reduce the reaction time. For exothermic reactions, the small dimensions of tube reactors enhance the performance of these reactions not only with better mixing but also with more efficient heat transfer. Regarding the temperature dependence of reactions, it is typically expressed using the Arrhenius rate law, derived from the observation that the reaction rate increases exponentially as the absolute temperature is increased. Because it is derived empirically, it is ignorant of mechanistic considerations and only takes into account the activation energy of the overall process. In contrast, transition-state theory gives the Eyring equation, which analyzes a single-step transformation and is useful in determining activation parameters such as $\Delta G^{\ddagger}$, $\Delta H^{\ddagger}$, and $\Delta S^{\ddagger}$. Although these equations describe two fundamentally different phenomena, they both illustrate a direct relationship between the absolute temperature and the rate constant of the reaction (k). Therefore, heating can speed up reactions that are prohibitively slow at room temperature [11].

Advantageously, the flow reactor configuration can also be readily customized to meet the specific demands of the reaction and the continuous processing requirements. The construction of the reactor is often modular being assembled from several specialized yet easily integrated components such as heating and cooling zones, micro-mixers, residence tubing coils, separators, and diagnostic/analysis units. This workflow not only allows for facile automation and continuous operation of such processes but also enables the chemist to perform more potentially hazardous and otherwise forbidden transformations in a safer and more reliable fashion [10, 12].

---

1 The **Reynolds number** (***Re***) is an important dimensionless quantity in fluid mechanics used to help predict flow patterns in different fluid flow situations.

The main advantages cited for improved operational safety are principally the reduced inventories of reactive chemicals, the small-contained reactor units, and the ability to install real-time monitoring of the system, leading to rapid identification of problems and the instigation of automated safe shutdown protocols. Furthermore, the use of direct in-line purification and analysis techniques can be implemented, thus generating a more streamlined and information-enriched reaction sequence [2, 7, 13–15].

11.2 Practical Approach for Biocatalysis Under Continuous Flow Conditions

Biocatalytic transformations have been traditionally carried out in batch reactors. However, performing those transformations in a continuous flow fashion has been shown to be a more efficient method capable of producing the desired product in higher yield and over shorter periods of time. Process security and integration of downstream processes such as in-line liquid–liquid extraction, solid adsorption, quenching, membrane separation, and solvent evaporation have also been important differentials from the traditional batch reactions [16].

Some recent reviews have presented important advances in this area. Turner and coworkers recently reviewed biotransformations in flow, giving many examples involving fine chemical synthesis [17]. Weiss and Tamborini have published extensive studies focusing on new technologies in the area, besides presenting appropriate terminology and metrics for biocatalytic reactions in flow [18, 19]. Furthermore, Woodley and coworkers had published a systematic approach to select the most suitable reactor for continuous pharmaceutical production [20]. Here, we would like to present some classes of reactions that have been applied in biocatalytic continuous flow setups and introduce a more practical approach of how to apply them in a laboratory routine.

11.2.1 Esterification

Esterification is widely used for the synthesis of many products of industrial interest as emulsifiers, lubricants, paints, emollients, and flavor and fragrance compounds [21, 22]. The chemical route to manufacture esters involves reacting an acid with an alcohol at high temperature in the presence of a metal or strong acid/basic catalyst. These process conditions can lead to ester degradation and formation of undesired side products, besides demanding a high energy cost [23]. In contrast, biotechnological routes use mild operating conditions and can be highly selective and efficient.

In biocatalytic routes, those reactions are mainly performed by lipases (see also Chapter 5), which are enzymes that catalyze the hydrolysis of carboxylic acids, and in low water concentration conditions, the opposite reaction [24]. Currently, they are the class of enzymes mostly used in biocatalytic continuous flow experiments, most likely because of their commercial availability, promiscuity, usage in industry and academia, and ability to operate under a wide variety of conditions such as organic solvents [18].

Among the commercially available enzymes, Novozym® 435 (a preparation of lipase B from *Candida antarctica* – CALB) is the most applied in esterification reactions because of its advanced properties, such as a wide range of operating temperature 20–110 °C, good recyclability, and high activity [18]. In 2008, Woodcock and coworkers applied Novozym® 435 to the synthesis of alkyl esters using fatty acids as substrates in a continuous flow packed bed system and compared the results with batch mode operation. Using the butyl hexanoate as a model reaction, they achieved after 10 minutes in the batch reaction only 3.3% ester conversion, compared to 92.7% in the continuous setup. Although comparable productivity has been obtained, the continuous flow system presented better stability, easier product recovery, and a potentially easier scale up [23].

Different ways to intensify the processes by applying different types of reactors have been recently studied. Besides the most used plug flow reactors, some papers have described promising results performing esterification reactions in microfluid devices and in different systems configuration (Table 11.1) [26, 27]. Condoret and coworkers performed the esterification of oleic acid with n-butanol in a biphasic enzymatic system in continuous mode using a centrifugal partition reactor (CPR) (Figure 11.1). They were able to achieve higher productivity than a conventional batch reactor with the additional benefit of working in a system with integrated separation of free catalyst, reducing separation steps [25].

Another key factor in the design of biocatalytic systems is to make a wise choice of the solvent. Some challenging esterification reactions involve substrates with different polarities and choosing the right medium can potentially improve the results. It also important to select solvents that comply with green chemistry postulates, biocatalyst compatible, cost-effective, with low viscosity, and reusable. One example of esterification that involves substrates with different polarities is the esterification of glycerol and benzoic acid to afford α-monobenzoate glycerol (α-MBG). In a recent work, Domínguez de María and coworkers performed this reaction in continuous flow using Novozym® 435 in a packed bed reactor.

Table 11.1 Reactors used for continuous flow synthesis using biocatalysts. The prices given are estimates from recent experience.

Reactor type cost per reactor				
Type	PFA	Microfluidic chip	Packed bed reactor	Cartridges
Cost	$ 150 per 50 ft.	~$ 100 per chip	~$ 150 (6 ft. Tubing ~$ 10	~$ 5–20 each

Source: Nioi et al. [25].

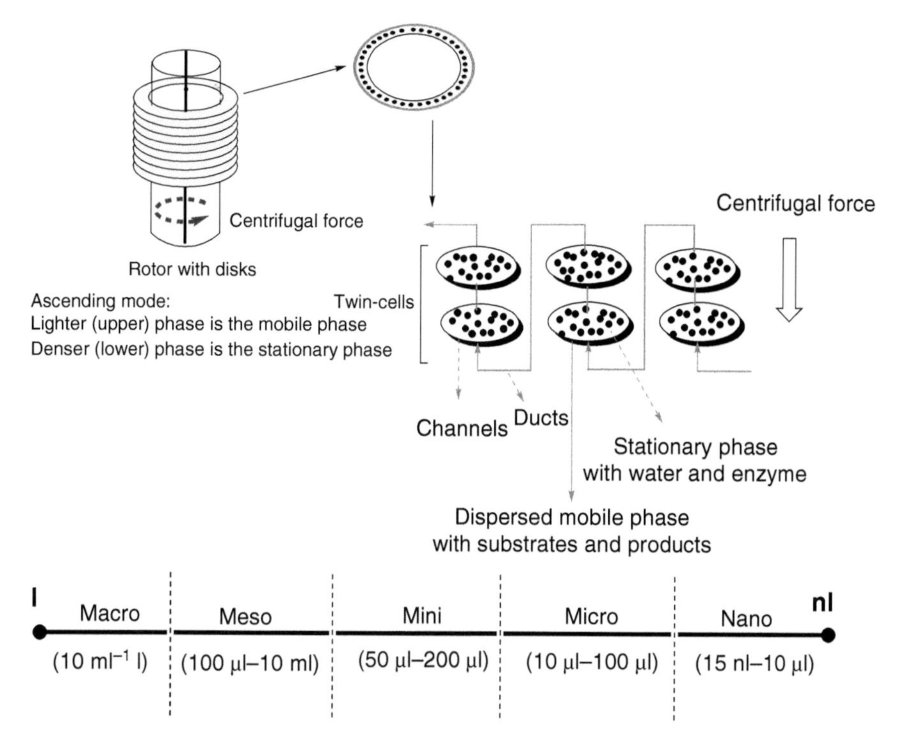

Figure 11.1 Diagram of a CPR reactor. The rotor consists of 21 disks that are composed of 90 twin cells linked by ducts. Source: Modified from Nioi et al. [25].

To overcome solubility problems, they used deep eutectic solvents (DESs) and phosphate buffer 100 mM pH 7.0 as a cosolvent (10% v/v), achieving a system with lower viscosity, efficient substrate solubility, high enzyme compatibility, and conversions 59% higher compared to the batch bioreactors [28].

11.2.1.1 Experimental Procedure

The biosynthesis of monostearin (**3**) from natural oil and fats is one example of reaction that can be performed in continuous flow. Using the commercial immobilized lipase from *Rhizomucor miehei* (RM IM), de Souza and coworkers synthesized the proposed molecule using the system represented in Scheme 11.1 [29].

In this work, authors utilized a X-Cube (ThalesNano) system that is composed of a packed bed reactor that was filled with RM IM, a temperature controller, and a continuous pump. For each experiment, heptane was pumped through the system before and after the experiment in order to clean the system from any remaining reactant or starting material. During the initial procedure, the required temperature was settled in order to condition the system before the beginning of the experiment. To proceed the reaction, a 35 mM stock solution containing stearic acid (**2**) and racemic 1,2-isopropylidene glycerol (**1**) equimolar proportion in heptane was pumped through the reactor in a continuous manner to yield a stream of product. The best conversion was obtained by heating the reactor at 60 °C with a flow rate of 0.6 ml min^{-1} (residence time of three minutes). Under these conditions, 87% of

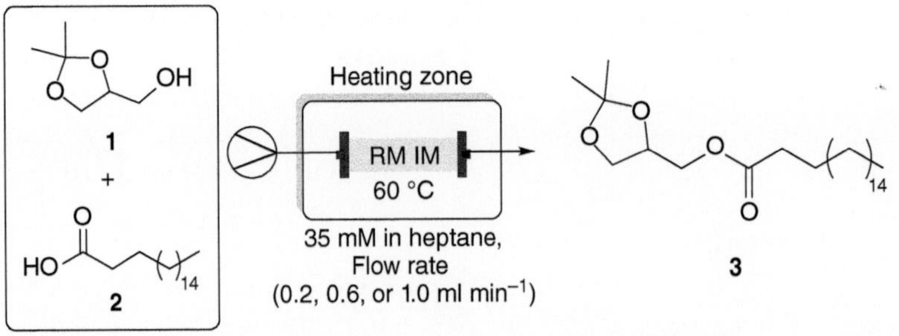

Scheme 11.1 Esterification reaction between racemic 1,2-isopropylidene glycerol and stearic acid catalyzed by immobilized lipase from *Rhizomucor miehei* (RM IM) under continuous flow conditions.

conversion was achieved. The reactor utilized had a catalytic bed of 0.6 ml (h = 7 cm, i.d. = 4 mm). The conversions were determined by collecting samples from the outlet stream and monitoring the fatty acid depletion by using a modification of the Lowry and Tinsley assay.

11.2.2 Transesterification

Besides the esterification of carboxylic acids, there is also the possibility to synthesize esters by transesterification reactions. In this type of reaction, it happens in the chemical reaction between an ester and an alcohol, resulting in a new ester and a new alcohol. Similar to esterification, transesterifications are also performed mainly by lipases, but instead of a carboxylic acid, an ester is used as substrate. Compared to the chemical route, that is performed at severe conditions and can present low selectivity, the transesterification performed by lipases can be conducted at milder conditions and are in general highly selectivity.

Because of its wide range of applications, lipases can be used in cascade processes. One example is the synthesis of glycerin carbonate under continuous flow conditions. De Souza and coworkers proposed a glycerol carbonate (**7**) production from vegetable oils by a cascade triacylglycerol (**4**) hydrolysis, biodiesel (**6**) synthesis, and esterification of the remaining glycerol with dimethyl carbonate (**5**) toward. This reaction was performed in a packed bed reactor using Novozym® 435 as a biocatalyst and achieved 85% conversion and high selectivity [30]. This protocol was further optimized by applying a handmade catalyst with the co-immobilization of lipases from *Porcine pancreas* (PPL) and CALB on epoxy resins, leading to a total conversion of vegetable oil with >99% selectivity (Scheme 11.2) [31].

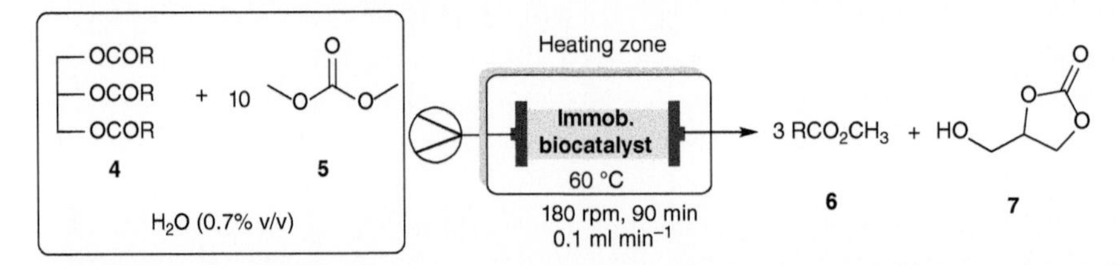

Scheme 11.2 Glycerol carbonate synthesis by applying a combination of the biocatalysts PPL and CALB. Reaction conditions: 1 : 1, 1 : 2, or 1 : 3 (w/w) mixtures of each biocatalyst; vegetable oil and dimethyl carbonate in a ratio of 1 : 10 (v/v); water (0.7% v/v) in *tert*-butyl methyl ether (MTBE) and 60 °C for 90 minutes. Source: Modified from do Nascimento et al. [31].

Transesterification in continuous flow can also be a strategy to synthesize APIs. A recent work from Wang and coworkers [32] reported the synthesis of 1-caffeoylglycerol from methyl caffeate and glycerin with the DES choline chloride urea as a cosolvent in a packed reactor containing Novozym® 435. In this setting, they were able to increase the yield to 97% and reduce the reaction time by 75% besides guarantee the reusability for until 20 times.

11.2.2.1 Experimental Procedure

Short-chain esters are commonly flavor and fragrance compounds used in the food, cosmetic, and pharmaceutical industries. In the industry, they are synthesized by the chemical esterification of monoterpenic alcohols, employing strong acid or basic catalysis under severe conditions. Alternatively, these molecules can be produced by a biocatalytic transesterification in continuous flow as reported by de Souza and coworkers (**10** and **11**) [22]. They performed this synthesis by following the reaction setup represented in Scheme 11.3 and using the commercial immobilized enzyme Novozym® 435.

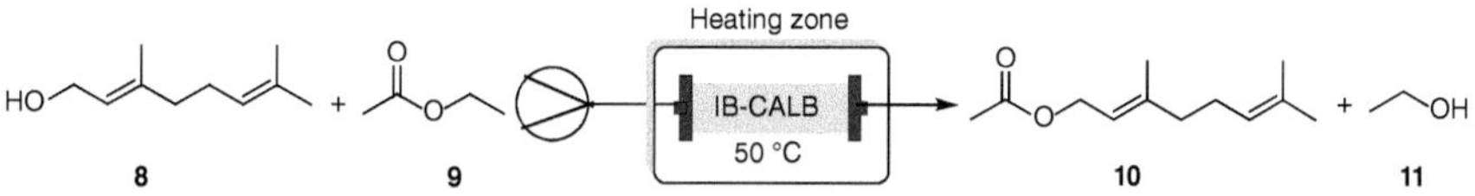

Heating zone

HO … 8 + 9 → IB-CALB 50 °C → 10 + OH 11

Scheme 11.3 Representation of the continuous flow reaction enzymatic synthesis of monoterpenic esters.

The system was composed by an Asia flow system (Syrris) equipped with a stainless steel column filled with the enzyme (packed bed reactor) and a continuous pump. For each experiment, ethyl acetate (**9**) was initially pumped through the system for three minutes. The same procedure was made after the experiment in order to clean the system from any remaining reactant or starting material. With the aid of a thermal bath, the reaction was performed in the desired temperature. To perform the reaction, a solution containing 150 mg of geraniol (**8**) in ethyl acetate was pumped through a packed bed reactor containing 1 g of immobilized enzyme. The best conversion was obtained by heating the reactor at 50 °C and applying a residence time of six minutes. Under these conditions, 88% of conversion was achieved. The column dimension was i.d. = 8.0 mm and h = 4.9 cm. The conversions were determined by collecting samples from the outlet stream and analyzing them by GC-FID.

11.2.3 Kinetic Resolutions

The synthesis of enantiomerically pure chiral compounds via cost-effective methods has become an important goal in the fine chemical and pharmaceutical industries. A conceptually simple and well-known method for the separation of enantiomers is the kinetic resolution, where the racemate is subjected to a selective reaction using a chiral agent and one of the enantiomers reacts faster than the other, thus enabling its separation [33]. Among these chiral mediators, the use of enzymes is particularly advantageous [34]. These naturally developed catalysts are biodegradable and have high chemo-, regio-, and stereoselectivity, leading to cleaner reactions. Another important factor is the increased commercial availability of a wide range of enzymes, especially lipases, proteases, esterases, acylases, and ω-transaminases, among others [35].

Kinetic resolution is based on the difference of reaction rates (k_R, k_S) of substrate enantiomers (S_R, S_S) during the transformation to products (P_R and P_S) by a chiral catalyst via diastereomeric transition states [36]. Because of the fact that two enantiomeric species are reacting simultaneously at different rates, the relative concentration of S_R/S_S and P_R/P_S is changing as the reaction proceeds, and as a consequence, the enantiomeric composition of S and P becomes a function of conversion (c). The mathematical basis for the treatment of the kinetics resolution was proposed by Fajans and coworkers [37] and later further developed for biocatalyzed resolutions by Sih and coworkers [38].

11.2.3.1 Kinetic Resolution of Amines Employing Lipases

Optically pure amines are highly valuable products or key intermediates for a vast number of APIs. Chemical synthesis of enantiopure amines is a challenging task as it requires harsh reaction conditions and toxic transition metal catalysts. To overcome those challenges, biocatalysis has been applied, developing efficient methods for racemate separation as an area of great interest for chiral amines synthesis (see also Chapter 6) [39]. The most remarkable biocatalysts for the kinetic resolution of amines are lipases because of its catalytic performance, wide substrate tolerance, and ability to be used in organic media [40]. Most of the reported examples for kinetic resolution of primary and secondary amines were in organic solvents by selective *N*-acylation. A beneficial property of lipases is that amide bond formation is essentially irreversible because lipases are not reactive to amide bond hydrolysis [41, 42].

De Miranda and coworkers reported a continuous flow system for production of (*R*)-amide from *rac*-1-phenylethylamine and ethyl acetate as the acyl donor catalyzed by Novozym® 435 [40]. An enantiomeric excess higher than 99% was achieved for the product (*R*)-amide and 48% conversion at 40 minutes. Oláh and et al. [41] also reported the kinetic resolution of *rac*-1-phenylethylamine by acylation with different isopropyl esters in continuous flow mode by various immobilized forms of lipase CALB. In addition, many lipases are readily available and can be immobilized and reutilized, being suitable for industrial purposes [17, 23, 43].

In a kinetic resolution, the choice of the acylating agent is an important parameter. As stated by Bornscheuer and Kazlauskas [42], and reported by Bassut et al. [43], who highlighted the importance of the acyl donor group in their recently published work, an ideal acylating agent should be inexpensive, unreactive in the absence of the lipase, and acylate quick- and irreversibly in the presence of the biocatalyst. Therefore, we present here the kinetic resolution of amines using poly(ethylene glycol) (PEG)-carboxylates as an acylating agent under continuous flow conditions [43].

11.2.3.1.1 Experimental Procedure A recent work by Bassut and coworkers reported the following methodology for the resolution of racemic amines using *rac*-1-phenylethylamine (**12**) and PEG diester (**13**) as an acyl donor because it proved to be a better acylating agent than diacid PEG (Scheme 11.4).

A stainless steel packed bed, similar to an high-performance liquid chromatography (HPLC) column (internal volume of 0.83 ml), was filled with 0.3 g of Novozym® 435 and placed in an incubator at 70 °C. Toluene (3–4 volumes of the reactor) was pumped by a continuous pump (Asia Pump system) to wet the catalytic bed. A solution of

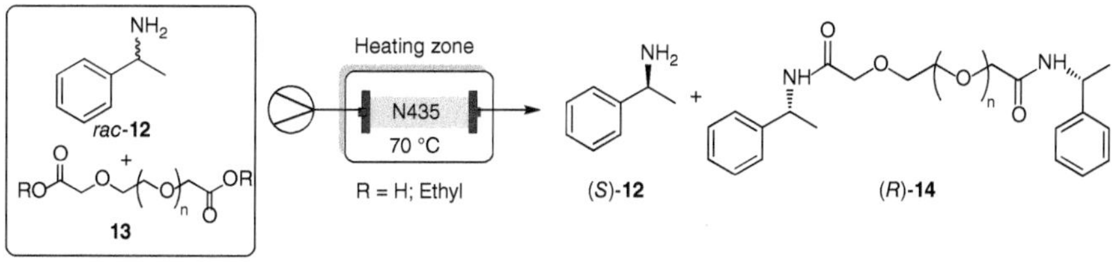

Scheme 11.4 Kinetic resolution of racemic 1-phenylethylamine using PEG600 diester as an acylating agent and Novozym® 435 as a biocatalyst. Source: Modified from Bassut et al. [43].

rac-1-phenylethylamine (0.15 M) and PEG diester (0.15 M) in toluene was then pumped at a flow rate of 27.6 μl min^{-1}, in order to achieve the desired residence time of 30 minutes. Further, 4.6 ml of the output stream was collected and a 2.0 ml sample was washed with a 1.0 M HCl aqueous solution to obtain (*S*)-**12** and (*R*)-**14**. Then, it was finally extracted with dichloromethane (3 × 1.0 ml), dried over anhydrous sodium sulfate, filtered, and derivatized before analysis by chiral gas chromatography (GC).

11.2.3.2 Kinetic Resolutions Employing ω-Transaminases

About ten years ago, ω-transaminases have emerged as the most useful and widely applied enzymes (see also Chapter 6) because they can be used in asymmetric synthesis of amino acids and asymmetric synthesis of chiral amines starting from prochiral ketones or kinetic resolution of racemic amines [44–47]. ω-Transaminases are pyridoxal-5′-phosphate (PLP)-dependent enzymes that produce α-chiral amines by transferring an amino group from the donor molecule to the carbonyl moiety of the acceptor molecule [45]. The cofactor PLP is extremely versatile, with its electron-dissipating nature allowing to react with a wide range of chemicals. This ability to displace the excess electron density around α-deprotonated carbon from the intermediate reaction allows PLP to act as a cofactor in transamination reactions [46].

Its application can be found in several recently published works, for example, in the synthesis of (*R*)-4-phenylbutan-2-amine as a precursor of antihypertensive Dilevalol [48], of 1-aminotetraline as a constituent of antidepressant Sertraline [49], and of (*R*)-*sec*-butylamine and cyclopropylethylamine [50] as important intermediates for corticotrophin-releasing factor (CRF-1) antagonist for the treatment of depression and anxiety [51].

In this context, the use of immobilized enzymes is gaining importance for continuous flow biocatalysis using isolated enzymes [52]. Recyclability of enzymes, along with increased stability, chemical selectivity, and operational window, can effectively be enabled by immobilization onto a support material, thereby making the enzyme perform as a heterogeneous catalyst [20]. Here, we report the application of immobilized ω-transaminases in the kinetic resolution of *rac*-1-phenylethylamine in continuous flow [53].

11.2.3.2.1 Experimental Procedure Commercial controlled pore glass carriers (EziG)[2] were used to immobilize the ω-transaminase from *Arthrobacter* sp. (AsR-ωTA) and applied in the kinetic resolution of the *rac*-1-phenylethylamine (**12**) (100 mM) in a continuous flow packed bed reactor (157 μl reactor volume), with conversion >49% and enantiomeric excess

2 The following EziG enzyme carrier materials were provided by EnginZyme AB (Stockholm, Sweden).

>99% of (*S*)-**12** (Scheme 11.5). The reaction was kept for 96 hours with no detectable loss of catalytic activity. The CPG-polymeric functionalized hybrid material creates a highly porous network for selective binding of enzymes with loadings up to 15% w/w.

In-flow Immobilization from Purified Enzyme Solution

ω-Transaminase from *Arthrobacter sp.* (15 mg, 403 nmol of purified enzyme) was diluted in 3-(*N*-morpholino)propanesulfonic acid (MOPS) buffer (10 ml, 100 mM, pH 7.5) supplemented with 0.1 mM PLP in an ice bath at 4 °C. A stainless steel column (h = 50 mm, i.d. = 2 mm) was filled with EziG material (100 mg) and hydrated by flowing MOPS buffer into the column (50 ml, 100 mM, pH 7.5, flow 0.5 ml min^{-1}). Then, the diluted stock solution of AsR-ωTA was loaded onto the column using a peristaltic pump (flow rate = 300 μl min^{-1}). MOPS buffer (30 ml, 100 mM, pH 7.5) was flowed through the column to wash out any possibly unbound protein. Buffer samples (20 μl) of the loading enzyme solution and of the flow-through obtained during washing were taken and the enzyme concentration was measured in both samples using the Bradford assay in order to calculate the immobilization yield. Immobilization was quantitative; thus, the final enzyme loading per unit of carrier material was 15% w/w enzyme loading. The column containing EziG-AsR was conditioned by flowing HEPES buffer (30 ml, 250 mM, pH 7, 25 μM PLP, flow 0.5 ml min^{-1}) and subsequently mounted on a Dionex P680 HPLC pump unit equipped with a flow controller. This setup was used for continuous flow kinetic resolution experiments.

Continuous Flow Kinetic Resolution with EziG-Immobilized ω-Transaminases

ω-Transaminase was immobilized on EziG carrier material (100 mg carrier plus enzyme, 15% w/w loading) in a stainless steel column (50 mm length × 2 mm diameter). A Dionex P680 HPLC pump unit was flushed with HEPES buffer (250 mM, pH 7.0, 25 μM PLP). The column containing EziG-ωTA was connected to the flow system and heated up to 30 °C in a water bath. The reaction mixture was prepared as follows: sodium pyruvate (5.4 g, 49.5 mmol, 55 mM concentration) was dissolved in HEPES buffer (900 ml final volume, 250 mM, pH 7.0) and *rac*-1-phenylethylamine (11.6 ml, 10.9 g, 90 mmol, 100 mM final concentration) was pre-dissolved in dimethyl sulfoxide (cosolvents concentration 10% v/v) before addition to the HEPES buffer containing sodium pyruvate. Then, the pH was adjusted to pH 7.0 and PLP (25 μM final concentration) was added. The solution was stirred

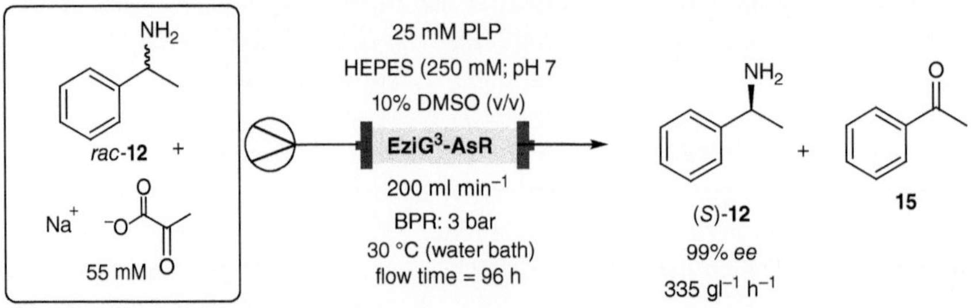

NH_2
rac-**12** +
Na^+ ^-O
55 mM
25 mM PLP
HEPES (250 mM; pH 7
10% DMSO (v/v)
EziG³-AsR
200 ml min^{-1}
BPR: 3 bar
30 °C (water bath)
flow time = 96 h
NH_2
(*S*)-**12**
99% *ee*
335 gl^{-1} h^{-1}
+
O
15

Scheme 11.5 Flow reactor setup in the application of EZiG³-AsR for the kinetic resolution of *rac*-1-phenylethylamine. Source: Böhmer et al. [53]. Licensed under CC-BY-4.0.

for one hour at room temperature in the dark. The reaction mixture was pumped through the column (average flow rate: 0.175 ml min^{-1}), and the product mixture was collected in fractions (ca. 8 ml each hour during the first day of operation and then 200 ml during the nighttime and 100 ml during the daytime). The column was operated for 96 consecutive hours without any detectable decrease of catalytic performance. A small aliquot of each fraction (0.5 ml) was basified with KOH (100 μl, 5 M) and extracted with EtOAc (2 × 500 μl). The organic layers were combined, dried over $MgSO_4$, and analyzed with GC equipped with an achiral column.

11.2.3.3 Kinetic Resolution of Alcohols Using Lipases

Nagy et al. [54] reported a continuous flow system using lipase from *Burkholderia cepacia* (Lipase PS) immobilized on hollow silica microspheres (M540) by bisepoxide activation to kinetic resolution of racemic 1-phenylethanol with vinyl acetate. Most examples using lipase use the commercially available CALB. However, with the new developments in enzyme immobilization, lipases from different organisms can be utilized. For example, sol–gel-immobilized lipase from *Burkholderia cepacia* (Amano PS) was utilized for the first time as a biocatalyst for kinetic resolution of 1,5-dihydroxy-1,2,3,4-tetrahydronaphthalene [55]. Various silane precursors for the sol–gel matrix such as methyl vinyl, octyl, phenyl-trimethoxysilane, and tetra-methoxysilane were tested as immobilization matrix and then tested for the kinetic resolution of chiral heteroaromatic secondary alcohols with benzofuran, benzothiophene, phenothiazine, and 2-phenylthiazol moieties. Here, we report a flow chemistry approach for the kinetic resolution of *rac*-1,2-propanediol protected by the trityl group using the packed bed reactor filled with different immobilized enzymes [56].

11.2.3.3.1 Experimental Procedure The acetylation of *rac*-1-(trityloxy) propan-2-ol (**16**) under continuous flow conditions was achieved in high yields and excellent selectivity in short reaction times (seven minutes), as shown in Scheme 11.6.

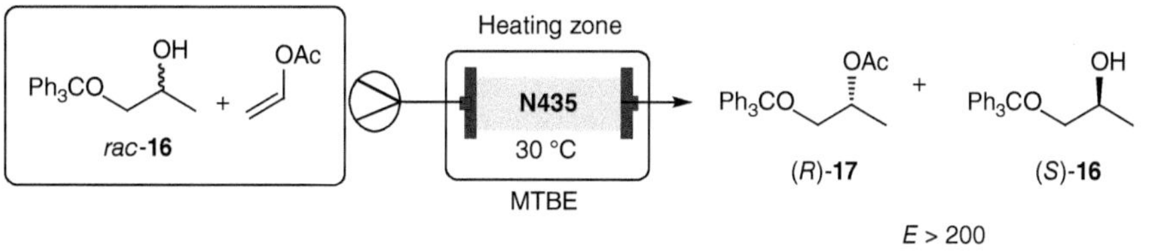

Scheme 11.6 Lipase Novozym® 435 in the resolution of *rac*-1-(trityloxy)propan-2-ol in flow. Source: Aguillón et al. [56]. © 2019, Elsevier.

Racemic 1-(trityloxy)propan-2-ol (6.78 mmol, 2.16 g) and vinyl acetate (8.14 mmol, 700 mg) were dissolved in 43.2 ml of *tert*-butyl methyl ether (MTBE). The starting mixture was pumped using the instrument Asia Flow Reactor equipped with different Omnifit borosilicate glass columns (h = 7 cm, i.d. = 1.5 cm) containing Novozym® 435 (0.937 g) at 30 °C. The reaction parameters (flow rate and concentration) were selected on the flow reactor, and processing was started, whereby only pure solvent was pumped through the system until the instrument had achieved the desired reaction parameters and stable processing was assured. Flow rates were selected in order to deliver reaction times in the range between 7 and 30 minutes [56].

Conversion was calculated by HPLC analysis with a column Zorbax Eclipse C18 (150 × 4.6 mm i.d. with particle size of 5 μm). The injection volume was 10.0 μl, and the mobile phase consisted of ethanol/water (85 : 15). The flow rate was 0.8 ml min^{-1} and data were acquired using a photodiode array detector, DAD, at 254 nm. The enantiomeric excess was determined by HPLC analysis, employing a chiral column Chiralpak® OD-H. The mobile phase consisted of 99 : 1 hexane:2-propanol; a flow rate of 0.8 ml min^{-1} and a gradient of 0–10 minutes 98%, followed by 13–15 min 97 : 3 hexane:2-propanol. Data were acquired using a photodiode array detector, DAD, at 254 nm.

11.2.4 Dynamic Kinetic Resolutions

The use of multiple biocatalysts or a combination of chemo- and biocatalysts in the same reaction vessel is an attractive method of synthesis because it can reduce the reaction time and product purification requirements (see also Chapters 13 and 14). Dynamic kinetic resolution (DKR) is an attractive method of increasing the yield of kinetic resolution reactions. In the classical enzymatic kinetic resolution, the theoretical maximum yield for one enantiomer is 50%. The conversion of the unreacted isomer back to the racemate, either actively or spontaneously, allows further resolution, producing the desired enantiomer. This process, known as DKR, is an example of deracemization and can lead to a theoretical yield of up to 100% of the desired enantiomer [57–59].

Several metal systems based on palladium (Pd), ruthenium (Ru), nickel (Ni), cobalt (Co), and iridium (Ir) have been employed as the racemization catalysts. Pd-based catalysts include Pd/C, Pd/$BaSO_4$, and Pd/AlO(OH) [60, 61]. Acids and bases, zeolites, and even light-activated catalysts have also been used as racemization catalysts for the DKR of alcohols [60]. An attractive option is the preparation of hybrid supports containing the metals for racemization and also the immobilized enzyme. Some lipases and proteases that have been employed as the resolution catalysts for DKR include the *Candida antarctica* lipase A (CALA), CALB, *Burkholderia* (formerly *Pseudomonas*) *cepacia* lipase (BCL), *Pseudomonas stutzeri* lipase (PSL), *Candida rugosa* lipase (CRL), and subtilisin Carlsberg (SC) from *Bacillus licheniformis* [60–63].

Akai and coworkers reported the lipase/metal-catalyzed DKR of racemic allylic alcohols using oxovanadium as a catalyst. They carried out racemization/isomerization using allylic alcohol as a starting material and applied the chiral product to the total synthesis of (−)-himbacine, obtaining a product with excellent enantioselectivity [60].

The Lozano group described for the first time the continuous DKR of *rac*-1-phenylethanol in different ionic liquid media/$scCO_2$ biphasic systems using a combination of the immobilized lipase (Novozym® 435) and the acid chromatographic support (silica modified with benzenesulfonic acid groups, SCX) as catalysts, providing a yield of 76% of (*R*)-1-phenylethyl propionate with excellent enantioselectivity (91–98% *ee*) [61].

Herein, we report the application of alcalase (subtilisin A) immobilized by simple hydrophobic adsorption onto various surface-grafted macroporous silica gels, resulting in easy-to-prepare and stable biocatalysts (Scheme 11.7). The procedure enabled the efficient DKR of racemic *N*-Boc-phenylalanine ethyl thioester (**18**) with benzylamine (**19**), producing

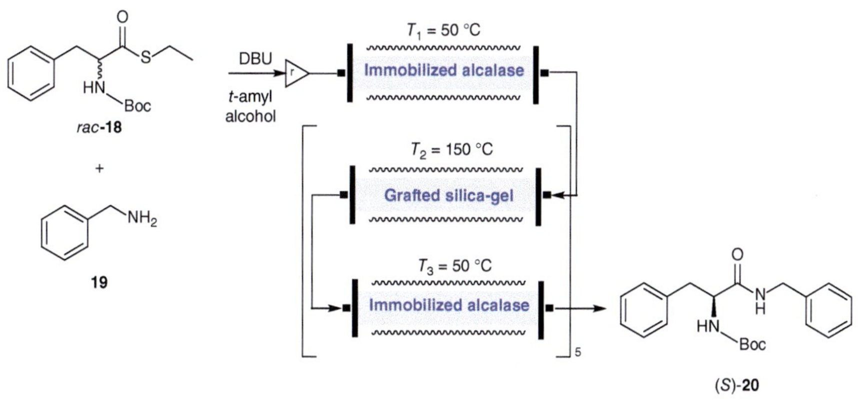

Scheme 11.7 Dynamic kinetic resolution of racemic *N*-Boc-phenylalanine ethyl thioester (*rac*-**18**) in continuous flow mode using an alternating cascade of packed bed enzyme reactors and racemization reactors kept at different temperatures. Source: Falus et al. [62]. © 2016, John Wiley & Sons.

(*S*)-benzylamide (*S*)-**20** with 79% conversion, 8.17 $g l^{-1} h^{-1}$ volumetric productivity, and 98% *ee* [62].

11.2.4.1 Experimental Procedure

Eleven CatCart™ columns (six columns filled with Alc-Dv250-Et [total filling weight: 1269.2 mg] and five columns filled with Dv250-Et [total filling weight: 1049.8 mg]) were connected in an alternating series starting with one filled with Alc-Dv250-Et. A solution of racemic *N*-Boc-Phe-SEt (*rac*-**18**, 5 $mg ml^{-1}$), benzylamine **17** (1.2 equiv), and 1,8-diazabicycloundec-7-ene (DBU, 3 equiv) in *tert*-amyl alcohol was pumped through the columns thermostated at 50 °C for the kinetic resolution process and 150 °C for the racemization step at a flow rate of 0.2 $ml min^{-1}$. Samples of size 200 ml diluted with hexane/2-propanol 98 : 2 to 500 ml) were collected after stationary operation had been established (three hours after starting the operation) and analyzed by HPLC [62].

The product solution emerging from the stationary phase of the DKR process was collected (24 ml, two hours) and the solvent was removed by a vacuum rotary evaporator. The residue was dissolved in MTBE (10 ml) and washed with 0.1 M HCl (2 × 5 ml). The organic phase was dried over $MgSO_4$ and the solvent was removed under vacuum. Further purification was performed by column chromatography on silica gel (eluent: ethyl acetate/*n*-hexane 1 : 3), yielding after removal of the solvent the pure amide (*S*)-**20** as white crystals with excellent optical purity.

11.2.5 Asymmetric Synthesis

Stereoselective transformations are among the many different classes of reactions that have been investigated using continuous flow technology. The development of new methods for the synthesis of enantiopure building blocks and intermediates is important. Asymmetric catalysis, in particular, has come to the forefront as a highly economical and efficient means for the generation of chiral compounds, whereby achiral starting materials are transformed directly into enantioenriched products using only minute amounts of a chiral component [64].

Asymmetric synthesis is referred to as enantioselective conversion of a prochiral substrate to an optically active product using a chiral addend via asymmetric step. Enzymes are frequently used as asymmetric catalysts for the preparation of enantiomerically pure organic compounds. Several studies of continuous flow biocatalysis can be found in the literature, for example: synthesis of chiral cyanohydrins [64], (*R*)-2-hydroxy-l-phenylpropanone with histidine-tagged benzaldehyde lyase (BAL–Ni–NTA) immobilized on Ni-NTA-sepharose as a heterogeneous catalyst [65], and (*S*)-1-phenylethylamine with cellulose immobilized transaminase [66].

Benítez-Mateos and coworkers reported the synthesis of chiral alcohols using ketoreductases (KREDs) and nicotinamide adenine dinucleotide phosphate (NADPH)) as cofactors (see also Chapter 7), in which both KRED and NADPH residues were co-immobilized and co-localized on the same surface and applied in continuous reactions in aqueous media [67], as shown in Scheme 11.8. The corresponding alcohols were obtained up to >99% yield and >99% *ee*, and this high performance could be maintained over five consecutive reaction cycles.

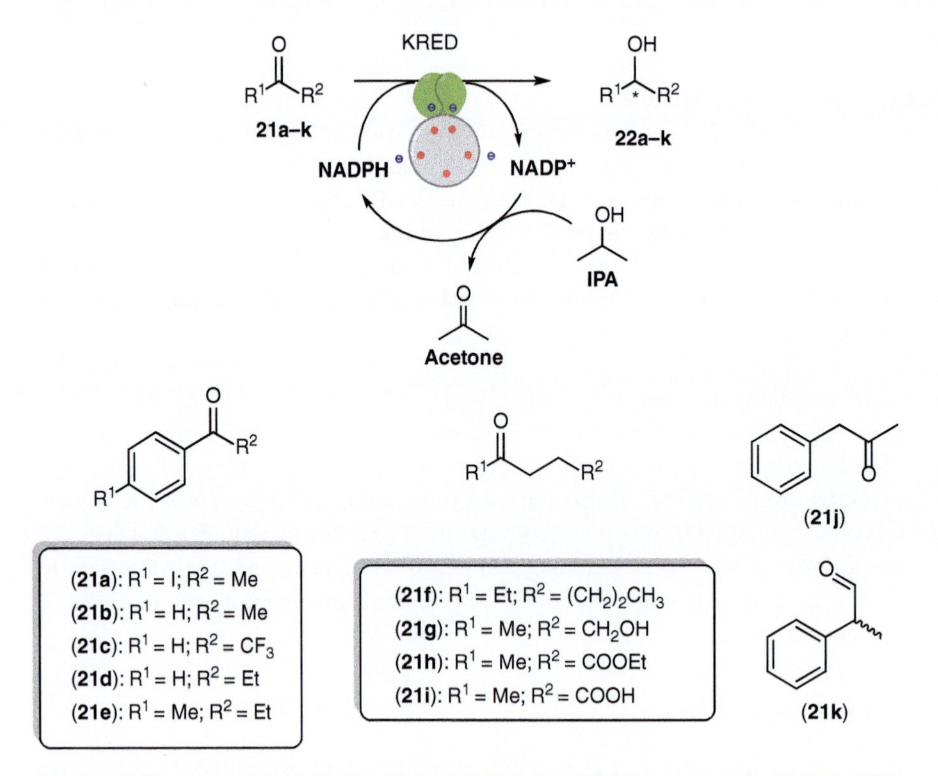

Scheme 11.8 Asymmetric reduction of prochiral ketones (**21a-j**) and racemic aldehyde **21k** to optically active secondary and primary alcohols (**22a-k**) catalyzed by a self-sufficient heterogeneous biocatalyst with co-immobilized KRED P1-A04 and NADPH. Source: Benítez-Mateos et al. [67]. © 2017, John Wiley & Sons.

11.2.5.1 Experimental Procedure

11.2.5.1.1 Protein Immobilization Typically, KRED (0.025–0.65 $mg\,ml^{-1}$, 1.0 ml) was incubated with the carrier (100 mg) for one hour at room temperature and under the optimized buffer conditions for each immobilization protocol. The immobilization on Ag-DVS, Ag-DEAE, Ag-PEI, Ag-DEAE/G, and Pu-A/E was carried with an enzyme solution prepared in sodium phosphate (10 mM) at pH 7.0. For the immobilization on Pu-G, the enzyme was prepared in sodium bicarbonate (100 mM) at pH 10.0 or in sodium phosphate buffer (100 mM) and dithiothreitol (10 mM) at pH 8.0. Upon the enzyme immobilization on a carrier that contained glyoxal groups, beads (100 mg) were incubated with $NaBH_4$ (1.0 $mg\,ml^{-1}$, 1.0 ml) in sodium bicarbonate (100 mM) at pH 10.0 and 4 °C for 30 minutes to irreversibly reduce the enzyme–carrier imine bonds into amine bonds [67].

11.2.5.1.2 Ion Exchange of NADPH on Ag-DEAE Beads immobilizing KRED (100 mg) were incubated with NADPH (1 mM, 1 ml) in Tris–HCl (10 mM) at pH 7.0 and room temperature. Samples of the supernatant were withdrawn after one hour and spectrophotometrically measured at 340 nm to quantify the adsorption of NADPH to the carrier. Following the adsorption process, the solid samples were washed with 10 column volumes of Tris–HCl (10 mM) at pH 7.0 and utilized in further experiments.

11.2.5.1.3 General Procedure for the Continuous Asymmetric Reduction: Self-sufficient heterogeneous biocatalysts (0.28 g, 6.5 mg KRED g^{-1} of carrier and 10 mmol NADPH g^{-1} carrier) were packed into a column (h = 2 cm, i.d. = 0.4 cm) and connected to a flow system driven by a semicontinuous syringe pump. A reaction mixture that contained the corresponding aryl ketone **21c** (10 mM), $MgCl_2$ (1 mM), and 17% v/v IPA in Tris–HCl (10 mM) at pH 7.0 was passed through the column at different flow rates (50–200 $ml\,min^{-1}$) and 258 °C. Different samples were collected from the outlet of the system and analyzed by HPLC [67]. A space–time yield of 97–112 $g\,l^{-1}\,day^{-1}$ was obtained, and additionally, the immobilized cofactor accumulated a total turnover number of 1076 for 120 hours.

11.3 Conclusions and Perspective

Continuous flow technology is still in early ages in terms of applying its potential toward biotechnology and biocatalysis, but we will only be able to advance in this research field if we change our way of designing a bioprocess. The lack of studies on continuous immobilization protocols turns this research field in a very interesting area where reduction on immobilization times and efficiency increase will be of great interest in industry. Further on, the continuous immobilization idea will be the possibility to run a cascade immobilization/reaction methodology in a *one-pot* strategy, allowing to maximize the efficiency and reduce the footprint of the process.

Cascade chemo-enzymatic or multi-enzymatic reactions have attracted great attention in the recent days and is an interesting field where a few authors have already started exploring the possibility of running continuous cascade reactions (see Chapters 13 and 14). One of the major benefits of performing cascade reactions under continuous conditions is compartmentalization (see also Chapter 4). This characteristic allows researchers to work with reactions under different reaction conditions of temperature, concentration, reagents, or reaction time, at the same cascade sequence. In this way, reagents or biocatalysts that are not compatible now can be placed in different compartments, under their specific reaction conditions, without the possibility of interacting with other chemical/biochemicals.

The possibility of performing cascade chemo-enzymatic or multi-enzymatic reactions under continuous conditions opens new process windows for biocatalyzed processes, which need to be explored in order to deliver new solutions for academia and industry.

References

1 Gutmann, B., Cantillo, D., and Kappe, C.O. (2015). Continuous-flow technology -a tool for the safe manufacturing of active pharmaceutical ingredients. *Angew. Chem. Int. Ed.* 54: 6688–6728.

2 Britton, J. and Raston, C.L. (2017). Multi-step continuous-flow synthesis. *Chem. Soc. Rev.* 46: 1250–1271.

3 Reizman, B.J. and Jensen, K.F. (2016). Feedback in flow for accelerated reaction development. *Acc. Chem. Res.* 49: 1786.1796.

4 Porta, R., Benaglia, M., and Puglisi, A. (2016). Flow chemistry: recent developments in the synthesis of pharmaceutical products. *Org. Process Res. Dev.* 20: 2–25.

5 Kobayashi, S. (2016). Flow "fine" synthesis: high yielding and selective organic synthesis by flow methods. *Chem. Asian J.* 11: 425–436.

6 Cambie, D., Bottecchia, C., Straathof, N.J.W. et al. (2016). Applications of continuous-flow photochemistry in organic synthesis, material science, and water treatment. *Chem. Rev.* 116: 10276–10341.

7 Mandity, I.M., Ötvös, S.B., and Fülöp, F. (2015). Strategic application of residence-time control in continuous-flow reactors. *ChemistryOpen* 4: 212–223.

8 Wiles, C. and Watts, P. (2014). Continuous process technology: a tool for sustainable production. *Green Chem.* 16: 55–62.

9 Vaccaro, L., Lanari, D., Marrocchi, A., and Strappaveccia, G. (2014). Flow approaches towards sustainability. *Green Chem.* 16: 3680–3704.

10 Movsisyan, M., Delbeke, E.I.P., Berton, J. et al. (2016). Taming hazardous chemistry by continuous flow technology. *Chem. Soc. Rev.* 45: 4892–4928.

11 Plutschack, M.B., Pieber, B., Gilmore, K., and Seeberger, P.H. (2017). The Hitchhiker's guide to flow chemistry. *Chem. Rev.* 117: 11796–11893.

12 Hessel, V., Kralisch, D., Kockmann, N. et al. (2013). Novel process windows for enabling, accelerating, and uplifting flow chemistry. *ChemsusChem.* 6: 746–789.

13 Munirathinam, R., Huskens, J., and Verboom, W. (2015). Supported catalysis in continuous-flow microreactors. *Adv. Synth. Catal.* 357: 1093–1123.

14 Baxendale, I.R. (2013). The integration of flow reactors into synthetic organic chemistry. *J. Chem. Technol. Biotecnol.* 88: 519–552.

15 Webb, D. and Jamison, T.F. (2010). Continuous flow multi-step organic synthesis. *Chem. Sci.* 1: 675–680.

16 Fitzpatrick, D.E. and Ley, S.V. (2016). Engineering chemistry: integrating batch and flow reactions on a single, automated reactor platform. *React. Chem. Eng.* 1: 629–635.

17 Thompson, M.P., Peñafiel, I., Cosgrove, S.C., and Turner, N.J. (2019). Biocatalysis using immobilized enzymes in continuous flow for the synthesis of fine chemicals. *Org. Process Res. Dev.* 23: 9–18.

18 Britton, J., Majumdar, S., and Weiss, G.A. (2018). Flow chemistry: integrated approaches for practical applications. *Chem. Soc. Rev.* 47: 5891–5918.

19 Tamborini, L., Fernandes, P., Paradisi, F., and Molinari, F. (2018). Flow bioreactors as complementary tools for biocatalytic process intensification. *Trends Biotechnol.* 36: 73–88.

20 Lindeque, R.M. and Woodley, J.M. (2019). Reactor selection for effective continuous biocatalytic production of pharmaceuticals. *Catalysts* 9: 262.

21 Lau, S.C., Lim, H.N., Basri, M. et al. (2014). Enhanced biocatalytic esterification with lipase-immobilized chitosan/graphene oxide beads. *PLoS One* 9: e104695.

22 Adarme, C.A.A., Leao, R.A.C., de Souza, S.P. et al. (2018). Continuous-flow chemo and enzymatic synthesis of monoterpenic esters with integrated purification. *Mol. Catal.* 453: 39–46.

23 Woodcock, L.L., Wiles, C., Greenway, G.M. et al. (2008). Enzymatic synthesis of a series of alkyl esters using novozyme 435 in a packed-bed, miniaturized, continuous flow reactor. *Biocatal. Biotrans.* 26: 466–472.

24 Itabaiana, I. Jr. and de Souza, R.O.M.A. (2018). *Biocatalysis at room temperature in Sustainable Catalysis: Energy Efficient Reactions and Applications* (eds. R. Luque and F.L.-K. Yun), 89–134. Wiley-VCH Verlag: Weinheim.

25 Nioi, C., Destrac, P., and Condoret, J.S. (2019). Lipase esterification in the Centrifugal Partition Reactor: Modelling and determination of the specific interfacial area. *Biochem. Eng. J.* 143: 179–184.

26 Swarts, J.W., Vossenberg, P., Meerman, M.H. et al. (2008). Comparison of two-phase lipase-catalyzed esterification on micro and bench scale. *Biotechnol. Bioeng.* 99: 855–861.

27 Bolivar, J.M., Wiesbauer, J., and Nidetzky, B. (2011). Biotransformations in microstructured reactors: more than flowing with the stream? *Trends Biotechnol.* 29: 333–342.

28 Guajardo, N., Schrebler, R.A., and Domínguez de María, P. (2019). From batch to fed-batch and to continuous packed-bed reactors: lipase-catalyzed esterifications in low viscous deep-eutectic-solvents with buffer as cosolvent. *Bioresour. Technol.* 273: 320–325.

29 Itabaina, I. Jr., Flores, M.C., Sutili, F.K. et al. (2012). Lipase-Catalyzed monostearin synthesis under continuous flow conditions. *Org. Process Res. Dev.* 16: 1098–1101.

30 Leão, R.A.C., de Souza, S.P., Nogueira, D.O. et al. (2016). Consecutive lipase immobilization and glycerol carbonate production under continuous-flow conditions. *Catal. Sci. Technol.* 6: 4743–4748.

31 do Nascimento, M.A., Gotardo, L.E., Leão, R.A.C. et al. (2019). Enhanced productivity in glycerol carbonate synthesis under continuous flow conditions: combination of immobilized lipases from Porcine Pancreas and *Candida antarctica* (CALB) on Epoxy Resins. *ACS Omega* 4: 860–869.

32 Liu, X., Meng, X.-Y., Xu, Y. et al. (2019). Enzymatic synthesis of 1-caffeoylglycerol with deep eutectic solvent under continuous microflow conditions. *Biochem. Eng. J.* 142: 41–49.

33 Todd, M. (2014). *Separation of Enantiomers*. Weinheim: Wiley-VCH Verlag.

34 de Miranda, A.S., Miranda, L.S.M., and de Souza, R.O.M.A. (2015). Lipases: valuable catalysts for dynamic kinetic resolutions. *Biotechnol. Adv.* 33: 372–393.

35 Faber, K. (2011). *Biotransformations in Organic Chemistry*, sixth ed. Berlin: Springer Science & Business Media.

36 Pellissier, H. (2011). *Chirality from Dynamic Kinetic Resolution*. Cambridge: Royal Society of Chemistry.

37 Faber, K. (2001). Non-sequential processes for the transformation of a racemate into a single stereoisomeric product: proposal for stereochemical classification. *Chem. Eur. J.* 23: 5004–5010.

38 Chen, C.-S., Fujimoto, Y., Girdaukas, G., and Sih, C.J. (1982). Quantitative analysis biochemical kinetic resolution of enantiomer. *J. Am. Chem. Soc.* 104: 7294–7299.

39 de Souza, R.O.M.A., Miranda, L.S.M., and Bornscheuer, U.T. (2017). A retrosynthesis approach for biocatalysis in organic synthesis. *Chem. Eur. J.* 23: 12040–12063.

40 de Miranda, A.S., Miranda, L.S.M., and De Souza, R.O.M.A. (2013). Ethyl acetate as an acyl donor in the continuous flow kinetic resolution of (±)-1-phenylethylamine catalyzed by lipases. *Org. Biomol. Chem.* 11: 3332–3336.

41 Oláh, M., Boros, Z., Hornyánszky, G., and Poppe, L. (2016). Isopropyl 2-ethoxyacetate—an efficient acylating agent for lipase-catalyzed kinetic resolution of amines in batch and continuous-flow modes. *Tetrahedron* 72: 7249–7255.

42 Bornscheuer, U.T. and Kazlauskas, R.J. (2006). *Hydrolases in Organic Synthesis: Regio- and Stereoselective Biotransformations*, second ed. Weinheim: Wiley-VCH Verlag.

43 Bassut, J., Rocha, A.M.R., da França, A. et al. (2018). PEG600-carboxylates as acylating agents for the continuous enzymatic kinetic resolution of alcohols and amines. *Mol. Catal.* 459: 89–96.

44 Patel, R. and Biocatalysis, N. (2011). Synthesis of key intermediates for development of pharmaceuticals. *ACS Catal.* 1: 1056–1074.

45 Kelly, S.A., Pohle, S., Wharry, S. et al. (2018). Application of ω-transaminases in the pharmaceutical industry. *Chem. Rev.* 118: 349–367.

46 Eliot, A.C. and Kirsch, J.F. (2004). Pyridoxal phosphate enzymes: mechanistic, structural, and evolutionary considerations. *Annu. Rev. Biochem.* 73: 383–415.

47 Malik, M.S., Park, E.-S., and Shin, J.-S. (2012). Features and technical applications of ω-transaminases. *Appl. Microbiol. Biotechnol.* 94: 1163–1171.

48 Koszelewski, D., Clay, D., Rozzell, D., and Kroutil, W. (2009). Deracemisation of α-chiral primary amines by a one-pot, two-step cascade reaction catalysed by ω-transaminases. *Eur. J. Org. Chem.* 2009: 2289–2292.

49 Pressnitz, D., Fuchs, C.S., Sattler, J.H. et al. (2013). Asymmetric amination of tetralone and chromanone derivatives employing ω-transaminases. *ACS Catal.* 3: 555–559.

50 Hanson, R.L., Davis, B.L., Chen, Y. et al. (2008). Preparation of (*R*)-amines from racemic amines with an (*S*)-amine transaminase from *Bacillus megaterium*. *Adv. Synth. Catal.* 350: 1367–1375.

51 Taché, Y., Martinez, V., Wang, L., and Million, M. (2004). CRF1 receptor signaling pathways are involved in stress-related alterations of colonic function and viscerosensitivity: implications for irritable bowel syndrome. *Br. J. Pharmacol.* 141: 1321–1330.

52 Britton, J., Majumdar, S., and Weiss, G.A. (2018). Continuous flow biocatalysis. *Chem. Soc. Rev.* 47: 5891–5918.

53 Böhmer, W., Knaus, T., Volkov, A. et al. (2019). Highly efficient production of chiral amines in batch and continuous flow by immobilized ω-transaminases on controlled porosity glass metal-ion affinity carrier. *J. Biotechnol.* 291: 52–60.

54 Nagy, F., Szabó, K., Bugovics, P., and Hornyánsky, G. (2019). Bisepoxide-activated hollow silica microspheres for covalent immobilization of lipase from *Burkholderia cepacia*. *Period. Polytech. Chem. Eng.* 63: 414–424.

55 Cimporescu, A., Todea, A., Badea, V. et al. (2016). Efficient kinetic resolution of 1,5-dihydroxy-1,2,3,4-tetrahydronaphthalene catalyzed by immobilized *Burkholderia cepacia* lipase in batch and continuous-flow system. *Process Biochem.* 51: 2076–2083.

56 Aguillón, A.R., Avelar, M.N., Gotardo, L.E. et al. (2019). Immobilized lipase screening towards continuous-flow kinetic resolution of (±)-1,2-propanediol. *Mol. Catal.* 467: 128–134.

57 Turner, N.J. (2010). Deracemisation methods. *Curr. Opin. Chem. Biol.* 14: 115–121.

58 Verho, O. and Baeckvall, J.-E. (2015). Chemoenzymatic dynamic kinetic resolution: a powerful tool for the preparation of enantiomerically pure alcohols and amines. *J. Am. Chem. Soc.* 137: 3996–4009.

59 Musa, M.M., Hollmann, F., and Mutti, F.G. (2019). Synthesis of enantiomerically pure alcohols and amines *via* biocatalytic deracemisation methods. *Catal. Sci. Technol.* 9: 5487–5503.

60 Akai, S., Hanada, R., Fujiwara, N. et al. (2010). One-pot synthesis of optically active allyl esters *via* lipase—vanadium combo catalysis. *Org. Lett.* 12: 4900–4903.

61 Lozano, P., De Diego, T., Larnicol, M. et al. (2006). Chemoenzymatic dynamic kinetic resolution of rac-1-phenylethanol in ionic liquids and ionic liquids/supercritical carbon dioxide systems. *Biotechnol. Lett.* 28: 1559–1565.

62 Falus, P., Cerioli, L., Bajnóczi, G. et al. (2016). A continuous-flow cascade reactor system for subtilisin a- catalyzed dynamic kinetic resolution of *N-tert*-butyloxycarbonylphenylalanine ethyl thioester with benzylamine. *Adv. Synth. Catal.* 358: 1608–1617.

63 Thalén, L.K. and Bäckvall, J.E. (2010). Development of dynamic kinetic resolution on large scale for (±)-1-phenylethylamine. *Beilstein J. Org. Chem.* 6: 823–829.

64 Chen, P., Han, S., Lin, G., and Li, Z. (2002). A practical high through-put continuous process for the synthesis of chiral cyanohydrins. *J. Org. Chem.* 67: 8251–8253.

65 Kurlemann, N. and Liese, A. (2004). Immobilization of benzaldehyde lyase and its application as a heterogeneous catalyst in the continuous synthesis of a chiral 2-hydroxy ketone. *Tetrahedron Asymmetry* 15: 2955–2958.

66 de Souza, S.P., Junior, I.I., Silva, G.M.A. et al. (2016). Cellulose as an efficient matrix for lipase and transaminase immobilization. *RSC Adv.* 6: 6665–6671.

67 Benítez-Mateos, A.I., San Sebastián, E., Ríos-Lombardía, N. et al. (2017). Asymmetric reduction of prochiral ketones by using self-sufficient heterogeneous biocatalysts based on NADPH-dependent ketoreductases. *Chem. Eur. J.* 23: 16843–16852.

Part IV

Recent Trends in Enzyme-Catalyzed Reactions

12

Photobiocatalysis

Martín G. López-Vidal, Guillermo Gamboa, Gabriela Oksdath-Mansilla, and Fabricio R. Bisogno

Universidad Nacional de Córdoba, Ciudad Universitaria, INFIQC-CONICET-UNC, Dpto. de Química Orgánica, Facultad de Ciencias Químicas, Haya de la Torre y Medina Allende, Córdoba, X5000HUA, Argentina

12.1 Introduction

The application of catalytic systems instead of stoichiometric ones supposed a change in paradigm in the way of thinking and planning chemical reactions. Catalysis enables the synthesis of useful compounds in a more economical, energy-saving fashion with the obvious advantage to be less detrimental to the environment. In this line, the so-called "practical elegance" in the design of a synthetic route entails a logic elegance along with a practical applicability. Scientists have spent tremendous efforts in setting up novel catalytic methodologies to achieve valuable synthetic targets [1].

Photocatalysis, in particular, photoredox catalysis, has witnessed an impressive burst thanks to the recent developments in collecting energy from the visible region of the electromagnetic spectrum. Indeed, this is possible by using transition metal complexes ($[Ru(bpy)_3]^{2+}$, *fac*-$Ir(ppy)_3$, etc.), organic dyes (rose Bengal [RB], methylene blue, eosin Y, rhodamine G, fluorescein, etc.), natural pigments (flavin, chlorophyll, etc.), and nanostructured semiconductors such as quantum dots (QDs), doped TiO_2, perovskites, carbon nitrides, etc. (Scheme 12.1). With this power in hands, chemists have set up a myriad of synthetically relevant reaction systems, enabling reduction, oxidation, and formation of C–C and C-heteroatom, among others, powered by visible light. It must be emphasized that a great number of these reactions are difficult (or too sluggish) under dark conditions [2].

Meanwhile, biocatalysis has gained momentum owing to the outstanding selectivity and negligible price compared with other types of catalysis, and therefore, chemical industry already adopted such technology even for large-scale processes (see also Chapter 15). The reason behind this success may be the simple modification of the enzyme by mutagenesis or directed evolution (see also Chapter 2), thus improving acceptance and selectivity toward non-natural substrates and activity in nonconventional media (see also Chapter 9),

Biocatalysis for Practitioners: Techniques, Reactions and Applications, First Edition.
Edited by Gonzalo de Gonzalo and Iván Lavandera.

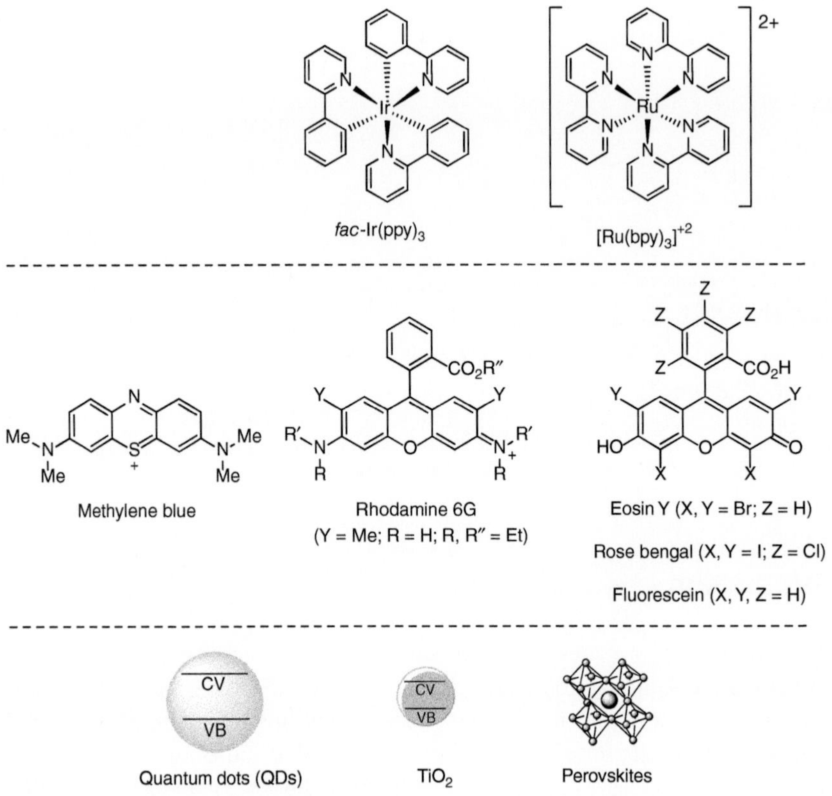

fac-$Ir(ppy)_3$ $[Ru(bpy)_3]^{+2}$

Methylene blue

Rhodamine 6G (Y = Me; R = H; R, R″ = Et)

Eosin Y (X, Y = Br; Z = H)

Rose bengal (X, Y = I; Z = Cl)

Fluorescein (X, Y, Z = H)

Quantum dots (QDs) TiO_2 Perovskites

Scheme 12.1 Common Photocatalysts (PCs).

such as organic solvents, ionic liquids, and deep eutectic solvents [3, 4]. There are several features that are exclusive to only one type of catalysis, for instance, the "innate" or "guided" reactivity achieved in metallocatalytic or photocatalytic processes, or the perfect enantioselectivity displayed by biocatalysis, or the robustness of and scalability of metal- and organo-catalyzed reactions [5]. Considering this, one may envision an ideal process comprising the advantages of several (or at least more than one) sort of catalysis. This might come true by a chemically reasonable and practically feasible combination of catalysts, either sequentially or in cascade/domino processes (see Chapter 14) [6–8]. In the past few years, scientists have successfully merged metal- and photocatalysis [9–11], metal- and biocatalysis [12–15], organo- and biocatalysis [16], organo- and photocatalysis [17–19], and of course a combination of photo- and biocatalysis [20–22]. In 2018 and 2019, comprehensive literature surveys collecting proof of concepts and applications of photobiocatalysis have appeared, illustrating the diverse underlying mechanisms. In this chapter, we will focus on examples of photobiocatalytic systems that can be useful in synthetic chemistry emphasizing practical aspects to be considered for the laboratory practitioner. Along it, several fundamental terms coming from radical chemistry and photochemistry will often be used. As many readers may not be familiar with some of them, the most common ones can be found in Scheme 12.2.

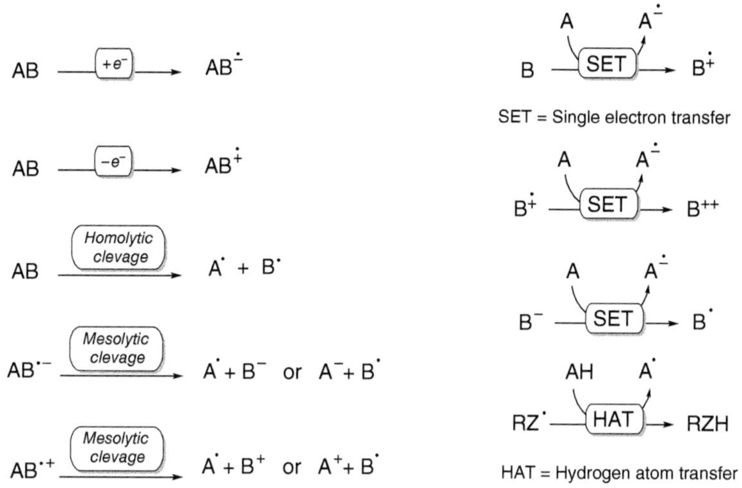

AB —(+e⁻)→ $AB^{\bullet-}$

AB —(−e⁻)→ $AB^{\bullet+}$

AB —(Homolytic clevage)→ $A^{\bullet}$ + $B^{\bullet}$

$AB^{\bullet-}$ —(Mesolytic clevage)→ $A^{\bullet} + B^-$ or $A^- + B^{\bullet}$

$AB^{\bullet+}$ —(Mesolytic clevage)→ $A^{\bullet} + B^+$ or $A^+ + B^{\bullet}$

B + A —(SET)→ $B^{\bullet+}$ + $A^{\bullet-}$

SET = Single electron transfer

$B^{\bullet+}$ + A —(SET)→ B^{++} + $A^{\bullet-}$

B^- + A —(SET)→ $B^{\bullet}$ + $A^{\bullet-}$

$RZ^{\bullet}$ + AH —(HAT)→ RZH + $A^{\bullet}$

HAT = Hydrogen atom transfer

Scheme 12.2 Common processes involved in radical chemistry.

12.2 Oxidative Processes

12.2.1 Baeyer–Villiger Oxidation

Enzymes can efficiently catalyze the selective formation of an ester (or lactone) from a (cyclic)ketone as substrate by the Baeyer–Villiger (BV) reaction [23]. In this process, a nucleophilic oxygen species is generated at the flavin-dependent monooxygenase active site by reductive activation of O_2, giving place to the corresponding ester or lactone product and water. In order to regenerate the flavin in the biocatalytic cycle, stoichiometric amounts of reducing equivalents are needed. These reducing equivalents (masked electrons) are delivered from a sacrificial substrate to the flavin by the mediation of a pyrimidine cofactor (NAD^+ or $NADP^+$). Considering the cost of the cofactors, different regeneration methods have been developed in order to use it in catalytic amounts. In this context, light activation of flavin in the presence of a sacrificial electron donor is a greener and simpler alternative to regenerate catalytic nicotinamide adenine dinucleotide (NADH) or nicotinamide adenine dinucleotide phosphate (NADPH) (Scheme 12.3). Even though flavin can be regenerated by photoredox catalysis, for BV-monooxygenases, the presence of nicotinamide cofactor is always needed because of its pivotal role during the catalysis (Scheme 12.4).

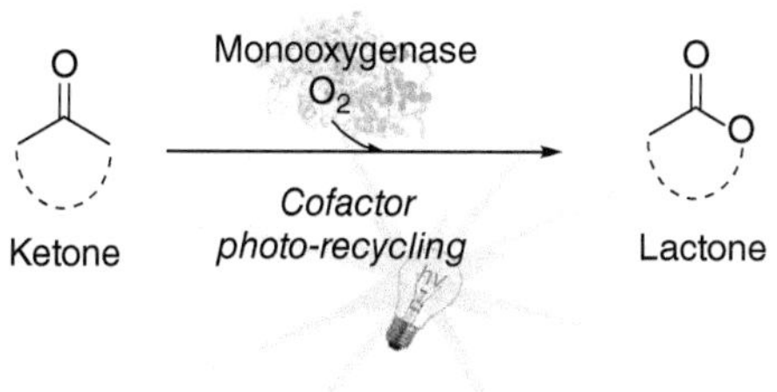

Ketone —(Monooxygenase, O_2; *Cofactor photo-recycling*, hν)→ Lactone

Scheme 12.3 General scheme for a Photobiocatalyzed Baeyer–Villiger oxidation.

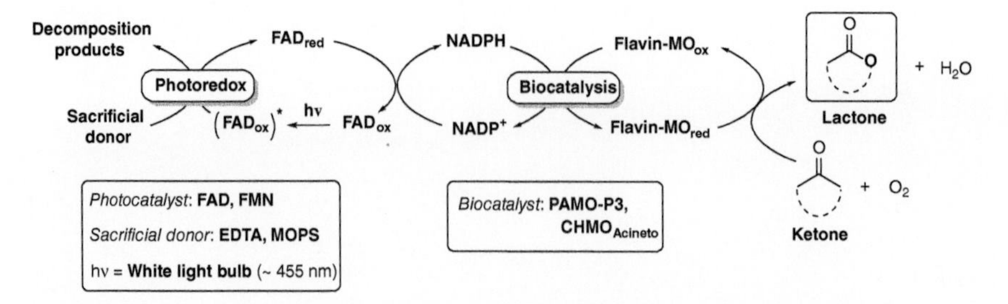

Scheme 12.4 Baeyer–Villiger oxidation with photocatalytic cofactor regeneration.

One of the first approaches was developed by Reetz's group in 2007, who evaluated the light-driven BV oxidation employing ethylenediaminetetraacetic acid (EDTA) as a sacrificial electron donor [24]. Specifically, the oxidation of cyclic ketones (1–2 μM) was performed using 10 μM of the mutant of phenylacetone monooxygenase from *Thermobifida fusca* (PAMO-P3) together with 25 μM EDTA, 100 μM additional flavin adenine dinucleotide (FAD), and 250 μM $NADP^+$. Under white light irradiation, the kinetic resolution of cyclic ketones was obtained with excellent conversion (40–50%) and high enantioselectivity (97% ee) after seven hours of reaction. Considering the FAD absorption in the visible region of the electromagnetic spectrum, sunlight irradiation can be used efficiently. In this case, even when the reaction rates were similar, the conversion depended on weather conditions. The turnover frequency and turnover number (TOF and TON, $10\,h^{-1}$ and 96, respectively) are lower in comparison of conventional regeneration systems (by electrolysis or by coupled enzyme systems). Another enzyme evaluated in a photobiocatalytic BV oxidation was cyclohexanone monooxygenase from *Acinetobacter calcoaceticus* NCIMB 9871 ($CHMO_{Acineto}$). In this case, perfect enantioselectivity was observed for similar substrates (>99% ee).

For these photobiocatalytic processes, some considerations have to be taken: on the one hand, the triplet excited state of flavin can produce singlet oxygen ($^1O_2$) by energy transfer to molecular oxygen. On the other hand, flavin radical anion intermediate can generate superoxide anion radical ($O_2^{\bullet-}$) by single electron transfer (SET) to molecular oxygen. These species may exert deleterious effects to substrates/products or to the enzyme, thus affecting the enzymatic activity/selectivity and reaction yield [25]. To avoid that, a new photoredox system was designed using FAD and MOPS (3-(*N*-morpholino)propanesulfonic acid)-based buffer, acting as an electron donor as well. Laser flash spectroscopy and steady-state absorption measurements support the fact that electron transfer process with MOPS competes with the O_2, giving at least a partial protection to active flavin species. However, the energy transfer process to the O_2 cannot be fully avoided, giving place to the formation of $^1O_2$ [26].

12.2.2 Alkane Hydroxylation

The regio- and stereo-specific hydroxylation of alkanes (C–H activation, Scheme 12.5) exhibits an enormous challenge in organic synthesis. In this context, the combination of biocatalysis and photocatalysis is a very elegant approach to get access to alcohols in a one-pot process.

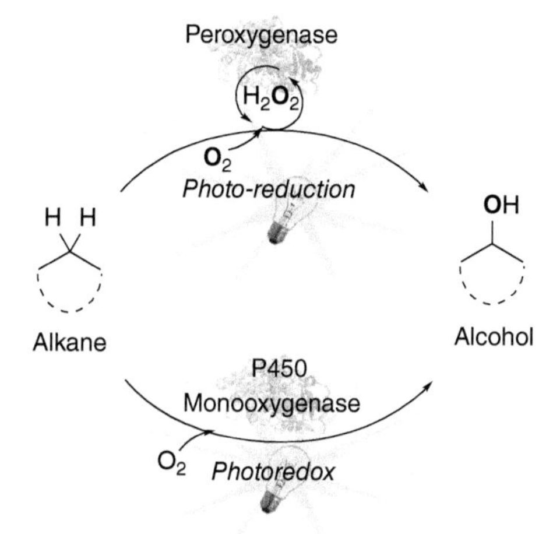

Scheme 12.5 General procedure for Photoenzymatic alkane hydroxylation.

Two different systems can be used to carry out the oxyfunctionalization of alkanes combining enzymes and light. Basically, the mechanism involved in the alcohol formation depends on the kind of enzyme used (Scheme 12.6). Particularly, cytochrome P450 monooxygenases can mediate hydroxylation through direct participation of a porphyrin π radical ferryl intermediate known as compound I (CpdI). By H-atom abstraction from the substrate, a new protonated hydroxoiron(IV) intermediate compound II (CpdII) is generated with the corresponding alkyl radical. This carbon-centered radical subsequently

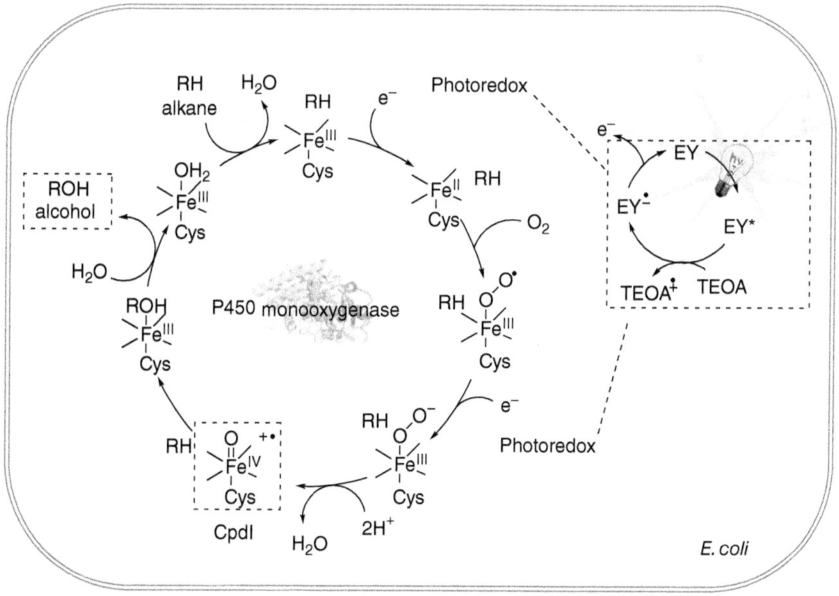

Scheme 12.6 Mechanism involved in the alkane hydroxylation reaction with *in vivo* photoreduction of P450 monooxygenase.

rebound onto the ferryl hydroxyl moiety, giving the hydroxylated substrate. One of the challenges of this system is the generation of the CpdI intermediate, which depends on a complicated electron transport chains from NADPH-P450 reductase (CPR) that contains two flavins (FAD and flavin mononucleotide [FMN]). To overcome this disadvantage, different strategies have been developed such as cofactor regeneration, alternative oxygen atom donor, electrochemistry, or light-driven reduction [27].

One may perform a photobiocatalytic version of this complex system. In this line, a very interesting cofactor-free system has been developed. A photoredox system was used to inject electrons (by electron transfer process) to the heme group of P450 monooxygenase [28]. Specifically, *Escherichia coli* cells producing whole *Bacillus megaterium* P450 variant (Y51F/F87A, BM3m2) system were incubated in a buffer solution with eosin Y (EY, the photosensitizer) and electron acceptor together with triethanolamine (TEOA, the electron donor) and the corresponding substrate. The BM3m2 system comprises two domains, the heme domain that performs the catalytic hydroxylation, and the diflavin-reductase domain, working as an electron shuttle to the heme group. Because of a specific interaction of EY with the P450 heme domain, an *in vivo* photoreduction of P450 via visible light irradiation can be applied to catalyze the alkane hydroxylation reaction (Scheme 12.6). Thus, to perform the photobiocatalytic process, a mixture of whole BM3m2-expressing cells, EY (20 μM), and 7-ethoxycoumarin (1 mM) in 0.2 ml of potassium phosphate (50 mM) and TEOA (100 mM) was irradiated using a white visible light source. The reaction showed a TON of 16 for BM3m2 at 18 hours.

On the other hand, for the photoredox system, other electron donors, such as EDTA, ascorbic acid, or formic acid, might be used, but they displayed poor solubility in this particular buffer solution at neutral pH and undesired side reactions with the protein. Therefore, this reaction is so far limited to the use of TEOA as a sacrificial electron donor.

Another system that can be considered to perform alkane oxyfunctionalization involves the use of peroxygenases. These enzymes can generate the catalytically active CpdI intermediate at the expense of hydrogen peroxide, avoiding the cumbersome electron transport chain. Nevertheless, even at low concentration of H_2O_2, this heme-enzyme is oxidatively deactivated. To avoid that, an *in situ* generation of H_2O_2 through reduction of O_2 can be efficiently achieved. In this sense, many approaches such as enzymatic, electrochemical, and photocatalytic alternatives have been developed for the H_2O_2 generation.

Nowadays, two different photobiocatalytic strategies to promote peroxygenase-catalyzed oxyfunctionalization reactions can be found in the literature. One of them involves the use of gold-loaded TiO_2 (Au-TiO_2) as a plasmonic photocatalyst (PC) for the oxidation of methanol (electron donor) and the reductive activation of molecular oxygen (Scheme 12.7) [29]. A mixture of 10 mM of substrate, 10 g l^{-1} of rutile Au-TiO_2, 150 nM of a recombinant unspecific peroxygenase from *Agrocybe aegerita* (r*Aae*UPO), and 250 mM of MeOH in phosphate buffer (pH 7.0, 60 mM), was irradiated with a white light bulb for 70 hours. A high conversion was observed for linear and cyclic alkanes as well as for benzylic substrate with a TON for the biocatalyst between 17 000 and 71 000. In all cases, traces of the corresponding ketone as a side product were detected via overoxidation of the alcohol product.

Recently, the same PC (Au-TiO_2) was evaluated as a heterogeneous water oxidation catalyst (WOC). Under photoexcitation, WOC can oxidize H_2O giving O_2 as a product. The extra electron into WOC is used to reduce O_2 to H_2O_2 that power up the enzymatic

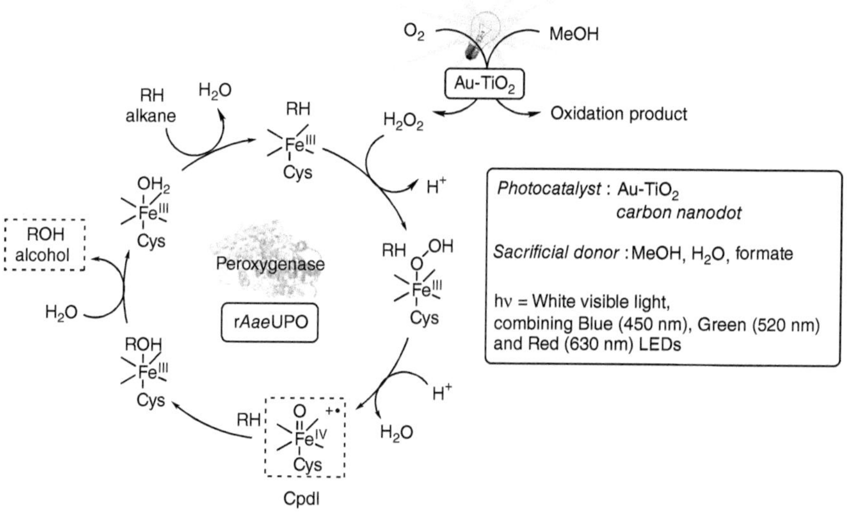

Scheme 12.7 Alkane hydroxylation catalyzed by peroxygenase involving H_2O_2 photogeneration.

oxyfunctionalization of alkanes. The same aforementioned conditions were evaluated using r*Aae*UPO peroxygenase as a biocatalyst, showing high conversion for the same alkanes with a TON for the biocatalyst between 17 000 and 38 800 [30].

As an extension of the use of heterogeneous PC, a carbon nanodot (CND) catalyst was evaluated in the photoenzymatic oxyfuntionalization. In this case, FMN was used as a cocatalyst to mediate the reduction of molecular oxygen to H_2O_2. Under anaerobic conditions, water is oxidized by the photoexcitated CND, and this CND with an additional electron can now reduce FMN, which is finally reoxidized forming H_2O_2 and promoting r*Aae*UPO-catalyzed hydroxylation. With this system, the overall reaction rates were significantly higher than using Au-TiO_2, showing a better performance, with a TON of 100 000 for r*Aae*UPO and more than 100 for FMN [30].

The other photobiocatalytic alternative involves the oxidation of NADH and reduction of O_2 to produce H_2O_2 through a photochemical process. Additionally, in order to recycle the expensive NADH cofactor, an enzymatic relay system was employed. Specifically, a formate dehydrogenase was used to mediate the hydride transfer from formate to NAD^+. The special feature of this system is the use of formate as a sacrificial donor, which produces CO_2 as a stoichiometric by-product, resulting in a cleaner enzymatic reaction. Therefore, connecting two different enzymatic processes through a photocatalytic system, H_2O_2 can be *in situ* generated to perform oxyfunctionalization of alkanes [31]. General conditions involve the use of 4.8 μM CbFDH formate dehydrogenase, 75 mM of $NaHCO_2$, 0.4 mM of NAD^+, 1 μM of PC together with 0.1 μM of r*Aae*UPO peroxygenase, and 10 mM of the corresponding substrate in buffer potassium phosphate 50 mM pH 7.0. Particularly, it was proposed that the simultaneous use of different PCs (such as FMN, phenosafranine, and/or methylene blue) in order to provide a more efficient use of the energy of polychromatic light (combining blue, green, and red light-emitting diodes (LEDs) with 450, 520, and 630 nm, respectively). Finally, The TON calculated for the catalytic components (r*Aae*UPO, PCs, NAD^+, and CbFDH) were 100 000, 500, 25, and 1785, respectively.

12.2.3 *O*-Dealkylation

Among different oxidative reactions catalyzed by heme-containing, nicotinamide-dependent cytochrome P450 monooxygenases, *O*-dealkylation is an interesting reaction, particularly for its application in group deprotection and synthesis of natural product derivatives [32]. Given that the enzyme requires reducing equivalents in the form of reduced cofactor NADPH, the Park's group developed in 2012 an artificial cofactor recycling photosystem (Scheme 12.8) [33]. This system involves the use of visible light, EY (PC), TEOA (sacrificial electron donor), and $[Cp^*Rh(bpy)H_2O]^{2+}$ (M, mediator). In the first step, TEOA is photooxidized by the excited EY that transfers this electron to M_{ox}. Afterward, the reduced mediator M_{red} regenerates NADPH by SET, returning to the initial form M_{ox}, being ready for a new photocatalytic cycle.

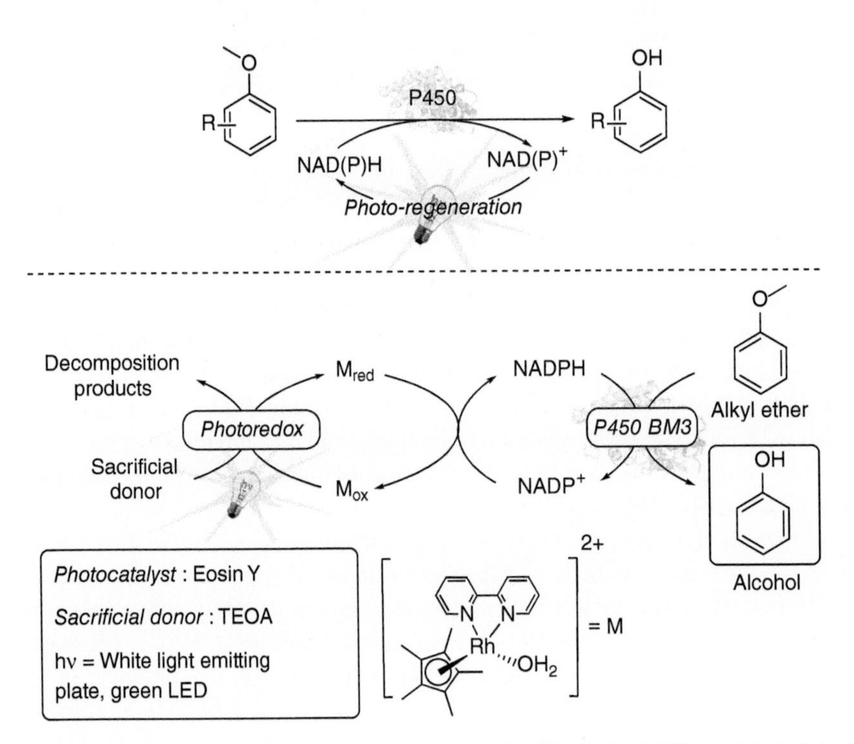

Scheme 12.8 Photocatalytic recycling system of NADPH for the P450-mediated *O*-dealkylation.

This light-driven cofactor recycling approach was tested along with P450 BM3 monooxygenase from a *B. megaterium* variant (Y51F/F87A, BM3m2) that consists of CPR (cytochrome P450 reductase) and P450. CPR transfers the electron required for enzymatic catalysis at P450 by oxidizing NADPH. Assays were performed using 7-ethoxycoumarin (1 mM) with cell-free cytochrome P450 solution (0.2 μM), EY (20 μM), M (0.1 mM), and

$NADP^+$ (0.1 mM) in potassium phosphate (50 mM)-triethanolamine (100 mM) buffer (pH 7.5). The reaction mixture was irradiated for 24 hours with visible light using a white light emitting plate, achieving low conversion (<1%). Nevertheless, this low conversion observed is similar to P450-mediated dealkylation without the cofactor recycling system, employing instead 50 mM of NADPH under dark conditions. Hence, low conversions are attributed to poor activity of the P450 enzyme toward 7-ethoxycoumarin. Considering the EY light absorption maximum at 525 nm, green LED (emission at ~535 nm) was tested. A similar low conversion was achieved after five hours of irradiation. In this direction, this light source becomes promising for future improvements of the reaction system. Some other factors have to be taken into account for P450-mediated *O*-dealkylation described. One factor is that EY concentration did not considerably affect the activity of the biocatalyst. For 1 mM of substrate, increasing EY concentration over 20 μM did not enhance the reaction yield. Another factor concerns mediator and TEOA concentrations: the higher these concentrations, either one or another, the less product formation. This is likely due to a deleterious effect on enzyme's activity. Finally, the last factor implies light intensity influence on the reaction rate: enzyme TOF at 5 hours increased from 2.77 to 8.37 h^{-1} switching the intensity from 15 to 60 $mW\,cm^{-2}$. In spite of this, using higher light intensity, the turnover slightly dropped probably because of photobleaching of EY.

12.2.4 Decarboxylation

12.2.4.1 Alkene Production

Oxidative enzymatic decarboxylation can be achieved by employing a P450 fatty acid decarboxylase ($OleT_{JE}$) from *Jeotgalicoccus* sp. ATCC 8456. This enzyme catalyzes the conversion of a range of natural fatty acids into the corresponding terminal alkenes (Scheme 12.9). For synthetic applications, a fusion variant of $OleT_{JE}$ was coupled with a reductase domain from *Rhodococcus* sp. that is able to directly accept NADPH as an electron donor [34]. Besides, a complex electron relay is needed, which involves ferredoxin as an electron mediator. Because electron transfer processes are sensitive to O_2, reactive species might be formed, thereby generating a misuse of the expensive cofactor NADPH, which affects enzymatic activity and reaction rate. Fortunately, this complex electron transport system can be circumvented because $OleT_{JE}$ can directly employ H_2O_2 in order to recycle the catalytically activate hydroperoxo-iron species, fundamental to the oxidative decarboxylation [35]. Considering the low stability of certain biocatalysts toward high concentration of H_2O_2, *in situ* light-driven continuous formation of H_2O_2 by flavin-mediated reduction of O_2 may avoid enzymatic loss of activity, as H_2O_2 would be consumed immediately after its generation.

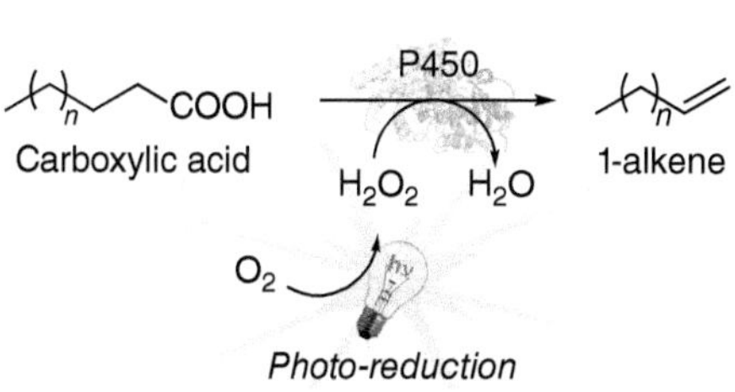

Scheme 12.9 Biodecarboxylation with photogeneration of hydrogen peroxide.

First studies were carried out with stearic acid (0.5 mM) as a model substrate in Tris–HCl buffer (50 mM Tris, 200 mM NaCl, pH 7.5), with FMN (10 μM), $OleT_{JE}$ (100 μg ml^{-1}) crude cell extracts, and EDTA (50 mM) as a sacrificial electron donor, upon illumination with white light at 25 °C. This procedure achieved an excellent conversion (99%) after two hours of irradiation (Scheme 12.10). In spite of this excellent result, lower conversions were observed with shorter chain length fatty acids (up to 42%), as well as a decrease in ratio 1-alkene:β-hydroxy acid (side product). Nevertheless, these short-chain fatty acids produce volatile alkenes that can be easily removed from the reaction mixture. On the other hand, long-chain fatty acids also showed lower conversions. Differently, an increase in 1-alkene:β-hydroxy acid ratio was observed. In the case of unsaturated fatty acids with *cis* configuration, the corresponding terminal alkene was not detected. Some considerations have to be taken for the described oxidative biodecarboxylation with photogeneration of H_2O_2. Firstly, enzyme purification is highly recommendable because of the presence of catalases in cell extracts that will consume H_2O_2. Secondly, the enzyme $OleT_{JE}$ may promote the undesired β-hydroxylation of substrates, reducing the yield of the corresponding 1-alkene. Finally, oleic acid inhibits the activity of $OleT_{JE}$. Thus, alkene production is limited to mixtures lacking of oleic acid.

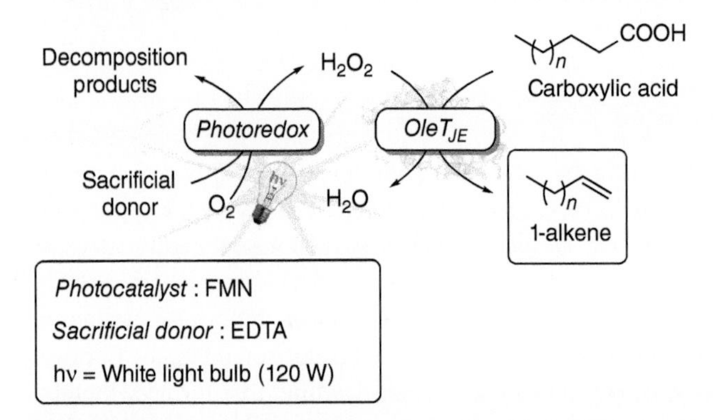

Scheme 12.10 Biodecarboxylation of fatty acids to obtain 1-alkenes by means of photogeneration of H_2O_2.

12.2.4.2 Alkane Production

A group of long-chain fatty acids can be irreversibly transformed into the corresponding C1-shortened alkanes and CO_2 by a recently discovered algal fatty acid photodecarboxylase from *Chlorella variabilis* NC64A (*Cv*FAP) (Scheme 12.11) [36]. This approach for the

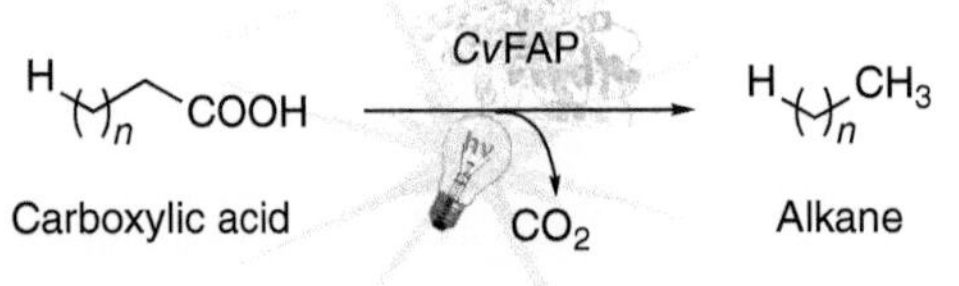

H COOH
Carboxylic acid
CvFAP
CO_2
H CH3
Alkane

Scheme 12.11 General scheme for enzymatic decarboxylation.

specific synthesis of alkanes comparatively results in a more environmentally friendly process than traditional protocols, as the reaction conditions are milder and free from transition metals [37–39]. Given that flavoenzyme *Cv*FAP is a photoenzyme, a continuous flux of photons is needed by the biocatalyst to work. The strongly bound FAD in its active site, which absorbs blue light (maximum absorption at 467 nm), is the responsible of the one-electron oxidation of the carboxylate substrate that triggers the whole process with the advantage of not needing additional cofactors [36, 40]. The ability of the catalyst to form hydrocarbons (alkanes) was evaluated and best performance was displayed at pH 8.5. Additionally, the enzyme loses its activity when the temperature surpasses 35 °C. Moreover, greater catalytic efficiency was observed toward long fatty acids (C16 and C17) [40]. In a recent study, crude cell-free extracts (CFEs) of *E. coli* overexpressing a shortened version of *Cv*FAP (comprising the residues 62–654) were used to perform the decarboxylation of some long-chain fatty acids [36]. The reason for the use of the CFE instead of the purified enzyme is that CFE presents higher efficiency and robustness in this reaction medium. The reactions were carried out in sealed vials at 37 °C using different fatty acids (30 mM), *Cv*FAP (6.0 μM) in Tris–HCl buffer (pH 8.5) with dimethylsulfoxide (DMSO) (30%) as a cosolvent, irradiating with blue LED (~ 450 nm) under gentle magnetic stirring for a time of 14 hours. Excellent conversions (≥90%) and respectable TONs (TON ≥ 4350) were achieved for palmitic (C16), margaric (C17), stearic (C18), and arachidic (C20) acids. Besides, unsaturated fatty acids – oleic (C18; Δ9) and linoleic (C18; Δ9,12) acid – could be transformed with modest conversions and TONs (65% and 49%, 2950 and 2600, respectively). Nonetheless, lauric (C12) and myristic (C14) acids were poorly converted by this photobiocatalytic process (11% and 25% of conversion, respectively). With the admirable results obtained in this study for the reaction using palmitic acid as a substrate, a preparative-scale experiment was performed under similar conditions. In this case, 155 mg of pentadecene (61% isolated yield) was obtained.

Furthermore, a two-step bienzymatic cascade to decarboxylate the fatty acid chains from triolein (a triglyceride) was performed in the same study. As depicted in the scheme below (Scheme 12.12), triolein is first hydrolyzed by a lipase from *Candida rugosa* (*Cr*Lip) into glycerol and three molecules of oleic acid. In the next step, *Cv*FAP is added into the system to mediate the light-driven decarboxylation of the released fatty acids molecules to afford (*Z*)-heptadec-8-ene and CO_2. A promising 83% of conversion and a TON of 8280 were observed for the photoenzyme. Further efforts have to be made yet to optimize the reaction conditions leading to higher conversions.

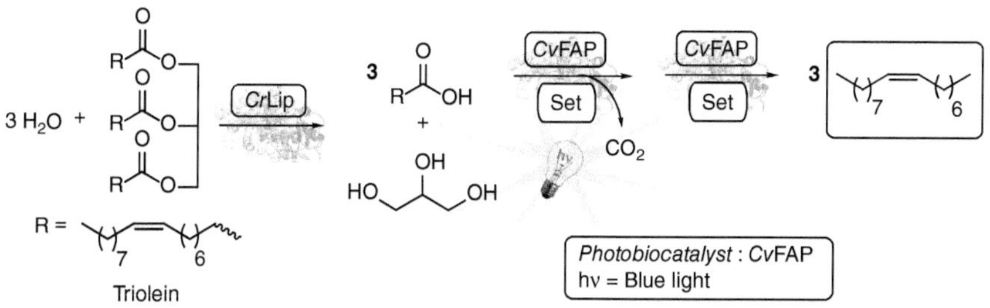

Scheme 12.12 Bienzymatic cascade to convert triolein into (*Z*)-heptadec-8-ene.

This cascade involving two enzymes offers the biotechnological possibility of transforming waste fatty acids and oils into alkanes, which can be employed as biofuels given their higher specific heat of combustion (c. 9%) in comparison with other biofuels obtained also from fatty acids and oils, such as methyl esters (biodiesel) [41]. The latter are obtained nowadays by a largely employed method of transesterification. Nevertheless, a transesterification process has the disadvantages of being reversible and the requirement of elevated amounts of methanol, issues that are absent in the cascade transformation involving light and enzymes.

Structurally, the enzyme *Cv*FAP presents a narrow hydrophobic tunnel from the exterior of the enzyme to its active site. When a carboxylic acid enters in this tunnel, the carboxylic group is finally oriented close to the FAD prosthetic group. Under irradiation with blue light, this cofactor triggers the decarboxylation process through two consecutive SET [40, 42].

It is thought that because of an insufficient stabilization of short- and medium-chain carboxylic acids in the tunnel of the biocatalyst *Cv*FAP, these substrates are not efficiently decarboxylated [40]. To solve this limitation and increase the reaction rate, a simple approach has been proposed [43]. An alkane is employed as a decoy molecule to fill up the space in the tunnel when short- or medium-chain carboxylic acids are used as substrates. The general rule is that the sum of the carbon atoms of the carboxylic acid and the decoy molecule has to be 16 in order to the reaction to take place. In this context, a wide range of carboxylic acids (from 1 to 5 carbon atoms) (150 mM) were transformed utilizing an appropriate decoy molecule (7.5 mM) [40]. It is important to notice that the group of acids tested was varied, involving linear and branched saturated acids as well as linear unsaturated ones. The reactions were performed at 30 °C in a Tris–HCl buffer solution (pH 8.5) with DMSO (20%) as a cosolvent, using *Cv*FAP (6 μM) as a photobiocatalyst. After three hours under irradiation with blue light (450 nm) and gentle magnetic stirring, excellent selectivity was achieved in most cases and acceptable TONs were obtained for certain substrates. Therefore, such an approach expands the substrates accepted by the enzyme toward the non-natural ones. More importantly, this experimental evidence will be useful to expand the potential biotechnological application of *Cv*FAP in the production of alkanes as biofuels.

12.2.5 Epoxidation

Two-component, diffusible flavin monooxygenases are enzymes able to catalyze different types of organic reactions, such as halogenations, hydroxylations, and epoxidations [44]. As their names indicate, these biocatalysts bear a reduced flavin cofactor (most commonly $FADH_2$) responsible for the catalytic activity. Precisely, the bound cofactor executes the reductive activation of molecular oxygen.

When these flavin monooxygenases are used to perform a chemical reaction, the reduced flavin cofactor is consequently oxidized. Regenerating the reduced cofactor when flavin-containing monooxygenases are used in isolated form requires two other enzymes: a NAD(P)H-reductase, which recycles the corresponding cofactor at the expense of reduced NAD(P)H, and another enzyme for regenerating the latter using an appropriate co-substrate. Overall, this approach seems too complicated, as too many reaction components are needed. In view of these limitations, some of the strategies being used to directly regenerate the flavin-reduced cofactor involve expensive transition metal catalysts or electrochemical methods [45–48]. Nevertheless, studies are being carried out on the use of light in cofactor recycling, leading to simpler, greener, and more inexpensive approaches [22, 49].

Styrenes can be enantioselectively oxidized into the corresponding epoxides by styrene monooxygenase (StyM) from *Pseudomonas* sp. VLB120 (Scheme 12.13). The enzyme consumes O_2, introducing one oxygen atom in the organic molecule and generating H_2O as a second product. In this approach, styrene and a wide range of its substituted derivatives (5 mM), both in the terminal double bond and in the aromatic ring, were evaluated in this epoxidation process. Thus, using StyM (5.3 μM) as a catalyst, FAD (200 μM), EDTA (20 mM) as a sacrificial electron donor, and catalase (600 U ml^{-1}). This latter enzyme is added to the system in order to consume any H_2O_2 formed by uncoupling that can exert deleterious effects on the biocatalytic system [25]. The reactions were conducted in KPi buffer (pH 7.0) with DMSO (1.25% v/v) as a cosolvent at 35 °C under magnetic stirring at 300 rpm and white light irradiation (400–700 nm) intensity of 40%. After one hour of reaction, excellent enantiomeric excesses were achieved (>95%). However, utilizing this reaction setup, the substrate conversion stopped after one hour of irradiation. The photostability of FAD is likely limiting the photoenzymatic process. Nevertheless, further investigation should be done on this reaction system to be applied to preparative scale.

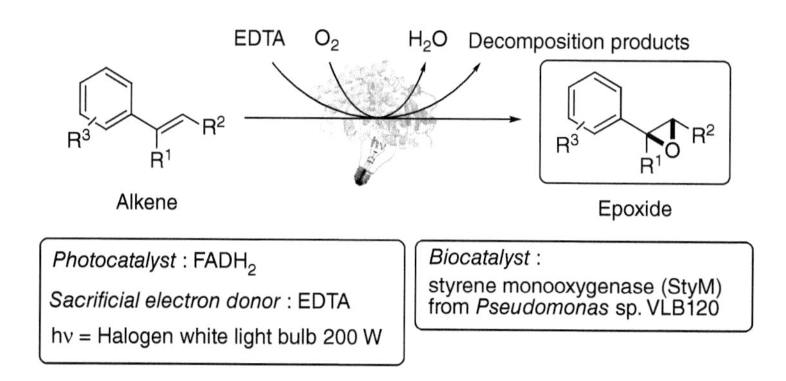

Scheme 12.13 Photobiocatalytic epoxidation of styrenes.

Another interesting system to photocatalytically epoxidize styrene derivatives and other olefins (cyclohexene and cyclooctene) was recently reported [50]. Nonetheless, for certain substrates, a mixture of oxidation products (diols and aldehydes) was observed, along with the corresponding epoxide. This approach involves a coupled system of the PC $[Ru(bpy)_3]^{2+}$ and the enzyme laccase LAC3 from *Trametes* sp. C30. In this process, the PC performs the epoxidation of the substrate introducing an oxygen atom from molecular oxygen. The reduced PC is oxidized again to the active Ru(III) complex by the action of the laccase. Thus, the biocatalyst transfers electrons received from the photocatalytic cycle to molecular oxygen, producing water as a product. Using molecular oxygen as a final electron acceptor is a highly promising feature of this approach, paving the way to greener oxidation processes of other organic molecules.

12.3 Reductive Processes

12.3.1 Carbonyl Reduction

Regarding stereoselective synthesis of chiral alcohols, enzymes are definitely an attractive option. Particularly, when it comes to reductions of carbonyl groups, nicotinamide-dependent dehydrogenases are often the preferred catalysts. Among these, alcohol dehydrogenases (ADHs, also called ketoreductases, KREDs) have been extensively studied for synthetic applications [48]. In the reduction process, these enzymes catalyze a reversible hydride transfer from a reduced nicotinamide cofactor NAD(P)H to the carbon atom in the carbonyl group, affording the corresponding alcohol and the oxidized cofactor NAD(P)$^+$ (Scheme 12.14) [51]. Stoichiometric amounts of the reduced cofactor would be required for the ADH-catalyzed reduction, thus representing a high economic cost. In this context, efficient strategies for *in situ* cofactor regeneration have been developed over the years [51]. One method implies direct cofactor regeneration using visible light. However, this involves two single-electron-transfer processes that generate biocatalytically inactive NAD(P)H isomers and dimers [52, 53].

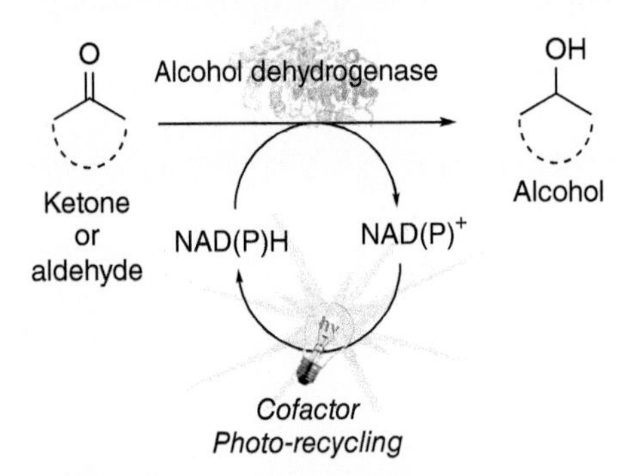

Scheme 12.14 General scheme of a photobiocatalytic carbonyl reduction.

To circumvent this, an effective relay system able to take advantage of these single-electron events for the selective hydride transfer is needed. In this line, a biocatalytic approach can be carried out using flavins, catalyzing pyrimidine cofactor reduction via a hydride transfer mechanism, whereas oxidized flavin is reduced by consecutive SET processes. Thus, the reduced flavin prosthetic group is photocatalytically regenerated using visible light at expenses of a sacrificial donor (Scheme 12.15).

In 2016, Brown and coworkers developed a photobiochemical strategy for NADPH regeneration employing NADPH-specific ferredoxin-NADP$^+$ reductase (FNR) from *Chlamydomonas reinhardtii*, CdSe QDs as a PC, and ascorbic acid as a sacrificial donor [54]. In this study, reduction of aldehydes (10 mM) was evaluated employing QDs (1.2 μM), FNR (0.6 μM), NADP$^+$ (0.25 mM), *Thermoanaerobium brockii* ADH (*Tb*ADH, 1 unit), and ascorbic acid (100 mM). Under 405 nm LED light irradiation, the corresponding alcohol was obtained with ~20% of conversion after three hours (Scheme 12.15). Although QDs capped with mercaptopropionic acid have a preference for ferredoxin-binding site in FNR,

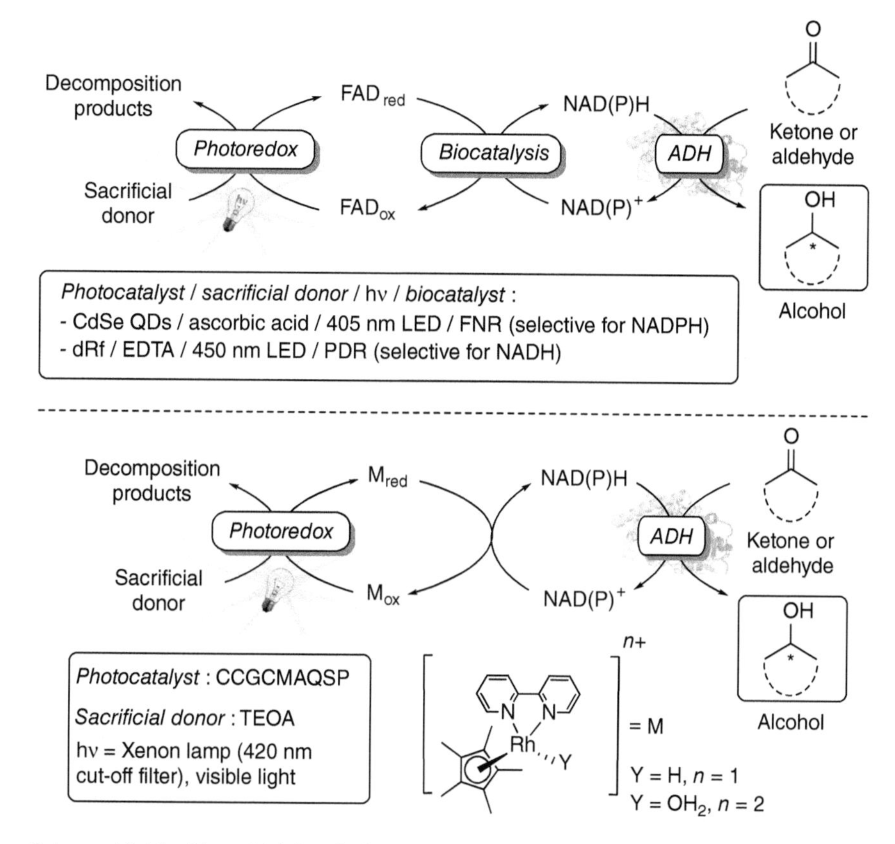

Scheme 12.15 Photo(bio)chemical regeneration of reduced cofactor for nicotinamide-dependent alcohol dehydrogenases.

TOF for NADPH regeneration ($1440\,h^{-1}$) is significantly lower than the TOF for FNR with natural mediator ferredoxin ($3.2\times10^5\,h^{-1}$), indicating a deficiency in electron transfer process into FNR. Complementary to that, and in order to expand the use of this photobiochemical regeneration system, the synthesis for isobutanol was targeted. This was achieved by a bienzymatic cascade. First, 2-ketoisovalerate was decarboxylated in a biocatalytic process employing a keto acid decarboxylase (KDC) from *Lactococcus lactis*. Then, the intermediate (isobutyraldehyde) was reduced to isobutanol by ADH, with a similar photobiochemical recycling of reduced cofactor NAD(P)H. For this enzymatic cascade, similar results were obtained, demonstrating the compatibility of the light-driven cofactor regeneration with multienzymatic systems. For this strategy, some practical considerations have to be taken into account. As far as QDs are concerned, both photooxidation and passivation of the catalyst might be promoted by oxygen and water; thus, inert atmosphere is mandatory [55]. However, aqueous medium, convenient for enzymatic reactions, may also influence photoinactivation of QDs. Because of this instability of QD-FNR complex (~3 hours), longer reaction times did not improve the conversion. Besides, QDs could produce singlet oxygen ($^1O_2$) from energy transfer to molecular oxygen, which has a deleterious effect in enzyme activity. Even traces of O_2 are able to diminish significantly the overall reaction, requiring the use of a glove box to prevent it. In order to achieve better results, further efforts have to be made for improving QD stability.

In addition to light-driven NADPH regeneration for carbonyl reductions with NADPH-dependent ADHs, in 2018, Hollmann's group developed an analog system for NADH regeneration in ADH-mediated bioreduction of ketones [56]. In this case, NAD-dependent putidaredoxin-NAD^+ reductase (PDR) from *Pseudomonas putida* was the biocatalyst of choice for cofactor regeneration and deazariboflavin (dRf) as a PC using EDTA as a sacrificial electron donor (Scheme 12.15). Under optimized conditions, ketone (10 mM) was tested in Tris–HCl buffer (50 mM, pH 8.0) with EDTA (20 mM), dRf (60 μM), PDR (5 mM), NAD^+ (0.2 mM), and ADH from *Rhodococcus ruber* (0.115 μM), employing a 450 nm LED in strict anaerobic conditions (glove box). After five hours of irradiation, less than 10% of conversion was obtained in the best case. In order to improve the electron relay into PDR-flavin, methyl viologen (MV^{2+}) was tested, considering its previous use as a mediator for FNR. In this way, conversion went up to 31% after five hours of irradiation. Remarkably, excellent enantiomeric excesses were obtained (96–99%), showing that the stereochemical outcome of the bioreduction by the ADH was not affected by the cofactor recycling promoted by light.

It is worth mentioning some considerations on this system. Particularly, when traces of O_2 are present, even in anaerobic conditions, studies on cofactor photoreduction showed that there is a significant decrease in NADH concentration as well as an irreversible inhibition of the PC. This drawback can be avoided by employing a glove box, in which case, stable NADH accumulation was observed at 24 hours. Furthermore, NADH concentration is dependent on PDR and dRf concentrations. Nevertheless, either in the absence of PDR or high quantities of dRf, significant nonselective reduction of the cofactor is observed. This may cause significant formation of biocatalytically inactive NADH isomers and/or NAD dimers.

Another consideration is the likely photoinactivation of PDR under blue light irradiation. Increasing photoexcitation wavelength seems to improve the stability for photoenzymatic processes. The reaction using 514 nm LED excitation was tested, employing formate dehydrogenase from *Methylobacterium extorquens* and Safranine O instead of PDR and dRf, respectively. Although conversion was lower than the PDR-dRF system (~4% after 5 hours), after 140 hours, continuous product formation was still observed. Hence, further efforts need to be spent on tuning up PCs in order to avoid high energy photoexcitation wavelengths deleterious for enzymatic activity.

The above-discussed examples comprise the use of specific enzymes for cofactor regeneration. The effectivity of this process partly relies on the selectivity over the oxidized cofactor. This is crucial for the prevention of undesired and inactive cofactor species that limit and inhibit the regeneration system. In addition to an enzymatic approach, selective formal hydride transfers to $NAD(P)^+$ can be achieved using metal complexes, which in turn could be regenerated by light (Scheme 12.15). In this line, Yadav's group has developed an artificial cofactor regeneration system involving pentamethylcyclopentadienyl rhodium bipyridine, $[Cp^*Rh(bpy)(H_2O)]^{2+}$ (M) as an electron mediator and a hydride transfer catalyst, along with substituted graphene as a PC and TEOA as a sacrificial electron donor. This system was coupled with a NADPH-dependent ADH from *Lactobacillus kefir* ADH (LKADH) for the enantioselective reduction of prochiral ketones (Scheme 12.15) [57, 58].

Reactions were carried out employing acetophenone derivatives (20 mM) in buffer (pH 7.0, 0.1 M) with $NADP^+$ (1 mM), M (0.2 mM), LKADH (1–2 U), TEOA (0.4 M), and 0.5 mg of chemically converted graphene coupled with multi-anthraquinone substituted porphyrin (CCGCMAQSP) as a PC. The mixtures were placed in quartz reactors using visible light irradiation (with a 420 nm cutoff filter) under argon atmosphere (Scheme 12.15). Moreover, addition of an immiscible cosolvent improved the conversion, resulting in a biphasic medium. The best results were obtained with 20% of *n*-hexane or *n*-heptane. The corresponding alcohols were obtained with moderate conversion (20–42%) and excellent enantioselectivity (up to >99% ee) after 50 hours of irradiation. In addition, heteroaromatic and aliphatic ketones were tested changing the PC to a chemically converted graphene (CCG) substituted with triazine-linked 4,4′-difluoro-4-bora-3*a*,4*a*-diaza-*s*-indacene (BODIPY) moieties. Similar results were observed, affording moderate conversion for chiral alcohols (30–45%) and also excellent enantioselectivity (up to >99% ee), showing that light-driven regeneration under these conditions did not affect the stereochemical outcome of the reaction.

In both cases, similar TOFs for cofactor recycling (15.3 h^{-1} for CCGCMAQSP and 7.4 h^{-1} for CCG-BODIPY) were observed. For this strategy, because of long reaction times, deleterious effects on enzymatic efficiency and PC stability may occur, diminishing conversions. In consequence, either further addition of LKADH or the PC may increase alcohol production.

On the other hand, flavin-dependent ene-reductases (EREDs) are enzymes that catalyze the reduction of activated C=C bonds with high enantioselectivity. However, a promiscuous catalytic activity can be achieved by merging with photoredox catalysis, as has been demonstrated by Hyster's group in 2019 [59]. By means of this strategy, aromatic ketones could be asymmetrically reduced into the corresponding chiral alcohols. Another interesting feature is that the reaction occurs through a radical pathway, and not through a hydride transfer mechanism from the flavin hydroquinone (FMN_{hq}) cofactor [60–62]. The biocatalyst employed in this study was morphinone reductase from *P. putida* (MorB), whereas $Ru(bpy)_3Cl_2$ was used as a PC (Scheme 12.16).

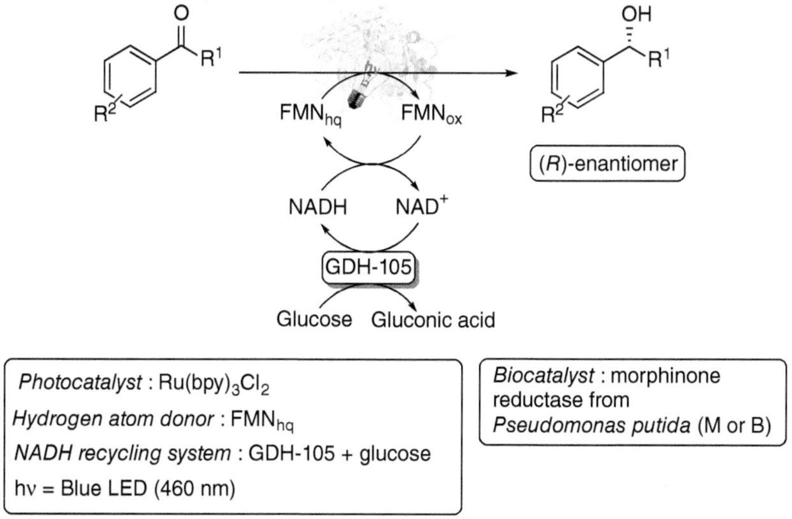

FMN_{hq} FMN_{ox} (*R*)-enantiomer NADH NAD^+ GDH-105 Glucose Gluconic acid

Photocatalyst : $Ru(bpy)_3Cl_2$
Hydrogen atom donor : FMN_{hq}
NADH recycling system : GDH-105 + glucose
hν = Blue LED (460 nm)

Biocatalyst : morphinone reductase from *Pseudomonas putida* (M or B)

Scheme 12.16 Promiscuous ERED photobiocatalytic enantioselective reduction of aromatic ketones.

Particularly, this reduction is feasible because of substrate activation in the enzyme's active site. This activation implies a hydrogen bonding between the carbonyl oxygen atom and two enzyme residues (hystidine/asparagine pair), reducing the redox potential of the substrate. For this approach, the reaction mixtures were prepared in an anaerobic chamber. The vials contained the aromatic ketones (0.01 mmol, introduced from a degassed DMSO solution), $Ru(bpy)_3Cl_2$ (0.08 mg), MorB (1 mol%), NAD^+ (2 mol%), glucose dehydrogenase (GDH)-105 (0.25 mg), and glucose (9 mg) in a degassed phosphate buffer solution (pH 8.0). After being sealed, the vials were taken out from the chamber and then placed on a shaker and irradiated with blue LEDs (460 nm) for 24 hours. This strategy permitted to transform acetophenone and *meta* and *para* substituted derivatives attaining moderate to high yields (50–99%) and good to very good enantiomeric excess, favoring the (*R*)-enantiomer (from 50% to 76% ee). Nevertheless, when *ortho*-methylacetophenone was tested, a low yield (15%) and poor enantiomeric excess were achieved (34% ee). Replacing the benzene ring by a naphthalene ring, high yield (90%), and good enantiomeric excess (76%) was obtained, although longer reaction time was required (48 hours instead of 24 hours). Likewise, replacing the methyl group of the original starting materials by bulkier groups, moderate yields, and poor enantiomeric excess were obtained.

A fundamental feature of this system is that it can be employed to reduce both the double bond and the carbonyl group in α,β-unsaturated acetophenones. Under light irradiation, MorB shows both the promiscuous activity (C=O reduction) and its natural one (C=C reduction). For instance, when 2-methyl-1-phenylprop-2-en-1-one was evaluated under the detailed reactions conditions, enantioenriched 2-methyl-1-phenylpropan-1-ol was obtained, achieving very good yield and good enantiomeric excess (89%, 64% ee). Mechanistically, the catalytic cycle goes through a SET-HAT-SET sequence.

12.3.2 Olefin Reduction

EREDs have proven to be reliable and effective enzymes for the production of valuable building blocks by means of the asymmetric bioreduction of alkenes bearing an electron-withdrawing group (EWG) (Scheme 12.17) [61]. These enzymes act by transferring a hydride – from a flavin group in the active site – and one proton in a Michael-type hydride addition/enolate protonation sequence onto the activated double bond. Because of its nature, as nicotinamide-dependent flavoproteins, stoichiometric amounts of the expensive reduced cofactor [NAD(P)H] are required for the regeneration of the catalytically active enzyme-bound flavin ($FMNH_2$). In addition to traditional approaches for nicotinamide

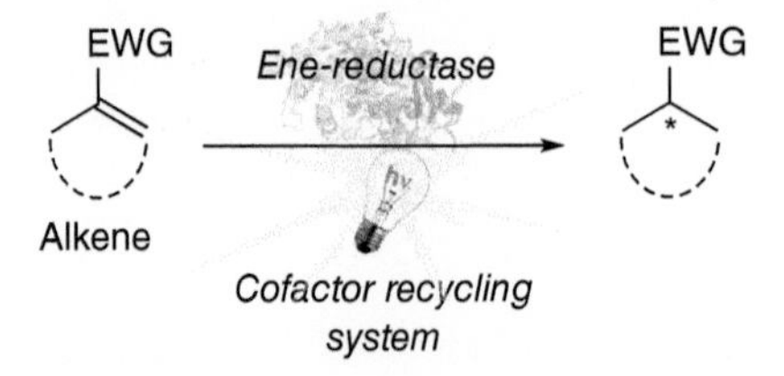

Scheme 12.17 Asymmetric bioreduction of activated alkenes by ene-reductases under photocatalytic cofactor recycling.

cofactor regeneration, some NAD(P)H-free systems have been studied. In this context, photochemical methods that might produce two efficient, consecutive SET processes toward the enzyme-bound flavin, are promising strategies for ER-$FMNH_2$ regeneration.

In 2009, Hollmann's group studied one of the first examples for the enzyme-bound flavin regeneration promoted by light, involving ERs [63]. In this study, the activated alkene (10 mM) was added to a potassium phosphate buffer solution (10 ml, 100 mM, pH 7.0) along with FMN (100 μM), YqjM (0.914 μM, the old yellow enzyme from *Bacillus subtilis* E.C. 1.6.99.1), and 25 mM EDTA as a sacrificial electron donor. Under irradiation with a 250 W halogen bulb under nitrogen atmosphere, tested alkenes were fully converted (Scheme 12.18).

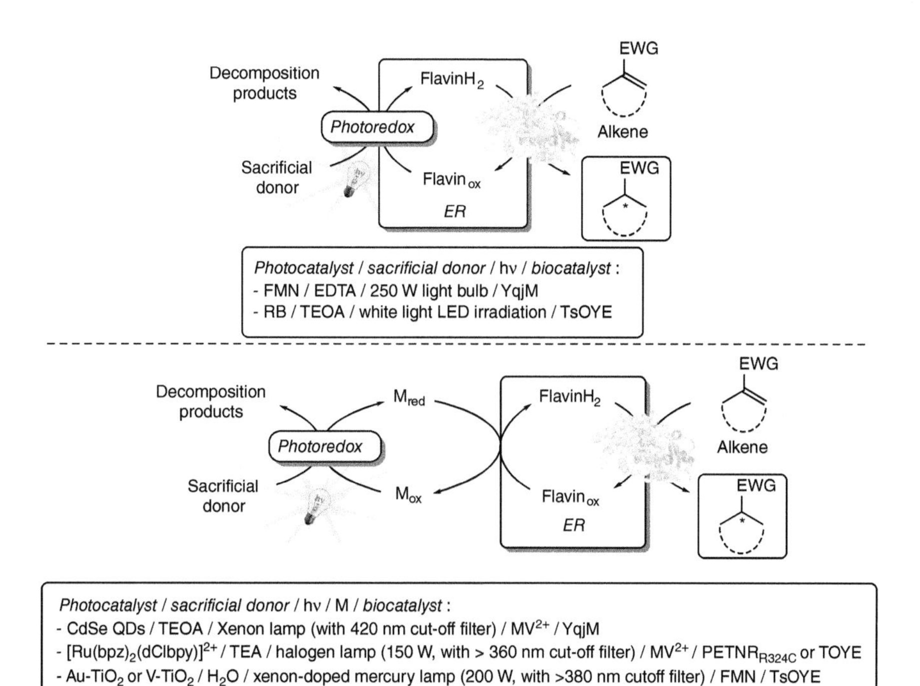

Scheme 12.18 Photoregeneration of reduced flavin in ene-reductase-catalyzed alkene reductions.

Lower reaction rates (almost half in comparison with a 250 W bulb) were obtained using a 40 W lamp. On the other hand, although the stereochemical outcome was not excellent (69% ee), practically the same results were observed with stoichiometric amounts of NAD(P)H (72% ee). This indicates that the observed selectivity is totally imparted by the enzyme and is not affected by the photochemical reestablishment of the ER-flavin. It is worth mentioning some considerations on this method. First of all, anaerobic conditions are mandatory for a correct functioning of the PC. Small amounts of O_2 lead to quenching of reduced flavin species generating H_2O_2. Therefore, degassing the reaction mixture is needed to enhance reaction rates. Secondly, employing other flavin-containing PCs, such as FAD or riboflavin, slightly changes both conversion and enantioselectivity. Interestingly, using an O_2-tolerant deazaflavin, a complete loss of enantioselectivity was detected, which suggests a nonenzymatic reduction of the alkene. Thirdly, protein purification is recommended because of the presence of endogenous ERED in *E. coli*, which affect negatively the ee with YqjM. However, this spurious ERED influences at a considerable extent, only at low concentrations of FMN. Last but not least, EDTA is shown to be the best sacrificial donor for this system. Other sacrificial donors give place to lower reaction rates and enantioselectivity.

Another approach to regenerate the flavin in the active site was carried out using organic dyes. These inexpensive PCs exhibit good photochemical properties because of the efficiency of intersystem crossing, which involves a transition between the excited singlet state and the triplet state. In 2017, Park's group studied the light-driven regeneration of the enzyme-bound flavin of EREDs with RB [64]. Specifically, bioreductions employing an ER from *Thermus scotoductus* (TsOYE) were promoted with RB under white light LED irradiation and TEOA as an electron source. Besides the observed association of RB with TsOYE by UV–vis spectroscopy, the formation of a flavin-semiquinone intermediate because of the first SET was also confirmed, as well as the reduced flavin.

Furthermore, the addition of RB produced a negative shift of 60 mV on the reduction potential of TsOYE. This result suggests that the association of RB and TsOYE presents a co-reducing property acting as an electrochemically hybridized species. For the light-driven enzymatic reduction, studies were carried out employing 2-methylcyclohex-2-enone (8 mM) in TEOA buffer (200 mM, pH 7.5), along with RB (50 μM), TsOYE (18 μM), and $CaCl_2$ (10 mM) under white light LED irradiation (Scheme 12.18). RB-sensitized bioreduction of 2-methylcyclohex-2-enone resulted in a 90% yield of (*R*)-2-methylcyclohexanone, with an excellent enantioselectivity (>99% ee) and TsOYE TOF of 256 h^{-1}. Regarding the PC, erythrosine B and phloxine B also showed activity toward the substrate, although lower TOF values for these organic dyes were observed (100 and 78 h^{-1} respectively). When changing the PC to FMN, only a TOF of 4.8 h^{-1} and 4.1% yield were obtained. Concerning amounts of enzyme and RB, reaction rates reached a saturation at a RB/TsOYE ratio of ~5. Remarkably, this system was extended to the ER from *B. subtilis* YqjM, whereas comparable results to TsOYE were obtained.

In 2012, Lutz's group explored the photoenzymatic regeneration of the FMN in the ERED of *B. subtilis* YqjM employing QDs, visible light, and TEOA as a sacrificial electron donor [65]. By photoexcitation of the QDs, an electron is promoted from the valence band toward the conduction band. This excited electron can be transferred directly to the enzyme-FMN or a mediator, which subsequently will transfer this electron toward the flavin. In fact, a formal two-electron transfer process is required for generating the catalytically active species in the ERED.

First studies were aimed to generate reduced FMN under inert atmosphere. Besides confirmation of the QD-flavin adduct, it did not produce any accumulation of the reduced cofactor. Thus, a mediator acting as an electron relay was needed. In this context, employing methyl viologen (MV^{2+}), complete photochemical reduction of FMN was obtained, which indicates that two consecutive SET processes successfully occur toward the oxidized cofactor. Coupling with the enzymatic reduction was carried out using ketoisophorone (200 μM), YqjM (3 μM), MV^{2+} (250 μM), CdSe QDs (10 μM), and TEOA (200 mM) in a phosphate buffer solution (200 mM, pH 7.2). After 60 min of irradiation with a Xe lamp (with 420 nm cutoff filter), 11% of conversion was detected, and no further conversion was achieved beyond this point (Scheme 12.18). Enantiomeric excess for the corresponding (*R*)-levodione was not perfect (66%); however, authors attribute this value to the presence of both enantiomers of levodione in the starting material.

Although conversion resulted fivefold lower than GDH-mediated NADPH-regeneration system, these conversions were slightly better than $[Ru(bpy)_3]^{2+}/MV^{2+}$-mediated processes. Further studies were carried out to elucidate the rate-limiting step of the reaction. Transient absorption spectroscopy showed that TEOA was an inefficient sacrificial donor, leading to small amounts of the active MV^{+},·thus limiting the reaction rate. Noteworthy, EDTA/$[Ru(bpy)_3]^{2+}$ produced MV^{+} 40% more than TEOA/$[Ru(bpy)_3]^{2+}$, however, EDTA was incompatible with QDs because of precipitation of this PC.

In 2016, improvement of a Ru-catalyst for light-driven bioreduction using ERs was explored by Scrutton's group [66]. For the enzymatic bioreduction, a series of activated alkenes were subjected to a mutant of pentaerythritol tetranitrate reductase ($PETNR_{R324C}$) from *Enterobacter cloacae* and the thermophilic Old Yellow Enzyme (TOYE) from *Thermoanaerobacter pseudoethanolicus*. Then, the alkene (25 mM) was directly added from a stock solution in an organic solvent (20% v/v) to a trimethylamine (TEA) buffer solution (50 mM, pH 8.0), along with the ER (10 μM), $[Ru(bpz)_2(dClbpy)]^{2+}$ (20 μM), and MV^{2+} (0.1 mM). After irradiation with a 150 W halogen lamp (360 nm cutoff filter) for 24 hours at room temperature under inert atmosphere, excellent results were obtained concerning conversion (up to >99%) and enantioselectivity (>99% ee) (Scheme 12.18). Remarkably, in every case, photochemical regeneration of the cofactor achieved the same or even better results regarding conversion than traditional G6PDH-mediated NADPH regeneration, although in detriment of the enantioselectivity for some substrates. It is worth noticing that the organic co-solvent has an important influence on the outcome of the reaction: higher yields were obtained in biphasic systems for some alkenes and even an improvement in enantiomeric excess. In addition, because of activity loss for $PETNR_{R324C}$, the use of *tert*-butyl methyl ether has to be avoided.

In another study, to circumvent the issues related to sacrificial electron donors (i.e. additional waste by-products, formation of harmful species for the enzymatic activity), reducing equivalents obtained by light-driven water splitting were coupled to ERs. Through photoexcitation of a doped TiO_2, electrons from the conduction band can reduce FMN by SETs. This reduced mediator acts as an electron relay, resulting in a hydride transfer toward the enzyme-bound flavin [10]. For the proof-of-concept, the reaction was carried out employing ketoisophorone (10 mM) in MOPS buffer (50 mM, pH 7.5, 10 mM $CaCl_2$), Au-TiO_2 (10 g l^{-1}), FMN (200 μM) as a mediator, and TsOYE (15.2 μM) under UV light irradiation (Xe-doped Hg lamp, 200 W) at 50 °C under argon atmosphere (Scheme 12.18). It has to be noted that the addition of the enzyme and FMN was performed after purging with argon the remaining components. After six hours of irradiation, 66% of conversion and 86% ee (up to 94%) to (*R*)-levodione was obtained.

Confirmation of the recycling system was carried out by monitoring the production of $^{18}O_2$ starting from $H_2{}^{18}O$ as a sacrificial electron donor. Noteworthy, in the absence of FMN and TsOYE, identical conversion was detected but the racemic product was formed. These results suggested that, on the one hand, a background nonenzymatic reduction competed with the bioreduction, being the reaction rate for TsOYE reduction significantly higher. On the other hand, Au-TiO_2-catalyzed water splitting was the overall rate-limiting step.

Changing to V-TiO_2, a more efficient catalyst for water oxidation, the reaction rate increased almost twice, although, in hand with a slight erosion of enantioselectivity (80% ee). Also, V-TiO_2 absorbs light in the visible light range, allowing water oxidation under less energetic irradiation. By the addition of a cutoff filter ($\lambda > 380$ nm) to the light source, after nine hours of irradiation, 50% yield and >85% ee were detected, paving the way for further applications (Scheme 12.18). It should be noted that the production of O_2 by water splitting did not affect to a significant extent the enzymatic activity because during nine hours of irradiation, continuous production of (*R*)-levodione was observed.

As mentioned, the use of natural nicotinamide cofactors in enzymatic reactions is often avoided because of its high cost and chemical stability issues. From this perspective, synthetic nicotinamide cofactor analogs (mNADHs) have emerged as an encouraging replacement by virtues such as lower cost, enhanced stability, and promising biocatalytic efficiency [67]. With this in mind, in 2018, Park's group developed a photocatalytic method for light-driven regeneration of mNADHs coupled with an ERED-mediated bioreduction, employing TEOA as a sacrificial electron donor [68]. In this work, upon light irradiation, a nitrogen-doped carbon nanodot (N-CD) was employed as a PC, delivering electrons to a mediator. This mediator subsequently transfers the electrons by SET toward the oxidized $mNAD^+$, forming the required mNADH for the enzymatic reduction (Scheme 12.19).

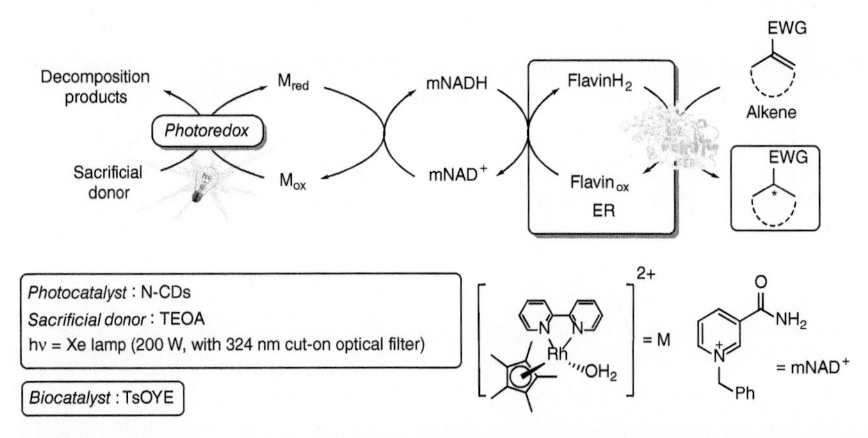

Scheme 12.19 Photoregeneration of reduced flavin using metal complexes and cofactor analogs as electron relays.

In this direction, in the presence of 1-benzyl-3-carbamoylpyridinium bromide ($mNAD^+$), regioselective reduction was accomplished using $[Cp^*Rh(bpy)H_2O]^{2+}$ as a mediator (M), N-CD as a PC, and TEOA as a sacrificial electron donor upon 200 W Xe lamp equipped with a 324 nm cut-on optical filter. With this result in hand, the light-driven reduction of 2-methylcyclohex-2-en-1-one (7 mM) was catalyzed by the ERED from *T. scotoductus* (9 μM, TsOYE), along with $(Cp^*Rh\text{-}(bpy)H_2O)^{2+}$ (M, 25 μM), $mNAD^+$ (2 mM), N-CDs (0.10 mg ml^{-1}), and $CaCl_2$ (5 mM) in a TEOA buffer (500 mM, pH 7.5) at 45 °C. The reaction was irradiated for 180 min, achieving excellent yield and high enantioselectivity (86% and 93% ee, respectively), and by doubling the quantity of N-CDs, a perfect yield was obtained (>99%, TsOYE TOF 576.3 h^{-1}). It worth mentioning that the same ee was obtained with the natural cofactor NADH. Besides, *trans*-cinnamaldehyde was also tested and good results were obtained after 75 min of irradiation (44% yield).

Previously commented examples depend on the capability of an external mediator to act as an electron shuttle, transferring two electrons to the flavin prosthetic group in the enzyme. In this context, some flavocytochromes bear certain specific mediators for the external injection of electrons onto the catalytically active flavin. These enzymes have heme groups that work as electron acceptors, subsequently transferring those electrons to the enzyme-bound flavin [69]. Taking advantage of this feature, direct irradiation may lead to the reduction of the flavin in the active site through SET processes between different parts of the enzyme.

In 2014, Armstrong's group reported a proof of concept for a photocatalytic regeneration system for the flavocytochrome c_3 (fcc3), a fumarate reductase from *Shewanella frigidimarina* NCIMB400 (Scheme 12.20) [70].

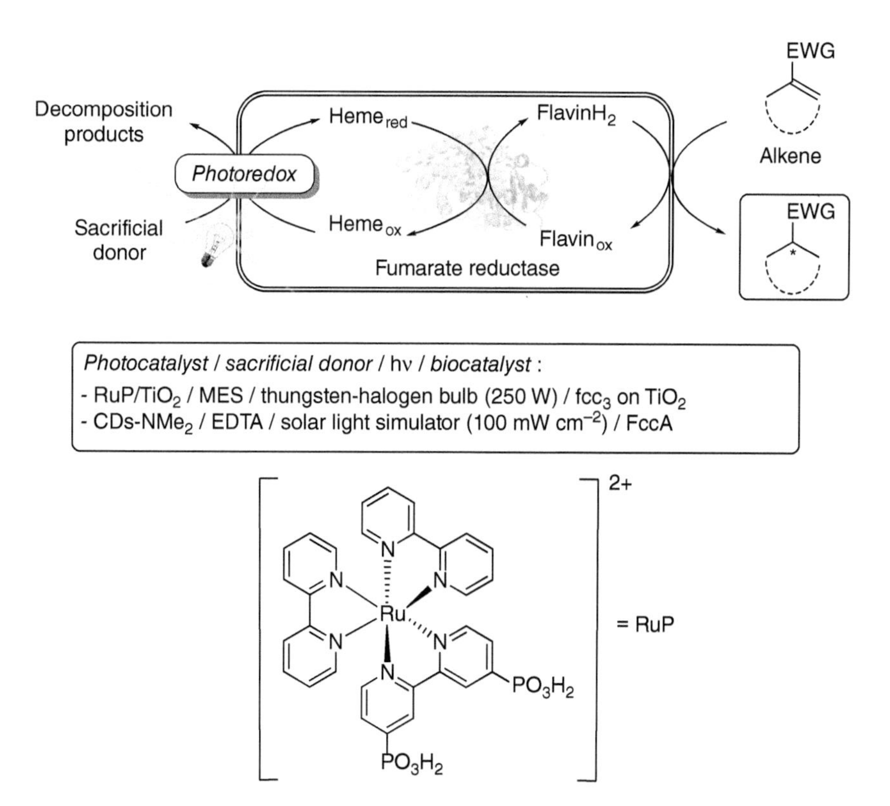

Scheme 12.20 Flavocytochrome c3-catalyzed alkene reduction employing photoredox approach for cofactor regeneration.

This enzyme contains four heme groups and a flavin (FAD) in the catalytic site. If an electron transfer from an excited PC to a semiconductor takes place, the injected electron in the conduction band can then be transferred to the heme group of the enzyme. A sacrificial electron donor is also needed for reestablishment of the PC. In this study, fcc_3 was adsorbed on a TiO_2 surface, along with a Ru-based PC $[Ru^{II}(bpy)_2(4,4'\text{-}(PO_3H_2)_2bpy)]$

Br_2 (RuP). In this line, fcc_3 (1.95 nmol) and RuP (56 nmol) were co-adsorbed on TiO_2 (5 mg). This solid phase was next suspended in 5 ml of 2-(*N*-morpholino)ethanesulfonic acid (MES) buffer (0.2 M), also working as a sacrificial electron donor, also containing fumarate (6 mM). It must be noted that this procedure was performed in an anaerobic glove box. After four hours of illumination with a 250 W tungsten-halogen bulb, 37.5% of conversion was observed (enzyme TOF 720 h^{-1}) (Scheme 12.20). Furthermore, additional irradiation for four hours (eight hours total) did not produce any significant increase in conversion. Loss of pink color (related to fcc_3) after illumination demonstrates enzyme photodegradation.

Through an analogous approach, Reisner's group studied in 2016 a simpler photoregenerating system for a fumarate reductase from *Shewanella oneidensis* MR-1 (FccA) [71]. This work comprises the photoexcitation and SET processes from functionalized carbon dots (CDs) toward FccA, at the expense of EDTA as a sacrificial electron donor. The best results were obtained using tertiary amine-capped CDs (CD-NMe_2). Conditions for 10 mM of fumarate bioreduction were CDs (1 mg), FccA (0.22 nmol), EDTA (0.1 M) in D_2O, under N_2 atmosphere, and irradiated by a solar light simulator at 100 mW cm^{-2} (Scheme 12.20). Under this setup, FccA-TOF for succinate synthesis was 1700 h^{-1}. Remarkably, the photocatalytic system remained active after 24 hours of irradiation, although a decrease in reaction rate was also observed during the irradiation course.

12.3.3 Imine Reduction

Certain 2-substitued-1-pyrrolines (imines) can be transformed into the corresponding amines with high enantioselectivity using a two-step, one-pot photobiocatalytic cascade (Scheme 12.21). Initially, the imine is reduced in a light-driven process to afford a racemic mixture of amines. Then, only one of the enantiomers is reoxidized to the original imine. As both reactions occur simultaneously in a cyclic manner, an enantioenriched mixture of amines is achieved. This cyclic deracemization has been carried out in water for the first time, avoiding the use of organic solvents and being therefore greener [72]. This cascade is depicted in detail in Scheme 12.22. Firstly, a photoredox-catalyzed reduction of the substrate takes place (step **I**). The water-soluble PC used ($Na_3[Ir(sppy)_3]_3$) is excited under irradiation of violet light (405 nm), allowing SET to the imine substrate, forming after protonation, a C2-centered α-amino alkyl radical. This highly reactive intermediate reacts with ascorbic acid through hydrogen atom transfer (HAT), yielding a racemic mixture of amines

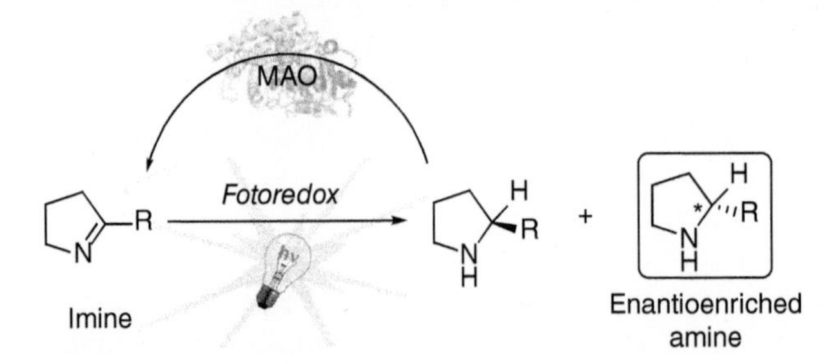

Scheme 12.21 Photobiocatalytic reduction of imines into enantioenriched amines featuring a cyclic deracemization.

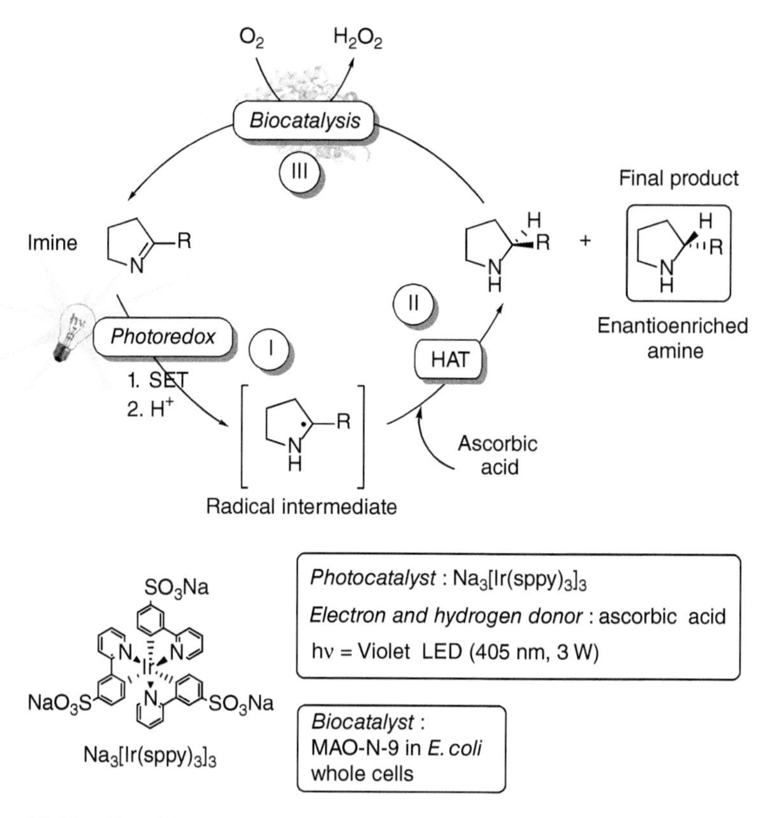

Scheme 12.22 Photobiocatalytic cascade for the asymmetric amine synthesis in a cyclic deracemization-like process.

(step **II**). It is worth mentioning that the hydrogen donor must be carefully selected, as it is fundamental to rapidly intercept the radical intermediate. Finally, a concurrent biocatalytic process oxidizes the (*S*)-enantiomer to the initial imine (step **III**). The selectivity is due to the use of *E. coli* whole cells overexpressing a monoamine oxidase (MAO-N-9). During this enantioselective oxidation, the enzyme consumes molecular oxygen, generating H_2O_2 as a product. It must be noted that the enzyme has to be maintained inside the bacteria cells because of a partial inactivation in the presence of the PC anion $Ir(sppy)_3^{3-}$, which is not able to cross the cell membrane because of its charge and high molecular weight. The cyclic reaction network was carried out in 1 ml of an aqueous phosphate buffer suspension (0.5 M, pH 8.0) containing the substrate (10 mM), $Na_3[Ir(sppy)_3]_3$ (0.1 mM), ascorbic acid (200 mM), and *E. coli* whole cells with MAO-N-9 (180 mg, wet weight).

1-Pyrrolines with a cyclohexyl or a butyl substituent in position 2 afforded an enantioenriched mixture of secondary amines with both excellent yield (≥95%) and enantiomeric excess (≥98% for the (*R*) or (*S*)-enantiomer – even though the enzyme always displays the same enantioselectivity, the product's configuration can vary according to the nature of substituents, following CIP priority rules). These results were obtained after 30 hours of reaction at room temperature and under air, using an orbital shaker (200 rpm). Moreover, when 1-pyrroline with a benzyl substituent was tested, an excellent yield was achieved

(>95%). Nevertheless, in this case, only 8% ee was obtained. Interestingly, this cyclic deracemization could be applied to an iminium salt even though it is a charged substrate. Using 2-cyclohexyl-1-methyl-1-pyrrolinium iodide as a substrate, the tertiary enantioenriched (*R*)-amine was as well formed with an excellent yield and enantiomeric excess (>95%, >99% ee). Additionally, by using a different setup (thiols as electron and hydrogen donors and $Ru(bpy)_3Cl_2$ as a PC), 1-methyl-3,4-dihydroisoquinoline could also be transformed, yielding the secondary (*R*)-amine with a very high yield (>95%) and a low-to-moderate enantiomeric excess (35% ee).

12.3.4 Reductive Amination

In 2008, Park's group developed one of the first approaches for the biosynthesis of L-glutamate employing a light-driven cofactor recycling (Scheme 12.23) [73]. The excitation of the PC $W_2Fe_4Ta_2O_{17}$ with visible light produces an electronic excitation toward the conduction band. Then, a Rh-based mediator ($[Cp^*Rh(bpy)H_2O]^{2+}$ = M) receives the excited electron (SET) and acts as an electron acceptor and then transfers the hydride to the oxidized NAD^+, generating the desired NADH. Afterward, this reduced cofactor is consumed by a glutamate dehydrogenase (GlDH), which converts α-ketoglutarate into L-glutamate using ammonium as a nitrogen source. The PC is reestablished at the expenses of a sacrificial electron donor such as EDTA. For this protocol, assays were performed employing $W_2Fe_4Ta_2O_{17}$ (5 mg), $[Cp^*Rh(bpy)H_2O]^{2+}$ (0.2 mM), NAD^+ (0.1 mM), and EDTA (5 mM) in

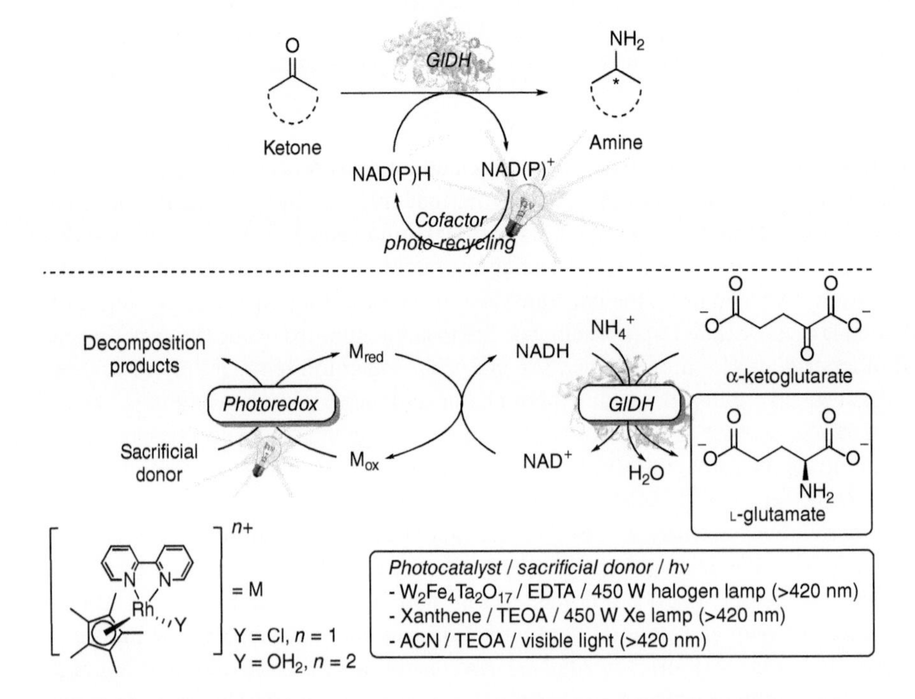

Scheme 12.23 Bioreductive amination of α-ketoglutarate coupled with different photorecycling cofactor systems.

a phosphate buffer (pH = 7.0, 100 mM) containing $(NH_4)_2SO_4$ (5 mM), α-ketoglutarate (0.1 mM), and GlDH (20 U). After 24 hours under irradiation with a 450 W halogen lamp ($\lambda \geq 420$ nm) and argon atmosphere, ~15% of conversion of α-ketoglutarate was achieved (Scheme 12.23).

Moreover, for a better SET from the PC toward the Rh mediator M, commercially available organic dyes were tested [74]. For this purpose, α-ketoglutarate (10 mM) was evaluated along with NAD^+ (50 mM), different xanthene dyes (20 mM), M (250 mM), ammonium sulfate (200 mM), and GDH (40 U), in phosphate buffer (0.1 M, pH 7.4) containing TEOA (15% w/v) as a sacrificial donor. Irradiation employing a 450 W Xe lamp equipped with a 420 nm cutoff filter, these TOFs were observed for the different PCs: phloxine B (1537.3 h^{-1}), RB (1270.8 h^{-1}), EY (1177.5 h^{-1}), and erythrosine B (1144.8 h^{-1}) (Scheme 12.23). In comparison with the previous work, α-ketoglutarate was fully converted in 1.5 hours using EY (in contrast to ~15% of conversion after 24 hours of irradiation using $W_2Fe_4Ta_2O_{17}$ as a PC). Lower conversions were found in the case of other xanthenes. The reaction time (1.5 hours) is convenient for EY because xanthene dyes are prone to photobleaching after long irradiation times. To improve the efficiency in this photocatalytic process, green LED should be tested, as EY absorbs strongly in this region of the visible spectrum.

Another example for the visible light-driven *in situ* regeneration of NADH was developed with amorphous carbon nitride (ACN, a heat-treated crystalline polymeric carbon nitride [CCN]) [75]. Coupling this recycling system with GlDH (30 U) was tested upon visible light irradiation ($\lambda > 420$ nm), using α-ketoglutarate (50 mM), ACN (2 mg ml^{-1}), M (250 μM), NAD^+ (1 mM), and $(NH_4)_2SO_4$ (100 mM) in a solution of TEOA 1 M (pH 7.5) (Scheme 12.23). Within 5 hours, 15.58 mM of L-glutamate and an enzyme TOF of 3637 h^{-1} were obtained. There are two very appealing features of such an approach: on the one side, even after eight cycles, ACN activity remained >90% as compared to the first cycle efficiency; on the other side, the PC remained active during dark periods in assays where light was alternatively switched on and off, increasing enzymatic conversion during light periods.

Although many alternatives for the reductive amination using GlDH and a photorecycling system for NADH were described here, it must be commented that, so far, the scope of this reaction is almost strictly limited to L-glutamate synthesis. Amines are very sensitive to one-electron oxidation processes. In fact, many sacrificial electron donors rely in this property (*i.e.* EDTA, TEA, and TEOA). Therefore, amines resulting from the reductive amination may suffer photodegradation under the reaction conditions. In the case of L-glutamate, product stability lies in the zwitterionic form, in which the protonation of the $-NH_2$ group protects the nitrogen atom from one-electron oxidation.

12.3.5 Dehalogenation

In 2016, Hyster's group has proven that some KREDs present a promiscuous activity under light irradiation, thus carrying out an asymmetric radical dehalogenation reaction of α-halolactones (Scheme 12.24) [76]. This is mainly based on the fact that NAD(P)H can act as a strong single-electron reductant upon photoexcitation [77–79]. Meanwhile, the enantioselectivity is due to a highly asymmetric HAT from radical cation NAD(P)H in the enzyme's active site. In this study, two enzymes were tested: KRED from *Ralstonia* sp. (*Ras*KRED) and KRED-P2D03 from *L. kefir* (*Lk*KRED). In both cases, the same wide range

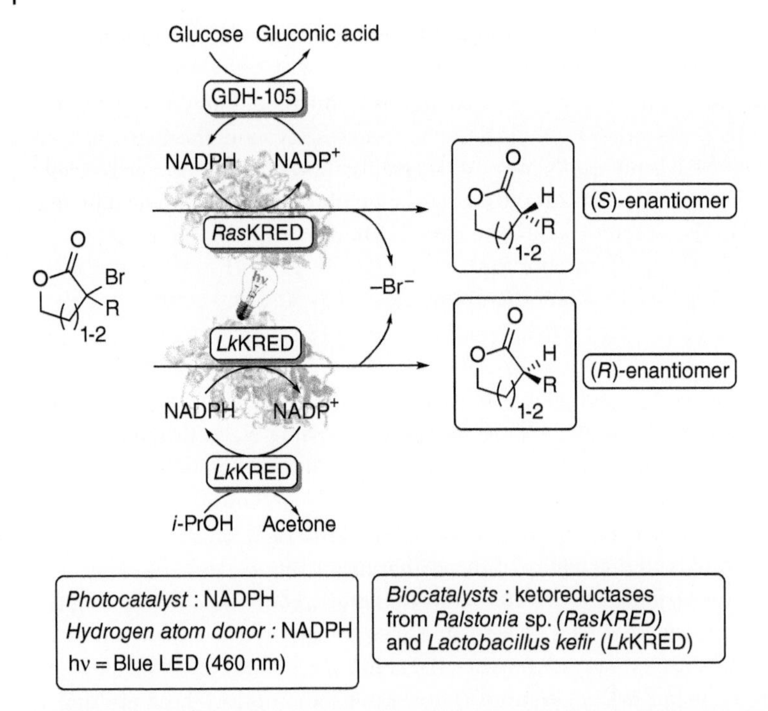

Scheme 12.24 Photobiocatalytic enantioselective radical dehalogenation of α-bromolactones.

of racemic substrates was tested, comprising mainly six-membered α-bromolactones bearing phenyl substituents at the α-position. On the one hand, *Ras*KRED yielded the corresponding (*S*)-enantiomer, whereas *Lk*KRED furnished the (*R*)-enantiomer (Scheme 12.24). When *Ras*KRED (1 mol%) was used as a biocatalyst, the reaction mixtures were prepared in an anaerobic chamber using the α-bromolactones (0.039 mmol, introduced from a degassed DMSO solution), $NADP^+$ (0.0312 μmol), commercially available GDH-105 (5 mg), and glucose (200 mM). The solvent employed was a 50 mM TRIS/$CaCl_2$/10% glycerol buffer solution (pH 7.0). Vials were sealed in the anaerobic chamber and then the reactions were carried out at 25 °C, stirring under irradiation with blue LEDs (460 nm) for 12 hours. In this case, the cofactor NADPH recycling system involved GDH-105 and glucose [51].

The same methodology was repeated employing *Lk*KRED (0.25 mol%) as a biocatalyst. The only difference was the reaction mixture composition: starting materials (0.039 mmol, introduced from a degassed DMSO solution) and $NADP^+$ (0.0156 μmol) in a 50 mM KPi/$MgSO_4$ buffer solution (pH 6.5) with isopropyl alcohol. Herein, NADPH was regenerated by the native ADH activity of the enzyme at the expense of isopropyl alcohol (coupled substrate process) [51]. Good to excellent enantiomeric excess (60–96% ee) was achieved for both enzymes (Scheme 12.24). It is worth mentioning that, among the evaluated starting materials, one chiral molecule β-methyl substituted afforded the *trans*-isomer with very high selectivity. Additionally, other five-membered α-bromolactones could also be converted, making this approach more interesting. However, for these five-membered substrates, a different enzyme from *L. kefir* was required to obtain good enantioselectivity

(73–82% ee with KRED-P1B02). From a mechanistic point of view, the dehalogenation takes place through a SET-HAT sequence.

Other carboxylic acid derivatives such as secondary α-bromoamides can be asymmetrically dehalogenated using a similar photobiocatalytic process involving *Lk*KRED variant (KRED-P2D03) and EY [80]. The promising results (yields: 58–71%; enantiomeric excess: from 80% to 94% ee) indicate that further efforts in developing these systems will enable chemists to perform reductive dehalogenation over a wide range of organic molecules using photobiocatalysis.

12.3.6 Deacetoxylation

In 2018, Hyster's group described a novel activity of C–C double-bond reductases using light (Scheme 12.25) [80]. In such an approach, the authors could perform an enantioselective radical deacetoxylation of ketones bearing an acetoxy group at the α-position, a non-natural (promiscuous) activity of the enzyme. In particular, the wild-type enzyme used was the C–C double-bond reductase from *Nicotiana tabacum* (*Nt*DBR) and RB was employed as a PC. In this strategy, racemic α-acetoxytetralones bearing other substituents at the α-position and the aromatic ring (both electron-donating and EWGs at the 6- and 7-positions) were evaluated (Scheme 12.25). The substrates were added into the reaction system dissolved in a degassed 2 : 1 THF-EtOH solution (10 μl mg^{-1} substrate). The radical deacetoxylation was performed in phosphate aqueous buffer solution (100 mM KPi, pH 8.0, 10% glycerol) using NtDBR (1.0 mol%), RB (0.5 mol%), $NADP^+$ (1.0 mol%), GDH-105 (0.2 mg mg^{-1} substrate), and glucose (6 eq). These two latter components account for the recycling system of NADPH. It is important to note that the preparation of the reaction mixture was done in an anaerobic chamber. Moreover, after sealing the vials, the interior atmosphere was replaced using a gentle N_2 stream while stirring the solution. Then, the N_2 stream was maintained through an inlet needle for 12 hours, under irradiation with green LEDs (530 nm) and magnetic stirring. Overall, modest to high yields (43–87%) and good to very good enantiomeric excesses were achieved (from 54% to 94% ee).

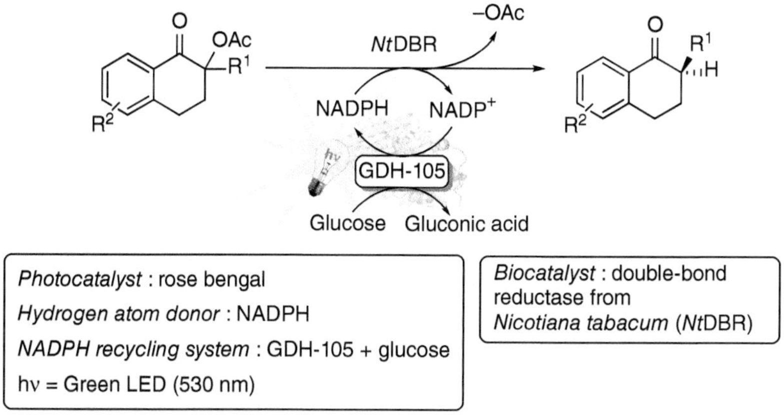

Photocatalyst : rose bengal
Hydrogen atom donor : NADPH
NADPH recycling system : GDH-105 + glucose
hv = Green LED (530 nm)

Biocatalyst : double-bond reductase from *Nicotiana tabacum* (*Nt*DBR)

Scheme 12.25 Photobiocatalytic enantioselective radical deacetoxylation of α-acetoxytetralones.

Interestingly, when acyclic ketones were tested, lower yields were obtained. Moreover, with sterically demanding substituents at the α-position (in comparison with a propyl substituent) of the cyclic starting materials, a decrease in the enantioselectivity was observed. Furthermore, even though the reaction occurs through a radical pathway, with a 4-pentynyl substituent at the α-position, no cyclization product was formed. It is worth noting that the substrate is activated inside the enzyme's active site by a hydrogen bond between the carbonyl and a tyrosine residue, making an SET feasible. Moreover, a highly enantioselective HAT from NADPH is responsible for the asymmetric deacetoxylation.

12.4 Combination of Photooxidation and Enzymatic Transformation

The development of cascade (*one-pot*) reactions is highly valuable in organic synthesis and industry (see also Chapters 13 and 14). In these processes, different catalytic steps are combined in order to perform chemical transformation in a sequential way and in a single reaction vessel, avoiding work-up steps, thus decreasing cost, wastes, and processing time (Scheme 12.26). Furthermore, the generation of unstable or toxic intermediates is controlled by their continuous consumption in the next reaction [6–8]. As aforementioned, one of the challenges in the development of cascade processes is to efficiently combine different catalyst working in the same vessel without cross-spoiling. In this sense, the field of photobiocatalysis has overcome this barrier exploring many transformations with the main goal to selectively obtain valuable compounds. Taking into account the possibility to carry out an alcohol oxidation mediated by visible light, a photoenzymatic *one-pot*, two-step cascade reaction has been designed for the synthesis of chiral amines (Scheme 12.27) [81]. In the first step, the direct conversion of alcohols to the corresponding carbonyl compounds was performed using a mixture of 0.75 mM of sodium anthraquinone-2-sulfonate (SAS) as a PC with 10 mM of substrate and irradiating with visible light for nine hours. After that, a reductive amination of the carbonyl intermediate was carried out using the corresponding pyridoxal-5-phosphate (PLP)-dependent ω-transaminase (ω-TA) enzyme according to the desired enantiomer: to obtain the (*R*)-antipode, the selective ω-TA from *Aspergillus terreus* (*At*ω-TA) was used, and the ω-TA from *B. megaterium* (*Bm*ω-TA) was used to get the (*S*)-enantiomer. In this transamination reaction, isopropylamine (IPA) is the favorite amine donor because of its low cost and more favorable thermodynamic equilibrium toward the expected product (Scheme 12.27). The general conditions for the second step were [ω-TA enzyme] = 10 mg ml^{-1}, [IPA] = 1 M, and [PLP] = 1 mM. The cascade

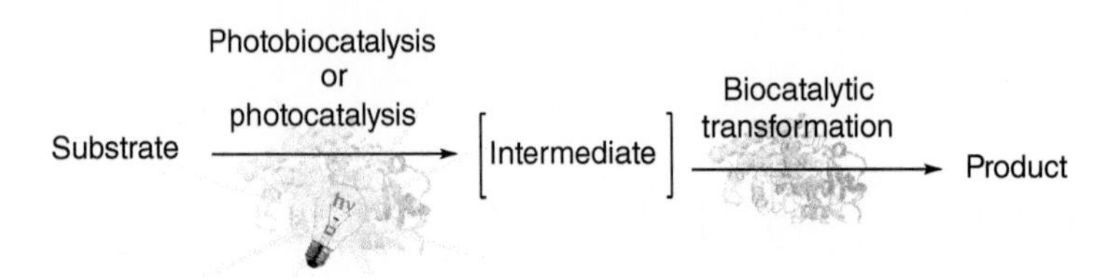

Scheme 12.26 Sequential strategies involving photocatalysis and biocatalysis.

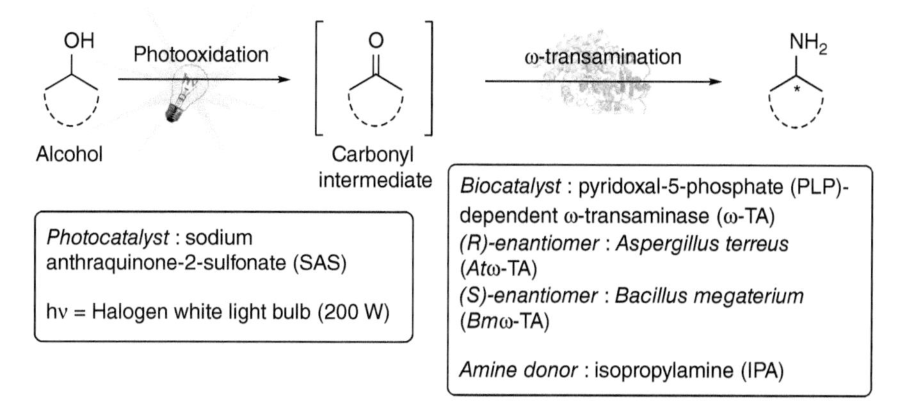

Scheme 12.27 Synthesis of chiral amines by photoenzymatic one-pot, two-step cascade reaction.

reaction shows a broader scope including aliphatic and aromatic substrates. In all cases, the methodology showed high enantioselectivity (>99% ee). However, higher conversions for aromatic derivatives were observed in comparison with aliphatic alcohols.

A similar approach has been developed by means of alkane photooxidation, thus forming carbonyl compounds (aldehydes and ketones) as a key intermediate of a subsequent enzymatic reaction. In order to explore various cascade reactions, photocatalysis was coupled with different asymmetric biotransformations to obtain chiral hydroxy nitriles, amines, acyloins, and α-chiral ketones as well as interconversion to alcohols, esters, and carboxylic acids (Scheme 12.28) [82]. Specifically, SAS was used as a PC and irradiated with white visible light (200 W halogen bulb) for 24 hours. Working in a biphasic system, an improvement of the reaction was observed, especially in the biocatalytic step. In order to get access to different chemical functionalities, several biocatalyzed transformations such as group transfer, oxidation, and reduction were explored. In this sense, amine transaminases (ATAs), hydroxynitrile lyases (HNLs), benzaldehyde lyases (BALs), aryl alcohol oxidases (AAOs), BV monooxygenases (BVMOs), KREDs, or EREDs (Scheme 12.28) were successfully applied. A similar approach to obtain chiral alcohol from α-alkyl carboxylic acid was developed by performing a photooxidative decarboxylation in a first step, leading to the formation of the corresponding carbonyl intermediate. Finally, by a stereoselective KRED-catalyzed carbonyl reduction, the enantioenriched alcohols were obtained with high conversion and selectivity [83].

Another photobiocatalytic cascade approach is based on the fact that an overoxidation reaction was observed during the oxyfunctionalization of alkanes with Au-TiO_2 and *A. aegerita* peroxygenase r*Aae*UPO. In particular, the ketone intermediates can be formed by photobiocatalytic oxidation of alkanes and subsequently used it in another enzymatic reaction [30]. With this in mind, the asymmetric synthesis of amines was performed by a cascade process in a one-pot, two-step reaction. Firstly, in order to obtain the ketone intermediate, an oxyfunctionalization of alkanes using higher concentration of Au-TiO_2 ($30\,g\,l^{-1}$) together with r*Aae*UPO peroxygenase (150 nM) for 96 hours under irradiation with visible light (150 W halogen white light bulb) was carried out. Following that, a ω-TA-catalyzed formal reductive amination was conducted using 105 μl of IPA, 130 μl of

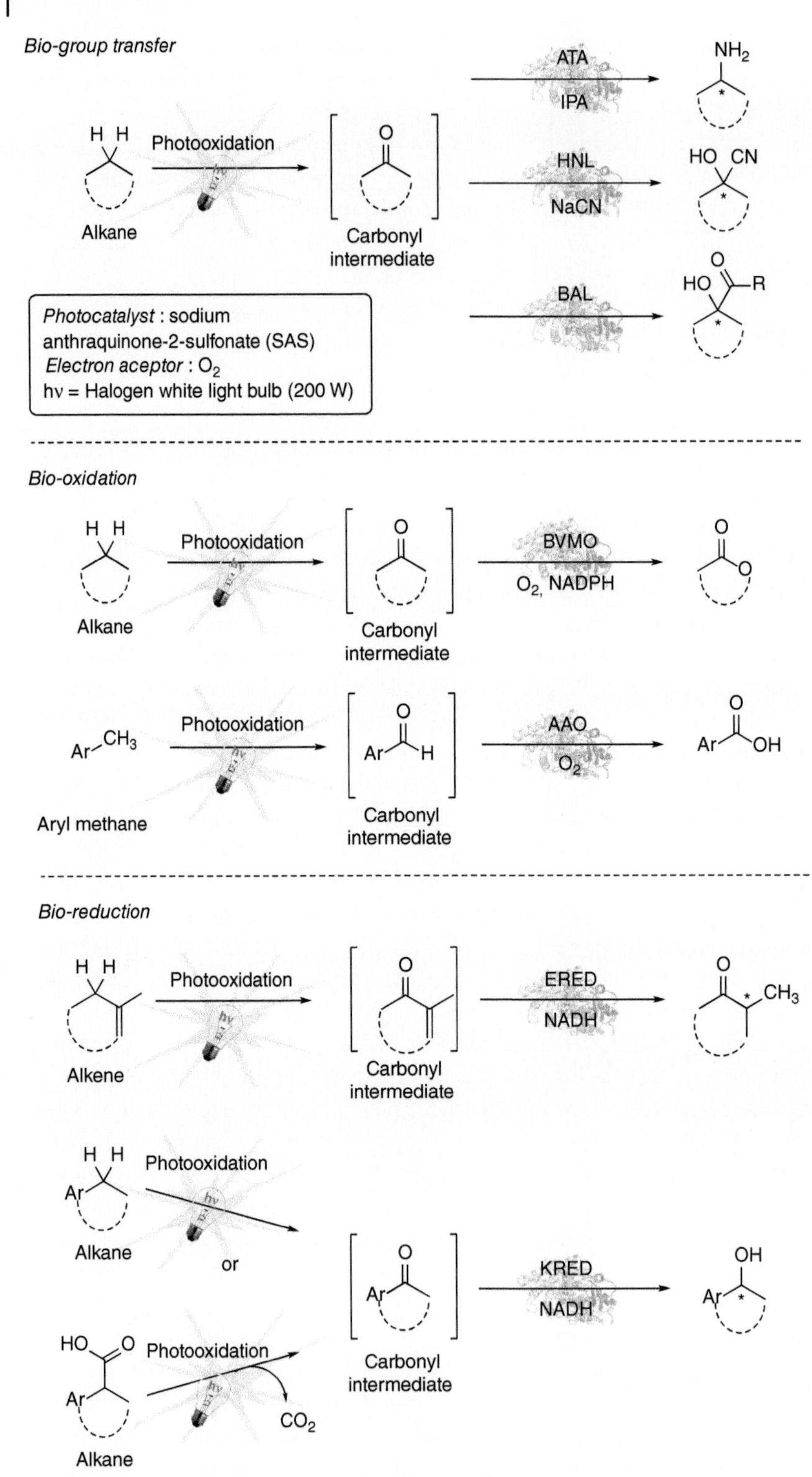

Scheme 12.28 Different combined approaches involving photocatalytic and enzymatic steps.

phosphoric acid (5 M), and 100 μl of pyridoxal phosphate (PLP, 10 mM) together with the ω-TAs enzymes from *A. terreus* (*R*-selective, *At*ω-TA) or *B. megaterium* (*S*-selective, *Bm*ω-TA). Both enantiomers were obtained with high selectivity (>99% ee).

In another approach, this photocatalyzed oxidation can be further combined with a subsequent enzymatic benzoin condensation. Specifically, benzaldehyde is generated by oxidation of toluene with the visible light/Au-TiO_2/r*Aae*UPO peroxygenase system. In the second step, BAL from *Pseudomonas fluorescens* (*Pf*BAL) was used to catalyze the corresponding condensation reaction.

Finally, another mild and efficient photobiocatalytic one-pot sequential approach has been developed by Castagnolo's group in 2018 for the synthesis of 1,3-mercaptoalkanols (Scheme 12.29) [84]. Volatile sulfur compounds (VSCs) constitute a large, diverse group of compounds thought to be responsible for different kinds of aromas and flavors of foods and beverages. Their olfactory features could vary depending on their chirality. Among these chemicals, 1,3-mercaptoalkanols are of great importance because they represent between 20% and 30% of VSCs [85].

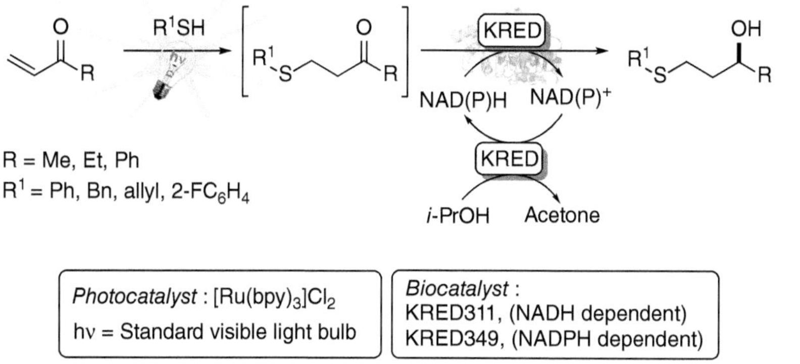

Scheme 12.29 Synthesis of chiral 1,3-mercaptoalkanols by photoenzymatic one-pot, two-step cascade reaction.

Firstly, a light-driven thio-Michael addition is performed to combine an α,β-unsaturated ketone (0.0455 mmol) and a thiol (1.0 equiv.). This photocatalytic process is carried out in a 200 mM phosphate buffer solution (pH 7.0, 700 μl) with DMSO as a cosolvent (50 μl) using visible light (angle poise lamp equipped with a standard visible light bulb) and $[Ru(bpy)_3]Cl_2$ (0.3 mol%) as a catalyst. After one minute of stirring, isopropanol (*i*-PrOH, 28.7 equiv.) and the required KRED cofactor (NADH or NADPH, 0.03 equiv., see Scheme 12.29) are added to the reaction mixture. Subsequently, a suspension of the purified, stabilized enzyme is added. Then, the mixture is placed on an orbital shaker at 37 °C and 160 rpm for 24 hours. During the biocatalytic process, the ketone intermediate is therefore consumed, transforming the carbonyl group into an alcohol. Depending on the chosen enzyme (KRED311 or KRED349, both novel enzymes identified and isolated by a metagenomic approach from the Prozomix library), opposite enantiomers can be achieved. Thus, after 24 hours of reaction, moderate to very good isolated yields (38–73%) and excellent enantiomeric excess values (≥95%) were achieved over a wide range of substrates.

12.5 Summary and Outlook

To date, the use of biocatalysis powered by light or photocatalysis using enzymes is still in its infancy. There is a burst of different reaction designs enabling better performances than those obtained under conventional conditions (dark or nonenzymatic) and some novel concepts are arising. Perhaps the critical points to be improved are the quantum yields (specifically when irradiation comes from a source different from sunlight) and the electron transport efficiency between the PC and the enzymatic cofactor/prosthetic group. Electrochemistry is another field that can be incorporated into more complex photobiocatalytic systems in order to accomplish energetically more demanding reactions. Immobilization of different (photo)catalyst in a controlled manner may engender assembly line-like architectures with enhanced performances in terms of selectivity and efficiency.

From a practical point of view, controlled atmosphere is critical in most cases because oxygen may deactivate exited triplet state of PC and, additionally, is easily reduced superoxide radical anion, which can exert deleterious effects in the overall reaction. Another aspect to be considered is the mass transfer issue. Most of the important organic substrates are hardly (if any) soluble in aqueous media at a synthetically relevant concentration, which usually hamper the application in more sophisticated systems. To tackle this, addition of cosolvents or additives, such as ionic liquids or deep eutectic solvents, has been extensively used. However, solvent compatibility of certain biocatalysts and that of PC often does not perfectly match. One may envision that continuous flow-based methodologies may improve this situation, mainly working under "segmented flow" regime. In this way, a sort of "compartmentalization" of the photo- and biocatalytic cycles takes place, thereby increasing the interfacial surface of the droplets, with concomitant increase in the mass transfer between organic-aqueous phase and, remarkably, avoiding the use of magnetic stirring that can impair the enzymatic activity due to mechanical deactivation of the biocatalyst(s).

It is expected that novel reactivities (different from those displayed by the isolated photo- and biocatalyzed processes) shall appear in the near future by exploring other combinations. Application of genetic code expansion in order to introduce noncanonical amino acids bearing photoactive moieties may engender novel photoprocesses. The introduction of artificial metalloenzymes into photobiocatalytic systems will expand the repertoire of feasible new-to-nature biocatalysis.

Abbreviations

AAO	aryl alcohol oxidase
ACN	amorphous carbon nitride
ADH	alcohol dehydrogenase
ATA	amine transaminase
BAL	benzaldehyde lyase
BODIPY	4,4′-difluoro-4-bora-3*a*,4*a*-diaza-*s*-indacene
bpy	bipyridine
bpz	4,4′-cyclohexane-1,1-diyldiphenol

BV	Baeyer–Villiger
BVMO	BV monooxygenase
CCG	chemically converted graphene
CCGCMAQSP	chemically converted graphene-coupled multi-anthraquinone-substituted porphyrin
CFE	cell-free extract
CHMO	cyclohexanone monooxygenase
CND	carbon nanodot
Cp*	pentamethylcyclopentadienyl
CPR	cytochrome P450 reductase
dClbpy	4,4′-dichloro-2,2′-bipyridine
DMSO	dimethylsulfoxide
dRf	deazariboflavin
EDTA	ethylenediaminetetraacetic acid
ERED	ene-reductase
EWG	electron-withdrawing group
EY	eosin Y
FAD	flavin adenine dinucleotide
FAP	fatty acid decarboxylase
FDH	formate dehydrogenase
FMN	flavin mononucleotide
FNR	ferredoxin-$NADP^+$ reductase
G6PDH	glucose-6-phosphate dehydrogenase
GDH	glucose dehydrogenase
GlDH	glutamate dehydrogenase
HAT	hydrogen atom transfer
HNL	hydroxynitrile lyase
IPA	*iso*-propylamine
ISC	intersystem crossing
KDC	keto acid decarboxylase
KPi	inorganic potassium phosphates
KRED	ketoreductase
LAC3	laccase from *Trametes* sp. C30
LED	light-emitting diode
MAO	monoamino oxidase
MeOH	methanol
MES	2-(*N*-morpholino)ethanesulfonic acid
MO	monooxygenase
MOPS	3-(*N*-morpholino)propanesulfonic acid
MV	methyl viologen
NAD^+/NADH	nicotinamide adenine dinucleotide
$NADP^+$/NADPH	nicotinamide adenine dinucleotide phosphate
N-CD	nitrogen-doped carbon nanodot
PAMO	phenylacetone monooxygenase
PDR	putidaredoxin-NAD^+ reductase

PETNR	pentaerythritol tetranitrate reductase
PLP	pyridoxal-5-phosphate
ppy	2-phenylpyridine
QD	quantum dot
RB	rose bengal
SAS	anthraquinone-2-sulfonate
SET	single-electron transfer
sppy	3-(pyridin-2-yl)benzenesulfonate
StyM	styrene monooxygenase
TA	transaminase
TEA	triethylamine
TEOA	triethanolamine
THF	tetrahydrofuran
TOF	turnover frequency
TON	turnover number
UPO	unspecific peroxygenase
VSC	volatile sulfur compounds
WOC	water oxidation catalyst

References

1 Noyori, R. (2009). The future of chemistry. *Nat. Chem.* 1: 5–6.

2 Stephenson, C.R.J., Yoon, T.P., and Macmillan, D.W.C. (2018). *Visible Light Photocatalysis in Organic Chemistry*. Wiley-VCH.

3 Gamenara, D., Seoane, G.A., Saenz-Méndez, P., and de María, P.D. (2012). *Redox Biocatalysis*. Hoboken, NJ: Wiley.

4 García-Urdiales, E., Alfonso, I., and Gotor, V. (2011). Update 1 of: Enantioselective enzymatic desymmetrizations in organic synthesis. *Chem. Rev.* 111 (5): PR110–PR180.

5 Michaudel, Q., Journot, G., Regueiro-Ren, A. et al. (2014). Improving physical properties via C-H oxidation: chemical and enzymatic approaches. *Angew. Chemie Int. Ed.* 53 (45): 12091–12096.

6 Tietze, L.F. (1996). Domino reactions in organic synthesis. *Chem. Rev.* 96 (1): 115–136.

7 Schrittwieser, J.H., Velikogne, S., and Kroutil, W. (2018). Artificial biocatalytic linear cascades for preparation of organic molecules. *Chem. Rev.* 118 (1): 270–348.

8 Bisogno, F.R., Lavandera, I., and Gotor, V. (2011). Biocatalytic concurrent processes. In: *Kirk-Othmer Encyclopedia of Chemical Technology*, vol. 9, 1–20. *John Wiley & Sons, Inc.*

9 Twilton, J., Le, C.C., Zhang, P. et al. (2017). The merger of transition metal and photocatalysis. *Nat. Rev. Chem.* 1: 0052.

10 Mifsud, M., Gargiulo, S., Iborra, S. et al. (2014). Photobiocatalytic chemistry of oxidoreductases using water as the electron donor. *Nat. Commun.* 5: 1–6.

11 Lang, S.B., Nele, K.M.O., Douglas, J.T., and Tunge, J.A. (2015). Dual catalytic decarboxylative allylations of α-amino acids and their divergent mechanisms. *Chem. Eur. J.* 21 (51): 18589–18593.

12 Wang, Y. and Zhao, H. (2016). Tandem reactions combining biocatalysts and chemical catalysts for asymmetric synthesis. *Catalysts* 6 (12): 2856–2864.
13 Martín-Matute, B. and Bäckvall, J.-E. (2007). Dynamic kinetic resolution catalyzed by enzymes and metals. *Curr. Opin. Chem. Biol.* 11 (2): 226–232.
14 Pàmies, O. and Bäckvall, J.E. (2003). Combination of enzymes and metal catalysts. A powerful approach in asymmetric catalysis. *Chem. Rev.* 103 (8): 3247–3261.
15 Ahn, Y., Ko, S.-B., Kim, M.-J., and Park, J. (2008). Racemization catalysts for the dynamic kinetic resolution of alcohols and amines. *Coord. Chem. Rev.* 252 (5–7): 647–658.
16 Bisogno, F.R., López-Vidal, M.G., and de Gonzalo, G. (2017). Organocatalysis and biocatalysis hand in hand: combining catalysts in one-pot procedures. *Adv. Synth. Catal.* 359 (12): 2026–2049.
17 Huo, H. and Meggersa, E. (2016). Cooperative photoredox and asymmetric catalysis. *Chimia (Aarau)* 70 (3): 186–191.
18 Nicewicz, D.A. and MacMillan, D.W.C. (2008). Merging photoredox catalysis with organocatalysis: the direct asymmetric alkylation of aldehydes. *Science* 322 (5898): 77–80.
19 Petronijević, F.R., Nappi, M., and MacMillan, D.W.C. (2013). Direct β-functionalization of cyclic ketones with aryl ketones via the merger of photoredox and organocatalysis. *J. Am. Chem. Soc.* 135 (49): 18323–18326.
20 Schmermund, L., Jurkaš, V., Özgen, F.F. et al. (2019). Photo-biocatalysis: biotransformations in the presence of light. *ACS Catal.* 9 (5): 4115–4144.
21 Maciá-Agulló, J.A., Corma, A., and Garcia, H. (2015). Photobiocatalysis: the power of combining photocatalysis and enzymes. *Chem. - A Eur. J.* 21 (31): 10940–10959.
22 Lee, S.H., Choi, D.S., Kuk, S.K., and Park, C.B. (2018). Photobiocatalysis: activating redox enzymes by direct or indirect transfer of photoinduced electrons. *Angew. Chem. Int. Ed.* 57 (27): 7958–7985.
23 Baeyer, A. and Villiger, V. (1898). Einwirkung den Caro'sohen reagens auf ketone. *Berichte der Dtsch. Chem. Gesellschaft* 32 (3): 3625–3632.
24 Hollmann, F., Taglieber, A., Schulz, F., and Reetz, M.T. (2007). A light-driven stereoselective biocatalytic oxidation. *Angew. Chem. Int. Ed.* 46 (16): 2903–2906.
25 Holtmann, D. and Hollmann, F. (2016). The oxygen dilemma: a severe challenge for the application of monooxygenases? *ChemBioChem* 17 (15): 1391–1398.
26 Gonçalves, L.C.P., Mansouri, H.R., Bastos, E.L. et al. (2019). Morpholine-based buffers activate aerobic photobiocatalysis via spin correlated ion pair formation. *Catal. Sci. Technol.* 9 (6): 1365–1371.
27 Munro, A.W., Mclean, K.J., Grant, J.L., and Makris, T.M. (2018). Structure and function of the cytochrome P450 peroxygenase enzymes. *Biochem. Soc. Trans.* 46 (1): 183–196.
28 Park, J.H., Lee, S.H., Cha, G.S. et al. (2014). Cofactor-free light-driven whole-cell cytochrome P450 catalysis. *Angew. Chem. Int. Ed.* 54 (3): 969–973.
29 Zhang, W., Burek, B.O., Fern, E. et al. (2017). Selective activation of C-H bonds in a cascade process combining photochemistry and biocatalysis. *Angew. Chem. Int. Ed.* 56 (48): 15451–15455.
30 Zhang, W., Fernández-Fueyo, E., Ni, Y. et al. (2018). Selective aerobic oxidation reactions using a combination of photocatalytic water oxidation and enzymatic oxyfunctionalizations. *Nat. Catal.* 1 (1): 55–62.

31 Willot, J., Ferna, E., Tieves, F. et al. (2019). Expanding the spectrum of light-driven peroxygenase reactions. *ACS Catal.* 9 (2): 890–894.

32 Zhang, M.X., Hu, X.H., Xu, Y.H., and Loh, T.P. (2015). Selective dealkylation of alkyl aryl ethers. *Asian J. Org. Chem.* 4 (10): 1047–1049.

33 Lee, S.H., Kwon, Y.C., Kim, D.M., and Park, C.B. (2013). Cytochrome P450-catalyzed O-dealkylation coupled with photochemical NADPH regeneration. *Biotechnol. Bioeng.* 110 (2): 383–390.

34 Liu, Y., Wang, C., Yan, J. et al. (2014). Hydrogen peroxide-independent production of α-alkenes by OleT JE P450 fatty acid decarboxylase. *Biotechnol. Biofuels* 7 (1): 1–12.

35 Zachos, I., Gaßmeyer, S.K., Bauer, D. et al. (2015). Photobiocatalytic decarboxylation for olefin synthesis. *Chem. Commun.* 51 (10): 1918–1921.

36 Huijbers, M.M.E., Zhang, W., Tonin, F., and Hollmann, F. (2018). Light-driven enzymatic decarboxylation of fatty acids. *Angew. Chem. Int. Ed.* 57 (41): 13648–13651.

37 Witsuthammakul, A. and Sooknoi, T. (2016). Selective hydrodeoxygenation of bio-oil derived products: acetic acid to propylene over hybrid CeO_2-Cu/zeolite catalysts. *Catal. Sci. Technol.* 6 (6): 1737–1745.

38 Sun, K., Schulz, T.C., Thompson, S.T., and Lamb, H.H. (2016). Catalytic deoxygenation of octanoic acid over silica-and carbon-supported palladium: support effects and reaction pathways. *Catal. Today* 269: 93–102.

39 Ford, J.P., Thapaliya, N., Kelly, M.J. et al. (2013). Semi-batch deoxygenation of canola- and lard-derived fatty acids to diesel-range hydrocarbons. *Energy Fuels* 27 (12): 7489–7496.

40 Sorigué, D., Légeret, B., Cuiné, S. et al. (2017). An algal photoenzyme converts fatty acids to hydrocarbons. *Science* 357 (6354): 903–907.

41 Aransiola, E.F., Ojumu, T.V., Oyekola, O.O. et al. (2014). A review of current technology for biodiesel production: state of the art. *Biomass Bioenergy* 61: 276–297.

42 Scrutton, N.S. (2017). Enzymes make light work of hydrocarbon production. *Science* 357 (6354): 872–873.

43 Zhang, W., Ma, M., Huijbers, M.M.E. et al. (2019). Hydrocarbon synthesis via photoenzymatic decarboxylation of carboxylic acids. *J. Am. Chem. Soc.* 141 (7): 3116–3120.

44 van Schie, M.M.C.H., Paul, C.E., Arends, I.W.C.E., and Hollmann, F. (2019). Photoenzymatic epoxidation of styrenes. *Chem. Commun.* 55 (12): 1790–1792.

45 Unversucht, S., Hollmann, F., Schmid, A., and Van Pée, K.H. (2005). $FADH_2$-dependence of tryptophan 7-halogenase. *Adv. Synth. Catal.* 347 (7–8): 1163–1167.

46 Hollmann, F., Lin, P.C., Witholt, B., and Schmid, A. (2003). Stereospecific biocatalytic epoxidation: the first example of direct regeneration of a FAD-dependent monooxygenase for catalysis. *J. Am. Chem. Soc.* 125 (27): 8209–8217.

47 Ruinatscha, R., Dusny, C., Buehler, K., and Schmid, A. (2009). Productive asymmetric styrene epoxidation based on a next generation electroenzymatic methodology. *Adv. Synth. Catal.* 351 (14–15): 2505–2515.

48 Hollmann, F., Hofstetter, K., Habicher, T. et al. (2005). Direct electrochemical regeneration of monooxygenase subunits for biocatalylic asymmetric epoxidation. *J. Am. Chem. Soc.* 127 (18): 6540–6541.

49 Zhang, W. and Hollmann, F. (2018). Nonconventional regeneration of redox enzymes-a practical approach for organic synthesis? *Chem. Commun.* 54 (53): 7281–7289.

50 Schneider, L., Mekmouche, Y., Rousselot-Pailley, P. et al. (2015). Visible-light-driven oxidation of organic substrates with dioxygen mediated by a $[Ru(bpy)_3]^{2+}$/laccase system. *ChemSusChem* 8 (18): 3048–3051.

51 Faber, K. (2011). *Biotransformation in Organic Chemistry*, 6 ed. Heidelberg: Springer-Verlag Berlin.

52 Duchstein, H.-J., Fenner, H., Hemmerich, P., and Knappe, W.R. (1979). (Photo)chemistry of 5-Deazaflavin. *Eur. J. Biochem.* 95 (1): 167–181.

53 Massey, V., Stankovich, M., and Hemmerich, P. (1978). Light-mediated reduction of flavoproteins with flavins as catalysts. *Biochemistry* 17 (1): 1–8.

54 Brown, K.A., Wilker, M.B., Boehm, M. et al. (2016). Photocatalytic regeneration of nicotinamide cofactors by quantum dot-enzyme biohybrid complexes. *ACS Catal.* 6 (4): 2201–2204.

55 Moon, H., Lee, C., Lee, W. et al. (2019). Stability of quantum dots, quantum dot films, and quantum dot light-emitting diodes for display applications. *Adv. Mater.* 31 (34): 1804294.

56 Höfler, G.T., Fernández-Fueyo, E., Pesic, M. et al. (2018). A photoenzymatic NADH regeneration system. *ChemBioChem* 19 (22): 2344–2347.

57 Choudhury, S., Baeg, J.-O., Park, N.-J., and Yadav, R.K. (2012). A photocatalyst/enzyme couple that uses solar energy in the asymmetric reduction of acetophenones. *Angew. Chem. Int. Ed.* 124 (46): 11792–11796.

58 Choudhury, S., Baeg, J.O., Park, N.J., and Yadav, R.K. (2014). A solar light-driven, eco-friendly protocol for highly enantioselective synthesis of chiral alcohols via photocatalytic/biocatalytic cascades. *Green Chem.* 16 (9): 4389–4400.

59 Sandoval, B.A., Kurtoic, S.I., Chung, M.M. et al. (2019). Photoenzymatic catalysis enables radical-mediated ketone reduction in ene-reductases. *Angew. Chem. Int. Ed.* 58 (26): 8714–8718.

60 Kohli, R.M. and Massey, V. (1998). The oxidative half-reaction of old yellow enzyme: the role of tyrosine 196. *J. Biol. Chem.* 273 (49): 32763–32770.

61 Winkler, C.K., Tasnádi, G., Clay, D., et al. (2012). Asymmetric bioreduction of activated alkenes to industrially relevant optically active compounds. *J. Biotechnol.* 162 (4): 381–389.

62 Winkler, C.K., Faber, K., and Hall, M. (2018). Biocatalytic reduction of activated C=C bonds and beyond: emerging trends. *Curr. Opin. Chem. Biol.* 43: 97–105.

63 Grau, M.M., Van Der Toorn, J.C., Otten, L.G. et al. (2009). Photoenzymatic reduction of C=C double bonds. *Adv. Synth. Catal.* 351 (18): 3279–3286.

64 Lee, S.H., Choi, D.S., Pesic, M. et al. (2017). Cofactor-free, direct photoactivation of enoate reductases for the asymmetric reduction of C=C bonds. *Angew. Chem. Int. Ed.* 56 (30): 8681–8685.

65 Burai, T.N., Panay, A.J., Zhu, H. et al. (2012). Light-driven, quantum dot-mediated regeneration of FMN to drive reduction of ketoisophorone by old yellow enzyme. *ACS Catal.* 2 (4): 667–670.

66 Peers, M.K., Toogood, H.S., Heyes, D.J. et al. (2016). Light-driven biocatalytic reduction of α,β-unsaturated compounds by ene reductases employing transition metal complexes as photosensitizers. *Catal. Sci. Technol.* 6 (1): 169–177.

67 Paul, C.E., Arends, I.W.C.E., and Hollmann, F. (2014). Is simpler better? Synthetic nicotinamide cofactor analogues for redox chemistry. *ACS Catal.* 4 (3): 788–797.

68 Kim, J., Lee, S.H., Tieves, F. et al. (2018). Biocatalytic C=C bond reduction through carbon nanodot-sensitized regeneration of NADH analogues. *Angew. Chem. Int. Ed.* 57 (42): 13825–13828.

69 Morris, C.J., Black, A.C., Pealing, S.L. et al. (1994). Purification and properties of a novel cytochrome: flavocytochrome c from *Shewanella putrefaciens*. *Biochem. J.* 302 (2): 587–593.

70 Bachmeier, A., Murphy, B.J., and Armstrong, F.A. (2014). A multi-heme flavoenzyme as a solar conversion catalyst. *J. Am. Chem. Soc.* 136 (37): 12876–12879.

71 Hutton, G.A.M., Reuillard, B., Martindale, B.C.M. et al. (2016). Carbon dots as versatile photosensitizers for solar-driven catalysis with redox enzymes. *J. Am. Chem. Soc.* 138 (51): 16722–16730.

72 Guo, X., Okamoto, Y., Schreier, M.R. et al. (2018). Enantioselective synthesis of amines by combining photoredox and enzymatic catalysis in a cyclic reaction network. *Chem. Sci.* 9 (22): 5052–5056.

73 Park, C.B., Lee, S.H., Subramanian, E. et al. (2008). Solar energy in production of L-glutamate through visible light active photocatalyst-redox enzyme coupled bioreactor. *Chem. Commun.* (42): 5423–5425.

74 Lee, S.H., Nam, D.H., and Park, C.B. (2009). Screening xanthene dyes for visible light-driven nicotinamide adenine dinucleotide regeneration and photoenzymatic synthesis. *Adv. Synth. Catal.* 351 (16): 2589–2594.

75 Son, E.J., Lee, Y.W., Ko, J.W., and Park, C.B. (2019). Amorphous carbon nitride as a robust photocatalyst for biocatalytic solar-to-chemical conversion. *ACS Sust. Chem. Eng.* 7 (2): 2545–2552.

76 Emmanuel, A., Greenberg, N.R., Oblinsky, G., and Hyster, T.K. (2016). Accessing non-natural reactivity by irradiating nicotinamide-dependent enzymes with light. *Nature* 540 (7633): 414–417.

77 Fukuzumi, S., Hironaka, K., and Tanaka, T. (1983). Photoreduction of alkyl halides by an NADH model compound. An electron-transfer chain mechanism. *J. Am. Chem. Soc.* 105 (14): 4722–4727.

78 Fukuzumi, S., Inada, O., and Suenobu, T. (2003). Mechanisms of electron-transfer oxidation of NADH analogues and chemiluminescence. Detection of the keto and enol radical cations. *J. Am. Chem. Soc.* 125 (16): 4808–4816.

79 Jung, J., Kim, J., Park, G. et al. (2016). Selective debromination and α-hydroxylation of α-bromo ketones using Hantzsch esters as photoreductants. *Adv. Synth. Catal.* 358 (1): 74–80.

80 Biegasiewicz, K.F., Cooper, S.J., Emmanuel, M.A. et al. (2018). Catalytic promiscuity enabled by photoredox catalysis in nicotinamide-dependent oxidoreductases. *Nat. Chem.* 10 (7): 770–775.

81 Zhang, W., Knaus, T., Mutti, F.G. et al. (2019). A photo-enzymatic cascade to transform racemic alcohols into enantiomerically pure amines. *Catalysts* 9: 1–11.

82 Zhang, W., Fueyo, F., Hollmann, F. et al. (2019). Combining photo-organo redox- and enzyme catalysis facilitates asymmetric C-H bond functionalization. *Eur. J. Org. Chem.* 2019 (1): 80–84.

83 Xu, J., Arkin, M., Peng, Y. et al. (2019). Enantiocomplementary decarboxylative hydroxylation combining photocatalysis and whole-cell biocatalysis in a one-pot cascade process. *Green Chem.* 21 (8): 1907–1911.

84 Lauder, K., Toscani, A., Qi, Y. et al. (2018). Photo-biocatalytic one-pot cascades for the enantioselective synthesis of 1,3-mercaptoalkanol volatile sulfur compounds. *Angew. Chem. Int. Ed.* 57 (20): 5803–5807.

85 Bentley, R. (2006). The nose as a stereochemist. Enantiomers and odor. *Chem. Rev.* 106 (9): 4099–4112.

13

Practical Multienzymatic Transformations: Combining Enzymes for the One-pot Synthesis of Organic Molecules in a Straightforward Manner

Jesús Albarrán-Velo, Sergio González-Granda, Marina López-Agudo, and Vicente Gotor-Fernández

University of Oviedo, Organic and Inorganic Chemistry Department, Avenida Julián Clavería s/n, 33006, Oviedo, Asturias, Spain

13.1 Introduction

Biocatalysis is currently considered a powerful technology for the production of organic molecules, generally under mild reaction conditions [1, 2]. To overcome synthetic challenges, a wide toolbox of biocatalysts is available, including the six enzyme classes (oxidoreductases (see also Chapters 6 and 7), transferases (see also Chapters 6 and 8), hydrolases (see also Chapter 5), lyases, isomerases, and ligases) [3]. Therefore, nature provides a wide number of possibilities as wild-type enzymes to be applied in synthetic selective transformations [4–7]. Interestingly, recent advances in immobilization techniques (see also Chapter 3) have allowed enzymes to work under different reaction media at the same time such that a higher protein stability can be conferred [8–10], while the irruption of advances in enzyme evolution techniques (see also Chapter 2) has led to the discovery of newly improved enzyme variants [11–14], broadening their substrate specificity and tailoring naturally occurring proteins for a wide number of biotechnological applications.

Overall, enzymes currently appear as excellent synthetic opportunities for the organic chemist, paving the way for the development of complex transformations under sustainable conditions and taking special advantage of the inherent selectivity of enzymes because of their natural amino acid composition. Obviously, the advances in biocatalysis have not gone unnoticed for the chemical industry (see also Chapter 15), and many enzyme applications have already been found [15, 16], especially highlighting the use of biotransformations for the synthesis of active pharmaceutical ingredients [17–19]. Remarkably, many biotechnological companies have flourished over the past decades, some of them gaining great maturity based on cooperations with the academic sector, providing to the market own enzyme portfolios obtained by applying different enzyme immobilization techniques or gene manipulation strategies to produce more stable, active, and versatile biocatalysts.

Biocatalysis for Practitioners: Techniques, Reactions and Applications, First Edition.
Edited by Gonzalo de Gonzalo and Iván Lavandera.

The development of chemoenzymatic and multienzymatic transformations presents great advantages over the classical step-by-step synthesis (stepwise transformations, Scheme 13.1a), avoiding the isolation and purification of reaction intermediates that can be unstable under determined conditions or even difficult to isolate from the reaction medium. Interestingly, these multistep one-pot processes can combine the productivity of chemical catalysis (i.e. metal catalysis and organocatalysis) with the stereoselectivity displayed by enzymes (see also Chapter 14) [20–22].

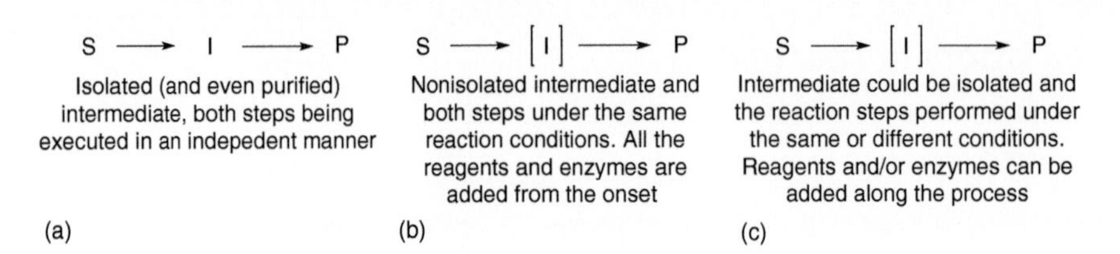

Scheme 13.1 (a) Stepwise, (b) cascade, and (c) sequential transformations: S, substrate; I, reaction intermediate; and P, product.

Two action modes can be selected for one-pot multicatalytic transformations, which are the design of cascade or sequential transformations (Scheme 13.1). The first ones are the most attractive approaches because of their easy manipulation consisting in the addition of all the reagents and enzymes from the first beginning, which work during the entire process under the same reaction conditions (Scheme 13.1b). The orchestral action of multiple catalysts, in this particular case, focusing in multienzymatic catalysis, depends on the enzyme requirements in terms of parameters that greatly modulate the enzyme activity such as the reaction medium, temperature, cofactor dependency, among others [23]. In fact, the chronological order in the addition of reagents and catalysts is undoubtedly a key issue, sometimes adding them in a sequential manner or even modifying the reaction conditions (pH or substrate concentrations) along the process results compulsory to achieve successful multicatalytic systems (sequential transformations, Scheme 13.1c). The importance of this hot topic has motivated prestigious scientists to perform extensive bibliographic revisions in the past seven years [24–35].

Overall, a cascade reaction, also called tandem or domino, can be defined as the combination of at least two chemical steps in a single recipient without requiring the isolation of reaction intermediates. Different cascade types have been reported in the literature (linear, orthogonal, parallel, or cyclic, Scheme 13.2), opening up multiple possibilities for the production of organic compounds [30].

In linear cascades, the product of one step is the substrate for the next transformation, increasing the product complexity with the development of subsequent modifications. Remarkably, a few trends have been identified for the effective development of one-pot multienzymatic transformations, which includes the rational design of cascades [36], the co-immobilization or co-expression of the catalysts [37], or the selection of the proper reactor for synthetic purposes [38]. Herein, this contribution compiles a representative selection of practical linear and cyclic multienzymatic cascades reported in recent years (2014–2020), trying to compare them in some cases with similar reported stepwise and

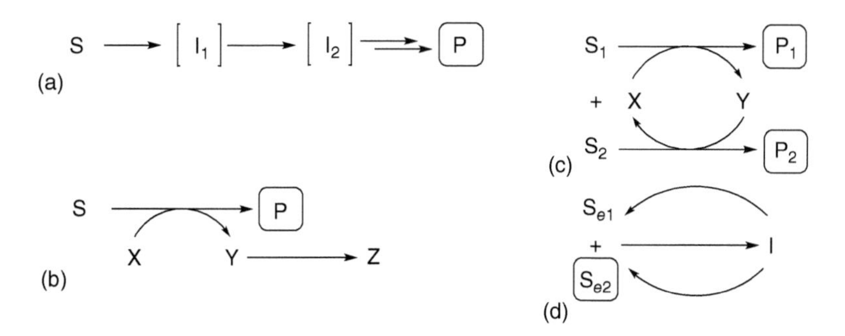

Scheme 13.2 Type of cascade processes: (a) linear; (b) orthogonal; (c) parallel; and (d) cyclic. S_n, substrate n (n = 1 or 2); S_{en}, enantiomer n (n = 1 or 2) of the substrate; P_n, product n (n = 1 or 2); I_x, reaction intermediate; and X, Y, and Z, co-substrate and coproduct(s).

sequential transformations. Special attention will be paid to those examples that include downstream process for product recovery rather than focusing in the description of examples summarizing conversion values. Scalability is a main issue when trying to implement a determined (sequence) transformation at industrial scale, so a clear emphasis will be made in reporting isolated yields to demonstrate the usefulness of multienzymatic synthesis for practical applications.

13.2 Non-stereoselective Bienzymatic Transformations

In this initial section, the description of efficient and scalable non-stereoselective bienzymatic systems is covered, describing first amine and amide syntheses, to later explore the possibility of combining different enzyme classes, mainly from the oxidoreductase family, for the preparation of a series of organic molecules.

13.2.1 Amine Synthesis

Amine synthesis has received particular attention by means of nonselective and stereoselective transformations because of the importance of amines as building blocks in numerous compounds with remarkable biological activities [39]. In this context, amine oxidases, amine transaminases (ATAs), imine reductases (IREDs), lipases, reductive aminases, among others have largely contributed to identify biocatalysis as a key technology for amine synthesis [40, 41]. In recent years, the combination of redox enzymes with transaminases has attracted great attention toward the preparation of non-chiral amines in a straightforward manner. For instance, Faber and coworkers have described the amination of nonactivated aliphatic fatty alcohols through a five-enzyme cascade, where the two main steps were developed in a linear mode consisting in the combined use of an alcohol oxidase (AOX) from *Aspergillus fumigatus* (*Af*LCAO) with the ω-transaminase from *Chromobacterium violaceum* (*Cv*TA), while an alanine dehydrogenase (AlaDH), a glucose dehydrogenase (GDH), and a catalase allow shifting the equilibrium to amine synthesis (Scheme 13.3a) [42]. Thus, from a series of 16 primary alcohols, the preparative amination

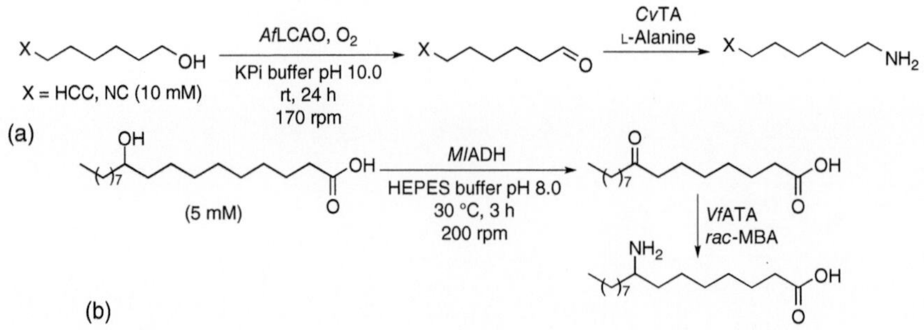

Scheme 13.3 Amination of (a) 7-hydroxyheptanitrile, oct-7-yn-1-ol and (b) ricinoleic acid using both an alcohol dehydrogenase and a transaminase.

of 7-hydroxyheptanitrile and oct-7-yn-1-ol (10 mM concentration) was described in moderate to complete conversions in the presence of oxygen and L-alanine as an amine donor, isolating the corresponding ethyl N-carbamates after chemical derivatization.

A similar strategy has been described by Park and coworkers for the amination of long-chain aliphatic hydroxy acids, obtaining the corresponding amino acids through a two-step *in vivo* transformation that employs a long-chain secondary alcohol dehydrogenase (ADH, also known as ketoreductase – KRED – or carbonyl reductase – CR) from *Micrococcus luteus* (*Ml*ADH) and an amine transaminase variant from *Vibrio fluvialis* (*Vf*ATA) using racemic methylbenzylamine (*rac*-MBA) as an amine donor [43]. Interestingly, the preparative cascade transformation of ricinoleic acid to (*Z*)-12-aminooctadec-9-enoic acid was carried out at 30 °C, yielding the amino acid in 70% isolated yield (54 mg) after liquid–liquid extraction and column chromatography purification (Scheme 13.3b).

The combination of an ADH from *Lactobacillus kefir* (*Lk*ADH) and a commercially available ATA from Codexis Inc. (ATA-200) has also been successfully applied for the transformation of cyclohexane-1,4-dione into 4-aminocyclohexanol (99 : 1 *cis-trans* ratio) at 1.5 mmol scale [44]. Trying to avoid the formation of the diol and diamine by-products, and after careful optimization of the reaction conditions, the amino alcohol was obtained as a unique product in 44% conversion through a cascade approach starting from 160 mg of the diketone (Scheme 13.4), while a higher 85% conversion was attained using a sequential strategy when starting from 80 mg of cyclohexane-1,4-dione, although with concomitant diol (9%) and diamine (6%) formation. Interestingly, the proper selection of ADH and ATA can favor the formation of the *trans*-amino alcohol instead of the *cis*-isomer (up to 20 : 80 ratio).

Very recently, Turner and coworkers have reported two independent strategies for the biocatalytic production of amines through the formation of a common aldehyde intermediate, which led to the *N*-alkylated amines by using a reductive aminase from *Aspergillus oryzae* (*Asp*RedAm) and 0.8–4.0 equiv of primary amines such as cyclopropylamine, benzylamine, propargylamine, or allylamine (Scheme 13.5) [45]. Depending on the starting material, the aldehyde can be produced either by primary alcohol oxidation using an engineered choline oxidase (AcCO$_6$, Scheme 13.5a) or carboxylic acid reduction employing a wild-type carboxylic acid reductase from *Segniliparus rugosus* (*Sr*CAR, Scheme 13.5b). High conversions (84–99%) and purities (>95%) were attained for all these preparative

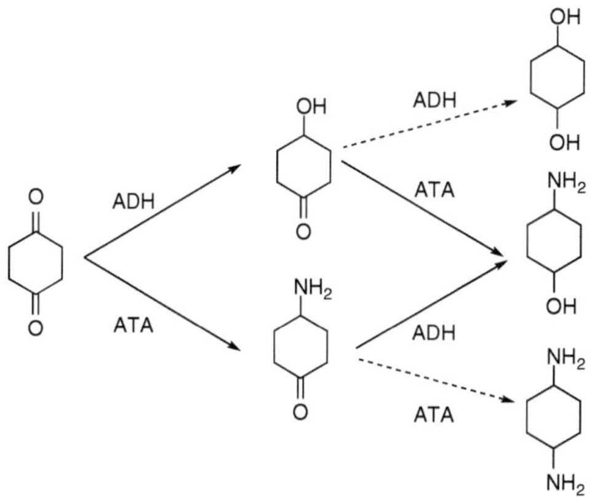

Scheme 13.4 Transformation of cyclohexane-1,4-dione into 4-aminocyclohexanol by combining an ADH and an ATA.

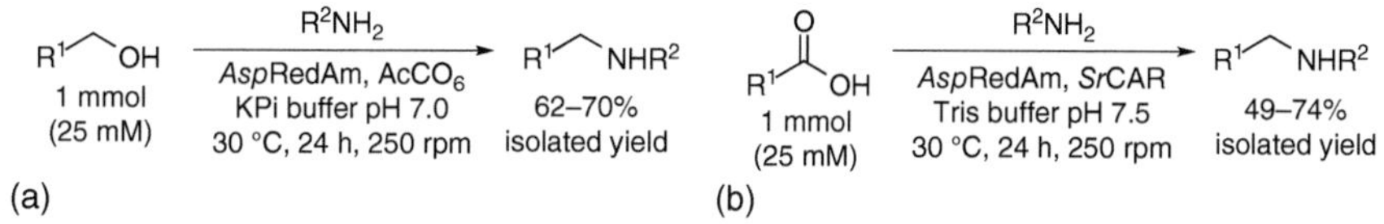

Scheme 13.5 Biocatalytic synthesis of N-alkylated amines using reductive aminases in combination with (a) alcohol oxidase and (b) carboxylic acid reductase.

transformations (1 mmol of alcohol), obtaining the amines after alkalinization of the reaction medium and extraction with ethyl acetate.

Finally, the combination of *Candida antarctica* lipase B (CAL-B) and α,ω-diamine transaminase (YgjG) have led to the synthesis of 2-substituted N-heterocyclic alkaloids including ruspolinone (71%), norsedaminone (60%), norhygrine (50%), and hygrine (75%) through a one-pot hydrolysis, transamination, and decarboxylative Mannich reaction after 18 hours at 30 °C. All these compounds were easily isolated after filtration through Celite® and acid–base work-up (Scheme 13.6) [46].

13.2.2 Bienzymatic Linear Cascades Toward the Production of Other Organic Compounds

The combination of transaminases and lipases has also resulted in an efficient strategy for the synthesis of other families of compounds such as amides. For instance, Anderson et al. reported a one-pot sequential strategy for the biotransamination of vanillin with 5 equiv of L-alanine using a mixture of dimethyl sulfoxide (DMSO): HEPES buffer (1 : 4.6 v/v) to obtain vanillylamine in 94% conversion (Scheme 13.7a) [47]. After freeze-drying the sample to remove the water, the amine was dissolved in 2-methyl-2-butanol and reacted with a fatty acid in the presence of molecular sieves to give nonivamide in 52% isolated after

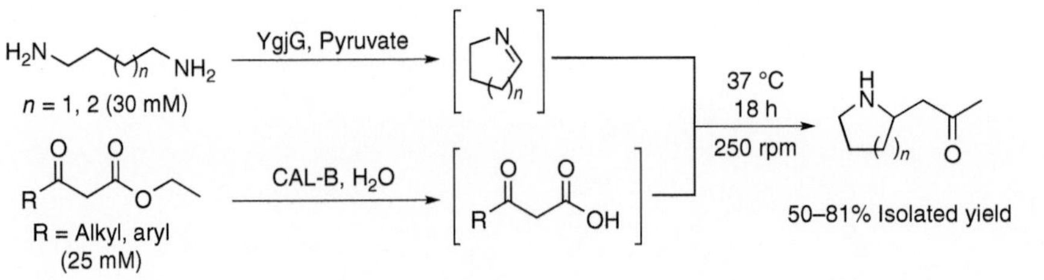

Scheme 13.6 One-pot hydrolysis, transamination, and decarboxylative Mannich reaction for the synthesis of 2-substituted N-heterocyclic alkaloids.

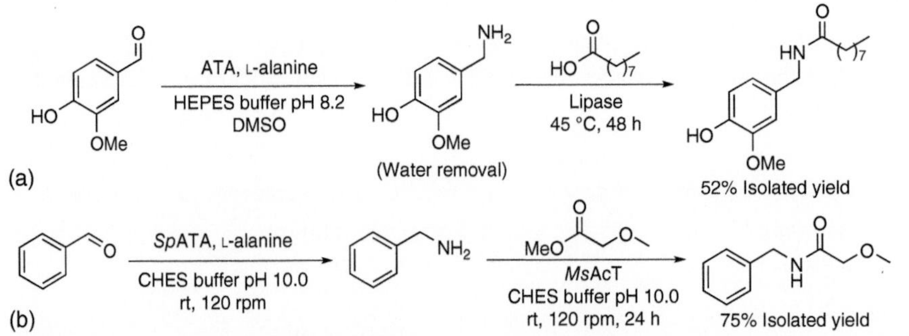

Scheme 13.7 Enzyme-catalyzed synthesis of (a) nonivamide from vanillin in a sequential manner and (b) *N*-benzyl-2-methoxyacetamide from benzaldehyde through a cascade.

column chromatography. Interestingly, in the same contribution, a bienzymatic cascade is reported for the transformation of vanillyl alcohol to vanillylamine employing an ADH and an ATA.

Berglund and coworkers have also reported the one-pot biocatalytic production of *N*-benzyl-2-methoxyacetamide in 92% conversion and 75% isolated yield through an ATA/acyl transferase cascade (Scheme 13.7b) [48]. Therefore, starting from benzaldehyde (1 mmol), L-alanine (25 equiv), and methyl methoxyacetate (10 equiv), the acyl transferase from *Mycobacterium smegmatis* (*Ms*AcT) and the ATA from *Silicibacter pomeroyi* (*Sp*ATA) were used under slow mixing conditions in a *N*-cyclohexyl-2-aminoethanesulfonic acid (CHES) buffer pH 10, yielding 134 mg of the desired amine after extraction at acidic pH.

The one-pot biocatalytic double oxidation of alkanes is also a recurrent strategy for the production of carbonyl compounds by the combination of an oxygenase and an oxidative enzyme. This is the case of the next two examples, which combined the use of P450 monooxygenases for the production of alcohols that are subsequently oxidized to the corresponding ketones (Scheme 13.8). For instance, Turner and coworkers have reported the double oxidation of α-isophorone to produce ketoisophorone by combining a variant of the chimeric self-sufficient P450cam-RhFRed with an ADH from *Candida magnoliae* (*Cm*ADH10, Scheme 13.8a) [49], while Faber and coworkers reported the transformation of octanoic acid into 2-oxooctanoic acid by regioselective α-hydroxylation catalyzed by P450 monooxygenase from *Clostridium acetobutylicum* (*CLA*P450) and next oxidation of

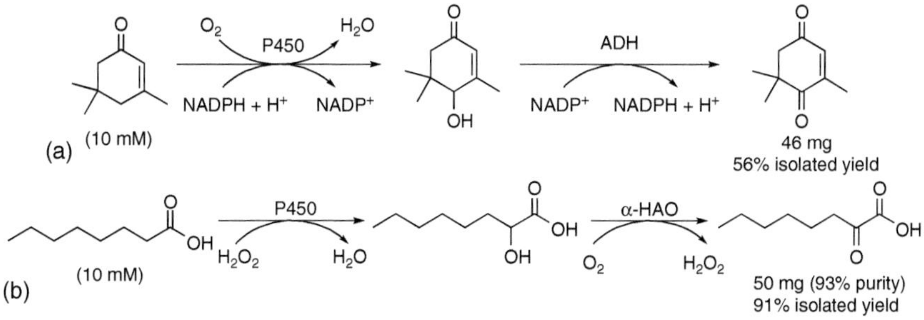

Scheme 13.8 One-pot combination of P450 monooxygenase-catalyzed hydroxylation with alcohol oxidation reactions for the formation of: (a) ketoisophorone; (b) 2-oxooctanoic acid.

the hydroxy acid intermediate using a α-hydroxy acid oxidase from *Aerococcus viridans* ((*S*)-α-HAO, Scheme 13.8b) [50].

Another example of the combination of two oxidoreductases is the one-pot conversion of 2,6-dimethoxy-4-allylphenol into syringaresinol using an engineered eugenol oxidase (EUGO I427A) and horseradish peroxidase (HRP) [51]. After optimization of the reaction conditions, the reaction was performed using 5% v/v of DMSO as a cosolvent, and after 22 hours at 30 °C, an 81% yield of the final product (870 mg) was obtained after extraction and column chromatography purification (Scheme 13.9).

Finally, in this section, the one-pot bienzymatic synthesis of ethyl (*R*)-3-hydroxyglutarate from ethyl (*S*)-3,4-chloro-3-hydroxybutyrate is described by using an engineered halohydrin dehalogenase from *Agrobacterium radiobacter* AD1 (HHDH) and a nitrilase from *Arabidopsis thaliana* (Scheme 13.10) [52]. After optimization of the reaction conditions, the cascade was performed at a significant 10 g substrate scale in the presence of sodium cyanide (NaCN), producing the desired cyano-hydroxy ester in 86.7% yield (9.16 g) after liquid–liquid extraction.

13.3 Stereoselective Bienzymatic Transformations

In this section, one-pot bienzymatic stereoselective synthesis will be covered paying special emphasis to describe procedures including product isolation. As discussed in the previous

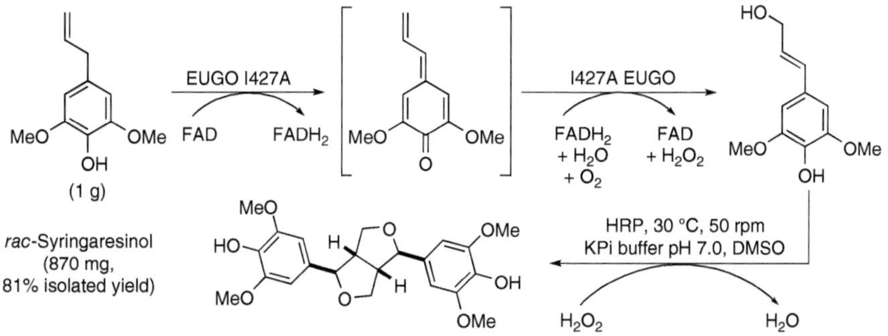

Scheme 13.9 Transformation of 2,6-dimethoxy-4-allylphenol into syringaresinol through the formation of the sinapyl alcohol intermediate.

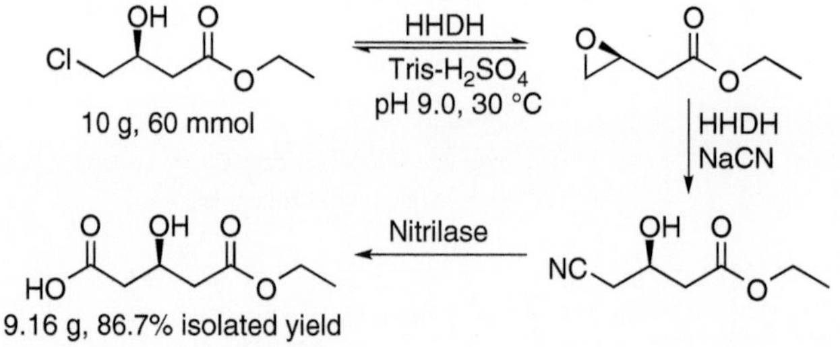

OH O
Cl O
10 g, 60 mmol
HHDH
Tris-H_2SO_4
pH 9.0, 30 °C
O O
O
HHDH
NaCN
O OH O
HO O
9.16 g, 86.7% isolated yield
Nitrilase
OH O
NC O

Scheme 13.10 Synthesis of ethyl (*S*)-3-4-chloro-3-hydroxybutyrate using a halohydrin dehalogenase and a nitrilase.

section related to the development of non-stereoselective processes, first amine syntheses will be discussed to later expand the potential of enzymes to different compound families.

13.3.1 Stereoselective Amine Synthesis Through Concurrent Processes

The preparation of optically active amines attracts great attention in organic synthesis, with biocatalysis providing elegant and sustainable solutions for the production of this class of nitrogen-containing molecules because of their possibilities as synthetic intermediates of a plethora of chemical compounds at industrial scale [40, 41]. To achieve this aim, many different enzyme families have been employed, so this section has been divided depending on the described strategies for chiral amine synthesis. First, examples regarding the amination of alcohols will be described because of the multiple existing possibilities to perform this task. Next, examples of deracemization of racemic and optically active amines, respectively, will be discussed to later introduce the production of amino alcohol, ATAs displaying a key role for the success of these types of transformations. Finally, a miscellaneous collection of other synthetic approaches will be summarized in an organized manner.

13.3.1.1 Amination of Alcohols

Turner and coworkers described in 2015 a pioneer work regarding multienzymatic amine synthesis, consisting in the combination of one or two ADHs such as the Prelog ADH from *Aromatoleum aromaticum* (*Aa*ADH) or the engineered anti-Prelog one from *Lactobacillus brevis* (*Lb*ADH) with an amine dehydrogenase (AmDH) from *Bacillus* species for the amination of (primary and secondary) aliphatic and aromatic alcohols with excellent conversion and selectivity levels (Scheme 13.11) [53]. Interestingly, a redox self-sufficient cascade is reported, working with an excellent atom efficiency as ammonium is used as a nitrogen source, releasing water as a sole by-product at 30 °C for 48 hours. Thus, it was possible to develop the hydrogen-borrowing conversion of either (*S*)-, (*R*)-, or racemic alcohols into the corresponding (*R*)-amines.

The same research group reported the hydrogen-borrowing cascade of 4-phenylbutan-2-ol and 1-(4-methoxyphenyl)butan-2-ol in a semi-preparative scale (100 mg) with a W110A/G198D variant from *Thermoanaerobacter ethanolicus* (*Tes*ADH) and a chimeric amine dehydrogenase (ChiAmDH), yielding the desired optically active (*R*)-amines in 69% and 84% yield, respectively [54]. Similarly, Mutti and coworkers described the amination of

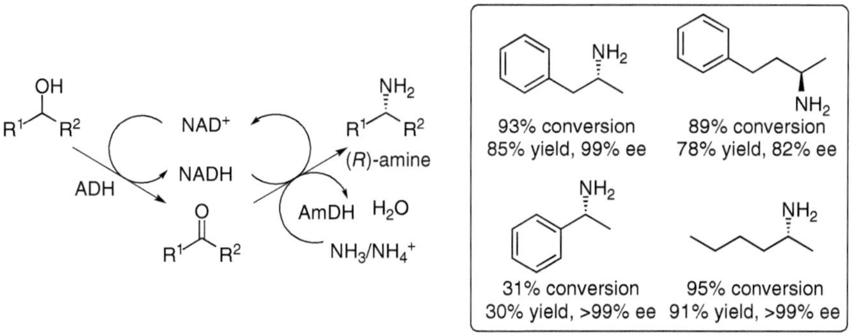

ADH NAD+ NADH AmDH H_2O NH_3/NH_4^+ (R)-amine

93% conversion 85% yield, 99% ee
89% conversion 78% yield, 82% ee
31% conversion 30% yield, >99% ee
95% conversion 91% yield, >99% ee

Scheme 13.11 Stereoselective hydrogen-borrowing amination of alcohols using an ADH and an AmDH, displaying below the results of the preparative transformations after 48 h at 30 °C and amine isolation after liquid–liquid extraction at basic pH.

2-heptanol (50 mg) using a I86A/W110A variant from *T. ethanolicus* (*Tes*ADH) and a ChiAmDH, yielding enantiopure (*R*)-2-heptanamine with 97% conversion and 90% yield (45 mg) [55]. The same research group has recently reported the co-immobilization of two dehydrogenases onto Fe(III) ion affinity beads (commercially available EziG Fe-amber), thus enhancing the efficiency of the hydrogen-borrowing amination and allowing the recyclability of the biocatalytic system composed by a ADH from *A. aromaticum* (*Aa*ADH) and a Chi1AmDH for five consecutive cycles [56]. From a series of (*S*)-alcohols, (*S*)-phenylpropan-2-ol was selected for the development of a preparative amination, leading to 90% conversion after 24 hours at 30 °C and 170 rpm, and 80% isolated yield of the (*R*)-amine.

Following with the stereoselective preparation of amines, Chen et al. reported the amination of aliphatic alcohols using an ADH from *Streptomyces coelicolor* and a NADH-dependent leucine dehydrogenase from *Exiguobactertium sibircum*, including the conversion of racemic 2-pentanol (44.1 mg, 0.5 mmol) into enantiopure (*R*)-2-pentanamine, isolated as a hydrochloride salt in a low 21% yield in spite of reaching a high conversion into the amine (94%) [57]. In an alternative approach, the reductive aminase from *A. oryzae* (*Asp*RedAm) was employed instead of an AmDH and combined with a single ADH (ADH-150, ADH from *Sphingobium yanoikuyae* or W110A *Tes*ADH) in a novel redox-neutral cascade releasing also water as a by-product (Scheme 13.12) [58]. Aliphatic and aromatic secondary amines were obtained including the nonselective preparative reaction between cyclohexanol (100 mg) and allylamine to produce *N*-allylcyclohexylamine in 98% conversion and 61% isolated yield. The synthesis of chiral amines was also reported, reaching moderate conversions and moderate to excellent selectivities from the corresponding racemic alcohol precursors.

Interestingly, the combination of a laccase mediator system with selected ATA has been properly addressed for the production of optically active benzylamines [59] and allylic amines [60]. The combined use of laccases and ATAs possesses some inherent limitations because of the enzyme characteristics. In this case, the laccase from *Trametes versicolor* (L*Tv*) was used in combination with the oxy-radical 2,2,6,6-tetramethylpiperidine-1-oxyl (TEMPO) (33 mol%), a system that requires a pH around 5 for the correct action of the enzyme, while ATAs usually work properly at slightly basic pHs. After the observation that

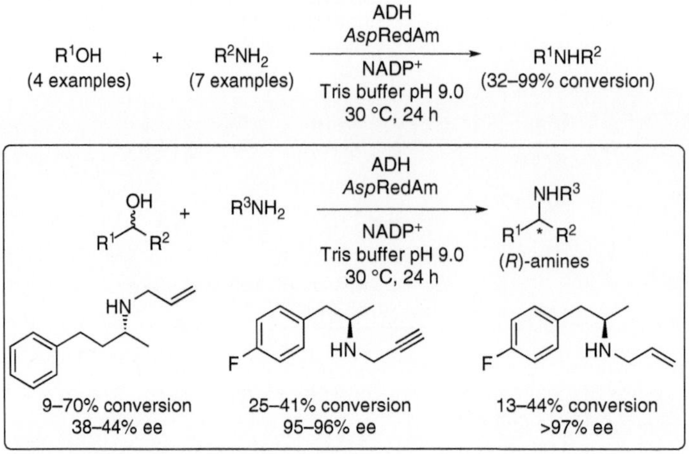

Scheme 13.12 Amination of primary and secondary alcohols combining the *Asp*RedAm with an ADH.

the laccase suffered a dramatic inactivation in the presence of the amine donor (isopropylamine, $^iPrNH_2$) required by the ATA, the necessity of different pH conditions motivate the development of a sequential approach rather than a cascade reaction (Scheme 13.13) [59, 60]. For the transformation of racemic 1-arylalkanols into benzylamines, a one-pot, two-step approach was developed by adding a large excess of $^iPrNH_2$ after complete oxidation of the alcohol into the corresponding ketone intermediate, demonstrating the scalability of the approach for two substrates (100 mg of alcohol, Scheme 13.13a) [59]. Interestingly, the use of a smart co-substrate, such as *cis*-but-2-ene-1,4-diamine, allowed the use of only 1.5 equiv of the amine donor instead of a large excess such as in the case of $^iPrNH_2$, obtaining enantiopure (*R*)-1-(3-bromophenyl)ethylamine in 91% isolated yield.

For the amination of allylic alcohols, a similar approach was adopted including aryl and heteroaryl substitutions for the development of analytical (0.08 mmol) and

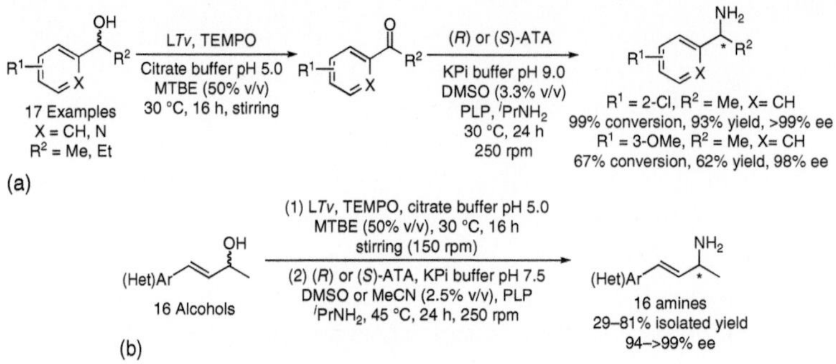

Scheme 13.13 Combination of laccase-mediator systems and selective amine transaminases for the sequential amination of (a) benzyl alcohols; (b) allylic alcohols.

semi-preparative transformations (0.16 mmol), isolating after extraction the corresponding (*R*)- or (*S*)-amines depending on the ATA selectivity (Scheme 13.13b) [60].

13.3.1.2 Deracemization of Amines

Deracemization strategies are highly appealing transformations for the stereoselective synthesis of alcohols and amines [61], overcoming the inherent limitation of the 50% yield associated with kinetic resolutions. In this context, several examples have been recently reported for the synthesis of chiral amines. For instance, Shin and coworkers have recently described the one-pot deracemization of various amines to exclusively obtain their (*S*)-enantiomers (85–99% yield) by using two complementary ATAs, one for the oxidative deamination of the (*R*)-enantiomer (*Arthrobacter* species) and the other for the biotransamination of the resulting ketone into the (*S*)-amine (engineered W58L/R417A ATA mutant from *Ochrobactrum anthropi*, Scheme 13.14) [62]. In particular, (*S*)-4-phenylbutan-2-amine was obtained in 79% isolated yield after ion exchange chromatography purification using a Tris buffer pH 7.0 and DMSO (15% v/v) as a cosolvent after 1.5 hours at 37 °C and 460 torr pressure. Alternatively, the combination of a (*S*)-ω-transaminase and an AmDH can lead to a panel of 13 enantiopure (*R*)- and (*S*)-amines with good to excellent conversions (80–>99%) [63]. The synthetic applicability of this approach has been demonstrated starting from 57.5 mg of racemic amine at 20 mM concentration (24 hours and 37 °C), obtaining after extraction (*R*)- and (*S*)-2-heptylamine in 53% and 75% isolated yield, respectively.

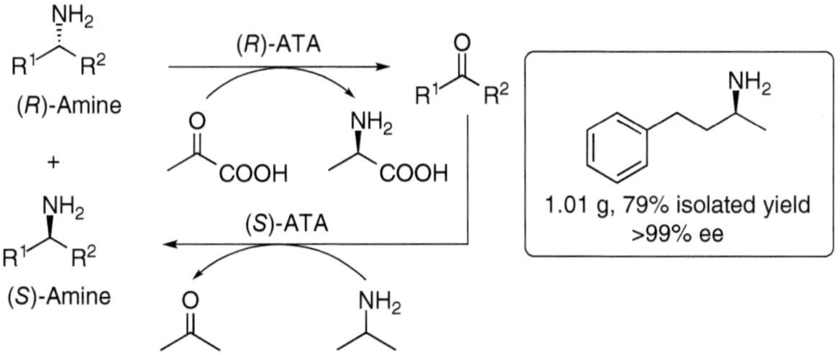

Scheme 13.14 One-pot deracemization of racemic amines combining two enantiocomplementary amine transaminases.

Monoamine oxidases (MAOs) are considered as very selective enzymes for the selective oxidation of amines to imines, a process that can be coupled with a nonselective chemical reduction for the production of chiral amines. Schrittwieser et al. reported a deracemization of racemic benzylisoquinolines using an engineered MAO from *Aspergillus niger* (MAO-N D11) in combination with the morpholine borane complex to later develop the synthesis of enantiopure (*S*)-berbines by berberine bridge enzyme (BBE)-catalyzed aerobic C–H activation of the N-methyl group, thus forming a new intramolecular C—C bond (Scheme 13.15) [64]. From the four studied amines, two preparative one-pot cascade transformations were developed (150–165 mg of amine, 0.5 mmol), yielding the corresponding enantiopure (>97% *ee*) (*S*)-berbines in 80–88% isolated yield (92–98% conversion) after extraction and column chromatography.

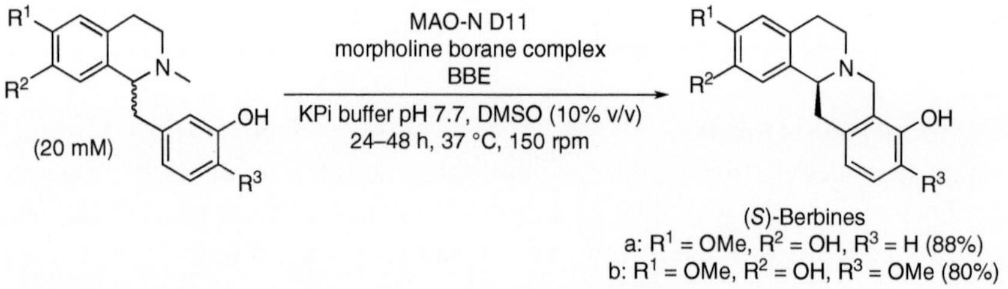

R^1, R^2, (20 mM), OH, R^3

MAO-N D11
morpholine borane complex
BBE
KPi buffer pH 7.7, DMSO (10% v/v)
24–48 h, 37 °C, 150 rpm

(*S*)-Berbines
a: R^1 = OMe, R^2 = OH, R^3 = H (88%)
b: R^1 = OMe, R^2 = OH, R^3 = OMe (80%)

Scheme 13.15 Deracemization of benzylisoquinolines by bio-oxidative kinetic resolution and stereoinversion using an MAO/reducing agent system and a BBE.

Finally, in this section, the stereoselective production of (*R*)-1-methyl-1,2,3,4-tetrahydroisoquinoline (MTQ) is described through a cascade approach involving the reduction of 1-methyl-3,4-dihydroisoquinoline (MDQ) and later aerobic oxidation (Scheme 13.16) [65]. For this purpose, the combination of the imine reduction step catalyzed by an artificial transfer hydrogenase (ATHase) coupled with a GDH to regenerate the reduced nicotinamide cofactor, with the selective amine oxidation using a MAO and a catalase, allows the accumulation of the desired (*R*)-MTQ from 89% yield and 75% ee of the amine after 12 hours to a complete conversion into the enantiopure amine after 24 hours.

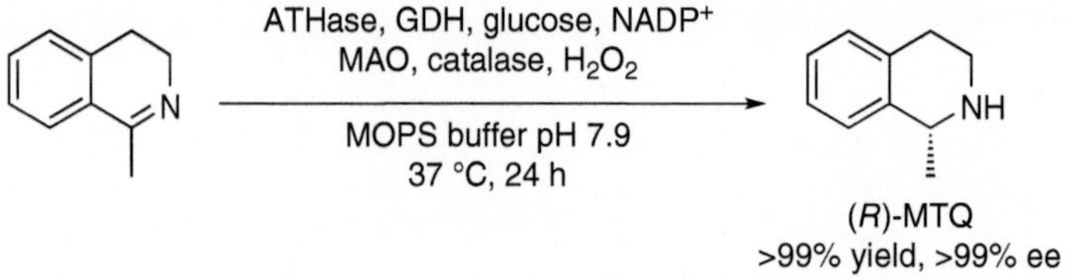

ATHase, GDH, glucose, $NADP^+$
MAO, catalase, H_2O_2
MOPS buffer pH 7.9
37 °C, 24 h

(*R*)-MTQ
>99% yield, >99% ee

Scheme 13.16 Combination of an artificial transfer hydrogenase with a monoamine oxidase for the production of (*R*)-1-methyl-1,2,3,4-tetrahydroisoquinoline.

13.3.1.3 Amino Alcohol Synthesis

Optically active amino alcohols are attractive organic molecules with many applications in medicinal chemistry and in asymmetric catalysis, mainly as chiral auxiliaries. Several practical bienzymatic cascades have been reported in recent years, which most of them combined the use of lyases and ATAs. For instance, Gefflaut and coworkers have described the use of a pyruvate aldolase (PyrAL) and an α-TA for the production of γ-hydroxy-α-amino acids [66]. Therefore, the preparative synthesis of D-*anti*-4,5-dihydroxynorvaline (DHNV) has been plausible using D-alanine (D-Ala, 1 mmol), pyruvic acid (PA, 0.1 mmol), and slow addition of glycoaldehyde (GL, 1.6 mmol) for four hours (Scheme 13.17), so after an additional five hours of stirring at room temperature, the remaining alanine was separated by ion exchange chromatography from the reaction mixture after DHNV lactonization and selective adsorption of the lactone on a cation exchange resin. Final elution of DHNV led to 124 mg of the compound in 83% yield and 7 : 93 *syn*:*anti* ratio.

More recently, Rother and coworkers reported the one-pot, two-step sequential synthesis of (1*S*,2*R*)-2-amino-1-(2,5-dimethoxyphenyl)propan-1-ol in 98% ee, which was isolated as

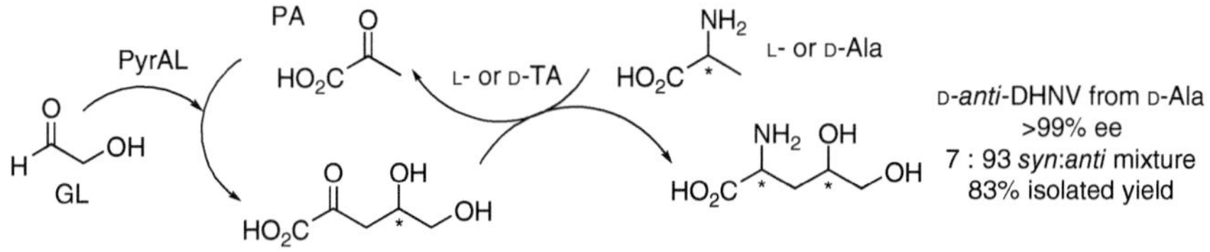

Scheme 13.17 Aldolase–transaminase recycling cascade for the synthesis of DHNV.

its hydrochloride salt with 94% purity (85 mg, 46% yield under non optimized conditions) [67]. The approach consists in the carboligation reaction between PA and 2,5-dimethoxybenzaldehyde (Scheme 13.18), followed by the transamination of the resulting carbonyl group using the *Bacillus megaterium* transaminase (*Bm*TA). The reaction was developed in a sequential mode by addition of the TA and the amine donor, once that an excellent conversion was reached in the carboligation step. Interestingly, depending on the carboligase and transaminase selection, it was possible accessing to all four amino alcohol stereoisomers.

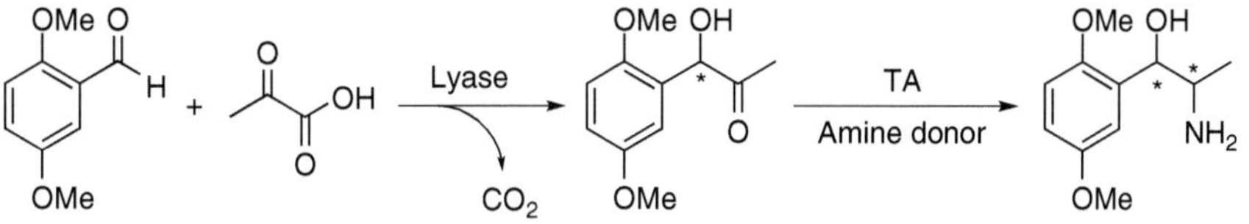

Scheme 13.18 Bienzymatic carboligation-transamination sequence for the one-pot, two-step synthesis of amino alcohol stereoisomers.

Also the synthesis of all four stereoisomers, in this case 4-amino-1-phenylpentane-1-ol, has been disclosed but in this case through the stepwise combination of a commercially available ketoreductase (KRED-P1-B10) and two complementary ATAs (*C. violaceum* or ATA-025) [68]. The approach depicted in Scheme 13.19 consists of (i) the oxidative kinetic resolution of 4-hydroxy-5-phenylpentan-2-one (321 mg, 1.8 mmol), obtaining (*R*)-4-hydroxy-5-phenylpentan-2-one (160 mg, 50% yield, 86% ee) and the diketone (85 mg, 27% yield), the latest being reduced to the (*S*)-hydroxy ketone (86% yield, 71% ee) and (ii) biotransamination of the corresponding hydroxy ketones, which led in analytical scale to

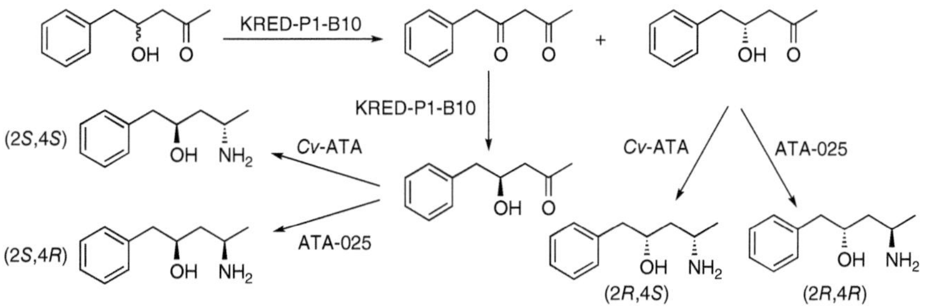

Scheme 13.19 Stepwise formation of all four 4-amino-1-phenylpentan-2-ol stereoisomers.

the four amino alcohol stereoisomers. The synthetic applicability of the approach was demonstrated with the scale-up of the biotransamination of the (*R*)-4-hydroxy-5-phenylpentan-2-one (90 mg, 0.5 mmol) with the ATA-025, leading to the (2*R*,4*R*)-amino alcohol (66 mg, 73% yield).

Finally, in this section, it is worth mentioning the use of firstly an ATA and later a KRED for the preparation of enantiopure amino alcohols and diols [69]. In this case, the kinetic resolution of 2-amino-2-phenylethan-1-ol is carried out by using a selective ATA and pyruvate as an amine acceptor to later develop in one-pot the bioreduction of the so-obtained keto acid intermediate using a CR to produce the 1-phenylethane-1,2-diol. In the work-up, the corresponding vicinal diol was obtained by extraction in an acidic medium to later basify the mixture and extract the amino alcohol (40–42% isolated yield).

13.3.1.4 Other Bienzymatic Stereoselective Synthesis of Amines

In this section, the development of miscellaneous approaches for the synthesis of chiral amines is compiled, combining mainly the use of ATAs with oxidoreductases such as ADHs, ene-reductases (EREDs), IREDs, laccases, MAOs, and synthases. The combination of ADHs and ATAs has been described in the previous section for the stereoselective synthesis of amino alcohols [68, 69] but is also possible to produce amine and alcohol enantiomers by co-immobilization of *Cv*TA and *Lodderomyces elongisporus* displaying KRED activity, converting 4-phenylbutan-2-amine or heptan-2-amine racemates under continuous flow conditions into the (*R*)-amines (44% and 30% yield, >99% ee) and the (*S*)-alcohols (46% and 43% yield, >99% ee) [70].

The use of protein engineering techniques has provided efficient solutions for the development of synthetically useful biotransformations in single and multienzymatic transformations [11–14]. This is the case for the conversion of 3-methylcyclohex-2-enone into the four stereoisomers of 1-amino-3-methylcyclohexane by the combination of an ERED and an ATA (Scheme 13.20) [71]. This approach consists in the reduction of an electron-deficient C=C bond catalyzed by an Old Yellow Enzyme (OYE) and the subsequent biotransamination of the saturated ketone using ATAs. The cascade reaction in preparative scale with the ERED and the Leu56Ile *Vf*ATA led to the (1*S*,3*S*)-1-amino-3-methylcyclohexane in 77% isolated yield (>99% ee and 70% de) after extraction.

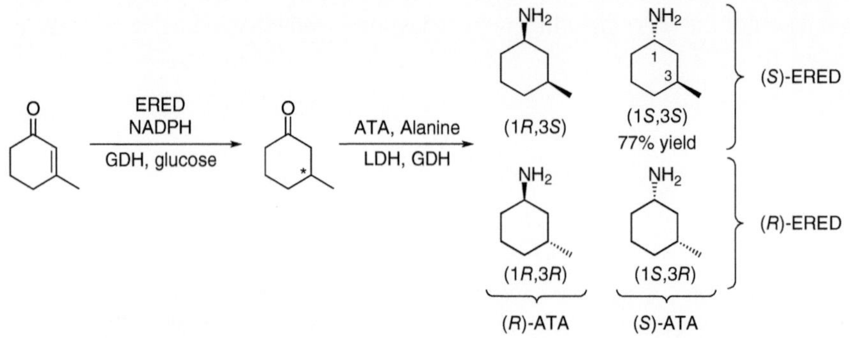

Scheme 13.20 Combination of ene-reductases and amine transaminases for the synthesis of all possible 1-amino-3-methylcyclohexane stereoisomers.

Home-made and commercially available (Prozomix and Johnson Matthey) EREDs and IREDs were tested in a telescopic sequence for the preparation of a variety diastereomerically enriched N-heterocycles [72]. In this manner, Turner and coworkers described for the first time the ERED-catalyzed α,β-unsaturated imines, probably through hydrolyzed ring open ω-amino enones, so later addition of the IRED ($R = CH_3$ after 4 hours, $R = CHCH_3$ after 16 hours) produces in semi-preparative scale the corresponding enantiopure cyclic secondary amines as hydrochloride salts after acidification (Scheme 13.21), (*S*)-2-isopropylpiperidine (40% yield, 14 mg) and (*R*)-2-[(*S*)-*sec*-butyl]piperidine (57% yield, 13 mg).

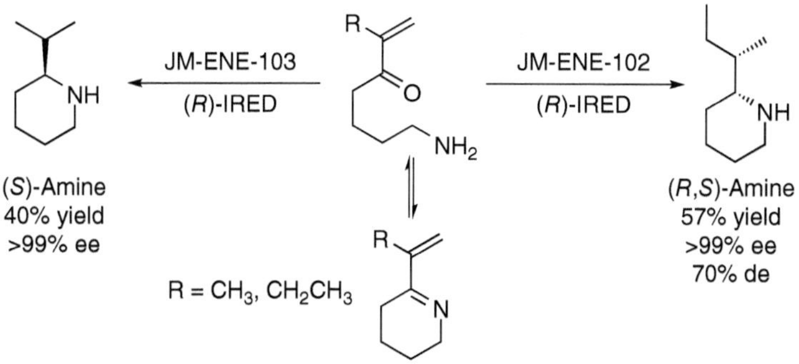

JM-ENE-103 (*R*)-IRED ← [R-substituted ω-amino enone, NH_2] → JM-ENE-102 (*R*)-IRED

(*S*)-Amine 40% yield >99% ee

(*R*,*S*)-Amine 57% yield >99% ee 70% de

$R = CH_3, CH_2CH_3$

Scheme 13.21 Telescopic sequence combining ERED and IRED for the production of diastereomerically enriched piperidines.

Another approach to gain stereoselective access to piperidines is the one described by Kroutil and coworkers by using now an IRED in combination with an ATA (Scheme 13.22) [73]. The authors proposed the biotransamination of nonane-2,6-dione (7.58 g) using the (*R*)-selective ATA from *Arthrobacter* (*ArR*-TA) and subsequent IRED-catalyzed reduction of the corresponding cyclic imine with the IRED-E preparation coupled with the *L. brevis* ADH for cofactor recycling purposes, obtaining the dihydropinidine as hydrochloride salt in 59% isolated yield (5.11 g).

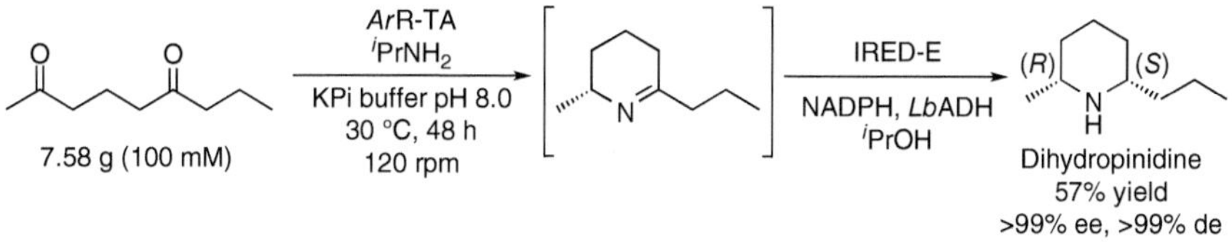

7.58 g (100 mM) → *ArR*-TA, $^{i}PrNH_2$, KPi buffer pH 8.0, 30 °C, 48 h, 120 rpm → [cyclic imine] → IRED-E, NADPH, *Lb*ADH, $^{i}PrOH$ → (*R*)(*S*) Dihydropinidine 57% yield >99% ee, >99% de

Scheme 13.22 Bienzymatic cascade production of dihydropinidine using the ATA-IRED catalytic system.

Moving to the asymmetric synthesis of 2,5-disubstituted pyrrolidines, Turner's group have reported to different approaches toward this class of heterocycles by combining a biotransamination process with the reduction of the corresponding imine either through a MAO-borane catalytic system or the use of a RedAm (Scheme 13.23). On the one hand, in 2014, after optimization of the individual steps, the commercially available ATA-113 from Codexis

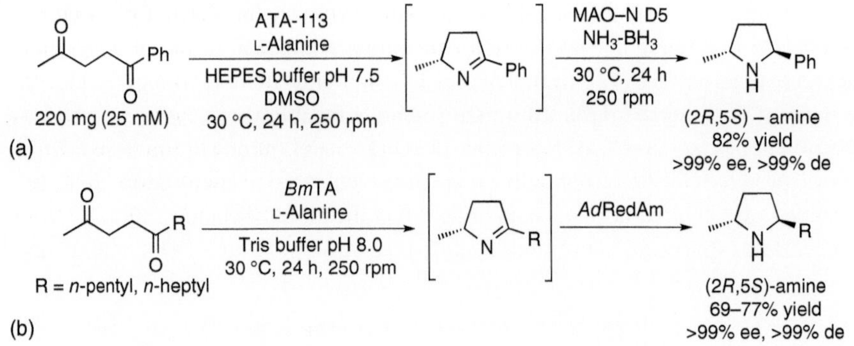

Scheme 13.23 Multienzymatic synthesis of chiral 2,5-substituted pyrrolidines using amine transaminases and (a) monoamine oxidases; (b) reductive aminases.

and the MAO-N D5 variant from *A. niger* were applied in a sequential one-pot strategy for the transformation of 1-phenylpentane-1,4-dione (220 mg, 1.25 mmol) into enantiopure (2*R*,5*S*)-2-methyl-5-phenylpyrrolidine, isolated in 82% yield after extraction and column chromatography purification (Scheme 13.23a) [74]. On the other hand, the combined use of *Bacillus megaterium* transaminase (*Bm*TA) with the reductive aminase from *Ajellomyces dermatitidis* (*Ad*RedAm) led to enantio- and diastereomerically pure (2*R*,5*S*)-2-methyl-5-alkylpyrrolidines in good yield and purity after extraction at basic pH (Scheme 13.23b) [75].

The synthesis of a series of alkaloids by means of the combined use of TAs and synthases is also an elegant example of the importance of biocatalysis in organic synthesis. For instance, (*S*)-1-(3,4-dihydroxybenzyl)-1,2,3,4-tetrahydroisoquinoline-6,7-diol was obtained starting from dopamine (0.5 mmol) using a *C. violaceum* TA variant (*Cv*2025 TA) and the norcoclaurine synthase Δ29*Tf*NCS, which can be later transformed into (*S*)-6,8,13,13*a*-tetrahydro-5*H*-isoquinolino[3,2-*a*]isoquinoline-2,3,10,11-tetraol by addition of formaldehyde to the reaction medium (Scheme 13.24a) [76]. Finally, strictosidine derivatives have been obtained by

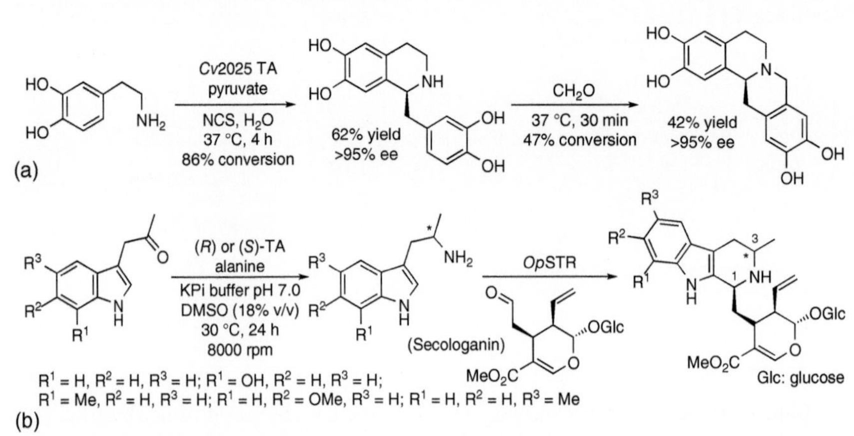

Scheme 13.24 Alkaloid syntheses by combination of amine transaminases and synthases for the production of (a) (*S*)-6,8,13,13a-tetrahydro-5*H*-isoquinolino[3,2-a]isoquinoline-2,3,10,11-tetraol; (b) strictosidine derivatives.

combining selective transaminases with the strictosidine synthase from *Ophiorrhiza pumila* (*Op*STR) through stepwise or sequential approaches by biotransamination of the corresponding ketones, and subsequent Pictet–Spengler condensation of α-methyltryptamine intermediates with secologanin (Scheme 13.24b) [77]. Interestingly, the use of enantiocomplementary TAs allowed the formation of the different possible epimers, developing in some cases semi-preparative transformations to isolate the C3-methylated strictosidine derivatives after purification by column chromatography and preparative high-performance liquid chromatography (HPLC) (12–15 mg, 70–80% yield, >98% de).

13.3.2 Stereoselective Bienzymatic Cascades Toward the Production of Other Organic Compounds

In this section, the focus will be on the development of linear bienzymatic transformations leading to optically active compounds different from amines, covering the synthesis of multiple families such as alcohols, amino acids, diols, epoxy alcohols, among many others. For that reason, the use of several classes of biocatalysts has been taken into account, trying to organize this section based on the type of employed enzymes.

13.3.2.1 Synthesis of Organic Compounds Other Than Amino Acids

ADHs have been largely applied in asymmetric synthesis for the production of chiral alcohols, their work under mild reaction conditions, and in a wide pH range making plausible their compatibility with other enzymes to carry out cascade transformations. This is the case of the lyase-ADH systems, which have been extensively exploited, especially for the synthesis of benzoins by benzaldehyde lyase (BAL)-catalyzed carboligation between two (different) aldehydes and later reduction of the remaining carbonyl group (Scheme 13.25a). Rother's group has made a series of interesting contributions in this field, such as the utilization of tea bags filled with whole cell enzymes, which possesses multiple advantages over traditional methods such as the easy enzyme handling and recovery, simplifying the downstream process and allowing the easy combination of different enzyme classes. This approach has been applied to the carboligation of benzaldehyde with acetaldehyde catalyzed by *Pseudomonas fluorescens* BAL (*Pf*BAL) and subsequent carbonyl reduction with the ADH from *Ralstonia* species (*Ras*ADH), yielding 1-phenylpropane-1,2-diol in complete enantio- and diastereoselectivity [78].

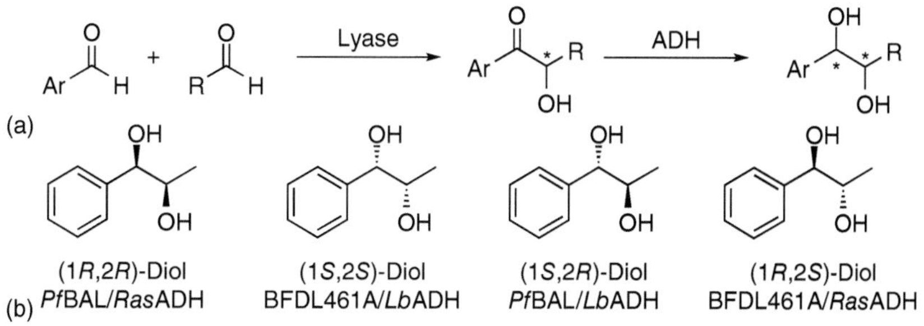

(a) Ar–CHO + R–CHO —Lyase→ Ar–CO–CH(OH)–R —ADH→ Ar–CH(OH)–CH(OH)–R

(b)
(1*R*,2*R*)-Diol *Pf*BAL/*Ras*ADH
(1*S*,2*S*)-Diol BFDL461A/*Lb*ADH
(1*S*,2*R*)-Diol *Pf*BAL/*Lb*ADH
(1*R*,2*S*)-Diol BFDL461A/*Ras*ADH

Scheme 13.25 Lyases and ADHs for the synthesis of optically active diols: (a) reaction scheme; (b) substrate scope.

The reaction setup was based in the use of eight tea bags of *Pf*BAL for 32 ml reactions, which was successfully achieved in a sequential mode under a microaqueous system (triethanolamine buffer pH 9.0 and 97.5% v/v MTBE). Therefore, the carboligation was developed in a high benzaldehyde concentration (500 mM), followed by feeding of acetaldehyde in portions and eight tea bags containing *Ras*ADH using cyclohexanol for cofactor recycling purposes, isolating 1.8 g of (1*R*,2*R*)-1-phenylpropane-1,2-diol (75.7% yield) after extraction and column chromatography. Interestingly, the enzyme catalytic system was used over four cycles without apparent activity loss. Two years later, the scale-up to 140 ml scale reaction was described using a SpinChem reactor [79]. This reaction has also been reported in a cascade approach by addition of lyophilized *Pf*BAL and *Ras*ADH cells to a screw-capped 23 ml glass vial (triethanolamine buffer pH 10.0 and 92% v/v MTBE), producing enantio- and diastereopure (1*R*,2*R*)-1-phenylpropane-1,2-diol (847.7 mg, 58% yield) [80]. Remarkably, combining two carboligases (*Pf*BAL or benzoylformate decarboxylase variant L461A from *Pseudomonas putida*) and two ADHs (*Ras*ADH or *Lb*ADH), it was possible to produce all four 1-phenylpropane-1,2-diol stereoisomers (Scheme 13.25b) [81].

An attractive strategy for the production of hydroxylated compounds is the *in situ* generation of aldehyde intermediates by either enzyme-catalyzed oxidation of primary alcohols or reduction of carboxylic acids to later react them with carbonyl compounds (Scheme 13.26). This is the case of the ADH from *P. putida* (AlkJ) or carboxyl acid reductase from *Nocardia iowensis* (*Ni*CAR) able to produce a series of five aldehydes, which can be accepted as substrates for the dihydroxyacetone (DHA)-dependent aldolase mutant Fsa1-A129S (Scheme 13.26a) [82]. By co-expression in *Escherichia coli* of the ADH and the fructose-6-phosphate aldolase (FSA), it was plausible to carry out the cascade reaction in a solution composed by the DHA and MeCN (5% v/v) at 25 °C and 250 rpm, obtaining (3*S*,4*R*)-1,3,4-trihydroxy-5-phenylpentan-2-one in 70% yield (147 mg) after six hours and solid-phase extraction protocol. Also generating an aldehyde *in situ* with an AOX, the stereoselective synthesis of benzoin derivatives is possible by later carboligation reaction with a lyase (Scheme 13.26b) [83]. This can be exemplified by the benzyl alcohol oxidation catalyzed by AOX from *Pichia pastoris* (*Pp*AOX) and subsequent condensation between two molecules of benzaldehyde, forming the (*R*)-benzoin in 83% isolated yield after extraction

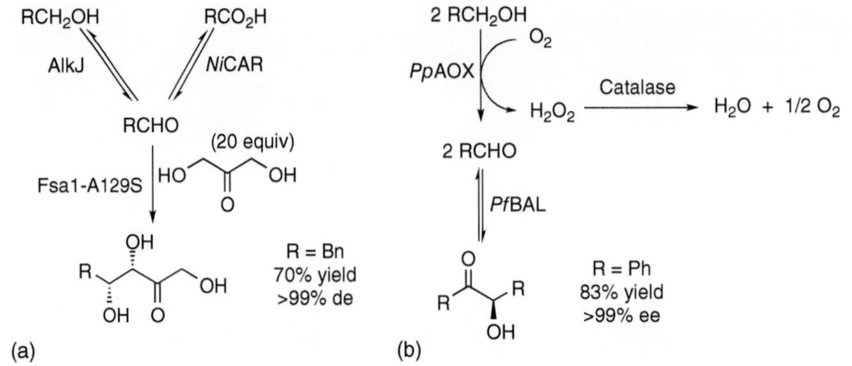

Scheme 13.26 *In situ* generation of aldehydes using an oxidoreductase for their subsequent C–C coupling with lyases to form (a) (3*S*,4*R*)-1,3,4-trihydroxy-5-phenylpentan-2-one; (b) (*R*)-benzoin.

(4.48 g). It must be highlighted that the use of a catalase is required to circumvent the enzyme inhibition caused by the hydrogen peroxide released in the oxidation reaction.

Taking advantage of the high reactivity of aldehydes, other examples can be here briefly commented, for instance, the sequential one-pot transformation of aliphatic aldehydes (glutaraldehyde $n = 1$, adipaldehyde $n = 2$) to optically active vicinal diols by intramolecular stereoselective carboligation using a pyruvate decarboxylase (PDC), followed by ADH-catalyzed carbonyl reduction of the corresponding 2-hydroxycycloalkanone intermediates (48–71% yield, Scheme 13.27). Interestingly, the screening of PDC and ADH led to the chiral *cis*- or *trans*-diols with excellent selectivity (>97% ee) and moderate to good yields (48–71%) after column chromatography [84].

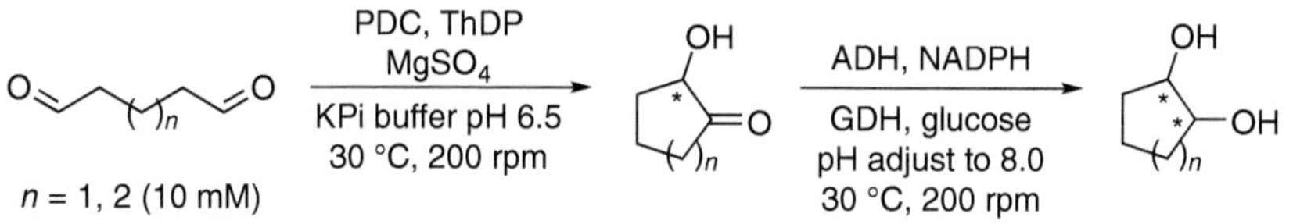

Scheme 13.27 Intramolecular C—C bond formation and carbonyl reduction for the synthesis of chiral cyclic vicinal diols.

α,β-Unsaturated aldehydes have turned out to be ideal substrates for the development of biocatalytic hydrogen-borrowing cascades combining EREDs and an aldehyde dehydrogenase (Ald-DH) to produce saturated carboxylic acids in a stereoselective manner (Scheme 13.28) [85]. Remarkably, the addition of external hydride source is not required because the hydride from the ERED-catalyzed reductive step is liberated in the CAR-catalyzed oxidative step. Preparative scale for the conversion of (*E*)-2-methyl-3-phenylacrylaldehyde (100 mg) into (*S*)-2-methyl-3-phenylpropanoic acid (>98% ee) was achieved with 96% chemoselectivity (88% isolated yield).

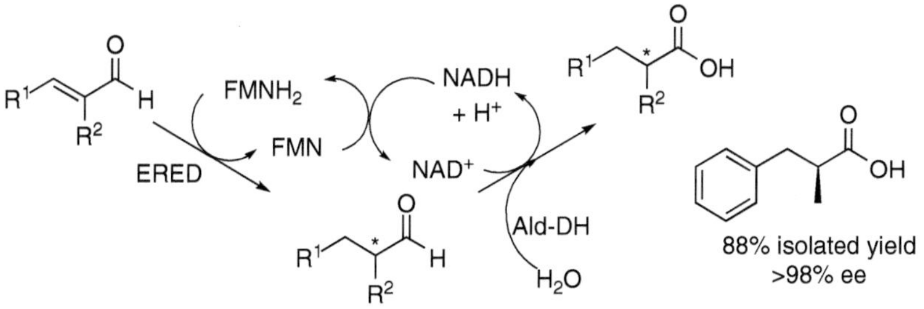

Scheme 13.28 Stereoselective hydrogen-borrowing conversion of α-substituted α,β-unsaturated aldehydes into chiral α-substituted carboxylic acids.

The combination of EREDs for the bioreduction of C=C bonds with the use of oxidoreductases have allowed the selective modification of carvone enantiomers by using ADHs or Baeyer–Villiger monooxygenases (BVMOs, Scheme 13.29). In the first case, You and coworkers have reported the synthesis of all four dihydrocarvone stereoisomers by ERED-catalyzed bioreduction of the carvone enantiomers, which can be followed by the carbonyl

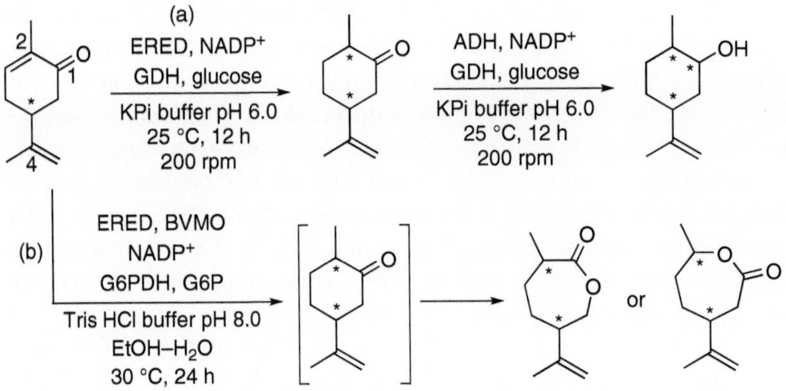

Scheme 13.29 Bienzymatic transformations over carvones involving the use of EREDs for the asymmetric C=C bond reduction and (a) ADHs; (b) BVMOs.

reduction using an ADH, leading to all eight stereoisomeric dihydrocarveols (>95% de, Scheme 13.29a) [86]. The synthesis of (1*R*,2*S*,4*R*)- (12.1 mg, 48% yield), (1*R*,2*S*,4*S*)- (9.9 mg, 40% yield), and (1*S*,2*S*,4*S*)-stereoisomers (12.8 mg, 52% yield) from 25 mg of the corresponding carvone enantiomer (5.6 mM) was described in a stepwise manner after extraction of the corresponding dihydrocarvone intermediate and column chromatography purification of the final dihydrocarveols. Interestingly, the combination of ERED and BVMO afford the one-pot cascade synthesis of carvolactones, four of them obtained after preparative biotransformation (30–76% yield after column chromatography, Scheme 13.29b) [87].

Also using EREDs and ADHs is possible for the performance of scalable processes to convert ethyl 4-oxo-pent-2-enoates into optically active γ-butyrolactones in high yield (171–656 mg, up to 90% after extraction and column chromatography) and excellent selectivity (98–99% ee) [88]. Pietruszka and coworkers have described a one-pot sequential approach starting with the bioreduction of the γ-keto ester with the *Bacillus subtilis* ERED (YqjM, wild type or mutants) and followed by carbonyl reduction using either Prelog-ADH from *Thermoanaerobacter* species or anti-Prelog-ADH from *L. brevis* (Scheme 13.30).

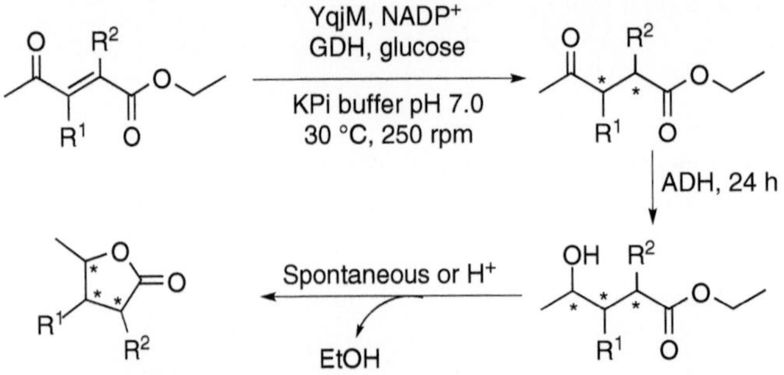

Scheme 13.30 ERED and ADH in sequential one-pot synthesis of γ-butyrolactones from γ-keto esters (R^1 = H and R^2 = Me; R^1 = Me and R^2 = H).

Next, the stereoselective synthesis of a wide variety of alcohols will be discussed, starting with the preparation of allylic alcohols, which are versatile building blocks in organic synthesis because they can be easily transformed into other chemical functionalities. For instance, Wu and coworkers have described the bioreduction of two families of α,β-unsaturated ketones, followed by their bioepoxidation, obtaining enantiopure epoxy alcohols in a straightforward manner (Scheme 13.31). In 2015, 14 α,β-unsaturated ketones including aryl, heteroaryl, and alkyl substituents were transformed into the corresponding enantiopure glycidol derivatives by bioreduction of the carbonyl group using a home-made (*S*)-specific oxidoreductase (*Ch*KRED03), and then addition of the styrene monooxygenase (SMO) from *P. putida* allowed the formation of the enantiopure epoxy alcohols in low to moderate yield after extraction and column chromatography (Scheme 13.31a) [89]. One year later, the same research group reported the co-expression of *Rhodococcus erythropolis* DSM 43297 ADH with a SMO for the formation of allylic epoxy alcohols, suppressing the concomitant oxidation of the alcohol by using isopropanol (2.5% v/v) as a co-substrate (Scheme 13.31b) [90].

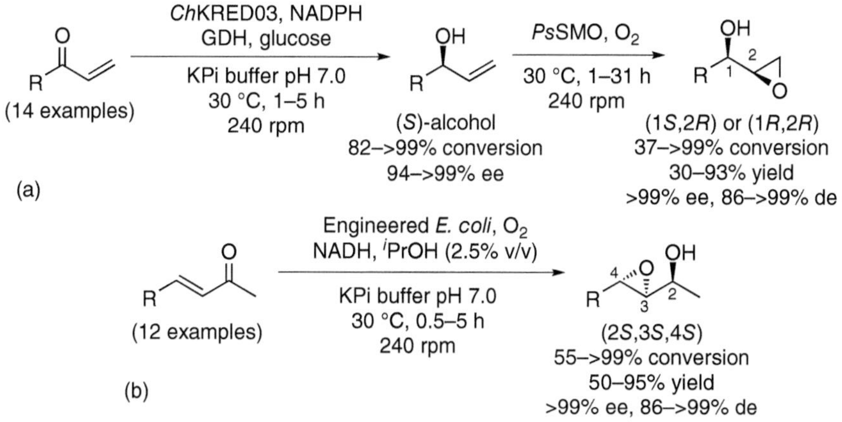

(a)
R (14 examples) → *Ch*KRED03, NADPH, GDH, glucose; KPi buffer pH 7.0, 30 °C, 1–5 h, 240 rpm → (*S*)-alcohol, 82–>99% conversion, 94–>99% ee → *Ps*SMO, O_2; 30 °C, 1–31 h, 240 rpm → (1*S*,2*R*) or (1*R*,2*R*), 37–>99% conversion, 30–93% yield, >99% ee, 86–>99% de

(b)
R (12 examples) → Engineered *E. coli*, O_2, NADH, ^{i}PrOH (2.5% v/v); KPi buffer pH 7.0, 30 °C, 0.5–5 h, 240 rpm → (2*S*,3*S*,4*S*), 55–>99% conversion, 50–95% yield, >99% ee, 86–>99% de

Scheme 13.31 ADHs and monooxygenases for the asymmetric synthesis of epoxy alcohols through (a) sequential approach; (b) cascade mode.

Allylic alcohols are also activated substrates for the development of chemoenzymatic oxidations through laccase mediator systems, which combined the use of the own laccase with an oxy-radical specie that is the real oxidizing agent. Recently, our research group has described a redox isomerization of allylic secondary alcohols involving two steps: (i) alcohol oxidation using the laccase from *T. versicolor* (L*Tv*) with the TEMPO for the formation of the corresponding α,β-unsaturated ketones and (ii) ERED-catalyzed bioreduction of the C=C bond using commercially available enzymes from Codexis Inc. (Scheme 13.32) [91]. The strategy was carried in a one-pot two-step sequential mode because of the incompatibilities when trying to develop it in a cascade manner: (i) the oxidation requires aerobic conditions, while the reduction needs to be developed in a closed vial because of ERED inactivation and (ii) TEMPO was not compatible with the GDH system required for the nicotinamide adenine dinucleotide phosphate (NADPH) recycling used in the bioreduction step as can

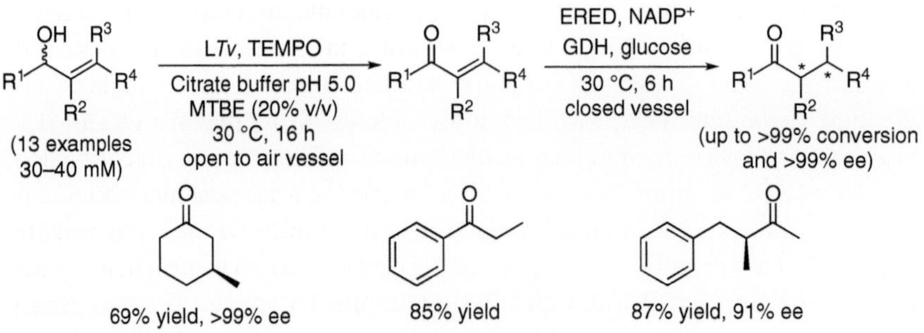

Scheme 13.32 Redox isomerization of secondary allylic alcohols.

oxidize the glucose. From a set of 13 allylic alcohols, the practical applicability of the redox isomerization was demonstrated for 1-phenylprop-2-en-1-ol (40 mM) and (*E*)-3-methyl-4-phenylbut-3-en-2-ol (30 mM), yielding (*S*)-3-methylcyclohexan-1-one (69% yield, >99% ee), propiophenone (85% yield), and (*S*)-3-methyl-4-phenylbutan-2-one (87% yield, 91% ee) after extraction.

Propargylic alcohols are also valuable organic molecules because they are present in a wide variety of biologically active molecules, possessing the alcohol and alkyne groups, which are susceptible to be easily modified for the formation of other classes of organic compounds. Recently, deracemization of 1-arylprop-2-yn-1-ols was described by combining their oxidation using the L*Tv*-TEMPO catalytic system, followed by selective ketone reduction with a (*S*)- (*Ras*ADH) or (*R*)-stereospecific enzyme (commercially available evo-1.1.200 ADH) [92]. Thus, deracemization of 1-phenylprop-2-yn-1-ol was demonstrated at 50 mg scale, yielding the (*S*)-alcohol in 83% yield and its antipode in 79% after extraction and column chromatography both in 98% ee (Scheme 13.33a). Similarly, deracemization of

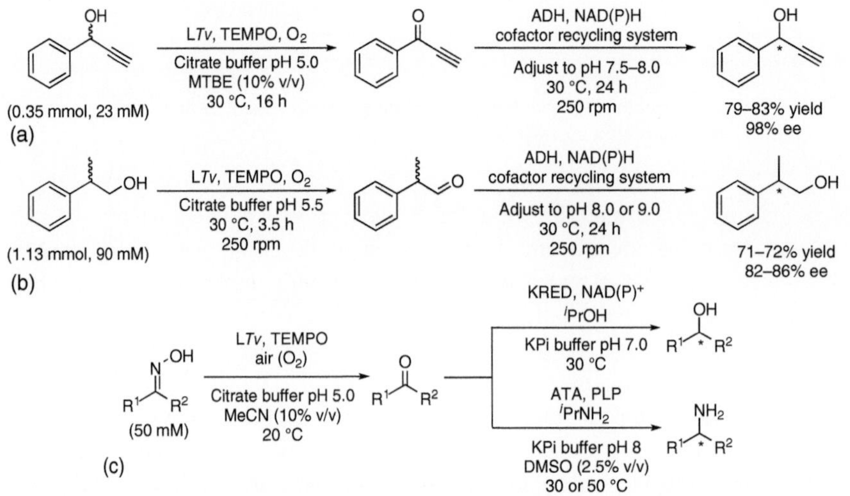

Scheme 13.33 Laccase mediator system and alcohol dehydrogenase-catalyzed bioreduction for the: (a) deracemization of 1-phenylprop-2-yn-1-ol; (b) deracemization of 2-phenylpropan-1-ol; and (c) transformation of ketoximes into optically active amines or alcohols.

2-phenylpropan-1-ol (150 mg, 90 mM) can be achieved under dynamic conditions by developing the oxidation for 3.5 hours under acidic conditions to later change the pH to 8 (ADH from horse liver, *Hl*ADH) or 9 (evo-1.1.2000), adding the cofactor and the co-substrate to yield the optically active alcohol in 71–72% yield and 82–86% ee (Scheme 13.33b) [93]. Different optically active amines (70–>99% conversion) or alcohols (83–>99% conversion) have been recently obtained in analytical scale by combining a laccase-mediated deoximation of the corresponding ketoximes to produce the corresponding ketone intermediates and subsequent biotransamination or bioreduction processes using ATAs or ADHs, respectively (Scheme 13.33c) [94]. Because of the different optimum pH of each class of enzymes, the approach is carried out in a sequential mode modifying until 25 and 17 mM ketone concentration for alcohol and amine syntheses, respectively, before adding the corresponding enzyme. For the deoximation reaction, generally 10% of acetonitrile (MeCN) was required as a cosolvent, but the use of only 1% v/v of Chemophor®, a polyethoxylated castor oil, enables to properly develop the reaction even at preparative scale, obtaining enantiopure (*S*)-1-(4-bromophenyl)propan-1-ol (80 mg, 85% isolated yield after filtration on silica gel) when using the ADH from *Rhodococcus ruber*.

The majority of the reported protocols developed for multienzymatic cascades are carried out in aqueous medium, the natural one for enzymes. However, it is also possible to perform multienzymatic transformations in organic solvents when using immobilized forms. This is the case for the synthesis of optically active cyanohydrin derivatives combining the use of *Manihot esculenta* hydroxynitrile lyase (*Me*HNL) and *C. antarctica* lipase type A (CAL-A). Thus, after a one-pot cascade involving the hydrocyanation of 4-anisaldehyde and subsequent benzoylation, (*S*)-4-methoxymandelonitrile benzoate was obtained in 81% yield and 98% ee (Scheme 13.34) [95], which is an immediate precursor of the natural compound (*S*)-tembamide after nickel Raney-catalyzed hydrogenation [96]. Best conditions for the HNL and lipase actions were found when adding the three equivalents of phenyl benzoate in equal portions during the time (0, 24, and 48 hours).

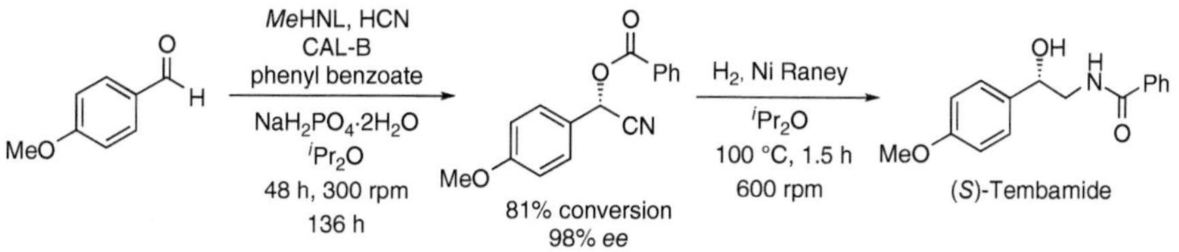

*Me*HNL, HCN
CAL-B
phenyl benzoate
$NaH_2PO_4{\cdot}2H_2O$
$^iPr_2O$
48 h, 300 rpm
136 h

81% conversion
98% *ee*

H_2, Ni Raney
$^iPr_2O$
100 °C, 1.5 h
600 rpm

(*S*)-Tembamide

Scheme 13.34 Chemoenzymatic synthesis of (*S*)-tembamide through bienzymatic hydrocyanation and benzoylation.

13.3.2.2 Amino Acid Synthesis

Amino acids represent a highly demanding class of compounds because they are essential chiral building blocks for the synthesis of multiple agrochemicals, pharmaceuticals, and fine chemicals. Multiple approaches have been exploited for their modification and production, so next, recent examples of bienzymatic linear and cyclic cascades will be discussed for their preparation in an asymmetric manner. For instance, L-tyrosine derivatives are key precursors for anticancer drugs and biochemical markers, and the stereoselective biocatalytic synthesis of L-tyrosine derivatives has been described in a one-pot cascade approach using arenes, pyruvate, and ammonia (NH_3) [97]. The strategy is based in two steps: (i) cytochrome P450-catalyzed hydroxylation of the substituted benzene at the *o*-position and (ii) C–C coupling and asymmetric amination of the corresponding phenol with a phenol lyase (Scheme 13.35). For this purpose, a series of P450 BM3 variants were tested in combination with the tyrosine phenol lyase from *Citrobacter freundii* (*Cf*TPL) transforming under optimized conditions a series of six arenes (40 mM) in low to moderate yield (14.9–124 mg, 6–49%) after cation exchange chromatography purification but with excellent selectivity (>97% ee).

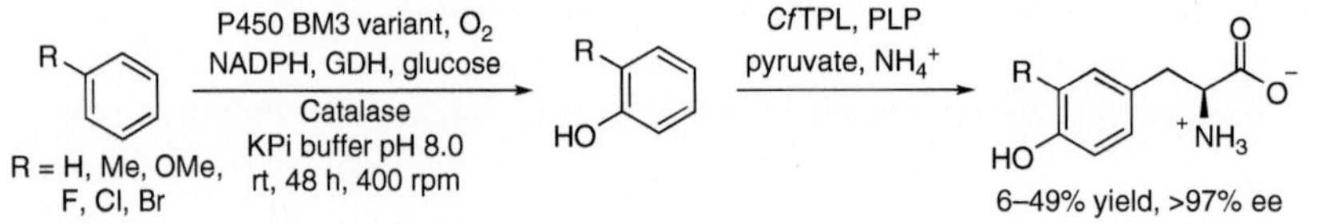

Scheme 13.35 One-pot cascade synthesis of L-tyrosine derivatives combining an engineered cytochrome P450 and a tyrosine phenol lyase.

Another interesting α-amino acid is serine, which can be prepared in its D-form from glycine and formaldehyde using a D-threonine aldolase from *Arthrobacter* species ATCC in 84% yield (22 g, >99% ee, Scheme 13.36) [98]. Interestingly, the later transformation of D-serine into a series of amino acids (13–84% yield, >99% ee), including L-tryptophan and L-cysteine, has been possible through dynamic kinetic resolution (DKR) using an alanine racemase from *B. subtilis* and a tryptophan synthase from *E. coli* k-12 MG1655 at 40 °C and 200 rpm.

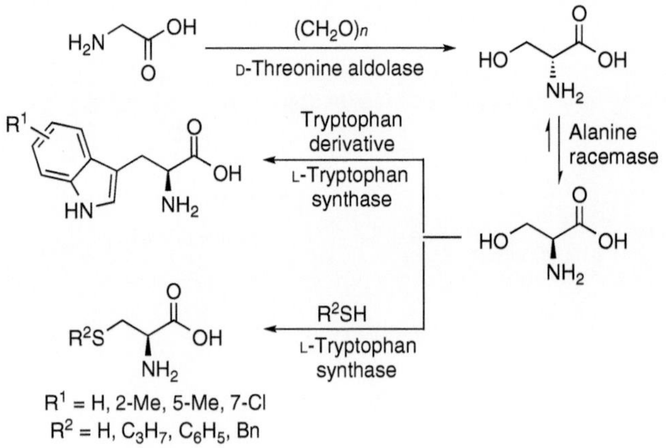

Scheme 13.36 Dynamic kinetic resolution of serine to produce L-noncanonical amino acids using a racemase and a synthase.

In a previous section, deracemization of amines has been discussed using a series of enzyme classes such as AmDHs, MAOs, or transaminases. Interestingly, this strategy has also been successfully exploited for the synthesis of optically active amino acids. For instance, the combined use of D-amino acid transaminase from *Bacillus sphaericus* (D-ATA) and a (*S*)-selective ω-TA from *O. anthropi* (*Oa*ATA) has been described by Park and Shin (Scheme 13.37), producing a series of L-amino acids including (*R*)-2-aminobutanoic acid and (*R*)-2-aminohexanedioic acid when starting from racemic 2-aminobutanoic (1.55 g, 300 mM) and 2-oxohexanedioic acid (1.24 g, 170 mM) [99].

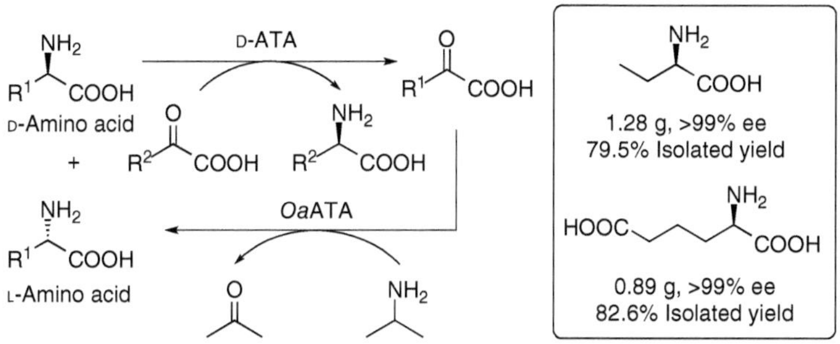

Scheme 13.37 Deracemization of amino acids using a d-amino acid transaminase and a (*S*)-selective ω-transaminase.

An alternative approach is the one described by Turner and coworkers consisting in the stereoinversion and deracemization of 18 L- and racemic phenyl alanine derivatives bearing electron-donating or -withdrawing substituents along the aromatic ring (Scheme 13.38) [100]. For that purpose, a L-amino acid deaminase (L-AAD from *Proteus mirabilis*) is coupled with a D-amino acid aminotransferase (D-AAT T242G from *Bacillus* species YM-1) both in whole cell preparations and reacted with D-glutamate as an amine donor for four hours at 37 °C in 10 mM substrate concentration. Preparative scale synthesis was developed for the production of D-4-fluoroalanine in 84% yield (76.9 mg, >99% ee) from its racemic form and using in this case D-aspartate as an amine donor. A similar strategy has been recently reported for the stereoinversion of L-amino acids involving a L-AAD from *P. mirabilis* with a D-amino acid dehydrogenase from *Symbiobacterium thermophilum*, leading to

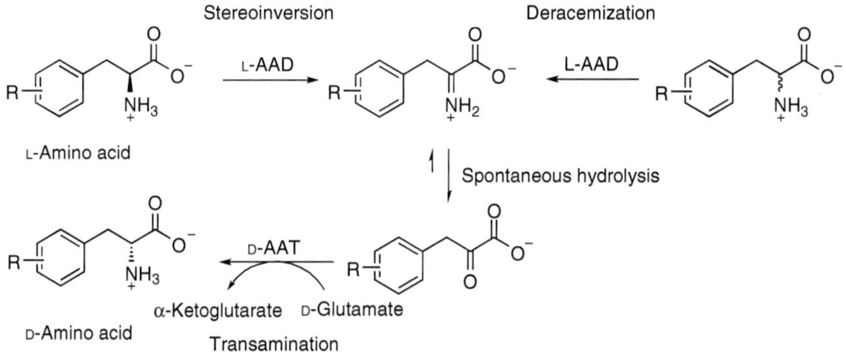

Scheme 13.38 Deracemization or stereoinversion of amino acids using l-amino acid deaminase with a d-amino acid aminotransferase.

a series of D-amino acids including the scale-up (80 mM, 50 ml reaction system) to produce D-phenyl alanine after six hours at 45 °C and 220 rpm with 91% isolated yield and enantiopure form after isolation by cation exchange resin column [101].

As a final example for the production of (*R*)-β-arylalanines, the combination of an enzyme displaying phenylalanine aminomutase (PAM) and phenylalanine ammonia lyase (PAL) activity such as the one from *Streptomyces maritimus* (EncP) and a strict lyase from *Anabaena variabilis* (*Av*PAL) has allowed the production and kinetic resolution of a panel of racemic amino acids, including the preparation of (*S*)-2-amino-3-(3-fluorophenyl)propanoic acid (40.5 mg) in enantiopure form after reaching 50% conversion in the kinetic resolution step (Scheme 13.39) [102].

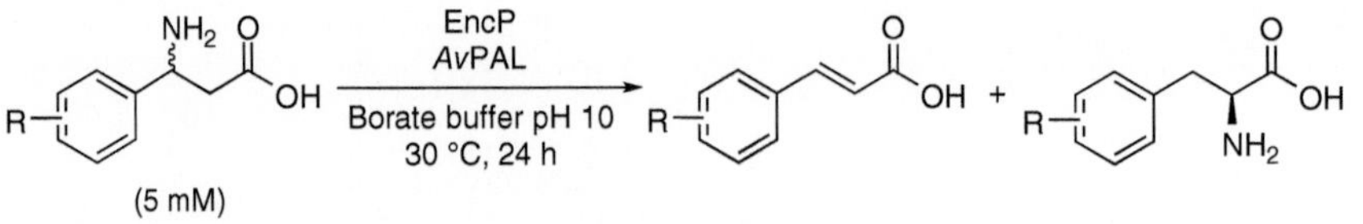

Scheme 13.39 Synthesis via kinetic resolution of β-arylalanines using enzymes displaying phenylalanine aminomutase and ammonia lyase activities.

13.4 Multienzymatic Transformations: Increasing Synthetic Complexity

To end this chapter, a series of multienzymatic reactions involving the linear use of more than two biocatalysts is disclosed, thus increasing the complexity level in the biotransformation. The keys for enzyme compatibility in these types of transformations will be analyzed toward the (stereo)selective production of a series of organic molecules. Brenna and coworkers have reported the three-step synthesis of chiral 3-oxoesters from alkyl cycloalkenecarboxylates (Scheme 13.40). The transformation was carried out in a sequential mode involving the chemoselective hydroxylation of cycloalkenecarboxylates, laccase-mediated oxidation of the corresponding allylic alcohol intermediate, and final ERED-catalyzed

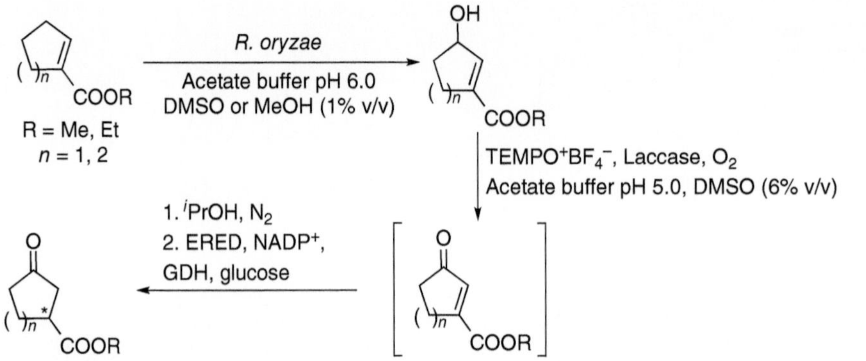

Scheme 13.40 Sequential synthesis of chiral cyclic γ-oxo esters involving enzymatic hydroxylation, oxidation, and reduction steps.

reduction of the C=C bond [103]. From four filamentous fungi, the best results in the hydroxylation reaction were found with the one from *Rhizopus oryzae*, which were entrapped in alginate beads employing minimum amount of cosolvent (1% v/v) such as MeOH for the methyl esters and DMSO for the ethyl esters to prevent the formation of by-products. The reactions took several days, and after beads filtration and extraction of the product, the oxidation was carried out without further purification using the catalytic system composed by $TEMPO^+BF_4^-$ and the laccase Amano M120 under oxygen atmosphere over 24 hours at 30 °C. After complete conversion of the hydroxyl ester, nitrogen was bubbled to remove dissolved oxygen and then the corresponding OYE and the required cofactor recycling system were added to obtain after additional 24 hours at 30 °C and 160 rpm, the corresponding (*S*) or (*R*)-oxo esters in 69–77% yield and 86–99% ee after extraction.

The same research group has reported a three-step synthesis of (1*S*,3*S*)- and (1*R*,3*S*)-3-methylcyclohexanol from 1-methylcyclohex-2-en-1-ol involving a laccase, an ERED, and an ADH (Scheme 13.41) [104]. The chemoenzymatic reaction has been carried out in a one-pot sequential mode involving (i) the [1,3]-oxidative rearrangement of the endocyclic tertiary alcohol with a laccase mediator system composed by the laccase from *T. versicolor* and the $TEMPO^+BF_4^-$ oxoammonium salt; (ii) stereoselective bioreduction of the endocyclic alkene; and (iii) asymmetric bioreduction of the carbonyl compound, thus creating a total of two stereocenters. The process was started in 50 mM substrate concentration at pH 5.2 under oxygen atmosphere for the oxidation step, and after 24 hours, nitrogen gas was bubbled and the reaction diluted to 2.5 mM, increasing the pH to a neutral value to enable the action of the OYE enzyme and the ADH, affording the *cis*- and *trans*-alcohol in 85% and 82% isolated yield (93–97 mg), respectively, after column chromatography purification.

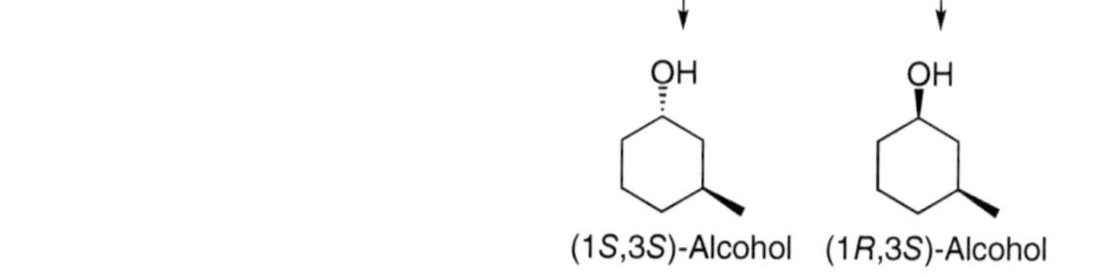

Scheme 13.41 Sequential synthesis of 3-methylcyclohexanol diastereoisomers involving enzymatic oxidation, followed by C=C and carbonyl reduction steps.

An ERED has also been involved in the synthesis of another alcohol such as 2-phenylethanol, which possesses a rose-like aroma, starting from cinnamaldehyde and involving the use of a BVMO and a formate dehydrogenase (FDH, Scheme 13.42a) [105]. After optimization of the reaction conditions, the cascade was developed in Tris–HCl buffer pH 8.0 at 25 °C and 200 rpm, reaching complete conversion after only 12 hours and recovering

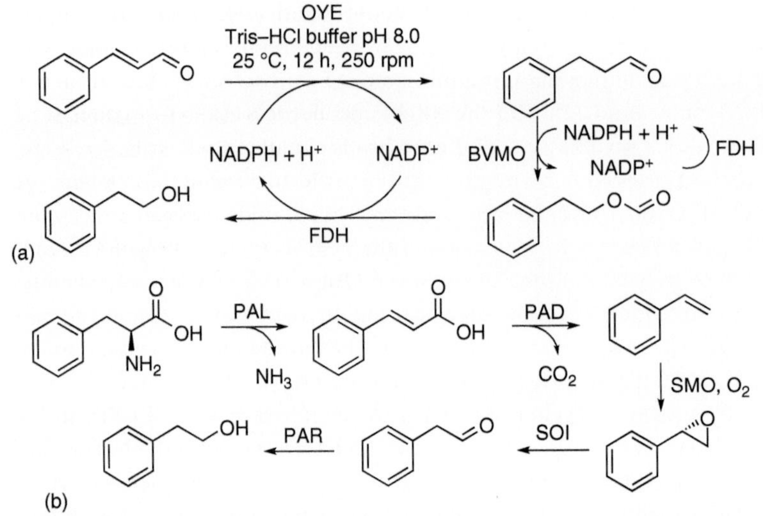

Scheme 13.42 One-pot syntheses of 2-phenylethanol from cinnamaldehyde (a) and l-phenylalanine (b).

2-phenylethanol through extraction in around 60% yield. The synthesis of 2-phenylethanol has also been reported in a fancy five enzyme cascade developed by engineering a recombinant *E. coli* strain expressing PAL, phenylacrylic acid decarboxylase (PAD), SMO, styrene oxide isomerase (SOI), and phenylacetaldehyde reductase (PAR) from L-phenyl alanine (248 mg, 1.5 mmol) in phosphate buffer pH 8.0 at 30 °C for six hours, obtaining a 89% conversion and recovering the alcohol in excellent purity and 72% yield after extraction and column chromatography (Scheme 13.42b) [106]. This study reports an extensive study focusing on the product recovery, showing the use of adsorbents for *in situ* product removal and a biodiesel-aqueous two-phase system to allow a better product recovery and avoid product inhibition, demonstrating the reusability of the catalytic system during various batches, although with continuous loss of the enzyme activity (87% of the original productivity remains after three batches but significantly decrease afterward).

Another strategy for the synthesis of 2-phenylethanol is the anti-Markovnikov hydration of styrene, an approach that can be applied to a series of alkenes and which can also be applied to the synthesis of amines through anti-Markovnikov alkene hydroamination (Scheme 13.43) [107]. The hydroamination and hydration of 12 alkenes consisted in the

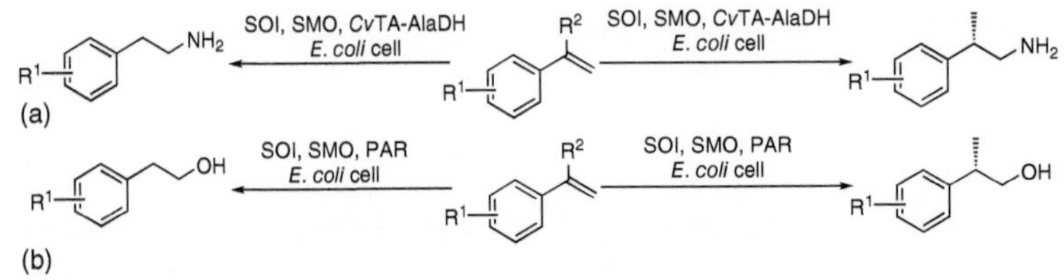

Scheme 13.43 One-pot syntheses of phenethylamines and 2-phenylethanols through (a) hydroamination and (b) hydration, respectively, of styrene derivatives.

epoxidation–isomerization–amination and epoxidation–isomerization–reduction, respectively, co-expressing the required enzymes in a *E. coli* strain composed by (i) SMO, SOI, transaminase from *C. violaceum* (*Cv*TA), and AlaDH for the amines (45–>99% conversion, >99% regioselectivity) and (ii) SMO, SOI, and PAR for the alcohols (60–>99% conversion, >99% regioselectivity). Preparative biotransformations over four alkenes (R^1 = H, 4-F, 4-Me, and 4-OMe and R^2 = H) were developed in 2.5 mmol scale to give the corresponding phenethylamines after extraction and column chromatography (236–309 mg, 83–95% conversion, 68–82% yield) and 2-phenylethanols (225–297 mg, 87–99% conversion, 66–83% yield).

Followed by the synthesis of amines, previously, it was shown that L-phenylalanine served as a valuable starting material for the synthesis of 2-phenylethanol when co-expressing 5 enzymes in *E. coli* [106], so the same amino acid can be transformed into benzylamine in 57% yield (71% conversion) after extraction and flash chromatography through a nine-step cascade involving the co-expression of 10 enzymes in *E. coli*: PAL, PAD, SMO, and epoxide hydrolase from *Sphingomonas* sp. HXN-200 (*Sp*EH), AlkJ, Ald-DH from *E. coli* (*Ec*ALDH), (*S*)-mandelate dehydrogenase (SMDH), benzoylformate decarboxylase from *P. putida* (*Pp*BFD), and *Cv*ATA-AlaDH (Scheme 13.44) [108]. Similarly, 4-fluorobenzylamine (147 mg, 49% yield) and 4-chlorobenzylamine (143 mg, 42% yield) were obtained.

Regarding amino alcohol synthesis, a series of multienzymatic cascades will be next described starting from epoxides, alkenes, or aldehydes. For instance, enantiopure (*R*)-phenylglycidol (R = Ph, Scheme 13.45) has been obtained from styrene oxide consisting in an epoxide hydrolase-catalyzed hydrolysis, alcohol oxidation catalyzed by a glycerol dehydrogenase, and final asymmetric amination using a ω-transaminase in combination with the Ala/AlaDH system for the cofactor NAD^+ recycling [109]. A cation resin adsorption was employed for *in situ* product removal, driving the equilibrium toward amino alcohol synthesis, which was obtained in 81.9% yield, much higher than in the absence of the resin (40.5%). The process can be extended toward the preparation of a series of chiral β-amino alcohols using lyophilized cell free extracts or resting cells of *Sp*EH, 2,3-butanediol dehydrogenase (BDHA) from *B. subtilis* and (*S*)-selective CR from *Gluconobacter oxydans*

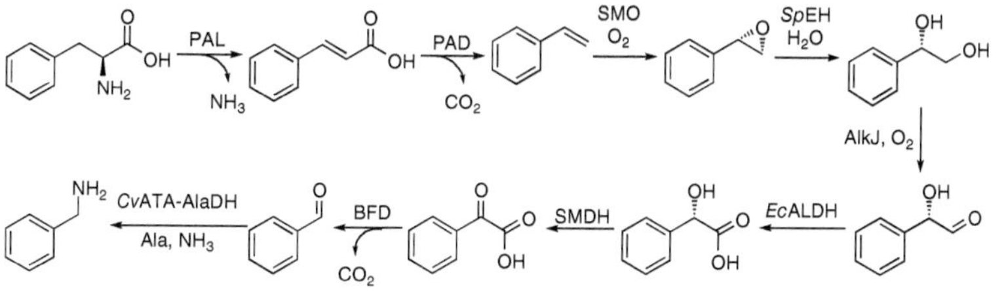

PAL, NH_3; PAD, CO_2; SMO, O_2; *Sp*EH, H_2O; AlkJ, O_2; *Ec*ALDH; SMDH; BFD, CO_2; *Cv*ATA-AlaDH, Ala, NH_3

Scheme 13.44 One-pot nine-step synthesis of benzylamine from l-phenylalanine co-expressing 10 enzymes in *E. coli*.

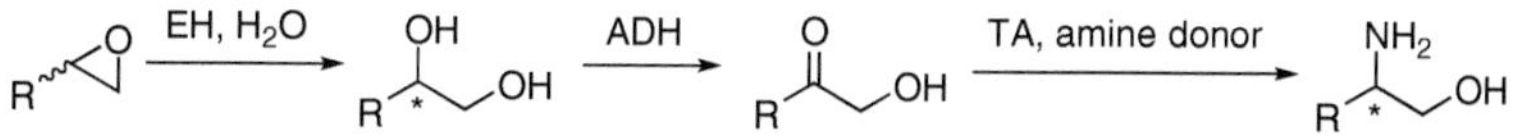

EH, H_2O; ADH; TA, amine donor

Scheme 13.45 Amino alcohol synthesis involving epoxide hydrolase, alcohol dehydrogenase, and transaminase-catalyzed steps.

(*Go*SCR), and (*R*)-ω-transaminase (*Mv*TA) from *Mycobacterium vanbaalenii*, including the preparative synthesis of (*S*)-phenylglycidol (R = Ph, 240 mg, Scheme 13.45) in 80% yield after extraction and flash chromatography [110].

A sequential aminohydroxylation of β-methylstyrene has allowed the production of phenylpropanolamines with high chemo-, regio-, and stereoselectivity through two consecutive bienzymatic cascades, requiring the isolation of the diol intermediates, consuming only molecular oxygen, one equivalent of ammonia and formate, and releasing only 1 equiv of carbonate as a by-product (Scheme 13.46) [111]. The first cascade consists in the epoxidation of *cis*- or *trans*-β-methyl styrene using a fused monooxygenase from *Pseudomonas species* (Fus-SMO) with either *Sphingomonas* species HXN200 (*Sp*EH) or *Solanum tuberosum* epoxide hydrolases (*St*EHs), which were expressed in *E. coli* and later lyophilized, leading to the four diol stereoisomers in 78–85% isolated yield and excellent selectivity (>99% ee and >99% de, Scheme 13.46a). Next, in a hydrogen borrowing cascade, the corresponding diol was oxidized at the C-2 position, finding from a set of reductases the ADHs from *Leifsonia* species, *A. aromaticum*, or *B. subtilis*, as the best enzymes, leading to the hydroxy ketones that were bioaminated, employing an (*R*)-selective chimeric AmDH (Ch1-AmDH) to obtain the (1*S*,2*R*)- and (1*R*,2*R*)-amino alcohols in 74% yield (>99% ee, >96% de, Scheme 13.46b).

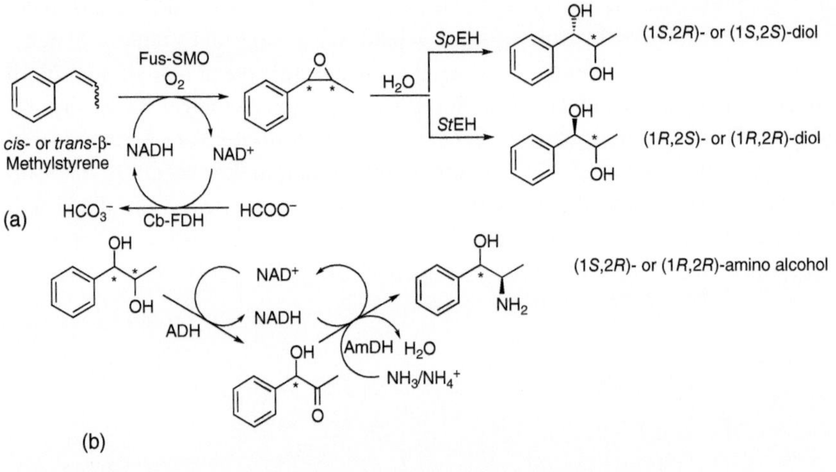

Scheme 13.46 Regio- and stereoselective aminohydroxylation of *cis*- and *trans*-β-methyl styrenes (a + b), involving the isolation of the corresponding diol intermediates (a).

Also, a three-step synthesis of trisubstituted tetrahydroisoquinolines is reported in a stepwise manner but without intermediate purification [112]. This approach involves the carboligation between pyruvate and 3-hydroxybenzaldehyde using acetohydroxy acid synthase I from *E. coli* (*Ec*AHAS-I), followed by biotransamination of the resulting hydroxy ketone using *Cv*2025 ATA and isopropylamine as the amine donor, and final norcoclaurine synthase-catalyzed Pictet–Spengler reaction with phenylacetaldehyde or 2-bromobenzaldehyde, obtaining the corresponding 1,2,3,4-tetrahydroisoquinolines with excellent selectivity in 92% and 54% yield, respectively (Scheme 13.47).

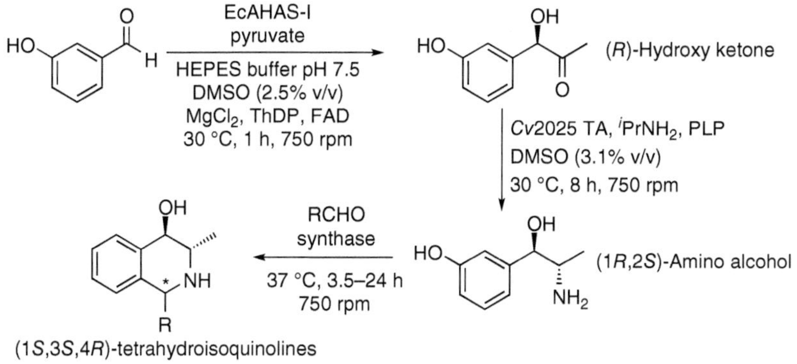

Scheme 13.47 Three-step stepwise synthesis of (1*S*,3*S*,4*R*)-1,2,3,4-tetrahydroisoquinolines.

Continuing with the synthesis of chiral amines, the combination of carboxylic acid reductase (CAR), ATA, and IRED has allowed the production of a series of piperidines and pyrrolidines (Scheme 13.48). For instance, these classes of substituted cyclic amines were obtained starting from keto acids or keto aldehydes after preparative scale synthesis (46–73 mg, 76–92% yield) employing the CAR from *Mycobacterium marinum* (*Mm*CAR), the commercially available (*S*)-selective ATA-113 and an (*R*)- or (*S*)-selective IRED from *Streptomyces* species (GF3546 or GF3587 variants) in a phosphate buffer pH 7.0 (Scheme 13.48a) [113]. Interestingly, the use of a whole cell enzyme where all these three enzyme activities are co-expressed has also been employed for the production of mono- and disubstituted chiral piperidines (Scheme 13.48b), including the preparative scale of racemic *cis*-4-methyl-2-phenylpiperidine (50 mg, 59% yield, >98% de) and (*S*)-2-phenylpiperidine with modest selectivity (70 mg, 58% yield, 30% ee) [114].

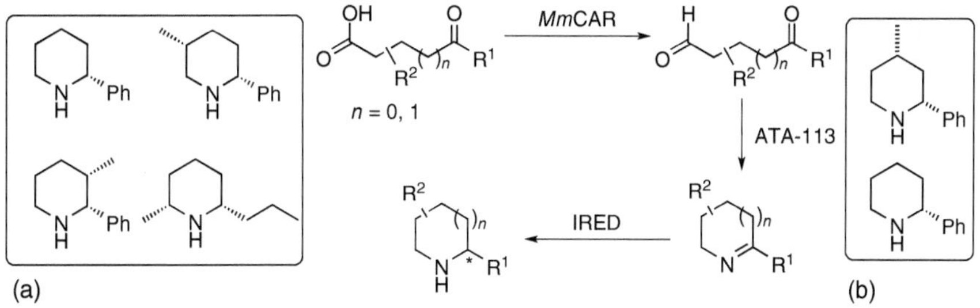

Scheme 13.48 Cascade synthesis of pyrrolidines and piperidines using CAR, ATA, and IRED enzymes as (a) individual biocatalysts; (b) co-expressing the enzyme activities in a whole cell unique biocatalyst.

Finally, the synthesis of a series of amino acids, hydroxy acids, and related compounds will be described by means of multienzymatic transformations. For instance, *p*-hydroxyphenyl lactic acid enantiomers were prepared through a one-pot, three-step sequential transformation designed as depicted in Scheme 13.49: (i) C–C coupling of the corresponding phenol and pyruvate in the presence of ammonia using M379V mutant of tyrosine

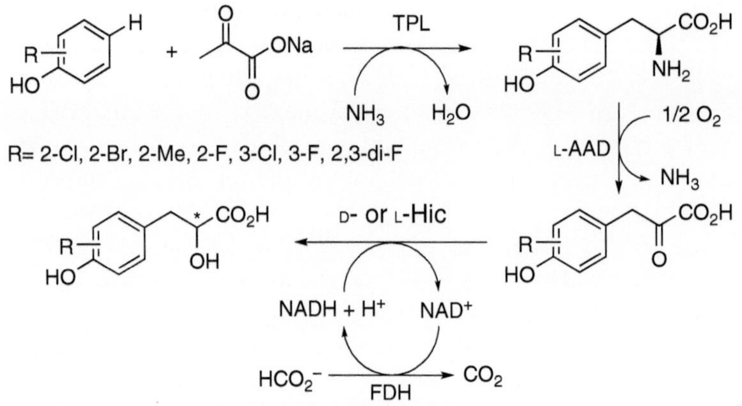

R= 2-Cl, 2-Br, 2-Me, 2-F, 3-Cl, 3-F, 2,3-di-F

Scheme 13.49 Synthesis of *p*-hydroxyphenyl lactic acid enantiomers starting from phenols in a one-pot, three-step sequential transformation.

phenol lyase from *Candida freundii* (*Cf*TPL); (ii) oxidative deamination of the resulting L-tyrosine intermediate with L-AAD from *Proteus myxofaciens*; (iii) and final stereoselective carbonyl reduction using complementary L- and D-isocaproate reductases from *Lactobacillus paracasei* (L-Hic) or *Lactobacillus confuses* DSM 20196 (D-Hic) [115]. The approach was conducted by reacting the corresponding phenol with pyruvate with tyrosine phenol lyase (TPL) in phosphate buffer for 24 hours at 21 °C and 170 rpm to later add NAD^+, ammonium formate, L-AAD, the selective isocaproate reductase, and FDH. Overall, the reaction was scaled-up using seven different phenols (23–92 mM), isolating after extraction the corresponding (*S*)-enantiomers in 60–81% yield (30.3–140.9 mg) and their antipodes in 58–85% yield (30.0–140.7 mg).

Park and coworkers described the conversion of long-chain fatty acids into medium-chain α,ω-dicarboxylic acids (21–46% yield) or ω-amino carboxylic acids (20–36% yield) using *E. coli* strains where a series of enzyme activities were co-expressed including the ones from KRED, BVMO, esterase, and ATA [116]. Followed by amino acid synthesis, Kroutil and coworkers reported two consecutive enzyme cascades for the transformation of cyclohexanol into 6-amino hexanoic acid (Scheme 13.50), also known as the nylon-6 monomer that is usually obtained through ε-caprolactam hydrolysis [117]. The first cascade consisted in the *Lb*ADH-catalyzed oxidation of cyclohexanol and subsequent

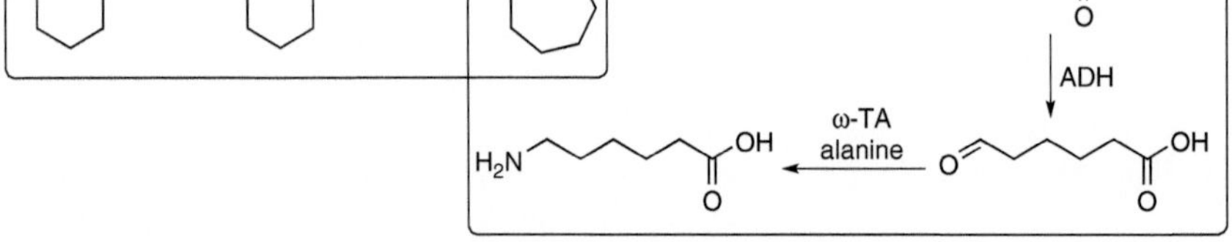

Scheme 13.50 Two consecutive multienzymatic cascades for the transformation of cyclohexanol into 6-amino hexanoic acid.

mono-oxygenation using the cyclohexanone monooxygenase from *Acinetobacter calcoaceticus* (*Ac*CHMO), yielding ε-caprolactone in 75% yield (85 mg) after extraction. Next, a second cascade was developed involving the lactone ring opening using the esterase from *B. subtilis*, the oxidation of the primary alcohol catalyzed by the ADH from *Bacillus stearothermophilus* and the biotransamination of the resulting aldehyde with the ω-TA from *Paracoccus denitrificans* using the AlaDH from *B. subtilis* for cofactor recycling purposes, producing 6-amino hexanoic acid.

Followed by the amino acid synthesis, a three-step cascade reaction was developed for the transformation of racemic mandelic acid (30.4 g) into enantiopure L-phenylglycine in 86.5% yield after column chromatography, consisting in the selective oxidation of mandelic acid using a D-mandelate dehydrogenase (DMDH) with *in situ* racemization of the unaltered enantiomer employing a mandelate racemase (MR), followed by L-amino acid dehydrogenase (L-AADH)-catalyzed reductive amination of the keto acid intermediate (Scheme 13.51) [118]. A novel DMDH was identified from *L. brevis* and purified, and the reaction conditions were optimized in terms of pH and temperature.

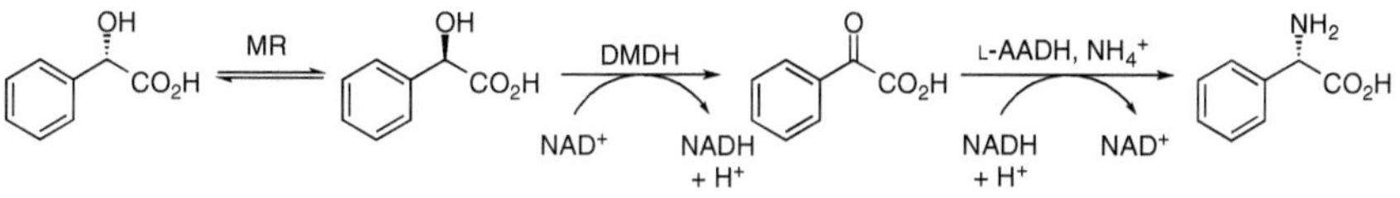

Scheme 13.51 Three-step biocascade for the conversion of the transformation of racemic mandelic acid into l-phenylglycine.

Li and coworkers reported three independent routes for the production of D-phenylglycines by co-expressing different enzyme activities in *E. coli* (Scheme 13.52) [119]. For

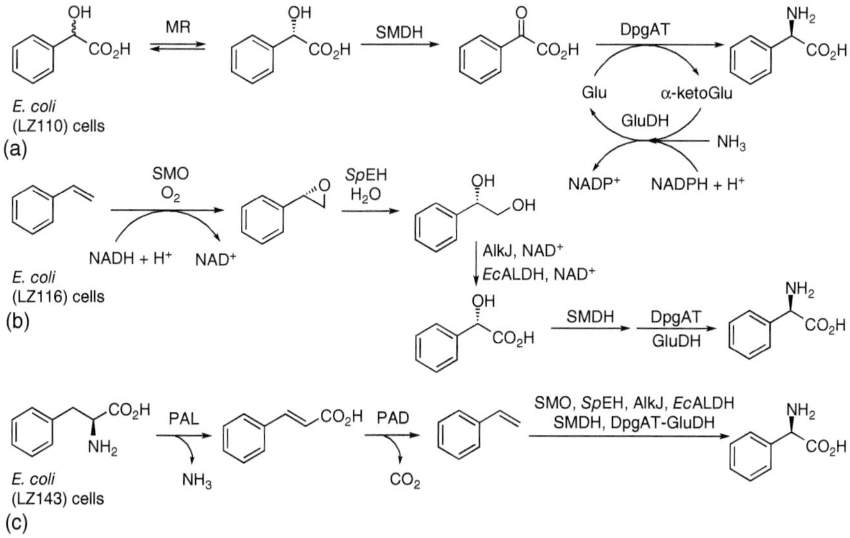

Scheme 13.52 Independent cascades for the synthesis of d-phenylglycine in *E. coli* recombinant multienzymatic catalytic systems.

instance, for the synthesis of D-phenylglycine as a model substrate, the following recombinant systems were successfully developed: (i) *E. coli* (LZ110) cells including mandelate racemase (MR), SMDH, and D-phenylglycine aminotransferase (DpgAT) and glutamate dehydrogenase (GluDH) from racemic mandelic acid; (ii) *E. coli* (LZ116) cells including SMO, epoxide hydrolase from *Sphingomonas* sp. HXN-200 (*Sp*EH), ADH from *P. putida* (AlkJ), Ald-DH from *E. coli* (*Ec*ALDH), SMDH, DpgAT, and GluDH from styrene; and (iii) *E. coli* (LZ143) cells including PAL, PAD, SMO, *Sp*EH, AlkJ, *Ec*ALDH, SMDH, DpgAT, and GluDH from L-phenylalanine. The extension of each approach for the synthesis of a series of D-phenylglycines was validated and preparative biotransformations were developed for the synthesis of D-phenylglycine and D-4-fluorophenylglycine using each recombinant *E. coli* system: (i) 71% and 52% isolated yield with LZ110; (ii) 62% and 46% isolated yield with LZ116; and (iii) 53% isolated yield for D-phenylglycine with LZ143.

The same research group successfully reported the functionalization of styrenes for the preparation of (*S*)-α-hydroxy acids, (*S*)-β-amino alcohols, and (*S*)-α-amino acids by constructing recombinant systems such as (i) *E. coli* (A-M1_R-M2) cells including SMO, *Sp*EH, AlkJ, and *Ec*ALDH; (ii) *E. coli* (A-M1_E-M3) cells including SMO, *Sp*EH, AlkJm, and *Cv*-ω-TA/AlaDH; (iii) *E. coli* (A-M1_R-M2_C-M4) cells including SMO, SpEH, AlkJ, *Ec*ALDH, and hydroxymandelate oxidase from *S. coelicolor*/catalase from *E. coli* and branch chain amino acid transaminase from *E. coli* (*Ec*-α-TA)/GluDH [120]. Each transformation was performed with 11 styrenes, demonstrating their practical applicability by executing preparative transformations for two compounds of each series for 24 hours: (i) (*S*)-hydroxy acids in 98% ee after extraction and recrystallization (R = H, 100 mM, 83% conversion and 72% yield; R = 4-F, 50 mM, 80% conversion and 61% yield); (ii) (*S*)-amino alcohols in 98% ee after extraction and column chromatography (R = H, 50 mM, 71% conversion and 62% yield R = 3-Me, 25 mM, 63% conversion and 55% yield; (iii) (*S*)-amino acids in >99% ee after extraction (R = H, 50 mM, 81% conversion and 70% yield; R = 4-F, 25 mM, 79% conversion and 59% yield).

Finally, the valorization of L-phenylalanine is also disclosed by Li's group, allowing the synthesis of five chiral molecules by developing a series of cascades in *E. coli* recombinant strains (Scheme 13.53) [121]: (i) LZ03 to synthesize the (*S*)-styrene oxide; (ii) LZ20 leading to the (*R*)-1-phenylethane-1,2-diol; (iii) LZ26 toward the antipode (*S*)-1-phenylethane-1,2-diol;

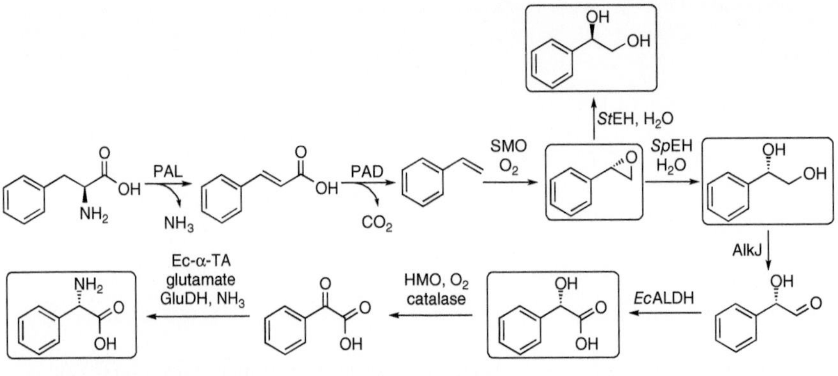

PAL NH_3 PAD CO_2 SMO O_2 *St*EH, H_2O *Sp*EH H_2O AlkJ *Ec*ALDH HMO, O_2 catalase Ec-α-TA glutamate GluDH, NH_3

Scheme 13.53 Biocatalytic cascades for the valorization of l-phenylalanine.

(iv) LZ37 until the (*S*)-mandelic acid; and (v) LZ76 for the preparation of (*S*)-phenylglycine, all of them being obtained in a range between 376 and 1082 mg (62–78% isolated yield and 96–>99% ee).

13.5 Summary and Outlook

Enzymes catalyze highly selective transformations in terms of chemo-, regio-, and stereoselectivity under mild reaction conditions. Traditionally, the use of only one enzyme for a determined reaction has required exhaustive optimization in terms of different parameters affecting the biocatalyst activity such as (i) solvent (aqueous ranging different pH and concentrations, organic, and neoteric solvents) but also when implying a cosolvent that modulates the substrate solubility and enzyme activity/selectivity; (ii) substrate scope and concentration; (iii) temperature and stirring speed/mode; (iv) cofactor requirements involving the action of co-substrates and coproducts; (v) immobilization types for enzyme activity/stability improvement; (vi) downstream process; and (vii) enzyme evolution for improving enzyme selectivity, change the biocatalysts stereopreference, accept different types of substrates (substrate promiscuity), or catalyze different reactions (enzyme promiscuity). Fortunately, the evolution of biocatalysis along the past four decades has facilitated the knowledge in developing one-step transformation, but how far we are from settling the development of multienzymatic transformations?

The performances of multicatalyzed chemoenzymatic and multienzymatic transformations have been largely spread in the past decade, allowing the design of complex transformations, including the stereoselective ones in multiple examples. Without any doubt, this area is open-up for a myriad of possibilities where enzyme compatibility represents the key issue, trying to favor cascade linear reactions over sequential and stepwise approaches.

This chapter has tried to collect representative examples in the use of multienzymatic synthesis involving linear and cyclic cascades for the production of a wide range of organic molecules, paying special attention to product recovery rather than reporting conversion values. Scientists must make strong efforts to demonstrate the applicability of reaction cascades in order to get the trust of chemical industry to implement this mature technology for practical applications [122].

References

1 Faber, K., Fessner, W.-D., and Turner, N.J. (2015). *Science of Synthesis, Biocatalysis in Organic Synthesis*. Stuttgart: Georg Thieme Verlag.

2 Drauz, K., Gröger, H., and May, O. (2012). *Enzyme Catalysis in Organic Synthesis*, 3 ed. Weinheim: Wiley-VCH.

3 Faber, K. (2018). *Biotransformations in Organic Chemistry*, 7 ed. Berlin: Springer-Verlag.

4 Sheldon, R.A. and Pereira, P.C. (2017). Biocatalysis engineering: the big picture. *Chem. Soc. Rev.* 46 (10): 2678–2691.

5 Sheldon, R.A. and Brady, D. (2018). The limits to biocatalysis: pushing the envelope. *Chem. Commun.* 54 (48): 6088–6104.

6 Foley, A.M. and Maguire, A.R. (2019). The impact of recent developments in technologies which enable the increased use of biocatalyst. *Eur. J. Org. Chem.* (23): 3713–3734.

7 Sheldon, R.A. and Brady, D. (2019). Broadening the scope of biocatalysis in sustainable organic synthesis. *ChemSusChem* 12 (13): 2859–2881.

8 DiCosimo, R., McAuliffe, J., Poulose, A.J., and Bohlmann, G. (2013). Industrial use of immobilized enzymes. *Chem. Soc. Rev.* 42 (15): 6437–6474.

9 Zhao, Z., Zhou, M.-C., and Liu, R.-L. (2019). Recent developments in carriers and non-aqueous solvents for enzyme immobilization. *Catalysts* 9 (8): 647.

10 Cen, Y.-K., Liu, Y.-X., Xue, Y.-P., and Zheng, Y.-G. (2019). Immobilization of enzymes in/on membranes and their applications. *Adv. Synth. Catal.* 361 (24): 5500–5515.

11 Bornscheuer, U.T., Huisman, G.W., Kazlauskas, R.J. et al. (2012). Engineering the third wave of biocatalysis. *Nature* 485 (7397): 185–194.

12 Arnold, F.H. (2018). Directed evolution: bringing new chemistry to life. *Angew. Chem. Int. Ed.* 57 (16): 4143–4148.

13 Bornscheuer, U.T., Hauer, B., Jaeger, K.E., and Schwaneberg, U. (2019). Directed evolution empowered redesign of natural proteins for the sustainable production of chemicals and pharmaceuticals. *Angew. Chem. Int. Ed.* 58 (1): 36–40.

14 Arnold, F.H. (2019). Innovation by evolution: bringing new chemistry to life (Nobel Lecture). *Angew. Chem. Int. Ed.* 58 (41): 14420–14426.

15 Choi, J.-M., Han, S.-S., and Kim, H.-S. (2015). Industrial applications of enzyme biocatalysis: current status and future aspects. *Biotechnol. Adv.* 33 (7): 1443–1454.

16 Chapman, J., Ismail, A.E., and Dinu, C.Z. (2018). Industrial applications of enzymes: recent advances, techniques, and outlooks. *Catalysts* 8 (6): 238.

17 Hughes, D.L. (2018). Biocatalysis in drug development-highlights of the recent patent literature. *Org. Process Res. Dev.* 22 (9): 1063–1080.

18 Adams, J.P., Brown, M.J.B., Diaz-Rodriguez, A. et al. (2019). Biocatalysis: a pharma perspective. *Adv. Synth. Catal.* 361 (11): 2421–2432.

19 Domínguez de María, P., de Gonzalo, G., and Alcántara, A.R. (2019). Biocatalysis as useful tool in asymmetric synthesis: an assessment of recently granted patents (2014–2019). *Catalysts* 9 (10): 802.

20 Denard, C.A., Hartwig, J.F., and Zhao, H. (2013). Multistep one-pot reactions combining biocatalysts and chemical catalysts for asymmetric synthesis. *ACS Catal.* 3 (12): 2856–2864.

21 Gröger, H. and Hummel, W. (2014). Combining the 'two worlds' of chemocatalysis and biocatalysis towards multi-step one-pot processes in aqueous media. *Curr. Opin. Chem. Biol.* 19: 171–179.

22 Filice, M. and Palomo, J.M. (2014). Cascade reactions catalyzed by bionanostructures. *ACS Catal.* 4 (5): 1588–1598.

23 Schmidt, S., Castiglione, K., and Kourist, R. (2018). Overcoming the incompatibility challenge in chemoenzymatic and multi-catalytic cascade reactions. *Chemistry* 24 (8): 1755–1768.

24 Oroz-Guinea, I. and García-Junceda, E. (2013). Enzyme catalysed tandem reactions. *Curr. Opin. Chem. Biol.* 17 (2): 236–249.

25 Ladkau, N., Schmid, A., and Bühler, B. (2014). The microbial cell-functional unit for energy dependent multistep biocatalysis. *Curr. Opin. Biotechnol.* 30: 178–189.

26 Otte, K.B. and Hauer, B. (2015). Enzyme engineering in the context of novel pathways and products. *Curr. Opin. Biotechnol.* 35: 16–22.

27 Köhler, V. and Turner, N.J. (2015). Artificial concurrent catalytic processes involving enzymes. *Chem. Commun.* 51 (3): 450–464.

28 Muschiol, J., Peters, C., Oberleitner, N. et al. (2015). Cascade catalysis-strategies and challenges *en route* to preparative synthetic biology. *Chem. Commun.* 51 (27): 5798–5811.

29 France, S.P., Hepworth, L.J., Turner, N.J., and Flitsch, S.L. (2017). Constructing biocatalytic cascades: in vitro and in vivo approaches to de novo multi-enzyme pathways. *ACS Catal.* 7 (1): 710–724.

30 Schrittwieser, J.H., Velikogne, S., Hall, M., and Kroutil, W. (2018). Artificial biocatalytic linear cascades for preparation of organic molecules. *Chem. Rev.* 118 (1): 270–348.

31 Sperl, J.M. and Sieber, V. (2018). Multienzyme cascade reactions: status and recent advances. *ACS Catal.* 8 (3): 2385–2396.

32 Schrittwieser, J.H., Velikogne, S., and Kroutil, W. (2018). Artificial biocatalytic linear cascades to access hydroxy acids, lactones, and α- and β-amino acids. *Catalysts* 8 (5): 205.

33 Wu, S. and Li, Z. (2018). Whole-cell cascade biotransformations for one-pot multistep organic synthesis. *ChemCatChem* 10 (10): 2164–2178.

34 Gandomkar, S., Żądło-Dobrowolska, A., and Kroutil, W. (2019). Extending designed linear biocatalytic cascades for organic synthesis. *ChemCatChem* 11 (1): 225–243.

35 Rudroff, F. (2019). Whole-cell based synthetic enzyme cascades-light and shadow of a promising technology. *Curr. Opin. Chem. Biol.* 49: 84–90.

36 Zhang, Y. and Hess, H. (2017). Toward rational design of high-efficiency enzyme cascades. *ACS Catal.* 7 (9): 6018–6027.

37 Hwang, E.T. and Lee, S. (2019). Multienzymatic cascade reactions via enzyme complex by immobilization. *ACS Catal.* 9 (5): 4402–4425.

38 Santacoloma, P.A., Sin, G., Gernaey, K.V., and Woodley, J.M. (2011). Multienzyme-catalyzed processes: next-generation biocatalysis. *Org. Process Res. Dev.* 15 (1): 203–212.

39 Breuer, M., Ditrich, K., Habicher, T. et al. (2004). Industrial methods for the production of optically active intermediates. *Angew. Chem. Int. Ed.* 43 (7): 788–824.

40 Ghislieri, D. and Turner, N.J. (2014). Biocatalytic approaches to the synthesis of enantiomerically pure chiral amines. *Top. Catal.* 57 (5): 284–300.

41 Patil, M.D., Grogan, G., Bommarius, A., and Yun, H. (2018). Oxidoreductase-catalyzed synthesis of chiral amines. *ACS Catal.* 8 (12): 10985–11015.

42 Pickl, M., Fuchs, M., Glueck, S.M., and Faber, K. (2015). Amination of ω-functionalized aliphatic primary alcohols by a biocatalytic oxidation-transamination cascade. *ChemCatChem* 7 (19): 3121–3124.

43 Lee, D.-S., Song, J.-W., Voß, M. et al. (2019). Enzyme cascade reactions for the biosynthesis of long chain aliphatic amines from renewable fatty acids. *Adv. Synth. Catal.* 361 (6): 1359–1367.

44 Sviatenko, O., Ríos-Lombardía, N., Morís, F. et al. (2019). One-pot synthesis of 4-aminocyclohexanol isomers by combining a keto reductase and an amine transaminase. *ChemCatChem* 11 (23): 5794–5799.

45 Ramsden, J.I., Heath, R.S., Derrington, S.R. et al. (2019). Biocatalytic N-alkylation of amines using either primary alcohols or carboxylic acids via reductive aminase cascades. *J. Am. Chem. Soc.* 141 (3): 1201–1206.

46 Galman, J.L., Slabu, I., Parmeggiani, F., and Turner, N.J. (2018). Biomimetic synthesis of 2-substituted N-heterocycle alkaloids by one-pot hydrolysis, transamination and decarboxylative Mannich reaction. *Chem. Commun.* 54 (80): 11316–11319.

47 Anderson, M., Afewerki, S., Berglund, P., and Córdova, A. (2014). Total synthesis of capsaicin analogues from lignin-derived compounds by combined heterogeneous metal, organocatalytic and enzymatic cascades in one pot. *Adv. Synth. Catal.* 356 (9): 2113–2118.

48 Land, H., Hendil-Forssell, P., Martinelle, M., and Berglund, P. (2016). One-pot biocatalytic amine transaminase/acyl transferase cascade for aqueous formation of amides from aldehydes or ketones. *Catal. Sci. Technol.* 6 (9): 2897–2900.

49 Tavanti, M., Parmeggiani, F., Castellanos, J.R.G. et al. (2017). One-pot biocatalytic double oxidation of α-isophorone for the synthesis of ketoisophorone. *ChemCatChem* 9 (17): 3338–3348.

50 Gandomkar, S., Dennig, A., Dordic, A. et al. (2018). Biocatalytic oxidative cascade for the conversion of fatty acids into α-ketoacids via internal H_2O_2 recycling. *Angew. Chem. Int. Ed.* 57 (2): 427–430.

51 Habib, M., Trajkovic, M., and Fraaije, M.W. (2018). The biocatalytic synthesis of syringaresinol from 2,6-dimethoxy-4-allylphenol in one-pot using a tailored oxidase/peroxidase system. *ACS Catal.* 8 (6): 5549–5552.

52 Yao, P., Wang, L., Yuan, J. et al. (2015). Efficient biosynthesis of ethyl (*R*)-3-hydroxyglutarate through a one-pot bienzymatic cascade of halohydrin dehalogenase and nitrilase. *ChemCatChem* 7 (9): 1438–1444.

53 Mutti, F.G., Knaus, T., Scrutton, N.S. et al. (2015). Conversion of alcohols to enantiopure amines through dual-enzyme hydrogen-borrowing cascades. *Science* 349 (6255): 1525–1529.

54 Thompson, M.P. and Turner, N.J. (2017). Two-enzyme hydrogen-borrowing amination of alcohols enabled by a cofactor-switched alcohol dehydrogenase. *ChemCatChem* 9 (20): 3833–3836.

55 Knaus, T., Cariati, L., Masman, M.F., and Mutti, F.G. (2017). In vitro biocatalytic pathway design: orthogonal network for the quantitative and stereospecific amination of alcohols. *Org. Biomol. Chem.* 15 (39): 8313–8325.

56 Böhmer, W., Knaus, T., and Mutti, F.G. (2018). Hydrogen-borrowing alcohol bioamination with coimmobilized dehydrogenases. *ChemCatChem* 10 (4): 731–735.

57 Chen, F.-F., Liu, Y.-Y., Zheng, G.-W., and Xu, J.-H. (2015). Asymmetric amination of secondary alcohols by using a redox-neutral two-enzyme cascade. *ChemCatChem* 7 (23): 3838–3841.

58 Montgomery, S.L., Mangas-Sanchez, J., Thompson, M.P. et al. (2017). Direct alkylation of amines with primary and secondary alcohols through biocatalytic hydrogen borrowing. *Angew. Chem. Int. Ed.* 56 (35): 10491–10494.

59 Martínez-Montero, L., Gotor, V., Gotor-Fernández, V., and Lavandera, I. (2017). Stereoselective amination of racemic *sec*-alcohols through sequential application of laccases and transaminases. *Green Chem.* 19 (2): 474–480.

60 Albarrán-Velo, J., Lavandera, I., and Gotor-Fernández, V. (2020). Sequential two-step stereoselective amination of allylic alcohols through the combination of laccases and amine transaminases. *ChemBioChem* 21 (1-2): 200–211.

61 Musa, M.M., Hollmann, F., and Mutti, F.G. (2019). Synthesis of enantiomerically pure alcohols and amines via biocatalytic deracemisation methods. *Catal. Sci. Technol.* 9 (20): 5487–5503.

62 Han, S.-W., Jang, Y., and Shin, J.-S. (2019). In vitro and in vivo one-pot deracemization of chiral amines by reaction pathway control of enantiocomplementary ω-transaminases. *ACS Catal.* 9 (8): 6945–6954.

63 Yoon, S., Patil, M.D., Sarak, S. et al. (2019). Deracemization of racemic amines to enantiopure (*R*)- and (*S*)-amines by biocatalytic cascade employing ω-transaminase and amine dehydrogenase. *ChemCatChem* 11 (7): 1898–1992.

64 Schrittwieser, J.H., Groenendaal, B., Resch, V. et al. (2014). Deracemization by simultaneous bio-oxidative kinetic resolution and stereoinversion. *Angew. Chem. Int. Ed.* 53 (14): 3731–3734.

65 Okamoto, Y., Köhler, V., and Ward, T.R. (2016). An NAD(P)H-dependent artificial transfer hydrogenase for multienzymatic cascades. *J. Am. Chem. Soc.* 138 (18): 5781–5784.

66 Guérard-Hélaine, C., Heuson, E., Ndiaye, M. et al. (2017). Stereoselective synthesis of γ-hydroxy-α-amino acids through aldolase-transaminase recycling cascades. *Chem. Commun.* 53 (39): 5465–5468.

67 Erdmann, V., Sehl, T., Frindi-Wosch, I. et al. (2019). Methoxamine synthesis in a biocatalytic 1-pot 2-step cascade approach. *ACS Catal.* 9 (8): 7380–7388.

68 Kohls, H., Anderson, M., Dickerhoff, J. et al. (2015). Selective access to all four diastereomers of a 1,3-amino alcohol by combination of a keto reductase- and an amine transaminase-catalysed reaction. *Adv. Synth. Catal.* 357 (8): 1808–1814.

69 Zhao, J.-W., Wu, H.-L., Zhang, J.-D. et al. (2018). One pot simultaneous preparation of both enantiomer of β-amino alcohol and vicinal diol via cascade biocatalysis. *Biotechnol. Lett.* 40 (2): 349–358.

70 Nagy-Győr, L., Abaházi, E., Bódai, V. et al. (2018). Co-immobilized whole cells with ω-transaminase and ketoreductase activities for continuous-flow cascade reactions. *ChemBioChem* 19 (17): 1845–1848.

71 Skalden, L., Peters, C., Dickerhoff, J. et al. (2015). Two subtle amino acid changes in a transaminase substantially enhance or invert enantiopreference in cascade syntheses. *ChemBioChem* 15 (7): 1041–1045.

72 Thorpe, T.W., France, S.P., Hussain, S. et al. (2019). One-pot biocatalytic cascade reduction of cyclic enimines for the preparation of diastereomerically enriched *N*-heterocycles. *J. Am. Chem. Soc.* 141 (49): 19208–19213.

73 Alvarenga, N., Payer, S.E., Petermeier, P. et al. (2020). Asymmetric synthesis of dihydropinidine enabled by concurrent multienzyme catalysis and a biocatalytic alternative to Krapcho dealkoxycarbonylation. *ACS Catal.* 10 (2): 1607–1620.

74 O'Reilly, E., Iglesias, C., Ghislieri, D. et al. (2014). A regio- and stereoselective ω-transaminase/monoamine oxidase cascade for the synthesis of chiral 2,5-disubstituted pyrrolidines. *Angew. Chem. Int. Ed.* 53 (9): 2447–2450.

75 Costa, B.Z., Galman, J.L., Slabu, I. et al. (2018). Synthesis of 2,5-disubstituted pyrrolidine alkaloids via a one-pot cascade using transaminase and reductive aminase biocatalysts. *ChemCatChem* 10 (20): 4733–4738.

76 Lichman, B.R., Lamming, E.D., Pesnot, T. et al. (2015). One-pot triangular chemoenzymatic cascades for the syntheses of chiral alkaloids from dopamine. *Green Chem.* 17 (2): 852–855.

77 Fischereder, E.-M., Pressnitz, D., and Kroutil, W. (2016). Stereoselective cascade to C3-methylated strictosidine derivatives employing transaminases and strictosidine synthases. *ACS Catal.* 6 (1): 23–30.

78 Wachtmeister, J., Jakoblinnert, A., Kulig, J. et al. (2014). Whole-cell teabag catalysis for the modularisation of synthetic enzyme cascades in micro-aqueous systems. *ChemCatChem* 6 (4): 1051–1058.

79 Wachtmeister, J., Mennicken, P., Hunold, A., and Rother, D. (2016). Modularized biocatalysis: immobilization of whole cells for preparative applications in microaqueous organic solvents. *ChemCatChem* 8 (3): 607–614.

80 Jakoblinnert, A. and Rother, D. (2014). A two-step biocatalytic cascade in micro-aqueous medium: using whole cells to obtain high concentrations of a vicinal diol. *Green Chem.* 16 (7): 3472–3482.

81 Wachtmeister, J., Jakoblinnert, A., and Rother, D. (2016). Stereoselective two-step biocatalysis in organic solvent: toward all stereoisomers of a 1,2-diol at high product concentrations. *Org. Process Res. Dev.* 20 (10): 1744–1753.

82 Bayer, T., Milker, S., Wiesinger, T. et al. (2017). In vivo synthesis of polyhydroxylated compounds from a "hidden reservoir" of toxic aldehyde species. *ChemCatChem* 9 (15): 2919–2923.

83 Schmidt, S., de Almeida, T.P., Rother, D., and Hollmann, F. (2017). Towards environmentally acceptable synthesis of chiral α-hydroxy ketones *via* oxidase-lyase cascades. *Green Chem.* 19 (5): 1226–1229.

84 Zhang, Y., Yao, P., Cui, Y. et al. (2018). One-pot enzymatic synthesis of cyclic vicinal diols from aliphatic dialdehydes via intramolecular C–C bond formation and carbonyl reduction using pyruvate decarboxylases and alcohol dehydrogenases. *Adv. Synth. Catal.* 360 (21): 4191–4196.

85 Knaus, T., Mutti, F.G., Humphreys, L.D. et al. (2015). Systematic methodology for the development of biocatalytic hydrogen-borrowing cascades: application to the synthesis of chiral α-substituted carboxylic acids from α-substituted α,β-unsaturated aldehydes. *Org. Biomol. Chem.* 13 (1): 223–233.

86 Guo, J., Zhang, R., Ouyang, J. et al. (2018). Stereodivergent synthesis of carveol and dihydrocarveol through ketoreductases/ene-reductases catalyzed asymmetric reduction. *ChemCatChem* 10 (23): 5496–5504.

87 Iqbal, N., Stewart, J.D., Macheroux, P. et al. (2018). Novel concurrent redox cascades of (*R*)- and (*S*)-carvones enables access to carvo-lactones with distinct regio- and enantioselectivity. *Tetrahedron* 74 (52): 7389–7392.

88 Classen, T., Korpak, M., Schölzel, M., and Pietruszka, J. (2014). Stereoselective enzyme cascades: an efficient synthesis of chiral γ-butyrolactones. *ACS Catal.* 4 (5): 1321–1331.

89 Liu, Y.-C., Liu, Y., and Wu, Z.-L. (2015). Synthesis of enantiopure glycidol derivatives *via* a one-pot two-step enzymatic cascade. *Org. Biomol. Chem.* 13 (7): 2146–2152.

90 Liu, Y.-C. and Wu, Z.-L. (2016). Switchable asymmetric bio-epoxidation of α,β-unsaturated ketones. *Chem. Commun.* 52 (6): 1158–1161.

91 Martínez-Montero, L., Gotor, V., Gotor-Fernández, V., and Lavandera, I. (2018). Mild chemoenzymatic oxidation of allylic *sec*-alcohols. Application to biocatalytic stereoselective redox isomerizations. *ACS Catal.* 8 (3): 2413–2419.

92 González-Granda, S., Méndez-Sánchez, D., Lavandera, I., and Gotor-Fernández, V. (2020). Laccase-mediated oxidations of propargylic alcohols. Application in the deracemization of 1-arylprop-2-yn-1-ols in combination with alcohol dehydrogenases. *ChemCatChem* 12 (2): 520–527.

93 Díaz-Rodríguez, A., Ríos-Lombardía, N., Sattler, J.H. et al. (2015). Deracemisation of profenol core by combining laccase/TEMPO-mediated oxidation and alcohol dehydrogenase-catalysed dynamic kinetic resolution. *Catal. Sci. Technol.* 5 (3): 1443–1446.

94 Cordeiro, R.S.C., Ríos-Lombardía, N., Morís, F. et al. (2019). One-pot transformation of ketoximes into optically active alcohols and amines by sequential action of laccases and ketoreductases or ω-transaminases. *ChemCatChem* 11 (4): 1272–1277.

95 Leemans, L., van Langen, L.M., Hollmann, F., and Schallmey, A. (2019). Bienzymatic-cascade for the synthesis of an optically active *O*-benzoyl cyanohydrin. *Catalysts* 9 (10): 822.

96 Leemans, L., Walter, M.D., Hollmann, F. et al. (2019). Multi-catalytic route for the synthesis of (*S*)-tembamide. *Catalysts* 9 (6): 522.

97 Dennig, A., Busto, E., Kroutil, W., and Faber, K. (2015). Biocatalytic one-pot synthesis of L-tyrosine derivatives from monosubstituted benzenes, pyruvate, and ammonia. *ACS Catal.* 5 (12): 7503–7506.

98 Yu, J., Li, J., Gao, X. et al. (2019). Dynamic kinetic resolution for asymmetric synthesis of L-noncanonical amino acids from D-ser using tryptophan synthase and alanine racemase. *Eur. J. Org. Chem.* (39): 6618–6625.

99 Park, E.-S. and Shin, J.-S. (2014). Deracemization of amino acids by coupling transaminases of opposite stereoselectivity. *Adv. Synth. Catal.* 356 (17): 3505–3509.

100 Walton, C.J.W., Parmeggiani, F., Barber, J.E.B. et al. (2018). Engineered aminotransferase for the production of D-phenylalanine derivatives using biocatalytic cascades. *ChemCatChem* 10 (2): 470–474.

101 Zhang, D., Jing, X., Zhang, W. et al. (2019). Highly selective synthesis of D-amino acids from readily available L-amino acids by a one-pot biocatalytic stereoinversion cascade. *RSC Adv.* 9 (51): 29927–29935.

102 Weise, N.J., Ahmed, S.T., Parmeggiani, F., and Turner, N.J. (2017). Kinetic resolution of aromatic β-amino acids using a combination of phenylalanine ammonia lyase and aminomutase biocatalysts. *Adv. Synth. Catal.* 359 (9): 1570–1576.

103 Brenna, E., Crotti, M., Gatti, F.G. et al. (2017). Biocatalytic synthesis of chiral cyclic γ-oxoesters by sequential C–H hydroxylation, alcohol oxidation and alkene reduction. *Green Chem.* 19 (21): 5122–5130.

104 Brenna, E., Crotti, M., De Pieri, M. et al. (2018). Chemo-enzymatic oxidative rearrangement of tertiary allylic alcohols: synthetic application and integration into a cascade process. *Adv. Synth. Catal.* 360 (19): 3677–3686.

105 Vorster, A., Smit, M.S., and Opperman, D.J. (2019). One-pot conversion of cinnamaldehyde to 2-phenylethanol via a biosynthetic cascade reaction. *Org. Lett.* 21 (17): 7024–7027.

106 Lukito, B.R., Wu, S., Saw, H.J.J., and Li, Z. (2019). One-pot production of natural 2-phenylethanol from l-phenylalanine via cascade biotransformations. *ChemCatChem* 11 (2): 831–840.

107 Wu, S., Liu, J., and Li, Z. (2017). Biocatalytic formal anti-Markovnikov hydroamination and hydration of aryl alkenes. *ACS Catal.* 7 (8): 5225–5233.

108 Zhou, Y., Wu, S., Mao, J., and Li, Z. (2018). Bioproduction of benzylamine from renewable feedstocks via a nine-step artificial enzyme cascade and engineered metabolic pathways. *ChemSusChem* 11 (13): 2221–2228.

109 Sun, Z.-B., Zhang, Z.-J., Li, F.-L. et al. (2019). One pot asymmetric synthesis of (*R*)-phenylglycinol from racemic styrene oxide via cascade biocatalysis. *ChemCatChem* 11 (16): 3802–3807.

110 Zhang, J.-D., Yang, X.-X., Jia, Q. et al. (2019). Asymmetric ring opening of racemic epoxides for enantioselective synthesis of (*S*)-β-amino alcohols by a cofactor self-sufficient cascade biocatalysis system. *Catal. Sci. Technol.* 9 (1): 70–74.

111 Corrado, M.L., Knaus, T., and Mutti, F.G. (2019). Regio- and stereoselective multi-enzymatic aminohydroxylation of β-methylstyrene using dioxygen, ammonia and formate. *Green Chem.* 21 (23): 6246–6251.

112 Erdmann, V., Lichman, B.R., Zhao, J. et al. (2017). Enzymatic and chemoenzymatic three-step cascades for the synthesis of stereochemically complementary trisubstituted tetrahydroisoquinolines. *Angew. Chem. Int. Ed.* 56 (41): 12503–12507.

113 France, S.P., Hussain, S., Hill, A.M. et al. (2016). One-pot cascade synthesis of mono- and disubstituted piperidines and pyrrolidines using carboxylic acid reductase (CAR), ω-transaminase (ω-TA), and imine reductase (IRED) biocatalysts. *ACS Catal.* 6 (6): 3753–3759.

114 Hepworth, L.J., France, S.P., Hussain, S. et al. (2017). Enzyme cascades in whole cells for the synthesis of chiral cyclic amines. *ACS Catal.* 7 (4): 2920–2925.

115 Busto, E., Simon, R.C., Richter, N., and Kroutil, W. (2016). One-pot, two-module three-step cascade to transform phenol derivatives to enantiomerically pure (*R*)- or (*S*)-*p*-hydroxyphenyl lactic acids. *ACS Catal.* 6 (4): 2393–2397.

116 Song, J.-W., Lee, J.-H., Bornscheuer, U.T., and Park, J.-B. (2014). Microbial synthesis of medium-chain α,ω-dicarboxylic acids and ω-aminocarboxylic acids from renewable long-chain fatty acids. *Adv. Synth. Catal.* 356 (8): 1782–1788.

117 Sattler, J.H., Fuchs, M., Mutti, F.G. et al. (2014). Introducing an *in situ* capping strategy in systems biocatalysis to access 6-aminohexanoic acid. *Angew. Chem. Int. Ed.* 53 (51): 14153–14157.

118 Fan, C.-W., Xu, G.-C., Ma, B.-D. et al. (2015). A novel D-mandelate dehydrogenase used in three-enzyme cascade reaction for highly efficient synthesis of non-natural chiral amino acids. *J. Biotechnol.* 195: 67–71.

119 Zhou, Y., Wu, S., and Li, Z. (2017). One-pot enantioselective synthesis of D-phenylglycines from racemic mandelic acids, styrenes, or biobased L-phenylalanine via cascade biocatalysis. *Adv. Synth. Catal.* 359 (24): 4305–4316.

120 Wu, S., Zhou, Y., Wang, T. et al. (2016). Highly regio- and enantioselective multiple oxy- and amino-functionalizations of alkenes by modular cascade biocatalysis. *Nat. Commun.* 7: 11917.

121 Zhou, Y., Wu, S., and Li, Z. (2016). Cascade biocatalysis for sustainable asymmetric synthesis: from biobased L-phenylalanine to high-value chiral chemicals. *Angew. Chem. Int. Ed.* 55 (28): 11647–116500.

122 Albarrán-Velo, J., González-Martínez, D., and Gotor-Fernández, V. (2018). Stereoselective biocatalysis: a mature technology for the asymmetric synthesis of pharmaceutical building blocks. *Biocatal. Biotransform.* 36 (2): 102–130.

14

Chemoenzymatic Sequential One-Pot Protocols

Harald Gröger

Bielefeld University, Chair of Industrial Organic Chemistry and Biotechnology, Faculty of Chemistry, Universitätsstr. 25, 33615, Bielefeld, Germany

14.1 Introduction: Theoretical Information and Conceptual Overview

The combination of various "classic" chemical as well as chemocatalytic reactions with biotransformations toward chemoenzymatic cascade-type one-pot processes has gained tremendous interest in recent years. A major "driving force" in this research area is the potential of such resulting one-pot processes to reduce the number of work-up and intermediate isolation steps, thus leading to a decreased amount of required solvent (utilized in, e.g., extraction steps) and consequently a lower amount of waste. Thus, the resulting one-pot processes are not only scientifically interesting but also attractive in terms of industrial applications because the improvement of E-factors (Kilogram of waste per Kilogram of product) goes hand in hand also with more favorable economic data.

When developing such one-pot processes, a key challenge is to achieve compatibility between the reaction steps with each other. This goal has been realized by various different concepts, which all contribute as valuable solutions to the "toolbox" of chemoenzymatic one-pot process development.

Such one-pot processes can be clustered (see also Chapter 13), for example, into cascades running

(1) in a sequential mode (in which one reaction is conducted and completed before the start of the subsequent reaction step), or
(2) in a tandem mode (in which the reactions proceed concurrently), or
(3) in a compartmentalized mode (in which the reactions are conducted in different reactor compartments), although this concept (see also Chapter 4) might also be considered as a modification of both concepts described in (1) and (2).

All of such concepts show advantages and disadvantages, and it is noteworthy that these tools are complementary to each other. Thus, case by case and depending on the specific

Biocatalysis for Practitioners: Techniques, Reactions and Applications, First Edition.
Edited by Gonzalo de Gonzalo and Iván Lavandera.

process demands of a multistep one-pot process, one of the described options might turn out as particularly advantageous for this specific case.

In this book chapter, the focus will be on the state of the art in the field of sequential one-pot cascades (described as option (1) above), which are visualized as "Concept I" of Figure 14.1. As briefly outlined above, in such a process, typically the first reaction is conducted upon completion before adding the catalyst for the second reaction step. In the field of chemoenzymatic one-pot syntheses according to this Concept I, two variants in terms of the sequence how to add the different types of catalysts are conceivable and both have been studied. In detail, an initial chemocatalytic reaction (or "classic" chemical reaction) followed by a subsequent enzymatic step can be conducted as well as the reverse option of an initial enzymatic step, followed by a chemocatalytic (or "classic" chemical) reaction step. It should be added that, however, in the latter case, the compatibility of the enzyme with the reaction components (including catalysts) from the following reaction steps is not relevant as such components are not added before the completion of the biotransformation.

For comparison, in Figure 14.1 also the basic concept of a tandem-like process is given (Concept II) and it should be added that such tandem-type chemoenzymatic processes according to Concept II in Figure 14.1 have already been reviewed elsewhere [1–4], as well as well as compartmentalization as a tool for merging chemo- and biocatalytic reactions [5].

Because for chemoenzymatic one-pot processes various options are conceivable (as outlined above), in this chapter, a brief overview about the drawbacks and advantages when applying a sequential one-pot concept according to the Concept I in Figure 14.1 will be given in comparison to other one-pot options. These scope and limitations are also visualized in the corresponding boxes in Table 14.1.

Among the major advantages (in addition to general advantages of one-pot processes such as reduced number of unit operations, decreased solvent consumption, and waste

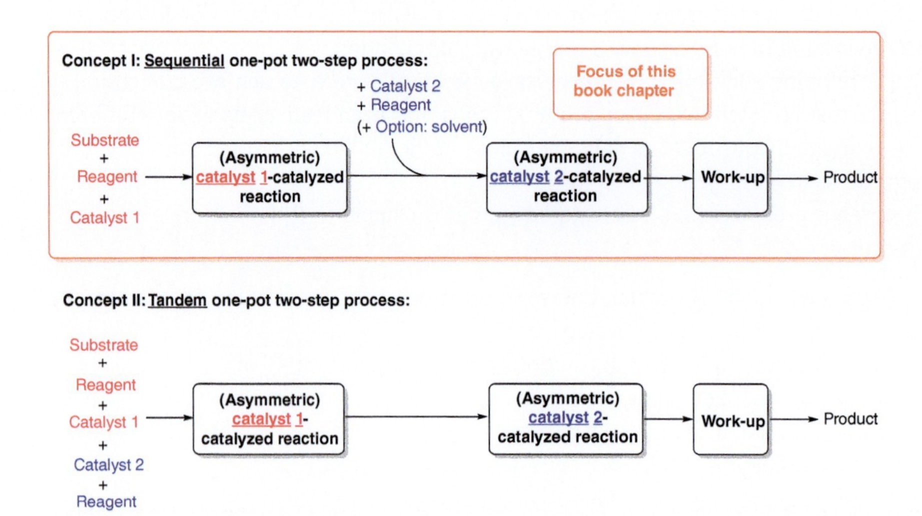

Figure 14.1 Conceptual overview about variants of chemoenzymatic one-pot syntheses with one chemocatalyst and one enzyme as catalysts 1 and 2.

Table 14.1 Advantages and disadvantages of the chemoenzymatic one-pot syntheses running in a sequential mode.

Advantages	Disadvantages
Avoiding cross-inhibitions	No "trapping" of labile intermediates (from first step)
Adjusting different "process windows"	Thermodynamically unfavored reactions difficult to integrate
Avoiding alternative, undesired reaction sequences	Low "steady-state" level of intermediate not possible

reduction) is the option to avoid cross-inhibitions. This option of a sequential one-pot process is of interest when reagents of the first step interact in a negative way with the catalyst or components being involved in the second step (since such reagents are then consumed prior to the addition of catalyst and components for the second step).

A further advantage is the opportunity to adjust suitable "process windows" with respect to temperature, solvent environment, and pH for both reaction steps individually. For example, if different pH optima exist, one can consider to adjust the pH from the optimal pH value for the first step after its completion to the optimal value of the second step before the start of this second step.

A third advantage is that "by definition" the desired reaction sequence can be "fixed" and other potential alternative reaction sequences can be excluded. This option is of importance if, for example, the catalyst for the second reaction step could also react with the substrate of the first reaction step. A synthetic example will be given below for the combination of the Suzuki reaction with a biotransformation.

Besides numerous advantages, however, the sequential chemoenzymatic cascade process technology also shows some drawbacks compared with the "tandem-mode" counterpart. For example, in contrast to tandem processes, labile intermediates (formed in the first step) cannot be "trapped" by direct *in situ* derivatization within such a cascade concept. Thus, typically thermodynamically unfavored reactions steps cannot be integrated into such a reaction cascade running in a sequential mode. Furthermore, in a sequential one-pot fashion, the product of the first step is enriched continuously up to a high final concentration, whereas in a tandem mode, the product of the first step is continuously converted, thus keeping its concentration at a low "steady-state" level. Keeping such a low level of the product of the first step serving as a substrate for the second step can be of advantage when such a molecule (i) has a short lifetime, thus being prone to decomposition or (ii) inhibits or deactivates any other component in the overall process, for example, the catalyst of the second step.

Thus, taking into account the "pro and cons" of the different variants of one-pot processes, it is of particular importance to analyze and define the challenges before the design of a one-pot cascade. In the following, some case studies will be presented, in which a sequential one-pot cascade has been turned out as a beneficial option and, thus, being realized for the combination of a chemocatalytic with a biocatalytic transformation.

14.2 State of the Art in Sequential Chemoenzymatic One-Pot Synthesis: Selected Examples and Historical Overview About Selected Contributions

The representative processes of sequential chemoenzymatic one-pot cascades, which are briefly summarized in the following sections, are subdivided by the type of the chemocatalyst utilized in the cascade (metal catalyst or organocatalyst). Although in principle both types of sequences (biocatalysis, followed by chemocatalysis or *vice versa*) are conceivable (and examples will be given for both concepts), the latter one has been in particular studied because only in this case the compatibility of the enzyme with the "chemical environment" of a chemocatalytic transformation is of relevance as the biotransformation represents the second step.

14.2.1 Sequential Chemoenzymatic One-Pot Synthesis Combining a Metal-Catalyzed Reaction with a Biotransformation

To start with the combination of a chemocatalytic reaction with a subsequent biocatalytic reaction when choosing a metal as a catalyst component, to the best of my knowledge, the first sequential one-pot cascade combining a chemocatalytic with an enzymatic transformation was reported by Hanefeld, Maschmeyer, Sheldon and coworkers [6]. The application is directed toward a chemoenzymatic synthesis of L-α-amino acids and consists of an initial enantioselective metal-catalyzed hydrogenation of *N*-acetyl amino acrylate **1** with a subsequent L-amino acylase-catalyzed hydrolytic resolution (Scheme 14.1). As a metal catalyst, a heterogeneous rhodium–phosphane complex was utilized. In the stepwise fashion, the enzyme was added after completion of the hydrogenation reaction. When using an L-amino acylase from *Aspergillus melleus*, formation of L-alanine, L-**3**, proceeds with excellent conversion and >98% ee.

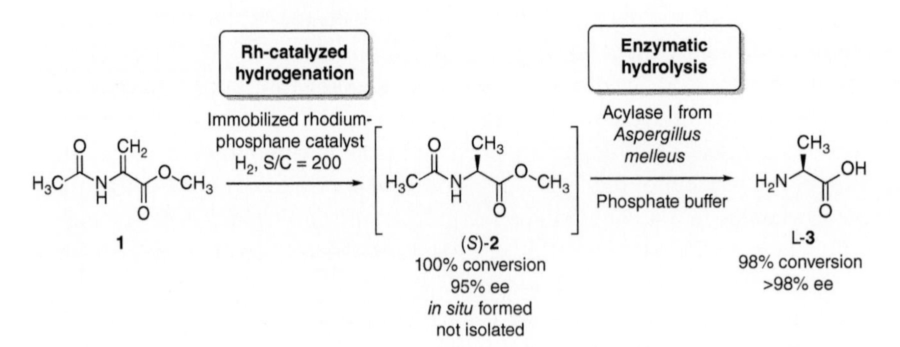

Scheme 14.1 Combination of a metal-catalyzed hydrogenation and an enzymatic hydrolytic resolution toward a one-pot process for preparing L-alanine from an acrylate derivative.

It should be added that in contrast to this successful one-pot process, the authors observed a negligible consumption of hydrogen when the enzyme was present from the beginning, thus indicating an interference of the enzyme with the metal catalyst leading to its deactivation and a low conversion when running the process in the tandem mode.

Besides hydrogenation reactions, palladium catalysis is a further prominent field of metal catalysis with applications even on industrial scale. Representative examples are the Suzuki cross-coupling and Heck reactions for the synthesis of biaryl and alkene compounds, which can serve as valuable substrates for enzymatic transformations. Accordingly, there has been a research interest to combine such fields by developing chemoenzymatic one-pot processes with an initial palladium-catalyzed transformation and a subsequent biocatalytic reaction, thus furnishing the desired target products. Addressing this concept, a one-pot process consisting of an initial Suzuki coupling reaction and a subsequent enzymatic reduction with both reactions running in an aqueous medium has been developed by the Gröger and Hummel groups (Scheme 14.2) [7]. As an intermediate, biaryl ketones are formed in the initial Suzuki reaction, which are then enantioselectively reduced in the presence of an alcohol dehydrogenase, thus furnishing the desired chiral biaryl alcohols with both high conversion and enantioselectivity after process development. However, a key challenge was to achieve compatibility of the components for the Suzuki reaction with the biocatalyst. It turned out that in particular the boronic acid as a reagent for the Suzuki reaction step has a negative impact on the enzyme. This problem could be solved by such a sequential one-pot concept because the boronic acid is consumed in the first step. Thus, a Suzuki reaction proceeding with just one equivalent of boronic acid (avoiding its use in excess) was developed, which then led to a full consumption before the addition of the enzyme. By means of such a sequential one-pot process, the biaryl alcohol (*S*)-**6** was formed with a conversion of 91% yield and >99% ee (Scheme 14.2). Thus, this example illustrates the advantage of applying a sequential one-pot process in the case of a negative interference of a reagent used in the initial step with the catalyst of the second step as in such a case, this reagent is consumed before adding the catalyst. It should also be mentioned that this methodology has been extended to the synthesis of analogous chiral C2-symmetric diols [8].

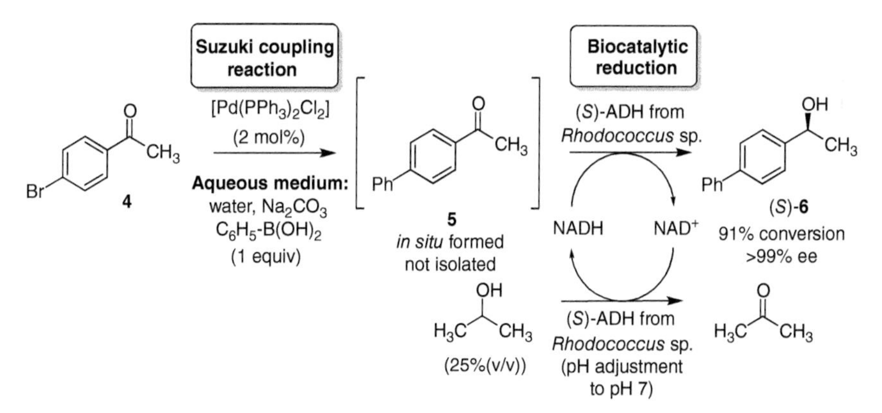

Scheme 14.2 Combination of a Pd-catalyzed Suzuki reaction with an enzymatic ketone reduction in a sequential one-pot process for the enantioselective synthesis of biaryl alcohols. Source: E. Burda et al. [7].

Furthermore, extensions and improvements of this type of sequential one-pot process have been achieved by various research groups. For example, water-soluble palladium nanoparticles stabilized within a protein cavity have been successfully applied by Cacchi and coworkers. This technology enables the option to use a phosphane-free palladium catalyst, and in the one-pot process, this catalyst for the Suzuki reaction turned out to be compatible with a subsequent ketone reduction, leading to the biaryl alcohol products in up to 91% yield and with excellent enantioselectivities of >99% ee [9]. A solvent engineering by means of ionic liquids for the Suzuki reaction was found by Schmitzer and Kroutil and coworkers to be beneficial for a recovery and reuse of the catalyst components in connection with this type of one-pot process [10]. In detail, the Suzuki reaction proceeds in an ionic liquid as a reaction medium, followed by an enzymatic ketone reduction, which then runs in a two-phase system consisting of ionic liquid and buffer. It is noteworthy that both phases can be reused, which has been demonstrated for four reaction cycles. The resulting biaryl alcohol was obtained in 94–98% yield and with >99% ee. The task to operate at lower temperature was addressed by Gröger, Schatz, Hummel and coworkers, who conducted the Suzuki reaction at room temperature by using a water-soluble palladium-TPPTS (tris(3-sulfonatophenyl)phosphine sodium salt) complex as a catalyst [11]. This type of palladium catalyst also showed high compatibility with the biotransformation, thus furnishing the desired biaryl alcohol efficiently.

This type of one-pot process technology for the enantioselective synthesis of biaryl-containing alcohols has also been conducted efficiently in a reaction media based on deep eutectic solvent (DES) as an alternative solvent system [12]. Furthermore, this process technology has been extended by the González-Sabín and Gröger groups to the combination of the Suzuki coupling reaction with a transaminase-catalyzed amine synthesis [13]. In this enantioselective one-pot process in a DES-based reaction medium, the desired biaryl-substituted amines have been obtained with overall conversions in the range of 78–92% and in 70–86% yields with excellent enantiomeric excess of >99% ee in all cases. This combination of Suzuki cross-coupling reaction and transamination has also been reported for other reaction media. The Bornscheuer and de Souza groups found suitable reaction conditions for combining these chemo- and biocatalysis reactions based on a water–*N*,*N*-dimethylformamide (DMF) reaction medium, thus furnishing the desired biaryl-containing amines with high conversions of up to 84% and excellent enantioselectivity (>99% ee) when utilizing an optimized mutant of an amine transaminase from *Aspergillus fumigatus* [14].

A further palladium-catalyzed cross-coupling reaction, which has been combined with a biotransformation toward a one-pot synthesis, is the Heck reaction. In their initial work, Cacchi and coworkers conducted a Heck reaction of an aryliodide with butenone in an organic reaction medium, followed by removal of the solvent (and other volatile components), leading to a crude product, which was then used directly without product purification in a subsequent enzymatic ketone reduction [15]. The applied enzyme turned out to be compatible with this crude product (and, thus, with the palladium catalyst), leading to the synthesis of the alcohol products in up to 85% yields and with excellent enantioselectivities of >99% ee. In a modified one-pot process version, Cacchi and coworkers combined a Heck reaction in aqueous medium utilizing a phosphine-free perfluoro-tagged palladium nanoparticle with a subsequent enzymatic reduction [16]. Again a high compatibility of the palladium catalyst with the biocatalyst was demonstrated, underlined by high yields for the desired alcohol products (*R*)-**9** of up to 92% and an enantioselectivity of >99% ee. Representative examples are shown in Scheme 14.3.

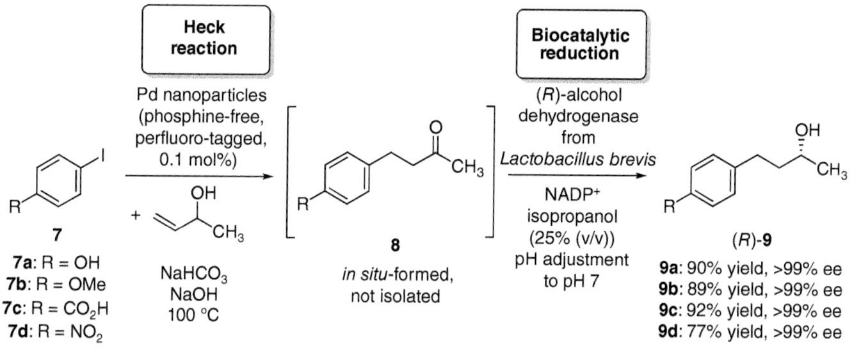

Heck reaction

Pd nanoparticles (phosphine-free, perfluoro-tagged, 0.1 mol%)

+ OH / CH_3

$NaHCO_3$
NaOH
100 °C

7
7a: R = OH
7b: R = OMe
7c: R = CO_2H
7d: R = NO_2

8
in situ-formed, not isolated

Biocatalytic reduction

(*R*)-alcohol dehydrogenase from *Lactobacillus brevis*
$NADP^+$
isopropanol (25% (v/v))
pH adjustment to pH 7

(*R*)-**9**
9a: 90% yield, >99% ee
9b: 89% yield, >99% ee
9c: 92% yield, >99% ee
9d: 77% yield, >99% ee

Scheme 14.3 Combination of a Pd-catalyzed Heck reaction with an enzymatic ketone reduction in a sequential one-pot process for the enantioselective synthesis of chiral alcohols.

However, compatibility of metal catalysts with enzymes is not limited to palladium and rhodium. A further metal having been demonstrated to be feasible for combination with enzymes is ruthenium. Ruthenium complexes are versatile catalysts being applied in numerous organic synthetic reactions. The first combination of a ruthenium-catalyzed reaction with a biotransformation was reported by Gröger and Schatz and coworkers, exemplified for a combination of a ring closure metathesis with an enzymatic hydrolysis [17]. In detail, in the initial step, a metathesis of 2,2-diallylmalonate **10** was conducted with the ruthenium catalyst **11**, followed by a selective ester-catalyzed hydrolysis (Scheme 14.4). The high yield of 94% for the resulting monoester **13** indicates a sufficient compatibility of the enzyme with the ruthenium catalyst as well as the advantage of this process concept of a sequential one-pot process as the enzyme is just added after completion of the metathesis, thus avoiding an ester hydrolysis of the diester substrate **10** (which would be disadvantageous because the metathesis runs much better when using the more hydrophobic diester substrate).

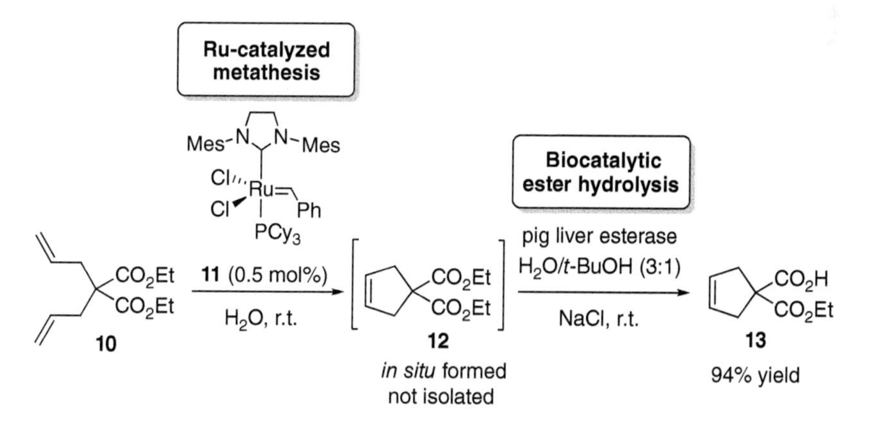

Scheme 14.4 Combination of a Ru-catalyzed metathesis reaction with an enzymatic ester hydrolysis in a sequential one-pot process for the synthesis of a cyclopentene carboxylic monoester.

Catalysis based on a Ru metal has also been applied by Gonzalez-Sabín and coworkers for a chemocatalytic rearrangement reaction in combination with a biotransformation in the presence of redox enzymes (Scheme 14.5) [18]. Among different concepts, also a sequential two-step one-pot process has been conducted. In detail, a Ru-catalyzed isomerization of racemic allylic alcohols to enones and subsequent asymmetric reduction in the presence of an alcohol dehydrogenase was carried out, thus forming the desired alcohols (*R*)-**17** in good to high yields and in enantiomerically pure form.

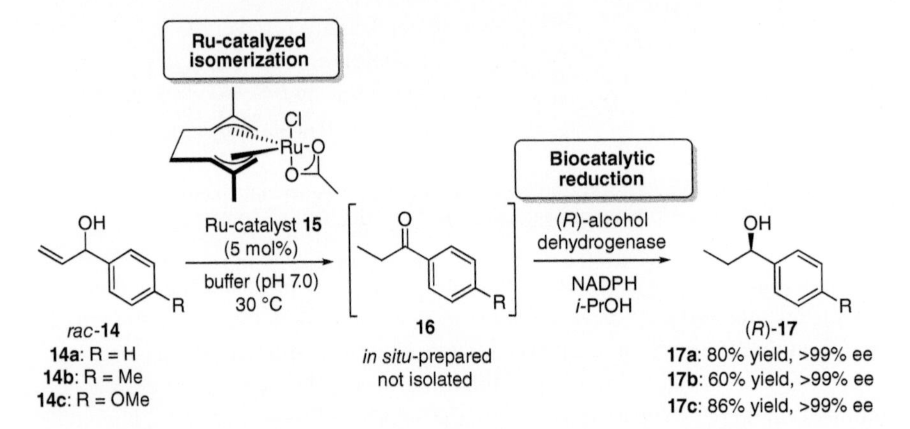

Scheme 14.5 Combination of a Rh-catalyzed isomerization of racemic allylic alcohols with an enzymatic C=O double bond reduction in a sequential one-pot process.

Another metal of wide interest for catalytic applications (in addition to hydrogenations) is rhodium, and a one-pot process based on the combination of this metal for a C—C bond formation with a biotransformation was reported by Zhao and coworkers (Scheme 14.6) [19]. In the initial step, a rhodium-catalyzed hetero-coupling of two diazoesters, **18** and **19**, led to the synthesis of an 1,4-dicarbonyl alkene **20**, which was then subsequently transformed by means of an ene reductase under formation of aryl-substituted 1,4-dicarbonyl compounds of type **21** in yields of up to 62% and with enantioselectivities of up to >99% ee. Also in this case, the concept of a sequential one-pot process offers advantages as the solvents can be adjusted according to the need of the individual steps. Thus, the rhodium-catalyzed

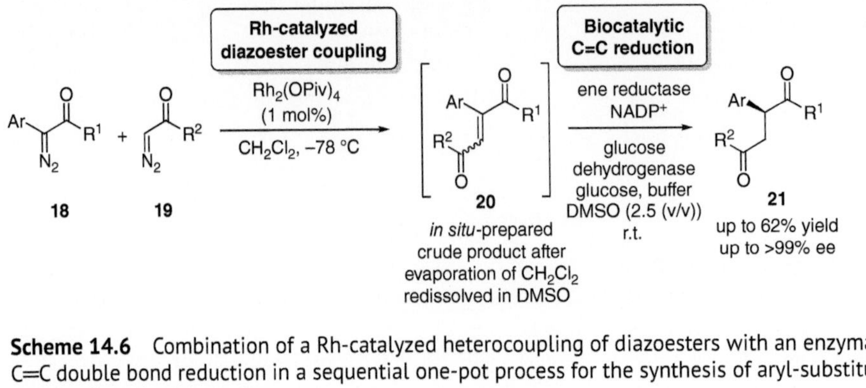

Rh-catalyzed diazoester coupling
Biocatalytic C=C reduction
$Rh_2(OPiv)_4$
(1 mol%)
CH_2Cl_2, −78 °C
ene reductase
$NADP^+$
glucose
dehydrogenase
glucose, buffer
DMSO (2.5 (v/v))
r.t.
18
19
20
in situ-prepared
crude product after
evaporation of CH_2Cl_2
redissolved in DMSO
21
up to 62% yield
up to >99% ee

Scheme 14.6 Combination of a Rh-catalyzed heterocoupling of diazoesters with an enzymatic C=C double bond reduction in a sequential one-pot process for the synthesis of aryl-substituted 1,4-dicarbonyl compounds.

reaction is conducted in methylene chloride, which is then removed (as it is an unfavorable solvent for the enzymatic step) and replaced by addition of dimethylsulfoxide (DMSO) before the subsequent enzymatic reduction of the C=C-double bond in the presence of an ene reductase. Thus, the enzymatic reaction runs in a favorable DMSO-containing aqueous medium with an *in situ* cofactor recycling in the presence of a glucose dehydrogenase and with D-glucose as a reducing agent.

Asymmetric carbon–carbon bond formation with metal catalysts have been combined as well with biotransformations. An example for this research area is the chemoenzymatic one-pot synthesis of 1,3-diol products developed by Aoki and coworkers, which consists of an asymmetric aldol reaction in the presence of a chiral Zn(II) catalyst and a subsequent enzymatic ketone reduction [20].

These selected examples illustrate the suitability of metal-catalyzed transformations to be combined with biotransformations toward one-pot processes running in a sequential mode. In this connection, it should be mentioned that this field of combined metal catalysis and biocatalysis toward one-pot processes (being conducted in a sequential as well as a tandem mode) has also been reviewed recently [1–5, 21].

14.2.2 Sequential Chemoenzymatic One-Pot Synthesis Combining an Organocatalytic Reaction with a Biotransformation

Stimulated by the tremendous achievements in the past three decades in the field of (asymmetric) organocatalysis [22–24], in recent years, the combination of organocatalytic reactions with biotransformations has also been studied with increased intensity. In the following, selected examples from this research area are summarized.

The first combination of an asymmetric organocatalytic reaction with a stereoselective biotransformation toward a one-pot process in aqueous medium has been realized by "merging" an asymmetric aldol reaction in the presence of a proline-derived organocatalyst with a subsequent enzymatic ketone reduction in various sequential one-pot setups [25, 26], leading, e.g., to 1,3-diols with a diastereomeric ratio of *d.r.* >25 : 1 and excellent enantiomeric excess of 99% ee [26]. The aldol reaction can be conducted under solvent-free conditions as well as in aqueous medium. Thus, by applying a sequential-mode one-pot concept, one can make use of both variants as the reaction medium for the enzymatic transformation can be adjusted after completion of the organocatalytic initial reaction step. This strategy (which is also reflected by one of the advantages in the box of Table 14.1) can be exemplified by the initial report in this field. Therein, the Gröger and Berkessel groups reported the combination of the asymmetric aldol reaction running under neat conditions with a high-space time yield with a subsequent enzymatic reduction of the *in situ*-formed aldol adduct (Scheme 14.7) [25].

After conducting the aldol reaction in acetone (4 equiv), which serves as both reagent and reaction medium, the reaction mixture is then directly passed into an aqueous reaction environment for the subsequent enzymatic step. In this medium, also isopropanol is added as a reducing agent. By means of this strategy enabling a one-pot-like process with both reactions running in different "process windows" being fine-tuned to their need, the desired 1,3-diol was formed with high conversion as well as diastereo- and enantioselecitvity as exemplified for 1,3-diol (1*R*,3*S*)-**25**, which was obtained with a conversion (related to the formation of this product) of 80% and with a diastereomeric ratio of *d.r.*=10 : 1 and an enantioselectivity of >99% ee (Scheme 14.7).

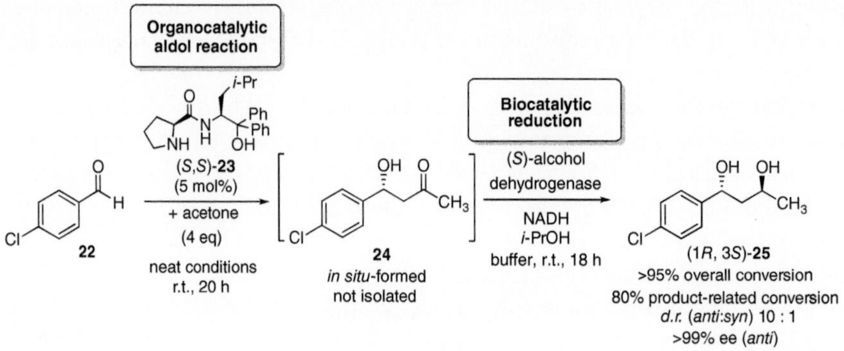

Scheme 14.7 Combination of an organocatalytic aldol reaction under neat conditions with a subsequent enzymatic reduction in aqueous medium.

An extension of this type of sequential one-pot cascade, in which both reactions proceed in aqueous medium, has also been reported by the Gröger and Berkessel groups [26]. Also in this sequential one-pot process, an adjustment of the reaction conditions has been conducted after completion of the initial organocatalytic reaction and before the biotransformation done as a second step. In addition, the concept of combining asymmetric organocatalytic aldol reactions and enzymatic reductions was also extended toward the synthesis of the industrial product (*R*)-pantolactone [27].

In a further version of a chemoenzymatic one-pot process based on an asymmetric aldol reaction as an organocatalytic step, Berry and Nelson and coworkers combined such an organocatalytic transformation with an enzymatic carbon–carbon bond [28]. The initial step consists of an asymmetric aldol reaction of acetaldehyde with a dialkyl glyoxylamide in the presence of a chiral diamine as an organocatalyst, and after dilution, the resulting aldol adduct was then further enzymatically transformed in the presence of an *N*-acetylneuraminic acid lyase and with sodium pyruvate, thus leading to the desired heterocyclic product.

A major benefit of asymmetric organocatalysis [22–24] is the opportunity to conduct a broad range of carbon–carbon bond formations in an asymmetric fashion (which is, interestingly, in contrast to preparative biocatalysis, for which in asymmetric organic synthesis only a few of such transformations are applied to a broader extent). Accordingly, also in the field of chemoenzymatic one-pot processes, there has been an interest to make use of the broad synthetic versatility of asymmetric organocatalysis in the field of carbon–carbon bond forming reactions. Addressing this research theme, Faber, Kroutil, and Pietruszka with coworkers realized a combination of an asymmetric organocatalytic Mannich-type reaction with a diastereoselective biocatalytic reduction (Scheme 14.8) [29]. Various synthetic strategies have been studied, and the optimized combination of the organocatalytic and enzymatic steps consists of a four-step, two-pot process. After the Mannich-type reaction, the solvent was evaporated, and the resulting crude product was then subjected to the

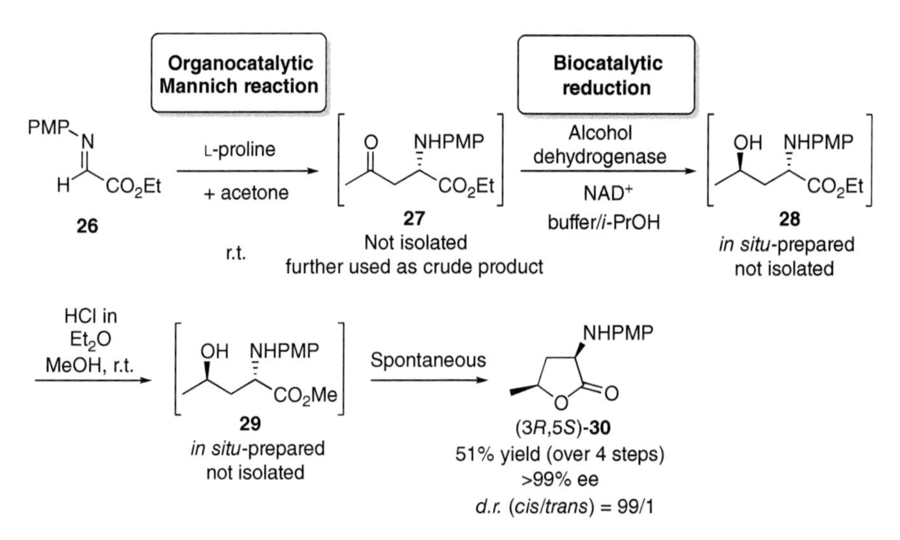

Scheme 14.8 Combination of an organocatalytic Mannich reaction with a subsequent enzymatic reduction and two further steps in a four-step two-pot process. PMP, *p*-methoxyphenyl.

biotransformation. The further reaction steps consisted of a transesterification and a lactone formation and lead to the formation of the desired *N*-protected α-amino γ-butyrolactone (3*R*,5*S*)-**30** in 51% yield and with an excellent diastereomeric ratio of *d.r.*= 99 : 1 and an enantiomeric excess of >99% ee (Scheme 14.8).

It is further noteworthy that this chemoenzymatic concept enables a highly enantio- and diastereoselective route to all four stereoisomers of the target molecule.

Furthermore, an organocatalytic reaction with an enol component has been combined with a biotransformation. Itoh and coworkers developed such a one-pot process based on an organocatalytic reaction of 2,3-dihydrofuran, which contains an enol ether moiety, with a hydroxyacetaldehyde precursor in the presence of a thiourea catalyst, followed by an enzymatic resolution of the racemic *in situ* formed hexahydrofuro[2,3-*b*]furan-3-ol by means of commercially available lipase PS and isolation of the remaining enantiomer in 30% yield and with an excellent enantiomeric excess of >99% ee [30]. This one-pot process also demonstrates the efficiency when applying lipase-catalyzed resolutions in such one-pot processes.

Furthermore, the concept of combining organo- and biocatalysis has been applied for a chemoenzymatic deracemization of secondary alcohols in a one-pot process. Interestingly, in such a process, which was developed by Lavandera, Gotor, Gotor-Fernández and coworkers, two redox reactions are combined, namely, an organocatalytic oxidation with an enzymatic reduction (Scheme 14.9) [31]. For the oxidation step, 2,2,6,6-tetramethylpiperidin-1-oxyl (TEMPO, 20 mol%) and iodine were chosen as "catalytic system." Together with a sonication, this chemocatalytic oxidation proceeds very efficiently, leading to the required ketone intermediates with full conversion within a short reaction time of only one hour. Subsequent enzymatic reduction under *in situ* cofactor recycling then furnished the desired alcohol products with both excellent conversions and enantioselectivities and representative examples are shown in Scheme 14.9.

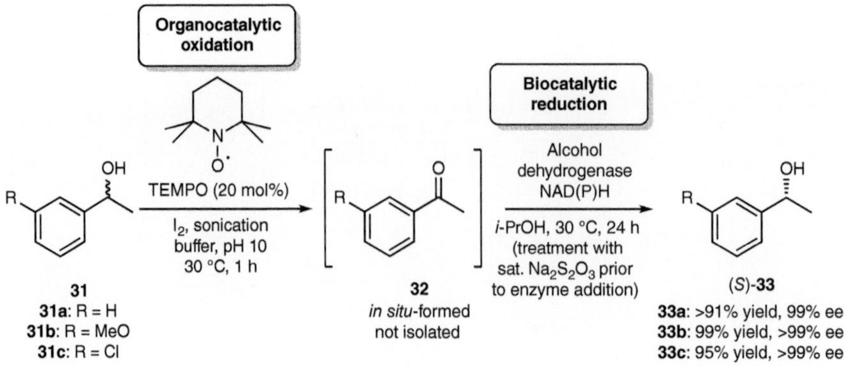

Organocatalytic oxidation

TEMPO (20 mol%)

I_2, sonication
buffer, pH 10
30 °C, 1 h

Biocatalytic reduction

Alcohol dehydrogenase
NAD(P)H

i-PrOH, 30 °C, 24 h
(treatment with sat. $Na_2S_2O_3$ prior to enzyme addition)

31
31a: R = H
31b: R = MeO
31c: R = Cl

32
in situ-formed
not isolated

(*S*)-**33**
33a: >91% yield, 99% ee
33b: 99% yield, >99% ee
33c: 95% yield, >99% ee

Scheme 14.9 Combination of an organocatalytic oxidation with a subsequent enzymatic reduction toward a one-pot deracemization process for secondary alcohols.

It should be added that the advantage of choosing a sequential one-pot strategy in this case is illustrated by the fact that this sequential strategy enables the beneficial destruction of iodine after the first reaction step and before the addition of the alcohol dehydrogenases. By means of this *in situ* decomposition of iodine by reduction to harmless iodide anions via addition of sodium thiosulfate, the required compatibility of the two types of reactions as a prerequisite to conduct both redox steps in one-pot is achieved in a very efficient way.

In addition, in this field of chemoenzymatic deracemization of secondary alcohols, other variants of TEMPO-based oxidation of secondary alcohols have been successfully combined with biocatalytic ketone reductions as demonstrated by González-Sabín and Rebolledo and coworkers [32]. The sequential mode of this one-pot process, which is based on the utilization of the organic radical 2-azaadamantane *N*-oxyl (AZADO) instead of TEMPO as an organocatalyst and sodium hypochlorite as an oxidation agent, turned out to be beneficial in this case as well as it enables to quench the remaining hypochlorite before the enzymatic step. The decomposition of hypochlorite can be done with isopropanol, which at the same time represents the reducing agent for the subsequent enzymatic reduction step.

It is noteworthy that this concept of coupling two redox reactions, namely, an organocatalytic oxidation and a biocatalytic reduction, has also been extended toward one-pot synthesis of chiral amines while still starting from secondary alcohols. The use of racemic secondary alcohols is beneficial as these substrates are readily accessible by various methods. Such a one-pot process running in a sequential mode by "merging" a TEMPO-catalyzed oxidation and a subsequent enzymatic transamination has been developed by Gotor-Fernández and Lavandera and coworkers (Scheme 14.10) [33]. In this case, for the oxidation of the racemic alcohols, a laccase/TEMPO-catalytic system was applied, and the enzymatic transamination reaction step was carried out with pyridoxal 5′-phosphate (PLP) as a cofactor and isopropylamine as a co-substrate. Although both steps run in aqueous medium, the laccase operates preferably at an acidic pH from 4.5 to 5.5, whereas the preferred pH for the enzymatic transamination step is pH 9. Thus, the sequential mode is an attractive option for conducting both steps in one-pot because it enables the adjustment of the pH to the specific needs of both steps and, thus, to conduct both reaction steps in its

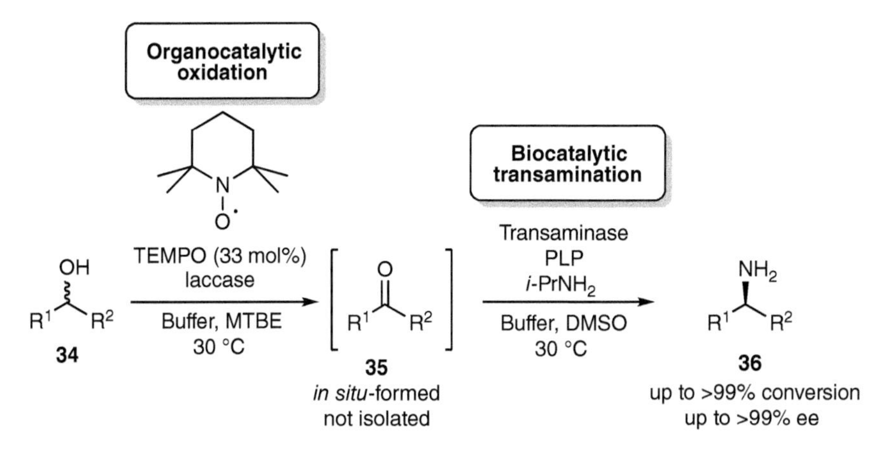

Scheme 14.10 Combination of an organocatalytic oxidation with a subsequent enzymatic transamination toward a one-pot transformation of racemic secondary alcohols to enantiomerically pure amines.

most suitable "process window." Accordingly, the resulting amines have been formed with both excellent conversions of >99% and enantioselectivities of >99% ee.

As for the related deracemization of secondary alcohols, also for this type of transformation, different versions of the oxidation step have been reported in combination with the enzymatic transamination. Rebolledo and González-Sabín applied the organic radical AZADO, which is sterically less hindered than TEMPO and enabled the oxidation step at lower catalyst loading (Scheme 14.11) [34]. Although both steps proceed in a basic aqueous reaction medium, different concentration requirements had to be taken into consideration during the development of the one-pot process. Thus, also here the sequential mode was chosen, as the organocatalytic step requires more concentrated conditions compared to the biotransformation. The oxidation step has been carried out at a substrate concentration of 250 mM with sodium hypochlorite as oxidation agent, followed by a dilution with buffer to adjust a concentration of 10 mM for the *in situ* formed ketone prior conducting the enzymatic step. By means of this approach, the resulting amines were formed with very high conversion and enantioselectivity, and a selected example is shown in Scheme 14.11.

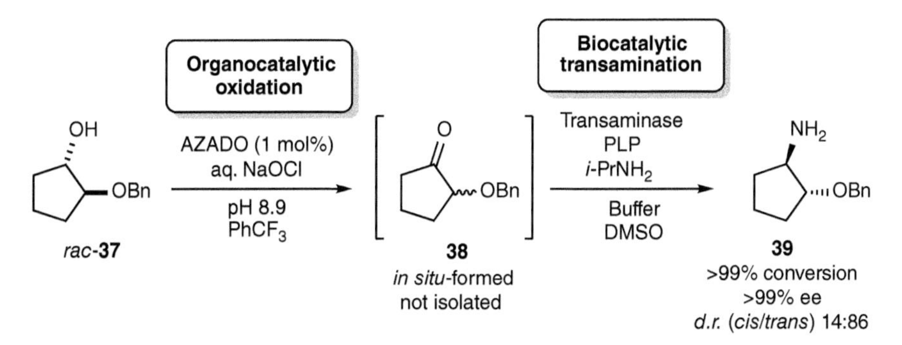

Scheme 14.11 Combination of an organocatalytic oxidation in the presence of AZADO as a catalyst with a subsequent enzymatic transamination toward a one-pot synthesis of chiral amines.

These selected examples underline the suitability to combine organo- and biocatalysts in one-pot processes running in a sequential mode. It should be added that this field of combined organocatalysis and biocatalysis toward one-pot processes (being conducted in a sequential as well as a tandem mode) has also been reviewed recently [35, 36].

14.2.3 Sequential Chemoenzymatic One-Pot Synthesis Combining a Reaction Catalyzed by a Heterogeneous Chemocatalyst with a Biotransformation

Although most of chemoenzymatic one-pot processes running in the sequential mode have been conducted in the presence of both homogeneous chemo- and biocatalysts, applications of heterogeneous catalysts are also known. Including heterogeneous catalysts in one-pot processes can be beneficial for various reasons, and in the following, some of them are briefly described.

A major advantage of utilizing heterogeneous catalysts is their simple separation from the product mixture by filtration, thus also making efficient reuse of the catalyst (assuming sufficient stability) easily possible. Heterogenization can also contribute to an increased catalyst stabilization and might represent a compartmentalization as it can suppress or avoid the contact with a homogeneous catalyst. Thus, in order to make use of such advantages, there has been consequently an interest to study the combination of heterogeneous with homogeneous catalysts in chemoenzymatic one-pot syntheses.

Starting with heterogenized chemocatalysts and their combination with enzymes, Monti and Guidotti and coworkers conducted an epoxidation of limonene with the heterogeneous titanium-grafted silica catalyst Ti/SiO_2, followed by hydrolytic ring-opening in the presence of a limonene epoxide hydrolase, which was added together with the buffer solution after completion of the initial step (Scheme 14.12) [37].

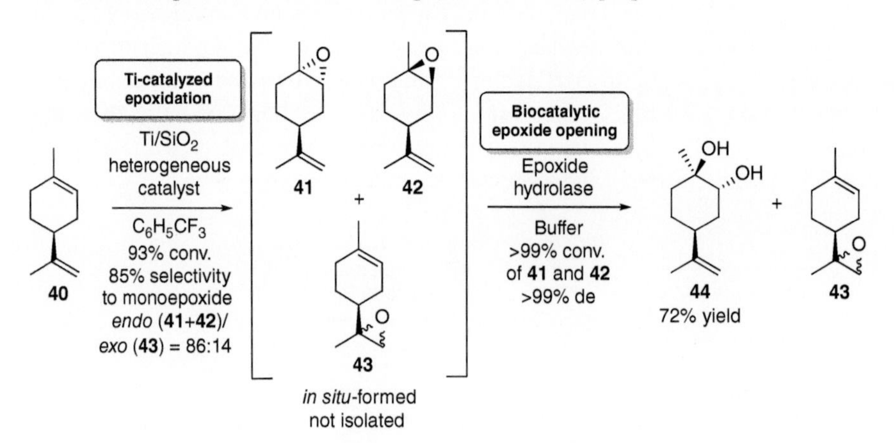

Scheme 14.12 Combination of a regio- and diastereoselective epoxidation with a heterogeneous chemocatalyst and an enzymatic epoxide opening reaction.

As a representative example, the (*R*,*R*,*S*)-diol **44** was obtained in 72% yield when starting from (−)-limonene, **40**. As the initial chemocatalytic step is carried out in pure organic medium (such as acetonitrile or α,α,α-trifluorotoluene) and under exclusion of water, this type of process also illustrates the advantage of sequential one-pot processes to utilize different media for the individual steps while still conducting both reactions in one-pot. In addition, a high compatibility of the catalytic systems was found.

14.2.4 Sequential Chemoenzymatic One-Pot Synthesis Combining a Reaction Catalyzed by a Heterogeneous Biocatalyst with a Chemocatalytic Transformation

As the reverse concept, one can consider the combination of a heterogenized biocatalyst with a chemocatalyst. Such a combination has been reported for the enantio- and diastereoselective synthesis of 1,3-diols (Scheme 14.13) [38]. Starting from acetone and an aromatic aldehyde, **45**, an asymmetric aldol reaction in the presence of a proline derivative as a homogeneous chiral organocatalyst gives the β-hydroxy ketone (*R*)-**46**, which was then diastereoselectively reduced in an organic solvent by means of a heterogenized alcohol dehydrogenase. Toward the heterogenization of the biocatalyst, the enzyme was entrapped in a superabsorber. It is noteworthy that the cofactor can also be retained by the aqueous superabsorber matrix as long as an organic reaction medium is used. Such a co-immobilization of an alcohol dehydrogenase and a cofactor in a superabsorber then enables to conduct the biotransformation also in pure organic medium. Furthermore, the superabsorber (and, thus, biocatalyst and cofactor) can be simply separated from the reaction medium bearing the organocatalyst and 1,3-diol product afterward. This sequential-type one-pot process consisting of an organocatalytic aldol reaction and a biocatalytic reduction gave the 1,3-diol (1*R*,3*S*)-**47** with high conversion as well as high enantioselectivity (>99% ee) and diastereoselectivity (diastereomeric ratio *d.r.* >35 : 1) [38].

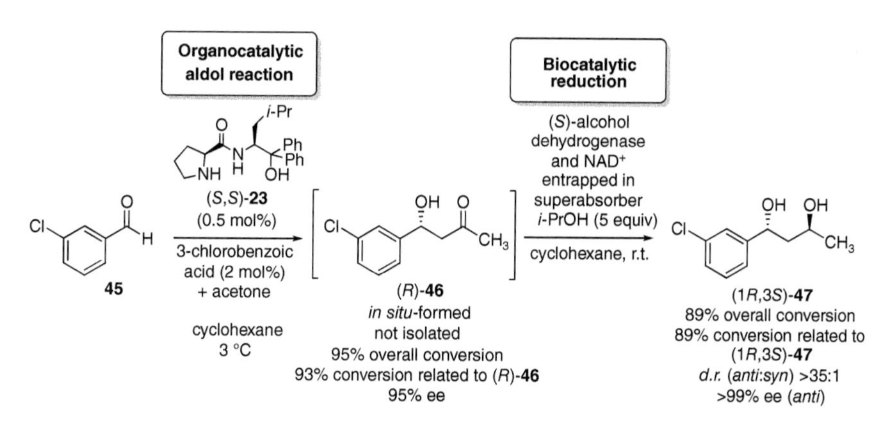

Scheme 14.13 Combination of an aldol reaction with a homogeneous organocatalyst and an enzymatic reduction with a heterogeneous biocatalytic system.

The heterogenization of enzymes also enables their use as immobilized biocatalysts in flow processes (see also Chapter 11). This has also been demonstrated for sequential-type one-pot processes as exemplified by the combination of a Suzuki cross-coupling reaction with a transamination in the presence of an immobilized biocatalyst developed by Bornscheuer and de Souza and coworkers (Scheme 14.14) [14]. At first, this type of process

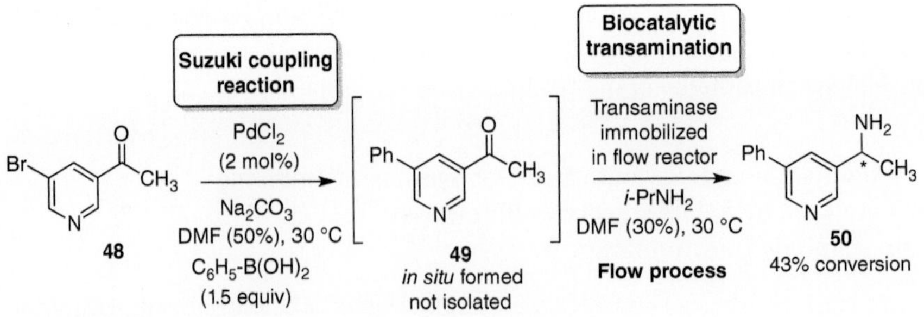

Suzuki coupling reaction
$PdCl_2$ (2 mol%)
Na_2CO_3
DMF (50%), 30 °C
C_6H_5-$B(OH)_2$ (1.5 equiv)
48
49 in situ formed not isolated
Biocatalytic transamination
Transaminase immobilized in flow reactor
i-$PrNH_2$
DMF (30%), 30 °C
Flow process
50 43% conversion

Scheme 14.14 Combination of a Suzuki cross coupling reaction with a heterogeneous biocatalyst for a one-pot-like process running in part in a flow mode.

has been demonstrated for a batch-type process using non-immobilized enzymes (see Section 14.2.1.) prior to the extension to the flow mode after immobilization of an optimized mutant of the transaminase [14]. Although this reaction is not a one-pot process, it can be considered as a one-pot-like process as intermediate purification is avoided, and as the resulting crude product mixture is directly applied without the need to work-up for the subsequent step running in a flow mode. The desired chiral amine product **50** was obtained with 43% conversion in this combined batch and flow process.

14.2.5 Sequential Chemoenzymatic One-Pot Synthesis Combining More than Two Reactions

Until now, most of the chemoenzymatic one-pot processes consists of the combination of a chemocatalytic or "classic" chemical reaction with a biotransformation. However, albeit to less extent it has also been demonstrated that more than two reactions can be combined toward a one-pot process. The pioneer work combining reactions catalyzed by all three types of catalysts, namely, a metal catalyst, an organocatalyst, and an enzyme, was reported by Kieboom and coworkers (Scheme 14.15) [39].

In this one-pot synthesis of methyl 4-deoxy-6-aldehydo-β-D-*xylo*-hexapyranoside, **54**, the first step is a biocatalytic oxidation of D-galactoside derivative **51** with air, followed by an organocatalytic dehydration with proline as an organocatalyst and a Pd/C-catalyzed hydrogenation under formation of product **54** in >95% yield.

A proof-of-concept work for combining two biocatalytic reactions and one chemocatalytic transformation has been recently demonstrated by Kroutil, Ward, Hailes and Rother and co-workers in a chemoenzymatic synthesis of 1,3,4-trisubstituted tetrahydroisoquinolines bearing three chiral centers (Scheme 14.16) [40]. The initial steps of this cascade are the two enzymatic transformations in the presence of a carboligase and a transaminase starting from 3-hydroxybenzaldehyde, **55**, and pyruvate as substrates, followed by a Pictet–Spengler reaction with a carbonyl moiety and catalyzed by phosphate as the third reaction step.

It is noteworthy that also here no intermediate purification is needed and that the advantage of the sequential one-pot strategy has been taken by removing the enzymes by ultrafiltration before the addition of the phosphate chemocatalyst as a final step, thus leading to the desired target product with high overall conversion and stereoselectivity.

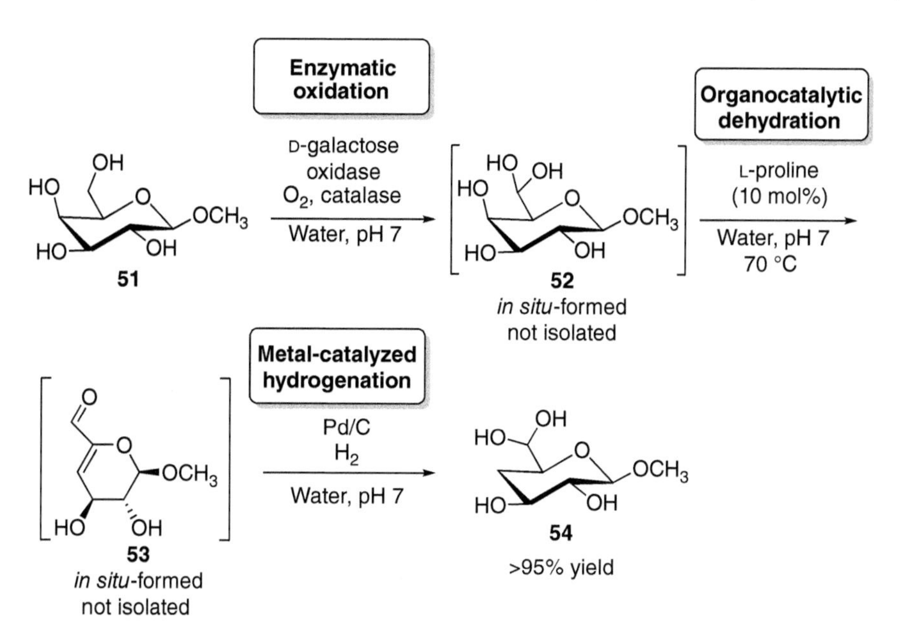

Scheme 14.15 Combination of biocatalytic, organocatalytic, and metal-catalyzed reactions toward a three-step one-pot process.

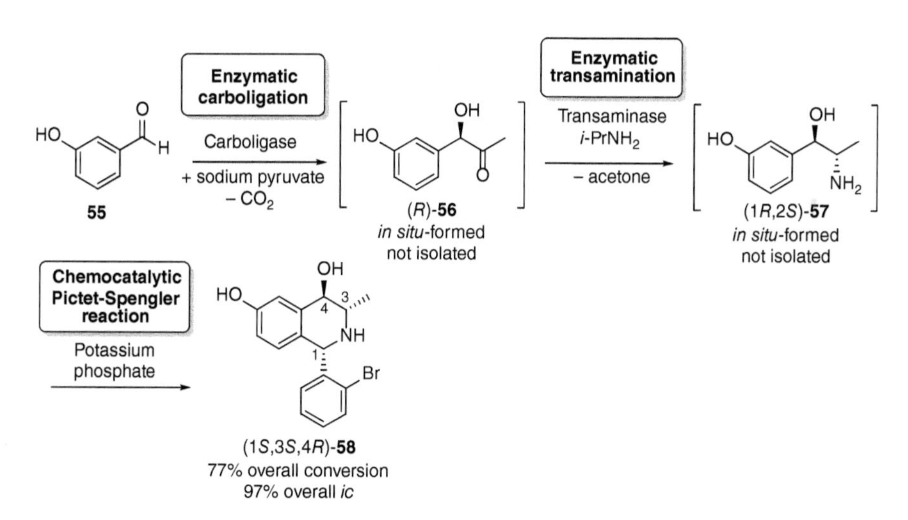

Scheme 14.16 Combination of two biocatalytic and a chemocatalytic reaction toward a three-step one-pot process. *ic*, isomeric content.

Very recently, a combination of three reactions comprising a metal-catalyzed alkene hydroformylation, a spontaneous condensation with hydroxylamine leading to an oxime, followed by a subsequent enzymatic dehydration under formation of an aliphatic nitrile has been reported by Gröger, Asano and Vorholt and co-workers [41]. This process being conducted in a one-pot-related fashion also does not require work-up and purification of the formed intermediates. However, as the enzyme is strongly deactivated by hydroxylamine used in the second step, decomposition of remaining traces of hydroxylamine before the addition of the enzyme is required.

14.3 Practical Aspects of the Development of Sequential Chemoenzymatic One-Pot Syntheses

In this chapter, some practical issues when developing sequential-type chemoenzymatic one-pot processes will also be discussed. In general, one particular issue to be considered when focusing on sequential multi-step, one-pot processes, compared to the "traditional" multistep approach utilizing purified product intermediates as substrates for the next steps, is how to achieve compatibility between the catalyst of the desired reaction step with the components of the previous reaction step(s) being still present in the reaction mixture or crude product used as a starting material. In order to gain insights into such compatibility issues, it appears to be beneficial to investigate the impact of each of the relevant process parameters on a catalyst in advance in individual experiments, before combining the reaction steps toward a one-pot process.

As the focus of this book is on biocatalysis, in the following, particularly the challenge of using an enzyme as a catalyst in the second step, thus raising the need of the biocatalyst to be compatible also with the components from the first reaction step, will be considered. The potential impact of the various reaction parameters on the enzyme (used in a second step of such a one-pot process), as well as the opportunities in terms of changing the "process window" after completion of the initial step and before the addition of the enzyme for the second (biotransformation) step of such a sequential-type chemoenzymatic one-pot process, is graphically summarized in Figure 14.2.

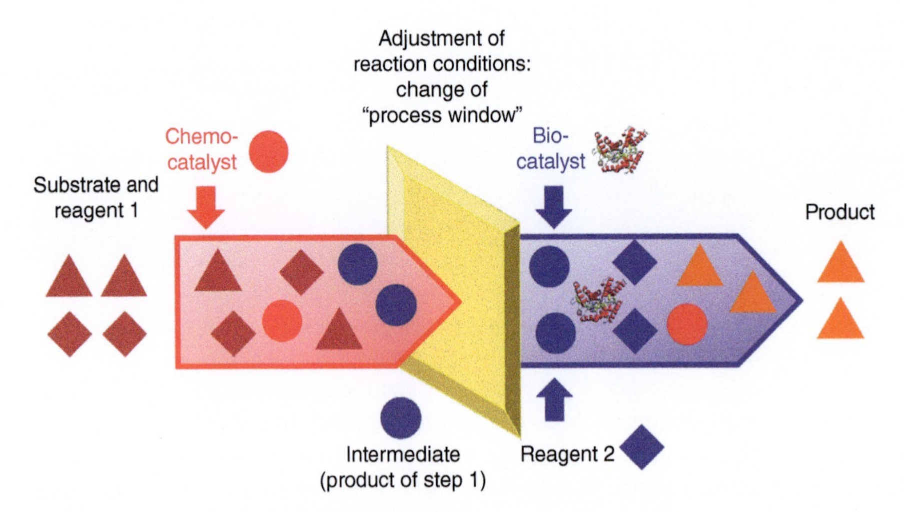

Figure 14.2 Overview about the impact of reaction parameters on the enzyme used in a second step of a sequential-type chemoenzymatic one-pot process.

The various influences of the reaction parameters on the biocatalyst used in the second step, also determine how a potential study in order to investigate in advance the impact of the different reaction parameters of reaction steps 1 and 2 on the enzyme as a catalyst for step 2 in such a sequential-type chemoenzymatic one-pot process could be conducted. Accordingly, such a list of relevant reaction parameters not only comprises the "typical" reaction parameters such as the solvent, temperature, substrate, and reagents of the reaction the (bio-)catalyst is catalyzing (which are shown in the blue-colored arrow in Figure 14.2), but also the reaction parameters being relevant for the previous step(s) such, for example, the reaction medium, the substrate(s), reagent(s), as well as (chemo-)catalysts being involved therein (which are shown in the red-colored arrow in Figure 14.2).

When studying the impact of various chemicals on the biocatalyst, such as the substrate and reagents of the biotransformation or the previous steps (as well as chemocatalyst in the latter case), it has to be differentiated between (bio-)catalyst inhibition and (bio-)catalyst deactivation. Although both effects led to a drop in the catalytic activity, they can be addressed and overcome by means of different solutions. For example, if it turns out that an enzyme is inhibited by the substrate of the enzymatic step or one of the components of the previous step from a certain concentration on (while maintaining its stability), dosing the substrate in order to keep the concentration at a low level (at which inhibition occurs to a less extent while still a high enzyme activity remains) represents a simple and efficient solution to overcome this hurdle. However, if in contrast a strong deactivation effect by a substrate or reagent occurs, even dosage might not be a suitable strategy to overcome this limitation as the enzyme is deactivated quickly also under those conditions. Furthermore, in case of a substrate inhibition, even a somewhat lower reaction velocity (caused by the inhibition) might be acceptable if the substrate inhibition is not too severe, whereas in the case of deactivation of an enzyme at a certain substrate concentration, the loss of enzyme activity over the time might be so severe that an efficient process cannot be realized. In such a case, other options such as, for example, enzyme engineering of the biocatalyst or (in case of deactivation by a component of a previous step) removal of the critical component before the addition of the biocatalyst, could be beneficial.

After studying the impact of substrates and reagents of steps 1 and 2 as well as the chemocatalyst from step 1 on the enzyme, one can get a "picture" if the enzyme will be compatible with such components and, thus, if a one-pot process could then become feasible in principle.

Further relevant criteria for developing a one-pot process are reaction parameters related to the "reaction environment" such as temperature, concentration, and pH value. However, in this case, a specific advantage of the sequential one-pot process is that to some extent, the reaction conditions can be re-adjusted after completion of the first step(s) and before the addition of the (bio-)catalyst for the subsequent step as indicated in Figure 14.2. Thus, this concept enables to adjust the "process window" to the specific needs of the second reaction while still having the option to run the first reaction under its own, optimized "process window." Such a re-adjustment of the "process window" can be in particular easily carried out, if just the reaction temperature or the substrate concentration has to be changed for the second step. At the same time, this option can be extremely helpful as often the reaction temperature for, e.g., an initial chemocatalytic step might differ strongly from the optimum temperature for, e.g., a second biotransformation step. An example described above, in which such a strategy of reaction temperature adjustment has been elegantly used, is the research work of Cacchi and coworkers on combining a Pd-catalyzed Heck reaction running at 100 °C with the subsequent enzymatic reduction running at ambient temperature (see Section 14.2.1 and Scheme 14.3 therein) [16].

In addition, the switch of the substrate concentration is an easy and at the same time very efficient option in order to achieve compatibility between the reaction steps as often chemo- and biocatalytic steps require very different substrate concentrations. Among the examples discussed above in the individual sections, the one-pot process developed by the Rebolledo and González-Sabín groups [34], which is shown in Section 14.2.2. and Scheme 14.11 therein, is such an example. In this combination of an organocatalytic oxidation and subsequent enzymatic reduction, the initial chemocatalytic reaction required a much higher substrate loading of 250 mM compared to the enzymatic process running at 10 mM. The needed compatibility could be achieved in an elegant manner by diluting the reaction medium of the first step with buffer before the start of the subsequent biotransformation reaction.

A further example of how a one-pot process can benefit from the option to adjust reaction parameters after completion of the first step and before the addition of the enzyme for the second step, is the switch of a solvent (and, thus, reaction medium). This has been demonstrated by Zhao and coworkers for the combination of the Rh-catalyzed diazoester coupling with an enzymatic C=C double bond reduction (see Section 14.2.1 and Scheme 14.6 therein) [19]. In this case, removal of methylene chloride after completion of the metal-catalyzed initial step enabled to conduct the biotransformation under optimized conditions.

Another case which benefits from conducting the chemoenzymatic one-pot process in a sequential mode is related to severe enzyme inhibitions (or deactivation) from components utilized in the initial chemocatalytic step. When being able to run such reactions with exactly stoichiometric amounts of such a problematic reagent, at the end of the chemocatalytic reaction, then the problematic component would be consumed. Thus, when adding the enzyme after completion of this initial step, such a negative interaction of the biocatalyst with such a component from the first step would not result in a problem for the one-pot process because of the lack of this problematic component at this stage. An example for such a case is the combination of the Pd-catalyzed Suzuki coupling reaction with a subsequent ketone reduction, which was reported by the Gröger and Hummel groups (see Section 14.2.1 and Scheme 14.2 therein) [7]. The alcohol dehydrogenase used for the biotransformation turned out to be strongly inhibited by the boronic acid. However, when using one equivalent of boronic acid only (instead of an excess of this compound) and adding the enzyme after completion of the metal-catalyzed step, then also the biotransformation (and, thus, the one-pot process) can be conducted efficiently.

These examples illustrate that (i) getting an insight by determining the impact of the various reaction parameters on the (bio-)catalyst used in the second step is very helpful when developing sequential-type chemoenzymatic one-pot processes, and (ii) that when choosing this sequential-type one-pot process option, the opportunity to modify the process window after completion of the initial reaction step and before the start of the second reaction steps represents a versatile tool for adjustment of the reaction parameters to the specific needs of the (bio-)catalyst used in the second step, while still benefiting from the avoidance of a work-up and purification step of the intermediate.

14.4 Conclusions and Outlook

In conclusion, various opportunities to combine chemocatalytic and biocatalytic reactions toward one-pot processes exist. All of those one-pot options provide advantages but also show drawbacks, and it is noteworthy that these methods thus complement each other in an advantageous fashion. At the same time, all of them enable to run multistep processes without the need for work-up of the intermediates formed in the initial step(s), thus making them very attractive for the development of sustainable processes with decreased waste formation, decreased solvent consumption, and reduced number of unit-operating steps (and, thus, improved space–time yield).

Among these chemoenzymatic one-pot process options, sequential-type one-pot processes have been increasingly developed in recent years. It has been demonstrated that a broad range of chemocatalytic transformations can be combined with biotransformations, thus also demonstrating a high compatibility of enzymes with respect to the components (substrates, reagents, and products), as well as catalysts from those chemocatalytic steps. It is noteworthy that these chemocatalytic reactions comprise transformations with metal catalysts as well as organocatalysts. Furthermore, these catalysts can be used in homogeneous as well heterogeneous form.

In spite of all the achievements made in the recent years in this research area of sequential-type chemoenzymatic one-pot processes and in general chemoenzymatic one-pot cascades, still many challenges are ahead of us. The design of novel cascades certainly belongs to those challenges and a particular goal can be seen in the future increase of the number of enzyme classes with proven applicability to be used in one-pot processes. This issue might also be connected with improved enzyme catalysts (as well as chemocatalysts) being robust also in such one-pot processes, in which more components are present compared to a single step transformation. For the identification of such more robust enzymes, protein engineering plays an important role. As so far most of the sequential-type one-pot processes have been conducted in the batch mode, also the evaluation of other reactor concepts, in particular the use of "flow reactors," will be of great interest. Last but not least, a further issue being of importance for implementation in industry is to increase space–time yield and decrease catalyst loading (of both enzyme and chemocatalyst). Thus, proof of concepts for sequential-type one-pot processes running at high substrate loading (of at least $50\,g\,l^{-1}$) and at low catalyst demand would also be of high interest.

In order to achieve such goals, expertise from different areas such as, for example, molecular biology and microbiology, organic chemistry, and chemical as well as biochemical engineering are required, thus also making this field of chemoenzymatic one-pot processes to a scientifically exciting interdisciplinary research area with a high potential for valuable future developments.

References

1 Schrittwieser, J.H., Velikogne, S., Hall, M., and Kroutil, W. (2018). Artificial biocatalytic linear cascades for preparation of organic molecules. *Chem. Rev.* 118: 270–348.

2 Rudroff, F., Mihovilovic, M.D., Gröger, H. et al. (2018). Opportunities and challenges for combining chemo- and biocatalysis. *Nat. Catal.* 1: 12–22.

3 Gröger, H. and Hummel, W. (2015, chapter 3.8.2). Merging of metal, organic, and enzyme catalysis. In: *Science of Synthesis, Biocatalysis in Organic Synthesis*, vol. 3 (eds. K. Faber, W.-D. Fessner and N.J. Turner), 491–514. Stuttgart: Thieme.

4 Gröger, H. and Hummel, W. (2014). Combining the 'two worlds' of chemocatalysis and biocatalysis towards multi-step one-pot processes in aqueous media. *Curr. Opin. Chem. Biol.* 19: 171–179.

5 Williams, G. and Hall, M. (eds.) (2018, chapter 15). Review with a focus on compartmentalization as a tool for combining chemo- and biocatalysis in one-pot processes: H. Gröger. In: *Modern Biocatalysis: Advances Towards Synthetic Biological Systems*, 439–472. The Royal Society of Chemistry.

6 Simons, C., Hanefeld, U., Arends, I.W.C.E. et al. (2006). Towards catalytic cascade reactions: asymmetric synthesis using combined chemo-enzymatic catalysts. *Top. Catal.* 40: 35–44.

7 Burda, E., Hummel, W., and Gröger, H. (2008). Modular chemoenzymatic one-pot syntheses in aqueous media: combination of a palladium-catalyzed cross-coupling with an asymmetric biotransformation. *Angew. Chem. Int. Ed.* 47: 9551–9554.

8 Burda, E., Bauer, W., Hummel, W., and Gröger, H. (2010). Enantio-and diastereoselective chemoenzymatic synthesis of C2-symmetric biaryl-containing diols. *ChemCatChem* 2: 67–72.

9 Prastaro, A., Ceci, P., Chiancone, E. et al. (2009). Suzuki-Miyaura cross-coupling catalyzed by protein-stabilized palladium nanoparticles under aerobic conditions in water: application to a one-pot chemoenzymatic enantioselective synthesis of chiral biaryl alcohols. *Green Chem.* 11: 1929–1932.

10 Gauchot, V., Kroutil, W., and Schmitzer, A.R. (2010). Highly recyclable chemo-/biocatalyzed cascade reactions with ionic liquids: one-pot synthesis of chiral biaryl alcohols. *Chem. Eur. J.* 16: 6748–6751.

11 Borchert, S., Burda, E., Schatz, J. et al. (2012). Combination of a Suzuki cross-coupling reaction using a water-soluble palladium catalyst with an asymmetric enzymatic reduction towards a one-pot process in aqueous medium at room temperature. *J. Mol. Catal. B: Enzym.* 84: 89–93.

12 Paris, J., Ríos-Lombardía, N., Morís, F. et al. (2018). Novel insights into the combination of metal- and miocatalysis: cascade one-pot synthesis of enantiomerically pure biaryl alcohols in Deep Eutectic Solvents. *ChemCatChem* 10: 4417–4423.

13 Paris, J., Telzerow, A., Ríos-Lombardía, N. et al. (2019). Enantioselective one-pot synthesis of biaryl-substituted amines by combining palladium and enzyme catalysis in deep eutectic solvents. *ACS Sustain. Chem. Eng.* 7: 5486–5493.

14 Dawood, A.W.H., Bassut, J., de Souza, R.O.M.A., and Bornscheuer, U.T. (2018). Combination of the Suzuki–Miyaura cross-coupling reaction with engineered transaminases. *Chem. Eur. J.* 24: 16009–16013.

15 Sgalla, S., Fabrizi, G., Cirilli, R. et al. (2007). Chiral (*R*)-and (*S*)-allylic alcohols via a one-pot chemoenzymatic synthesis. *Tetrahedron: Asymmetry* 18: 2791–2796.

16 Boffi, A., Cacchi, S., Ceci, P. et al. (2011). The Heck reaction of allylic alcohols catalyzed by palladium nanoparticles in water: chemoenzymatic synthesis of (*R*)-(−)-rhododendrol. *ChemCatChem* 3: 347–353.

17 Tenbrink, K., Seßler, M., Schatz, J., and Gröger, H. (2011). Combination of olefin metathesis and enzymatic ester hydrolysis in aqueous media in a one-pot synthesis. *Adv. Synth. Catal.* 353: 2363–2367.

18 Ríos-Lombardía, N., Vidal, C., Liardo, E. et al. (2016). From a sequential to a concurrent reaction in aqueous medium: ruthenium-catalyzed allylic alcohol isomerization and asymmetric bioreduction. *Angew. Chem. Int. Ed.* 55: 8691–8695.

19 Wang, Y., Bartlett, M.J., Denard, C.A. et al. (2017). Combining Rh-catalyzed diazocoupling and enzymatic reduction to efficiently synthesize enantioenriched 2-substituted succinate derivatives. *ACS Catal.* 7: 2548–2552.

20 Sonoike, S., Itakura, T., Kitamura, M., and Aoki, S. (2012). One-pot chemoenzymatic synthesis of chiral 1,3-diols using an enantioselective aldol reaction with chiral Zn^{2+} complex catalysts and enzymatic reduction. *Chem. Asian J.* 7: 64–74.

21 Gröger, H. (2015, chapter 11). Metals and metal complexes in cooperative catalysis with enzymes within organic-synthetic one-pot processes. In: *Cooperative Catalysis* (ed. R. Peters), 325–349. Weinheim: Wiley-VCH.

22 Berkessel, A. and Gröger, H. (2005). *Asymmetric Organocatalysis: From Biomimetic Concepts to Application in Asymmetric Synthesis*. Weinheim: Wiley-VCH.

23 Pellissier, H. (2007). Asymmetric organocatalysis. *Tetrahedron* 63: 9267–9331.

24 List, B. and Maruoka, K. (eds.) (2012). *Science of Synthesis: Asymmetric Organocatalysis*, vol. 1–2. Stuttgart: Thieme.

25 Baer, K., Kraußer, M., Burda, E. et al. (2009). Sequential and modular synthesis of chiral 1,3-diols with two stereogenic centers: access to all four stereoisomers by combination of organo-and biocatalysis. *Angew. Chem. Int. Ed.* 48: 9355–9358.

26 Rulli, G., Duangdee, N., Baer, K. et al. (2011). Direction of kinetically versus thermodynamically controlled organocatalysis and its application in chemoenzymatic synthesis. *Angew. Chem. Int. Ed.* 50: 7944–7947.

27 Heidlindemann, M., Hammel, M., Scheffler, U. et al. (2015). Chemoenzymatic synthesis of vitamin B5-intermediate (*R*)-Pantolactone via combined asymmetric organo- and biocatalysis. *J. Org. Chem.* 80: 3387–3396.

28 Kinnell, A., Harman, T., Bingham, M. et al. (2012). Development of an organo-and enzyme-catalysed one-pot, sequential three-component reaction. *Tetrahedron* 68: 7719–7722.

29 Simon, R.C., Busto, E., Schrittwieser, J.H. et al. (2014). Stereoselective synthesis of γ-hydroxynorvaline through combination of organo-and biocatalysis. *Chem. Commun.* 50: 15669–15672.

30 Kanemitsu, T., Inoue, M., Yoshimura, N. et al. (2016). A concise one-pot organo- and biocatalyzed preparation of enantiopure hexahydrofuro[2,3-*b*]furan-3-ol: an approach to the synthesis of HIV protease inhibitors. *Eur. J. Org. Chem.* 2016: 1874–1880.

31 Méndez-Sánchez, D., Mangas-Sánchez, J., Lavandera, I. et al. (2015). Chemoenzymatic deracemization of secondary alcohols by using a TEMPO–iodine–alcohol dehydrogenase system. *ChemCatChem* 7: 4016–4020.

32 Liardo, E., Ríos-Lombardía, N., Morís, F. et al. (2018). A straightforward deracemization of *sec*-alcohols combining organocatalytic oxidation and biocatalytic reduction. *Eur. J. Org. Chem.*: 3031–3035.

33 Martínez-Montero, L., Gotor, V., Gotor-Fernández, V., and Lavandera, I. (2017). Stereoselective amination of racemic *sec*-alcohols through sequential application of laccases and transaminases. *Green Chem.* 19: 474–480.

34 Liardo, E., Ríos-Lombardía, N., Morís, F. et al. (2017). Hybrid organo- and biocatalytic process for the asymmetric transformation of alcohols into amines in aqueous medium. *ACS Catal.* 7: 4768–4774.

35 Yamashita, Y. and Gröger, H. (2019, chapter 2.5). Organocatalyst/biocatalyst dual catalysis. In: *Science of Synthesis: Dual Catalysis in Organic Synthesis*, vol. 2 (ed. G.A. Molander), 339–364. Stuttgart: Thieme.

36 Aranda, C., Oksdath-Mansilla, G., Bisogno, F.R., and de Gonzalo, G. (2020). Deracemisation processes employing organocatalysis and enzyme catalysis. *Adv. Synth. Catal.* 362: 1233–1257.

37 Palumbo, C., Ferrandi, E.E., Marchesi, C. et al. (2016). One–pot selective dihydroxylation of limonene combining metal and enzyme catalysis. *Chem. Select* 1: 1795–1798.

38 Heidlindemann, M., Rulli, G., Berkessel, A. et al. (2014). Combination of asymmetric organo- and biocatalytic reactions in organic media using immobilized catalysts in different compartments. *ACS Catal.* 4: 1099–1103.

39 Schoevaart, R. and Kieboom, T. (2002). Combined catalytic conversion involving an enzyme, a homogeneous and a heterogeneous catalyst: one-pot preparation of 4-deoxy-D-glucose derivatives from D-galactose. *Tetrahedron Lett.* 43: 3399–3400.

40 Erdmann, V., Lichman, B.R., Zhao, J. et al. (2017). Enzymatic and chemoenzymatic three-step cascades for the synthesis of stereochemically complementary trisubstituted tetrahydroisoquinolines. *Angew. Chem. Int. Ed.* 56: 12503–12507.

41 Plass, C., Hinzmann, A., Terhorst, M. et al. (2019). Approaching bulk chemical nitriles from alkenes: a hydrogen cyanide-free approach through a combination of hydroformylation and biocatalysis. *ACS Catal.* 9: 5198–5203.

Part V

Industrial Biocatalysis

15

Industrial Processes Using Biocatalysts

Florian Kleinbeck, Marek Mahut, and Thierry Schlama

Novartis Pharma AG, Chemical and Analytical Development, 4056, Basel, Switzerland

15.1 Introduction

Biocatalytic transformations applying enzymes as catalysts for chemical reactions have been successfully used on commercial scale in the chemical and food industries for decades [1] (for reviews on the use of biocatalysis in the manufacturing of pharmaceuticals, see Refs [2–10]). Although early examples of biocatalytic transformations in the manufacturing of active pharmaceutical ingredients (APIs) date back almost 100 years [11], the number of examples have started to grow significantly only in the 1990s. This growth was triggered by the tremendous advances in genomics, molecular biology, and bioinformatics. These advances give access to a continuously increasing number and variety of enzymes, offer the possibility to modify enzymes to tailor their performance to specific needs, and enable large-scale fermentation to provide commercial quantities of enzymes [12–14].

The growing interest in the application of biocatalytic transformations for the large-scale manufacturing of APIs can be attributed to multiple prominent features of enzyme-catalyzed reactions that in many cases provide significant advantages over traditional synthetic methodology, thus rendering biocatalysis an attractive alternative, if not even the method of choice.

- *Mild reaction conditions.* Biocatalytic reactions generally operate at moderate temperatures and pH values, i.e. typically at ambient or slightly elevated temperatures of 40–60 °C and mildly basic or acidic conditions. The reaction conditions applied in biocatalytic transformations are defined by the stability limitations of enzymes with respect to temperature, pH value and environment, and often reflect the natural environment of (wild-type) enzymes. In the vast majority of cases, water is the only solvent or the primary component of the solvent system, with polar organic solvents sometimes used as cosolvents for hydrophobic substrates.
- *High chemoselectivity.* The strong substrate specificity of most enzymes ensures a chemoselective transformation of the substrate exclusively to the desired product. In combination

Biocatalysis for Practitioners: Techniques, Reactions and Applications, First Edition.
Edited by Gonzalo de Gonzalo and Iván Lavandera.

with the mild reaction conditions, the high chemoselectivity suppresses or significantly reduces the formation of by-products because of undesired side reactions or instability of substrate and product under the reaction conditions. Consequently, high yield and chemical purity of the product can be achieved in most cases.

- *Excellent stereoselectivity*. One of the most distinct advantages of enzymes is their ability to form stereogenic centers with high enantio- or diastereoselectivity because of the chiral environment provided by the enzyme around the active site. The mild reaction conditions facilitate to prevent epimerization after installation of the stereogenic center(s) and therefore allow access to products with high stereoisomeric purity.
- *Benign from an environmentally and process safety perspective*. Biocatalytic reactions apply reaction conditions that are – with organic cosolvents as the main exception – generally found in nature. Enzymes and cofactors are readily biodegradable, providing the option to dispose of aqueous wastewater effluent streams via wastewater treatment plants. Raw materials specifically associated with the biocatalytic reaction (i.e. excluding substrate and product) are usually of low toxicological concern. Because of the mild reaction conditions, the use of aqueous reaction media, and the absence of thermal safety hazards of the employed raw materials (excluding the substrate and product), biocatalytic transformations generally pose little challenges from a process safety perspective.
- *Operationally simple processes*. Many biocatalytic processes are operationally simple, essentially requiring consecutive addition of the reaction components and subsequent stirring. They can be readily executed in standard multipurpose equipment, and special precautions such as the operation under rigorously oxygen-free conditions or the exclusion of moisture, which are frequently required, e.g. for asymmetric metal-catalyzed reactions, do generally not have to be applied.
- *Tunability of biocatalysts*. Besides the option to improve the performance of enzymes via standard process optimization approaches (e.g. optimization of temperature and pH value), enzyme evolution offers the significant advantage to tune the active biocatalyst and modify its performance [12–14]. Compared to most approaches taken for the improvement of other catalysts, e.g. asymmetric ligands for metal catalysis or organocatalysts, enzyme evolution quickly provides access to a vast range of new enzyme mutants. The choice of the evolutionary pressure allows to specifically address individual performance aspects of a biocatalyst, e.g. activity, stereoselectivity, chemoselectivity, tolerance toward desired operating conditions (e.g. pH and temperature range), or stability of the enzyme.

When the advantages summarized above are combined with the specific conception of biocatalytic transformations at the design stage of the synthesis [15–17], ample opportunities arise for the development of synthetic routes to the complex structures of APIs. Full utilization of these opportunities then enables the implementation of efficient, safe, environmentally friendly, resource-conscious, and cost-attractive manufacturing processes for large-scale commercial manufacturing.

15.2 Biocatalysis in the Pharmaceutical Industry

The past two decades have brought many outstanding examples of ingenious applications of biocatalysis in the pharmaceutical industry [2–10]. Some of these applications will be discussed in more detail in this section as examples that highlight the power, efficiency, and conceptual beauty of biocatalytic strategies for the large-scale manufacture of APIs. Although not all of the examples were finally implemented for commercial supply, they provide a representative overview of the state-of-the-art and the breadth of biocatalytic applications. In the discussion of each example, a focus is placed on process development aspects rather than the development of the biocatalytically active enzyme itself by enzyme evolution, or structural details of the enzyme.

15.2.1 Pregabalin

The synthesis of pregabalin (**3**), marketed for the treatment of neurological indications such as epilepsy or anxiety, represents one of the early prominent cases in which the use of biocatalysis allowed to dramatically improve the efficiency and environmental sustainability of an existing synthetic route to a target compound (Scheme 15.1) [18] (for the synthesis of pregabalin using a classical non-biocatalytic approach, see Ref. [19]).

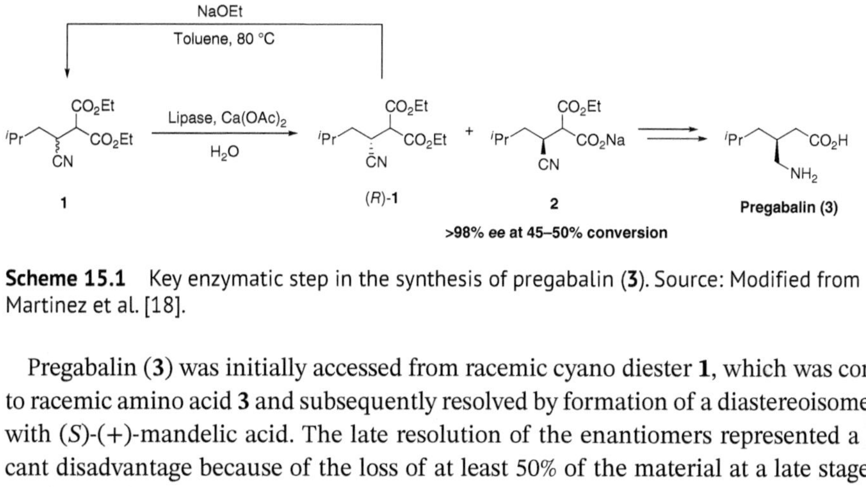

NaOEt
Toluene, 80 °C
Lipase, $Ca(OAc)_2$
H_2O
1 → (*R*)-1 + 2 → Pregabalin (3)
>98% *ee* at 45–50% conversion

Scheme 15.1 Key enzymatic step in the synthesis of pregabalin (**3**). Source: Modified from Martinez et al. [18].

Pregabalin (**3**) was initially accessed from racemic cyano diester **1**, which was converted to racemic amino acid **3** and subsequently resolved by formation of a diastereoisomeric salt with (*S*)-(+)-mandelic acid. The late resolution of the enantiomers represented a significant disadvantage because of the loss of at least 50% of the material at a late stage of the synthesis. The use of racemic cyano diester **1** provided the option to shift the resolution to the very beginning of the sequence, applying a lipase to convert the diester group into an ester acid functionality. The ability of the lipase to distinguish between the two enantiomers of the substrate then resulted in a mixture of enantiomerically enriched diester **1** and enantiomerically pure ester acid **2**. Interestingly, the differentiation between the enantiomers was not achieved by reaction at the existing stereogenic center to be resolved, but by generation of a second stereogenic center at the pendant diester group. Although the second stereogenic center in product **2** disappeared during a thermally induced decarboxylation in subsequent steps and was therefore inconsequential, the potential presence of diastereoisomers in **2** required the development of sophisticated analytical methods to determine the enantioselectivity of the resolution step as well as the diastereoselectivity of the transformation. While the first task, the determination of the enantioselectivity, was successfully achieved by *in situ* decarboxylation (i.e. conversion of the mixture of diastereoisomers into a mixture of enantiomers) under the conditions of gas chromatography (GC) analysis, determination of the diastereoselectivity was achieved by methylation to give the corresponding diastereoisomeric mixture of methyl ethyl diesters of compound **2**.

Screening of commercially available lipases led to the identification of lipolase as the superior enzyme in terms of activity and enantioselectivity, providing product (*R*)-**1** with an E value of more than 200 at 45–48% conversion after approximately 24 hours at 25 °C.[1] A pH value of pH 8.0 was identified as optimal for the reaction, and the pH value was kept constant over the course of the reaction by use of a phosphate buffer and pH-controlled addition of aqueous sodium hydroxide (NaOH). Optimization of the reaction conditions allowed reducing the enzyme loading from 5% (v/v) with respect to substrate **1** during screening to 1% (v/v) on commercial scale, resulting in total turnover numbers in the range of 10^5 for the saponification of diester **1**.

At concentrations of more than 1 M in substrate **1**, conversion was found to stall at less than 40% conversion. Comparison of the results from reactions with a single charge of substrate **1** vs fed batch conditions demonstrated no negative impact of the substrate concentration on conversion. However, addition of product **2** at the beginning of the reaction resulted in enzyme inhibition. The negative impact of product **2** on the reaction performance could be suppressed by addition of substoichiometric amounts of calcium acetate, potentially because of complex formation of product **2** with divalent calcium cations and enzyme stabilization. Suppression of product inhibition by addition of calcium acetate allowed raising the substrate concentration to 3 M, corresponding to 765 g of cyano diester **1** per liter of reaction medium.

As an additional benefit, recovery of the unreacted diester (*R*)-**1** from the reaction mixture was facilitated at increased substrate concentration. Phase separation between the aqueous product phase and the unreacted enantiomer as the organic phase occurred significantly faster than at lower substrate concentration as a result of a higher density of the more concentrated aqueous phase and a beneficial increase of the volume ratio between the organic and aqueous phases. The undesired enantiomer (*R*)-**1** was recycled in a separate step by racemization under basic conditions using NaOEt in toluene at 80 °C. Under these conditions, almost complete racemization was observed in approximately 16 hours.

The optimized reaction conditions were successfully confirmed in pilot trials and subsequently applied without additional modifications at commercial manufacturing scale of up to 3.5 ton of substrate **1** per batch using standard production equipment. Subsequent conversion of compound **2** to pregabalin (**3**) followed a closely related sequence of steps as in the established synthesis.

In summary, the change to a biocatalytic approach allowed the shift of the resolution step to an early point in the synthesis, recycling of the undesired enantiomer, and the replacement of organic solvents by mainly aqueous reaction media, resulting in an overall significant improvement of the environmental sustainability compared to the established conventional manufacturing process.

15.2.2 Vernakalant

The biocatalytic synthesis of vernakalant (**6**), used for the treatment of atrial fibrillation as a common cardiovascular disease, represents a compelling example for the application of

1 Product (*R*)-**1** was obtained in high diastereoselectivity of >99.5% *de*, as determined by GC analysis after derivatization as the corresponding methyl ester.

biocatalysis to a racemic substrate amenable to dynamic kinetic resolution, setting two stereogenic centers in a single transformation (Scheme 15.2) [20] (for the synthesis of vernakalant using a classical non-biocatalytic approach, see Refs [21–24]).

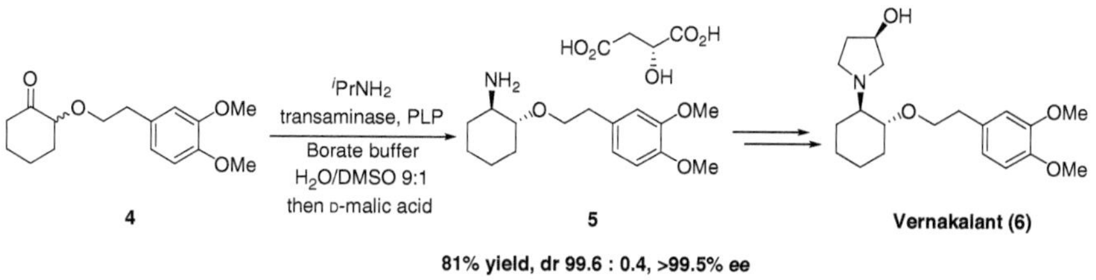

Scheme 15.2 Synthesis of a precursor of vernakalant (**6**) through an enzymatic dynamic kinetic resolution. Source: Modified from Limanto et al. [20].

Diastereoselective transamination of the (*R*)-enantiomer of racemic ketone **4** and concomitant racemization of the remaining (*S*)-enantiomer under the basic reaction conditions was demonstrated to be viable, providing access to amino ether **5** via dynamic kinetic resolution. Isopropylamine was used as amine donor in the presence of pyridoxal 5′-phosphate (PLP) as cofactor. Dimethyl sulfoxide (DMSO) was applied as organic cosolvent to increase the solubility of the substrate.

Initial screenings of the available transaminase libraries provided product **5** in high enantiomeric purity, although favoring the undesired *cis*-diastereoisomer. Enzyme evolution in combination with *in silico* design subsequently led to a complete reversal of the diastereoselectivity after only one round of evolution. With the desired diastereoselectivity established, further evolution rounds focused on improvements in activity, diastereoselectivity, and enzyme stability toward the high pH values required for epimerization of the undesired enantiomer under the reaction conditions.

Under optimized process conditions, the reaction was run at pH 10.5 to facilitate epimerization at a sufficient rate. The pH value was kept constant during the reaction by continuous addition of isopropylamine. Over a total reaction time of approximately 20 hours at 45 °C, conversion typically exceeded 94%. The final enzyme variant provided product **5** as a single enantiomer in a >99 : 1 ratio of *trans*- and *cis*-diastereoisomers and 81% yield from racemic ketone **4**, as demonstrated on multi-gram scale.

Isolation of amino ether **5** was achieved by salt formation, with various acids identified in screenings as potential salt forming agents. D-Malic acid was finally chosen as the preferred option, as the use of a chiral acid allowed additional improvement of the diastereoselectivity up to 99.6 : 0.4 dr during crystallization. Furthermore, the isolated salt **5** could be directly used in the downstream transformation toward vernakalant (**6**).

15.2.3 Sitagliptin

The development of a transaminase enzyme to access the diabetes mellitus type 2 medication sitagliptin (**8**) in a collaboration between Codexis and Merck elegantly demonstrated the power of enzyme evolution (Scheme 15.3) [25, 26] (for a process variant using an

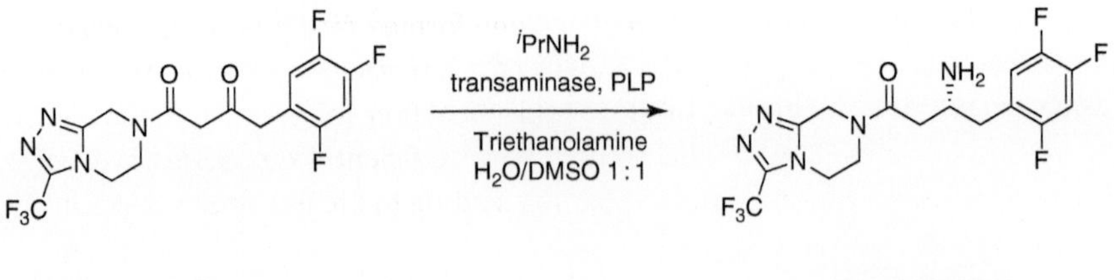

$^iPrNH_2$
transaminase, PLP
Triethanolamine
H_2O/DMSO 1 : 1

7

Sitagliptin (8)
92% yield, >99.9% ee

Scheme 15.3 Synthesis of sitagliptin (**8**) through a biocatalytic transamination reaction. Sources: Modified from Savile et al. [25].

immobilized transaminase, see Truppo et al. [27]; for the synthesis of sitagliptin using a classical non-biocatalytic approach, see Refs [28, 29]).

Starting from an enzyme with low activity on the target keto substrate **7**, the use of a substrate walking approach with initial enzyme evolution on a simplified truncated substrate and in total 11 rounds of evolution allowed to significantly improve the performance of the transaminase enzyme. Under optimized conditions, the process was run with enzyme loadings of 3–6% (w/w) with respect to substrate **7**, leading to reaction times of 12–24 hours at constant performance with respect to quality and conversion. Adjustment of the enzyme loading allowed adaptation of the cycle time to optimize the throughput of the overall manufacturing process.

During process development, a focus was placed on performance parameters that are crucial for large-scale commercial manufacturing. A suitable concentration of the hydrophobic substrate **7** in the aqueous reaction medium was achieved by addition of DMSO as a cosolvent. In the final reaction medium, consisting of a 1 : 1 mixture of water and DMSO, a substrate concentration of $200\,g\,l^{-1}$ could be achieved. Under these conditions, substrate **7** possesses a solubility of ca. $8\,g\,l^{-1}$ in the reaction medium. Besides DMSO, methanol was evaluated as a cosolvent but was finally not chosen because of increased dimerization via imine formation between substrate **7** and product **8** at higher concentrations.

As transamination reactions are reversible, with the equilibrium often favoring the ketone substrate over the amine product, tolerance of the transaminase enzyme to a significant excess of the amine donor isopropylamine (up to concentrations of 1 M isopropylamine) was crucial to drive the reaction toward the product **8**. Increasing the reaction temperature from 22 °C to 45 °C allowed to shorten the reaction time in line with Arrhenius' law. Although multiple enzyme variants tolerated temperatures of 60 °C or above, instability of substrate **7** above 50 °C limited the maximum reaction temperature. As the pH value was found to affect the reaction rate as well, adjustment of the pH value from originally pH 7.5 to pH 8.5 resulted in a 2.5-fold increase of the reaction rate. Because of a clear difference in the pK_a values of $^iPrNH_2$ and sitagliptin, with $^iPrNH_2$ being more basic than **8** (pK_a 10.6 vs pK_a 8–9), the reaction was buffered with 0.2 M triethanolamine. Control of the pH value in the desired range was ensured by monitoring and pH-controlled dosing of 4 M aqueous $^iPrNH_2$ solution.

Operational measures implemented during process development achieved a shift of the reaction equilibrium from an initial ratio of 87 : 13 between product **8** and substrate **7** under sealed conditions, i.e. preventing loss of acetone formed from the amine donor isopropylamine, to more than 90 : 10 by application of a low vacuum of 300–350 Torr and gentle nitrogen sweeping of the headspace over the reaction mixture. Dimerization by imine formation between substrate **7** and product **8** was efficiently suppressed by feeding the substrate as a solution in DMSO over 2–3 hours, leading to dimer levels of 2–5% in the reaction.

For work-up, the reaction mixture was quenched by acidification to pH 2 with concentrated aqueous HCl at 45 °C, leading to denaturation of the transaminase enzyme and formation of an insoluble precipitate. The precipitate could be removed by filtration, followed by adjustment of the pH value to pH 10.5 with NaOH and extraction of the product **8** with isopropyl acetate. In an alternative procedure, product **8** was extracted with a 1 : 1 mixture of isopropyl acetate and isopropanol directly after adjustment to pH 10.5 without prior removal of the precipitate by filtration. The denatured enzyme was in this case discarded together with the aqueous layer from the work-up. A protocol of the optimized process on a 1 kg scale is reported in Savile et al. [25].

15.2.4 Esomeprazole

A biocatalytic approach using a Baeyer–Villiger monooxygenase (BVMO) and oxygen as the terminal oxidant has been reported for the installation of the chiral sulfoxide moiety in esomeprazole (**10**), a medication to treat gastric reflux (Scheme 15.4) [30] (for the synthesis of esomeprazole using a classical non-biocatalytic approach, see Refs [31–33]; a similar approach to access a chiral sulfoxide using BVMO has been described for an intermediate in the synthesis of AZD6738, a lead candidate for an oncology project of AstraZeneca, see Goundry et al. [34]). In this case, oxidation of the sulfur atoms requires not only control over the enantioselectivity of the reaction, but also over the chemoselectivity to avoid formation of the corresponding sulfone,[2] a known by-product formed by overoxidation and found to be difficult to remove below acceptable levels in esomeprazole (**10**).

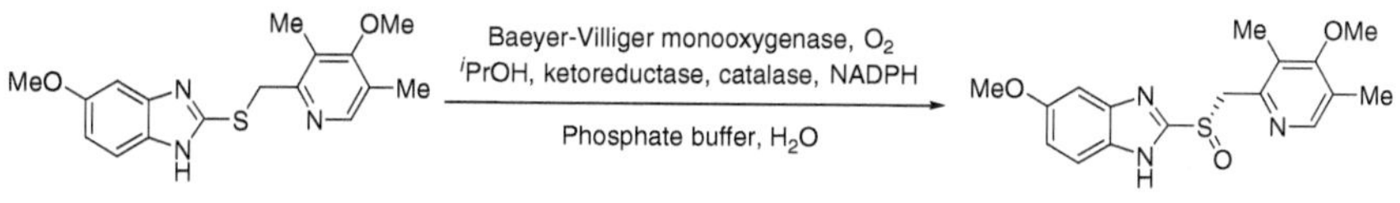

Baeyer-Villiger monooxygenase, O_2
^{i}PrOH, ketoreductase, catalase, NADPH
Phosphate buffer, H_2O

9

Esomeprazole (10)
87% yield, >99% *ee*

Scheme 15.4 Synthesis of esomeprazole (**10**) through BVMO-mediated oxidation. Source: Modified from Bong et al. [30].

In the catalytic cycle of BVMOs (see also Chapter 7), flavin adenine dinucleotide (FAD) acts as a cofactor that reacts with oxygen to form a hydroperoxide intermediate, which oxidizes sulfide **9** to the corresponding sulfoxide **10**, leaving oxidized FAD behind. Prior to initiation of the next cycle, oxidized FAD needs to be reduced by nicotinamide adenine

2 For the Kagan sulfoxidation, ratios of 35 : 1 to 79 : 1 between sulfoxide and the sulfone resulting from overoxidation are reported.

dinucleotide phosphate (NADPH) to provide the reduced form of FAD again. Because of the high cost of $NADP^+$ as a precursor to NADPH, the cofactor was recycled in the case at hand using a ketoreductase (KRED) and isopropanol, which is oxidized to acetone, as terminal reductant. The amount of isopropanol was reduced as much as possible to prevent formation of an explosive atmosphere in the presence of oxygen and allow operation under safe conditions below the flash point of 5% (v/v) isopropanol in water at 50 °C.

Starting from a wild-type BVMO with only trace levels of activity toward sulfide **9**, enzyme evolution led to the development of an enzyme variant with significantly improved performance. Enzymatic activity was increased by a factor of 140 000 compared to the original wild-type enzyme to achieve enzyme loadings as low as 2% (w/w) with respect to substrate **9**. At the same time, the enantioselectivity of the oxidation reaction improved to >99% *ee* for product **10**, sulfone formation was suppressed to levels below 0.1%, and the amount of $NADP^+$ needed as a cofactor was significantly reduced. A lab-scale protocol on multi-gram scale was reported for the optimized enzymatic process [30]. It is interesting to note that during the course of the evolution program, the enzyme lost its ability to catalyze Baeyer–Villiger oxidations, i.e. its original reactivity was completely changed toward sulfide oxidation, likely implying as well a change in the enzymatic mechanism.

Because of the low solubilities of oxygen (ca. 0.2 mM at 25 °C), sulfide **9** and esomeprazole in the reaction medium, process development primarily focused on optimization of mass transfer in the system. Pure oxygen was found to be superior to air, as mass transfer of oxygen from the gaseous to the aqueous phase was identified as a rate limiting factor under many investigated reaction conditions. As sulfide **9** only partially dissolved in the reaction medium, and **10** precipitated from solution, the reaction was run as a slurry-to-slurry process. Despite the low solubilities of substrate **9** and oxygen, resulting in low concentrations of both reaction partners, a practically useful reaction rate was observed, highlighting the excellent catalytic activity exhibited by the BVMO enzyme.

BVMOs are known to release hydrogen peroxide in the case of inefficient substrate binding [33], resulting in both non-enzymatic oxidation of sulfide **9** to racemic sulfoxide **10** and oxidation of the sulfoxide to the corresponding sulfone as potential background reactions. Although suppression of sulfoxide formation was a target parameter for optimization during enzyme evolution, addressed by increasing the binding affinity to sulfide **9** to avoid generation of hydrogen peroxide, catalase was added to the reaction mixture as a safeguard to reduce any hydrogen peroxide potentially present in the reaction medium to water and oxygen.

A relatively small pH range was tolerated by the reaction. Low pH values were not acceptable because of the instability of **10** at neutral or acidic pH. At high pH values, however, the extent of overoxidation to the sulfone increased. At pH 9, the pH value finally chosen for the reaction, crystallization of esomeprazole occurred during the course of the reaction. The polymorphism of **10** led to variations in the viscosity of the reaction mixture, negatively affecting the efficiency of the oxygen uptake for the chosen reactor design. These challenges could be addressed by adjustment of the agitation speed and by implementation of seeding.

15.2.5 Montelukast

Montelukast (**13**), marketed for the treatment of asthma, possesses one stereogenic center, derived from a chiral alcohol with (*S*)-configuration in intermediate **12**. In the original synthesis, the stereogenic center was accessed from the ketone precursor **11** by an asymmetric reduction using enantiomerically pure (−)-chlorodiisopinocampheylborane as a stoichiometric reductant at low temperature [35, 36]. The development of a KRED-catalyzed reduction of the keto functionality in substrate **11** to provide chiral alcohol **12** highlights the significant potential of biocatalytic processes for safer and environmentally more benign manufacturing processes (Scheme 15.5) [37].

The extremely low solubility of substrate **11** in aqueous media because of its hydrophobicity was identified as a key challenge to overcome in the development of the biocatalytic transformation. The results of solubility screenings for substrate **11** in binary and ternary mixtures of water and water-miscible organic solvents indicated the highest solubility of approximately $1\,g\,l^{-1}$ in a 1 : 5 : 3 mixture of tetrahydrofuran (THF), ^{i}PrOH, and water. The use of polar organic solvents as cosolvents was assumed to provide a viable approach, as enzymes can retain high catalytic activity even in hydrophobic reaction media [38].

In the case at hand, ^{i}PrOH was used as a terminal reductant, with the cofactor NADPH acting as a shuttle to transfer a hydride equivalent to the keto substrate under catalysis by the KRED enzyme (see also Chapters 7 and 10). Similar to a transfer hydrogenation, ^{i}PrOH as the hydride donor is concomitantly oxidized to acetone as the corresponding ketone. Whereas regeneration of the cofactor NADPH is in many cases performed by a second enzymatic system, e.g. oxidation of glucose to gluconic acid by glucose dehydrogenase, the acceptance of ^{i}PrOH as a reductant thus avoided the need for a second enzyme for the NADPH–$NADP^{+}$ cycle.

Reversibility of ketone reductions by KRED enzymes and the use of ^{i}PrOH as a reductant typically require removal of acetone by distillation to shift the equilibrium toward the product. However, because of the low solubility in the reaction medium, product **12** crystallized as the monohydrate during the reaction and was thus constantly removed from the system, driving the reaction to completion. Under optimized conditions, >99% conversion was obtained on large scale.

In order to reach the desired minimum activity, enzyme evolution was necessary, starting from available evolved variants that already showed an excellent (*S*)-selectivity of >99.9% *ee*. Furthermore, tolerance and stability of the KRED enzyme toward the high concentration of organic solvents, present in more than 70% in the reaction medium, as well as toward higher temperatures (increased from ambient temperature to 45 °C, with tolerance of higher temperatures up to 55 °C over short periods of time) had to be improved. Last, early variants of the enzyme showed inhibition by acetone, forming as a by-product in stoichiometric quantities. All defined goals were successfully achieved by in total five rounds of enzyme evolution. By serendipity, it was observed during evolution that 10% toluene as a cosolvent proved to be superior to THF, leading to an additional 2.5-fold increase in activity without any negative impact on the performance. Overall, an activity increase by a factor of 3000 was achieved compared to the starting enzyme variant.

Similar to sitagliptin (**8**), reaction time and enzyme loading could be adjusted to fit the needs of the commercial manufacturing process, with reaction times of 24 hours at 3% (w/w) enzyme loading or 10 hours at 5% (w/w) enzyme loading. The reaction was found to be robust against changes in the pH value, tolerating pH values over a wide range from pH 7.5 to pH 9.8. Under the final process conditions, the pH value was controlled at pH 8.0 by a triethanolamine buffer system. A concentration of $0.1\,g\,l^{-1}$ of the cofactor NADPH was

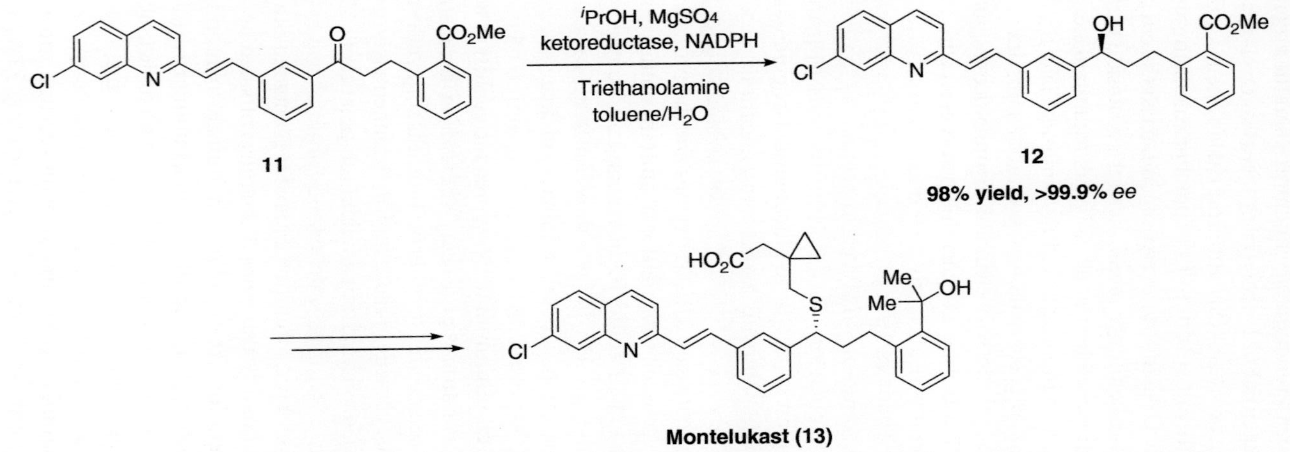

Scheme 15.5 Synthesis of a precursor of montelukast (**13**) via stereoselective biocatalytic reduction. Source: Modified from Liang et al. [37].

used, and $MgSO_4$, needed for optimal performance of the KRED, was added to the reaction mixture in small quantities.

Despite the presence of organic cosolvents, neither substrate **11** nor product **12** was completely soluble under the optimized reaction conditions, and the ketone reduction was run as a slurry-to-slurry process at a concentration of $100\,g\,l^{-1}$. On lab scale, product **12** was isolated by filtration of the reaction mixture after the requirement for the in-process control on conversion was met, followed by washing of the filter cake with water. For commercial manufacturing on a batch size of 230 kg, saturated aqueous NaCl solution and ethyl acetate were added after the reaction had reached full conversion. Residual enzyme was removed by filtration on a pressure nutsche filter, and subsequent standard extractive work-up, concentration, and recrystallization provided alcohol **12** in 97% yield and >99.9% *ee*.

Compared to the original process using (−)-chlorodiisopinocampheylborane for the transformation of ketone **11** to alcohol **12**, the biocatalytic process significantly reduced the process mass intensity (PMI) index of the reaction from previously 52 to 34 for the KRED-catalyzed reaction.

15.2.6 Boceprevir

An elegant biocatalytic desymmetrization approach of bicyclic *meso*-compound **14** forms the basis for the synthesis of key intermediate **18** in the synthesis of boceprevir (**19**), a protease inhibitor used for the treatment of hepatitis [39] (a similar approach using a desymmetrization approach based on the oxidation of *meso*-pyrrolidines by monoamine oxidase has been described for the structurally related drug substance telaprevir, see Refs [40–42]). The development of the biocatalytic route was inspired by the previous generation approach, which applied the same sequence of transformations, yet in a racemic fashion, with subsequent resolution to access enantiomerically pure ester **18** (Scheme 15.6) [43–46].

Desymmetrization of *meso*-compound **14** by a monoamine oxidase to enantioselectively provide Δ^1-pyrroline **15** was identified as an attractive alternative to the resolution of a racemate. Molecular oxygen serves as the oxidant in amine oxidations with monoamine oxygenases, which are enzymes that either contain copper or depend on FAD as a cofactor (see also Chapter 6). Oxygen is reduced to hydrogen peroxide, which itself can be further converted to oxygen and water by a catalase as a second enzyme in the catalytic cycle.

The need for molecular oxygen as a terminal oxidant precluded the use of organic solvents in a commercial process for boceprevir (**19**) because of safety aspects. Mass transfer between the gaseous and the liquid phase had to be optimized to improve oxygen uptake by the reaction medium. This task was rendered more challenging by the already low concentration of oxygen in the aqueous solution because of the high ionic strength of the reaction medium. A solution was identified by the application of a gas dispersion tube, which allowed maximizing the area for transfer of oxygen from the air into the liquid phase and thus increasing the mass transfer.[3] For large-scale manufacturing, pure oxygen was applied

3 Though addition of an anti-foaming agent is described for the lab-scale protocol in the Supporting Information, its use is not commented on in the manuscript. Addition of an anti-foaming agent likely suppresses foam formation due to introduction of air or oxygen via the gas dispersion tube.

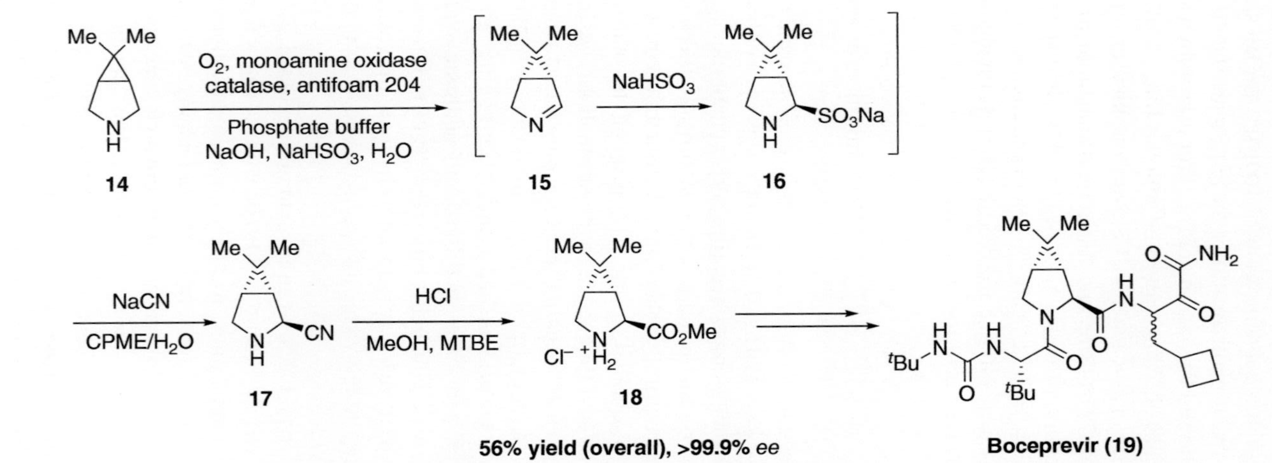

Scheme 15.6 Synthesis of boceprevir (**19**) involving an enzymatic desymmetrization process. Source: Modified from Li et al. [39].

instead of air, as the activity of the monoamine oxidase enzyme increased by a factor of up to 3 at higher oxygen pressure. As acidity increased over time because of the release of protons in the catalytic reaction, the pH value was kept constant at the target of pH 7.4 by pH-controlled addition of aqueous NaOH over the course of the reaction. The volatility of imine **15** required the off-gas to be passed through a cooled condenser in order to minimize product loss and removal of imine **15** from the reactor headspace to address a potential explosion risk in the presence of oxygen. The flow rate of air through the reaction mixture was fine-tuned to balance reaction rate and loss of product **15**.

Whereas catalase displays high activity and is readily available on commercial scale in sufficient quantities, thus not requiring any further optimization, enzyme evolution was necessary for the monoamine oxidase. The generation of hydrogen peroxide as a stoichiometric by-product enabled a straightforward readout of the enzyme activity using an established fluorescence-based assay. Following identification of a suitable wild-type enzyme as a starting point, enzyme evolution improved activity, solubility, and thermal stability of the enzyme – providing at the same time increased stability during storage – over in total four evolution rounds. As the starting wild-type variant of the monoamine oxygenase was found to already provide product **15** in >99% *ee*, further optimization of the enantioselectivity could be excluded from the enzyme evolution program.

In initial studies, the reaction stalled at approximately 50% conversion because of complete deactivation of the monoamine oxidase enzyme. Investigations into the correlation between reaction rate and the concentrations of substrate **14** and product **15** indicated enzyme inhibition by both reaction partners, with significant and irreversible inhibition displayed by product **15**. Options to generate imine **15** as a transient intermediate by direct conversion into the corresponding nitrile **17**, thus reducing its concentration in the reaction mixture, failed because of the inhibitory effect of cyanide on the catalase as well as the instability of the stereogenic center in nitrile **17** in the aqueous reaction medium. Formation of the bisulfite adduct of imine **15**, however, was found to work well under the reaction conditions, and monoamine oxygenase tolerated the bisulfite adduct **16** well. The high solubility of compound **16** in aqueous media effectively addressed the challenges arising from the volatility of imine **15**, significantly decreased the loss of material in the off-gas of the reaction, reduced the concentration of free imine **15** under the reaction conditions, and provided a suitable reservoir for the product until the end of the transformation, thus avoiding inhibition of the monoamine oxygenase enzyme.[4]

As sodium hydrogen sulfite is oxidized to sodium sulfate in the presence of hydrogen peroxide, the dosing regime was carefully optimized to ensure a sufficiently high concentration of bisulfite for formation of bisulfite adduct **16**, while retaining the concentration of free imine **15** as low as possible. Under the conditions for commercial manufacturing, a mixture of amine **14** and sodium hydrogen sulfite was added over 24 hours to the reaction medium containing both the monoamine oxidase and the catalase enzyme, providing mainly bisulfite adduct **16** and small amounts of free imine **15**.

The transformation of bisulfite adduct **16** to methyl ester **18** was achieved by sequential cyanation in a biphasic reaction mixture with cyclopentylmethyl ether (CPME) as an organic solvent to selectively give *trans*-nitrile **17**, followed by a subsequent reaction with

4 ^{1}H NMR studies indicated instantaneous reaction of imine **15** to bisulfite adduct **16**.

methanol under acidic conditions to provide methyl ester **18** as the corresponding hydrochloride salt. Starting from amine **14**, product **18** was obtained in overall 56% yield and excellent enantiomeric purity of >99% *ee*, as demonstrated on multi-gram scale [39].

15.3 Aspects to Consider for Development of a Biocatalytic Process on Commercial Scale – A Case Study

Chemists naturally focus on the optimization of standard chemical process parameters, e.g. isolated yield, product purity, or stereoselectivity, during process development. However, for late development stages and application on industrial scale, many additional aspects have to be taken into consideration. These aspects may not seem to be directly related to chemistry, yet play a crucial role to enable large-scale manufacturing of a drug substance in consistent quality and yield following a robust manufacturing process and efficient operations. Process development for industrial applications therefore requires the identification of the optimum of a multitude of aspects instead of maximization of individual chemistry-related parameters. Although these challenges are not specific to biocatalytic transformations, but universally apply to process development for any kind of process and any kind of industrial application, the use of enzymes on commercial scale is associated with certain unique challenges compared to standard non-enzymatic chemical processes. The enzyme-catalyzed transamination shown in Scheme 15.7 is used as a case study to exemplify these challenges [47] (for the synthesis of (*R*)-biphenylalanine (**21**) using alternative biocatalytic approaches, see Refs [48–50]; for the synthesis of chiral biphenylalanine using a classical non-biocatalytic approach, see Refs [51–53]).

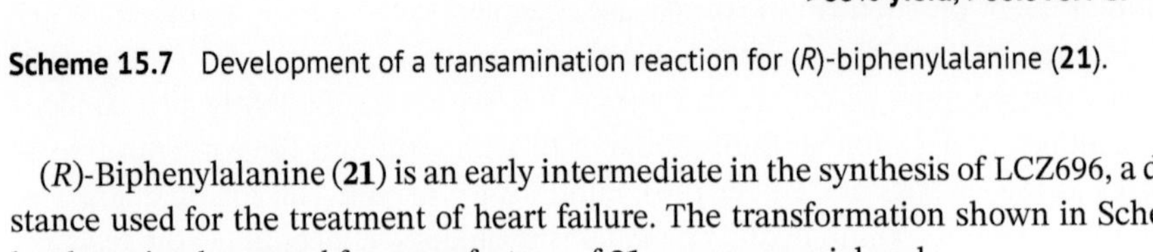

$^{i}PrNH_2$
transaminase, PLP
H_2SO_4, phosphate buffer
H_2O
20
21
>99% yield, >99.9 : 0.1 er

Scheme 15.7 Development of a transamination reaction for (*R*)-biphenylalanine (**21**).

(*R*)-Biphenylalanine (**21**) is an early intermediate in the synthesis of LCZ696, a drug substance used for the treatment of heart failure. The transformation shown in Scheme 15.7 has been implemented for manufacture of **21** on commercial scale.

The following topics are subsequently discussed in more detail:

- Identification of a suitable enzyme
- Process development
- Control strategy and regulatory considerations
- Health, process safety, and environmental aspects
- Equipment utilization and throughput time

- Equipment cleaning
- Enzyme release testing
- Transport and storage

As the product **21** from the enzyme-catalyzed transamination in this case study is used for the manufacturing of a drug substance, the aspects above are discussed against the applicable regulations for pharmaceutical products. Although the regulations in other areas, e.g. crop protection or fine chemicals, may be different, the underlying general principles are expected to be similar.

15.3.1 Identification of a Suitable Enzyme

The enzymatic transamination of ketones to the corresponding amines represents one of the most established classes of biocatalytic transformations in the synthesis of APIs. Numerous examples for industrial applications have been reported in the literature, even for structurally complex substrates (for a recent review on the application of transaminases in the pharmaceutical industry, see Tufvesson et al. [54]). Consequently, screening kits of transaminase enzymes are commercially available from various suppliers, covering a wide and diverse range of possible substrates.

A minimum productivity target has to be met by the enzyme in order to provide an economically attractive solution for industrial applications. For transamination reactions, state-of-the-art examples typically operate at product concentrations of $50\,g\,l^{-1}$ or above, with biocatalyst productivities of at least 10 g of product per gram of the biocatalyst [54]. In addition, minimum requirements for conversion and – in the case of chiral transformations – stereoselectivity have to be fulfilled. For non-natural substrates, the likelihood to identify a wild-type enzyme with suitable activity and stereoselectivity is not considered to be high. Screening of commercial libraries that contain already evolved transaminases increases the chances for substrates with low to moderate complexity, even though extensive screening of libraries targeting specific substitution patterns of the substrate may be required. For more complex substrates, screening usually provides a suitable starting point for subsequent enzyme evolution. In the case of keto acid **20**, a transaminase enzyme with adequate activity and enantioselectivity could be identified by multiple rounds of extensive library screening. Transaminase ATA-032 was found to provide optimum performance for substrate **20**.

15.3.2 Process Development

Process development for (*R*)-biphenylalanine initially focused on the definition of suitable set points for key process parameters, in particular pH value and temperature, under which the identified enzyme ATA-032 demonstrated best performance. The tolerance of the biocatalytic system toward parameter deviations was investigated, and stability tests for ATA-032 under target conditions as well as extreme process parameter settings were performed. These experiments demonstrated consistently high enzymatic activity under mildly basic conditions around pH 8.5 at temperatures of 40–45 °C, allowing low enzyme loadings of 1.0% (w/w) or below with respect to the substrate biphenylpyruvic acid. Interestingly, once set at the beginning of the reaction, the pH value was found to be stable with minor deviations from the set point over the entire reaction time, without the need for active pH adjustment by addition of acid or base over the course of the reaction.

As the biocatalytic transamination of ketones to amines is reversible, leading to establishment of an equilibrium between substrate and product with typically a preference for the ketone substrate, the impact of changes in the stoichiometry was investigated (for strategies to shift the equilibrium toward the amine product, see Refs [3, 54, 55]). In many transamination reactions, high conversion is only achieved by removing acetone from the reaction mixture. In the transformation of **20** to **21**, however, essentially complete conversion was reached without the need to actively shift the equilibrium toward the amine product. Differences in the solubility between the substrate and product, with a lower solubility of (*R*)-biphenylalanine in the reaction mixture, are assumed to be the likely reason for this behavior, leading to continuous removal of **21** from the reaction mixture and thus a shift of the equilibrium toward the product side. Although an excess of isopropylamine was required to drive the reaction to completion, the amount of the amine donor isopropylamine could be varied within a broad range without major impact on the conversion. Further optimization during later development stages led to a significant reduction of the amount of isopropylamine. Because of ease of handling, solid isopropylammonium chloride (or other isopropylammonium salts), initially used as amine donor, was replaced by a 70% aqueous isopropylamine solution for commercial manufacturing.

Under standard conditions, the transaminase enzyme ATA-032 was added to the reaction mixture as a suspension in an aqueous buffer solution. Alternatively, addition as a solution could be performed without major impact on the reaction performance. In all cases, addition of the enzyme occurred last in the sequence of operations to avoid extended exposure of ATA-032, e.g. because of unexpected hold times at commercial scale, to higher temperatures in the absence of the substrate **20** or any other reaction component. For analogous reasons, enzyme addition was designed such that any contact with the reactor wall – which is expected to have a higher temperature than the reaction mixture – was excluded.

After complete conversion, the product was isolated by filtration after acidification of the reaction mixture to pH 1–2. Under these conditions, the enzyme and potential degradation products such as peptide fragments or single amino acids remained dissolved in the aqueous reaction mixture and were removed with the mother liquor. Interestingly, despite the low pH value, (*R*)-biphenylalanine was isolated as the free amino acid (likely present as the corresponding zwitterion) instead of the corresponding ammonium salt. These findings were confirmed by determination of chloride levels by titration in the isolated product when the reaction was acidified with aqueous HCl, demonstrating the absence of chloride as the expected counterion in case formation of the hydrochloride salt by protonation with HCl occurred.

The process description that was applied for initial kilo lab trials and subsequently used as a basis for late-phase process development is described below.

Biphenylpyruvic acid (**20**) (1.00 kg, 4.16 mol, 1.00 equiv) was suspended in a preformed solution of isopropylamine hydrochloride (1.35 kg, 14.1 mol, 3.39 equiv) and 66 mM aqueous K_2HPO_4 solution (15.5 kg) at 20 °C. The pH value was adjusted to pH 6–8 by addition of isopropylamine (0.30 l, 3.50 mol, 0.84 equiv). The mixture was warmed to 45 °C, followed by control of the pH value and adjustment to pH 8.5–8.6 with 1 M aqueous NaOH or 2 M aqueous HCl. A suspension of PLP (8.30 g, 0.034 mol, 0.008 equiv) in 4.9 mM aqueous K_2HPO_4

solution (400 ml) was added, followed by a suspension of transaminase enzyme ATA-032 (10.0 g; 1.00% (w/w) with respect to biphenylpyruvic acid) in 4.9 mM aqueous K_2HPO_4 solution (800 ml). The reaction was stirred for 12 hours at pH 8.5–8.6, with pH adjustment using 1 M aqueous NaOH or 2 M aqueous HCl if required. A sample was taken to check conversion (target: residual biphenylpyruvic acid (**20**) ≤ 1.0% (area) by HPLC; stirring continued in case conversion target not met). After cooling to 25 °C, 2 M aqueous HCl (9.0 l) was added with temperature control to adjust the pH value to pH 1–2. The reaction mixture was stirred for one hour, then the suspension was filtered off, and the filter cake was washed with water (12.6 kg). The product was dried at 50 °C under vacuum to provide 1.00 kg (>99% yield; corrected for assay) of (*R*)-biphenylalanine (**21**) in >99.9% (area) HPLC purity and an enantiomeric ratio of >99.9 : 0.1.

In summary, with the exception of the identification of the enzyme described in Section 15.3.1, process development was performed in line with routine development practice for chemical transformations. No specific requirements that would not be encountered in the development of standard chemical reactions had to be specifically addressed for the biocatalytic transamination reaction described in Scheme 15.7.

15.3.3 Control Strategy and Regulatory Considerations

Insurance of consistent and high quality is eminent for commercial drug substances or drug substances that are under development and used in clinical trials. Regulatory guidelines explicitly outline the expectations of health authorities with respect to impurity identification, investigations of a potential quality impact due to parameter variations or carryover of impurities, and the definition of a robust strategy to reliably control the quality of the final drug substance.[5] Because biocatalytic reactions are not inherently different from other – catalyzed or non-catalyzed – reactions, the same general considerations for definition of regulatory control strategies apply as for non-biocatalytic chemical manufacturing processes.

Attention needs, however, to be paid to impurities that are specifically related to the use of enzymes, but typically not of relevance for non-biocatalytic processes. Excellent summaries on the key aspects to consider for a control strategy on impurities associated with biocatalytic transformations are provided in Refs [56, 57]. The definition of a suitable control strategy to address any toxicological concern requires information on the following aspects:

---

5 Though local regulations may differ, the regulatory guidelines issued by the International Council for Harmonisation of Technical Requirements for Pharmaceuticals for Human Use (ICH) generally serve as fundamental guidance to the pharmaceutical industry. The ICH was created in 1990 with the intention to harmonize regulatory requirements for the assessment of quality, efficacy and safety of drug substances and drug products between the regulatory authorities of the USA, the European Union and Japan. Guidelines are established in consultation with the pharmaceutical industry associations following a defined workflow and require consensus among the members. Basic quality expectations for pharmaceutical drug substances and drug products are summarized in the guidelines Q3A "Impurities in New Drug Substances", Q3B "Impurities in New Drug Products", Q3C "Guideline for Residual Solvents" and Q3D "Guideline for Elemental Impurities". Regulations for the development and manufacture of drug substances are outlined in the guideline Q11 "Development and Manufacture of Drug Substances (Chemical Entities and Biotechnological/Biological Entities)". Further information on the ICH and the various guidelines can be found at the ICH's official web site. https://www.ich.org/.

- Types of impurities associated with the biocatalytic transformation.
- Type of biocatalyst applied for the biocatalytic reaction.
- Type of expression system used for the preparation of the enzyme.
- Route of administration of the drug substance.
- Position of the biocatalytic step in the overall synthesis and nature of downstream transformations.

The toxicological concerns associated with impurities that result from the use of biocatalysis are strongly linked to the types of impurities that may potentially be present in the final drug substance and – importantly – to the route of administration of the drug substance. For drug substances administered orally, many of the impurities associated with biocatalytic transformations are degraded in the human gastrointestinal tract, leading to a noticeable reduction of toxicological concerns. In most cases, application of a generic control strategy for orally administered drug substances may therefore be appropriate. For non-oral routes of administration, however, the toxicological concerns for the same impurities may be significantly increased compared to oral administration, resulting in the need for a thorough case-by-case risk assessment [56, 57].

15.3.3.1 Impurities

Relevant impurity classes that are typically associated with the use of enzymes in chemical transformations and therefore need to be assessed are discussed below. The scientific rationale outlined in this chapter targets orally administered drug substances and may not be directly applicable to non-oral routes of administration.

Depending on the position of the biocatalytic step in the overall synthesis and the nature of the downstream transformations up to the drug substance, residual enzyme could potentially be carried over to the final drug substance. The carryover of the catalytically active enzyme itself is in particular of relevance if the biocatalytic step occurs late in the synthesis, and if the transformations between the biocatalytic reaction and the drug substance apply mild reaction conditions (e.g. do not involve strongly acidic or basic conditions) that do not lead to breakage of the enzyme into smaller fragments (either peptides or single amino acids). Residual enzyme may be carried over either as a denatured enzyme without enzymatic activity or with its catalytic activity preserved. Levels of residual enzyme can be determined using chromatographic (LC–UV, LC–MS), electrophoretic (sodium dodecyl sulfate–polyacrylamide gel electrophoresis, SDS–PAGE), or immunochemical techniques such as enzyme-linked immunosorbent assay (ELISA).

The enzyme used in a biocatalytic step may be degraded into smaller fragments by breakage of the peptidic bonds as a result of the applied work-up conditions in the biocatalytic step or the reaction conditions used in downstream operations. Reaction conditions that typically lead to fragmentation into either smaller peptides (of various size and amino acid sequence) or single amino acids employ strongly acidic or basic conditions, sometimes in combination with elevated temperatures. The determination of specific peptide fragments or single amino acids is generally neither feasible nor required. Hence, peptide fragments and amino acids may be determined together by amino acid quantification after acidic protein hydrolysis. Because all proteins are hydrolyzed during degradative treatment (e.g. with hot concentrated HCl), the determination covers any potential residual enzyme or peptide fragments, including any host cell proteins that may be present from the fermentation process used to manufacture the enzyme. The amino acids resulting after digestion are quantified using chromatographic techniques, either with (LC–UV, LC–FL, LC–MS, GC–FID, GC–MS) or without (LC–MS, LC–MS/MS) derivatization, by comparison against a reference hydrolysate of the parent enzyme. Because any proteinaceous contamination is most likely associated with the biocatalyst itself, the biocatalyst is also used as a quantitation reference (i.e. for preparation of comparison and spiking solutions).

The specific conditions applied for analysis of the sum of residual enzyme, peptide fragments, and single amino acids in the synthesis of (*R*)-biphenylalanine (**21**) are outlined in Table 15.1. A chromatogram of the amino acid separation after derivatization is shown in

Table 15.1 Analysis of residual enzyme, peptide fragments, and single amino acids in the synthesis of (*R*)-biphenylalanine (**21**).

Item	Description
Analysis principle	A digested sample is analyzed using a limit test approach by comparison against standard solutions of derivatized amino acids (as HPLC retention time standard) and digested parent enzyme (as quantification standard).
Sample preparation	• Sample digestion is performed in 6 M aqueous HCl in a closed vial at 100 °C for 18 h or alternatively for 20 min at 160 °C in a microwave. • The resulting solution is concentrated to dryness on a hot plate at 55 °C by passing a stream of nitrogen over the solution. • The residue is dissolved in borate buffer and acetonitrile. • Amino acids present in the sample are subsequently derivatized by treatment with 6-aminoquinolyl-*N*-hydroxysuccinimidyl carbamate at 55 °C for 10 min. • The resulting solution is concentrated to dryness on a hot plate at 55 °C by passing a stream of nitrogen over the solution. • The residue is dissolved in 1 M aqueous HCl and acetonitrile. • Remaining solids are removed by centrifugation or settling to obtain a clear supernatant. • The supernatant is used for analysis after appropriate dilution.
Chromatographic conditions	Column: Waters AccQ Tag Ultra, 1.7 μm; length 100 mm, internal diameter 2.1 mm Mobile phase: • Phase A: 50 ml AccQ Tag Eluent A (Waters) diluted with 950 ml water • Phase B: acetonitrile Gradient (amount of phase B given in %): • 0 min: 0.1% • 0.5 min: 0.1% • 5.7 min: 9.0% • 7.7 min: 21.0% • 8.5 min: 60.0% • 8.6 min: 95.0% • 11.6 min: 95.0% • 11.7 min: 0.1% • 14.0 min: 0.1% (end of acquisition) Flow rate: 0.7 ml min^{-1} Detection: 260 nm (UV) or 250 nm/395 nm (FL, excitation/emission) Column temperature: 50 °C

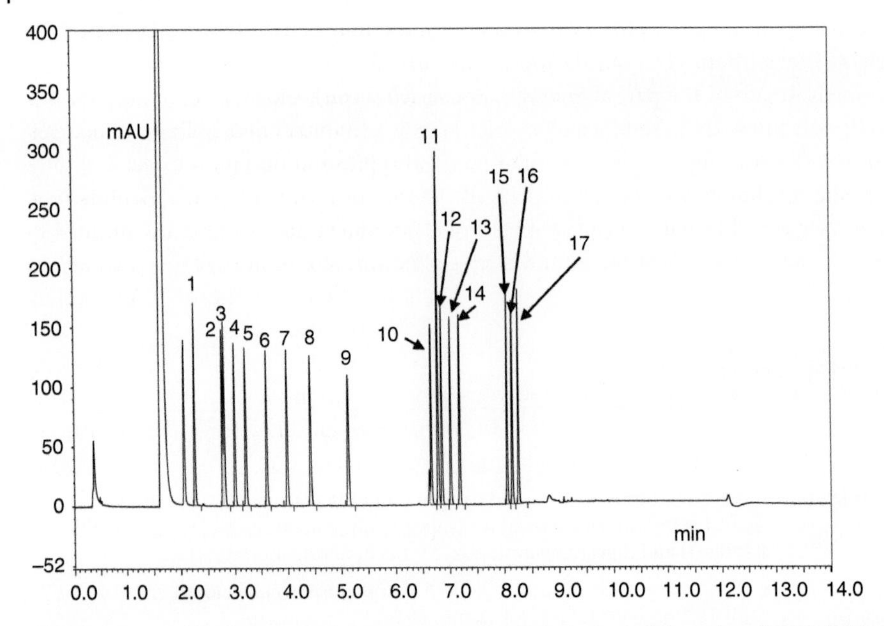

Figure 15.1 Separation of amino acids under the analytical conditions described in Table 15.1. Peak list: (1) His; (2) Arg; (3) Ser; (4) Gly; (5) Asp; (6) Glu; (7) Thr; (8) Ala; (9) Pro; (10) Cys; (11) Lys; (12) Tyr; (13) Met; (14) Val; (15) Ile; (16) Leu; and (17) Phe.

Figure 15.1 [58] (for a review, see Ref. [59]). Note that methionine (Met), tyrosine (Tyr), tryptophan (Trp), cysteine (Cys), asparagine (Asn), glutamine (Gln), and histidine (His) are not stable under acid hydrolysis conditions and hence are not evaluated.

Residual DNA from the host organism used for fermentation may be an additional potential impurity to consider. The likelihood for the presence of host cell DNA depends on the type of the biocatalyst. Typically, it is highest when the biocatalytic transformation is carried out with cell lysate after fermentation (i.e. the crude mixture after rupture of the host cells) and lower in the case of lyophilized (isolated) enzymes due to several purification operations that are typically part of the isolation process. Enzyme immobilization further reduces the likelihood of host cell DNA presence. Residual host cell DNA may be determined using quantitative real-time PCR (qPCR). Determination is recommended to be conducted as a limit test with spiking of DNA of the host organism, which is commercially available for all typical microorganisms. Similarly, primers specific to certain host strain gene(s) can be obtained from commercial sources.

Similar to the presence of host cell DNA, proteins from the host cell used for the fermentation process may be present as impurities. In analogy to host cell DNA, the likelihood for the presence of host cell proteins is highest when the cell lysate after fermentation is used and lower in case lyophilized (isolated) enzymes are applied. Host cell proteins may be specifically determined using immunochemical techniques such as ELISA. As for qPCR, kits for all typical organisms are commercially available. Immunoassays are very sensitive and allow detection of host cell proteins in ppm amounts or lower. However, determination of specific host cell proteins is usually not required, particularly not for orally administered drug substances, because host cell protein quantification may be covered by the amino acid determination after hydrolysis (see discussion above).

In case bacteria are used as host cells (e.g. *Escherichia coli*), the DNA encoding for the enzyme to be expressed is located on plasmids, which are small circular DNA molecules outside of the main bacterial DNA. Depending on the plasmid design, the plasmid may contain genes encoding for a specific antibiotic resistance. This approach ensures that addition of the corresponding antibiotic to the seed bacterial culture suppresses the growth of all bacteria that do not carry the plasmid and hence would also not express the desired biocatalytic enzyme encoded on the plasmid. If an antibiotic is used in the fermentation process, the potential for carryover of the antibiotic to the drug substance needs to be assessed considering the permitted daily exposure (PDE) of the antibiotic. Evaluation of worst-case exposure based on the amount of antibiotic present in the fermentation and the amount of enzyme produced is recommended. Examples of antibiotics applied in fermentation processes are, e.g. kanamycin or chloramphenicol. In addition, microbial count testing for the presence of genetically modified microorganisms (GMMs), applicable in case whole cells are applied, makes use of the antibiotic resistance located on the plasmid.

Endotoxins (lipopolysaccharides and lipoglycans) are large molecules that consist of a lipid and a polysaccharide part. These molecules are found in the membranes of some bacteria and may trigger strong immune responses in humans. Endotoxins are not stable to the conditions in the gastrointestinal tract and therefore are of no significant toxicological concern for orally administered drugs, but may have to be controlled in line with applicable guidelines for parenteral APIs. Standard compendial analytical test methods for the determination of endotoxin levels are available[6] (for a review on endotoxin removal, see Magalhães et al. [60]).

Many biocatalytic transformations require the presence of one or multiple cofactors for the catalytic reaction to occur. Typical examples are PLP (vitamin B6) for transaminase enzymes or nicotinamide adenine dinucleotide (NADH/NADPH) for dehydrogenase enzymes. As these cofactors are organic small molecules, their levels are controlled according to the ICH Q3A guideline [61], and their carryover, fate, and depletion have to be investigated in analogy to other organic small-molecule impurities. Metals used as cofactors are controlled as elemental impurities according to the ICH Q3D guideline [62].

Viruses that are harmful to humans could potentially be present as impurities under specific circumstances if an enzyme used in a biocatalytic reaction was produced in host systems of animal or human origin [63]. Analytical methods to reliably quantify a "total virus count" do not exist because available methods are either too generic (e.g. determination of total proteins), too specific (i.e. targeting a specific virus or virus family only), or have low sensitivity (e.g. transmission electron microscopy). Because of the complexity of the test methods, the need for specialized laboratories, and straightforward ways to prevent virus contamination, it is recommended to mitigate any risk of virus contamination by the request for TSE/BSE-free manufacturing of the enzyme (i.e. materials of human or animal origin have not been used for manufacturing, and contamination with materials of human or animal origin can be excluded) and confirmation by the enzyme supplier via a

6 For compendial methods, see United States Pharmacopoeia (USP) <85>, United States Pharmacopoeia (USP) <161>, European Pharmacopoeia (EP) 2.6.14 or Japanese Pharmacopoeia (JP) 4.01.

certificate [64]. The risk of contamination with viruses that are infectious for humans is not relevant in case of enzyme expression in bacteria, fungi, or yeasts as host systems. Contrary to mammalian or human cells, these host systems cannot support the growth of viruses that are of toxicological concern to humans, but only of virus types (e.g. bacteriophages and yeast viruses) not infectious to humans. In this case, viral clearance is implicitly covered by the TSE/BSE certificate. In general, standard chemical process conditions and operations, such as exposure to elevated temperatures, exposure to organic solvents, and crystallization, result in either inactivation or removal of any viral particles from chemical manufacturing processes, e.g. by denaturation of the protein envelope surrounding the viral particles. If required, deliberate inclusion of process conditions known to inactivate or remove viral particles would be an additional measure to further mitigate any risk from viral contamination.

In case the biocatalytically active enzyme is immobilized on a solid support (e.g. by covalent linkage, adsorption, or encapsulation), the potential for impurity introduction or formation due to leaching from the solid support needs to be assessed. The chemical linkers applied for immobilization, such as glutaraldehyde, are usually the main components of concern from a toxicological perspective. Furthermore, as most solid supports are organic polymer-based materials, the risk for the presence of residual monomers needs to be evaluated as well.

15.3.3.2 Types of Biocatalysts

Depending on the specific transformation to perform, various types of biocatalysts may be used. The actual and potential impurities present and their levels depend on the type of the biocatalyst. Common types of biocatalysts are summarized below.

- *Whole cells*. In these systems, complete host cells are used to catalyze a biocatalytic reaction. Consequently, multiple other enzymes and cell components besides the target enzyme that is specifically expressed by the host cell are present, but confined within the cell wall. As a result, other (undesired) biocatalytic reactions could occur, leading to formation of by-products.
- *Cell lysates*. Following fermentation, the cell wall of the host cells is broken up to release all cell components. The target enzyme that is specifically expressed by the host cell is in consequence mixed with other cell components (e.g. other enzymes) in the cell lysate. As a result, other (undesired) biocatalytic reactions could occur, leading to formation of by-products. In addition, components typically not found in the exterior of the host cells are present in the cell lysate.
- *Isolated enzyme*. After fermentation and cell lysis, the target enzyme that is specifically expressed by the host cell in high copy numbers is separated from other cell components (e.g. other enzymes) by purification. Isolated enzymes are often obtained as a dry powder by lyophilization, less commonly aqueous solutions are employed.
- *Immobilized enzyme*. Isolated enzymes may be attached (covalently or non-covalently) to a solid support to allow reuse of the immobilized enzyme by facile separation from the reaction mixture, or to facilitate removal of the enzyme from the reaction mixture by filtration.

15.3.3.3 Type of Expression System

The type of expression system chosen for fermentation of the target enzyme is a key factor for the definition of a suitable control strategy for impurities associated with a biocatalytic reaction. The fermentation processes applied for most commercial enzymes use established host systems that are derived from non-pathogenic and non-toxic microbial strains and possess a long history of safe use. Host systems of animal or human origin are generally not applied for fermentation, and manufacturing of isolated enzymes can therefore be conducted in TSE/BSE-free conditions.[7] Examples of host systems known to be safe to use are bacteria such as *E. coli* (safe strain lineage, e.g. K12 or B), yeast strains such as *Pichia pastoris* and *Saccharomyces cerevisiae*, and the fungal strain *Aspergillus oryzae* [56, 57]. According to the "Classification of infective microorganisms by risk group" by the World Health Organization (WHO) from 2003 (for associated regulations, see ICH [65]), these microorganisms belong to Risk Group I, which comprises microorganisms that are unlikely to cause human disease or animal disease of veterinary importance. In case non-standard expression systems are applied that have no history of safe use or belong to a risk group other than Risk Group I, an individual case-specific assessment needs to be performed.

15.3.3.4 Route of Administration

The route of administration has a strong influence on the definition of a suitable control strategy for impurities associated with biocatalytic transformations. For drug substances administered orally, toxicological concerns with regard to the potential presence of residual enzyme, residual peptides and amino acids, host cell DNA and proteins, as well as endotoxins in the drug substance are considerably reduced, as the human gastrointestinal tract is capable of degrading these potential impurities into non-toxic components. For other routes of administration, in particular intravenous administration, a detailed case-specific risk assessment is required.

15.3.3.5 Position of the Biocatalytic Step in the Synthesis and Downstream Transformations

Similar principles as outlined in the ICH Q11 guideline for the definition of regulatory starting materials can be applied to assess the likelihood for carryover of impurities related to biocatalytic transformations and the associated risk of an impact on drug substance quality [65] (see footnote 5 for general information on the ICH). The ICH Q11 guideline states that "*Impurities introduced or created early in the manufacturing process typically have more opportunities to be removed in purification operations (e.g. washing, crystallization of isolated intermediates) than impurities generated late in the manufacturing process, and are therefore less likely to be carried into the drug substance.*" In analogy, the likelihood of a biocatalytic step to affect the quality of the drug substance due to carryover of impurities related to the biocatalytic transformation decreases with the number of chemical steps between the biocatalytic step and the final drug substance.

---

7 Materials used in the synthesis of pharmaceuticals are required to be TSE/BSE-free in order to exclude the presence of prions. Prions are misfolded proteins of high physico-chemical stability that are the cause of neurodegenerative diseases like the Creutzfeldt–Jakob disease in humans.

Standard chemical processing operations such as the use of chemical reagents and organic solvents, salt formation, crystallization, and filtration or distillation generally lead to degradation and inactivation of enzymes, and to degradation of most of the impurities associated with biocatalytic transformations (e.g. host cell proteins or host cell DNA) because of their physicochemical properties. In the Food and Drug Administration (FDA) Guidance for Industry No. 216 on fermentation-derived intermediates, drug substances, and related drug products for veterinary medicinal use [66], the statement is made that "*For most fermentation products (e.g. antibiotics), it is expected that purification and downstream processing effectively remove process-related impurities, such as residual media components, residual protein and nucleic acid derived from microbial cells, and other processing reagents.*" The same scientific rationale can be applied to biocatalytic reactions, in which the fermentation product itself – the enzyme – is not even incorporated into the drug substance, but only used in catalytic quantities to enable a chemical transformation. In all cases, a case-specific assessment to evaluate the impact of the downstream process conditions and operations on the fate and depletion of impurities associated with the biocatalytic transformation is recommended as part of the control strategy definition.

15.3.3.6 Summary of the Case Study

The key aspects and elements of the control strategy applied for impurities associated with the transamination in the synthesis of (*R*)-biphenylalanine (**21**) are summarized in Table 15.2. As no general regulatory guidelines are available for biocatalytic transformations in chemical reactions, the definition of the control strategy was inspired by the available literature precedence for other APIs [56, 57].

Table 15.2 Elements of the control strategy for impurities associated with the transamination in the synthesis of (*R*)-biphenylalanine (**21**).

Route of administration	Oral
Compliance level	Non-GMP (before regulatory starting material)
Stage in synthesis	12 steps before API
Type of biocatalytic transformation	Transamination
Product phase	Marketed
Type of biocatalyst	Dry enzyme powder, produced from *E. coli* fermentation
Control strategy	Supportive data generated in support of health authority requests; no routine testing for any impurities associated with the transamination step performed in starting materials, intermediates, and/or the final drug substance Generated supportive data: • Residual amino acids (after digestion) determined in multiple batches of regulatory starting material; absence at levels below LOQ of 0.10% applicable for the final drug substance demonstrated (in line with ICH Q3A guideline) • Fate and depletion of enzyme along the synthetic route up to the regulatory starting material investigated in spiking studies (spiking level above the charging quantity of the enzyme applied)

15.3.4 Health, Process Safety and Environmental Aspects

Health, process safety and environmental aspects take a prominent place in process development for large-scale manufacturing processes, as the severity of any incident inherently grows with increasing scale of the transformation. The associated risks therefore need to be avoided, appropriately reduced or – if risk avoidance or reduction to an acceptable level cannot be achieved – suitably mitigated and controlled, e.g. by engineering means or analytical controls.

15.3.4.1 Health

Information on the toxicity of specific biocatalytic enzymes will likely be absent in most cases, in particular during (early) development stages. Different from many standard chemical reagents, however, enzymes used as biocatalysts in chemical transformations are generally of limited toxicological concern for acute or chronic exposure [67]. The lack of a significant toxicological potential is mainly a result of the ready biodegradability of enzymes and their low bioavailability. In addition, biocatalytic enzymes are usually not expected to be active toward natural targets (e.g. in the human body), specifically if they have been evolved from wild-type enzymes to improve performance on a specific target substrate in a chemical transformation. Based on the available data from genotoxicity testing, e.g. Ames tests or chromosome aberration tests, enzymes are not expected to be mutagenic, carcinogenic, or reprotoxic, reflecting the intrinsic properties of enzymes.

Most enzymes are not expected to be skin irritants based on the available data, with the exception of proteases, which are regarded as mildly irritant. The use of enzymes is, however, generally associated with the risk of respiratory sensitization. Studies triggered by incidents in the detergent industry in the early 1960s showed that enzymes may lead to respiratory allergies in case of exposure to unsafe high levels [67–69]. Dust formation during handling of solid (lyophilized) enzyme powders or aerosol formation during handling of solutions are of highest concern. Exposure to enzymes by inhalation must therefore be stringently controlled both during research and development work on lab scale, e.g. by use of appropriate fume hoods and weighing cabinets that ensure fulfillment of suitably low occupational exposure levels, as well as during handling of lyophilized enzyme powders on pilot and production scale, e.g. by charging of enzyme powders to reactors via glove boxes.

15.3.4.2 Process Safety

Biocatalytic reactions are typically performed in water or mixtures of water and a suitable organic cosolvent (e.g. DMSO), using reagents with only low to moderate reactivity. Because of the limited thermal stability of enzymes, the majority of reactions is run at ambient or slightly elevated temperatures, only in rare cases exceeding 60 °C. Consequently, the potential for thermal events, e.g. exothermicity during reagent dosing or the need for high reaction temperatures to achieve the target conversion, is inherently reduced for biocatalytic transformations. Furthermore, enzymatic reactions generally operate at pH values that are only moderately basic or acidic, thereby reducing the risks associated with corrosivity and material incompatibility. Considering the importance of thermal process safety in the development of manufacturing processes for large-scale commercial manufacturing, biocatalytic transformations provide essential benefits in this respect compared to many traditional chemical processes.

In the case of the transamination to access (*R*)-biphenylalanine (21), all standard safety tests and studies (e.g. reaction calorimetric investigations) were performed in analogy to traditional chemical steps, indicating no thermal safety risks associated with the biocatalytic reaction.

15.3.4.3 Environmental Aspects

Enzymes – including enzymes used as biocatalysts – are in most cases readily biodegradable. Work-up and further downstream processing usually result in denaturation of the active enzyme with concomitant loss of its catalytic activity, or even degradation into peptide fragments, single amino acids, or amino acid derivatives such as esters or amides. In most biocatalytic transformations, denaturation is deliberately performed, e.g. by application of extreme temperatures or pH values, high salt concentration, or the use of organic solvents (e.g. ethanol), and constitutes an inherent part of the manufacturing process. Residual enzyme and associated degradation products are then removed with the waste streams and treated according to the applicable regulations, e.g. by treatment in wastewater treatment plants for aqueous waste streams or by incineration for organic waste streams. Cofactors such as PLP (vitamin B6), thiamine (vitamin B1), or nicotinamide adenine dinucleotide (NADH/NADPH), that are required for some biocatalytic transformations, are generally well degradable and universally present in the environment, therefore not causing any environmental concern.

Although for most biocatalytic applications, the components of the biocatalytic system itself could in theory be disposed of into the environment as part of the aqueous waste streams after treatment in wastewater treatment plants, practical implementation may be more challenging because of the following aspects:

- Aqueous waste streams often contain residual substrate and product of the biocatalytic transformation. For pharmaceutical products, these materials are, however, in many cases not readily biodegradable. Specific assessments for the tolerable levels in the aqueous waste streams are therefore required.
- Aqueous waste streams may contain residual amounts of organic solvents, used either as cosolvents in the biocatalytic step or during aqueous work-up and product isolation. Disposal of aqueous waste streams via wastewater treatment plants may be restricted depending on the applicable limits for residual organic solvents in aqueous waste streams and the biodegradability of the specific organic solvents.
- In order to avoid overfertilization in rivers and lakes, the total nitrogen content in aqueous waste streams may be limited in line with applicable local requirements. Such restrictions are to be taken into consideration for biocatalytic transformations that introduce nitrogen into a substrate, e.g. transformations catalyzed by transaminases or phenylalanine lyases [70]. The reagents used in these reactions – typically isopropylamine for transaminations and ammonium salts such as ammonium carbamate for phenylalanine lyases – eventually lead to generation of ammonia, which may end up in the effluent streams after wastewater treatment.

It is important to point out that the challenges listed above are not peculiar to biocatalytic transformations, but are equally applicable to chemical transformations using standard chemical reaction conditions.

15.3.5 Equipment Utilization and Throughput Time

Because of the biphenyl moiety, both biphenylpyruvic acid (**20**) and (*R*)-biphenylalanine (**21**) are hydrophobic, resulting in low solubilities in the aqueous reaction medium. In order to achieve a sufficiently high productivity of the step, defined as yield normalized against reactor volume and reaction time [71], the transformation was run as a slurry-to-slurry process, with a suspension of substrate **20** converted to a suspension of product **21**. Under the equilibrium conditions of the transamination reaction, the small fraction of substrate **20** dissolved in the reaction medium is converted to product **21**, which itself precipitates from solution. The concentration of substrate **20** remains unchanged, as additional substrate is constantly dissolved over the course of the reaction. Mass transfer between solid state and dissolved state is sufficiently fast to achieve complete conversion within a reasonable reaction time, even in the absence of a polar organic cosolvent. It is interesting to note that the observed reaction rate was found to increase with increasing scale, i.e. from lab scale to kilo lab scale to commercial scale, presumably because of improved mass transfer as a result of higher mixing efficiency.

Productivity can also be positively influenced by modification of the reaction rate for the chemical transformation. In order to maximize the reaction rate for the conversion of biphenylpyruvic acid **20** to (*R*)-biphenylalanine **21**, the temperature was raised to profit from higher rates at higher temperatures in line with Arrhenius' law. Transaminase ATA-032 showed sufficient temperature stability, tolerating temperatures of 40–45 °C over multiple hours, and both biphenylpyruvic acid and (*R*)-biphenylalanine were found to be stable toward elevated temperatures.

15.3.6 Equipment Cleaning

Equipment cleaning is a vital component in the manufacturing of pharmaceuticals to ensure appropriate product quality and avoid cross-contamination of other products when various drug substances are manufactured in the same multipurpose equipment. For a biocatalytic transformation that uses catalytic amounts of an isolated enzyme (i.e. not cell lysates or whole cells), which itself is not therapeutically active (i.e. different from peptidic drug substances such as cyclosporine) and is manufactured by fermentation using a host system that has an established history of safe use, a two-stage approach consisting of the elements below may be applied:

1) Decontamination and denaturation to address potential residual enzyme and enzyme degradants, applying one of the cleaning conditions defined in the current version of the European Medicines Agency (EMA) "Note for guidance EMA/410/01: Minimizing the risk of transmitting animal spongiform encephalopathy via human and veterinary medicinal products" for denaturation of prions to address TSE/BSE risks for manufacturing equipment [72].

    - Sodium hydroxide (NaOH) treatment: use of 1 M aqueous NaOH solution (corresponding to 4% (w/w) concentration) with at least two hours of contact duration at not less than 25 °C.
    - Sodium hypochlorite (NaOCl) treatment: use of aqueous NaOCl solution with at least 20 000 ppm active chlorine (corresponding to 2% active chlorine or 5.25% NaOCl) for one hour.

2) Standard equipment cleaning procedure in line with applicable operating procedures for equipment cleaning after chemical reactions.

The conditions defined in the EMA guidance to address TSE/BSE risks for manufacturing equipment have been demonstrated to be suitable for the denaturation of prions, i.e. a class of proteins that exhibit significant stability toward standard denaturation conditions. Consequently, these conditions are also expected to ensure complete denaturation of any residual enzyme or peptide fragments following a biocatalytic process step. As enzymes are generally well soluble in aqueous conditions, the use of aqueous NaOH or aqueous NaOCl is expected to lead to retention of residual enzyme in the aqueous cleaning media.

Subsequent use of a standard equipment cleaning protocol, e.g. by boiling out the equipment with an organic solvent, provides an additional safeguard to ensure denaturation and removal of the enzyme. Like the exposure to pH extremes, application of heat to degrade and inactivate protein-based components is accepted by health authorities and defined as a measure for equipment cleaning in regulatory guidance. CVMP [73] states that "*Therapeutic macromolecules and peptides are known to degrade and denature when exposed to pH extremes and/or heat, and may become pharmacologically inactive. The cleaning of biopharmaceutical manufacturing equipment is typically performed under conditions which expose equipment surfaces to pH extremes and/or heat, which would lead to the degradation and inactivation of protein-based products. In view of this, the determination of health based exposure limits using PDE limits of the active and intact product may not be required.*" A similar rationale applies for the denaturation and removal of residual peptide fragments, single amino acids, or host cell DNA.

15.3.7 Enzyme Release Testing

Although enzymes are to be considered as standard chemical raw materials and should thus be treated like any other homogeneous or heterogeneous catalyst (e.g. palladium catalysts such as $PdCl_2$ or Pd/C), various aspects specific to biocatalysts need to be taken into consideration for the release testing strategy.

- Activity of the enzyme toward a specific substrate and selectivity in the biocatalytic transformation is usually addressed by an activity test (e.g. use test with defined criteria for conversion and selectivity).
- Enzyme powders may be hygroscopic. Hygroscopicity is generally addressed by a test item for water determination.
- Elevated temperatures as well as pH extremes will in many cases lead to denaturation and loss of catalytic activity. Potential denaturation is usually addressed by a (generic or specific) activity test (e.g. use test with defined criteria for conversion and selectivity).
- Enzymes may require cofactors for their activity, e.g. metal salts (e.g. Zn^{2+}, K^+, or Mg^{2+}) or specific organic small molecules (e.g. NADH/NADPH or PLP). Cofactors are typically covered by a (generic or specific) activity test (e.g. use test with defined criteria for conversion and selectivity) or a separate release test like for other raw materials.
- Enzymes are produced by fermentation employing microorganisms. Potential impurity classes inherent to this manufacturing principle are addressed primarily by the TSE/BSE assessment (including viral clearance) and may be further addressed by specific test items, e.g. for purity.

The selection of analytical tests should reflect the position of the biocatalytic step in the overall synthesis and the route of administration. For lyophilized enzyme powders, the following test items are recommended to be part of the specifications for the biocatalytic enzyme as a minimum:

- Appearance
- Activity (use test)
- Water content

In addition, a TSE/BSE statement should be requested from the supplier or should be available in case of in-house fermentation. A separate identity test is generally not required when the activity test is performed with the substrate of the actual biocatalytic transformation. Since this transformation will in most cases – in particular for structurally complex substrates and if enzyme evolution has been performed – be catalyzed only by the appropriate ("correct") enzyme, identity is intrinsically tested by the use test and sufficiently proven. In cases where activity is tested with a different substrate than the actual substrate for the transformation, confirmation of the identity may be recommended, e.g. by SDS-PAGE. Lyophilized enzyme powders usually contain water, which is typically included in the reaction medium as a solvent. Consequently, the water content only affects the enzyme assay and thus is not considered as a critical quality attribute.

The standard test items listed above are generally considered sufficient for release of enzymes when the biocatalytic step is not located at the end of the synthesis, i.e. when denaturation and degradation are likely to occur in downstream operations (e.g. because of work-up and reaction conditions in subsequent transformations), and when the drug substance is orally administered. If one or both of these requirements are not met, additional test items to confirm identity (e.g. by SDS-PAGE or ELISA) and purity are recommended.

In the case of the transamination to access **21**, transaminase enzyme ATA-032 was released for manufacture following the standard approach outlined above. Appearance, water content, and activity using the actual biphenyl keto acid substrate **20** were tested for release.

15.3.8 Transport and Storage

Although optimal conditions for storage are distinctive to each protein, application of some general guidelines for protein storage and stability is recommended.

Lyophilized enzyme powders should be stored at low temperatures for long-term storage (i.e. over several years), ideally at 2–8 °C or at −20 °C. Storage at −20 °C is recommended in particular if no stability data are available for a specific enzyme. Stability data generated for a similar biocatalyst, e.g. a closely related enzyme, may provide indications about the temperature stability. Other options include the storage of non-lyophilized enzymes at −20 °C or lower and storage in solution with a cryoprotectant such as 50% glycerol. Enzyme solutions should be stored at 2–8 °C to avoid the freeze/thaw cycles in case storage is required. Biocatalyst suppliers generally provide enzymes in suitable packaging types. Use of protecting gas is not necessary, and desiccants or drying agents should not be employed. In addition, smaller quantities (up to several 100 g) might be packaged into laboratory plastic (e.g. Nalgene) containers.

Transport conditions for biocatalysts typically correspond to the storage conditions. If desired, the temperature may be tracked during transport to detect significant temperature excursions. For transport of smaller amounts, e.g. during early development stages, lower transport temperatures than defined as storage conditions should be considered (e.g. transport at −78 °C) to mitigate the impact of temperature deviations.

Changes in the physical environment of enzymes lead to stress that can affect both stability and activity. Freezing/thawing cycles may cause instability and damage to enzymes through several mechanisms, most of them attributed to ice crystal formation. Ice crystals can concentrate (buffer) salts and enzymes ("freeze concentration"), causing significant stress, e.g. by protein unfolding at the ice/water interface. Based on literature reports, variation in the cooling and freezing rates of enzyme solutions during freezing/thawing cycles may in addition affect enzyme activity and stability. Consequently, freezing/thawing cycles should be avoided whenever possible to reduce enzyme degradation and concomitant activity loss. Bulk shipments of lyophilized enzyme powders for lab development should therefore always be dispensed into smaller quantities to circumvent that for every day work, the bulk of the material needs to be brought from cold storage to room temperature. For pilot campaigns and commercial production, bulk quantities may be split into smaller packaging sizes – ideally adjusted to the intended batch size – by the enzyme supplier to avoid the need for later dispensing from bulk quantities at the production site.

15.4 Conclusions, Expectations, and Prospects

Over the past 20 years, applications of biocatalysis in the pharmaceutical industry have seen a tremendous growth. Although biocatalytic transformations remained limited to selected examples comprising only few reaction classes at the beginning, biocatalysis has turned into a routine tool for most pharmaceutical companies by now. For some reaction and product classes, e.g. the asymmetric reduction of ketones to chiral secondary alcohols, or the synthesis of chiral amines from the corresponding ketone precursors, biocatalytic approaches have even become the new state-of-the-art methodologies, replacing well-established traditional chemical methods such as stoichiometric or metal-catalyzed asymmetric ketone reductions. With the ever-growing number of available enzyme variants and enzyme classes, the range of products accessible by biocatalysis will be amplified, and more and more traditional chemical reactions come into scope, leading eventually to their replacement by biocatalytic approaches.

Commercial enzyme kits for straightforward and fast screening are available today from many established vendors, offering high performance in terms of activity and selectivity for a diverse range of substrates. A large selection of enzymes is available "off-the-shelf" and can be accessed in sufficient quantities to support any scale from lab up to commercial manufacturing. With only a handful of well-performing enzyme classes, an impressive range of substrates and building blocks has become readily and efficiently accessible even on pilot and commercial scale. This trend is expected to continue with new enzyme variants and enzyme classes being continuously added to the chemist's toolbox.

Although explicit regulatory guidance for the use of enzymes as biocatalysts in chemical reactions has not yet been implemented by health authorities worldwide, the regulatory landscape has been shaped by the growing number of examples of biocatalytic reactions applied to the manufacturing of drug substances. Suitable control strategies that combine analytical and process measures have been devised and are mutually accepted by industry and leading health authorities [56, 57].

Until now, the development of biocatalytic processes most of the time focuses on the introduction of chirality in target molecules. Enzymes are perfectly suited for this task, but their abilities significantly expand beyond the control of enantioselectivity and diastereoselectivity.

- Enzymes enable novel classes of reactions and thus have to be considered as an enabling technology that allows to reach out to parts of the chemical space that were until now not accessible by traditional chemical means.
- Enzymes are prime tools to provide the synthesis of products in increased yields, purities, and selectivities They allow the implementation of shorter synthesis routes and simplify manufacturing operations for reaction, work-up and isolation. For these reasons, enzymatic processes offer significant productivity and cost-saving potential.
- Enzymes help to "green" manufacturing processes by the substitution of stoichiometric reactions with catalytic approaches, reduction of waste streams (e.g. by switching from organic solvents to aqueous reaction media), or removal of transition-metal catalysts or hazardous reagents from the manufacturing process.
- Enzymes operate under mild reaction conditions, thereby significantly reducing safety hazards that are often associated with traditional chemical processes.

So far, many enzyme-catalyzed transformations targeted bond formation or bond modification between carbon atoms and heteroatoms such as oxygen or nitrogen. Over time, examples of transformations that address C—C bond formations or belong to reaction classes that are difficult to implement today on large scale, such as P450-mediated oxidations, are expected to increase in number and diversify [6].

The growth in biocatalytic applications runs in parallel to the maturation of the foundational technologies. Directed enzyme evolution and other protein engineering technologies have opened up tremendous possibilities to modify enzymes in a targeted way [12–14] (see also Chapter 2). Enzymes can now be tailored to perform specific tasks in a highly selective manner, designed to meet reaction and process needs, or even to expand into new reactivity space. The fantastic ability of enzymes to adapt to various substrates, conditions, and media when subjected to evolution seems to unleash an unlimited potential for enzymatic catalysis. Research in academia and industry aims at shortened cycle times for enzyme evolution [74]. Technologies under development comprise fast protein engineering methods in combination with new analytical and process engineering tools, e.g. fast mass spectroscopy, fluorescence analysis, and microfluidic or *in vitro* transcription methods. Enzyme evolution is therefore expected to become even more accessible and impactful than it currently is.

Besides the technical implications, these new technologies revolutionize the synthetic strategy paradigms in place and commonly accepted for the manufacturing of drug substances over decades [15–17]. Although enzymatic transformations were initially applied mainly in lifecycle management projects to optimize process performance with the implementation of second- or third-generation manufacturing processes after commercialization, focusing on the replacement of individual transformations by biocatalytic alternatives to address specific deficiencies, a shift toward the use of biocatalytic reactions already in early development stages can be observed. In addition, initial examples of complete cascades that combine various biocatalytic transformations into a single sequence – in some cases run as one-pot processes [75, 76] (see also Chapter 13) or in combination with other technologies such as photocatalysis [77] (see also Chapters 12 and 14) – have appeared, and more examples are expected to be reported in the near future. As such, biocatalysis has started to change the way we think about synthetic routes to complex drug substances and will more and more turn into a strategic tool rather than a (often opportunistic) tactical mean.

We can already anticipate today that the speed of enzyme evolution will further increase, benefiting from advances in areas such as analytics or statistical methods, machine learning and artificial intelligence. In combination with new enzyme classes, either derived from natural sources or artificially generated, the field of biocatalysis will keep its high dynamics, and many more enzymatic steps will be utilized and integrated in synthetic strategies right from the very beginning. The "Holy Grail" could be full *in silico* enzyme identification and evolution [78–81]. If this target is ever reached, synergies with AI-driven retrosynthetic software and automated synthesis will pave the way to unprecedented opportunities.

Acknowledgments

We would like to acknowledge our colleagues Daniel Gschwend, Yunzhong Li, Thomas Ruch, Kurt Laumen, and Claude Haby for their contributions to the development of the biocatalytic synthesis of (*R*)-biphenylalanine, as well as Sounia Touchene, Régine Wicky, Xu Su, and Liang Lu for their contributions to the establishment of the analytical method for residual amino acid determination. We thank Codexis for helpful discussions during the development of the transamination reaction of (*R*)-biphenylalanine.

List of Abbreviations

The following abbreviations are used in the manuscript:

API	Active pharmaceutical ingredient
ATA	(amine) transaminase
BVMO	Baeyer–Villiger monooxygenase
ELISA	enzyme-linked immunosorbent assay
EMA	European Medicines Agency
FAD	flavin adenine dinucleotide
FDA	Food and Drug Administration
GC-FID	gas chromatography–flame ionization detector
GC-MS	gas chromatography–mass spectrometry
GMM	genetically modified microorganism
ICH	International Council for Harmonisation of Technical Requirements for Pharmaceuticals for Human Use

KRED	ketoreductase
LC-FL	liquid chromatography–fluorescence detection
LC-MS	liquid chromatography–mass spectrometry
LC-MS/MS	liquid chromatography–tandem mass spectrometry
LC-UV	liquid chromatography–ultraviolet detection
NADPH	nicotinamide adenine dinucleotide phosphate
PCR	polymerase chain reaction
PDE	permitted daily exposure
PLP	pyridoxal 5'-phosphate
SDS-PAGE	sodium dodecyl sulfate–polyacrylamide gel electrophoresis
TSE/BSE	transmissible spongiform encephalopathies/bovine spongiform encephalopathy

References

1 Huges, G. and Lewis, J.C. (2018). Introduction: biocatalysis in industry. *Chem. Rev.* 118: 1–3.

2 Panke, S. and Wubbolts, M. (2005). Advances in biocatalytic synthesis of pharmaceutical intermediates. *Curr. Opin. Chem. Biol.* 9: 188–194.

3 Pollard, D.J. and Woodley, J.M. (2006). Biocatalysis for pharmaceutical intermediates: the future is now. *Trends Biotechnol.* 25: 66–73.

4 Huisman, G.W. and Collier, S.J. (2013). On the development of new biocatalytic processes for practical pharmaceutical synthesis. *Curr. Opin. Chem. Biol.* 17: 284–292.

5 Choi, J.M., Han, S.S., and Kim, H.S. (2015). Industrial applications of enzyme biocatalysis: current status and future aspects. *Biotechnol. Adv.* 33: 1443–1454.

6 Patel, R.N. (2018). Biocatalysis for synthesis of pharmaceuticals. *Bioorg. Med. Chem.* 26: 1252–1274.

7 Abdelraheem, E.M.M., Busch, H., Hanefeld, U., and Tonin, F. (2019). Biocatalysis explained: from pharmaceutical to bulk chemical production. *React. Chem. Eng.* 4: 1878–1894.

8 Adams, J.P., Brown, M.J.B., Diaz-Rodriguez, A. et al. (2019). Biocatalysis: a pharma perspective. *Adv. Synth. Catal.* 361: 2421–2432.

9 Sheldon, R.A. and Reymond, J.L. (2010). Biocatalysis and biotransformation. *Curr. Opin. Chem. Biol.* 14 (2): 113–284.

10 Hughes, G. and Lewis, J.C. (2018). Biocatalysis in industry. *Chem. Rev.* 118 (1): 1–368.

11 Bornscheuer, U.T. and Buchholz, K. (2005). Highlights in biocatalysis – historical landmarks and current trends. *Eng. Life. Sci.* 5: 309–323.

12 Arnold, F.H. (2018). Directed evolution: bringing new chemistry to life. *Angew. Chem. Int. Ed.* 57: 4143–4148.

13 Davids, T., Schmidt, M., Böttcher, D., and Bornscheuer, U.T. (2013). Strategies for the discovery and engineering of enzymes for biocatalysis. *Curr. Opin. Chem. Biol.* 17: 215–220.

14 Woodley, J.M. (2013). Protein engineering of enzymes for process applications. *Curr. Opin. Chem. Biol.* 17: 310–316.

15 Turner, N.J. and O'Reilly, E. (2013). Biocatalytic retrosynthesis. *Nat. Chem. Biol.* 9: 285–288.

16 de Souza, R.O.M.A., Miranda, L.S.M., and Bornscheuer, U.T. (2017). A retrosynthesis approach for biocatalysis in organic synthesis. *Chem. Eur. J.* 23: 12040–12063.

17 Hönig, M., Sondermann, P., Turner, N.J., and Carreira, E.M. (2017). Enantioselective chemo- and biocatalysis: partners in retrosynthesis. *Angew. Chem. Int. Ed.* 56: 8942–8973.

18 Martinez, C.A., Hu, S., Dumond, Y. et al. (2008). Development of a chemoenzymatic manufacturing process for pregabalin. *Org. Process Res. Dev.* 12: 392–398.

19 Hoekstra, M.S., Sobieray, D.M., Schwindt, M.A. et al. (1997). Chemical development of CI-1008, an enantiomerically pure anticonvulsant. *Org. Process Res. Dev.* 1: 26–38.

20 Limanto, J., Ashley, E.R., Yin, J. et al. (2014). A highly efficient asymmetric synthesis of vernakalant. *Org. Lett.* 16: 2716–2719.

21 Plouvier, B.M., Chou, D.T.H., Jung, G. et al. (2006). Synthetic process for aminocyclohexyl ether compounds. WO2006088525.

22 Machiya, K., Ike, K., Watanabe, M. et al. (2006). Production method of optically active cyclohexane ether compounds. WO2006075778.

23 Jung, G., Yee, J.G.K., Chou, D.T.H., and Plouvier, B.M.C. (2006). Synthetic processes for the preparation of aminocyclohexyl ether compounds. WO2006138673.

24 Ye, H., Yu, C., and Zhong, W. (2012). New procedure for the preparation of (1*R*,2*R*)-2-[(*R*)-3-(benzyloxy)pyrrolidin-1-yl]cyclohexanol. *Synthesis* 44: 51–56.

25 Savile, C.K., Janey, J.M., Mundorff, E.C. et al. (2010). Biocatalytic asymmetric synthesis of chiral amines from ketones applied to sitagliptin manufacture. *Science* 329: 305–309.

26 Desai, A.A. (2011). Sitagliptin manufacture: a compelling tale of green chemistry, process intensification, and industrial asymmetric catalysis. *Angew. Chem. Int. Ed.* 50: 1974–1976.

27 Truppo, M.D., Strotman, H., and Hughes, G. (2012). Development of an immobilized transaminase capable of operating in organic solvent. *ChemCatChem* 4: 1071–1074.

28 Hansen, K.B., Balsells, J., Dreher, S. et al. (2005). First generation process for the preparation of the DPP-IV inhibitor sitagliptin. *Org. Process Res. Dev.* 9: 634–639.

29 Hansen, K.B., Hsiao, Y., Xu, F. et al. (2009). Highly efficient asymmetric synthesis of sitagliptin. *J. Am. Chem. Soc.* 131: 8798–8804.

30 Bong, Y.K., Song, S., Nazor, J. et al. (2018). Baeyer–Villiger monooxygenase-mediated synthesis of esomeprazole as an alternative for Kagan sulfoxidation. *J. Org. Chem.* 83: 7453–7458.

31 Larsson, M.E., Stenhede, U.J., Sorensen, H. et al. (1999). Process for synthesis of substituted sulfoxides. US Patent 5,948,789.

32 Cotton, H., Elebring, T., Larsson, M. et al. (2000). Asymmetric synthesis of esomeprazole. *Tetrahedron Asymmetry* 11: 3819–3825.

33 Balke, K., Beier, A., and Bornscheuer, U.T. (2018). Hot spots for the protein engineering of Baeyer–Villiger monooxygenases. *Biotechnol. Adv.* 36: 247–263.

34 Goundry, W.R.F., Adams, B., Benson, H. et al. (2017). Development and scale-up of a biocatalytic process to form a chiral sulfoxide. *Org. Process Res. Dev.* 21: 107–113.

35 Belley, M.L., Leger, S., Labelle, M. et al. (1996). Unsaturated hydroxyalkylquinoline acids as leukotriene antagonists. US Patent 5,565,473.

36 King, A.O., Corley, E.G., Anderson, R.K. et al. (1993). An efficient synthesis of LTD4 antagonist L-699,392. *J. Org. Chem.* 58: 3731–3735.

37 Liang, J., Lalonde, J., Borup, B. et al. (2010). Development of a biocatalytic process as an alternative to the (−)-DIP-Cl-mediated asymmetric reduction of a key intermediate of montelukast. *Org. Process Res. Dev.* 14: 193–198.

38 Klibanov, A.M. (2001). Improving enzymes by using them in organic solvents. *Nature* 409: 241–246.

39 Li, T., Liang, J., Ambrogelly, A. et al. (2012). Efficient, chemoenzymatic process for manufacture of the boceprevir bicyclic [3.1.0]proline intermediate based on amine oxidase-catalyzed desymmetrization. *J. Am. Chem. Soc.* 134: 6467–6472.

40 Köhler, V., Bailey, K.R., Znabet, A. et al. (2010). Enantioselective biocatalytic oxidative desymmetrization of substituted pyrrolidines. *Angew. Chem. Int. Ed.* 49: 2182–2184.

41 Znabet, A., Polak, M.M., Janssen, E. et al. (2010). A highly efficient synthesis of telaprevir by strategic use of biocatalysis and multicomponent reactions. *Chem. Commun.* 46: 7918–7920.

42 Mijts, B., Muley, S. and Liang J. et al. (2010). Biocatalytic processes for the preparation of substantially stereomerically pure fused bicyclic proline compounds. WO2010008828.

43 For the synthesis of boceprevir using a classical non-biocatalytic approach, see: Park, J., Sudhakar, A., Wong, G.S. et al. (2004). Process and intermediates for the preparation of (1*R*,2*S*,5*S*)-6,6-dimethyl-3-azabicyclo[3.1.0]hexane-2-carboxylates or salts therefore. WO2004113295.

44 Kwok, D.L., Lee, H.C., and Zavialov, I.A. (2009). Dehydrohalogenation process for the preparation of intermediates useful in providing 6,6-dimethyl-3-azabicyclo[3.1.0]hexane compounds. WO2009073380.

45 Berranger, T. and Demonchaux, P. (2008). Process for the preparation of 6,6-dimethyl-3-azabicyclo[3.1.0]hexane compounds utilizing bisulfite intermediate. WO2008082508.

46 Wu, G., Chen, F.X., Rashatasakhon, P. et al. (2007). Process for the preparation of 6,6-dimethyl-3-azabicyclo[3.1.0]hexane compounds and enantiomeric salts therefore. WO2007075790.

47 Kleinbeck-Riniker, F.K., Kapferer, T., Laumen, K. et al. (2018). New process for early sacubitril intermediates. WO2018116203.

48 Ahmed, S.T., Parmeggiani, F., Weise, N.J. et al. (2015). Chemoenzymatic synthesis of optically pure L- and D-biarylalanines through biocatalytic asymmetric amination and palladium-catalyzed arylation. *ACS Catal.* 5: 5410–5413.

49 Parmeggiani, F., Ahmed, S.T., Thompson, M.P. et al. (2016). Single-biocatalyst synthesis of enantiopure D-arylalanines exploiting an engineered D-amino acid dehydrogenase. *Adv. Synth. Catal.* 358: 3298–3306.

50 Filip, A., Nagy, E.Z.A., Tork, S.D. et al. (2018). Tailored mutants of phenylalanine ammonia-lyase from *Petroselinum crispum* for the synthesis of bulky L- and D-arylalanines. *ChemCatChem* 10: 2627–2633.

51 Ksander, G.M., Ghai, R.D., deJesus, R. et al. (1995). Dicarboxylic acid dipeptide neutral endopeptidase inhibitors. *J. Med. Chem.* 38: 1689–1700.

52 van den Berg, M., Minnaard, A.J., Haak, R.M. et al. (2003). Monodentate phosphoramidites: a breakthrough in rhodium-catalysed asymmetric hydrogenation of olefins. *Adv. Synth. Catal.* 345: 308–323.

53 Willemse, T., Van Imp, K., Goss, R.J.M. et al. (2015). Suzuki–Miyaura diversification of amino acids and dipeptides in aqueous media. *ChemCatChem* 7: 2055–2070.

54 Tufvesson, P., Lima-Ramos, J., Jensen, J.S. et al. (2011). Process considerations for the asymmetric synthesis of chiral amines using transaminases. *Biotechnol. Bioeng.* 108: 1479–1493.

55 Kelly, S.A., Pohle, S., Wharry, S. et al. (2018). Application of ω-transaminases in the pharmaceutical industry. *Chem. Rev.* 118: 349–367.

56 Wells, A.S., Finch, G.L., Michels, P.C., and Wong, J.W. (2012). Use of enzymes in the manufacture of active pharmaceutical ingredients – a science and safety-based approach to ensure patient safety and drug quality. *Org. Process Res. Dev.* 16: 1986–1993.

57 Wells, A.S., Wong, J.W., Michels, P.C. et al. (2016). Case studies illustrating a science and risk-based approach to ensuring drug quality when using enzymes in the manufacture of active pharmaceuticals ingredients for oral dosage form. *Org. Process Res. Dev.* 20: 594–601.

58 Cohen, S.A. and Michaud, D.P. (1993). Synthesis of a fluorescent derivatizing reagent, 6-aminoquinolyl-*N*-hydroxysuccinimidyl carbamate, and its application for the analysis of hydrolysate amino acids via high-performance liquid chromatography. *Anal. Biochem.* 211: 279–287.

59 Fountoulakis, M. and Lahm, H.W. (1998). Hydrolysis and amino acid composition analysis of proteins. *J. Chromatogr. A* 826: 109–134.

60 Magalhães, P.O., Lopes, A.M., Mazzola, P.G. et al. (2007). Methods of endotoxin removal from biological preparations: a review. *J. Pharm. Pharm. Sci.* 10: 388–404.

61 ICH (2006). International Conference on Harmonisation of Technical Requirements for Registration of Pharmaceuticals for Human Use. ICH harmonized tripartite guideline. Impurities in new drug substances Q3A(R2). Current step 4 version [internet]. https://database.ich.org/sites/default/files/Q3A%28R2%29%20Guideline.pdf (accessed 18 April 2020).

62 ICH (2019). International Council on Harmonisation of Technical Requirements for Registration of Pharmaceuticals for Human Use. ICH harmonized guideline. Guideline for elemental impurities Q3D(R1). Final version [internet]. https://database.ich.org/sites/default/files/Q3D-R1EWG_Document_Step4_Guideline_2019_0322.pdf (accessed 18 April 2020).

63 World Health Organization (WHO) (2004). *Laboratory Biosafety Manual*, 3rde. Geneva: World Health Organization (WHO). https://www.who.int/csr/resources/publications/biosafety/en/Biosafety7.pdf (accessed 18 April 2020).

64 European Medicines Agency (EMA) (2008). *CHMP/BWP (Committee Abbreviation). Guideline on Virus Safety Evaluation of Biotechnological Investigational Medicinal Products. EMEA/CHMP/BWP/398498/2005 [internet].* London: European Medicines Agency (EMA). https://www.ema.europa.eu/en/documents/scientific-guideline/guideline-virus-safety-evaluation-biotechnological-investigational-medicinal-products_en.pdf (accessed 18 April 2020).

65 ICH (2012). International Conference on Harmonisation of Technical Requirements for Registration of Pharmaceuticals for Human Use. ICH harmonized tripartite guideline. Development and manufacture of drug substances (chemical entities and biotechnological/biological entities) Q11. Current step 4 version [internet]. https://database.ich.org/sites/default/files/Q11_Guideline.pdf (accessed 18 April 2020).

66 U.S. Department of Health and Human Services, Food and Drug Administration, Center for Veterinary Medicine (2012). Chemistry, manufacturing, and controls (CMC) information – fermentation-derived intermediates, drug substances, and related drug products for veterinary medicinal use. Guidance for industry no. 216 [internet]. FDA. https://www.fda.gov/media/79873/download (accessed 18 April 2020).

67 Association Internationale de la Savonnerie, de la Détergence et des Produits d'Entretien (AISE), Enzymes Occupational Exposure Working Group (2018). Guidelines for the safe handling of enzymes in detergent manufacturing. Version 2.2 [internet]. AISE. https://www.aise.eu/documents/document/20180405111438-aise-enzymes_safe_handling-v2-2-march_2018.pdf (accessed 18 April 2020).

68 Basketter, D.A., Broekhuizen, C., Fieldsend, M. et al. (2010). Defining occupational and consumer exposure limits for enzyme protein respiratory allergens under REACH. *Toxicology* 268: 165–170.

69 Budnik, L.T., Scheer, E., Burge, P.S., and Baur, X. (2017). Sensitising effects of genetically modified enzymes used in flavour, fragrance, detergence and pharmaceutical production: cross-sectional study. *Occup. Environ. Med.* 74: 39–45.

70 Parmeggiani, F., Weise, N.J., Ahmed, S.T., and Turner, N.J. (2018). Synthetic and therapeutic applications of ammonia-lyases and aminomutases. *Chem. Rev.* 118: 73–118.

71 Dach, R., Song, J.J., Roschangar, F. et al. (2012). The eight criteria defining a good chemical manufacturing process. *Org. Process Res. Dev.* 16: 1697–1706.

72 European Commission, European Medicines Agency (EMA) (2011). Minimizing the risk of transmitting animal spongiform encephalopathy via human and veterinary medicinal products. Note for guidance EMA/410/01 [internet]. EMA. https://www.ema.europa.eu/en/documents/scientific-guideline/minimising-risk-transmitting-animal-spongiform-encephalopathy-agents-human-veterinary-medicinal_en.pdf (accessed 18 April 2020).

73 European Medicines Agency, Committee for Medicinal Products for Human Use (CHMP) and Committee for Medicinal Products for Veterinary Use (CVMP) (2014). Guideline on setting health based exposure limits for use in risk identification in the manufacture of different medicinal products in shared facilities. EMA/CHMP/CVMP/SWP/169430/2012 [internet]. EMA. https://www.ema.europa.eu/en/documents/scientific-guideline/guideline-setting-health-based-exposure-limits-use-risk-identification-manufacture-different_en.pdf (accessed 18 April 2020).

74 Truppo, M.D. (2017). Biocatalysis in the pharmaceutical industry: the need for speed. *ACS Med. Chem. Lett.* 8: 476–480.

75 Sperl, J.M. and Sieber, V. (2018). Multienzyme cascade reactions – status and recent advances. *ACS Catal.* 8: 2385–2396.

76 Huffman, M.A., Fryszkowska, A., Alvizo, O. et al. (2019). Design of an in vitro biocatalytic cascade for the manufacture of islatravir. *Science* 366: 1255–1259.

77 Seel, C.J. and Gulder, T. (2019). Biocatalysis fueled by light: on the versatile combination of photocatalysis and enzymes. *ChemBioChem* 20: 1871–1897.

78 Kries, H., Blomberg, R., and Hilvert, D. (2013). *De novo* enzymes by computational design. *Curr. Opin. Chem. Biol.* 17: 221–228.

79 Wijma, H.J., Floor, R.J., Bjelic, S. et al. (2015). Enantioselective enzymes by computational design and *in silico* screening. *Angew. Chem. Int. Ed.* 54: 3726–3730.

80 Garrabou, X., Verez, R., and Hilvert, D. (2017). Enantiocomplementary synthesis of nitroketones using designed and evolved carboligases. *J. Am. Chem. Soc.* 139: 103–106.

81 Osuna, S., Jimenez-Oses, G., Noey, E.L., and Houk, N.K. (2015). Molecular dynamics explorations of active site structure in designed and evolved enzymes. *Acc. Chem. Res.* 48: 1080–1089.

16

Enzymatic Commercial Sources

Gonzalo de Gonzalo[1] *and Iván Lavandera*[2]

[1] *Universidad de Sevilla, Departamento de Química Orgánica, c/ Profesor García González 1, 41012, Sevilla, Spain*
[2] *Universidad de Oviedo, Departamento de Química Orgánica e Inorgánica, Avda. Julián Clavería 8, 33006, Oviedo, Spain*

16.1 Introduction

Biocatalysts are nowadays essential tools in the preparation of valuable compounds, particularly in the synthesis of pharmaceutical and fine chemical targets [1–3]. As highly selective catalysts, they offer a direct and simple way to synthesize complex achiral and chiral compounds. These biological catalysts also offer a high efficiency and are able to perform their activity under mild and environmentally friendly conditions. Depending on the reaction they catalyze, biocatalysts can be divided into seven main groups: oxidoreductases, transferases, hydrolases, lyases, isomerases, ligases, and translocases, the first three groups being the most employed with synthetic purposes. Biocatalysts can be employed in different industrial sectors (see Chapter 15), the most demanding being food and beverages, cleaning reagents, biofuel production, agriculture, and (bio)pharmaceuticals.

Enzymes can be extracted from any living organism. Most of the biocatalysts with industrial and laboratory applications are obtained from microorganisms, from eukaryotic systems as yeasts and fungi to prokaryotic bacteria from both the Gram-positive and Gram-negative families. In contrast, a small percentage of them can be obtained from plants and animals. In general, the microbial enzymes are preferred to plant and animal biocatalysts because of some advantages:

- Microbial enzymes are cheaper to obtain.
- Enzyme contents in microorganisms are more predictable and controllable.
- Plant and animal tissues can contain more potentially harmful materials than microbes.

There are several companies focused on the preparation biocatalysts, presenting different sizes. Some of them are only focused on the synthesis of enzymes, whereas others developed this activity as a part of their global business. Most of these companies produce enzyme formulations for large-scale applications, including food and beverage production, household care, and bioenergy or feed, but in the past few years, there is an increasing

Biocatalysis for Practitioners: Techniques, Reactions and Applications, First Edition.
Edited by Gonzalo de Gonzalo and Iván Lavandera.

production of biocatalysts for the pharmaceutical industry and biotechnology [4]. By this reason, carbohydrases, proteases, and lipases are the types of enzymes most widely employed on large-scale applications.

In this chapter, we are going to describe some of the most important companies focusing mainly on the production of biocatalysts with applications in organic synthesis. These companies have been divided depending on the location of their headquarters between Europe, North America, and Asia. Apart from their portfolio of biocatalytic preparations, many of these companies are able to offer other types of services, which include the following:

- Screening of novel biocatalysts as well as improved variants of them.
- Production of the biocatalysts at high scale (fermentation procedures).
- Purification of the biocatalysts and preparation of enzymatic formulations ready to be employed.
- Development of novel chemoenzymatic procedures for the synthesis of highly valuable compounds.

Finally, it has to be taken into consideration that the biocatalyst supplier market is very dynamic. Thus, the information shown herein is updated to July 2020.

16.2 European Companies

16.2.1 AB Enzymes

This is a German company located at Darmstadt which supplies enzymes for industrial applications, covering a huge range of sectors including baking, pulp and paper, grain and oilseed processing, detergents, protein modification, textiles, and animal feed [5].

16.2.2 Almac

This company, located in Northern Ireland, was founded in 1968. Its initial activity was devoted to provide clinical services to the pharmaceutical industry. After some time of activity, the company expanded including a biocatalysis section that offers services to different clients worldwide [6].

Almac also offers services in process development for scaling-up reactions, employing as biocatalysts hydrolases, alcohol dehydrogenases (ADHs), ω-transaminases (amine transaminases [ATAs]), Baeyer–Villiger monooxygenases (BVMOs), nitrilases, cytochrome P450s, and cyanide dihydratases. These biocatalysts can be acquired individually, but biocatalysts' libraries are also supplied in the selectAZymes™ kits, in which a number of enzymes with the same activity can be purchased.

Almac can offer to its customers the specific enzymes in different preparations including concentrated biomass (cell paste), cell-free extracts, or lyophilized enzyme powders. The company provides services in fermentation development and scaling-up from 1 l to 63 m^3 to deliver the desired biocatalysts. Enzymes can also be acquired as immobilized preparations, as this company presents a high expertise in immobilization techniques for both whole cells and isolated enzymes.

16.2.3 Biocatalysts

This company was first registered in 1983, Cardiff (Great Britain). Biocatalysts develop and manufacture specialty enzymes from small- to large-scale quantities for a variety of industries [7]. The most important applications of these biocatalysts are the food, flavor, and fragrance and life sciences sectors, also being employed in pharma and fine chemicals. Some examples of the products offered by this supplier with applications in food sciences are shown in Table 16.1.

16.2.4 c-Lecta GmbH

This German biotechnological supplier, with headquarters in Leipzig, is a relatively young company. c-Lecta offers different types of biocatalysts including NuCLEANase, a cost-effective tech-grade endonuclease for industrial applications, able to catalyze the removal of nucleic acids from biotechnological products [8]. This company also commercialized ENARASE®, a genetically engineered endonuclease from *Serratia marcescens* expressed in *Bacillus* sp., which catalyzes the cleavage of all forms of DNA and RNA very efficiently.

Lipase B from *Candida antarctica* (CAL-B) is one of the most employed biocatalysts for industrial applications because of its usually high chemo-, regio-, and/or stereoselectivity, combined with its high stability, as this enzyme is able to work in a wide set of different nonconventional media [9]. CAL-B is commercialized under different preparations, being supplied by c-Lecta as CalB Immo Plus, a CAL-B immobilized onto a Purolite Corporation polymeric support.

Finally, c-Lecta supplies Customized Enzyme c-Lections, pools of industrially relevant and so far unavailable enzymes that are individually assorted upon request according to the custom-specific needs. Different types of biocatalysts, including ADHs, ω-transaminases, nitrilases and esterases, lipases, and cytochrome P450s, are available in these screening kits at the quantities desired by the customers.

Table 16.1 Enzymes supplied by Biocatalysts with applications in food science.

Enzyme	Activity
Promod™ 950L	Protease
Promod™ 192P	Microbial acid protease
Flavopro™ 954MDP	Microbial deaminase
Glucose oxidase 789L	Oxidase
Cellulase 13L/MDP	Degradation of cellulose
Beta glucosidase 16L	Exo-carbohydrase
Amylase AD11MDP	Starch digestion

Source: Based on [7].

16.2.5 Enzymicals

Enzymicals is a German company (Greifswald), founded in 2009, which offers a set of recombinant enzymes suitable for research and development [10]. This company also supplies chemicals as well as their expertise in process development, optimization, and piloting of novel chemoenzymatic reactions for the production of fine chemicals. Enzymicals present a portfolio of several biocatalysts of different activity, including ω-transaminases, BVMOs, imine reductases, halohydrin dehalogenases, phosphotransferases, aminoacylases, carboxyl esterases, and lipases, as shown in Table 16.2. Apart from the individual enzymes, some types of biocatalysts can be acquired as enzyme kits on demand by the customers.

16.2.6 Evoxx Technologies GmbH

Evoxx Technologies GmBH is a company located at Monheim am Rhein (Germany), being the European headquarter of the global enzyme manufacturer AETL group, which supplies different types of biocatalysts and probiotics [11]. Evoxx Technologies presents a

Table 16.2 Some selected examples of the biocatalysts supplied by Enzymicals .

Enzyme	Activity	Quantity
ATA-01 *Aspergillus fumigatus*	ω-transaminase	On demand
ATA04 *Aspergillus oryzae*	ω-transaminase	On demand
ATA07 *Mycobacterium vanbaalenii*	ω-transaminase	On demand
BVMO 01 *Acinetobacter calcoaceticus*	Baeyer–Villiger monooxygenase	50 mg (0.5 ml) or 500 mg (5.0 ml)
BVMO 02 *Thermobifida fusca*	Baeyer–Villiger monooxygenase	0.5 ml or 5.0 ml
BVMO 04 *Pseudomonas putida*	Baeyer–Villiger monooxygenase	0.5 ml or 5.0 ml
IRED01 *Streptomyces* sp. GF3587	Imine reductase	On demand
IRED04 *Paenibacillus elgii* B69	Imine reductase	On demand
HHDH01 *Tistrella mobilis*	Halohydrin dehalogenase	50 mg or 500 mg
HHDH03 *Methylibium petroleiphilum*	Halohydrin dehalogenase	50 mg or 500 mg
PTF01 *Limulus polyphemus*	Phosphotransferase	50 mg or 500 mg
PTF02 *Brucella abortus*	Phosphotransferase	50 mg or 500 mg
PLE isozymes 1 to 6 from pig liver	Esterase	50 mg or 500 mg
Esterase 01 *Bacillus subtilis*	Esterase	50 mg or 500 mg
Esterase 04 *Pseudomonas fluorescens*	Esterase	50 mg or 500 mg
LIP01 *Candida antarctica*	Lipase	0.5 ml or 5.0 ml
LIP02 *Fusarium solani pisi*	Lipase	0.5 ml or 5.0 ml
LIP03 *Geobacillus thermoleovorans* IHI-91	Lipase	0.5 ml or 5.0 ml

Source: Based on [10].

broad range of enzymes employed in food applications, as in fruit and vegetable processing, baking, brewing and malting, starch and grain processing, oil and fats treatments, and dairy and cheese processing. However, apart from this line of business, this company also supplies several biocatalysts to be employed in chemical synthesis and in pharmaceutical technology, focusing mainly on three types of enzymes: ω-transaminases, ADHs, and lipases. Regarding the first enzymes, Evoxx supplies two screening kits, one with 9 lyophilized (*R*)-transaminases from different microorganisms and one with 10 lyophilized (*S*)-selective transaminases. It is also possible to acquire the comprehensive screening kit including the 19 enzymes. Table 16.3 shows some of the ADHs and lipases that can be purchased on demand, taking into account that ADHs can also be supplied as a screening kit, whereas the lipases can be obtained as different preparations, including immobilized forms, powders, or liquid preparations.

16.2.7 GECCO

The Groningen Enzyme and Coenzymes Company is a dynamic Dutch spin-off company born from the University of Groningen [12]. GECOO was created in 2018 and presents a wide selection of valuable enzymes to be employed with synthetic purposes. The main portfolio consists in redox enzymes, including ADHs and oxidative biocatalysts as copper-dependent oxidases (laccases), flavin-containing monooxygenases (FMOs), BVMOs, and dye-decolorizing peroxidases (DyP), as indicated in Table 16.4. GECCO also provides other types of enzymes such as dehalogenases or epoxide hydrolases (EHs). Finally, this company offers various cofactors and cofactor analogs, such as deazaflavin cofactors F0 and F420.

Table 16.3 ADHs and lipases supplied by Evoxx Technologies.

Enzyme	Activity
ADH 030, lyophilized powder, NAD-dependent enzyme	(*S*)-alcohol dehydrogenase
ADH 040, lyophilized powder, NAD-dependent enzyme	(*S*)-alcohol dehydrogenase
ADH 200, lyophilized powder, NAD-dependent enzyme	(*R*)-alcohol dehydrogenase
ADH 270, lyophilized powder, NADP-dependent enzyme	(*R*)-alcohol dehydrogenase
ADH 440, lyophilized powder, NADP-dependent enzyme	(*S*)-alcohol dehydrogenase
Addzyme CalB 165G, immobilized *Candida antarctica* B	Lipase
Addzyme TL 165G, immobilized *Thermomyces lanuginosus*	Lipase
Addzyme RD 165G, immobilized *Rhizopus delemar*	Lipase
Addzyme CalB 5L, *Candida antarctica* B liquid preparation	Lipase
Addzyme TL 100P, *Thermomyces lanuginosus* powder	Lipase
Addzyme RD 50P, *Rhizopus delemar* powder	Lipase

Source: Based on [11].

Table 16.4 Some of the enzymes commercialized by GECCO.

Enzyme	Activity
ADH A *Pyrococcus furiosus*	Alcohol dehydrogenase
*Rj*ADH *Rhodococcus jostii*	Alcohol dehydrogenase
*Lb*ADH *Lactobacillus brevis*	Alcohol dehydrogenase
Bacillus licheniformis ATCC 9945a	Laccase
Human FMO5	Flavin-containing monooxygenase
*Me*FMO *Methylophaga aminisulfidivorans*	Flavin-containing monooxygenase
FMOs *Rhodococcus jostii* RHA1 (8 enzymes)	Flavin-containing monooxygenases
STMO *Rhodococcus rhodochrous*	Baeyer–Villiger monooxygenase
PAMO *Thermobifida fusca*	Baeyer–Villiger monooxygenase
CPMO *Comamonas* sp. NCIMB 9871	Baeyer–Villiger monooxygenase
*Tm*CHMO *Thermocrispum municipale*	Baeyer–Villiger monooxygenase
*Tfu*DyP *Thermobifida fusca*	Dye-decolorizing peroxidase
*Pf*DyP *Pseudomonas fluorescens*	Dye-decolorizing peroxidase

Source: Based on [11].

16.2.8 Inofea AG

This Swiss company founded in 2014 is an enzyme producer for different applications including biocatalysis, bioconjugation processes in order to obtain antibody–drug conjugates, and bioanalysis [13]. Inofea offers the Enzzen® Technology, a system capable of immobilizing almost all types of biocatalysts or enzyme cocktails onto safe silica particles. This immobilization procedure is able to protect the biocatalyst(s) by growing a structured shield on the outer surface of the particle, improving the properties and the reusability of the enzymatic systems.

16.2.9 Johnson-Matthey

This British company with headquarters located at London is a multinational specialty chemicals and sustainable technologies supplier [14]. Its biocatalytic section offers different enzymatic products to be employed in organic synthesis (Table 16.5).

Johnson-Matthey supplies a wide set of ADHs, as well as other dehydrogenases such as alanine dehydrogenase, able to catalyze the conversion of pyruvate to L-alanine, amine dehydrogenases for the synthesis of primary amines, and lactate dehydrogenases, which catalyze the reduction of pyruvate to lactate. This company also offers a set of ene-reductases with different activities for the reduction of double bonds, as well as (*R*)- and (*S*)-selective ω-transaminases for the synthesis of aromatic and aliphatic primary amines. These biocatalysts can be supplied individually or as biocatalyst kits. Thus, a C=C bond reduction kit containing seven ene-reductases and a ketoreductase kit consisting in 17 different ADHs are available. Finally, the company offers a chiral amine kit, containing different enzymes able to catalyze the synthesis of primary, secondary, and tertiary amines.

Table 16.5 Enzymes supplied by Johnson-Matthey.

Enzymes	Activity
ADH-105, ADH-110, ADH-150, ADH-153, ADH-159, ADH-160, ADH-171, ADH-19, ADH-20, ADH-220, ADH-230, ADH-244, ADH-27, ADH-61, ADH-62	Alcohol dehydrogenases
AlaDH-6	Alanine dehydrogenase
AmDH-1, AmDH-2, AmDH-3, AmDH-4, Am-DH-8	Amine dehydrogenases
LDH-4, LDH-12	Lactate dehydrogenases
ENE-101, ENE-102, ENE-103, ENE-105, ENE-107, ENE-108, ENE-109	Ene-reductases
RTA-102, RTA-103, RTA-104, RTA-105, RTA-25, RTA-40, RTA-45, RTA-57, RTA-58, STA-1, STA-113, STA-118, STA-120, STA-121, STA-13, STA-14, STA-2	ω-transaminases

Source: Based on [14].

16.2.10 Metgen Oy

Metgen Oy is a Finnish company with headquarters in Kaarina founded in 2008, which performs its activity in the development and production of industrial enzymes [15]. MetGen offers a range of biocatalysts to solve critical industrial issues that address needs in the markets of pulp and paper, biofuels, and wastewater treatment. Thus, since its origin, this company has been mainly focused on the development of enzymes useful for biomass treatment in pulp and paper industry. Some of the products that Metgen commercializes are summarized in Table 16.6.

Table 16.6 Examples of the enzymatic products supplied by Metgen Oy in the fields of paper and pulps and wastewater treatment.

Products	Application
MetZyme® Brila™ Enzyme product family	Recycle fiber applications
MetZyme® Ligno™ Enzyme product family	Mechanical pulping processes
MetZyme® Povon™ Enzyme product family	Chemical pulp bleaching and extracts control
MetZyme® Forico™ Enzyme product family	Industrial water treatment and effluent control
MetZyme® Sekalo™ (α-amylase)	Starch conversion solutions
MetZyme® Plata™ Enzyme product family	Recycled fiber deinking and wood pitch removal
MetZyme® Suno™ Enzyme product family	Lignocellulosic chemicals production

Source: Based on [15].

16.2.11 Novozymes

This Danish biotechnological company, with headquarters close to Copenhagen, has been expanded to the rest of Europe, Asia, and America, representing nowadays one of the most important enzyme suppliers worldwide [16]. Novozymes offers biocatalysts for several applications, including pharma, organic synthesis, food processing, laundry, flavors and fragrance, etc.

Regarding their use in organic synthesis and pharma, most of the enzymes commercialized by this company are lipases and other types of hydrolases, being widely employed in several hydrolytic and synthetic procedures. Table 16.7 shows some examples of the enzymes from Novozymes with synthetically valuable purposes, including Novozym® 435, an immobilized *C. antarctica* B preparation that has been extensively employed in organic synthesis.

16.2.12 Prozomix

This British biotechnological company is located at Haltwhistle [17]. Since its creation, Prozomix develops, produces, and supplies a huge range of biocatalysts with several applications in organic chemistry. Thus, this company offers hydrolases, ADHs, aldolases, alkyl transferases, BVMOs, decarboxylases, ene-reductases, EHs, glycosyl transferases, halogenases, halohydrin dehalogenases, imine reductases, nitroreductases, nitrile hydratases, cytochrome P450s, ammonia lyases, monoamine oxidases, and transaminases, some examples being shown in Table 16.8. Most of the enzymes can be purchased in Screening Kits, whereas other enzymes can be obtained individually. Prozomix also supplies different cofactors required for the activity or certain enzymes, as well as enzymes for the regeneration of these cofactors, such as formate- or glucose-dehydrogenases and nicotinamide adenine dinucleotide phosphate (NADPH)-oxidases. Finally, this company has an extensive portfolio of enzymes with industrial application in different sectors, including amylases, catalases, pectinases, and different proteases.

Table 16.7 Some hydrolytic enzymes commercialized by Novozymes.

Enzyme	Activity	Specific activity
Novozym® 435, immobilized	Nonspecific lipase	10 000 PLU g^{-1}
Lypozime® CalB L, liquid	Nonspecific lipase	5 KLU g^{-1}
Resinase® HT from *Aspergillus oryzae*, liquid	Lipase	50 KLU g^{-1}
Alcalase® 2.4LFG from *Bacillus licheniformis*, liquid	Endo-peptidase	2.4 AU g^{-1}
Esperase® 8.0L from *Bacillus licheniformis*, liquid	Endo-peptidase	8 KNPU g^{-1}
rTrypsin® 8.0L from *Fusarium venenatum*, granulate	Serine protease	800 USP mg^{-1}

Source: Based on [16].

Table 16.8 Some of the biocatalysts supplied by Prozomix.

Biocatalysts	Quantity
PRO-AKR(001-203), Aldo-ketoreductases from different sources	50 mg-bulk enquires
PRO-450(001-010), Cytochrome P450s from different sources	100 mg crude
PRO-ERED(001-0040), Ene-reductases from different sources	50 mg-Bulk enquires
Nitroreductase, *Escherichia coli* K12	1–1000 mg
PRO-E0260, Nitrilase from *Bradyrhizobium japonicum*	1–1000 mg
PRO-NITR(001-018), Nitrilases from different sources	50 mg-bulk enquires
PRO-AldP, Aldolases screening kit	100–200 mg
PRO-CoE-BVMOs screening kit	1 mg or 1–10 g
PRO-DACARBP, Decarboxylases screening kit	100–200 mg
PRO-GHP(MTP), Glycosyl hydrolases screening kit	1 mg
PRO-CoE-IREDs, Imine reductases screening kit	1 mg or 1–10 g
PRO-NHASEP, Nitrile hydratases screening kit	2 ml
PRO-CoE-AMLP, Ammonia lyases screening kit	1 mg or 1–10 g
PRO-CoE-RAMO, Monoamine oxidases screening kit	10 mg or 2–20 g

Source: Based on [17].

16.2.13 Royal DSM

This Dutch multinational is involved in the manufacturing and supplying of food and dietary supplements, personal care products, bio-based materials, as well as yeast and enzymes [18]. DSM is one of the major suppliers of enzymes and yeasts in the world, producing enzymes with different applications in food and beverage production, from boosting baking and brewing, to empowering egg and fruit processors, to optimizing dairy, oils, and fats.

16.3 American Companies

16.3.1 Codexis Inc.

Codexis, Inc. is a leading protein engineering company applying technologies to use enzymes at laboratory and industrial scale [19]. Their headquarters are located in Redwood City (USA), and since 2002, they are focused on the development of new proteins through molecular biology tools and also to their production, in order to find applications in the pharmaceuticals and fine chemicals, biotherapeutics, food and beverage, and agriculture fields.

Guided by proprietary software and bioinformatic systems, Codexis powers existing knowledge and expertise to improve protein stability, activity, specificity, and other properties to enhance clinical function or process performance. With this approach to protein engineering starts with an understanding of the desired performance specifications and optimizes a protein specifically to fit the desired requirements.

Table 16.9 Screening kits provided by Codexix Inc.

Screening kits	Quantity
Codex® Ketoreductase (KRED) Screening Kit	24 enzymes × 250 mg
Codex® Amine Transaminase (ATA) Screening Kit	24 enzymes × 250 mg
Codex® Nitrilase (NIT) Screening Kit	12 enzymes × 250 mg
Codex® Ene Reductase (ERED) Screening Kit	7 enzymes × 250 mg
Codex® Standard MicroCyp® Screening Kit	23 enzymes (1 screen per kit)

Source: Based on [19].

Enzymes can be produced from grams to kilos to tons, developing optimized proteins for specific applications in weeks and can scale up to full commercial production in short periods such as three months. Codexis can provide screening kits (Table 16.9) to rapidly determine the feasibility of using the engineered enzymes. Each kit contains a selected set of enzymes that have been engineered for enhanced activity, selectivity, substrate range, solvent, and temperature stability. Conveniently packaged in individual vials and with enough enzyme for several screens and hit verification, kits include protocols and co-reagents for use. Follow-on quantities of all kit enzymes are stocked in 50–100 g quantities to enable rapid delivery.

Furthermore, other available enzyme kits are BVMOs, imine reductases, monoamine oxidases, halohydrin dehalogenases, and acylases.

16.3.2 Dupont Nutrition and Biosciences

This transnational company is a world leader of innovative and sustainable solutions across food, health, pharma, and biotech industries [20]. Among the different products that are provided by this company are antimicrobials, antioxidants, dairy cultures, emulsifiers, probiotics, and food enzymes. DuPont™ Danisco® food enzymes can help to improve food freshness, optimize production, add texture, ensure consistent quality, and therefore reduce costs.

Among the different enzymes that are available from this company, lipases to modify the flavor of cheeses, lactases for reducing the lactose content in milk, chymosin and/or pepsin coagulants for traditional cheese making, alkaline proteases for enhancing the quality of fish-processing by-products, or α-amylases, β-glucanases, and xylanases for brewing applications, can be mentioned.

16.3.3 IBEX Technologies

This company is based in Montreal (Canada) and manufactures and markets reagents for biomedical use through its wholly owned subsidiary IBEX Pharmaceuticals Inc. The company product lines comprise high-purity enzymes for *in vitro* diagnostics (IVD) and research, as well ELISA kits for osteoarthritis research. IBEX also provides custom production services for *in vitro* diagnostic market, including custom fermentation as well as filling and lyophilization for disposable diagnostic components.

Among the different enzymes they trade, several heparinases, chondroitinases, collagen antibodies, and a diamine oxidase can be mentioned. They can be acquired from few enzymatic units to bulk quantities [21].

16.3.4 MP Biomedical

MP Biomedicals is a worldwide corporation developing, manufacturing, and distributing products for the life science, IVD, fine chemicals, and dosimetry markets from small-scale research to large-scale manufacturing. They are located in Irvine (USA), having more than 50 years of experience in research products, also introducing innovative new products of high quality [22]. Apart from many diagnostics reagents, they can also provide with many different fine chemicals for organic synthesis. Among these, several cofactors and enzymes (Table 16.10) can be found in their catalog.

16.3.5 Sigma-Aldrich

This transnational chemical company was founded in 1975 and is headquartered in St. Louis (USA). In 2014, it was acquired by Merck KGaA and it is located all over the world. In the past decades, this chemical company has significantly expanded its business lines,

Table 16.10 Selection of biocatalysts supplied by MP Biomedical.

Enzyme	Activity	Quantity
Lyticase from *Arthrobacter luteus*	Protease	Up to 50 kU
Trypsin	Peptidase	Up to 500 mg
Leucine aminopeptidase from porcine kidney	Peptidase	Up to 100 U
Lipase from porcine pancreas	Lipase	Up to 500 g
Lipase from *Candida rugosa*	Lipase	100 g
Cholesterol esterase from *Pseudomonas* sp.	Esterase	Up to 200 kU
Amylase from *Aspergillus oryzae*	Amylase	Up to 5 MU
β-Glucuronidase from *Helix pomatia*	Glycosidase	Up to 50 ml
β-Galactosidase from *Escherichia coli*	Glycosidase	Up to 3 kU
Alcohol oxidase from *Pichia pastoris*	Oxidase	Up to 5 kU
Horseradish peroxidase	Peroxidase	Up to 25 kU
Glucose oxidase from *Aspergillus niger*	Oxidase	Up to 500 kU
Alcohol dehydrogenase from yeast	Dehydrogenase	Up to 150 kU
Glucose-6-P dehydrogenase from *Leuconostoc mesenteroides*	Dehydrogenase	Up to 1 kU
Lactate dehydrogenase from bovine heart	Dehydrogenase	10 kU

Source: Based on [22].

Table 16.11 Selection of biocatalysts provided by Sigma-Aldrich divided by enzyme class.

Enzyme class	Total number
Oxidoreductases (EC 1.x.x.x): alcohol dehydrogenases, alcohol oxidases, amino acid oxidases, monoamine oxidases, cytochrome P450s, laccases, catalases, peroxidases, dioxygenases, monooxygenases, superoxide dismutases	177
Transferases (EC 2.x.x.x): transaldolase, acyltransferases, phosphorylases, transaminases, kinases	79
Hydrolases (EC 3.x.x.x): esterases, lipases, (deoxy)ribonucleases, phosphatases, sulfatases, glycosidases, epoxide hydrolases, peptidases, proteases, acylases, amidases, β-lactamases, deaminases	542
Lyases (EC 4.x.x.x): decarboxylases, cyanohydrin lyases, aldolases, carbonic anhydrases, phenylalanine ammonia lyases, phospholipases	46
Isomerases (EC 5.x.x.x): sugar-converting isomerases, topoisomerases	18
Ligases (EC 6.x.x.x): synthetases, carboxylases	8

Source: Based on [23].

serving customers focused on identifying and developing medicines, delivering end-to-end products and expertise to customers who take what is developed in labs and manufacture it, and applying analytical tools to ensure that drugs, food, and beverages are safe for consumption.

Among the many different chemical products they can provide, an important number of reagents, catalysts, inhibitors, and cofactors related to enzyme catalysis can be mentioned. For instance, a summary of the different biocatalysts that are accessible through its webpage [23] is displayed in Table 16.11.

16.3.6 Strem Chemicals, Inc.

This chemical company was established in 1964, located in Newburyport (USA), and manufactures and markets specialty chemicals of high purity. Its clients include academic, industrial, and government research and development laboratories as well as commercial-scale businesses in the pharmaceutical, microelectronic, and chemical or petrochemical industries. Strem also provides custom synthesis (including high-pressure synthesis) and current good manufacturing practice regulations (cGMP) manufacturing services.

Strem offers over 5000 specialty products in the area of metals, inorganics, organometallics, and nanomaterials. Its first commercial product was cobalt carbonyl, manufactured in Strem's high-pressure autoclaves. The first catalog also contained (pre)catalysts and phosphorus ligands, which continue to be important classes of products offered by Strem.

More recently, they are focused on the commercialization of biocatalysts, mainly on lipases. In its catalog, different immobilization forms from CAL-B and also a number of immobilized proteases are accessible. Also, various enzyme kits including different lipases, immobilized forms of CAL-B, and proteases can be found through its webpage [24].

16.3.7 Worthington Biochemical Corp

This biotech company was founded in 1947, is located in New Jersey (USA), and manufactures enzymes as the main products [25]. Worthington does not actually make enzymes: they extract them from various animal and plant tissues and various microbial sources such as bacteria, fungi, and molds. A starting material for a particular enzyme is selected according to the prevalence of that enzyme in the material.

Some of the animal tissues used at Worthington include beef pancreas, hog kidney, cow eyes, horse liver, rabbit muscle, beef liver, pig heart, horse blood, and calf intestine. Various proteins are also extracted from horseradish roots, sweet potatoes, almonds, and pokeweed. Some products are isolated from the bacteria *Escherichia coli*, different species of *Clostridia*, and various other bacteria. Yeast, mushrooms, whole milk, and eggs are also used for this purpose. All enzymatic products are materials extracted from living cells. Table 16.12 shows a selection of enzymes that are accessible at this company.

Table 16.12 Selection of biocatalysts that are provided by Worthington Biochemical Corp.

Enzyme	Activity
Alcohol dehydrogenase from yeast	>300 U mg^{-1}
Lactate dehydrogenase from rabbit muscle	>250 U mg^{-1}
Glucose-6-phosphate dehydrogenase from *Leuconostoc mesenteroides*	>600 U mg^{-1}
Galactose oxidase from *Dactylium dendroides*	>30 U mg^{-1}
Catalase from bovine liver	>40 000 U mg^{-1}
Peroxidase from horseradish	>500 U mg^{-1}
L-Amino acid oxidase from *Crotalus adamanteus* venom	>4 U mg^{-1}
Hexokinase from yeast	>150 U mg^{-1}
Cholesterol esterase from porcine pancreas	>300 U g^{-1}
Deoxyribonuclease I from bovine pancreas	>2000 U mg^{-1}
Alkaline phosphatase from calf intestine	>3000 U mg^{-1}
β-Galactosidase from *Escherichia coli*	>300 U mg^{-1}
Cellulase from *Trichoderma reesei*	>45 U mg^{-1}
α-Chimotrypsin from bovine pancreas	>45 U mg^{-1}
Trypsin from bovine pancreas	>180 U mg^{-1}
Pepsin A from porcine stomach	>2500 U mg^{-1}
Collagenase from *Clostridium histolyticum*	>500 U mg^{-1}
Adenosine deaminase from calf spleen	>15 U mg^{-1}
Tyrosine decarboxylase from *Streptococcus faecalis*	>0.2 U mg^{-1}
Aldolase from rabbit muscle	>10 U mg^{-1}

Source: Based on [25].

16.4 Asian Enzyme Suppliers

16.4.1 Advanced Enzymes Technologies, Ltd.

An Indian company located at Louiswadi Thane is a research-driven company manufacturing enzymes and probiotics, marketing more than 400 proprietary products developed from over 65 indigenous enzymes and probiotics [26]. They commercialize different biocatalysts for human nutrition (e.g. peptidases, lipases, and proteases) and food processing (e.g. pectinases, amylases, and esterases), among others.

16.4.2 Amano Enzyme Co., Ltd.

This company was founded in 1899 in Japan as a pharmaceutical business. In 1948, this business expanded to use the process of koji fermentation (traditionally used to create soy sauce, miso, and sake) to produce specialty enzymes, beginning with Malt Diastase. This company, with headquarters in Nagoya (Japan), is also located in Elgin (USA) and supplies high-quality microbial enzymes for the food, dietary supplement, industrial, diagnostic, and regenerative medical industries [27]. They feature multiton manufacturing capabilities, safe quality food (SQF) Level 2 certification, and technical expertise to provide service along with guaranteed quality products. Some of the products that Amano commercializes are summarized in Table 16.13.

Table 16.13 Examples of the enzymes supplied by Amano and their possible applications.

Enzymes	Application
KLEISTASE E5NC, KLEISTASE SD80, KLEISTASE T10S, KLEISTASE PLF3	Starch processing
Protease A "Amano" 2 SD, Protease M "Amano" SD, Protease P "Amano" 6 SD, ProteAX, Peptidase R, THERMOASE PC10F, THERMOASE GL30, PROTIN SD-AY10, PROTIN SD-NY10	Flavor enhancement
Lipase A "Amano" 12, Lipase AY "Amano" 30SD, Lipase G "Amano" 50, Lipase R "Amano", Lipase MH "Amano" 10SD, Lipase DF "Amano" 15, Lipase MER "Amano"	Oils and fats refining
Glucose dehydrogenase "Amano" 8 [GDH-8], Glucose dehydrogenase "Amano" 2A [GDH-2A], Glucose dehydrogenase "Amano" NA [GDH-NA], Glucose oxidase "Amano" AM [GO-AM], Glucose oxidase "Amano" NA [GO-NA], Glucose oxidase "Amano" M [GO-M], Glucose oxidase "Amano" 2 [GO-2]	Glucose sensor
Cholesterol dehydrogenase "Amano" 6 [CHDH-6], Cholesterol oxidase "Amano" 6E [CHO-6E], Cholesterol oxidase "Amano" 7 [CHO-7], Cholesterol esterase "Amano" 2A [CHE-2A], EST "Amano" 2 [EST-2]	Cholesterol determination
Lipase PS "Amano" SD, Lipase PS "Amano" IM, Lipase AK "Amano", Lipase AH "Amano" SD, Lipase AYS "Amano", Lipase AS "Amano", Acylase H "Amano", D-Aminoacylase "Amano"	Chiral resolution/ synthesis

Source: Based on [27].

16.4.3 Aumgene Biosciences

This biotech company is based in Surat, India. Founded in 2004, the company undertakes development, manufacturing, and marketing of enzymes, probiotics, and agribiotech products [28]. The company has state-of-the-art research and development facilities for high-throughput screening of microbial strains, fermentation process development, and purification of enzymes and biologics. Among the different enzymes they commercialize, laccases, cellulases, and amylases for large-scale production can be mentioned.

16.4.4 EnzymeWorks

EnzymeWorks is a company that supplies enzymes, cofactors, and biochemicals and also provides contract manufacturing organization (CMO) services. The company is located in Jiangsu (China). EnzymeWorks develop bio-based green processes for chemical manufacturing through retrosynthetic integration of chemical and biological transformation toolboxes [29]. They are focused on biosynthesis, enzyme development, protein engineering, high-throughput screening, metabolic engineering, and fermentation from the lab to pilot scale, manufacturing enzymes and fermented chemicals from 3000 l to 10 000 l scale.

They provide different enzyme family kits such as ketoreductases, transaminases, cytochrome P450s, hydrolases, nitrilases, ene-reductases, monoamine oxidases, nitrile hydratases, EHs, imine reductases, glycosyltransferases, BVMOs, aldolases, nitroreductases, decarboxylases, and dehydrogenases.

16.4.5 Meito Sangyo Co., Ltd.

This biotech company is located in Nagoya (Japan). They produce different sweets and beverages by using fermentation technology from microorganisms. Meito Sangyo Co., Ltd. Fine chemicals Division produces unique products, such as polysaccharide and enzymes. The different biocatalysts they offer are described in Table 16.14, all of them with application in the food industry [30].

Table 16.14 Hydrolytic enzymes commercialized by Meito Sangyo Co., Ltd.

Enzyme	Purpose	Specific activity
Lipase MY from *Candida cylindracea*	Flavor development	$30\,000\,\mathrm{U\,g^{-1}}$
Lipase OF from *Candida cylindracea*	Flavor development Hydrolysis of oils and fats	$360\,000\,\mathrm{U\,g^{-1}}$
Lipase PL from *Alcaligenes* sp.	Modification of oil and fats	*ca.* $100\,000\,\mathrm{U\,g^{-1}}$
Lipase QLM from *Alcaligenes* sp.	Modification of oil and fats	*ca.* $60\,000\,\mathrm{U\,g^{-1}}$
Lipase SL from *Burkholderia cepacia*	Hydrolysis of oils and fats	*ca.* $45\,000\,\mathrm{U\,g^{-1}}$
Lipase TL from *Pseudomonas stutzeri*	Hydrolysis of oils and fats	*ca.* $50\,000\,\mathrm{U\,g^{-1}}$
Phospholipase D from *Actinomadura* sp.	Modification of phospholipids	$1500\,\mathrm{U\,g^{-1}}$

Source: Based on [30].

16.4.6 Oriental Yeast Co., Ltd.

Oriental Yeast is a manufacturer with headquarters in Tokyo (Japan), with a long-established history of technological innovations originally derived from baker's yeast production. In its food business, the company develops a variety of food ingredients in addition to yeast while focusing on yeast-related biotechnology applications. Among the different companies that belong to Oriental Yeast, OYC Americas is based in Vista (CA, USA), and its laboratory of yeast and fermentation has over 13 000 strains of yeast as stock cultures and works to develop applications for high-performing, high-functioning yeasts, and fermented products.

Among the different products they commercialize [31], various enzymes (e.g. dehydrogenases, phosphatases, dioxygenases, oxidases, peroxidases, and glucosidases), coenzymes, and recombinant enzymes from human source can be mentioned. These enzymes can be acquired at different quantities (from few enzyme units or mg to kU or bulk).

16.4.7 Takabio

Takabio is a global company that supplies a series of enzymes for food industry applications produced by Shin Nihon Chemical in Anjo (Japan). With more than 20 years of expertise in the field of enzymes and biotechnologies behind them, the Takabio teams support food sector manufacturers all over the world by providing enzymes for a wide variety of industrial applications [32]. Some of the different biocatalysts that are commercialized appear in Table 16.15.

16.4.8 Toyobo Co., Ltd.

This company has been a leading and prominent manufacture of overall textiles in Japan. After diversification, they are expanding their product lines in other fields such films,

Table 16.15 Selected enzymes commercialized by Takabio company.

Enzyme	Application
α-Galactosidase from *Aspergillus niger*	Nutrition (human and animal), dietary supplements
Amylases from *Aspergillus niger* and *Aspergillus oryzae*	Liquefaction and saccharification of starch, dietary supplements
Catalase from *Aspergillus niger*	Multiple food applications
Dipeptidyl-aminopeptidase from *Aspergillus niger*	Dipeptide production
Glucose oxidase from *Penicillium chrysogenum*	Multiple food applications
Lipases from *Aspergillus niger*, *Candida rugosa*, and *Rhizomucor miehei*	Hydrolysis of oils and fats
Proteases and peptidases from *Aspergillus oryzae*, *Aspergillus niger*, *Bacillus subtilis*, and *Aspergillus melleus*	Dairy industry, dietary supplements

Source: Based on [32].

plastics, electronics-based, biochemical-related, and medical-related business [33]. One of those product line-ups is Toyobo Enzymes, which comes out of the technologies in fermentation and purification accumulated over the past several decades. The Biochemical Department is located in Osaka (Japan), manufacturing enzymes since two decades (Table 16.16).

16.5 Outlook

In the past few years, the development of biocatalytic methodologies has experienced a great development. As it has been shown in the present book, biocatalysts offer a set of advantages that make them really useful in the preparation of highly valuable compounds. Most of the enzymatic preparations are employed in sectors such as food and beverages, for cleaning purposes, for biofuel production, and in agriculture, and an important number of processes have also been performed applying (tailored) biocatalysts in the (bio)pharmaceutical industrial sector and also applied to organic synthesis in both industry and academia. By this reason, nowadays several chemical and biotechnological companies provide different types of enzymes in their portfolio of products, from SMEs devoted mainly to enzyme's preparation, modification and immobilization, to huge transnational companies in which biocatalysts are part of their business interest.

With this final contribution, authors want to stress the relevance that biocatalysis field is gaining at commercial level and also to facilitate to researchers interested in developing

Table 16.16 Selected biocatalysts commercialized by Toyobo Co., Ltd.

Enzyme	Microorganism	Specific activity
N-Acetylneuraminic acid aldolase	Microorganism not specified	$>15\,U\,mg^{-1}$
Cholesterol esterase	*Pseudomonas* sp.	$>100\,U\,mg^{-1}$
Diaphorase	*Clostridium* sp.	$>30\,U\,mg^{-1}$
β-Galactosidase	*Escherichia coli*	$>500\,U\,mg^{-1}$
Glucose oxidase	*Aspergillus* sp.	$>180\,U\,mg^{-1}$
Glucose-6-phosphate dehydrogenase	*Leuconostoc* sp.	$>400\,U\,mg^{-1}$
Glutamate dehydrogenase	*Proteus* sp.	$>300\,U\,mg^{-1}$
Glycerol dehydrogenase	*Cellulomonas* sp.	$>50\,U\,mg^{-1}$
D-3-Hydroxybutyrate dehydrogenase	*Pseudomonas* sp.	$>100\,U\,mg^{-1}$
Leucine dehydrogenase	*Bacillus* sp.	$>20\,U\,mg^{-1}$
Peroxidase	Horseradish	$>250\,U\,mg^{-1}$
Purine-nucleoside phosphorylase	Microorganism not specified	$>15\,U\,mg^{-1}$
Pyruvate oxidase	*Pseudomonas* sp.	$>1.5\,U\,mg^{-1}$

Source: Based on [33].

experiments in this area, some of the most important supplier enzyme companies at different locations. Many of them have recently appeared, emphasizing the larger impact that Biotechnology, and more specifically biocatalysis, is taking place at industrial level and, in a broader sense, in society. With the outstanding developments made in the recent decades in molecular biology, bioinformatics, microbiology, material science, and process engineering, more robust, active, and selective enzymatic forms (as pure or partially purified enzymes, as whole cell biocatalysts or in immobilized forms) are expected to appear in the next years not only to fulfill the desired requirements to synthesize a target molecule at industrial level but also in order to find novel reactivities that nowadays seem to be inaccessible [34].

References

1 de Gonzalo, G. and Domínguez de María, P. (eds.) (2018). *Biocatalysis: An Industrial Perspective*. Cambridge: Royal Society of Chemistry.

2 Adams, J.P., Brown, M.J.B., Díaz-Rodríguez, A. et al. (2019). Biocatalysis: a pharma perspective. *Adv. Synth. Catal.* 361: 2421–2432.

3 Hauer, B. (2020). Embracing Nature's catalysts: a viewpoint on the future of biocatalysis. *ACS Catal.* 10: 8418–8427.

4 Sheldon, R.A., Brady, D., and Bode, M.L. (2020). The Hitchhiker's guide to biocatalysis: recent advances in the use of enzymes in organic synthesis. *Chem. Sci.* 11: 2587–2605.

5 https://www.abenzymes.com/en/ (accessed June 2020).

6 https://www.almacgroup.com (accessed June 2020).

7 https://www.biocatalysts.com/ (accessed June 2020).

8 http://www.c-lecta.com/ (accessed June 2020).

9 Gotor-Fernández, V., Busto, E., and Gotor, V. (2006). *Candida antarctica* lipase B: an ideal biocatalyst for the preparation of nitrogenated organic compounds. *Adv. Synth. Catal.* 348: 797–812.

10 http://www.enzymicals.com/index_en.html (accessed June 2020).

11 https://evoxx.com/ (accessed June 2020).

12 http://www.gecco-biotech.com/ (accessed June 2020).

13 https://inofea.com/ (accessed June 2020).

14 https://matthey.com/en (accessed June 2020).

15 https://www.metgen.com/ (accessed June 2020).

16 https://www.novozymes.com/ (accessed June 2020).

17 http://www.prozomix.com/cap/home (accessed June 2020).

18 https://www.dsm.com/ (accessed June 2020).

19 https://www.codexis.com/ (accessed July 2020).

20 https://www.food.dupont.com/ (accessed July 2020).

21 https://www.ibex.ca/ (accessed July 2020).

22 https://www.mpbio.com/ (accessed July 2020).

23 https://www.sigmaaldrich.com (accessed July 2020).

24 https://www.strem.com/index.php (accessed July 2020).

25 http://www.worthington-biochem.com/default.html (accessed July 2020).

26 https://www.advancedenzymes.com/ (accessed July 2020).

27 https://www.amano-enzyme.com/ja/ (accessed July 2020).

28 http://www.aumgene.com/ (accessed July 2020).

29 http://www.enzymeworking.com/ (accessed July 2020).

30 https://www.meito-sangyo.co.jp/kaseihin/index_e.html (accessed July 2020).

31 https://oyc.co.jp/en/ (accessed July 2020).

32 https://www.takabio.com/en/ (accessed July 2020).

33 https://www.toyobo-global.com/ (accessed July 2020).

34 Arnold, F.H. (2019). Innovation by evolution: bringing new chemistry to life (Nobel lecture). *Angew. Chem. Int. Ed.* 58: 14420–14426.

Index

Biocatalysis for Practitioners: Techniques, Reactions and Applications, First Edition.
Edited by Gonzalo de Gonzalo and Iván Lavandera.

b

c

f

i

m

n

o

p

q

r

S